Springer Proceedings in Advanced Robotics 11

The Springer Proceedings in Advanced Robotics (SPAR) publishes new developments and advances in the fields of robotics research, rapidly and informally but with a high quality.

The intent is to cover all the technical contents, applications, and multidisciplinary aspects of robotics, embedded in the fields of Mechanical Engineering, Computer Science, Electrical Engineering, Mechatronics, Control, and Life Sciences, as well as the methodologies behind them.

The publications within the "Springer Proceedings in Advanced Robotics" are primarily proceedings and post-proceedings of important conferences, symposia and congresses. They cover significant recent developments in the field, both of a foundational and applicable character. Also considered for publication are edited monographs, contributed volumes and lecture notes of exceptionally high quality and interest.

An important characteristic feature of the series is the short publication time and world-wide distribution. This permits a rapid and broad dissemination of research results.

More information about this series at http://www.springer.com/series/15556

spar

Jing Xiao · Torsten Kröger · Oussama Khatib
Editors

Proceedings of the 2018 International Symposium on Experimental Robotics

Springer

Editors
Jing Xiao
Robotics Engineering
Worcester Polytechnic Institute
Worcester, MA, USA

Torsten Kröger
Karlsruhe Institute of Technology
Karlsruhe, Baden-Württemberg, Germany

Oussama Khatib
Department of Computer Science
Stanford University
Stanford, CA, USA

ISSN 2511-1256 ISSN 2511-1264 (electronic)
Springer Proceedings in Advanced Robotics
ISBN 978-3-030-33949-4 ISBN 978-3-030-33950-0 (eBook)
https://doi.org/10.1007/978-3-030-33950-0

This Springer imprint is published by the registered company Springer Nature Switzerland AG
The registered company address is: Gewerbestrasse 11, 6330 Cham, Switzerland

Foreword

At the dawn of the century's third decade, robotics is reaching an elevated level of maturity and continues to benefit from the advances and innovations in its enabling technologies. These all are contributing to an unprecedented effort to bringing robots to human environment in hospitals and homes, factories and schools, in the field for robots fighting fires, making goods and products, picking fruits and watering the farmland, saving time and lives. Robots today hold the promise for making a considerable impact in a wide range of real-world applications from industrial manufacturing to health care, transportation, and exploration of the deep space and sea. Tomorrow, robots will become pervasive and touch upon many aspects of modern life.

The *Springer Tracts in Advanced Robotics (STAR)* was launched in 2002 with the goal of bringing to the research community the latest advances in the robotics field based on their significance and quality. During the latest fifteen years, the STAR series has featured publication of both monographs and edited collections. Among the latter, the proceedings of thematic symposia devoted to excellence in robotics research, such as ISRR, ISER, FSR, and WAFR, have been regularly included in STAR.

The expansion of our field as well as the emergence of new research areas has motivated us to enlarge the pool of proceedings in the STAR series in the past few years. This has ultimately led to launching a sister series in parallel to STAR. The *Springer Proceedings in Advanced Robotics (SPAR)* is dedicated to the timely dissemination of the latest research results presented in selected symposia and workshops.

This volume of the SPAR series brings the proceedings of the sixteenth edition of the International Symposium on Experimental Robotics (ISER). This symposium took place in Buenos Aires, Argentina, from November 5 to 8, 2018. The seven-part volume edited by Jing Xiao, Torsten Kröger, and Oussama Khatib is a collection of 67 contributions spanning a wide range of topics in robotics, including medical robotics, unmanned aerial vehicles, mobile robot navigation, mapping and localization, field robotics, robot learning, manipulation, human–robot interaction, and design and prototyping. Experimental validation of algorithms, concepts, or techniques is the common thread running through this large research collection.

From its classical venue to its excellent program, the fifteenth edition of ISER culminates with this unique reference on the current developments and new directions of experimental robotics—a genuine tribute to its contributors and organizers!

July 2019

Bruno Siciliano
SPAR Editor

Preface

Experimental Robotics XVI is the collection of contributions presented at the 2018 International Symposium on Experimental Robotics (ISER 2018), November 5–8, 2018, Buenos Aires, Argentina. A total of 67 contributions were selected via peer review out of 120 submissions. The contributions cover a wide range of topics in robotics, including medical robotics, unmanned aerial vehicles, mobile robot navigation, mapping and localization, field robotics, robot learning, manipulation, human-robot interaction, and design and prototyping. The common emphasis is on experimental testing and validation of new ideas and methodologies, and this collection of contributions represents the latest in experimental research in robotics.

ISER is a series of biennial international symposia. Sponsored by the International Foundation of Robotics Research (IFRR), the goal of ISER symposia is providing a single-track forum for research in robotics that focuses on novelty of theoretical contributions validated by experimental results. Taking advantage of the small group setting, ISER 2018, for the first time, paired a discussion session with each regular session to better facilitate the exchange of ideas related to the session theme.

ISER 2018 also featured a pre-symposium workshop for outreach to robotics researchers and students in Argentina and surrounding countries and a post-symposium industrial forum to facilitate discussion of robotics applications and technology transfer. In doing so, ISER 2018 continued the unique tradition of ISER as an ambassador of robotics in the host country and region.

We thank all the reviewers for their important work in maintaining the high quality of the symposium. We also thank Franka Emika and the US National Science Foundation for providing financial support to the symposium or its participants. Most of all, we thank all the authors and participants for their technical contributions.

Jing Xiao
Torsten Kröger
Oussama Khatib

Contents

Contacts and Manipulation

Unmanned Aerial Vehicles III

Medical Robotics

Medical Robotics

Torsten Kroeger

All five papers of this session cover advancements for hardware and software components as well as algorithms that will improve or enable medical robotics applications. Niyaz *et al.* presented a motion planning concept validated by physical and simulated experiments, which allowed tentacle-like robots composed of pre-curved telescoping tubes to navigate through tissue. This would enable novel minimally invasive surgery applications requiring robot navigation in highly constraint spaces. The work of Kim *et al.* offered a comparable hardware setup and focused on non-surgical approaches to open narrowed coronary arteries and restore arterial blood flow to the heart tissue. Compared to the tentacle-like robots, the robots were wire-guided. The wires were tracked through fluoroscopy. Chitalia *et al.* introduced a new hardware design for robot joints for robotic pediatric neuroendoscopes – a tendon-driven continuum robot, which was feasible for use in the tip of neuroendoscopes. Wartenberg *et al.* presented a novel design of a six-degree-of-freedom system that could be used for needle insertion tasks inside of MRI scanning system. With the goal of modeling, understanding, and transferring human skills to robot devices, Chaudhari *et al.* discussed parameterizing musculoskeletal models with deep learning methods – an approach that offered the potential for future human-robot rehabilitation applications.

Given the current interest and growth in machine learning algorithms, most discussions during the discussion panel of this session were geared to data-driven modeling and engineering approaches. For instance, the panel and the audience discussed opportunities for using data-driven and model-free approaches for kinematic models of complex systems such as continuum robots.

Following Surgical Trajectories with Concentric Tube Robots via Nearest-Neighbor Graphs

Sherdil Niyaz[1(✉)], Alan Kuntz[2], Oren Salzman[3], Ron Alterovitz[2], and Siddhartha Srinivasa[1]

[1] School of Computer Science and Engineering, University of Washington, Seattle, USA
sniyaz@cs.washington.edu
[2] Department of Computer Science, University of North Carolina at Chapel Hill, Chapel Hill, USA
[3] The Robotics Institute, Carnegie Mellon University, Pittsburgh, USA

1 Introduction

Concentric tube robots, or CTRs, are tentacle-like robots composed of precurved telescoping tubes (Fig. 1a) and are controlled by rotating and translating each individual tube [6]. Their dexterity and small diameter enable minimally-invasive surgery in constrained areas, such as accessing the pituitary gland via the sinuses. Unfortunately, their unintuitive kinematics make manually guiding the tip while also avoiding obstacles with the entire tentacle-like shape extremely difficult [19]. This motivates a need for new user interfaces and planning algorithms.

Although existing planners [19] enable CTRs to reach specified points in task-space, this is often insufficient. For example, cutting a window in the skull during brain surgery (Fig. 1b) requires specifying an entire *path* R for the robot's tip. We have, to the best of our knowledge, implemented the first planner in this domain able to compute a trajectory that closely follows such a task-space path.

Algorithms that are able to follow some task-space path R by computing a constrained path in configuration space ($\mathcal{C}$-space) [2,14,17,22] have two main challenges: (i) some (e.g., a vector-field planner [17]) are fast but myopic, often falling into local minima and (ii) all are oblivious to how closely the generated paths follow R. In contrast, Holladay et al. [9] address both concerns by constructing a layered graph L in $\mathcal{C}$-space and searching it for a path that *directly* minimizes the Fréchet distance [21] from R.

Key to this algorithm's efficiency are the graph's ordered layers (Fig. 2a), each consisting of multiple Inverse Kinematics (IK) solutions for a specific point on R. Directed edges connect each configuration to others in the same and adjacent layers, ensuring monotonic progress. Unfortunately, this requires any path through

---

Electronic supplementary material The online version of this chapter (https://doi.org/10.1007/978-3-030-33950-0_1) contains supplementary material, which is available to authorized users.

J. Xiao et al. (Eds.): ISER 2018, SPAR 11, pp. 3–13, 2020.
https://doi.org/10.1007/978-3-030-33950-0_1

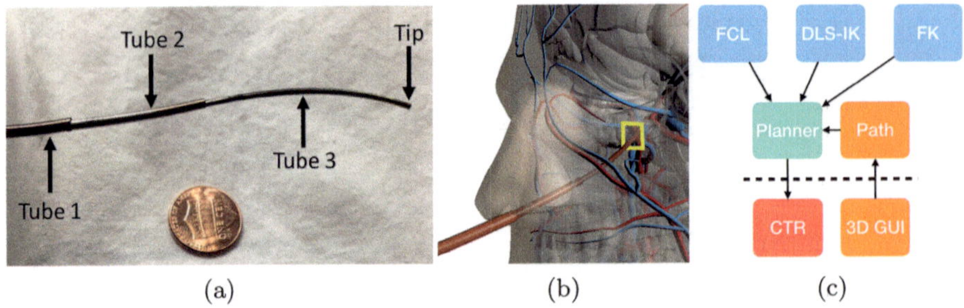

Fig. 1. (a) CTR with coin for scale. (b) Depiction of a CTR deployed via the sinus. A reference path R, used to cut a window in the skull with the CTR's tip, is depicted in yellow. (c) An example of a surgical system that uses our planner as a core component.

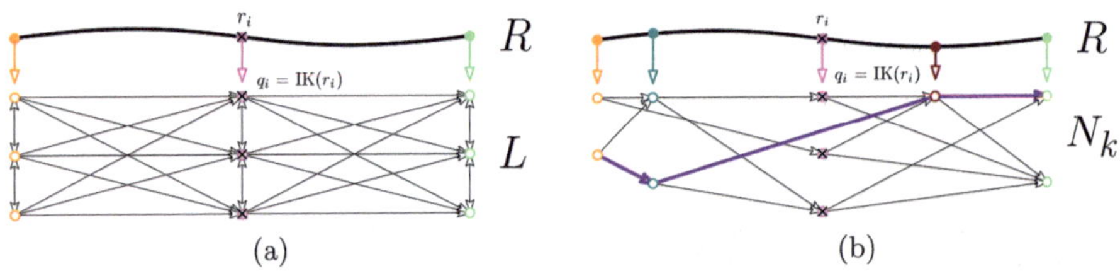

Fig. 2. (a) Layered graph L and (b) nearest-neighbor graph N_k constructed in $\mathcal{C}$-space to follow a reference path R in task-space. Note that a layer in L is depicted as a vertical stack of nodes. If reaching r_i is impossible, all paths in L become blocked. The structure of N_k, however, can bypass such blockages.

L to touch R at *every* layer, leading to failure in constrained environments such as ours. For example, a layer mapping to a point $r_i \in R$ made unreachable due to collisions causes no collision-free path to exist in the graph. The CTR's kinematics make it difficult for a surgeon specifying a path to perfectly distinguish feasible locations in task space, accentuating this failure mode.

Our key insight is to generalize this algorithm by searching over a *nearest-neighbor* graph, with less restrictive connections that allow deviation from infeasible portions of R (Fig. 2b). This makes planning in constrained anatomical environments possible, and thus enables our key contribution—a system implemented on a CTR, presented with a set of simulated and physical experiments. The former reveal that our planner produces higher-quality solutions more reliably than alternative algorithms, while the latter inform algorithmic next-steps necessary to bridge the simulation and real-robot gap.

2 Technical Approach

2.1 Notation and Definitions

The Fréchet distance is a metric of path similarity that has been extensively studied in computational geometry, with varied applications such as speech [12] and handwriting [18] recognition. It can be explained intuitively via an analogy where a dog on a leash traverses one path while its owner traverses another, each with independent speed parameterizations α and β. In this case, the Fréchet distance is the shortest leash length required for the two to stay connected, assuming the dog and its owner are selecting optimal values of α and β.

Motion plans for surgical tasks, such as tissue manipulation, must respect the physician's intent by (i) minimizing the tip's deviation from the specified path at each point in time and (ii) following the "flow" of the path by ensuring that consecutive points along the two paths are traversed in the same order. Because the Fréchet distance captures both objectives, it provides an excellent optimization criteria for our domain.

Our planner searches for these motion plans in $\mathcal{C}$-space, the set of all possible configurations of the robot. Each configuration is a d-dimensional point that uniquely defines the robot's shape. Similarly, the task-space of the robot is defined as the space of all possible end-effector positions. We use the Forward Kinematics (FK) operator to map points in $\mathcal{C}$-space to points in task-space, and use the Inverse Kinematics (IK) operator to perform the reverse mapping.

For a CTR, a configuration must describe the translation ℓ_i and rotation θ_i of each individual tube. Thus, for a CTR comprised of k tubes (typically three or four), each configuration is a $2k$-length tuple $q = (\ell_1, \theta_1, \ldots, \ell_k, \theta_k)$. We define the robot's task-space as the $\mathbb{R}^3$ location of its tip, which is sufficient for many procedures such as cutting with heat or a laser. However, we note that our planner can easily be used with other definitions of task-space that also account for the tips's orientation.

In our setting, the physician specifies a reference path R in task-space as a sequence of $\mathbb{R}^3$ waypoints. Let Γ denote the set of collision-free $\mathcal{C}$-space paths and $\mathcal{F} : \mathbb{R}^3 \times \mathbb{R}^3 \rightarrow \mathbb{R}$ denote the discrete[1] Fréchet distance between two task-space paths. Our problem now calls for computing[2]

$$\arg\min_{\gamma \in \Gamma} \mathcal{F}(\mathrm{FK}(\gamma), R).$$

2.2 Algorithmic Approach

Recall that Holladay et al. [9] construct a layered graph L in $\mathcal{C}$-space. This is done by evenly sampling waypoints along the reference path R, and computing a "layer" of distinct IK solutions for each one. Each of these configurations has out-edges to all configurations in both the same and next immediate layer. Following Har-Peled and Raichel [8], the cross-product graph $\Phi = L \times R^3$ is

[1] Computing the continuous Fréchet distance, is notoriously difficult [16]. Thus, we use the *discrete* variant, easily computed using dynamic programming [5]. Here, the "leash" between points on the two paths is computed only for a discrete set of points and serves as an approximation of the (continous) Fréchet distance.

[2] By a slight abuse of notation we use FK to map both points as well as paths in $\mathcal{C}$-space to points and paths in task space, respectively.

[3] By a slight abuse of notation we treat R both as the discrete reference path as well as the one-dimensional graph defined by this sequence of waypoints.

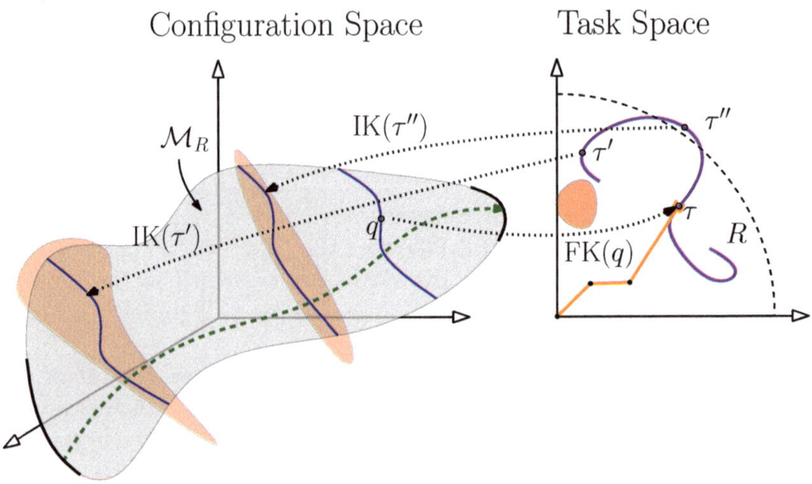

Fig. 3. A reference path R (purple, right) in the task-space of a robot (yellow, right) induces a manifold $\mathcal{M}_R$ of IK solutions in $\mathcal{C}$-space (grey, left). Following R via a layered graph requires traversing $\mathcal{M}_R$ from one end to the other, such as by taking the path in green. However, even if R is not directly in collision with some obstacle (red, right), its presence may cause a point $\tau' \in R$ to map to a self-manifold [3] IK(τ') that is completely in collision (red, left) given the kinematics of the robot. These in-collision self-manifolds require deviating from $\mathcal{M}_R$ to approximate R, and make traversal via the layered graph impossible.

then constructed and searched for a minimal-bottleneck path, which induces the path $\xi \in L$ such that FK(ξ) minimizes the discrete Fréchet distance with R.

Algorithmic Enhancement: Our early experiments, however, revealed that this planner often fails in surgical scenarios. In these constrained environments, some sampled waypoint $r_i \in R$ may be unreachable, often due to a combination of collisions and robot kinematics (Fig. 3). Even the existence of *one* such waypoint prevents the algorithm from finding any solutions, as all paths are blocked by a layer of configurations in collision (Fig. 2a). To prevent this, we construct a nearest-neighbor graph N_k by uniformly sampling IK solutions from R and connecting each to its k-nearest neighbors (NN) in $\mathcal{C}$-space. This less-rigid structure allows the planner to avoid IK solutions sampled from infeasible waypoints (Fig. 2b).

Specifically, each configuration q is connected to its k-NN in $Q_s(q)$, the set containing all $q_s \in N_k$ such that FK(q_s) lies after FK(q) along R. This ensures that any path through N_k monotonically follows R. As in [9], we generate the path $\xi \in N_k$ that minimizes the Fréchet distance with R by searching the cross-product graph $\Phi = N_k \times R$.

Densification: This graph-based approach allows solutions to be improved on-demand by sampling additional IK solutions from R and densifying N_k. To connect each newly-sampled solution q to N_k, we add out-edges from q to its

k-NN in $Q_s(q)$. In addition, we add in-edges to q from its k-NN in $Q_p(q)$, the set containing all $q_p \in N_k$ such that $\mathrm{FK}(q_p)$ lies before $\mathrm{FK}(q)$ along R. These additional connections increase the set of paths the planner can search over, while still constraining it to monotonically follow R.

Implementation details: We collision check edges lazily [4,7], using LPA* [11] for efficient replanning. Our CTR has three tubes, each of which can be rotated and translated, inducing a 6-dimensional $\mathcal{C}$-space. We use the damped least squares (DLS) IK algorithm [20], which has previously been used with the CTR [19]. We represent obstacles as polygonal meshes, and use FCL [13] for collision checks. We define the distance between two configurations q_1 and q_2 as:

$$\sum_k |\ell_{k1} - \ell_{k2}| + \rho(\theta_{k1}, \theta_{k2})$$

where ρ denotes the $SO(2)$ difference between two angles. We note that the choice of distance metric can drastically affect the algorithm's performance [1].

3 Experiments and Results

We consider four surgically-motivated reference paths in our experimental evaluation (Fig. 4), all within a human skull base. The first (i) moves the CTR's tip in a straight line towards the back of the sinuses, ending near a region where the second (ii) traces a trapezoid. These two paths are inspired by a clinical scenario where the CTR is deployed via the first path, and then uses the second to cut a window allowing deeper access into the skull. We also test (iii) a different version of the second path, as well as (iv) a W-shaped path that could be used for wound irrigation. While not in collision with the skull, these last two paths reside in constrained areas and thus exhibit the phenomenon depicted in Fig. 3.

3.1 Simulation Comparison

We evaluate our algorithm, which we term NN Graph, in simulation and compare its performance with the following alternatives: (a) VFP which follows a vector-field along R by using the Jacobian to integrate a path in $\mathcal{C}$-space [17], (b) GreedyIK which adaptively samples a set of ordered IK solutions from R

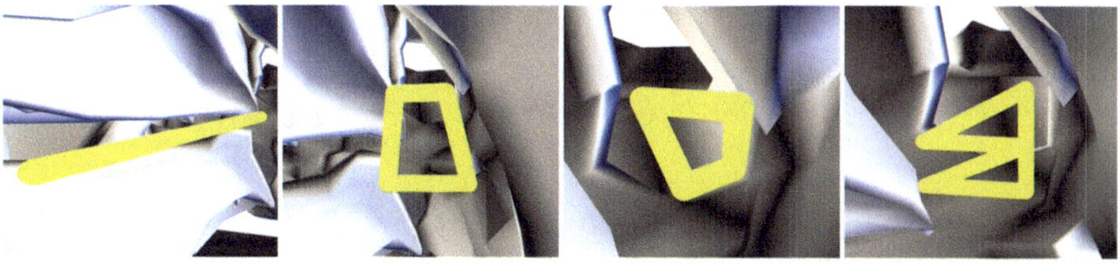

Fig. 4. From left to right, reference paths (i) through (iv) used in our experiments.

and attempts to interpolate a collision-free path in $\mathcal{C}$-space between them [17], (c) CBiRRT which enables planning on constraint manifolds [2] and (d) Layered Graph—the approach presented by Holladay et al. [9]. It is worth noting that planners (a) and (b) are myopic but efficient and that all but (d) were not designed to minimize the Fréchet distance from a reference path.

Simple Paths: We first compare the planners using paths (i) and (ii), both without regions made unreachable by the collision geometry or CTR kinematics. We report the Fréchet distances of the produced trajectories from R for a wide set of parameters (Fig. 5) and average results over ten random seeds.

We see that for both paths, all planners successfully produce solutions. When using path (i), a straight line, performance for all planners remains fairly stable as parameters vary. However, both paths reveal weaknesses in VFP and GreedyIK. With the former, optimal performance on path (ii) is only achieved under two sets of parameters. Similarly, GreedyIK produces results of markedly lower quality than NN Graph across most parameters. CBiRRT, meanwhile, is used to pin the CTR's tip within a tolerance of R and succeeds when this tolerance is made sufficiently tight. This is especially evident on path (i), the straight line.

Constrained Paths: We also compare the planners using paths (iii) and (iv), as seen in Fig. 6. Due to being located in constrained areas, these paths contain small regions infeasible for the CTR to follow perfectly. We note that these paths were not hand-engineered adversarially, but were generated by a human familiar with the CTR and real-world surgical procedures.

Due to the infeasible regions, Layered Graph was unable to find any solutions with either path, confirming the failure mode we observed in our early experiments. While CBiRRT still constrains the CTR's tip to lie within a tolerance of R, it lacks any notion of path following, causing it to violate the flow of these more constrained paths. This oblivious nature also hinders GreedyIK, causing it to display unstable behavior where sampling *more* IK solutions can cause a sharp drop in solution quality.

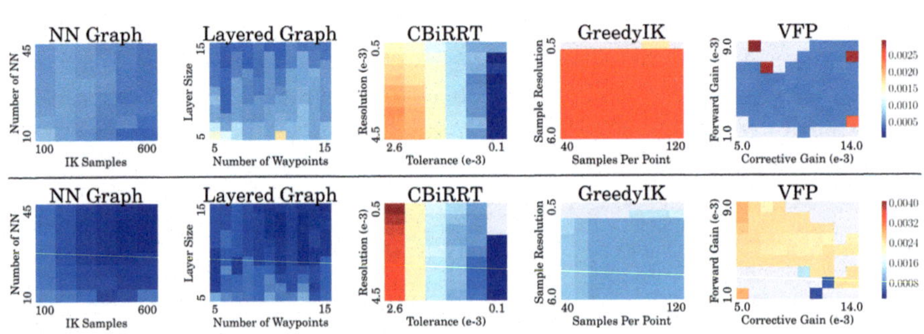

Fig. 5. Fréchet distances over a range of parameters for path (i) at top and path (ii) at bottom. A grey square indicates the algorithm failed to find a solution with that combination of parameters.

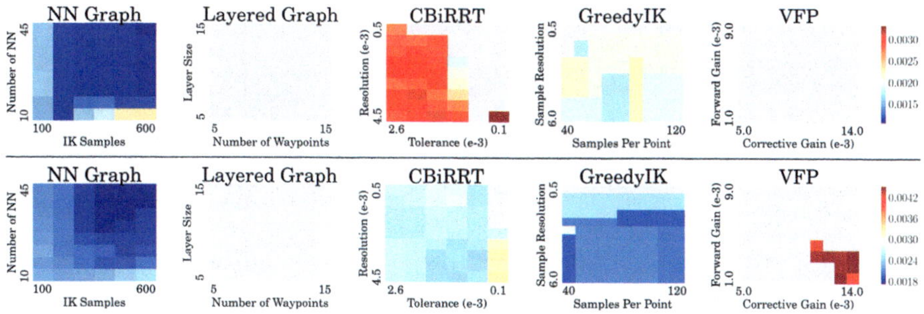

Fig. 6. Fréchet distances for path (iii) at top and path (iv) at bottom. We note that Layered Graph failed on both paths, and that VFP failed on path (iii).

In our experiments with all four paths, NN Graph displayed stable behavior under parameter variation and was the only planner able to consistently produce high-quality solutions. In fact, for each of the four other planners, at least one path led to either very low quality solutions or no solutions being found.

Additional Comparisons: We use our results from path (iii) to perform additional comparisons between planners. In the first of these, we correlate Fréchet distance and runtime for each planner by averaging results across all parameters over ten-second windows (Fig. 7a). This confirms that GreedyIK produces solutions quickly due to its simple nature. However, our planner outperforms the others, which display fluctuating behavior over time.

In addition to evaluating our planner over a range of parameters, we are interested in how a *single* set of parameters generalizes across a range of reference paths. We compare against GreedyIK using its optimal parameters, and select a set for NN Graph that lead to similar runtime on path (iii). Generalization is evaluated by adding independent Gaussian noise to the corners defining this path and recording planner success rates (each taken over 20 trials) under increasing levels of noise (Fig. 7b). We see that as the level of noise increases, our planner has a higher success rate than GreedyIK.

3.2 Algorithmic Behavior

Runtime Analysis: A decomposition of NN Graph's runtime on path (iii) as we linearly scale both the number of nearest-neighbors and IK solutions sampled is shown in Fig. 7c. FK, used in constructing the nodes of the cross-product graph Φ [8,9], is notably the most expensive component. This cost arises from modeling complex interactions between tubes, which requires solving a series of differential equations rather than the typical matrix multiplication.

Densification: We also explore the ability of densification to improve existing NN Graph solutions. This provides our planner an advantage over alternatives, as it enables a physician to improve a solution online until they are satisfied. Our

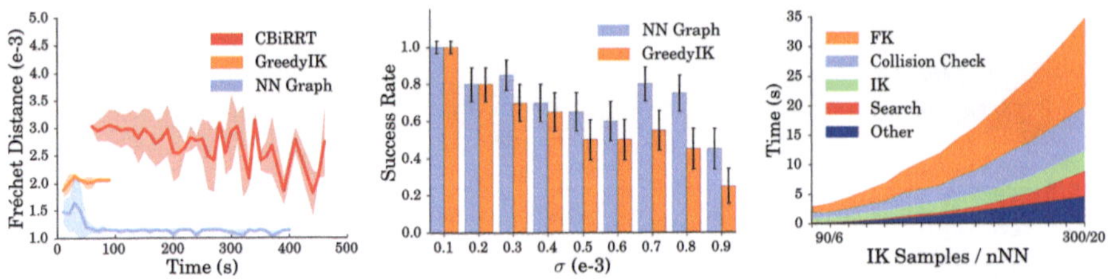

(a) Fréchet vs. Time (b) Robustness to Noise (c) Runtime Components

Fig. 7. (a) Fréchet distance as a function of the runtime for each planner. (b) Success rates for NN Graph and GreedyIK as we inject more noise into R. (c) A decomposition of the runtime of our NN Graph. All results use path (iii)

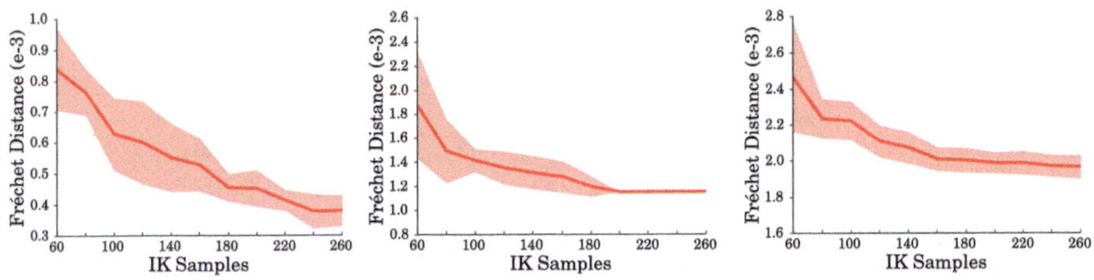

(a) Path (ii) (b) Path (iii) (c) Path (iv)

Fig. 8. Densification results.

trials use 20 nearest-neighbors and augment N_k with an additional 20 configurations per densification iteration. We report the Fréchet distances of solutions as the graph increases in size.

Densification provides very little improvement with path (i), likely because this simple straight-line path makes the initial solution already close to optimal. The other three paths display notable improvement as the size of N_k increases (Fig. 8). However, paths (iii) and (iv) both exhibit diminishing returns past a given iteration.

3.3 Physical Experiments

We evaluate our algorithm on the real robot, testing all four paths inside a 3D-printed model of the skull base anatomy. Parameters are selected that create high-quality solutions in simulation. To measure Fréchet distance, we use a magnetic tracking system to localize the tip during execution. Because this system is noisy, we execute each trajectory ten times and average the results.

Imperfections in our kinematic model (discussed in the next section) prevent us from scoring path (iii) due to collisions. While this path was not executed successfully, the other three were. Paths (iv), (ii), and (i) were each executed with average real Fréchet errors of 0.00806, 0.00747, and 0.00744 m respectively. These are all significantly higher than the respective simulation errors of 0.00211,

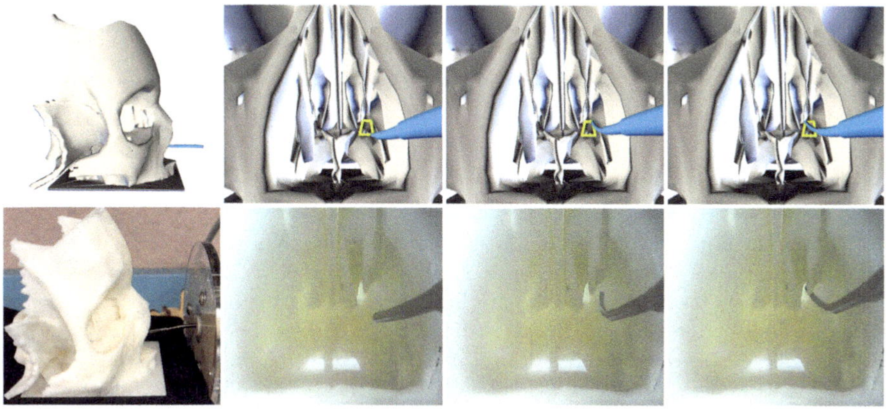

Fig. 9. Simulated (top) and real (bottom) execution of path (ii). The reference path is colored yellow in simulation.

0.00031, and 0.00034 m. This discrepancy arises primarily from the inaccuracies in our kinematic model. A comparison of simulated and real execution using path (ii) is shown in Fig. 9.

4 Experimental Insights and Future Work

Our experiments support our use of a nearest-neighbor graph and demonstrate the advantages of our planner over alternatives. They also reveal previously unconsidered domain-specific challenges, which we can only now begin to address. We propose solutions to each of these, which we will explore in future work.

Our planner's high runtime is mainly due to our robot's expensive FK (Fig. 7c), motivating a paradigm shift in planner design for all robots that incur high cost in this operation. To address this, we propose constructing the cross-product graph Φ implicitly, which avoids unnecessarily invoking FK to create nodes never expanded during the search.

While we demonstrate a successful planner, our physical experiments reveal the kinematic model has now become the limiting factor. Although this model is state-of-the-art [15], it fails to perfectly model certain physical phenomenon, such as friction between tubes. This causes the shape of the robot in simulation to only approximate reality. Naturally, improving this model is a solution. We also propose taking its uncertainty into account via a bicriteria optimization problem of (i) minimizing Fréchet and (ii) maximizing the distance of the robot from obstacles to prevent collisions. This has the added benefit of mitigating error in the collision geometry, constructed from a preoperative volumetric scan [10].

Additionally, successful execution of path (ii) on the real CTR required manually tuning the insertion point of the robot into the skull. The kinematics of the robot induce a trumpet-shaped workspace, leading to greater range of motion farther from this insertion point. One of the insertion points was too close to the reference path, placing it outside the robot's workspace and causing a planning failure. Likewise, the successful insertion point placed the shaft of the robot in a less-constrained area, preventing a collision that otherwise occurred due to modeling error. Because it proved so consequential, we propose incorporating this selection of insertion point as a planning subproblem.

Acknowledgments. We thank Bob Webster and his group at Vanderbilt University for numerous discussions on CTRs and for creating the CTR used here. We thank Rachel Holladay for her invaluable insight in discussing her previous work. This work was (partially) funded by the National Institute of Health R01 (#R01EB019335), National Science Foundation CPS (#1544797), National Science Foundation NRI (#1637748), National Science Foundation RI Award 1149965, the Office of Naval Research, the RCTA, Amazon, and Honda.

References

1. Atias, A., Solovey, K., Salzman, O., Halperin, D.: Effective metrics for multi-robot motion-planning. CoRR, abs/1705.10300 (2017)
2. Berenson, D., Srinivasa, S., Ferguson, D., Kuffner, J.: Manipulation planning on constraint manifolds. In: ICRA, pp. 625–632 (2009)
3. Burdick, J.: On the inverse kinematics of redundant manipulators: characterization of the self-motion manifolds. In: ICRA, pp. 264–270 (1989)
4. Dellin, C., Srinivasa, S.: A unifying formalism for shortest path problems with expensive edge evaluations via lazy best-first search over paths with edge selectors. In: ICAPS, pp. 459–467 (2016)
5. Eiter, T., Mannila, H.: Computing discrete Fréchet distance. Technical report, Citeseer (1994)
6. Gilbert, H., Rucker, D., Webster III, R.J.: Concentric tube robots: the state of the art and future directions. In: ISRR, pp. 253–269 (2013)
7. Haghtalab, N., Mackenzie, S., Procaccia, A., Salzman, O., Srinivasa, S.: The provable virtue of laziness in motion planning. In: ICAPS, pp. 106–113 (2018)
8. Har-Peled, S., Raichel, B.: The Frećhet distance revisited and extended. TALG **10**(1), 3 (2014)
9. Holladay, R., Salzman, O., Srinivasa, S.: Minimizing task space Frećhet error via efficient incremental graph search. CoRR, abs/1710.06738 (2017)
10. Johnson, H., McCormick, M., Ibanez, L.: The ITK Software Guide Book 1: Introduction and Development Guidelines, vol. 1. Kitware Inc., New York (2015)
11. Koenig, S., Likhachev, M., Furcy, D.: Lifelong planning A*. Artif. Intell. **155**(1–2), 93–146 (2004)
12. Kwong, S., He, Q.H., Man, K.F., Tang, K.S., Chau, C.W.: Parallel genetic-based hybrid pattern matching algorithm for isolated word recognition. Int. J. Pattern Recogn. Artif. Intell. **12**(05), 573–594 (1998)
13. Pan, J., Chitta, S., Manocha, D.: FCL: a general purpose library for collision and proximity queries. In: ICRA, pp. 3859–3866 (2012)
14. ROS Industrial. Descartes (2015). Accessed 20 Apr 2018
15. Rucker, D.: The mechanics of continuum robots: model-based sensing and control. Ph.D. thesis, Vanderbilt University (2011)

16. Solovey, K., Halperin, D.: Sampling-based bottleneck pathfinding with applications to Fréchet matching. In: European Symposium on Algorithms, ESA, pp. 76:1–76:16 (2016)
17. Srinivasa, S., Johnson, A., Lee, G., Koval, M., Choudhury, S., King, J., Dellin, C., Harding, M., Butterworth, D., Velagapudi, P., Thackston, A.: A system for multi-step mobile manipulation: architecture, algorithms, and experiments. In: ISER, pp. 254–265 (2016)
18. Sriraghavendra, E., Karthik, K., Bhattacharyya, C.: Fréchet distance based approach for searching online handwritten documents. In: Document Analysis and Recognition, vol. 1, pp. 461–465. IEEE (2007)
19. Torres, L., Kuntz, A., Gilbert, H., Swaney, P., Hendrick, R., Webster III, R.J., Alterovitz, R.: A motion planning approach to automatic obstacle avoidance during concentric tube robot teleoperation. In: ICRA, pp. 2361–2367 (2015)
20. Wampler, C.: Manipulator inverse kinematic solutions based on vector formulations and damped least-squares methods. IEEE Trans. Syst. Man Cybern. **16**, 93–101 (1986)
21. Wylie, T., et al.: The discrete Fréchet distance with applications. Ph.D. thesis, Montana State University-Bozeman, College of Engineering (2013)
22. Yao, Z., Gupta, K.: Path planning with general end-effector constraints: using task space to guide configuration space search. In: IROS, pp. 1875–1880 (2005)

An Experimental Validation of Behavior-Based Motions for Robotic Coronary Guidewire Crossing Techniques

Young-Ho Kim(✉), Ankur Kapoor, Rodolfo Finocchi, and Erin Girard

Siemens Healthineers, Medical Imaging Technologies, Princeton, NJ, USA
{young-ho.kim,ankur.kapoor,rodolfo.finocchi,
erin.girard}@siemens-healthineers.com

Abstract. Percutaneous Coronary Intervention (PCI) is a non-surgical approach used to open narrowed coronary arteries and restore arterial blood flow to heart tissue. During a PCI procedure, the clinician uses X-ray fluoroscopy to visualize and guide the catheter, guidewire, and other devices (e.g. angioplasty balloons and stents). Robot-assisted PCI could potentially reduce the radiation exposure of operators and provide a more ergonomic workflow. However, modeling and controlling of the guidewire remains challenging because of the interplay of guidewire motions, the tip properties (e.g., loads, coating), and the local cross-sectional area of the vessel lumen (e.g, stenosis) and results in a highly non-linear system. Thus, robot-assisted PCI devices are still passively controlled by human operators at the cockpit. In this paper, we introduce methods to generate distal guidewire motions that take advantage of the fast response of a robotic system and which may be difficult to generate by a human hand. The fundamental motions that a robot can use to control the movement and direction of the guidewire are rotation and pushing/retracting, from the proximal end of the guidewire outside the insertion point on the patient's body. We begin by investigating combinations of these fundamental motions under structured environmental settings and conduct a systematic empirical comparison of task completion time for a given setting. We then demonstrate improved dynamic behavior motions for a soft guidewire, which shows a promising speed-up by 33% and 44% for two difficult stenosis cases.

Keywords: Robot-assisted PCI · Guidewire navigation · Behavior-based control · High-speed dynamic motion · Medical robots

---

Electronic supplementary material The online version of this chapter (https://doi.org/10.1007/978-3-030-33950-0_2) contains supplementary material, which is available to authorized users.

J. Xiao et al. (Eds.): ISER 2018, SPAR 11, pp. 14–23, 2020.
https://doi.org/10.1007/978-3-030-33950-0_2

1 Introduction

Percutaneous Coronary Intervention (PCI) [1] is a minimally invasive approach used to open coronary arteries that have been narrowed or blocked by atherosclerosis. During a PCI procedure, X-ray fluoroscopy is the primary imaging modality used to visualize and guide the catheter, guidewire, and other devices (*e.g.* angioplasty balloons and stents). The clinician is continuously exposed to X-rays while manipulating a guidewire through the catheter, into the coronary, and across the stenosis or blockage. Moreover, in order to reduce the degree of longterm radiation exposure, the interventional cardiologist wears a heavy lead apron during the entire procedure. To alleviate these workflow conditions and allow the clinician to sit on an ergonomic console, robots have been introduced [2] to remotely manipulate the guidewire and the catheter. However, they are still passively controlled by human operators at the cockpit. Modeling and controlling of the guidewire remains challenging because of the interplay of guidewire motions, the tip properties (*e.g.*, loads, coating), and the local cross-sectional area of the vessel lumen (*e.g*, stenosis) and results in a highly non-linear system. For the task of controlling guidewires to cross stenosed lesions, referred to as *wiring a vessel*, it is ideal to use as soft a wire as possible. A softer wire is likely to cause less damage to the vessel lumen and is less likely to result in adverse events, such a vessel perforations. During the procedure, clinicians often change to stiffer wires to cross a difficult stenosis and then switch back to a softer wire. A system that can combine the benefits of environmental modeling and *optimal* guidewire manipulation with human operator control could potentially reduce task completion time and allow the clinician to cross the stenosis without switching wires.

In this paper, we introduce methods to generate distal guidewire motions that take advantage of the fast response of a robotic system and which may be difficult to generate by a human hand. The fundamental motions that a robot can use to control the movement and direction of the guidewire are limited to rotation about its axis and pushing/retracting at the proximal end of the guidewire. We begin by investigating combinations of these fundamental motions under structured environmental settings and conduct a systematic empirical comparison of task completion time for a given setting. We then demonstrate improved dynamic behavior motions for soft guidewires on difficult stenosis cases.

1.1 Related Work

A detailed assessment of the mechanical properties of wires used in PCI is described by Schröder [3]. Guidance on various crossing techniques, wire selection and manipulation based on environmental conditions for the manual operator are described in [1,4,5]. Prior efforts at kinematics and dynamics modeling, including simulating wires [6–11], were directed at developing training simulators for physicians. In this work we focus on dynamic behavior motions and their interactions with the environment to assist the operator in crossing a stenosis with the softest wire possible.

Behavior motions are well known methods to model complex systems and have merits of simplicity to achieve a specific task [12]. There exist many applications using behavior motions to overcome the difficulties of modeling and planning in different settings [13–15]. In our prior work, we also employed a set of behavior motions to show a possible usage for controlling guidewires in tortuous vessels [16]. In this paper, we focus on how to combine behavior motions under structured environmental interactions for the task of *wiring a vessel*.

Currently, several commercial robotic-assisted catheter systems (*e.g.*, Magellan of Hansen Medical, CorPath of Corindus Vascular Robotics) are available. The common features existing robotic-assisted systems offer are positional control of devices with reduced radiation exposure to the primary operator (but not necessarily the laboratory staff). However, they are still passively controlled by human operators at the cockpit [2]. Recently, Madder et al. [17] proposed automatic wire retraction motions with a rotation of 180° for a coronary artery bifurcation. However, it is specifically tailored for crossing coronary bifurcations and is of limited use as a general purpose crossing technique. Furthermore, for a general purpose crossing system, it is essential to consider the relationships between wire properties, wire motions, and environments. Here we empirically establish behavior motions for crossing a range of stenoses types for different guidewires.

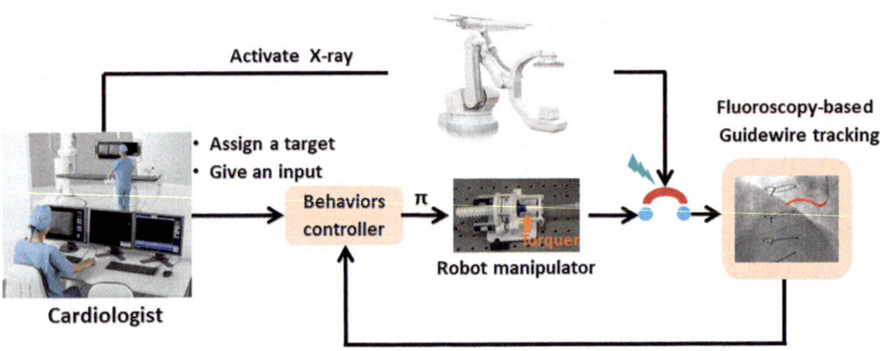

Fig. 1. An overview of the proposed robotic-assisted system.

1.2 Preliminaries and Problem Description

Figure 1 shows an overview of a PCI workflow using our robotic system. The operator identifies the target stenosis that is to be crossed, which activates our proposed behavior controller for the guidewire. The behavior controller provides outputs to the positional controller of the robot manipulator. The user can also provide additional input to our controller based on the visual cues observed in the image.

The proximal end of the wire is held by a small plastic device, the *torquer*, which in turn is rigidly attached to the robot. At the torquer's attachment point to the robot, the wire's state, $\boldsymbol{x}_k \equiv [d, \phi]$, has a translation, d and rotation, ϕ. At the tip of the wire, the state, $\boldsymbol{x}_t$ is defined by its position and orientation in $SE(3)$. Let the physical parameters of the wire be defined as a tuple $\boldsymbol{p} = \langle p_d, p_s, p_c, p_l \rangle$ representing the wire diameter, the tip shape, the coating type, and the loads, respectively. Let tuple, $\boldsymbol{e} \equiv \langle \boldsymbol{r}, l, \boldsymbol{\alpha}, \theta \rangle \in \mathscr{E}$ represent the spatial area of interest to cross, from the set of environments, $\mathscr{E}$. We define the environment model in detail in Sect. 2.1. Then our manipulation system can be treated as a transition function, $f(\boldsymbol{\pi}; \boldsymbol{p}, \boldsymbol{e})$ that uses a control, $\boldsymbol{\pi} \in \Pi$ to transition from state $\boldsymbol{x}(t)$ to another state $\boldsymbol{x}(t+1)$ given $\boldsymbol{p}$ and $\boldsymbol{e}$:

$$f(\boldsymbol{\pi}; \boldsymbol{p}, \boldsymbol{e}) : \boldsymbol{x}(t) \rightarrow \boldsymbol{x}(t+1), \tag{1}$$

where $\boldsymbol{x}$ consists of states $\boldsymbol{x}_k$ and $\boldsymbol{x}_t$. We define motion models in Sect. 2.2

Problem: Given $\boldsymbol{p}$ and $\boldsymbol{e}$, manipulate the wire to cross a stenosis, starting from location $\boldsymbol{q}_s$ to the end location $\boldsymbol{q}_e$ of the stenosis, making all state transitions via control of the torquer, finding a motion that minimizes cost C, defined as the completion time to cross target area. Then, the problem follows:

$$\pi^* = \underset{\pi \in \Pi}{\arg\min}\ C(\pi | \boldsymbol{p}, \boldsymbol{e}). \tag{2}$$

2 Technical Approach

To understand the interactions between motions and environment, we model a small set of highly stenosed structures by varying the parameters of tuple, $\boldsymbol{e}$. Without loss of generality, we can assume a small section of the vessel to have a cylindrical cross section and a cylindrical lumen. A vessel can then be composed by joining multiple such cross-sections end-to-end. Next we propose a small set of motions designed to cross these structured environments. We take a data driven approach to solve Eq. (2) because the system and transition function in Eq. (1) is highly non-linear. We present a formal model for the environment below followed by that of motion behaviors.

2.1 Environment Modeling

A representation of the environment for a short segment of an occluded vessel is shown in Fig. 2. The wire entry is assumed to be from the left. As previously mentioned, the tuple, $\boldsymbol{e} \equiv \langle \boldsymbol{r}, l, \boldsymbol{\alpha}, \theta \rangle$ defines the environment. The length of the longitudinal section of the open-ended cylinder at the center of the hole gives $l = \beta_1$, the length of the true lumen. The entry angles made by the true lumen with the vessel wall along any longitudinal cross section are θ_1 and θ_2. The lengths of the same longitudinal section along the vessel wall are β_2 and β_3. Since the generated motions can symmetrically rotate and sweep on the cross-sectional area, we find the critical angle with which to rotate the guidewire,

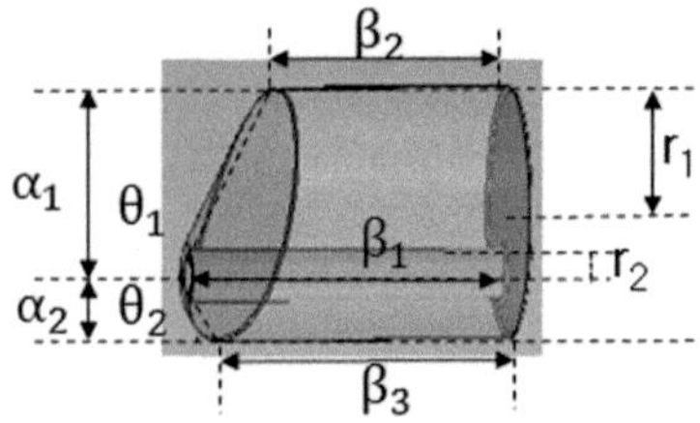

Fig. 2. Close up of an exemplary vessel: The progression of the guidewire is from the left to the right. The open-ended cylinder is represented by the diameter of the through holes and the slope of the conic face surrounding the hole.

which is $\theta = \arg\min_{\forall(\beta_2,\beta_3)} \min(\theta_1, \theta_2)$. The vessel radius and true lumen radius are r_1 and r_2, respectively. We define the relative location of the true lumen, α as α_2/α_1 if $\alpha_1 \geq \alpha_2$, and α_1/α_2 otherwise.

2.2 Motion Modeling

We design control output, π as a special case of the discrete-time model [18] referred to as "motion behaviors". In our problem, we have two such basic discrete motions (translational, d and rotational, ϕ) each consisting of acceleration and time as (a, t) where d (or ϕ) $= \frac{1}{2} \cdot a \cdot t^2$. For *translational motions*, the robot generates a forward (or backward) motion of i mm with an acceleration of j mm/s^2, whereas, a *rotational motion* comprises a clockwise (or counter-clockwise) rotation of n cycle(s) with an angular acceleration of k rad/s^2. Using these two basic independent functions, we generate eight combinatorial "behavior motions" as follows:

Table 1. Set of permutations for "motion behaviors". *Slow* is designed as one half the velocity and acceleration of *fast* motions. For rotation, numbers in parenthesis are the number of cycles with signs indicating the cycle direction.

Behaviors	M_1	M_2	M_3	M_4	M_5	M_6	M_7	M_8
Translation	slow	slow	slow	slow	fast	fast	fast	fast
Rotation	slow(1,−1)	slow(3,−3)	slow(3,3)	fast(3,3)	slow(1,−1)	fast(3,−3)	fast(3,3)	fast(1,−1)

After experimentation, we have narrowed the possible permutations of behaviors to use one variable for translational motions: acceleration (slow and fast), and three variables for rotational motions: a direction (clockwise, counter-clockwise), a speed (slow and fast), the number of cycles (one cycle, three cycles). In this paper, we use a fixed displacement of 3 mm for translational motions. Linear accelerations of 20 mm/s^2 and 40 mm/s^2 and angular accelerations of 30 rad/s^2 and 60 rad/s^2, were used for *slow* and *fast* motions respectively. We selected eight illustrative motion combinations for experimentation, as shown in

Table 1. From our initial experiments, we observed that certain motions, such as slow clockwise (slow, slow(1,−1)) and slow counter-clockwise (slow, slow(1,1)) rotational motions, gave similar results. Only one of such similar motions were considered in our experiments.

3 Experimental Setup

To validate our motions, we used 3D vessel phantoms constructed using acrylic rods. For environmental settings, we use values of $r_1 = 4$ mm, $r_2 = 1$ mm (75% stenosis) and $l = 10$ mm for all phantoms. Further details of the test environments are illustrated in Fig. 3.

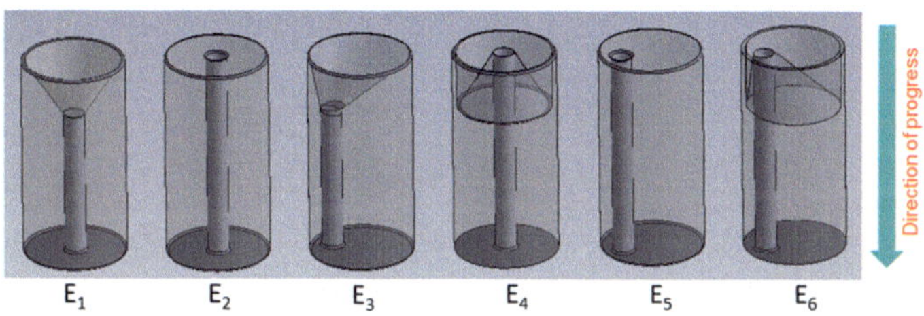

	θ	α
E_1	45°	1
E_2	90°	1
E_3	60°	0.3
E_4	135°	1
E_5	90°	0.3
E_6	135°	0.3

Fig. 3. A set of structured environments representing typical coronary stenosis.

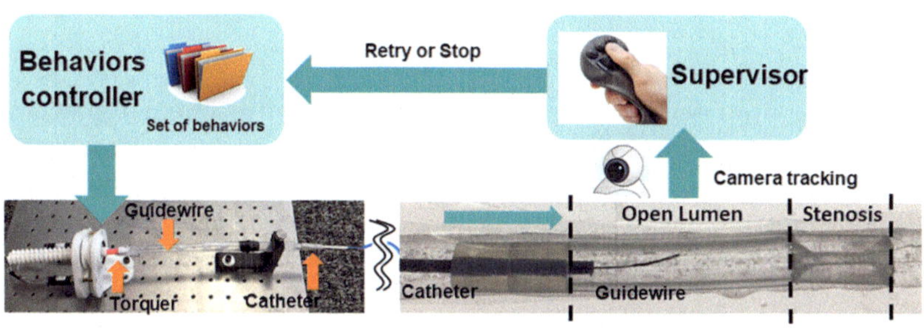

Fig. 4. Experimental setup: The camera provides image feedback of the guidewire tip position in real-time. The user can also advise and enter a command into the controller for collaborative manipulation.

The experimental setup shown in Fig. 4 was built to demonstrate our conceptual idea of *wiring a vessel* during PCI. The robot was fixed to a table along with a camera tracking system to simulate an X-ray system (gray-scale projection imaging). We use two geared motors (3000 rpm, 15 mNm) to provide high-speed motions. We use a JL4 catheter with .014 inch diameter straight coronary guidewires with three tip loads: *soft* (0.4–0.6 gf), *intermediate* (1.5–2.5 gf), and *stiff* (4–6 gf), which is determined by measuring the amount of force required to buckle the distal tip of the guidewires [19]. The distal tip (2 *cm*) has a hydrophilic coating. The guidewire is initially located at the center of the open lumen as shown in Fig. 4. A "motion behavior" is selected and repeated until the wire crosses the occlusion in the phantom. During this process, the user can request a retraction of the wire if no progress is observed or if the guidewire buckles. This event triggers the controller to execute a retraction behavior once before re-executing the given motion behavior. The retraction is a backwards motion of 5 mm at 40 mm/s^2.

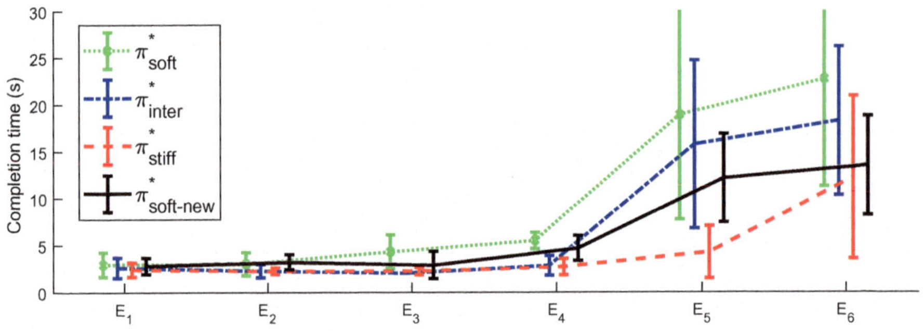

Fig. 5. Completion times for optimal motions for *soft*, *intermediate*, *stiff* wires, and our proposed motion $\pi^*_{soft-new}$ to cross difficult stenoses with a *soft* wire.

4 Results

We tested our proposed motions (M_1 through M_8) with six environmental settings (E_1 through E_6) for 15 trials each. The times reported include retraction time. The complete results are in the supplementary materials due to space constraints; we summarize the key results in this section. As seen from Fig. 5 we can categorize these environments based on completion time T, as easy (E_1, E_2, E_3, E_4; $T < 5s$) and difficult (E_5, E_6; $T > 10s$) tasks. Motion M_5 (fast translation with one-cycle of the slow rotation) gives the best results for easy environments. However, M_4 works well for more difficult tasks because it has more dwell time at the entrance of the stenosis, thus giving the guidewire tip a higher chance of entering the true lumen. Additionally, M_7 and M_8 are the best options for stiffer wires. For easy tasks, slow motions (M_1, M_3, and M_4) do not help to save time. In fact, these motions caused the wires to buckle along the walls, while the high speed motions help to get the wire through.

Our optimal motion solutions for each wire type are shown in Fig. 5. The optimal motion for the difficult cases are: M_4 for E_5 and M_6 for E_6 [π^*_{soft} and π^*_{inter}] and M_6 for E_5 and E_6 [π^*_{stiff}]. It is clear that the *stiff* wire gives better controllability and better results. However, the *stiff* wire is the clinically least preferred option due to safety issues such as dislodgement of plaque, perforation of vessels, or tissue trauma. The ideal situation would be to use the *soft* wire and have the same results as with the *stiff* wire. We seek to experimentally establish a new behavior motion that could reduce completion times for *soft* wires in difficult cases.

We found that M_4 which has slower acceleration provides a better solution for *soft* wires. Using this intuition, we setup a new motion, $\pi^*_{soft-new}$, which has a similar setup to M_4, but provides a steady push with a slower linear acceleration of 10 mm/s^2. This new motion, $\pi^*_{soft-new}$, shows a promising speed-up by 33% for E_5 and 41% for E_6 when compared to π^*_{soft} (Fig. 5).

5 Discussion

Most cardiologists have to control guidewires/catheters repeatedly every day. There is a limit to the continuous generation of stable motions by hand. The best of our knowledge, under stenosis cases, human operators usually use motions with accelerations less than 20 rad/s^2 for rotational motions and 20 mm/s^2 for translational motions. So, we used these observation as our base-line parameters of "slow".

As a long term goal, we are interested in maximizing the robot's performance while ensuring safety issues (which we believe is possible if we use a soft wire). So in this paper, we are not focused on capturing the human operator's motions (*e.g.*,Rafii-Tari et al. [20]) but exploring new simple motions that might not be easily consistently (or even possibly) generated by human hands.

Due to practical constraints, this paper cannot provide a comprehensive review of realistic settings and all possible guidewires: Firstly, we use acrylic stenosis environments for the proof of our concept. In practice, it is hard to generate real stenosis models. So we consider an animal test and a cadaver test for proper tissue interactions in the near future. Secondly, here we only employed general physical properties of guidewires (a straight tip and 2 mm hydrophilic coating). In the future, we will test more varied physical properties such as tip shapes and coating types with various environmental settings.

To our knowledge, the present study is the first to study the effects of high-speed dynamic motions in robotic PCI. The use of such motions could be safe and efficient when integrated with user feedback. Further, such robotic behavior motions provide another set of tools for lesion crossing, delivering motions that a human operator could not. This could be useful in Chronic Total Occlusion (CTO) PCI to support crossing difficult lesions and used in combination with new guidewire technology that applies energy at the distal tip to help cross fibrous or heavily calcified regions.

Behavior-based motions need to be characterized by parameters which may be difficult to estimate for each guidewire and varying environment. However, the relevant workspace of guidewires, defined as the length of stenoses, is very small. Thus, a data-driven approach whereby sub-optimal parameters are updated with repeated motion trials based on previous motion feedback can be used.

Acknowledgement. This feature is based on research, and is not commercially available. Due to regulatory reasons its future availability cannot be guaranteed.

References

1. Kini, A., Sharma, S., Narula, J.: Practical Manual of Interventional Cardiology. Springer, New York (2014)
2. Maor, E., Eleid, M.F., Gulati, R., Lerman, A., Sandhu, G.S.: Current and future use of robotic devices to perform percutaneous coronary interventions: a review. J. Am. Heart Assoc. **6**(7), e006239 (2017)
3. Schröder, J.: The mechanical properties of guidewires. Part I: stiffness and torsional strength. Cardiovasc. Intervent. Radiol. **16**, 43–46 (1993)
4. Liau, C.-S., Ho, S.-G., Chen, S.-N., Yang, L.-F.: A new guidewire technique for difficult cardiac catheterization. Cardiology **97**, 24–28 (2001)
5. Dash, D.: Guidewire crossing techniques in coronary chronic total occlusion intervention: A to Z. Indian Heart J. **68**, 410–420 (2016)
6. Konings, M.K., van de Kraats, E.B., Alderliesten, T., Niessen, W.J.: Analytical guide wire motion algorithm for simulation of endovascular interventions. Med. Biol. Eng. Comput. **41**(6), 689–700 (2003)
7. Luboz, V., Blazewski, R., Gould, D., Bello, F.: Real-time guidewire simulation in complex vascular models. Vis. Comput. **25**(9), 827–834 (2009)
8. Tang, W., Wan, T.R., Gould, D.A., How, T., John, N.W.: A stable and real-time nonlinear elastic approach to simulating guidewire and catheter insertions based on cosserat rod. IEEE Trans. Biomed. Eng. **59**(8), 2211–2218 (2012)
9. Wang, H., Wu, J., Wei, M., Ma, X.: A robust and fast approach to simulating the behavior of guidewire in vascular interventional radiology. Comput. Med. Imag. Graph. **40**, 160–169 (2015)
10. He, B., Xu, S., Ding, A., Zhou, Y.: Analysis of rotation angles and motions of the flexible mechanisms in bifurcated blood vessels. Proc. Inst. Mech. Eng. Part H J. Eng. Med. **231**(8), 747–757 (2017)
11. Sharei, H., Alderliesten, T., van den Dobbelsteen, J.J., Dankelman, J.: Navigation of guidewires and catheters in the body during intervention procedures: a review of computer-based models. J. Med. Imag. **5**(1), 010902 (2018)
12. Shell, D.A., Matarić, M.J.: Behavior-based methods for modeling and structuring control of social robots. In: Sun, R. (ed.) Cognition and Multi-Agent Interaction: From Cognitive Modeling to Social Simulation, Chap. 11. Cambridge University Press, Cambridge (2005)
13. P.C. for Agile Semi-Autonomous Ground Vehicles using Motion Primitives, "Predictive Control for Agile Semi-Autonomous Ground Vehicles using Motion Primitives". In: Proceedings of American Control Conference, Montréal, Canada, June 2012
14. Paranjape, A.A., Meier, K.C., Shi, X., Chung, S.-J., Hutchinson, S.: Motion primitives and 3D path planning for fast flight through a forest. Int. J. Robot. Res. **34**(3), 357–377 (2015)
15. Kim, Y.-H., Shell, D.A.: Using a compliant, unactuated tail to manipulate objects. IEEE Robot. Autom. Lett. **2**(1), 223–230 (2017)
16. Kim, Y.-H., Kapoor, A., Finocchi, R., Girard, E.: Evaluation of high-speed dynamic motions for robotic guidewire crossing techniques. In: Proceedings of Hamlyn Symposium on Medical Robotics, London, United Kingdom, June 2018
17. Madder, R., Lombardi, W., Parikh, M., Kandzari, D., Grantham, J., Rao, S.: Impact of a novel advanced robotic wiring algorithm on time to wire a coronary artery bifurcation in a porcine model. J. Am. Coll. Cardiol. **70**(80) (2017)
18. LaValle, S.M.: Planning Algorithms. Cambridge University Press, New York (2006)

19. Boston Scientific: Guidewire Family of Products (2018). http://www.bostonscientific.com/content/gwc/en-US/products/guidewires/choice-floppy-guide-wire/_jcr_content/maincontent-par/image_1.img.MTC_Guidewire-chart.jpg
20. Rafii-Tari, H., Liu, J., Payne, C.J., Bicknell, C., Yang, G.-Z.: Hierarchical HMM based learning of navigation primitives for cooperative robotic endovascular catheterization. In: Proceedings of Medical Image Computing and Computer-Assisted Intervention (MICCAI), Boston, USA, September 2014

Design and Analysis of a Bidirectional Notch Joint for a Robotic Pediatric Neuroendoscope

Yash Chitalia[1(✉)], Xuefeng Wang[2], Vinh Nguyen[3], Shreyes Melkote[3], Joshua Chern[4], and Jaydev P. Desai[1]

[1] Medical Robotics and Automation (RoboMed) Laboratory, Wallace H. Coulter Department of Biomedical Engineering, Georgia Institute of Technology, Atlanta, GA, USA
{yashc,jaydev}@gatech.edu

[2] Department of Mechanical Engineering, University of Alabama, Tuscaloosa, AL, USA
xwang201@eng.ua.edu

[3] George W. Woodruff School of Mechanical Engineering, Georgia Institute of Technology, Atlanta, GA, USA
{vnguyen43,shreyes.melkote}@gatech.edu

[4] Children's Hospital of Atlanta, Atlanta, GA, USA
joshua.chern@choa.org

Abstract. In this paper, we propose the design of a novel joint particularly for a tendon-driven continuum robot, which is feasible for use in the tip of a robotic neuroendoscope. We detail the functionality of this joint, and by varying geometric parameters of the joint, we analyze its bending capabilities in a simulated environment. Analysis, simulation and experiment show that the joint achieves higher bending flexibility in comparison to other types of joints used in this application, while retaining its stiffness towards external forces. A prototype of the joint has been developed using a superelastic material by milling, in order to validate the claims. The results we obtained are satisfactory and match with the proposed analysis and simulation.

1 Motivation, Problem Statement and Related Work

Pediatric hydrocephalus is a lethal condition [1] occurring more commonly in infants, where a blockage in the flow of cerebrospinal fluid (CSF) in the brain expands the ventricles and imparts pressure on the patient's brain. This condition is a common one in developed countries affecting 0.7 live births over a thousand, with even higher rates in the developing world [2]. Endoscopic third

Disclosure Statement: The development of the robotic pediatric neuroendoscope is based on the steerable guidewire technology. Dr. Jaydev P. Desai and Mr. Yash Chitalia are co-inventors of the steerable guidewire technology and the intent is to commercialize it.

J. Xiao et al. (Eds.): ISER 2018, SPAR 11, pp. 24–33, 2020.
https://doi.org/10.1007/978-3-030-33950-0_3

ventriculostomy (ETV) is increasingly becoming a common procedure used to handle this condition [2]. In this procedure, the clinician makes a small opening in the floor of the third ventricle thus allowing the CSF to flow out of the blocked ventricle, and complete its circuit through the brain normally. ETV tends to be a more attractive option especially in developing countries, due to its success rate without the use of expensive solutions involving shunts. In several cases, the access to a conducive spot on the floor of the third ventricle may not be linear, which requires the need for a flexible robotic tool for this procedure [3]. Due to the tortuous nature of the path to the target region, this endoscopic robot must have at least two joints so as to be able to navigate around sensitive blood vessels and reach a larger workspace inside the ventricle. Furthermore, the tools required for conducting these procedures must pass through the narrow working channels of commonly used neuroendoscopes such as the OI HandyPro endoscope (1.3 mm working channel) or the Minop endoscope from Aesculap (2.2 mm working channel) [4]. Robotic endoscope designs usually include joints that involve rolling or sliding motion between two rigid members [5,6], or exploit the physical compliance of the material to form bending flexural joints [7]. Bending flexural joints are constructed by etching away material from a tubular shaft, thus enabling flexion of the tube essentially creating directional springs [8–11]. Figure 1 shows the design of three such commonly used flexural joints which are often termed *notch joints*, that are actuated using tendons attached to the distal ends of the joints (shown by blue lines in Fig. 1). A bending member is created by the removal of material in each of the notch joints (shown by the dotted lines in Fig. 1 (i)–(iii)), and actuation of the tendon with a force F_t causes this bending member to deflect thereby causing the joint to bend. Due to the absence of moving parts, such as in rolling or sliding joints, notch joints are suitable for miniaturization, making them appropriate for the neuroendoscopes used in the ETV procedures.

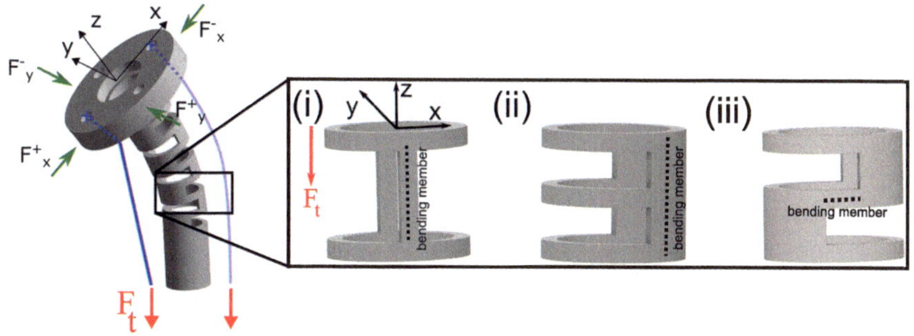

Fig. 1. Bending flexural joints are designed by creating notches into the body of a tubular shaft. The notch geometries allow us to classify these joints into three types, (i) symmetric bidirectional notch joint, (ii) unidirectional asymmetric notch joint, and the (iii) asymmetric bidirectional notch joint.

The bidirectional symmetric notch joint [12–15] (see Fig. 1 (i)) is obtained by removing material symmetrically on either side of the Y-Z plane creating a bending member symmetrically about this plane. This allows the joint to be deflected on both sides of the Y-Z plane. However, due to the bending member running centrally along the Y-Z plane, the moment arm achieved by the actuating force F_t is low, thereby making this joint very stiff. As a result, the authors in [16,17] suggest a unidirectional asymmetric notch design (see Fig. 1 (ii)) which can achieve a higher moment arm due to the bending member being shifted to one side. Note that both the above notch geometries, while being stiff to forces applied transverse to the bending plane (Y-Z plane in Fig. 1), are complaint in the bending plane (X-Z plane in Fig. 1). Therefore, they are susceptible to external forces F_x^+ and F_x^- which may result in undesired deflections. However, unidirectional notch joints cannot resist forces F_x^- by design, while the bidirectional joints can resist these forces by tensioning the appropriate tendons. While the authors in [18] propose the addition of contact-aids to reduce this problem in unidirectional geometries, it cannot be eliminated completely.

In this work, we focus on the design-parameters of a type of a flexural notch joint we call the bidirectional asymmetric notch joint (see Fig. 1 (iii)) [19–22]. In comparison to the unidirectional notch joints [18], this joint has notches on both sides of the Y-Z plane (see Fig. 1 (iii)). These notches are arranged asymmetrically about this plane, unlike symmetric notch joints [11], where material is etched out on both sides of the Y-Z plane in a symmetric manner. We find that the bending flexibility of the bidirectional asymmetric notch joint is much higher than that of a unidirectional asymmetric notch joint, given the same amount of material removal. This is due to the fact that the bending is dominated by the member created between the two notches of this joint. This phenomenon has been demonstrated in Sect. 5. We note that this additional improvement in flexibility does not excessively affect the stiffness of the beam to external forces applied.

We will first generate an Euler beam based bending model for the same, using a Finite Element Analysis (FEA) to guide this model. Finally, we analyze the bending, transverse and axial stiffness of this joint in comparison to the other types of notch joints, since these are the primary forces observed on such joints in a surgical setting.

2 Technical Approach

In this section, we provide a model to estimate the bending stiffness of the bidirectional asymmetric notch joint. A simplified schematic is shown in Fig. 2 (a), where $L = 2d - \mathrm{OD}$ is effective beam length from the left side of the bottom notch to the right side of the top notch, and $l = l_{arm} + (OD - d)$ is the distance from the tendon to the notch. The deformation of the beam, v, due to the tendon tension force, F_t, is governed by the following equation:

$$EIv'' = \frac{F_t}{2}(x + l) \tag{1}$$

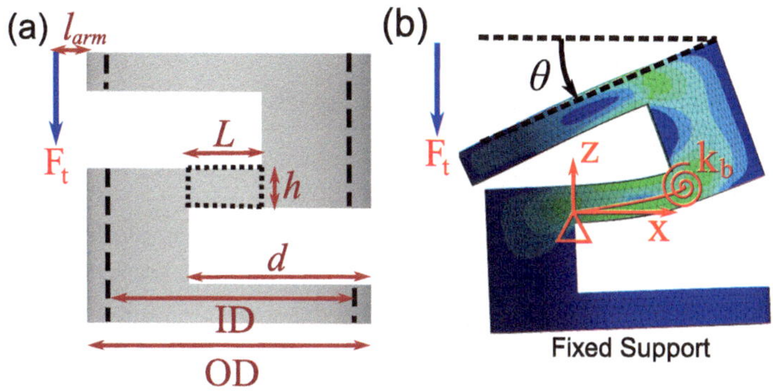

Fig. 2. (a) Euler beam model used to model forces applied in the bending plane, (b) Finite element model of the bidirectional asymmetric joint bending primarily due to deflection of intermediate beam.

where E is the elasticity of the tube, $I = \frac{1}{24}(\mathrm{OD} - \mathrm{ID})h^3$ is the second moment of area on the beam, x is the horizontal coordinate origining from the left side of the notch and half of F_t results from the two beams between the notches. By solving (1), we get

$$v = \frac{F_t}{2EI}(\frac{l}{2}x^2 + \frac{x^3}{6} + c_1 x + c_2) \tag{2}$$

From the FEA, it has been found that the left side of the beam has negligible deflection. Thus, the boundary condition on the left side can be assumed as a fixed end, which yields $c_1 = c_2 = 0$ and

$$v = \frac{F_t}{2EI}(\frac{l}{2}x^2 + \frac{x^3}{6}) \tag{3}$$

The bending angle on the right side of the beam is

$$\theta_1 = v(L)' = \frac{F_t}{2EI}(Ll + \frac{1}{2}L^2) \tag{4}$$

However, by observation, there is an additional bending angle θ_2 at the connecting position of the beam and the right-side curved wall. A torsional spring model with a stiffness k_b is used here and

$$\theta_2 = \frac{F_t}{2k_b}(L + l) \tag{5}$$

Thus, the bending angle of a notch is

$$\theta = \theta_1 + \theta_2 = F_t(\frac{2Ll + L^2}{4EI} + \frac{L + l}{2k_b}) \tag{6}$$

3 Experiments Performed

In this work, we analyze a finite element model of a single bidirectional asymmetric notch pair, and will use the information gathered from FEM simulations to guide the design of the final robot. All FEM simulations are conducted in ANSYS 18.2 Research Version (ANSYS Inc., Canonsburg, PA, United States). Material properties for Nitinol were taken from literature [18]. We ran a series of simulations, varying the notch geometry varying notch depth d and the height of the bending member h to estimate joint stiffness and torsional stiffness parameters defined in the previous section. A tetrahedral mesh with a mesh density of approximately 0.25×10^6 elements/mm^3 was used to ensure high mesh quality, especially at the notched regions.

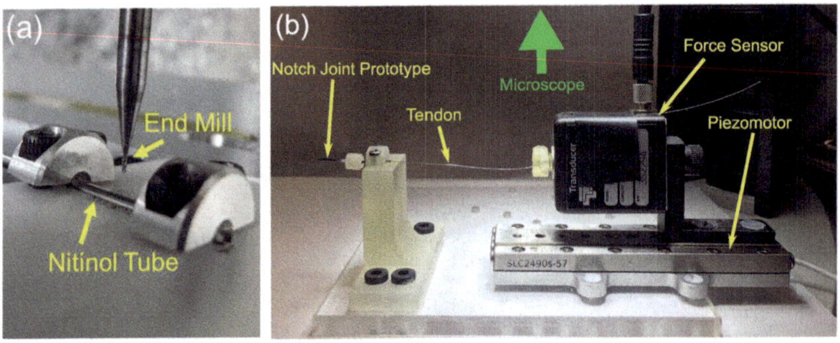

Fig. 3. (a) Machining setup to manufacture the bidirectional asymmetric joints, (b) Experimental setup to measure joint deflection under the application of various forces.

Nitinol tubes having outer diameter of 0.8 mm and inner diameter of 0.62 mm (Confluent Medical, CA, United States) were machined on a 3-axis CNC milling machine (Okuma MC-V4020, Okuma America Corporation, NC, United States) with a 200 micron diameter 4 flute end mill (Accupro 62773387, MSC, NY, United States) (see Fig. 3 (a)). The nominal cutting speed was 157 mm/min and feed rate was 1.27 mm/sec. In order to validate our FEA models, we constructed a compact setup, designed such as to fit on the observation surface of a standard stereo microscope (Leica Camera AG, Wetzlar, Germany) (see Fig. 3 (b)). A single tendon from the joint to be tested is connected to a piezo-based linear actuator (SmarAct GmbH, Oldenburg, Germany), via a load cell with a maximum load capacity of 5 lbs (Transducer Techniques, CA, United States). Data from the load cell and the microscope are acquired via a 16-bit Analog to Digital Converter (Model 826, Sensoray, OR, United States) and USB respectively. The motor is controlled to provide a step input of 50 μm, while force and encoder data is sampled continuously throughout this time. An image from the microscope is captured periodically once the joint has moved and force data has reached steady state for each step input. A Hough transform computed for each image automatically will provide us with the ground truth for the angle achieved by each joint at each point of time.

In the next section, we validate our Euler beam model for a set of three notch geometries where the dominating bending member is the beam between two notches (see Fig. 1 (iii)), and we follow this with an analysis of a set of unidirectional and bidirectional asymmetric joints to understand the effects of tendon forces in the bending and transverse planes, and axial external forces on each type of joint (see Sect. 5).

Table 1. Set of notch samples tested to validate the FEA and Euler beam models for the bending stiffness of the joint, as well as the torsional stiffness values obtained from each of the FEA simulations.

Sample	Parameters			
	d mm	h mm	k_b	# Trials
S1	0.5	0.15	0.0021	3
S2	0.55	0.125	0.0023	3
S3	0.6	0.15	0.0017	3

4 Results

Table 1 shows a set of notch samples selected to verify our FEA simulations and euler beam model. These dimensions were selected such that the intermediate beam was the dominating cause of joint deflection, as assumed in our model in Sect. 2. We can see in Fig. 4, that the FE model is able to satisfactorily model the deflection of the bidirectional notch joint for small forces and small deflections. Since the total deflection in a robotic tool is caused due to a serial chain of several such joints, this model may be capable of addressing the overall deflection for multiple bidirectional notches.

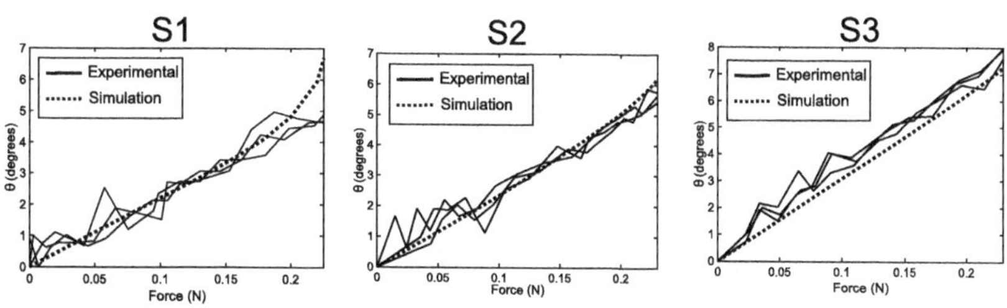

Fig. 4. Experimental results for each of the dimensions specified in Table 1, compared to the FEA simulations for the specific dimensions. Solid lines represent each trial.

Next, the ability of the notch joint to be relatively stiff in the plane orthogonal to the bending plane makes it very desirable for the construction of multi-degree-of-freedom (DoF) robots, because distal tendons can then be routed along the plane orthogonal to the bending plane of the proximal joint. These tendons, when tensioned, will affect any proximal joint only minimally, thereby achieving inter-joint decoupling by design. To this end, we arrive experimentally at a value of notch depth, maximizing flexibility in the bending plane, minimizing flexibility in the transverse plane, without exceeding the ultimate stress that can be applied to the joint. A plot of the tendon tension vs. equivalent (von-Mises) stress in order to achieve complete notch bending, for differing notch depths is shown in Fig. 5 (a). It can be seen that the ultimate stress of 1070 MPa used in our model is exceeded by notch depths under 500 microns. Therefore, while the usage of these geometries will result in high orthogonal stiffness, they will not be able to achieve complete bending of the notch without breakage.

The tip deflection in the plane orthogonal to the bending plane for various notch geometries is shown in Fig. 5 (b). Here, it can be seen that joint stiffness decreases with increasing notch depth. We find that the largest ratio of the bending angle vs the tip deflection happens for the depth of 500 microns without any breakage.

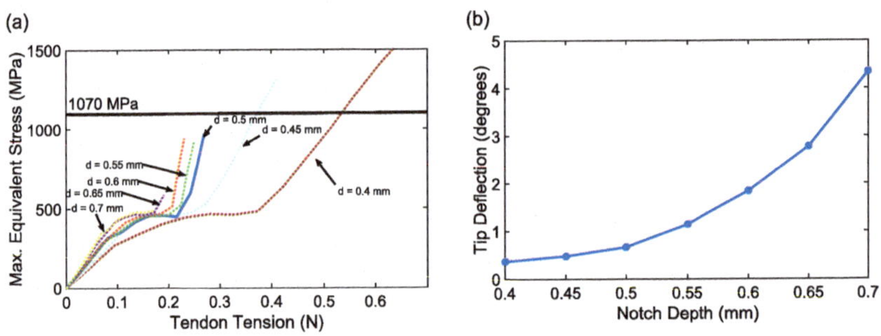

Fig. 5. (a) Maximum equivalent (von-Mises) stresses seen in a single notch for different tendon tension values and varying geometries, (b) Notch depths vs. tip deflection in the plane orthogonal to the bending plane.

5 Main Experimental Insights

One of the major insights from the finite element analysis of the bidirectional asymmetric notch joint can be observed in Fig. 6. In this figure, we analyze the deformation of each of the three types of joints that can be used in designing robotic neuroendoscopic tools, under loading from a moment created by tendon tension force, F_t, at a moment arm, $l_{arm} + \left(\frac{\mathrm{OD}}{2}\right)$. Note that in each case, the bending member is different. In the case of the symmetric bidirectional joint (see Fig. 6 (a)), these members are the beams created by the remaining wall of the tube after cutting the joint. In the case of the unidirectional joint (see Fig. 6 (b)),

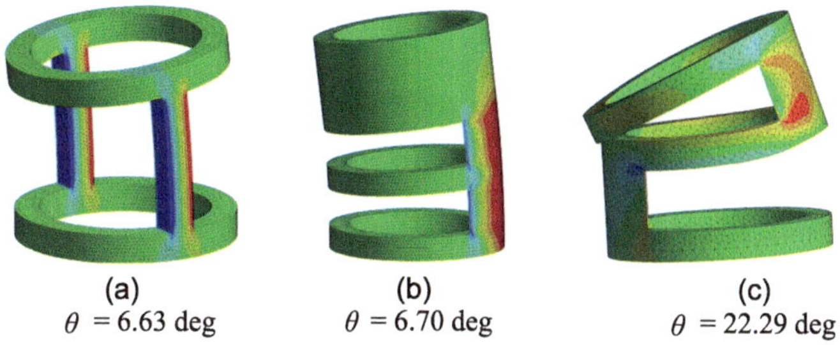

Fig. 6. The deflection observed in a (a) symmetric bidirectional notch joint [11], (b) unidirectional asymmetric notch joint [18], and the (c) asymmetric bidirectional notch joint, due to a tendon tension force, $F_t = 0.2$ N, applied at distance, $l_{arm} = 0.19$ mm, from the outer wall of each joint.

the bending member is the backbone of the joint, seen by the reddened region to the right of the notches. In our case, the bending portion is both the walls of the uncut portion of each notch as well as the horizontal member that connects the two notches. As a result, the beam that bends in the case of the asymmetric bidirectional joint (see Fig. 6 (c)), is longer than the ones in the symmetric bidirectional joint or the asymmetric unidirectional joint. This results in the higher bending flexibility seen in this type of a joint with the same amount of material removal.

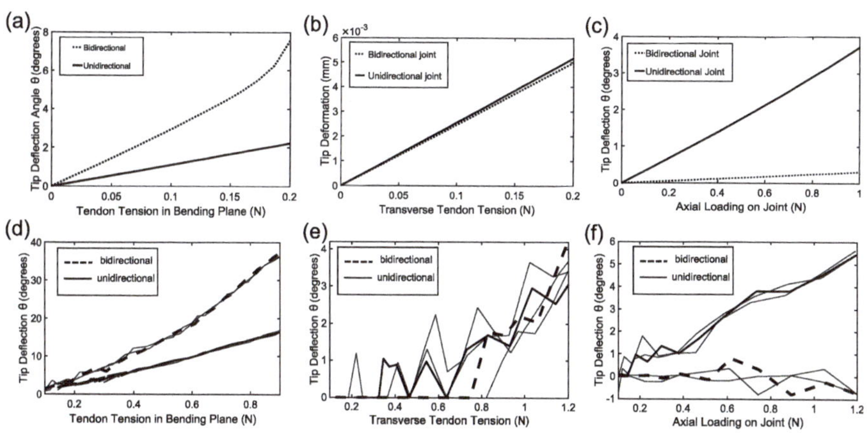

Fig. 7. Forces applied along the bending plane ((a), (d)), orthogonal to the bending plane ((b), (e)), and axially ((c), (f)) on bidirectional notch joint (d = 0.5 mm, h = 0.2 mm) and unidirectional asymmetric joint (d = 0.6 mm, h = 0.2 mm). Figures (a)–(c) indicate FEM results for small loads and small deformations and Figures (d)–(f) indicate experimental results.

Next, we perform an FE analysis for the bidirectional and unidirectional asymmetric notch joints for small deformations under small loads. These simulations allow us to design a bidirectional notch joint to have higher flexibility than the unidirectional joint in the bending plane, but equivalent stiffness in the direction orthogonal to the bending plane (see Fig. 7 (a)–(c)). Using the experimental setup described in Sect. 3, we confirm these simulated results for larger forces and deformations (see Fig. 7 (d)–(f)) . More importantly, the depth of the cut made to achieve these numbers is much lower ($d_{bidirectional} = 500\ \mu m$) than that seen for the unidirectional asymmetric joint ($d_{unidirectional} = 600\ \mu m$). As a result, a better performance can be obtained using a bidirectional asymmetric notch joint, while improving resistance to external axial forces due to the presence of more material in the joint than in the case of the unidirectional asymmetric joint (see Fig. 7 (c), (f)).

Acknowledgements. This work is based on the steerable guidewire technology, which was supported in part by the Georgia Research Alliance.

References

1. Vinchon, M., Rekate, H., Kulkarni, A.V.: Pediatric hydrocephalus outcomes: a review. Fluids Barriers CNS **9**(1), 18 (2012)
2. Kulkarni, A.V., et al.: Endoscopic third ventriculostomy in the treatment of childhood hydrocephalus. J. pediatr. **155**(2), 254–259 (2009)
3. Eastwood, K.W., Bodani, V.P., Drake, J.M.: Three-dimensional simulation of collision-free paths for combined endoscopic third ventriculostomy and pineal region tumor biopsy: implications for the design specifications of future flexible endoscopic instruments. Operative Neurosurg. **12**(3), 231–238 (2015)
4. Gaab, M.R.: Instrumentation: endoscopes and equipment. World Neurosurg. **79**(2), S14–e11 (2013)
5. Cepolina, F., Michelini, R.C.: Review of robotic fixtures for minimally invasive surgery. Int. J. Med. Robot. Comput. Assisted Surg. **1**(1), 43–63 (2004)
6. Catherine, J., Rotinat-Libersa, C., Micaelli, A.: Comparative review of endoscopic devices articulations technologies developed for minimally invasive medical procedures. Appl. Bionics Biomech. **8**(2), 151–171 (2011)
7. Jelínek, F., et al.: Classification of joints used in steerable instruments for minimally invasive surgery–a review of the state of the art. J. Med. Dev. **9**(1), 010801 (2015)
8. Haga, Y., et al.: Development of minimally invasive medical tools using laser processing on cylindrical substrates. Electric. Eng. Jpn. **176**(1), 65–74 (2011)
9. Dewaele, F., et al.: A novel design for steerable instruments based on laser-cut nitinol. Surg. Innov. **21**(3), 303–311 (2014)
10. Ryu, S.C., et al.: Feasibility study of an optically actuated MR-compatible active needle. In: 2011 IEEE/RSJ International Conference on Intelligent Robots and Systems (IROS), pp. 2564–2569. IEEE (2011)
11. Kato, T., et al.: Extended kinematic mapping of tendon-driven continuum robot for neuroendoscopy. In: 2014 IEEE/RSJ International Conference on Intelligent Robots and Systems (IROS 2014), pp. 1997–2002. IEEE (2014)
12. Peirs, J., et al.: A flexible distal tip with two degrees of freedom for enhanced dexterity in endoscopic robot surgery. In: Proceedings of the 13th Micromechanics Europe Workshop, pp. 271–274 (2002)
13. Peirs, J., et al.: Design of an advanced tool guiding system for robotic surgery. In: IEEE International Conference on Robotics and Automation. Proceedings, ICRA 2003, vol. 2, pp. 2651–2656. IEEE (2003)

14. Yoneyama, T., et al.: Force detecting gripper and flexible micro manipulator for neurosurgery. In: 2011 Annual International Conference of the IEEE Engineering in Medicine and Biology Society, EMBC, pp. 6695–6699. IEEE (2011)
15. Liu, J., et al.: Compliant articulation structure using superelastic NiTi-NOL. Smart Mater. Struct. **22**(9), 094018 (2013)
16. York, P.A., et al.: A wrist for needle-sized surgical robots. In: 2015 IEEE International Conference on Robotics and Automation (ICRA), pp. 1776–1781. IEEE (2015)
17. Eastwood, K.W., et al.: Kinetostatic design of asymmetric notch joints for surgical robots. In: 2016 IEEE/RSJ International Conference on Intelligent Robots and Systems (IROS), pp. 2381–2387. IEEE (2016)
18. Eastwood, K.W., et al.: Design of a contact-aided compliant notched-tube joint for surgical manipulation in confined workspaces. J. Mech. Robot. **10**(1), 015001 (2018)
19. Chitalia, Y., Wang, X., Desai, J.P.: Design, modeling and control of a 2-DoF robotic guidewire. In: 2018 IEEE International Conference on Robotics and Automation (ICRA). IEEE (2018), accepted
20. Barnett, A.C., et al.: Compliant needle vibration cutting of soft tissue. J. Manuf. Sci. Eng. **138**(11), 111011 (2016)
21. Gao, A., et al.: Mechanical model of dexterous continuum manipulators with compliant joints and tendon/external force interactions. IEEE/ASME Trans. Mechatron. **22**(1), 465–475 (2017)
22. Jones, J.A., Lee, Y.-S., Moore, J.Z.: Parametric study for asymmetric flexure hinge design for tissue cutting. Proc. Inst. Mech. Eng. Part B J. Eng. Manuf. **233**(4), 1302–1309 (2018). https://doi.org/10.1177/0954405418774587

In-Bore Experimental Validation of Active Compensation and Membrane Puncture Detection for Targeted MRI-Guided Robotic Prostate Biopsy

Marek Wartenberg[1(✉)], Katie Gandomi[1], Paulo Carvalho[1], Joseph Schornak[1], Niravkumar Patel[2], Iulian Iordachita[2], Clare Tempany[3], Nobuhiko Hata[3], Junichi Tokuda[3], and Gregory S. Fischer[1]

[1] Worcester Polytechnic Institute, Worcester, USA
{marek.wartenberg,kygandomi,pacarvalho,jgschornak,gfischer}@wpi.edu
[2] Johns Hopkins University, Baltimore, USA
{npatel89,iordachita}@jhu.edu
[3] Brigham and Women's Hospital, Harvard Medical School, Boston, USA
{ctempany,hata,tokuda}@bwh.harvard.edu

1 Motivation, Problem Statement, Related Work

It is estimated that in the United States there will be 164,690 new cases and 29,430 deaths from prostate cancer in 2018 [1]. Trans-Rectal Ultrasound (TRUS) has typically been used to facilitate sampling of up to twenty biopsy cores, but due to variable prostate size this technique often still misses clinically significant cancers [2]. Instead, MRI provides higher image quality and multiparametric imaging, allowing for procedures with fewer needle insertions via direct targeting of suspicious lesions.

Regardless of imaging modality, accuracy of needle tip placement remains an issue. Targeting in open loop can be performed by predicting curvature of a bevel tipped needle through waypoints along the insertion path [3], but his often leads to errors due to unmodeled needle deflection and target shift. Instead, deflection of an asymmetric bevel tipped needle can be exploited for closed-loop active compensation, through rotational positioning of the bevel location [4]. Additionally, targeting accuracy of a system controlling robotic needle rotation would benefit from robotic control of needle insertion as well. This has been implemented through teleoperated systems to keep the physician loop [5], but still distances the physician from the patient. Instead, a cooperatively controlled system would keep the physician directly at the surgical site with full control of the procedure, while maintaining the advantages of robotic accuracy. In a cooperative control paradigm, the user initiates all motions and the robotic system provides haptic cues that can be used to improve physician's performance [6], which may have wider acceptance by the medical community than a teleoperated counterpart [7].

Motivation for this work came as our research group performed clinical trials of MRI-guided robot-assisted prostate biopsy at Brigham and Women's Hospital

J. Xiao et al. (Eds.): ISER 2018, SPAR 11, pp. 34–44, 2020.
https://doi.org/10.1007/978-3-030-33950-0_4

in Boston, MA [8]. This system employed robot assistance for target alignment but was still dependent on manual needle insertion along the robotically aligned axis. As a result, the manual insertion suffered from loss of targeting accuracy due to unmodeled needle deflection and target shift, with targeting error across all clinical trials of 6.5 ± 5.1 mm [9]. The contribution of this work is to increase targeting accuracy seen in clinical trials through robotic active compensation for deviations from an initial straight-line trajectory using a typical clinically used 18G biopsy needle. More so, to experimentally validate this function during cooperatively controlled needle insertions inside the MRI bore, to maintain a procedure directed by the physician at the procedure site.

2 Technical Approach

A 6 degree of freedom (DoF) robotic system suitable for use in the MRI environment was used. This system employs piezoelectric actuation and consists of a 4-DoF robotic manipulator that was previously used in clinical trials [8,9], and a 2-DoF needle driver configured to collect user input and needle forces for cooperatively controlled needle insertions during autonomous rotational bevel tip positioning [10]. An annotated image of the robot along with the system integrated into the MRI room shown in Fig. 1. The two functional components of this robotic system are cooperative needle insertion and closed-loop rotational control of the bevel tipped needle, where this work aims on experimentally validating these functions inside the MRI. The functionality of these subsystems is introduced here in context of experimental validation, with more detailed explanations of function in Wartenberg et al. [11].

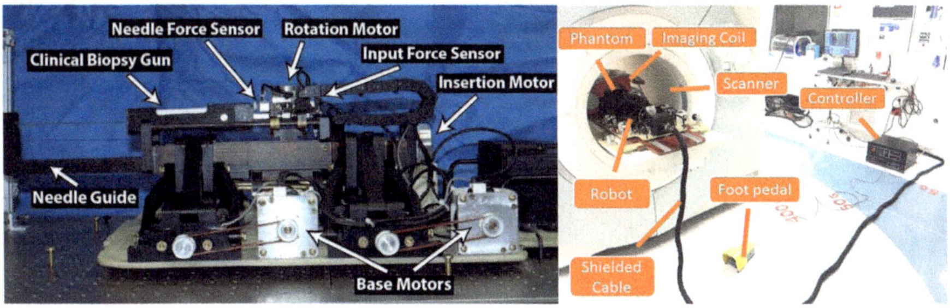

Fig. 1. (Left) An annotated figure of the robotic needle placement manipulator suitable for used in the MR environment, (Right) Hardware integration of the robot system into the clinical environment. The robot sits inside the MR scanner and is connect to the controller via a shielded cable.

2.1 Active Compensation

Closed-loop image-guided active compensation can be performed during a cooperative needle insertion. By localizing the needle tip and target, continuous rotational bevel tip position can be updated to induce deflection in a desired direction to account for deviations from an initial straight-line trajectory. With insertion defined along the Z-direction, a desired rotation position of compensatory effort is determined by projecting the current X-Y needle tip position onto the target plane. The angle of the two-dimensional straight line from the needle tip projection to the target on the target plane is the direction of compensatory effort used to position the rotational bevel tip, described with figures in Wartenberg et al. [11].

2.2 Cooperative Insertion

To keep the physician in full control while maintaining robotic accuracy, active compensation was implemented to autonomously rotate the needle while the physician performs a cooperatively controlled insertion by applying an input force directly on the robot. The system collects both user input force, F_{input}, and the sum of all needle forces (cutting, stiffness, friction), $\sum F_{needle}$, where the velocity in based on a difference of the two. The velocity curve is a decreasing exponential, where the difference in forces in the normal operating range ($F_{input} > F_{needle}$) is scaled by the exponential decay constant λ. Increasing the magnitude of λ leads to an increased insertion velocity during initial insertion with higher sensitivity between F_{input} and F_{needle} as their values approach each other. The goal is augmented feedback, not transparent haptic sensation. This algorithm gives the user control of insertion speed through a force input, while increasing the sensitivity of tactile feedback as the needle force approaches this input. The cooperative control algorithm is shown in Eq. 1, where the second equation of motion along with the subtraction of 0.9 times the exponential decay are included to ensure an insertion velocity is always present if there is a user applied input force.

$$v_{insertion} = \begin{cases} v_{max}(1 - 0.9e^{-\lambda(F_{input} - \sum F_{needle})}, & F_{input} > \sum F_{needle} \\ 0.1v_{max}, & F_{input} < \sum F_{needle}, F_{input} > 0 \\ 0, & \text{otherwise} \end{cases} \quad (1)$$

This admittance based closed loop velocity control provides an insertion velocity output to a user force input. A block diagram describing cooperative insertion is shown below in Fig. 2, and originally shown in Wartenberg et al. [11]. The two inputs are a target depth and the user applied input force. The current needle tip position, determined via real-time imaging, is subtracted from the target depth to determine if the target depth has been reached. If it has then insertion stops, and if it hasn't than the cooperative control equation determines the closed loop insertion velocity based on the measured forces. An increased insertion velocity increases insertion position, which is fed back to make the supervisory control loop decision. As the needle inserts deeper into the tissue the force on the needle increases and this real-time needle force is fed back to the beginning of the control loop to determine the cooperative insertion velocity.

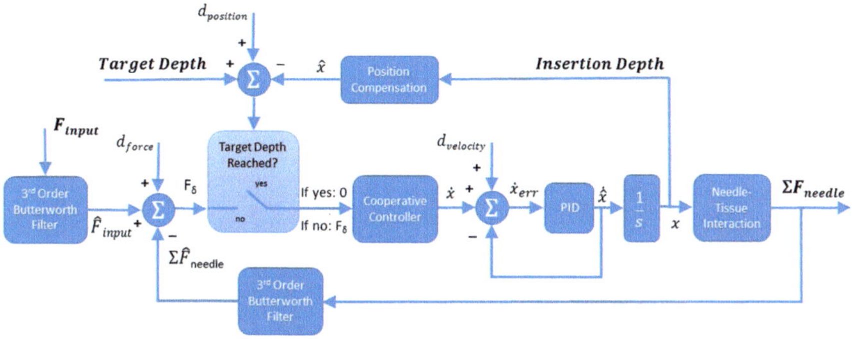

Fig. 2. A block diagram showing the implementation of cooperative needle insertion. Closed-loop velocity control maintains a setpoint based on force values with a supervisory position loop to cease insertion at the target depth [11].

While providing compensation during cooperative insertion via a robotic system, the tactile forces typically used by a physician for anatomical localization are lost. By implementing membrane puncture detection, an augmented haptic feedback can be provided during the cooperatively controlled insertion to return this information back to the physician. Due to tissue deformation and needle deflection, kinematic motion of the robot cannot accurately describe the needle tip position, and puncture of a membrane cannot be guaranteed. Instead, using preoperative imaging to localize the depth of a membrane, a window (5 mm in this case) can be provided to monitor real-time needle force data for the puncture. Within the expected window, membrane puncture was determined by thresholding the derivative of needle force to determine the rapid increase of needle force seen at boundary displacement followed by a sudden drop when puncture occurs, similarly to the algorithm implemented by Barbe et al. [12]. Once puncture is detected by the system a haptic response can be provided to the physician, returning tactile forces that were lost.

3 Experiments Completed

For clinical translation, the subsystems developed in the laboratory needed to be integrated with the standard protocols in the operating room. Three in-bore experiments were performed using a Verio 3T MRI (Siemans Healthineers, Erlangen, Germany) in the Advanced Multi-Modality Image Guided Operating (AMIGO) Suite at Brigham and Women's Hospital, with scanner room integration seen previously on the right of Fig. 1.

Table 1. MRI parameters for the image sequences used to test SNR

Sequence	ACRT1	ACRT2
Repetition Time (ms)	500	2000
Echo Time (ms)	20	20/80
Field of View (mm)	250 × 250	250 × 250
Matrix	256 × 256	256 × 256
Number of Slices	11	11
Slice Thickness (mm)	5	5
Flip Angle (degrees)	90	90
Receiver Bandwidth (kHz)	16.6	16.6

The first experiment was MRI compatibility validation through assessing changes in image quality by measuring signal-to-noise ratio (SNR). Mechatronic systems within the MRI can cause significant image degradation. Poor image quality may result in incorrect localization of needle tip and target as well as planning locations for puncture detection. The effect of the system on the SNR of images was evaluated in four conditions: Baseline, Robot Present, Robot Powered and Robot Moving, with a standardized phantom similar to that in Stoianovici et al. [13]. Two standard imaging protocols were used, T1-weighted spin echo and T2-weighted dual spin echo sequences designed by the American College of Radiology for quality control tests. Table 1 shows the parameters for both image sequences where signal intensity was calculated using Eq. 2, signal noise with Eq. 3, with SNR calculated as shown in Eq. 4, based on the National Electronics Manufacturing Association (NEMA) standard [14]. Paired t-tests were used to determine statistical significance of any signal changes seen across conditions.

$$S_s = \frac{\underset{u,v:ROI}{mean}\left(p_{u,v}^{Is}\right) + \underset{u,v:ROI}{mean}\left(p_{u,v}^{Js}\right)}{2} \tag{2}$$

$$N_s = \frac{\underset{u,v:ROI}{std}\left(p_{u,v}^{Is} - p_{u,v}^{Js}\right)}{\sqrt{2}} \tag{3}$$

$$SNR_s = \frac{S_s}{N_s} \tag{4}$$

The second experiment was to validate puncture detection with augmented haptic feedback, performing ten trials of continuous cooperative needle insertion through layered phantoms as seen in Fig. 3. Upon detection, several options are available for providing an enhanced feedback response to the user at the puncture event. Most simply, the robot could be made to stop at a detection, or perhaps withdrawn slightly to account for tissue relaxation. Vibrotactile feedback could be implemented with the use of an additional actuator, or during cooperative insertion a short freeze of the insertion axis can provide a haptic response related to the puncture. In this case. the insertions were performed continuously under cooperative control and if the system detected a puncture using the paired imaging and force threshold technique, the user was informed via a short (100 ms) freeze of insertion.

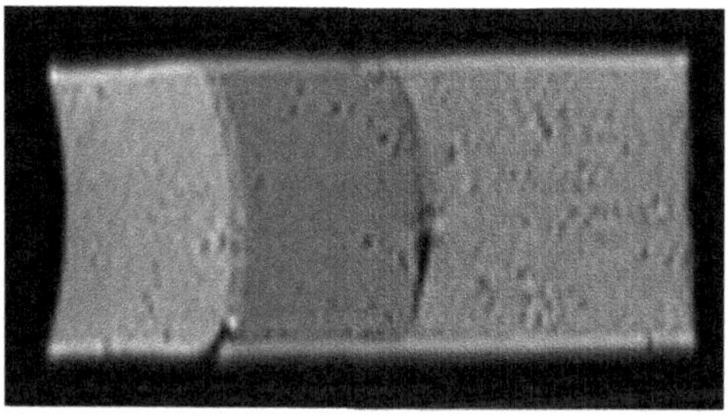

Fig. 3. An MR image of the multi-layered phantom used for in-bore validation of membrane puncture detection

The third experiment was validation of autonomous rotational active-compensation during cooperatively controlled insertion. A custom phantom was manufactured using anonymized MR images, representing realistic anatomical and mechanical properties of the pelvic region. The needle was aligned to an initial straight-line trajectory towards a target selected in planning images, and an input force was applied to cooperatively insert the needle approximately 5–10 mm. An intraoperative scan was taken and localizations of the needle tip and target were performed in 3DSlicer (www.slicer.org) to update the rotational bevel tip position at each step. Insertion continued in this manner until the target depth was reached where the in-plane targeting error (plane normal to needle insertion axis) was determined.

4 Results

The results of SNR evaluation for the T1 and T2 (80 ms echo) sequences are shown in Fig. 4. In the T1-weighted scans there was no statistically significance between two baseline images taken ($p = 0.518$), followed by a statistically significant drop of approximately 10% of signal in the Robot Present case ($p = 0.006$). For the remaining conditions there was no statistical significance of signal loss between the Robot Present-Robot On, Robot Present-Robot Moving, and Robot On-Robot Moving cases ($p = 0.355, p = 0.174, p = 0.242$, respectively).

In the T2-weighted scans there was also no statistical significance between baseline images ($p = 0.872, p = 0.832$), for the 20 ms echo and 80ms echo, respectively. Similarly to T1, for each echo time there was an approximate 15% drop in signal for the Robot Present case with statistical significance, ($p = 0.003$) for 20 ms, ($p = 0.004$) for 80 ms, followed by no statistical significance between the three cases thereafter ($p = 0.540, p = 0.426, p = 0.457$) for the 20 ms echo, and ($p = 0.182, p = 0.725, p = 0.375$) for the 80 ms echo. These results show that the presence of the robot in the MR bore degrades the image SNR between 10% and 15%, but there is no further statistically significant loss by powering on the robot or moving the motors during a scan sequence.

The results of the second experiment showed the system detected all but a single puncture, and a haptic response in the form of a 100 ms freeze of insertion was felt by the user in each detected case. Figure 5 shows a graph of the needle force throughout insertion with each puncture labeled at the point it was detected. The length of each insertion was between 10 and 15 s.

Finally, the results of the third experiment showed the in-plane targeting error during in-bore actively-compensated cooperative needle insertions was 2.88 ± 0.54 mm across all trials, and puncture was detected in each case providing the same 100 ms freeze of insertion to the user. On the left of Fig. 6 are two MR images, the first showing the needle during insertion and the second showing final needle placement. On the right of Fig. 6 are 3D visualization of the segmented anatomical structures and each insertion trajectory towards its target.

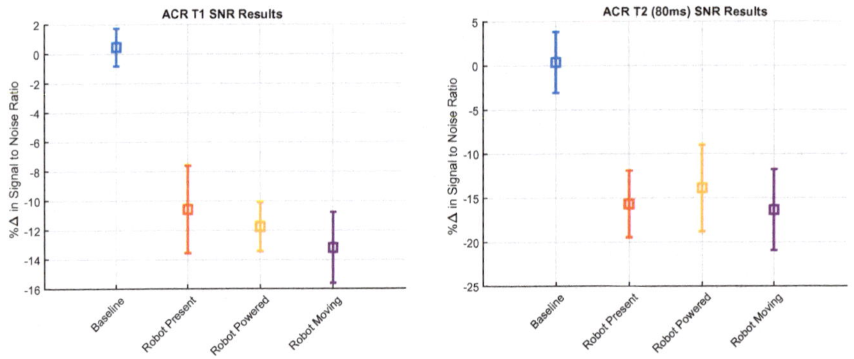

Fig. 4. The results of the SNR analysis. There was a statistically significant drop in image signal from the baseline to the robot present case, with no statistically significant difference seen between the robot present case and either of the robot on or robot moving cases.

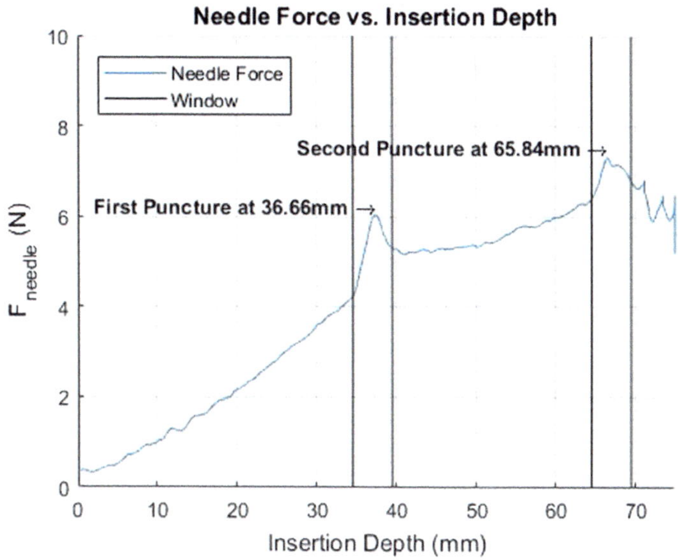

Fig. 5. Needle force during cooperative insertion through two membrane transitions. The two puncture locations are labeled at their locations, where the detection window was 5 mm.

5 Main Experimental Insights

Full robotic control of needle insertion is expected to be the most accurate way to place a needle tip for biopsy, but this leap from the clinical standard of practice would likely be met with caution from both acting physicians as well as regulatory bodies. Instead, through user directed cooperative control, a physician would maintain full control of the procedure directly at the surgical site, while adding the benefit of robotic accuracy.

Outcomes from this research build upon clinical trials of robot assisted prostate biopsy, where the insertion was performed manually and suffered from decreased targeting accuracy due to unmodeled needle deflection and target shift. Preliminary tests for both puncture detection and active compensation under cooperative control were performed in the lab [11], and experimental system validation took place in the AMIGO suite at Brigham and Women's Hospital.

Using paired t-tests, the results of SNR evaluation showed a 10% to 15% drop in signal quality between the baseline and robot present cases, but no statistically significant change ($p > 0.05$) in signal quality between the robot present and either the robot on or robot moving cases. This drop is acceptable and less than the 80% seen in previous literature without shielding, and the 40%–60% degradation that has been previously seen even with shielding [15]. The authors would like to note that the standardized phantom and pulse sequences used in SNR analysis are different than either sequences used to collect the MRI images shown in this paper. Figure 3 was collected using a spin-echo sequence, while the

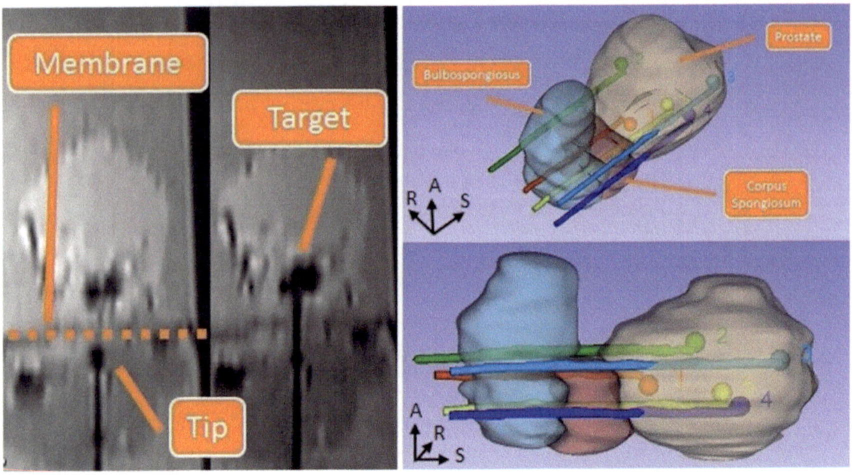

Fig. 6. (Right) Coronal images of a needle insertion into a heterogeneous phantom showing the needle tip and target, (Left) In-bore needle insertion trajectories. Each of the five needle trajectories performed during in-bore validation testing along with segmented pelvic structures.

image in Fig. 6 was from a real-time pulse sequence which provides images at a much faster rate, at the expense of resolution.

Force sensing of needle interaction with tissue can return the tactile forces used by the physician for anatomical localization. If an augmented haptic response can be provided at detection of membrane puncture during cooperative insertion, perceived anatomical localization of the needle tip can be restored to the physician. It has been shown that probabilistic models can perform better than simple thresholding to separate a true puncture event from other needle-tissue interactions [12], but a more complex model may not be necessary when searching for a puncture event in real-time, if the location of expected puncture can be narrowed as is done here. Using this idea, a window was created based on pre-operative imaging wherein the real-time needle force data could be used to identify a membrane puncture. In this experiment the user was able to identify all but one membrane puncture from the provided feedback, where the missed detection was found to be the algorithm not detecting puncture, rather than the user not feeling the feedback.

Cooperative needle insertions were performed by applying an input force directly on the needle driver. This was done with autonomous rotational positioning for active compensation of the needle trajectory into heterogeneous phantoms characterized to mimic periprostatic tissue. The final targeting error across all insertions was 2.88 ± 0.54 mm, comparable to the 3.24 ± 0.99 mm targeting error found during in-lab tests and improved over 50% compared to the 6.5 ± 5.1 mm targeting error seen in clinical trials with manual insertion through the robotically aligned axis. In these results, similarly to the clinical trials, in-plane error at the target plane was used to quantify accuracy as opposed to 3D tip-to-target error. This is because the biopsy core taken is 1 cm long, therefore accuracy in-plane has much more effect on the outcome of clinical biopsy than insertion position.

This work was performed with respect to clinical translation and future work continues this path. The authors suggest to move forward towards pre-clinical trials a clutch should be added to the insertion axis to easily allow for disengagement of the insertion motor. This would increase safety to release the insertion axis in the case of emergency, in addition to creating a system for the biopsy to be performed with hands-on the sliding carriage but without robotic control of insertion. This could be a first step in regulatory approval, an encoded hands-on insertion pushed along the linear guides with autonomous rotational compensation. With this modification the system could be ready for cadaver and pre-clinical animal trials within one year.

6 Disclosures and Acknowledgment

NH has a financial interest in Harmonus, a company developing Image Guided Therapy products. NH's interests were reviewed and are managed by Brigham and Women's Hospital and Partners HealthCare in accordance with their conflict of interest policies. This research was funded by NIH R01 CA111288, NIH R01 CA166379 and NIH R01 EB020667.

References

1. Society, A.C.: Cancer facts and figures 2018. American Cancer Society (2018)
2. Djavan, B., Margreiter, M.: Biopsy standards for detection of prostate cancer. World J. Urol. **25**(1), 11–17 (2007)
3. Pacchierotti, C., Abayazid, M., Misra, S., Prattichizzo, D.: Teleoperation of steerable flexible needles by combining kinesthetic and vibratory feedback. IEEE Trans. Haptics **7**(4), 551–556 (2014)
4. Abayazid, M., Roesthuis, R.J., Reilink, R., Misra, S.: Integrating deflection models and image feedback for real-time flexible needle steering. IEEE Trans. Rob. **29**(2), 542–553 (2013)
5. Tse, Z.T.H., Elhawary, H., Rea, M., Davies, B., Young, I., Lamperth, M., et al.: Haptic needle unit for mr-guided biopsy and its control. IEEE/ASME Trans. Mechatron. **17**(1), 183–187 (2012)
6. Coles, T.R., Meglan, D., John, N.W.: The role of haptics in medical training simulators: a survey of the state of the art. IEEE Trans. Haptics **4**(1), 51–66 (2011)
7. Troccaz, J., Peshkin, M., Davies, B.: Guiding systems for computeraided surgery: introducing synergistic devices and discussing the different approaches. Med. Image Anal. **2**(2), 101–119 (1988)
8. Patel, N., Li, G., Shang, W., Wartenberg, M., Heffter, T., Burdette, E., Iordachita, I., Tokuda, J., Hata, N., Tempany, C., Fischer, G.: System integration and preliminary clinical evaluation of a robotic system for mri guided transperineal prostate biopsy (accepted). J. Med. Robot. Res. **4**(02), 1950001 (2018)

9. Tokuda, J., Tuncali, K., Li, N., G an Patel, Heffter, T., Fischer, G., Iordachita, I., Burdette, E., Hata, N., Tempany, C.: In-bore mri-guided transperineal prostate biopsy using 4-dof needle-guide manipulator. In: International Society for Magnetic Resonance in Medicine, Annual Meeting (2016)
10. Wartenberg, M., Schornak, J., Carvalho, P., Patel, N., Iordachita, I., Tempany, C., Hata, N., Tokuda, J., Fischer, G.: Closed-loop autonomous needle steering during cooperatively controlled needle insertions for MRI-guided pelvic interventions. In: The 10th Hamlyn Symposium on Medical Robotics, pp. 33–34 (2017)
11. Wartenberg, M., Schornak, J., Gandomi, K., Carvalho, P., Nycz, C., Patel, N., Iordachita, I., Tempany, C., Hata, N., Tokuda, J., Fischer, G.S.: Closed-loop active compensation for needle deflection and target shift during cooperatively controlled robotic needle insertion. Ann. Biomed. Eng. **46**(10), 1582–1594 (2018)
12. Barbe, L., Bayle, B., de Mathelin, M., Gangi, A.: In vivo model estimation and haptic characterization of needle insertions. Int. J. Robot. Res. **26**(11–12), 1283–1301 (2007)
13. Stoianovici, D., Jun, C., Lim, S., Li, P., Petrisor, D., Fricke, S., Sharma, K., Cleary, K.: Multi-imager compatible, mr safe, remote center of motion needle-guide robot. IEEE Trans. Biomed. Eng. **65**(1), 165–177 (2017)
14. National Electrical Manufacturers Association Std. Determination of signal-to-noise ratio (SNR) in diagnostic magnetic resonance imaging.
15. Krieger, A., Song, S.E., Cho, N.B., Iordachita, I.I., Guion, P., Fichtinger, G., Whitcomb, L.L.: Development and evaluation of an actuated mri-compatible robotic system for mri-guided prostate intervention. IEEE/ASME Trans. Mechatron. **18**(1), 273–284 (2013)

Understanding and Transfer of Human Skills in Robotics Using Deep Learning and Musculoskeletal Modeling

Dipti Chaudhari[1(✉)], Kunj Bhagat[2], and Emel Demircan[2,3]

[1] Department of Computer Science, CSULB, Long Beach, USA
Dipti.Chaudhari.24@gmail.com

[2] Department of Mechanical and Aerospace Engineering, CSULB, Long Beach, USA
Kunj.Bhagat@student.csulb.edu

[3] Department of Biomedical Engineering, CSULB, Long Beach, USA
Emel.Demircan@csulb.edu

Abstract. With the application of deep learning, prosthetic rehabilitation can be carried out in a manner that not only emulates human manipulation skills and performance, but can also work more efficiently. In this study, we introduced computer vision capability for a rehabilitation robot using a convolutional neural network (CNN). The human skill of scooping was studied by dividing it into four motion primitives or sub-tasks. For each primitive, optimum human posture was identified in terms of muscular effort. Human motion skills were analyzed in terms of physiological parameters, including wrist pronation-supination angle, elbow flexion angle, shoulder rotation/abduction/flexion angles, and hand accelerations by three dimensional musculoskeletal modeling. This analysis identified how humans execute the same activity for eight different materials. Optimum human motion for each material was mapped to a robotic arm with six degrees-of-freedom (DOFs), which was equipped with a camera. The success ratio while examining the scooping motion over all trials was found to be 85%. Consequently, the activity can be performed efficiently based on human intuition in a dynamic environment.

1 Introduction

Robots have transformed the way in which an artificial device can replace a missing body part, which may have been lost through trauma, disease, or congenital conditions. These artificial devices are called prosthetic devices. Early static prosthetics are replaced by more advanced robotic prosthetics with greater degrees of freedom and capability of performing more complex tasks [1]. Such robots can also be programmed to perform specific tasks, but robot control is

Electronic supplementary material The online version of this chapter (https://doi.org/10.1007/978-3-030-33950-0_5) contains supplementary material, which is available to authorized users.

J. Xiao et al. (Eds.): ISER 2018, SPAR 11, pp. 45–55, 2020.
https://doi.org/10.1007/978-3-030-33950-0_5

still a complicated task. Biosensors such as electromyography (EMG) sensors or needle electrode can be used to identify the muscle activities and fully automate the interaction [2]. This research proposes an efficient way of transferring human skills to rehabilitation robots. The interaction between a human subject and a rehabilitation robot can be reduced to a remarkable level [3], if the prosthetics are made smart, i.e. if they can take decisions on their own. The aim of the research is to train the robot by solving intelligence on various levels and use it for the improvement of rehabilitation robots [4]. To achieve that, we considered upper extremity (human arm) rehabilitation [5]. We taught a robotic arm to perform the task of scooping, which is very trivial for human beings but extremely difficult for robots to perform. Work related to motion primitives in human-robot interaction is discussed in [6]. We trained the robotic manipulator to perform scooping on various materials. Thus, we investigated human scooping activity in detail and analyzed motion primitives to generate natural movements [7]. This investigation was done using the Optitrack [8] motion capture system and electromyography (EMG). Muscle activations were investigated for the biceps and extensor carpi radialis longus (ECRL), which are primarily responsible for scooping tasks. Based on the activation signals of these methods, we determined optimum human posture which corresponds to minimum muscular effort [9] to perform the task. We studied human motion for eight different types of materials such as water, sand, soil, pebbles, rock, honey, stones, and cereals. Using human musculoskeletal modeling [10,11], the captured human motion of scooping was investigated in terms of physiological parameters such as joint angles and angular velocities for each DOF of the human arm. Based on the angular data, the task of scooping is divided into sub-tasks or motion primitives: Approach, Scooping, Dragging, and Boomerang. Subdivision of the task into these primitives helped in focusing on key events during the motion, and studying different techniques that humans use to scoop a variety of materials. The understanding of human skill is then used for robot motion planning. There are different types of robot motion planning techniques based on motion prediction and constraints [12,13]. Optimum human motion was mapped to equivalent robotic motion and the robot was programmed to execute the motion for each category of material. Work related to imitating and mapping human motion to robot is discussed in [14]. The robotic arm is equipped with the computer vision ability, which identifies which of the eight studied materials it is being asked to scoop.

2 Methodology

This research was completed in three significant stages: human motion data analysis, computer vision, and robot motion planning. For upper extremity rehabilitation, we selected the human activity of scooping different materials because it involves physiological manipulation skills [15] that are natural to humans but extremely difficult for robots to learn. The experiment was performed with a 23-year-old healthy male subject, with a height of 6 ft 2 in., and a body weight of 170 pounds.

2.1 Human Musculoskeletal Model

The human arm is a highly complex system. Thus, to understand the motion of scooping, we observed the Right Arm Flexion, Right Arm Adduction, Right Arm Rotation, Right Elbow Flexion, Right Wrist Flexion, Right Wrist Deviation, Right Wrist Pronation and Supination joint angles of the human arm (Fig. 1). We also monitored the biceps and ECRL muscles which are primarily responsible for the action of scooping.

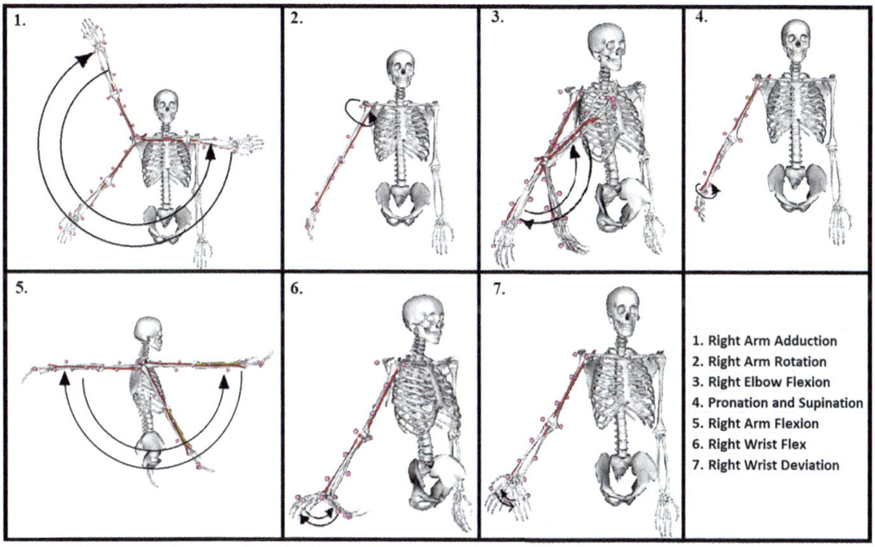

Fig. 1. Degrees of Freedom of human right arm considered for the experiment

We used OpenSim's upper extremity kinematic model [16], which includes 15 DOFs, representing the shoulder, elbow, forearm, wrist, thumb, and index finger, and 50 muscle compartments crossing these joints. The scooping motion was reconstructed by placing virtual markers on the model exactly as they were positioned on the human subject. We studied the joint angles and muscles that were discussed earlier. We inferred the relation between the materials and the joint angles for the human right arm, and the relation between different joint angles for the same material. Based on the joint space data, the task of scooping is divided into sub-tasks or motion primitives: Approach, Scooping, Dragging, and Boomerang. The motion primitives were determined manually and offline based on factors including human arm's joint angles and velocities. Approach is the first sub-task in the activity of scooping, in which the arm moves from a start position to the surface of the material. Scooping sub-task involves getting the material into the scooping equipment. Dragging sub-task is responsible for maximizing the material in the scooping equipment and ensuring that the material remains in the equipment during motion. The final sub-task is Boomerang, in which the arm carefully takes the scooped material to the final position. Table 1 shows the corresponding DOFs for each motion primitive during scooping.

Table 1. Human activity scooping motion primitives

No	Name	Focused joint-angles
1	Approach	Arm Flexion, Elbow Flexion
2	Scooping	Wrist Pronation/Supination
3	Dragging	Elbow Flexion, Wrist Pronation/Supination
4	Boomrange	Arm Flexion, Elbow Flexion

2.2 Human Motion Capture

The experiment was performed for 8 different materials-cereal, honey, rock, pebble, sand, soil, stone, and water. We covered materials that are significantly different in appearance, nature, and force feedback upon being scooped. The Optitrack motion tracking system with five Prime-41 cameras was used for this experiment. Each camera had 170 LEDs and 4.1MP resolution. For tracking the motion, a 120 Hz sampling frequency was used. From the captured marker-data, we reconstructed the human motion in real-time through consistent tracking of marker trajectories [17]. In addition to motion tracking, we also collected data on linear acceleration of the wrist and EMG signal activation in the biceps and ECRL using the EEGO [18] system. EMG data were post-processed using high-pass filtering at 20 Hz to remove motion artifacts, full-wave rectifying, and then low-pass filtering at 5 Hz to generate a linear envelope. These data were then normalized to a maximum isometric muscle contraction. Out of the ten trials for each material, the one with least muscular effort [9] was selected as the optimal posture, and was investigated further.

2.3 Training the Robot

The very first stage of the image identification module pipeline is the convolutional neural network, which represents the input (image of the material to be scooped) in the program in an understandable format. For image classification we have selected to use transfer learning to extract image features from pre-trained convolutional neural network, and building classifier on small sample of the training data. Transfer learning is a technique in which the already trained model is used for solving similar problems [19]. For representing the material images in the proposed approach, we used the transfer learning method [20] and a pre-trained neural network model with VGG architecture [21]. With transfer learning, the model can be extended to a greater number of materials with very small sample for each material to train the model. We have collected enough data to make the background information like bowl colors invariant. Using transfer learning, the model can understand more features other than visual representation which could be difficult to identify when features are hand crafted. We used the Visual Geometry Group (VGG) convolutional neural network model [22,23]. The model achieved 92.7% top-5 test accuracy in ImageNet [24,25], which is a dataset of over 14 million images belonging to 1000 classes. The last layer off the VGG16 network is SoftMax function, which classifies the input in one of the 1000 ImageNet classes depending upon the 4096 features learned by the previous layer (fully connected + ReLu layer output). For this research, the bottleneck features were captured after the fully connected + ReLu layer and the SoftMax function was replaced by the model trained using k-NN algorithm [26]. For each material, 15 images were taken by the USB camera and resized to an image of size $224 \times 224 \times 3$. The feature vector representation from VGG network was used to build the k-NN classifier using 'knn' package of R programming language for value of $k = 3$.

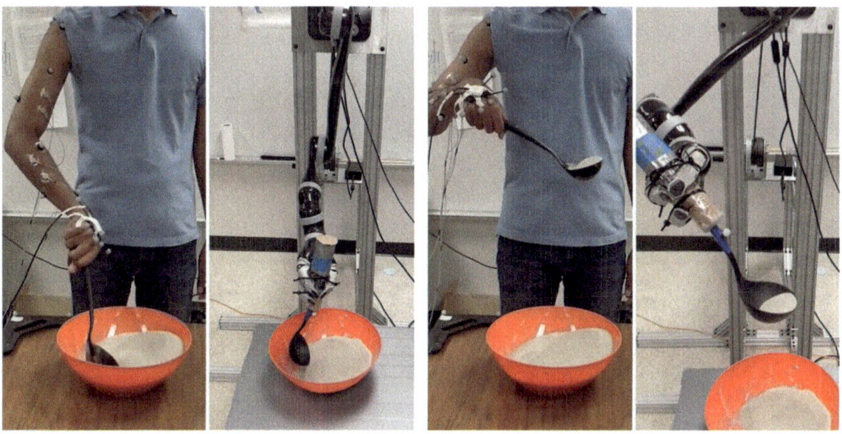

Fig. 2. Initial and Final optimal posture (i.e minimum muscular effort) of the human subject and the robot

2.4 Robot Motion Planning

To perform tasks in real-time, probabilistic approach is considered where there is need for fast algorithms for motion planning of a high DOF robot. For this research, a Kinova JACO robotic arm with 6 DOFs was used. The robot was equipped with a generic USB camera for computer vision ability. The executable code to manipulate and control the robot was developed with the API exposed by Jaco SDK and related libraries using C++ programming language. The robot model could be controlled and manipulated using two types of gains: velocity and position. We selected velocity gain over position gain, as in this case accuracy is of greater importance than efficiency. This relative importance is due to the robot's proximity to human subject, especially in applications such as rehabilitation. Velocity gain has more accuracy while dealing with angular control.

We performed human motion capture to identify the natural human posture while executing the scooping activity. Human motion capture was performed using Optitrack system [8]. This system requires the human subject to place passive motion-capture markers on the body-part whose motion is under concern. In our study, subject's dominant arm which performed the scooping action was tracked. The markers were placed on the right arm of our subject such that the resulting motion of the arm would produce smooth and continuous trajectories in the captured data. We obtained motion data by placing the markers on all fingers, back of the hand, wrist, forearm, elbow, biceps, triceps and the shoulder. The obtained motion data was then manually processed to generate readable data files. The data files are then imported in Opensim and inverse kinematics is applied on the data to convert the motion data into joint-angles of the subject's arm. These joint-angles is critical in robot motion planning. Average joint-angles were derived for each material from the converted files. These average joint-angles were then fed to the robot's API to parameterize and generate the robot motion based on the identified material.

Alongside motion capture data, human subject's wrist linear acceleration data was also captured during the experiment and used for robot motion planning. The average speed of a wrist during the scooping of each material was derived using this data. This study helped us to set a sustainable speed for the end-effector of the robot. We preset the velocity gain of the arm to a constant value of 0.20. The robot motion control executed the trajectory defined for scooping of the material type, which it received from the image classification module (Fig. 2).

3 Results

The objective of the research was to develop the decision-making capability in assistive robots to execute a task with human intuition in dynamic situations. We used machine learning algorithms CNN and k-NN for real-time material identification. We achieved 100% image classification accuracy for material image classification algorithm for k = 3. The performance of image classification algorithm for different values of k is shown in the Table 2.

By performing human motion analysis using musculoskeletal modeling, we observed how every joint angle changed according to the nature of the material. The joint angles of human arm for two extremely different materials, water and rock, can be observed in Fig. 3.

In the dragging motion primitive, we can see that the plot for pronation and supination is very smooth for water and has very prominent spikes for rock. Similarly, the average angular velocities for elbow flexion, wrist pronation-supination, and wrist flexion were calculated in each motion primitive. Figure 4 shows graphs to represent this information for water and rock. These angular velocities were used in robot motion planning for respective materials.

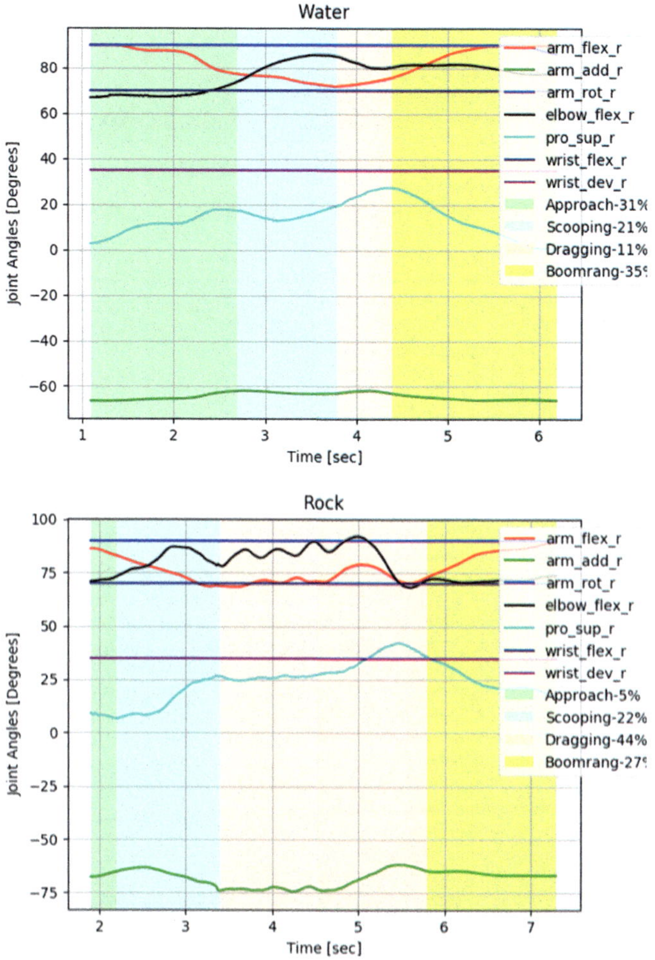

Fig. 3. Joint angles in different motion primitives for scooping water and rock

Table 2. k-NN model performance for different values of k

K-value	Classification accuracy
1	78%
2	82%
3	100%
4	87%
5	92%
6	88%

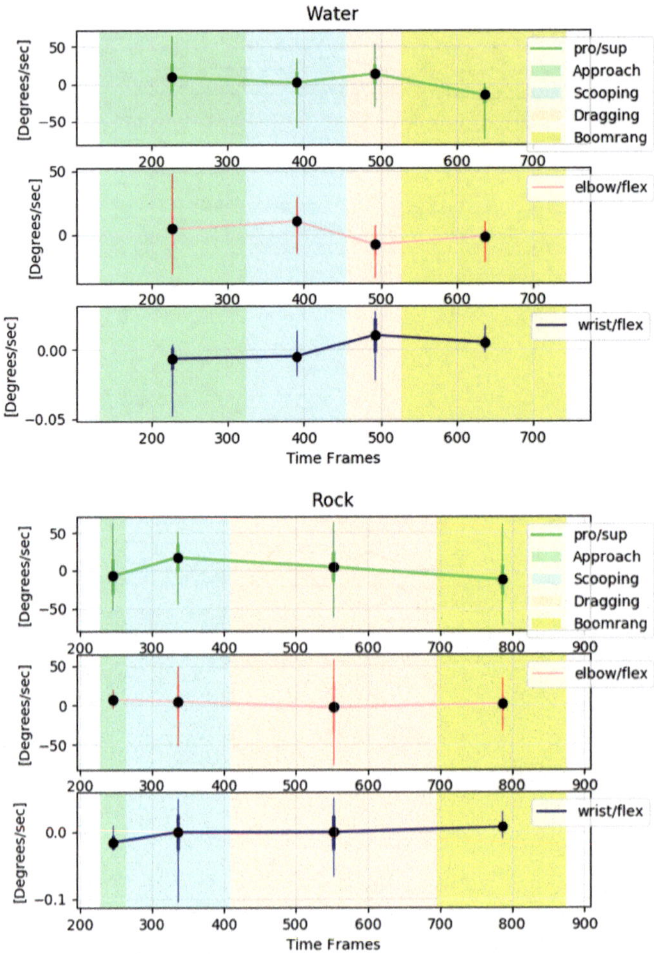

Fig. 4. Average angular velocity for scooping water and rock.

The robot was programmed and could execute the scooping action for each material with high efficiency and was comparatively efficient than human subject in terms of execution time. The scooping execution time for the human and the robot was measured manually and is compared in the graph represented by Fig. 5. The success ratio while examining the scooping motion over all trials was found to be 85%.

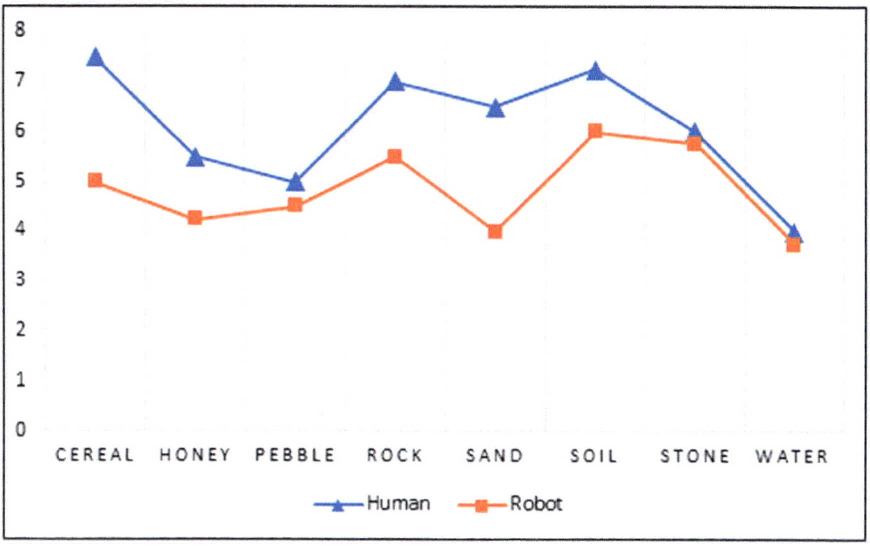

Fig. 5. Scooping activity time (in seconds) by human subject and trained robot.

4 Conclusions

In the proposed method, we recorded the dynamics of human motion for different situations by using techniques of motion capture, EMG signal monitoring and musculoskeletal modeling. We investigated the motion by simulating a human musculoskeletal model and deriving the essential information for robot motion planning. The robot was programmed to perform the desired tasks by using an angular control method, which provides more accuracy. Moreover, we provided computer vision capability to the robot, which enabled it to decide which program to execute based on the materials that it identified. As a result of these innovations, the resulting robotic motion was highly improved as the robot had human intuition (computer vision) and skills (learned from human motion).

The experimental evaluation presented in this paper yields new insights into human movement understanding and motion characterization using deep learning and biomechanical modeling. We plan to use numerical optimization methods (i.e. EMG-informed computed muscle control) to reconstruct muscle forces [2]. Since we are analyzing a non-complex upper-extremity movement, simultaneous data acquisition from EMG and mocap is not complex. The physical environment in rehabilitation robotics prioritizes accuracy over efficiency because the robot is near the human subject who is likely to have limited mobility. We carefully selected the parameters, which assured the safety of the human subject in case the motion was not executed in a desired way. Using machine learning techniques, robots can be made more capable. Additional human skills can be incorporated in robots, thereby making the motion more human like. This idea can be used to design robotic systems with different mechanical potentials and different environmental constraints, thereby adapting the methods to a wide variety of practical purposes.

We plan to extend the work by integrating force feedback (haptics) into the robot model by using artificial intelligence. By using deep learning techniques such as reinforcement learning, the robot can learn to behave like humans and make decisions in a dynamic environment. We also plan to inculcate the intelligence into the application, which will help the robot to perform its programmed activities in such scenarios. In these future experiments, we plan to make robots capable of executing actions in dynamic situations, such as mixing two different materials, changing the elevation level or the quantity of the material to scoop, or changing the robots' range of motion and degrees of freedom during the motion. In addition, we plan to analyze motion data from more than one human to investigate differences among the subjects. This study can be extended to multiple environments with different mechanical constraints, thereby enabling individuals with musculoskeletal disorders to gain independence in their daily activities.

References

1. Orin, D.E.: Application of robotics to prosthetic control. Ann. Biomed. Eng. **8**, 305 (1980). https://doi.org/10.1007/BF02363434
2. Demircan, E., Wheeler, J., Anderson, F.C., Besier, T., Delp, S.: EMG-informed computed muscle control for dynamic simulations of movement. In: Proceedings of the XXII Congress of the International Society of Biomechanics (ISB), Cape Town, South Africa (2009)
3. Riener, R., Frey, M., Bernhardt, M., Nef, T., Colombo, G.: Human-centered rehabilitation robotics. In: 9th International Conference on Rehabilitation Robotics 2005. ICORR 2005, pp. 319-322 (2005)
4. Zhang, X.Y., Wang, K.X.: Robot assisted rehabilitation technology robotics and intelligent devices. Chin. J. Rehabil. **28**(4), 246–248 (2013)
5. Gonzalez, R.V., Buchanan, T.S., Delp, S.L.: How muscle architecture and moment arms affect wrist flexion-extension moments. J. Biomech. **30**(7), 705–712 (1997)
6. Grigore, E.C., Scassellati, B.: Discovering action primitive granularity from human motion for human-robot collaboration. In: Proceedings of Robotics: Science and Systems, July 2017
7. Lim, B., Ra, S., Park, F.: Movement primitives principal component analysis and the efficient generation of natural motions. In: Proceedings IEEE International Conference on Robotics Automation, pp. 4630-4635 (2005)
8. Optitrack system. https://optitrack.com/
9. Khatib, O., Demircan, E., De Sapio, V., Sentis, L., Besier, T., Delp, S.: Robotics-based synthesis of human motion. J. Physiol. Paris **103**, 211–219 (2009)
10. Seth, A., Sherman, M., Reinbolt, J.A., Delp, S.L.: OpenSim: a musculoskeletal modeling and simulation framework for in silico investigations and exchange. PMC, April 2015
11. Thelen, D.G., Anderson, F.C., Delp, S.L.: Generating dynamic simulations of movement using computed muscle control. J. Biomech. **36**(3), 321–328 (2003)
12. Park, J.S., Park, C., Manocha, D.: Intention-aware motion planning using learning based human motion prediction. In: Proceedings of Robotics: Science and Systems, July 2017

13. Dantam, N.T., Kingston, Z.K., Chaudhuri, S., Kavraki, L.E.: Incremental task and motion planning: a constraint-based approach. In: Robotics: Science and Systems (2016)
14. Albrecht, S., Ramırez-Amaro, K., Ruiz-Ugalde, F., Weikersdorfer, D., Leibold, M., Ulbrich, M., Beetz, M.: Imitating human reaching motions using physically inspired optimization principles. In: IEEE-RAS International Conference on Humanoid Robots, Bled, Slovenia (2011)
15. Demircan, E., Besier, T., Menon, S., Khatib, O.: Human motion reconstruction and synthesis of human skills. In: Lenarcic, J., Stanisic, M.M. (Eds.) Advances in Robot Kinematics, pp. 283-292. Springer, Heidelberg (2010)
16. Holzbaur, K.R., Murray, W.M., Delp, S.L.: A model of the upper extremity for simulating musculoskeletal surgery and analyzing neuromuscular control. Ann. Biomed. Eng. **33**(6), 829–840 (2005)
17. Demircan, E., Sentis, L., De Sapio, V., Khatib, O.: Human motion reconstruction by direct control of marker trajectories. In: Lenarcic, J., Stanisic, M.M. (Eds.) Advances in Robot Kinematics, pp. 263-272. Springer, Heidelberg (2008)
18. EEGO system. https://www.ant-neuro.com/products/eego_sports
19. Pan, S.J., Yang, Q.: A survey on transfer learning. IEEE Trans. Knowl. Data Eng. **22**, 1345–1359 (2010)
20. Oquab, M., Bottou, L., Laptev, I., Sivic, J.: Learning and transferring mid-level image representations using convolutional neural networks. In: Proceedings of the 2014 IEEE Conference on Computer Vision and Pattern Recognition, Columbus, OH, USA, pp. 1717-1724, 23–28 June 2014
21. Abadi, M., Barham, P., Chen, J., Chen, Z., Davis, A., Dean, J., Devin, M., Ghemawat, S., Irving, G., Isard, M., Kudlur, M., Levenberg, J., Monga, R., Moore, S., Murray, D.G., Steiner, B., Tucker, P., Vasudevan, V., Warden, P., Wicke, M., Yu, Y., Zheng, X.: TensorFlow: a system for large-scale machine learning. Google Brain. In: 12th USENIX Symposium on OSDI, (2016)
22. Simonyan, K., Zisserman, A.: Very deep convolutional networks for large-scale image recognition. arXiv:1409.1556v6 [cs.CV], 10 April 2015
23. Caffe Model. https://gist.github.com/ksimonyan/211839e770f7b538e2d8#file-readme-md
24. Imagenet. http://image-net.org/
25. Krizhevsky, A., Sutskever, I., Hinton, G.: ImageNet classification with deep convolutional neural networks. In: Advances in Neural Information Processing Systems (NIPS) (2012)
26. K-nearest neighbor. https://en.wikipedia.org/wiki/K-nearest_neighbors_algorithm

Unmanned Aerial Vehicles I

Unmanned Aerial Vehicles (UAV) I

Nick Roy

The session focused on aerial systems, with particular consideration of multiple air or mixed air and ground systems. In the first paper, Wang *et al.* proposed a novel exploration algorithm for a team of air and ground robots working together. Their paper showed that by explicitly modelling the differential capabilities of air and ground vehicles, a more efficient exploration strategy could be computed. In the second paper, Ebadi and Agha-Mohammadi showed that similar consideration of air and ground vehicles could enable air to ground localization on Mars. The paper described compelling experimental results produced in a Martian analog. In the third paper, Brodie *et al.* described their VoliroX platform, a 12 rotor MAV with 6 tiltable arms. Their results showed that their new rotary wing morphology achieves both force- and pose-omnidirectionality with highly dynamic capabilities, while maintaining high efficiency in hover. In the fourth paper, Gautam *et al.* examined the problem of autonomous landing, which is notoriously challenging due to external disturbances and estimation errors. The paper presents and assesses a novel vision based guidance approach for landing on a moving platform. Finally, in the last paper, Klein *et al.* showed how a UAV could learn to classify terrain during flight, and presented results of tests in the Calaveras River in Stockton, CA. This set of papers represent substantial progress in pushing the capabilities and level of autonomy in complex aerial systems.

A Collaborative Aerial-Ground Robotic System for Fast Exploration

Luqi Wang[1(✉)], Daqian Cheng[2], Fei Gao[1], Fengyu Cai[1], Jixin Guo[1], Mengxiang Lin[2], and Shaojie Shen[1]

[1] The Hong Kong University of Science and Technology, Kowloon, Hong Kong
{lwangax,fgaoaa,fcaiaa,jguoaj,eeshaojie}@connect.ust.hk
[2] Beihang University, Beijing, China
{chengdaqian,linmx}@buaa.edu.cn

1 Introduction

Exploration of unknown environments using autonomous robots has been considered a fundamental problem in robotics applications such as search and rescue [10], industrial inspection and 3D modelling. For exploration, the basic requirement for robots is to scan an unknown space or detect free space as fast as possible. Unmanned ground vehicles (UGVs) [12] and unmanned aerial vehicles (UAVs) [2] have both been employed for such tasks, with each having different advantages: UGVs are more payload-capable, so that they can carry heavy, long-range laser scanners which are inapplicable for weight-constrained UAVs. Meanwhile, UAVs have superiors mobility and agility, since they can fly above obstacles and cover areas that are inaccessible to UGVs, like obstacles' top surfaces. Consequently, a UGV often enjoys a larger sensor coverage, yet cluttered and view-blocking environments hamper its performance. On the other hand, a UAV may deliver inferior exploration efficiency due to its short-range sensor, but enjoys an unblocked downward-looking view. Therefore, UGVs are more favorable for open areas, while UAVs are preferable in cluttered environments.

There are numerous works conducted on robotic exploration. [12] first proposed the concept of frontiers, which are defined as unknown grid-map cells that are adjacent to free ones and thus represent accessible new information. Harmonic function, the solution to *Laplace's Equation*, shown as Eq. 1, is used to plan the path to frontiers [5]. This method generates a scalar field in free-space based on its surrounding boundary conditions (occupied cells and frontier cells) and obtains the path using gradient-descent. These works are only conducted on

L. Wang and D. Cheng—contributed equally to this work.
This work was supported by HKUST project R9341.

Electronic supplementary material The online version of this chapter (https://doi.org/10.1007/978-3-030-33950-0_6) contains supplementary material, which is available to authorized users.

J. Xiao et al. (Eds.): ISER 2018, SPAR 11, pp. 59–71, 2020.
https://doi.org/10.1007/978-3-030-33950-0_6

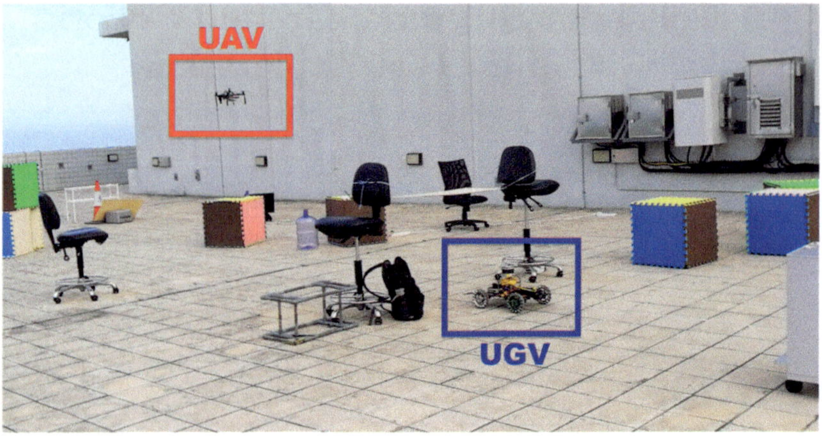

Fig. 1. Outdoor exploration in progress.

ground platforms. For aerial-ground exploration, [1] deploys a UAV as a back-up instead of as an independent explorer, only when the UGV encounters high, invisible areas. The system in [4] is also proposed based on the same idea that one vehicle helps another, failing to exploit both vehicles' full potential.

In this paper, considering the different environmental advantages of UAVs and UGVs, we propose an autonomous collaborative framework which utilizes their complementary characteristics to achieve higher efficiency and robustness in exploration applications. We summarize our contribution as follows:

1. An efficient exploration framework that combines UAVs' and UGVs' advantages.
2. A more efficient computation method for the harmonic function for robotic exploration tasks.
3. Integration of the proposed collaborative exploration framework with state estimation, sensor fusion and trajectory optimization modules. Extensive field experiments and simulations are presented to validate the efficiency and robustness of the proposed method.

2 Environment Representation

The information about the exploration environment is represented in a global grid map. Besides obstacles and free space, other information is required to be represented in the map, including frontiers and harmonic fields, to facilitate the exploration.

2.1 Frontier

The original definition of a frontier, the boundary that separates unknown space and free space, is proposed in [12]. This definition of a frontier can be applied to UGVs directly. However, it is not suitable to directly apply to UAVs in some cases as they hover above obstacles. To facilitate the aerial-ground collaborative case, we define two types of frontier, as shown in Fig. 5:

1. Frontier A: The conventional definition of frontier proposed by [12], which is the boundary between free space and unknown space. This kind of frontier is visible to both ground and aerial vehicles.
2. Frontier B: A new type of frontier, which is the boundary between obstacles and unknown space. This kind of frontier usually occurs on the top surfaces of obstacles, and can, therefore, only be seen by an aerial vehicle.

According to the advantages introduced stated in Sect. 1, we can intuitively conclude that UGVs pursue longer Frontier A lines, while UAVs are more favorable for Frontier B over Frontier A. The exploration planner is designed based on this intuition. In practice, we are using a grip map to store map information. Hence, the frontiers and other objects are all formed by grid cells in the exploration map.

2.2 Harmonic Field

To facilitate the UGV path planning, we employ a harmonic function, formulating the exploration space as a potential field. The governing equation of potential ϕ in 2D is

$$\nabla^2 \phi = \frac{\partial^2 \phi}{\partial x^2} + \frac{\partial^2 \phi}{\partial y^2} = 0. \tag{1}$$

The boundary conditions are set as below:

1. Occupied cells and the ground vehicle: zero potential value.
2. Frontier cells: the negative value of its frontier lines' lengths.

3 Path Planning

To set up an effective exploration planner, we first formulate the proposed exploration-with-preference problem using a frontier method. Then a harmonic field method is adopted to generate a path towards one of the frontiers for the ground vehicle. Meanwhile, motion primitives are generated for evaluation and the primitive with the highest score is chosen as the target of the aerial vehicle. In the case that no primitive has a positive score, a target on the frontiers with the highest score is chosen. Finally, the paths are optimized and executed by the vehicles.

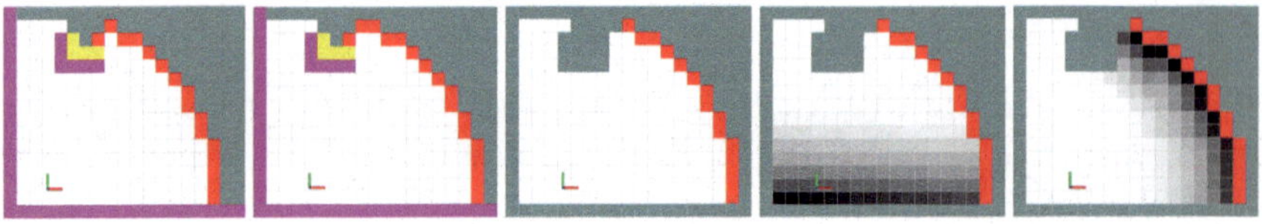

(a) original map (b) updated map (c) new boundary (d) conventional sequence (e) spreading sequence

Fig. 2. Illustration of harmonic field's update sequence. As the UGV moves forward, the map updates and new boundary cells are generated. In (c), white cells are free and need updating, gray cells are non-free and fixed-value, and red cells are new boundary cells. (d) and (e) show the conventional-sequence and spreading-sequence respectively; each cell's darkness represents its update sequence.

3.1 Terrestrial Planning

To utilize the formulated harmonic field mentioned in Sect. 2.2 for UGV path planning, We first discretize the governing equation, Eq. 1, and adopt the Successive over-relaxation (SOR) method [3] with relaxation parameter w, as shown in Eq. 2, to iteratively compute the potential value until convergence, and in each iteration, cells' potential values are updated sequentially. Such an update-sequence is conventionally the map-array's sequence (row-by-row), but we propose an update-sequence that spreads out from the new boundary conditions as shown in Fig. 2(e). This sequence is able to spread their influence faster, and the computation can converge within fewer iterations.

$$\phi_{x,y}^{k+1} = (1-w)\phi_{x,y}^{k} + \frac{w}{4}(\phi_{x+1,y}^{k} + \phi_{x-1,y}^{k} + \phi_{x,y+1}^{k} + \phi_{x,y-1}^{k}) \quad (2)$$

When the field converges, a simple gradient-descent path can be obtained from the location of the ground vehicle.

Figure 5 gives an example of a harmonic function field and its gradient field, and the path obtained in this method has the following desirable features:

1. Being repelled from obstacles; thus the sensor's range is more utilized.
2. Being inclined to lead to longer and nearer frontier lines.

3.2 Aerial Planning

To achieve rapid exploration for the UAV, we employ a motion primitive method [7] to generate path candidates for evaluation. The state space model is modeled as Eq. 3 and the input is the body linear acceleration a_{body}:

$$\dot{x} = Ax + Ba_{body}. \quad (3)$$

With the state space model, the state transition equation can be derived to calculate the primitives:

$$x(t) = e^{At}x_0 + \int_0^t e^{A(t-\sigma)} B a_{body} d\sigma. \tag{4}$$

The evaluation of the a primitive P is based on its score, which is formulated as the information gain along the path to the primitive divided by the cost to the primitive:

$$P.Score = \frac{P.InfoGain}{P.Cost}. \tag{5}$$

The cost is modeled as a linear quadratic minimum-time problem [11]:

$$P.Cost = \int_0^t (a_{body}^T R a_{body} + \rho) d\tau, \tag{6}$$

where ρ is the weight parameter tuning time and input and R is the weight parameter of the acceleration in different directions. Larger ρ implies a larger weight of time.

The information gain of a grid cell C is modeled as

$$C.InfoGain = \sum_{f_i}^{Fv_A^{UAV}} w_A + \sum_{f_j}^{Fv_B^{UAV}} w_B + \sum_{u_k}^{Uv^{UAV}} w_U, \tag{7}$$

where Fv_A^{UAV}, Fv_B^{UAV} and Uv^{UAV} are the set of visible Frontier A cells, Frontier B cells and unknown cells. As Frontier B cells indicate the top surface of the obstacles, Frontier B cell clusters indicate obstacle-cluttered areas, preferable for the aerial vehicle. Therefore, a larger weight should be assigned to Frontier B cells. In practice, w_B, w_A and w_U are set to 0.89, 0.1 and 0.01 respectively. The information gain along the path to the primitive is calculated generally by the summation of the information gain of the cells along the path, except that the visible cells are not counted repetitively.

An extensive primitive-tree is acquired by iteratively enforcing a discretized set of accelerations, as shown in Fig. 3. Meanwhile, the primitives that exceed the speed limit are trimmed to ensure dynamic feasibility. Finally, the path candidate with the highest score is selected.

Nevertheless, restricted by the field of view (FOV) and the dynamic state of the aerial vehicle, it is possible that the paths to the primitives do not possess positive information gains or scores, since the planning strategy stated earlier in this section is just locally optimal. In this situation, a global target selector is adopted. The candidate target set is assigned to be all the frontier cells and the information gain at a cell is computed utilizing Eq. 7. The costs to each of the frontier cells are set to be the Euclidean norm between the UAV's position and the frontier cells. The scores are calculated in the same way as mentioned in Eq. 5 and the position with the highest score is selected as the target.

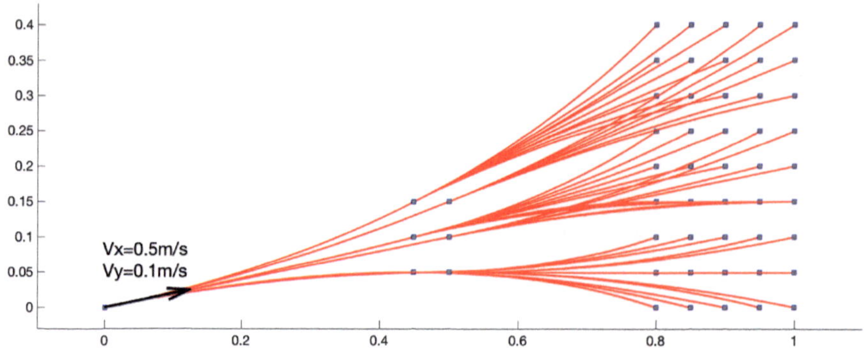

Fig. 3. Illustration of motion primitives with discretization of linear acceleration. The arrow at the initial position represents the initial velocity of 0.5 m/s and 0.1 m/s in x and y directions. The squares indicate the primitives, while the red curve represents the path to the primitives. Some of the primitives are trimmed since they exceed the speed limit of 0.5 m/s on any of the axes.

3.3 Trajectory Optimization

With the target path or position selected for UGV and UAV in Sects. 3.1 and 3.2, the trajectories along the paths or towards the positions are required to be optimized against the energy cost while ensuring dynamic consistency for the vehicles to follow.

For the ground vehicle, we adopt the method described in our previous work [6] for optimization. With a gradient-descent path generated, an obstacle-free travel corridor along it can be produced. Then, with the waypoints, continuity, safety, and dynamically feasibility constraints enforced, the energy cost, which is set to be the squared jerk of the trajectory, is optimized inside the corridor to generate piece-wise polynomial trajectories using the Bernstein polynomial basis. The ith-order Bernstein polynomial basis is defined as

$$b_n^i(t) = \binom{n}{i} \cdot t^i \cdot (1-t)^{n-i}, \tag{8}$$

where n is the degree of the Bernstein polynomial basis, $\binom{n}{i}$ is the binomial coefficient and t is the variable parameterizing the trajectory. The waypoints constraints are enforced by assigning the initial and final positions, velocities and accelerations; the continuity constraints restrict the continuity of the derivatives between the curve pieces; the safety constraints add additional boundary limits to the corridor to ensure safety; and the dynamical feasibility constraints bound the velocity and the acceleration along the trajectories.

For the aerial vehicle, we adopt a minimum jerk trajectory generator [8] to minimize the squared jerk with waypoint and continuity constraints enforced, which is similar to the ground vehicle, using polynomial trajectories.

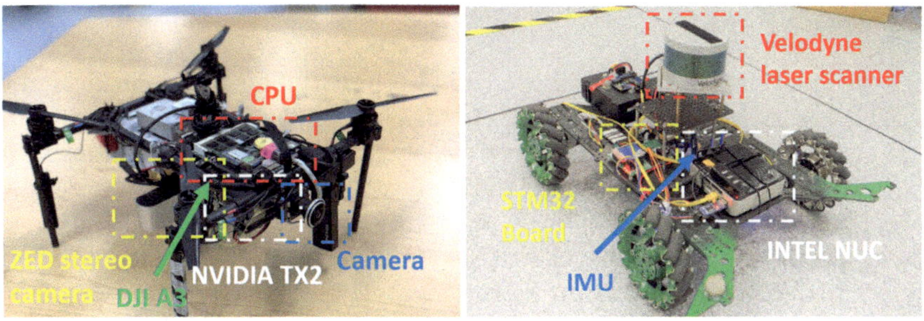

(a) Aerial platform (b) Ground platform

Fig. 4. Experiment platforms.

4 System Description

The collaborative system contains an aerial platform and a ground platform. Both of the platforms shown in Fig. 4 are custom-built vehicles, which can localize themselves and perform mapping in dense and complex environments. The exploration planner collects the real-time dynamic states of the vehicles to plan trajectories for them to follow. There is a wireless network set up for the communication between the vehicles.

4.1 Ground Platform

Our ground platform is an omni-directional vehicle with four Mecanum wheels, driven by four DJI RM 3510 brushless motor, which are installed on a carbon fiber chassis. A Velodyne VLP-16 laser scanner with a range up to 100 m is mounted above the chassis to serve both localization with an inertial measurement unit (IMU) and mapping purposes. The laser scan data are collected by an on-board Intel NUC with an i7-5500U processor, running Ubuntu 16.04 and ROS Kinetic, through Ethernet. The NUC is also responsible for other computation, including state estimation, mapping and high-level control. The velocity commands are sent to a STM32F407 microcontroller board through a serial port to directly control the motors on the chassis.

Localization, mapping and control programs are implemented on the NUC. An Extended Kalman Filter (EKF) is implemented to fuse IMU data and odometry data from the SLAM program LOAM [13] for real-time localization. The scanned laser point cloud is collected and transformed into the world frame to integrate into the globally maintained map. Meanwhile, the fused odometries are sent to the exploration planner to generate trajectories for the vehicle to follow. The pose controller receives trajectory commands and sends velocity commands to the STM32 microcontroller to control the motors.

4.2 Aerial Platform

Our aerial platform is a self-developed quadrotor with four DJI Snail 2305 motors installed to serve as actuators. A DJI A3 flight controller is utilized as the autopilot to control the attitude of the drone. Meanwhile, the IMU inside together with the forward-looking MatrixVision mvBlueFox-MLC200w global shutter camera with a 235°-FOV fish-eye lens performs state estimation to localize the aerial vehicle. A downward-looking Stereolabs ZED stereo camera with an FOV of $110^\circ \times 80^\circ$ and a range of 20 m is mounted below the vehicle to produce depth images for the mapping. The computations are all performed onboard on an Intel i7-5500U processor and an Nvidia Jetson TX2, both of which are running Ubuntu 16.04 and ROS Kinetic.

The software running onboard is responsible for localization, mapping and exploration planning. The VINS-Mono from our previous work [9], which fuses IMU data and image data from the forward-looking camera together, is adopted for localization of the aerial vehicle. Besides the control purpose, the fused odometries are also sent to the exploration planner for planning. The SDK of the ZED stereo camera provides depth images and point clouds for dense mapping and the point clouds are transformed and merged into the global map.

5 Experiments and Results

5.1 Outdoor Autonomous Exploration

The outdoor experiment site is an $17\,\text{m} \times 8\,\text{m}$ area, with various obstacles, as shown in Fig. 1. In Fig. 7, the quadrotor and the ground vehicle both start from the bottom to perform exploration. The downward-looking camera is configured to have a $2\,\text{m} \times 2\,\text{m}$ FOV. The UAV is flying at a constant height of 2 m with maximum speed 1.4 m/s, while the UGV is configured to have a laser scanned range of 6 m and travels at a maximum speed of 0.5 m/s. The exploration results are shown in Figs. 7 and 8(b).

5.2 Simulative Exploration

We also perform simulations on distinctive maps. All environments are $20\,\text{m} \times 20\,\text{m}$ in size and have a 0.1-m resolution. Exploration starts with both robots placed at a corner. The UGV travels at a maximum speed of 1m/s, while the UAV has a higher maximum speed of 1.4 m/s. The simulated UAV is equipped with a downward-looking depth camera with a FOV of $60^\circ \times 60^\circ$ in the vertical and horizontal directions, and the UGV carries a 360° laser scanner with a 6-m range. The UAV hovers at a constant height of 2 m, resulting in a 4.6-m^2 sensor coverage. The exploration results are shown in Figs. 6 and 8(a).

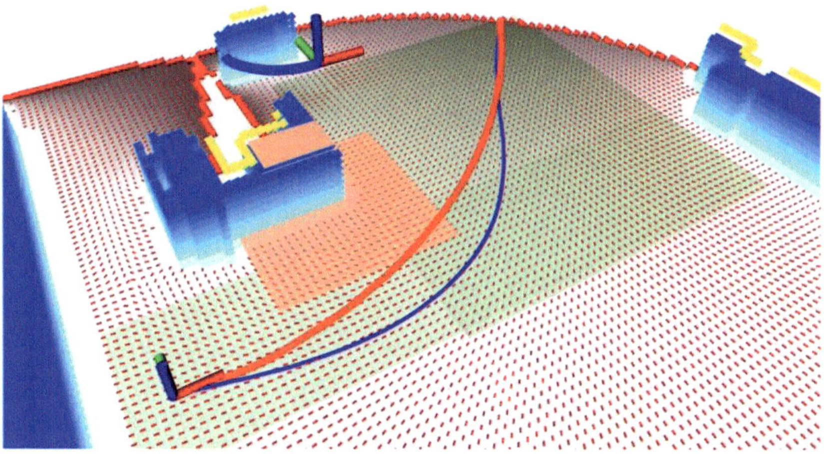

Fig. 5. Red and yellow points are Frontier A/B cells respectively. The orange point cloud is from the UAV's sensor. Darkness represents the harmonic function value, and red vectors represent the cells' gradient. The field guides the UGV to a further but longer frontier line. Green squares are the travel corridor, and the blue and red curve on the ground are the original gradient-descent and the optimized trajectory respectively. The blue curve above is the trajectory planned for the UAV.

5.3 Comparisons and Analyses

Figure 5 shows a visualization-capture, as an illustration of the framework. To examine the framework, we compare the performance (progress against time) of the UAV-UGV team, a single UAV, and a single UGV. Figures 6, 7 and 8 show the exploration performance in simulations and outdoor experiments. We also test the sequences mentioned in Sect. 3.1, and the results are shown in Table 2.

As shown in Fig. 8, the collaborative team has significantly higher exploration efficiency than individual vehicles. The time for exploration of the system is significantly reduced, as is the trajectory length of each vehicle, as shown in Figs. 6 and 7. According to the results shown in Table 1, the collaborative system has at least 23% higher exploration speed compared with UGV alone, at least 145% higher exploration speed compared with UAV alone and the advantages can reach up to 163% and 235% respectively. From the definition of the collaborative efficiency η_{col} in Eq. 9, the variable can reflect the effectiveness of the collaboration. For the worst of the testing cases, uniform random, the collaborative efficiency can still reach 0.90, even though the exploration site is the least suitable for utilizing the advantages of both vehicles. For the intuitively most favorable case, the maze, the efficiency can reach 1.45, indicating the proper modeling of the system. The high collaborative efficiency reflects the effectiveness of the work distribution of the vehicles. Meanwhile, the computation times of the trajectories can all be restricted to below 80 ms, meaning that the method can be adopted for real-time online navigation.

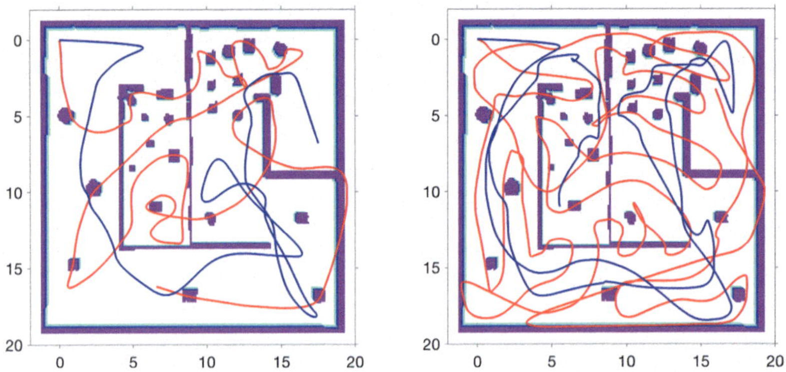

(a) Collaborative trajectory (b) Individual trajectory

Fig. 6. Exploration results of the maze map. Red curve represents UAV; blue curve represents UGV.

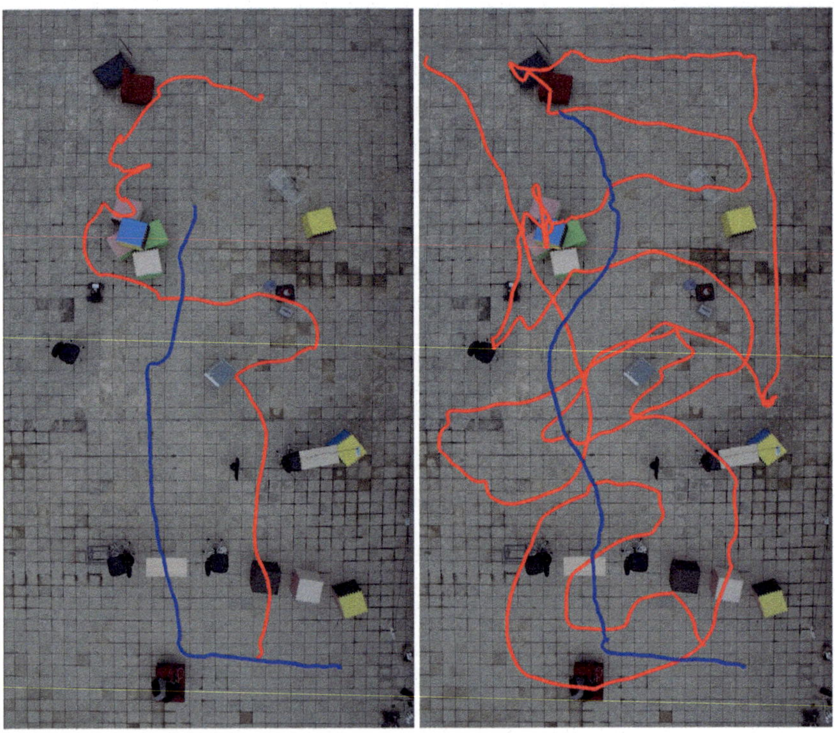

(a) Collaborative trajectory (b) Individual Trajectory

Fig. 7. Exploration results of the experiment site. Red curve represents UAV; blue curve represents UGV.

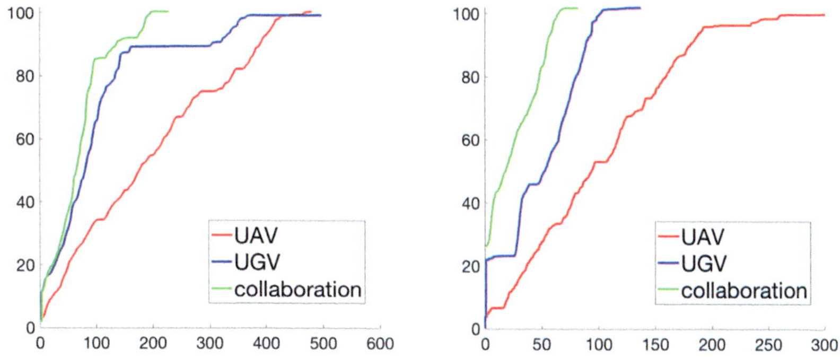

(a) Maze (b) Outdoor Experiment

Fig. 8. Exploration progress in simulation and outdoor experiment.

Table 1. Comparison of exploration time and trajectory generation time

Environment type	T_{exp}^{UGV}[a] (s)	T_{exp}^{UAV}[a] (s)	T_{exp}^{col}[a] (s)	η_{col}[b]	UAV Avg. $t_{traj.gen.}$[c] (ms)	UGV Avg. $t_{traj.gen.}$ (ms)
Maze	366.25	448.75	**139.50**	1.45	27.3	21.5
Fence and clusters	278.75	448.75	**178.00**	0.97	28.5	23.9
Uniform random	146.80	401.50	**119.75**	0.90	28.9	17.4
Outdoor experiment site[c]	94	191	**59**	1.07	73.9	45.0

[a] T_{exp}^{UGV}, T_{exp}^{UAV} and T_{exp}^{col} denote the time duration from the start of the exploration to exploration progress of 95% of the UGV alone, the UAV alone and the collaborative system respectively, since the interior of the obstacles cannot be fully scanned by the UGV.

[b] η_{col} denotes the collaborative efficiency, which is defined as:

$$\eta_{col} = \frac{\frac{1}{T_{exp}^{col}}}{\frac{1}{T_{exp}^{UGV}} + \frac{1}{T_{exp}^{UAV}}} \quad (9)$$

[c] $t_{traj.gen.}$ denotes the computation time for trajectory generation.

[d] Note that the simulation and the outdoor experiment platforms are different. The trajectory generation time section of the outdoor experiment is just for reference.

Table 2. Comparison of the different update sequences of the harmonic field

Environment type	Convergence iterations of conventional sequence	Convergence iterations of BFS sequence	Iteration reduction of BFS in percentage
Maze	268.837	**250.068**	6.98%
Fence and clusters	347.648	**331.658**	4.60%
Uniform random	396.084	**365.863**	7.63%

In Table 2, the proposed BFS update sequence of the harmonic field mentioned in Sect. 3.1 is compared with the normal row-by-row update sequence. It is clearly shown that the BFS sequence consumes fewer iterations for each of the cases, hence reducing the computation power by 4.60% to 7.63%.

6 Experimental Insights

Control is always the most basic issue in robotics experiment. The control of a quadrotor system can be affected by multiple factors. The error from the actuators can significantly deteriorate the performance, affecting other parts of the system. On our aerial platform, the original DJI 2312E motors with a maximum thrust of 850-g/rotor was substituted with DJI 2305 Snail motors with a maximum thrust of 1.32-kg/rotor. Reserving a larger margin on the actuators can reduce the uncertainty in the control segment, enhancing the control performance, and it can further increase the reliability in field applications.

In this experiment, the re-planning strategy is one of the most crucial issues. Time allocation is still a significant problem in path planning and it can produce serious effects on the consistency of the generated trajectories. In our system, to avoid the inconsistency caused by improper time allocation, we take trajectory length, time, and initial and end state into consideration to produce an effective re-planning strategy.

7 Conclusion

In this paper, we propose an autonomous aerial-ground collaborative exploration system. The system consists of a UGV and a UAV and utilizes their complementary characteristics to achieve high efficiency in the exploration process. The ground vehicle carries a laser scanner to localize itself with the IMU and construct the map while traveling amid obstacles; the UAV utilizes a forward-looking camera to localize itself with the installed IMU, while constructing the map through a downward-looking stereo camera when flying above obstacles. The proposed exploration planning method combines a frontier method, a field method and a motion primitive method to plan efficient and consistent motions for the vehicles.

Extensive simulated experiments are performed to validate the proposed method. Furthermore, the aerial and ground platforms, which are integrated with state estimation, dense mapping and exploration planning modules, are implemented to verify the practicability of the method in an outdoor environment. The results show that the collaborative system has superior exploration speed against each vehicle alone and has a high collaborative efficiency.

References

1. Butzkey, J., Dornbushy, A., Likhachevy, M.: 3-D exploration with an air-ground robotic system. In: 2015 IEEE/RSJ International Conference on Intelligent Robots and Systems (IROS), pp. 3241–3248 (2015)
2. Cieslewski, T., Kaufmann, E., Scaramuzza, D.: Rapid exploration with multi-rotors: a frontier selection method for high speed flight. In: 2017 IEEE/RSJ International Conference on Intelligent Robots and Systems (IROS), pp. 2135–2142 (2017)

3. Connolly, C.: On the application of harmonic functions to robotics. J. Robot. Syst. **10**, 931–946 (1993)
4. Delmerico, J., Mueggler, E., Nitsch, J., Scaramuzza, D.: Active autonomous aerial exploration for ground robot path planning. IEEE Robot. Autom. Lett. **2**(2), 664–671 (2017)
5. Silva, E.P., Engel, P.M., Trevisan, M., Idiart, M.A.P.: Exploration method using harmonic functions. Robot. Auton. Syst. **40**(1), 25–42 (2002)
6. Gao, F., Wu, W., Lin, Y., Shen, S.: Online safe trajectory generation for quadrotors using fast marching method and Bernstein basis polynomial. In: Proceedings of the IEEE International Conference on Robotics and Automation (ICRA) (2018)
7. Likhachev, M., Ferguson, D.: Planning long dynamically feasible maneuvers for autonomous vehicles. Int. J. Robot. Res. **28**(8), 933–945 (2009)
8. Mellinger, D., Kumar, V.: Minimum snap trajectory generation and control for quadrotors. In: International Conference on Robotics and Automation (ICRA), pp. 2520–2525. IEEE (2011)
9. Qin, T., Li, P., Shen, S.: VINS-Mono: a robust and versatile monocular visual-inertial state estimator. arXiv preprint arXiv:1708.03852 (2017)
10. Shen, C., Zhang, Y., Li, Z., Gao, F., Shen, S.: Collaborative air-ground target searching in complex environments. In: IEEE International Symposium on Safety, Security and Rescue Robotics (SSRR), pp. 230–237 (2017)
11. Verriest, E.I., Lewis, F.L.: On the linear quadratic minimum-time problem. IEEE Trans. Autom. Control **36**(7), 859–863 (1991)
12. Yamauchi, B.: A frontier-based approach for autonomous exploration. In: Proceedings of 1997 IEEE International Symposium on Computational Intelligence in Robotics and Automation, CIRA 1997, pp. 146–151 (1997)
13. Zhang, J., Singh, S.: LOAM: lidar odometry and mapping in real-time. In: Robotics: Science and Systems Conference (2014)

Rover Localization in Mars Helicopter Aerial Maps: Experimental Results in a Mars-Analogue Environment

Kamak Ebadi(✉) and Ali-Akbar Agha-Mohammadi

NASA Jet Propulsion Laboratory, California Institute of Technology, Pasadena, CA, USA
{ebadi,aliagha}@jpl.nasa.gov

Abstract. In a potential Mars sample return mission, a Mars rover is required to visit previously explored and mapped environments in order to retrieve previously collected samples for subsequent return to Earth. In such a mission, the rover needs to establish its position within a provided map to safely and efficiently plan a path toward the goal locations. In this work, we study the feasibility and performance of aerial-to-ground (A2G) localization of the Mars rover by registering rover's ground imagery to an aerial map of a Mars analogue environment. Through empirical experiments at the Jet Propulsion Laboratory's Mars Yard, we present performance, robustness and sensitivity analysis for A2G localization in varying lighting conditions, viewing angles, terrain types and using different image feature detectors and descriptors.

1 Introduction

As depicted in Fig. 1, NASA will deploy a low-flying rotorcraft scout [1] to collaborate with agency's Mars 2020 rover mission that is scheduled to launch in July

Fig. 1. A Mars helicopter scouts the Martian terrain, and provides high resolution aerial maps of the terrain. The rover needs to establish its position within a provided map in order to safely and efficiently navigate toward the goal locations. Image credit: NASA/JPL-Caltech

J. Xiao et al. (Eds.): ISER 2018, SPAR 11, pp. 72–84, 2020.
https://doi.org/10.1007/978-3-030-33950-0_7

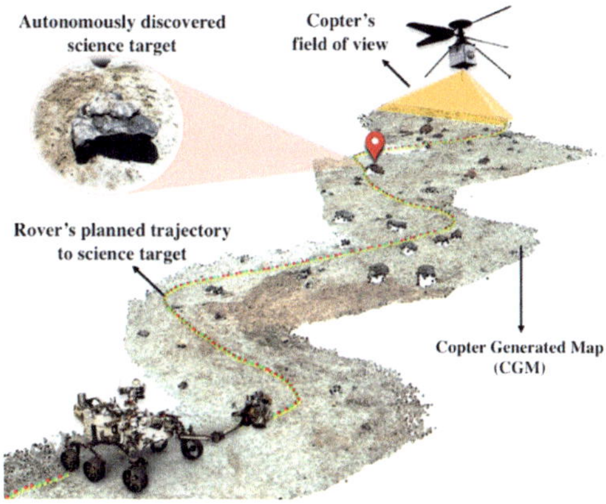

Fig. 2. Using copter's aerial imagery in JPL's Mars Yard, a 3D map is created that encodes information about the traversability of the terrain, as well as the location of autonomously discovered science targets. Rover's safe and efficient traverse toward the science targets is contingent upon establishing localization in the provided map.

2020. The Mars helicopter would fly at altitudes ranging from 3 m to 10 m above the surface, and provide high resolution aerial imagery that could be used to create high-fidelity and information-rich aerial maps of the local terrain. Rover's safe and efficient traverse toward goal locations in a mapped environment is contingent upon establishing accurate localization in the provided copter-generated map (CGM). Figure 2 illustrates a collaborative rover-copter mission in a Mars-analogue environment, where the CGM is created from a sequence of copter's aerial imagery in a Mars-analogue environment and encodes information about traversability of the terrain and perceptual features with known locations that can serve as information sources for rover's localization.

Related Work: There is a broad literature on rover localization based on satellite imagery. Authors in [2], warp images captured by the rover to obtain a bird's eye view of the ground before registering the warped images to a grid of satellite

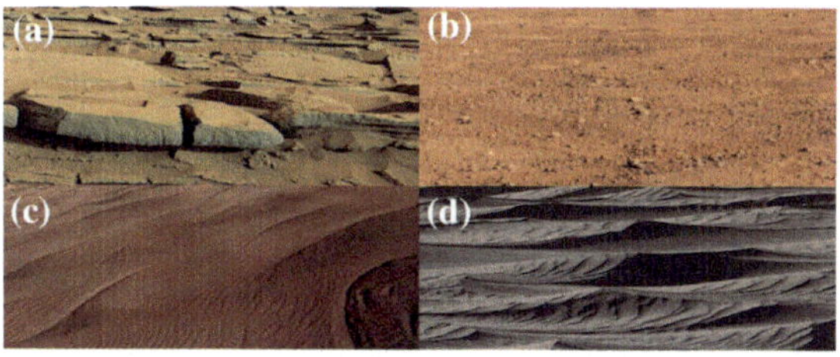

Fig. 3. (a) represents a feature-rich environment. (b) and (c) represent a feature-poor environment with sparse perceptual features. (d) represents an ambiguous environment with repetitive features. Image credit: NASA/JPL-Caltech

imagery. Similarly in [3–5], a localization approach using HiRISE [6] imagery is introduced, where outstanding landmarks such as craters, cliffs, and hills are used for establishing orbital to ground localization. The low spatial resolution of HiRISE imagery limits the use of such methods to environments with large distinctive features.

Localization methods that rely on A2G image matching [2,7–9], have proven to be successful where sufficient salient features can be extracted from aerial and ground imagery. Upon establishing a successful A2G correspondence, the location of the ground image is inferred from its corresponding geotagged aerial images.

As presented in Fig. 3, distinctive landmarks and rich perceptual features are often sparse in a Mars-like environment. In addition, variations in illumination and viewing angles between aerial and ground imagery could further exacerbate the challenge of establishing accurate A2G localization.

Objective: A Mars-like environment with poor and ambiguous perceptual features poses a challenge to establishing accurate A2G localization. In this work, we study the feasibility and performance of localization in a Mars-analogue environment by relying on aerial-to-ground image correspondence integrated with a Bayesian filtering scheme.

2 A2G Scene Correspondence

A Mars rover's safe and efficient traverse in a previously mapped environment is contingent upon establishing the pose of the rover on the provided map. In a GPS-denied Martian environment, the rover needs to rely on A2G image correspondence to estimate its position in the map. However, A2G image correspondence in a feature-poor Mars-like environment can be ambiguous and lead to perceptual aliasing. In this section, as depicted in Fig. 4 we present an A2G image similarity evaluation pipeline that takes the rover's ground observation as an input and outputs a similarity distribution that represent the likelihood of the rover's observation at all known locations in the aerial map. In the rest of this section we explain different components of the A2G similarity evaluation pipeline.

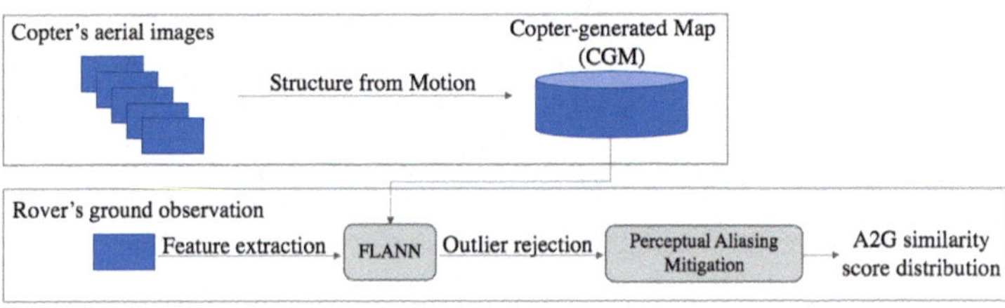

Fig. 4. A2G image similarity evaluation pipeline

Copter-Generated Map (CGM): We consider a dataset $A = \{a^i\}_{i=1}^n$ where a^i is the i-th aerial image and n is the total number of aerial images obtained by the Mars helicopter. In order to extract distinctive image features and construct a map by estimating their 3D pose, we use the iterative structure from motion (SfM) method presented in [10]. SfM involves image feature extraction and matching, camera pose estimation, and recovery of the 3D structure of the terrain. We denote the copter-generated map by $CGM = \{f^i_{a\in A}, l^i_{a\in A}\}_{i=1}^n$, where $f^i_a \in A$ is the i-th feature extracted from an aerial image and $l^i_a \in A$ is the estimated 3D pose of the extracted feature in CGM. We denote the set of all estimated camera positions obtained from SfM with Γ and define a mapping function $\psi(\gamma^{(i)}) = a^i$, where a^i is the i-th aerial image and $\gamma^{(i)}$ is the estimated camera pose in CGM that corresponds to aerial image a^i.

Vision-Based Ground Observation: In order to localize the rover in the CGM we acquire ground observations (referred to as query images) using rover's onboard camera. We denote rover's query image at the k-th time-step by z_k followed from observation model $z_k = h(x_k, v_k)$, where x_k is the rover's position at the k-th time step and $v_k \sim p_v(\cdot|x_k)$ is the state-dependent observation noise.

A2G Feature Matching: As depicted in Fig. 4, for every keypoint extracted from rover's query image, we perform the search for its closest neighbor in CGM by using the Fast Library for Approximate Nearest Neighbors (FLANN) [11,12] which is specifically efficient and fast when working on high dimensional spaces. FLANN uses multiple randomized KD-tree or K-means tree forests for search in high dimensional spaces which results in significant improvement in computation and speed compared to methods that rely solely on the nearest neighbor search.

Weak Correspondence Rejection: Once features in the rover's query image are matched with the keypoints in the CGM, the spurious corresponding points (outliers) are classified and rejected using the Random Sample Consensus (RANSAC) algorithm [13]. we denote the remaining set of corresponding keypoints in CGM by C. Since each point in C can be traced back to an aerial image, we define a group $C^i_a \subset C$ as the set of features in C that correspond to the aerial image a^i. Due to high level of perceptual similarity between the scenes in a repetitive and ambiguous Mars-like environment, a query image can be matched with multiple reference images that correspond to different locations in the CGM. Reference images with low number of features are considered *weak* correspondences and are removed from consideration. Through experiments we observed that it is efficient to reject weak correspondences with number of matched features less than $m^{-1}n$, where n and m are the total number of keypoints and groups in C, respectively.

Data Association Ambiguit: Perceptual aliasing is caused by the ambiguity in data association where a rover's query image is registered to multiple aerial images where each image corresponds to a different location in the CGM. Figure 5 presents a typical example of perceptual aliasing in a Mars-analogue environment, where after applying RANSAC and weak correspondence rejection, keypoints in the query image are registered to four aerial images that each represents a different location in the map.

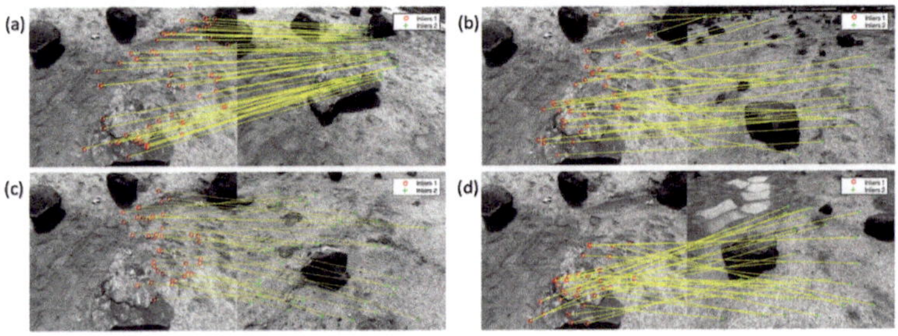

Fig. 5. Due to sparsity of salient features and perceptual similarity between scenes in a Mars-analogue environment, after outlier and weak correspondence rejection, there are four corresponding images to the rover's query image that each represents a different location in CGM.

After weak correspondence rejection, for the remaining corresponding aerial images, an affine 2D geometric transformation T is estimated that maps the points in a query image denoted by C_z to the matching points in an aerial image a denoted by C_a. We apply transformation T to all the points in C_z and measure the average residual between the reprojected points C'_z and C_a by

$$R(C_a, C'_z) = |C_a|^{-1} \sum_{j=1}^{|C_a|} ||c^j_a - c'^j_z|| \quad (1)$$

where $|C_a|$ is the cardinality of C_a. The similarity score of z_k to the i-th aerial image is given by,

$$s(z_k, a^i) = (1 + R(C^i_a, C'_{z_k}))^{-1}. \quad (2)$$

and the similarity score for images with corresponding points less than the set threshold is set to zero.

Observation Likelihood: By relying on Eq. 2 and using the mapping function $\psi(\gamma^{(i)}) = a^i$ that associates an aerial image to its corresponding estimated camera position in the CGM, we obtain a distribution that represents the likelihood of rover's measurement z_k for an estimated camera location in the CGM as

$$p(z_k|\gamma^{(i)}) = s(z_k, \psi(\gamma^{(i)}))(\sum_{\gamma \in \Gamma} s(z_k, \psi(\gamma))^{-1} \quad (3)$$

3 Bayesian Filtering for Location Estimation

Due to ambiguity and high degree of perceptual similarity between scenes at different locations in a Mars-like environment, the obtained similarity distribution is non-Gaussian and multimodal. In this section, we rely on a Bayesian filtering scheme [14] to cope with the non-Gaussianity of A2G similarity distributions that pose a challenge to accurate data association and rover localization.

State and Dynamics Mode: Let $x_{k+1} = f(x_k, u_k, q_k)$ denote the system dynamics, where x_k and u_k encode rover's location in CGM and action at the k-th time step, and $q_k \sim p_q(\cdot|x_k)$ is the state-dependent process noise.

Rover's Belief: Let $b_k = p(x_k|z_{0:k}) \in \mathbb{B}$ denote the probability distribution over all possible locations of the rover in CGM at the k-th time step. $\mathbb{B}$ denotes the space of all possible beliefs. We approximate the belief with a particle-based representation, i.e., $b_k(x) = \sum_{i=1}^{n} w_k^{(i)} \delta(x - p_k^{(i)})$ where $s_k = (p_k^{(i)}, w_k^{(i)})_{(i=1:n)}$ is a list of n weighted particles, with $p^{(i)}$ and $w^{(i)}$ denote the pose in the CGM and the importance weight of i-th particle respectively. $\delta(\cdot)$ is the Kronecker delta function that is one at zero and zero otherwise.

Bayesian Recursion: We will rely on particle filter to recursively evolve the belief b based on the last action and observation of the rover. The algorithm is composed of three steps of prediction τ^-, update $\tau^+$and resampling τ^s, as detailed further below,

$$b_{k+1} = \tau(b_k, u_k, z_{k+1}) = \tau^s(\tau^+(\tau^-(b_k, u_k), z_{k+1})). \tag{4}$$

Prediction: At each filtering step, we will first predict the belief b_{k+1}^- based on the rover action. At time $t = 0$, the rover's prior belief is given by $b_0^- = p(x_0|z_0) \equiv p(x_0)$, and it can be presented by a finite set of n particles with importance weight of n^{-1}, distributed uniformly over the CGM. According to the rover's action u_k, obtained from visual or wheel odometry the belief is predicted as

$$b_{k+1}^-(x) = conv(b_k(x), \mathcal{N}(u_k, Q\|u_k\|)) =: \tau^-(b_k(x), u_k), \tag{5}$$

which can be implemented as

$$w_{k+1}^{(i)-} = w_k^{(i)}, \quad p_{k+1}^{(i)-} = f(p_k^{(i)-}, u_k, q_k^{(i)})_{(i=1:n)}, \quad q_k^{(i)} \sim \mathcal{N}(0, Q\|u_k\|). \tag{6}$$

We capture the noise accumulation over longer traverses by modeling the noise intensity as the nominal covariance Q scaled by the magnitude of control effort u_k.

Importance Weight Update: After each observation z_k, a new likelihood distribution (Eq. 3) is obtained and we can update the propagated belief as given by

$$\grave{b}_{k+1}(x) = b_{k+1}^-(x) p(z_{k+1}|x) =: \tau^+(b_{k+1}^-(x), z_{k+1}) \tag{7}$$

where $\grave{b}_{k+1}(x)$ is the updated belief based on the latest observation $z_k + 1$. To implement this in the particle representation, we need to compute the importance weight of each particle by searching for the likelihood assigned to the closest estimated camera position to the particle's current position in the CGM and show this by

$$\gamma^{(i)} = \arg\min_{\Gamma} \|p^{(i)} - \gamma\|, \quad i = 1 \cdots n. \tag{8}$$

Using the likelihood $p(z_k|\gamma^{(j)})$ each particle is updated as,

$$\grave{w}_{k+1}^{(i)} = p(z_k|\gamma^{(i)}), \quad \grave{p}_{k+1}^{(i)} = p_{k+1}^{(i)-}, \quad i = 1 \cdots n. \tag{9}$$

Importance Resampling: In the importance resampling phase, we denote the resampled belief by,

$$b_{k+1} = \tau^s(\grave{b}_{k+1}(x)), \tag{10}$$

where τ^s is the resampling function. The probability of a particle being selected is proportional to its importance weight that is the likelihood of the particle being the correct representation of rover's location in the CGM. The resampled particles with normalized importance weights represent the posterior belief at the $k+1$ time step as given by

$$w_{k+1}^{(i)} = \grave{w}_{k+1}^{(i)}\Big(\sum_{i=1:n} \grave{w}_{k+1}^{i}\Big)^{-1}, \quad p_{k+1}^{(i)} \sim b_{k+1}(x), \quad i = 1 \cdots n. \tag{11}$$

4 Experimental Evaluation

In this section we study the feasibility, performance, and robustness of A2G localization in varying illumination conditions, viewing angles, environments, and using different methods for image feature detection and description.

Mars-Analogue Environment: Experiments are carried out at the Jet Propulsion Laboratory's Mars Yard which is an outdoor simulated Mars-analogue environment that provides a large test area to test different robotic applications under various natural lighting conditions, soil characteristics, and rock colors. As depicted in Fig. 6, three environments with rich, poor, and ambiguous and repetitive features are created for our experiments at the Mars Yard.

Aerial Image Dataset: A Sony ICX445 camera with a 2.7 mm FUJINON Fish-eye lens is used for acquiring the aerial imagery at $640X480$ resolution. In each environment, the terrain is mapped by taking a sequence of 200 RGB aerial images in a rectangular area with the size of 42 m^2. Images are obtained between the hours of 9am to 10am on the same day under clear skies and using a drone flying with a downward looking camera at the fixed altitude of 1.5 m above the ground.

Ground Image Dataset: A Sony ICX445 camera onbaord the Athena rover testbed (Fig. 7) with a 2.7 mm FUJINON Fish-eye lens is used to collect the

Fig. 6. Feature-rich terrain with distinctive landmarks (top row), repetitive and ambiguous features (middle row), and feature-poor terrain (bottom row). On the right, an aerial view of JPL's Mars Yard with markings that correspond to the locations of created environments.

Fig. 7. Athena rover is a prototype rover for earth-based testing of Mars Rover in the Jet Propulsion Laboratory's Mars Yard.

query images at $640X480$ resolution in each environment. Ground images are acquired at three different times of the same day (9am, 1pm and 5pm) and under clear skies, while keeping the camera pointed at 65° viewing angle with respect to the ground.

A2G Viewing Angle Variations: To study the performance and sensitivity of A2G image matching under varying viewing angles, we study different image

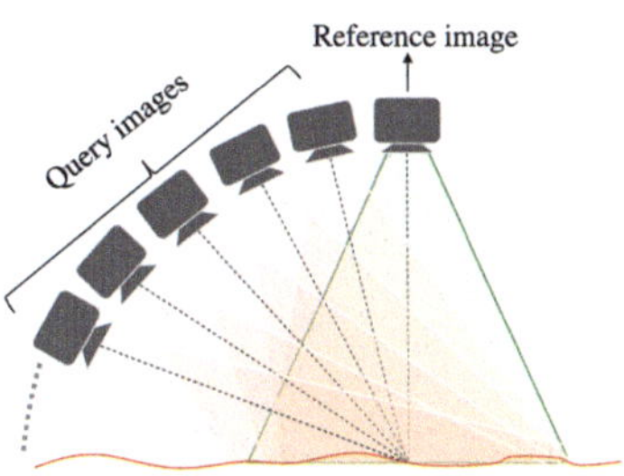

Fig. 8. A reference image is obtained by pointing the camera prependicular to the ground. Query images are obtained in different viewing angles ranging from 15 to 75° with respect to the ground.

feature detection and description methods for A2G image matching. As depicted in Fig. 8, we obtain a reference image of a feature-rich terrain by pointing the camera perpendicular to the ground. We then obtain query images from the same part of the terrain at different viewing angles. Image feature matching is performed between the reference image and each query image and the total number of extracted features, matched features and correct matches (inliers) are used as metrics to evaluate and compare the performance of each method.

In this experiment we rely on SIFT [15], SURF [16], Harris [17], FAST [18] and BRISK [19] for image feature detection and description. As presented in Fig. 9, SIFT and Harris yield the largest number of extracted features, however, after outlier rejection only SIFT and SURF yield the largest number of correct matches. It can be observed that once the viewing angle between aerial and ground images exceeds 35°, there is a sharp decline such that the remaining number of correct matches is not sufficient for a reliable A2G similarity evaluation.

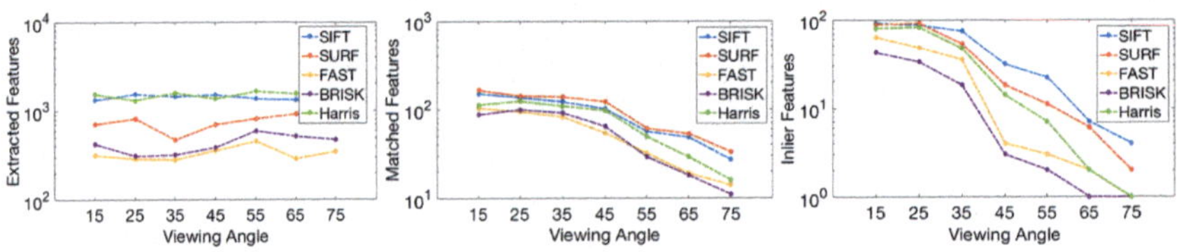

Fig. 9. Total number of extracted, matched and inlier features in A2G Image matching under different viewing angles and using different feature detection/description methods

Evaluation of Localization Accuracy. To study the effect of different feature detection and description methods on the localization accuracy, we use the aerial and ground imagery collected in the feature-rich terrain. From the set of aerial imagery the copter-generated map (CGM) is constructed (Fig. 10).

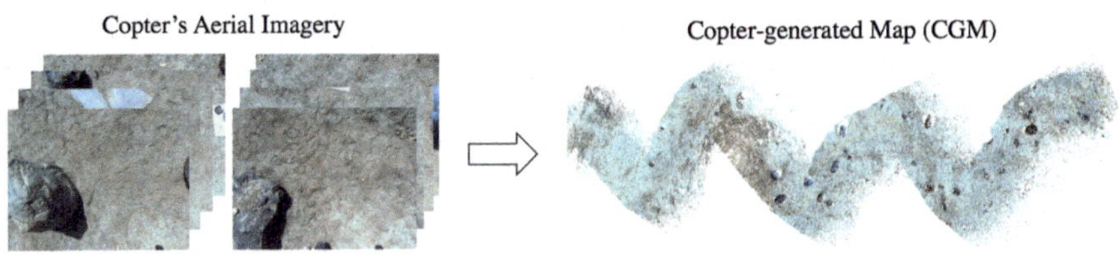

Fig. 10. The copter-generated map (CGM) of a Mars-analogue environment is feature-based representation of the environment from the set aerial imagery.

In the next step, the sequence of ground imagery is used to establish rover's position in the CGM. As shown in Fig. 11(a), at time step $t = 0$ with no prior observations, particles (blue dots) that represent the initial belief for rover's position in CGM are distributed uniformly over the map. The localization error at each time step is defined as the euclidean distance between the average location of all particles and the ground truth position of the rover. As rover traverses the terrain to acquire new ground observations, particles are propagated by relying on rover's motion model obtained from visual odometry. After acquiring a new observation, the similarity distribution obtained from A2G image matching is used to update the importance weight of particles. In the resampling step, particles are then resampled where the probability of a particle being selected is proportional to its importance weight. As shown in Fig. 11(c) through this recursive filtering process, particles are eventually converged on the true position of the rover in the CGM.

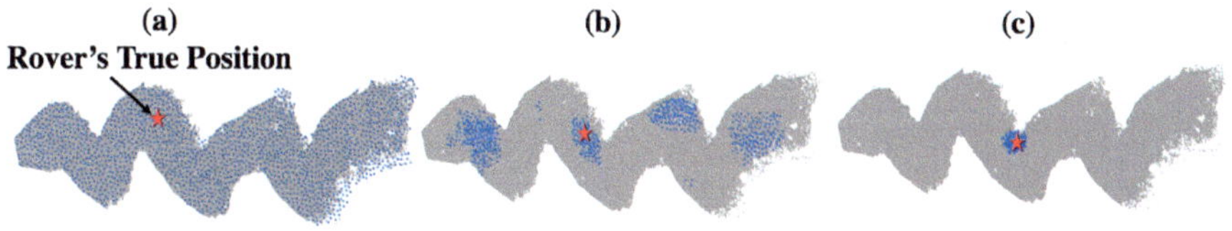

Fig. 11. (a): The initial distribution of particles at time t = 0, (b): distribution of particles after acquiring multiple ground observations, (c): convergence of particles on rover's true position.

The performance of A2G localization using different feature detection and description methods is presented in Fig. 12. We can observe that regardless of the feature detection and description method used, the localization is achieved in all cases. However, because SIFT and SURF yield the largest number of correct A2G matches compared to other methods, this enables a faster and more accurate A2G localization from a smaller number of query images. Due to higher computational speed of SURF and its high performance, for the rest of our experiments we exploit SURF for feature detection and description of aerial and ground imagery.

A2G Localization in Different Terrain Types: In a Mars-like environment, terrain types can be classified into three categories of feature-rich, feature-poor, and ambiguous and repetitive. In this experiment, we use the aerial and ground imagery collected in the morning to study the A2G localization performance in each terrain type. As presented in Fig. 13, in the feature-rich terrain localization accuracy is significantly improved after processing 10 ground observations. In the feature-poor terrain, accurate localization is not established as there are not sufficient landmarks that can be exploited for successful image matching. In the repetitive and ambiguous environment reduced number of distinctive landmarks and high similarity between multiple scenes leads to perceptual aliasing which in turn diminishes the chances of establishing an accurate correspondence.

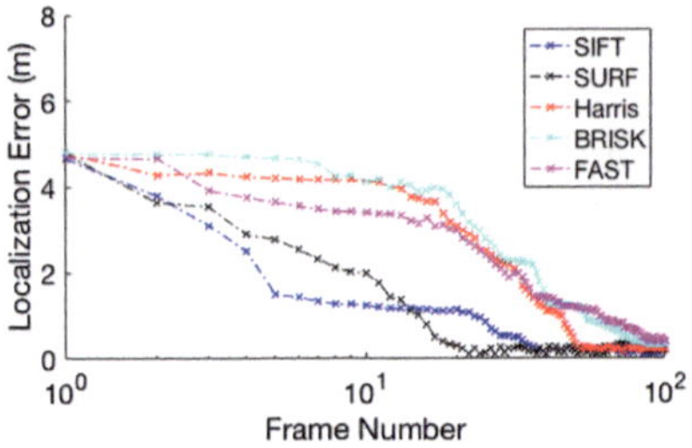

Fig. 12. A2G localization performance evaluation in a feature-rich environment using different feature detection and description methods.

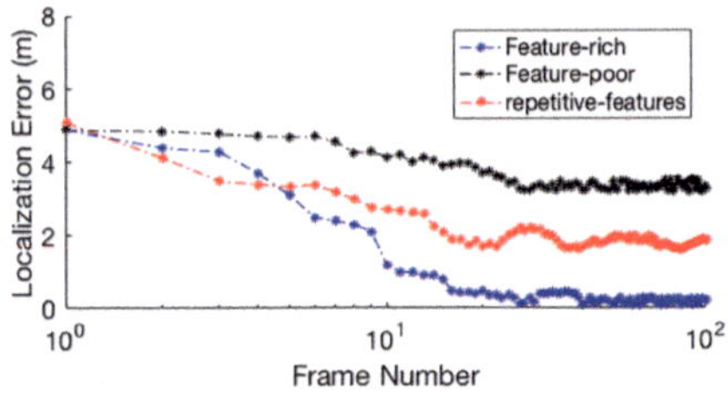

Fig. 13. A2G localization performance evaluation in feature-rich, feature-poor and terrain with repetitive perceptual features.

Localization Robustness in Varying Illuminations: In this experiment, in order to study the effect of illumination variations on A2G localization performance, we use ground image datasets that are collected at different hours of the same day. As presented in Fig. 14, despite dramatic illumination variations the A2G localization performs well in the feature-rich terrain. In feature-poor and repetitive features environments accurate localization is not achieved as salient features are sparse, and illumination variations have a bigger impact on robust and reliable data association.

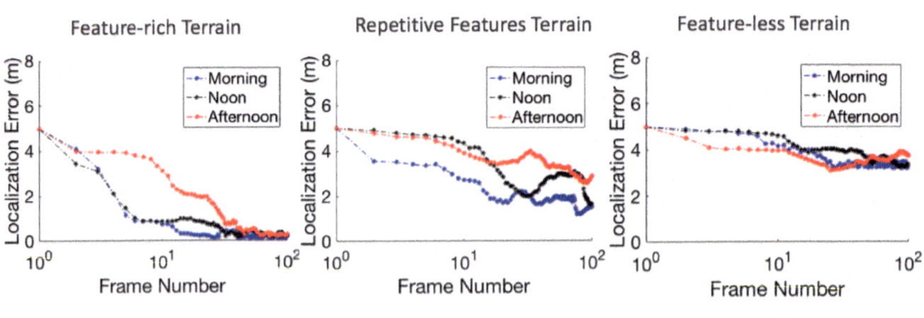

Fig. 14. A2G localization performance evaluation in different terrain types and under varying illumination conditions.

5 Conclusion and Future Work

In this paper, we studied the feasibility and performance of rover localization in copter-generated maps in a Mars-analogue environment. Various challenges including varying viewpoints, scale, and illumination as well as variations in the level of perceptual features were investigated. The experimental results demonstrated that accurate A2G localization is feasible in a feature-rich Mars-like environment. However, in feature-poor and ambiguous environments the A2G localization does not have sufficient accuracy. We also investigated different feature detectors and descriptors for A2G image matching, and concluded that SIFT and SURF provide the largest number of correct matches, and yield a faster and more efficient A2G localization compared to Harris, FAST and BRISK features.

For future work, we plan to conduct extensive experiments in a larger Mars-analogue environment and using aerial imagery that are obtained over a longer time period. In addition we plan to study how active perception methods (e.g. POMDP-like methods [20,21]) can enhance the localization performance by actively taking actions that lead to higher perceptual information gain and ultimately a higher situational awareness. We will also investigate the graph-based methods for distributed collaborative mapping using a team of rover and helicopter in a Mars-like environment.

Acknowledgement. The research was carried out at the Jet Propulsion Laboratory, California Institute of Technology, under a contract with the National Aeronautics and Space Administration.

References

1. Balaram, B., et al.: Mars helicopter technology demonstrator. In: 2018 AIAA Atmospheric Flight Mechanics Conference (2018)
2. Viswanathan, A., Pires, B.R., Huber, D.: Vision based robot localization by ground to satellite matching in GPS-denied situations. In: 2014 IEEE/RSJ International Conference on Intelligent Robots and Systems (IROS 2014). IEEE (2014)
3. Tao, Y., Muller, J.-P., Poole, W.: Automated localisation of Mars rovers using co-registered HiRISE-CTX-HRSC orthorectified images and wide baseline Navcam orthorectified mosaics. Icarus **280**, 139–157 (2016)
4. Di, K., et al.: Mars rover localization based on feature matching between ground and orbital imagery. Photogram. Eng. Remote Sens. **77**(8), 781–791 (2011)
5. Li, R., et al.: MER spirit rover localization: comparison of ground image–and orbital image–based methods and science applications. J. Geophys. Res. Planets **116** (2011)
6. Kirk, R.L., et al.: Ultrahigh resolution topographic mapping of Mars with MRO HiRISE stereo images: meter-scale slopes of candidate Phoenix landing sites. J. Geophys. Res. Planets. **113**(E3) (2008)
7. Forster, C., et al.: Air-ground localization and map augmentation using monocular dense reconstruction. In: 2013 IEEE/RSJ International Conference on Intelligent Robots and Systems (IROS). IEEE (2013)
8. Majdik, A.L., et al.: Air-ground matching: appearance-based GPS-denied urban localization of micro aerial vehicles. J. Field Robot. **32**(7), 1015–1039 (2015)

9. Majdik, A.L., Albers-Schoenberg, Y., Scaramuzza, D.: Mav urban localization from google street view data. In: 2013 IEEE/RSJ International Conference on Intelligent Robots and Systems (IROS). IEEE (2013)
10. Snavely, N., Seitz, S.M., Szeliski, R.: Photo tourism: exploring photo collections in 3D. ACM Trans. Graph. (TOG) **25**(3), 835–846 (2006)
11. Muja, M., Lowe, D.G.: Scalable nearest neighbor algorithms for high dimensional data. IEEE Trans. Pattern Anal. Mach. Intell. **36**(11), 2227–2240 (2014)
12. Silpa-Anan, C., Hartley, R.: Optimised KD-trees for fast image descriptor matching. In: IEEE Conference on Computer Vision and Pattern Recognition, CVPR 2008. IEEE (2008)
13. Fischler, M.A., Bolles, R.C.: Random sample consensus: a paradigm for model fitting with applications to image analysis and automated cartography. In: Readings in Computer Vision, pp. 726–740 (1987)
14. Fox, D., et al.: Particle filters for mobile robot localization. In: Doucet, A., de Freitas, N., Gordon, N. (eds.) Sequential Monte Carlo Methods in Practice, pp. 401–428. Springer, New York (2001)
15. Lowe, D.: Distinctive image features from scale-invariant keypoints. Int. J. Comput. Vis. **60**(2), 91–110 (2004)
16. Bay, H., Tuytelaars, T., Van Gool, L.: SURF: speeded up robust features. In: Leonardis, A., Bischof, H., Pinz, A. (eds.) European Conference on Computer Vision, vol. 3951, pp. 404–417. Springer, Heidelberg (2006)
17. Harris, C., Stephens, M.: A combined corner and edge detector. In: Alvey Vision Conference, vol. 15, no. 50 (1988)
18. Rosten, E., Drummond, T.: Fusing points and lines for high performance tracking. In: Tenth IEEE International Conference on Computer Vision, ICCV 2005, vol. 2 (2005)
19. Leutenegger, S., Chli, M., Siegwart, R.Y.: BRISK: binary robust invariant scalable keypoints. In: 2011 IEEE International Conference on Computer Vision (ICCV). IEEE (2011)
20. Agha-Mohammadi, A., Chakravorty, S., Amato, N.M.: FIRM: sampling-based feedback motion-planning under motion uncertainty and imperfect measurements. Int. J. Robot. Res. **33**(2), 268–304 (2014)
21. Otsu, K., Agha-Mohammadi, A., Paton, M.: Where to look? Predictive perception with applications to planetary exploration. IEEE Robot. Autom. Lett. **3**(2), 635–642 (2018)

Towards Efficient Full Pose Omnidirectionality with Overactuated MAVs

Karen Bodie(✉), Zachary Taylor, Mina Kamel, and Roland Siegwart

Autonomous Systems Lab, ETH Zürich, Zürich, Switzerland
kbodie@ethz.ch

Abstract. Omnidirectional MAVs are a growing field, with demonstrated advantages for aerial interaction and uninhibited observation. While systems with complete pose omnidirectionality and high hover efficiency have been developed independently, a robust system that combines the two has not been demonstrated to date. This paper presents VoliroX: a novel omnidirectional vehicle that can exert a wrench in any orientation while maintaining efficient flight configurations. The system design is presented, and a 6 DOF geometric control that is robust to singularities. Flight experiments further demonstrate and verify its capabilities.

Keywords: Overactuated · Omnidirectional MAV · Geometric control

1 Introduction

Traditional rotary micro aerial vehicles (MAVs) are underactuated, providing only 4 controllable degrees of freedom (DOFs) by nature of their aligned propeller axes. This can lead to unstable behavior when a force and torque are applied to such a system in flight. Omnidirectional MAVs allow for decoupling of the translational and rotational dynamics, permitting stable interaction with the environment in any pose. With the addition of complete pose omnidirectionality, the system can achieve uninhibited aerial movement and robust tracking of 6 DOF trajectories, a unique boon for aerial filming and 3D mapping, as well as configuration-based navigation in constrained environments.

A dominant struggle in the development of MAVs is achieving a good compromise between performance and efficiency. For aerial interaction and high dynamic control authority in omnidirectional flight, performance can be represented by the force and torque control volumes of a system. For pose-omnidirectional platforms, the force envelope must exceed gravity in all directions with an additional buffer to maintain dynamic movement. Countering this performance goal is the

Electronic supplementary material The online version of this chapter (https://doi.org/10.1007/978-3-030-33950-0_8) contains supplementary material, which is available to authorized users.

J. Xiao et al. (Eds.): ISER 2018, SPAR 11, pp. 85–95, 2020.
https://doi.org/10.1007/978-3-030-33950-0_8

desire for high efficiency and longer flight times, which is compromised in systems that generate high internal forces (thrust forces which counteract each other), or add additional weight for actuation.

Problem Statement: This paper presents a novel platform that achieves both force- and pose-omnidirectionality with highly dynamic capabilities, while maintaining high efficiency in hover, by nature of its six actuated tilt arms. Experimental results demonstrate that the additional complexity and weight of the tilt-arm configuration are justified by the system's performance.

Related Work: Within the past 5 years, huge growth has occurred in the field of fully actuated omnidirectional MAVs. These works can be generally categorized as fixedly tilted rotor platforms, and tilt-arm rotor unit platforms. By its design, a platform with fixedly tilted rotors that is able to generate a significant wrench on the environment creates a proportionately significant amount of internal force, which directly detracts from flight efficiency [1–4].

Tilt-arm platforms can achieve optimal hover efficiency in the absence of external disturbances, when all propeller thrust vectors are aligned against gravity. This has been demonstrated in the form of a quadrotor [5,6] with limited roll and pitch, and also in the form of a hexarotor, with our platform's predecessor Voliro [7], which is limited by low thrust and singularity issues in certain configurations. These platforms achieve force omnidirectionality with high hover efficiency at the cost of additional complexity, weight, and inertia. An intersection of the two concepts which couples the tilt axes to a single motor has also been evaluated in simulation [8], enabling efficient flight, but without pose omnidirectionality.

2 Technical Approach

2.1 System Design

This paper presents the VoliroX platform: a 12 rotor MAV with 6 tiltable arms, shown in Fig. 1(a). The system layout consists of equally spaced arms along the z_b-plane, with two rotors per arm to balance the rotational inertia. Two KDE2315XF-885 motors per arm with 9in propellers also provide high thrust

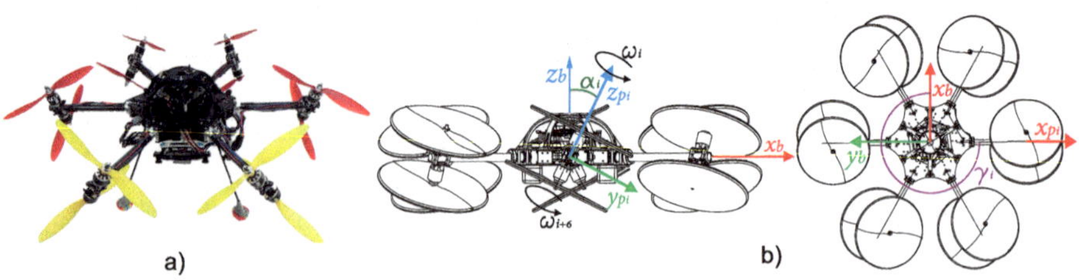

Fig. 1. (a) Prototype system, (b) coordinate frames and variables

(13.7N per motor), enabling dynamic flight even in the least efficient configurations. The arm is tilted by a Dynamixel XL430 servo actuator, located in the base to reduce system inertia. Upper and lower propellers counter-rotate such that drag torque from the propellers is canceled in the nominal case, and can be controlled by varying the relative inputs to the rotors. Moreover, the counter-rotating propeller configuration minimizes the gyroscopic moment on the tilting mechanism, allowing the use of a small Dynamixel servo. An onboard Intel NUC i7 manages higher level computation and communication tasks, while the combined position and attitude controller operates on a PX4 flight controller. The system is powered by two 3800 mAh 6s LiPo batteries, and can be tethered for the case of near-horizontal endurance flights. The total system mass is 4 kg, and it can generate 150N of force at maximum thrust in the horizontal hover configuration.

Coordinate systems for the platform are described in Fig. 1(b), with the body-fixed coordinate frame x_b, y_b, z_b, and definitions of the fixed arm spacing angles γ_i and tilting angles α_i.

2.2 System Dynamics and Control

The following assumptions are adopted to simplify the model:

- The body structure is rigid and symmetric.
- Thrust and drag torques are proportional to the square of rotor's speed, and rotors are able to achieve desired speeds ω_i with negligible transients.
- The primary axes of the system correspond with the principal axes of inertia.
- The dynamics of the tilt motors are independent of the rotational speed of rotors.
- The thrust and drag torques produced by each rotor are independent, i.e. there is no airflow interference.

The system dynamics are derived by the Newton-Euler approach under the stated assumptions, resulting in the following equations of motion expressed in the body-fixed frame:

$$\begin{bmatrix} \boldsymbol{F}_b \\ \boldsymbol{\tau}_b \end{bmatrix} = \begin{bmatrix} m\boldsymbol{I}_3 & 0 \\ 0 & \boldsymbol{J}_b \end{bmatrix} \begin{bmatrix} \ddot{\boldsymbol{x}}_b \\ \dot{\boldsymbol{\omega}}_b \end{bmatrix} + \begin{bmatrix} \boldsymbol{\omega}_b \times m\dot{\boldsymbol{x}}_b \\ \boldsymbol{\omega}_b \times \boldsymbol{J}_b\boldsymbol{\omega}_b \end{bmatrix} \tag{1}$$

where $\boldsymbol{F}_b$ and $\boldsymbol{\tau}_b$ are the total forces and torques on the system, $\dot{\boldsymbol{x}}_b$ and $\dot{\boldsymbol{x}}_b$ are velocity and acceleration of the origin, and $\boldsymbol{\omega}_b$ is the angular body velocity, and m and $\boldsymbol{J}_b$ are the mass and inertia, where subscript $_b$ denotes quantities expressed in the body-fixed frame.

Trajectory tracking on SE(3) is used to guarantee local exponential stability [9], and has been extended to the omnidirectional case [10]. The control allocation presented here is an extension of [5,7], a nonlinear function of tilt angles α_i. The force allocation matrix mapping propeller thrust and torques to the net body force and torque is defined as $\boldsymbol{A}_\alpha$, and can be thought of as the instantaneous allocation matrix, since it does not depend on any delay in response of the tilting motors. One can take advantage of the fact that all entries in $\boldsymbol{A}_\alpha$ are linear combinations of $\sin(\alpha_i)$ and $\cos(\alpha_i)$, and extract them to give a simplified allocation matrix $\boldsymbol{A}$. The result vector $\boldsymbol{u}$ has 4 values for each tilt arm, with upper rotors $j \in \{1...6\}$ and lower rotors $\{7...12\}$. For a system with 6 arms spaced evenly around the z_b-axis at rotation angles γ_i, with arm length l_x, rotor spin direction s_j, and thrust and drag coefficients c_f and c_d:

$$\begin{bmatrix} \boldsymbol{F}_b \\ \boldsymbol{\tau}_b \end{bmatrix} = \boldsymbol{A}_\alpha \boldsymbol{\Omega} = \boldsymbol{A}\boldsymbol{u} \qquad \boldsymbol{\Omega} = \begin{bmatrix} \Omega_1 \\ \vdots \\ \Omega_{12} \end{bmatrix} = \begin{bmatrix} \omega_1^2 \\ \vdots \\ \omega_{12}^2 \end{bmatrix} \tag{2}$$

$$\boldsymbol{A} = c_f \begin{bmatrix} \sin(\gamma_i) & 0 & \dots \\ -\cos(\gamma_i) & 0 & \dots \\ 0 & 1 & \dots \\ -s_j c_d \sin(\gamma_i) & l_x \cos(\gamma_i) & \dots \\ s_j c_d \cos(\gamma_i) & -l_x \sin(\gamma_i) & \dots \\ -l_x & -s_j c_d & \dots \end{bmatrix} \quad \boldsymbol{u} = \begin{bmatrix} \sin(\alpha_i)\Omega_{i,} \\ \cos(\alpha_i)\Omega_i \\ \vdots \\ \sin(\alpha_i)\Omega_{i+6} \\ \cos(\alpha_i)\Omega_{i+6} \\ \vdots \end{bmatrix} \begin{matrix} \forall i \in \{1...6\} \\ \forall j \in \{1...12\} \end{matrix}$$

where Ω_i is the square of the rotor velocity ω_i, and is always chosen to be positive with the proposed allocation method.

A control law for the system is derived from (2) using the Moore-Penrose pseudo-inverse of the force allocation matrix $\boldsymbol{A}$ to compute the actuator commands from a desired 6 DOF wrench $[\boldsymbol{F}_d \ \boldsymbol{\tau}_d]^\top$ expressed in the body-fixed frame:

$$\boldsymbol{u} = \boldsymbol{A}^\dagger \begin{bmatrix} \boldsymbol{F}_d \\ \boldsymbol{\tau}_d \end{bmatrix} \tag{3}$$

The desired forces and torques on the system are expressed as

$$\begin{aligned} \boldsymbol{F}_d &= m(\boldsymbol{R}^\top(-k_p \boldsymbol{e}_p - k_v \boldsymbol{e}_v + \ddot{\boldsymbol{x}}_{sp} + \boldsymbol{g}) + (\boldsymbol{\omega}_b \times \boldsymbol{R}^\top \dot{\boldsymbol{x}})) \\ \boldsymbol{\tau}_d &= \boldsymbol{J}_b(-k_R \boldsymbol{e}_R - k_\omega \boldsymbol{e}_\omega) + (\boldsymbol{\omega}_b \times \boldsymbol{J}_b \boldsymbol{\omega}_b) + (\boldsymbol{x}_{com} \times \boldsymbol{F}_d) \end{aligned} \tag{4}$$

and are based on the equations of motion described in (1), with accelerations taken as a combination of position and velocity errors multiplied by tuned gain values. To compensate for a center of mass offset $\boldsymbol{x}_{com}$, a counter-torque is added to the desired torque equation. In (4), $\boldsymbol{R} \in \mathrm{SO}(3)$ is the rotation matrix from the body-fixed frame to the inertial frame, $\boldsymbol{g}$ is the gravity vector along the inertial frame z-axis, and k_p, k_v, k_R, and k_ω, are positive constant gains on position, velocity, rotation, and angular rate errors. Errors are expressed in the inertial frame relative to set point values with subscript $_{sp}$, with angular error terms defined in (5) according to [9].

$$\begin{aligned} \boldsymbol{e}_p &= \boldsymbol{x} - \boldsymbol{x}_{sp} \\ \boldsymbol{e}_v &= \dot{\boldsymbol{x}} - \dot{\boldsymbol{x}}_{sp} \\ \boldsymbol{e}_R &= \frac{1}{2}(\boldsymbol{R}_{sp}^\top \boldsymbol{R} - \boldsymbol{R}^\top \boldsymbol{R}_{sp})^\vee \\ \boldsymbol{e}_\omega &= \boldsymbol{\omega}_b - \boldsymbol{R}^\top \boldsymbol{R}_{sp} \boldsymbol{\omega}_{sp} \end{aligned} \tag{5}$$

To extract α_i values from the result vector $\boldsymbol{u}$, each pair of rows is summed, and the trigonomic identity $\theta = atan2(\sin(\theta), \cos(\theta))$ is used. The resulting α_i values are then limited according to $\dot{\alpha}_{\max}$. To extract Ω_i values, consider the identity $\sin^2(\theta) + \cos^2(\theta) = 1$, to obtain the solution as follows:

$$\begin{aligned} \Omega_i &= \Omega_i \sin^2(\alpha_i) + \Omega_i \cos^2(\alpha_i) = \sin(\alpha_i) u_{2i} + \cos(\alpha_i) u_{2i+1} \\ \Omega_{i+6} &= \Omega_{i+6} \sin^2(\alpha_i) + \Omega_i \cos^2(\alpha_i) = \sin(\alpha_i) u_{2(i+6)} + \cos(\alpha_i) u_{2(i+6)+1} \end{aligned} \tag{6}$$

Notice that the solution in (6) projects the desired force onto the best achievable tilt angle. If the direction of desired force would require an inversion of the propeller, the corresponding Ω_i is then set to zero, permitting only positive thrust values.

Thrust values for individual motors have been computed based on test results of a double-rotor propeller group on an RCbenchmark test stand, to measure the rotor z_{p_i}-axis thrust and torque. Thrust and torque values are correlated linearly with pulse-width modulation (PWM) commands sent to the rotors' electronic speed controllers (ESCs). To remain in a thrust region that maintains a good linear approximation, and to reduce the risk of overheating, the maximum output of the rotors is limited to 80%, which corresponds with a thrust of 20 N for each double rotor unit. Using this test data, maximum achievable force and torque envelopes of the system can be determined by feeding commands forward through the allocation matrix (2). The resulting plots for maximum force and torque envelopes are shown in Fig. 2(a). The force envelope is computed in the absence of torque, and vice versa.

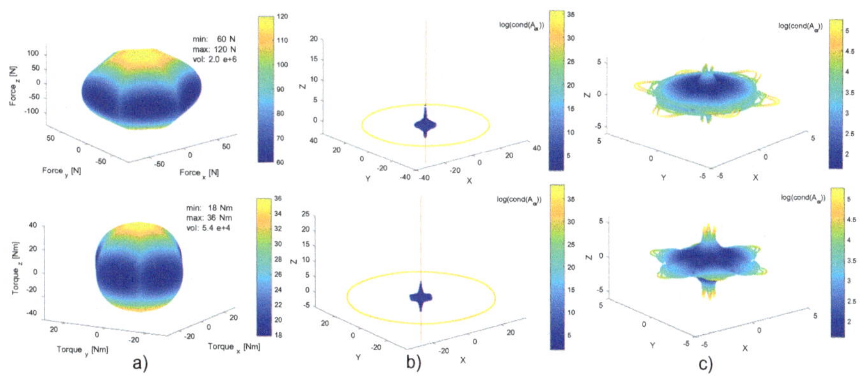

Fig. 2. (a) Maximum force and torque envelopes, (b) log of condition numbers of $\boldsymbol{A}_\alpha(\boldsymbol{\alpha})$, and (b) log of condition number for modified $\boldsymbol{A}_\alpha(\tilde{\boldsymbol{\alpha}})$, with scale significantly zoomed in.

2.3 Singularity Handling

Omnidirectional wrench generation for the described vehicle encounters two types of singularity cases. The first case is where alignment of the propellers results in a rank reduction of the instantaneous allocation matrix $\boldsymbol{A}_\alpha$, which corresponds to an ill-conditioned matrix (or high condition number). This rank reduction occurs when the desired force is either orthogonal to the body z_b-frame (rank is reduced to 4, no instantaneous force exertion along x_b or y_b axes is possible), and when the desired force lies in the z_b-plane (rank is reduced to 5, no instantaneous force exertion is possible along z_b). This phenomenon is shown in Fig. 2(b), where the condition number asymptotically extends along the z_b-axis and z_b-plane.

The above mentioned kinematic singularities can be handled by identifying the case where the desired force vector comes within an angle φ_t of the z_b-axis or the z_b-plane. A multiplier k_t then quadratically scales an additional tilt component up to a value of c_t when the misalignment angle ϕ goes to zero. Bias tilt directions b_i alternate for neighboring tilt-arms. The modified tilt angle $\tilde{\alpha}_i$ can be expressed as

$$\tilde{\alpha}_i = \alpha_{i,prev} + \text{sign}(\delta\tilde{\alpha}_i)\text{min}(|\delta\tilde{\alpha}_i|, \dot{\alpha}_{max}\delta t) \quad \delta\tilde{\alpha}_i = \delta\alpha_i + k_t b_i c_t$$

$$k_t = \begin{cases} 0 & \forall\, \phi \geq \varphi_t \\ (1 - \dfrac{\phi}{\varphi_t})^2 & \forall\, \phi < \varphi_t \end{cases} \qquad b_i = (-1)^i \forall i \in \{1...6\}$$

The resulting condition number envelope for the modified allocation matrix $\boldsymbol{A}_\alpha(\tilde{\alpha})$ is significantly reduced in the singularity cases, as seen when comparing Fig. 2(b) and (c), where the scale of (c) is magnified for a better view of the new condition number envelope. Figure 3(b) shows a visual interpretation of the thresholding approach described above, with the quadratic tilt bias multiplier k_t shown in Fig. 3(c).

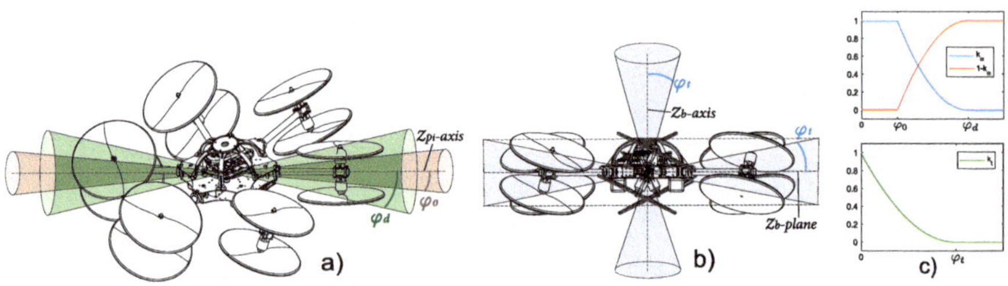

Fig. 3. Visual depiction singularity handling for (a) tilt angle singularities, (b) kinematic singularties, and (c) associated multipliers k_α and k_t.

The second singularity case is when the force aligns with an arm axis, causing a kinematic singularity where the associated thrusters cannot contribute to the desired force. Since tilt angle dynamic limitations are not considered in the allocation, when the desired force crosses this axis directly, it results in an angular velocity that approaches infinity as the time step goes to zero. To expose the tilt motor angular velocities and evaluate this singularity condition, the allocation is taken to a derivative level, as proposed by Ryll et al. To do this, tilt angle velocities are first expressed in the body wrench expression by including a structural null matrix:

$$\begin{bmatrix} \boldsymbol{F}_b \\ \boldsymbol{\tau}_b \end{bmatrix} = \begin{bmatrix} \boldsymbol{A}_\alpha(\boldsymbol{\alpha}) \ \boldsymbol{0} \end{bmatrix} \begin{bmatrix} \boldsymbol{\Omega} \\ \dot{\boldsymbol{\alpha}} \end{bmatrix} \tag{7}$$

Now, the differential of $\boldsymbol{A}_\alpha(\boldsymbol{\alpha})$ may be expressed by expanding the terms as

$$\begin{aligned} \boldsymbol{A}_\alpha(\boldsymbol{\alpha})\boldsymbol{\Omega} &= \sum_{i=1}^{12} \boldsymbol{a}_i(\boldsymbol{\alpha})\Omega_i \\ \frac{\mathrm{d}\boldsymbol{A}_\alpha(\boldsymbol{\alpha})\boldsymbol{\Omega}}{\mathrm{d}t} &= \boldsymbol{A}_\alpha(\boldsymbol{\alpha})\dot{\boldsymbol{\Omega}} + \sum_{i=1}^{12} \frac{\partial \boldsymbol{a}_i(\boldsymbol{\alpha})}{\partial \alpha}\Omega_i\dot{\boldsymbol{\alpha}} \end{aligned}$$

and expressing the time derivative of (7) in the body frame as

$$\begin{bmatrix} \dot{\boldsymbol{F}}_b \\ \dot{\boldsymbol{\tau}}_b \end{bmatrix} = \begin{bmatrix} \boldsymbol{A}_\alpha(\boldsymbol{\alpha}) & \sum_{i=1}^{12} \frac{\partial \boldsymbol{a}_i(\boldsymbol{\alpha})}{\partial \alpha}\Omega_i \end{bmatrix} \begin{bmatrix} \dot{\boldsymbol{\Omega}} \\ \dot{\boldsymbol{\alpha}} \end{bmatrix} \tag{8}$$

From expression (8) it is clear that the system experiences a singular condition in the tilt angle velocities when an Ω_i term is zero, which occurs when the force and torque capabilities of the given propeller unit cannot contribute to the desired wrench. This corresponds with a case where the desired force and torque vectors align with the propeller group x_{p_i}-axis. An approach commonly used in robot manipulators can be invoked to reduce this issue, by instead using the damped pseudo-inverse of the allocation matrix [11]. This approach, however, requires the inversion of a 6×6 matrix in real time, which currently due to high use of CPU and memory make it a poor choice for use on a micro-controller. Taking inspiration from this approach, a solution that approximates a damping effect is derived, requiring significantly less computational power.

When the angle between the desired force vector comes within the damping threshold angle φ_d of an arm axis, the tilt-angle speeds associated with that axis are quadratically scaled to zero by multiplier $k_{\alpha,i}$. As the alignment angle η_i approaches a zero threshold, φ_0, the tilt angle freezes in position. A visual interpretation of the alignment geometry is shown in Fig. 3(a), with multiplier curves depicted in Fig. 3(c). The change in α_i can then be expressed as follows:

$$\delta\alpha_i' = \delta\alpha_i(1 - k_{\alpha,i})$$

$$k_{\alpha,i} = \begin{cases} 0 & \forall \quad \eta_i > \varphi_d \\ (1 - \dfrac{\eta_i - \varphi_0}{\varphi_d - \varphi_0})^2 & \forall \ \varphi_d > \eta_i > \varphi_0 \\ 1 & \forall \quad \eta_i < \varphi_0 \end{cases}$$

This approach is justified due to the fact that at small angles of alignment deviation, the force contribution of the thrusters in question is negligible when compared to the remaining thrust contributions, and the error can be treated as a disturbance. When the new angles are computed based on these values, and limited according to $\dot{\alpha}_{max}$, the rotor speeds are then calculated as in (6). A force from the limited propeller group exists only if the thrust axis can be projected as a positive component onto the desired force vector.

Since the contribution of propeller groups in the tilt-angle singularity region is so small, one can take advantage of this situation to address another issue of the physical system, notably the windup of cables around an axis. As the tilt velocity is ramped down, it is balanced with an unwinding velocity ω_u, that is calculated based on a maximum unwinding rate and direction to the zero position, resulting the new tilt angle change $\delta\alpha_i^* = \delta\alpha_i' - \text{sign}(\alpha_i)k_{\alpha,i}\omega_u\delta t$.

3 Experimental Results

The presented results are derived from flight tests in an indoor space, using a VICON external camera system for state estimation. Parameter values used for the experiments are shown in Table 1.

Table 1. Parameters used for experimental flight tests

φ_0[deg]	φ_d[deg]	φ_t[deg]	c_t[deg]	$\omega_u\,[\frac{\text{rad}}{s}]$
5	15	10	10	8

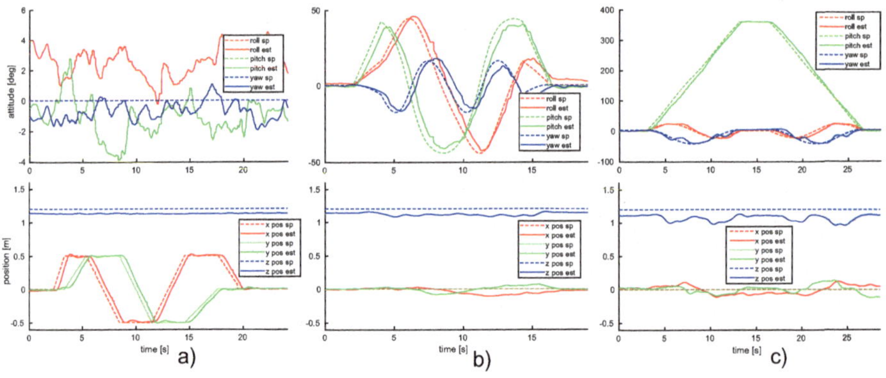

Fig. 4. Experimental trajectory tracking (a) translation without rotation, (b) rotation without translation (c) 360° flip about the y_b-axis

Omnidirectional Flight: To validate the controller, three trajectories were tested to demonstrate decoupling of the force and torque. Tracking results are shown in Fig. 4. The first two trajectories demonstrate the decoupled translation and rotation of the system, and the third trajectory is a 360° flip about the $\boldsymbol{y}_b$-axis. The system is able to track these trajectories with position errors below 5 cm and attitude errors below 4°.

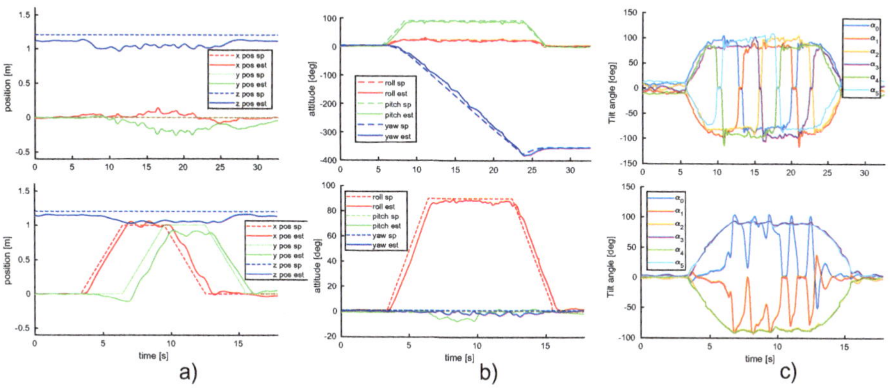

Fig. 5. Experimental (a) position tracking, (b) attitude tracking, and (c) commanded tilt angles for translation in a singular configuration (upper row) and a cartwheel through 6 singular positions (lower row)

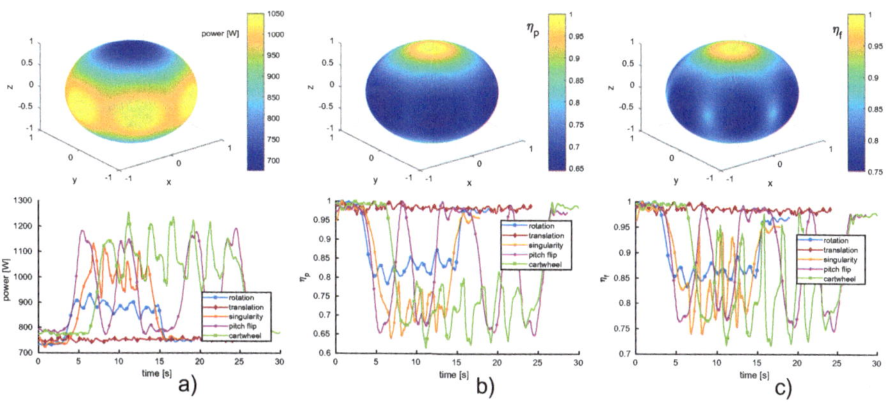

Fig. 6. (a) Power consumption, (b) power efficiency and (c) wasted force index for theoretical hover, and measured values from tested trajectories.

Robustness to Singular Configuration: In order to demonstrate robustness to the singularity cases presented above, two additional trajectory tracking experiments have been conducted, with results shown in Fig. 5. The first trajectory rotates the vehicle such that one arm is aligned with gravity, then tracks a translation trajectory. The second trajectory moves the vehicle to 90° pitch, then completes a 360° rotation about the body frame $\boldsymbol{z}$-axis. In both cases which pass through known singularity regions, experimental results confirm stable tracking of the desired trajectory.

Efficiency Evaluation: Efficiency is defined here in terms of relative power consumption, and evaluated relative to the best case power consumption p_h of horizontal hover, $\eta_P = \frac{v_{total}\Sigma_{i=1}^{12} i_i}{p_h} \in [0, 1]$. A wasted force index is also computed, as proposed in [8], $\eta_f = \frac{\|\boldsymbol{F}_b\|}{\Sigma_{i=1}^{12} f_i} \in [0, 1]$, where f_i are the thrust values for each rotor. Estimated total power, power efficiency and wasted force for each hover configuration are shown in Fig. 6, projected onto the force envelope. Below the theoretical envelopes, measured values from experimental results are displayed, using measured PWM-to-current correlation and flight-test voltage logs to find the power consumed. Experimental results agree well with the ranges predicted on the theoretical envelopes, showing that efficient flight configurations are achieved in specific orientations, while the system remains pose-omnidirectional with varying degrees of efficiency.

4 Conclusion

Results presented in this paper demonstrate that although actuated tilting rotor units have added complexity and weight compared to other fully actuated MAVs, the additional weight can be justified by improved hover efficiency of the platform in particular configurations, and additional complexity by the reliability of the prototype system, VoliroX.

As a result of the assumption that rotor thrust and drag torques are independent, any airflow interference between rotor groups is treated as an unknown external disturbance to be rejected. This is a reasonable assumption for some but not all body orientations, such as when flying the cartwheel maneuver, where lower rotors are affected by airflow from those above. As a result there is some loss in tracking performance, solutions to which are being developed in ongoing work. Moving to more dynamic flight trajectories, Further techniques for singularity avoidance will be investigated, and additional benefits that can be yielded from an overactuated system.

References

1. Brescianini, D., D'Andrea, R.: Design, modeling and control of an omni-directional aerial vehicle. In: ICRA 2016, pp. 3261–3266. IEEE (2016)
2. Park, S., Lee, J., Ahn, J., Kim, M., Her, J., Yang, G.-H., Lee, D.: ODAR: aerial manipulation platform enabling omni-directional wrench generation. IEEE/ASME Trans. Mechatron. **23**, 1907–1918 (2018)
3. Park, S., Her, J., Kim, J., Lee, D.: Design, modeling and control of omni-directional aerial robot. In: IROS 2016, pp. 1570–1575. IEEE (2016)
4. Staub, N., Bicego, D., Sablé, Q., Arellano, V., Mishra, S., Franchi, A.: Towards a flying assistant paradigm: the OTHex. In: ICRA 2018. IEEE (2018)
5. Falconi, R., Melchiorri, C.: Dynamic model and control of an over-actuated quadrotor UAV. IFAC Proc. Volumes **45**(22), 192–197 (2012)
6. Ryll, M., Bülthoff, H.H., Giordano, P.R.: A novel overactuated quadrotor unmanned aerial vehicle: modeling, control, and experimental validation. IEEE Trans. Control Syst. Technol. **23**(2), 540–556 (2015)
7. Kamel, M., Verling, S., Elkhatib, O., Sprecher, C., Wulkop, P., Taylor, Z., Siegwart, R., Gilitschenski, I.: Voliro: an omnidirectional hexacopter with tiltable rotors. arXiv preprint arXiv:1801.04581 (2018)

8. Ryll, M., Bicego, D., Franchi, A.: Modeling and control of fast-hex: a fully-actuated by synchronized-tilting hexarotor. In: IROS 2016, pp. 1689–1694. IEEE (2016)
9. Lee, T., Leoky, M., McClamroch, N. H.: Geometric tracking control of a quadrotor UAV on SE(3). In: Decision and Control (CDC), pp. 5420–5425. IEEE (2010)
10. Invernizzi, D., Lovera, M.: Geometric tracking control of a quadcopter tiltrotor UAV. IFAC-PapersOnLine **50**(1), 11565–11570 (2017)
11. Wampler, C.W.: Manipulator inverse kinematic solutions based on vector formulations and damped least-squares methods. IEEE Trans. Syst. Man Cybern. **16**(1), 93–101 (1986)

Vision Based Robust Autonomous Landing of a Quadrotor on a Moving Target

Alvika Gautam[1(✉)], P. B. Sujit[1], and Srikanth Saripalli[2]

[1] IIIT-Delhi, Delhi, India
{alvikag,sujit}@iiitd.ac.in
[2] Department of Mechanical Engineering, Texas A & M, College Station, TX, USA
sri@fastmail.fm

Abstract. Autonomous landing is the most difficult maneuver in a quadrotor mission because of challenges such as external disturbances and localization error, both of which cause deviation from the desired trajectory and may result in a crash. In this paper, we present a guidance framework that allows a multirotor to land on a moving target accurately. The framework utilizes onboard vision to detect and estimate the landing target parameters. We then analyze the effects of environmental disturbances, and abrupt changes in the motion of the landing target to allow us to investigate them from a computer vision perspective. This will aid in development of a robust autonomous landing strategy for a moving target. We present robustness results through outdoor hardware experiments.

Keywords: Autonomous landing · Guidance · Computer vision · Robustness

1 Introduction

Unmanned aerial vehicles, particularly rotary wing ones like quadrotors are ubiquitous in civilian and defense applications [1,2], owing to their agile maneuvering and hovering capability. Autonomous quadrotor landing on a desired target in these applications has gained considerable attention during the past decades. The two essential components of an autonomous landing process are target detection, and guidance & control. Typically, autonomous landing is attempted using sensor-based guidance strategies, which use a combination of vision sensors and range finders, that generate velocity commands to minimize lateral, longitudinal error and descend on the target based on it's estimated state. The condition for landing on a moving target, from a guidance perspective can be described as; Given an aerial vehicle (quadrotor), with an initial velocity of V_{u0} and at an altitude h, and an initial distance R_0 from a moving target (velocity V_t), necessary condition for a successful landing is:

$$\lim_{t \to t_f} R = \lim_{t \to t_f} -\dot{R} = 0, \text{For a final time } t_f \quad (1)$$

J. Xiao et al. (Eds.): ISER 2018, SPAR 11, pp. 96–105, 2020.
https://doi.org/10.1007/978-3-030-33950-0_9

where, $-\dot{R}$ is the closing velocity between the vehicle and the landing target. Thus, prospective landing guidance strategies desire a smooth velocity variation with zero closing velocity at touchdown [3]. Guidance strategies along with vision based landing techniques have been studied in the literature from different perspectives [4–6]. However, robust, precise and time efficient autonomous landing on a moving target remains a fundamental, yet challenging problem due to environmental disturbances along with constraints imposed by localization errors, motion of the target and time critical nature of applications. Thus, it becomes imperative to thoroughly test the landing setup in order to ensure that the controller degrades gracefully even in unforeseen conditions, and a decision to abort and/or reinitialize the landing process can be made accordingly.

In this work, we study the problem of autonomous landing on a moving target from a robustness perspective. We discuss our vision based guidance approach for landing and present outdoor experiment results for tracking and landing with perturbations in target speed/trajectory and presence of environmental disturbances.

2 Technical Approach

We have previously developed a vision based guidance system which used pure-pursuit guidance for aligning the UAV with the target, and a speed controller based on log of polynomial function. This was validated using outdoor experiments for a stationary target [7]. In this work we extend and evaluate the approach for landing on a moving target. The proposed velocity profile is given as,

$$V_u = V_t N^{\log_n\left(1+ax+bx^2\right)} \tag{2}$$

where, x represents the ratio of current and initial distance to landing target,

$$x = \frac{R}{R_0}, \quad 0 \leq x \leq 1 \tag{3}$$

V_u and V_t represent the UAV and target velocity and the constants a and b are related to base of the log n by,

$$a + b = n - 1, \quad n > 1, \tag{4}$$

and N represents the ratio of initial UAV to target velocity.

Fig. 1. Hardware setup consisting of the quadrotor attempting to land on the target mounted on a ground rover.

Initially at $t = t_0$, $R = R_0$, which implies $x = 1$ and at $t = t_f$, $x = 0$ which leads to a successful landing condition ($V_u = V_t$) as governed by Eq. (1). The log polynomial controller was analyzed in [8] and closed-form design parameter solutions were derived for different nature of velocity profiles as shown in Table 1. These solutions were derived by analysing the rate of change of V_u, with the variation in (a, b). In addition to this, approximate landing time and maximum velocity expressions were derived as a function of the design parameters of the log polynomial controller function. We leverage this analysis in order to extend the approach for autonomous landing on a moving target and evaluate the robustness of the same. We use an off the shelf 3DR solo quadcopter for our outdoor experiment setup.

Table 1. Intervals of (a, b) for various flight profiles

(a, b)	Velocity profile
$0 \leq a \leq 2(n-1)$, $-(n-1) \leq b \leq (n-1)$	Continuous decelerating landing profile
$a > 2(n-1), b < -(n-1)$	Initially accelerating landing profile
$a < 0, b > (n-1)$	Continuous decelerating hovering profile

2.1 Target Detection and Tracking

The landing algorithm is implemented on a linux based embedded computer: (Odroid U3) which is mounted on a quadrotor and connected to a downward facing Ptgrey USB camera. The hardware setup along with the landing target is shown in Fig. 1. The target is chosen in the form of a red colored blob mounted on a ground rover. The landing algorithm is implemented using ROS and target detection is performed, using color based blob detection approach using OpenCV.

In case of a moving target tracking and landing, velocity of the UAV must be controlled according to the velocity of the target. Thus, it is important to compute the target velocity correctly. We use a KF for a robust estimate of the landing pad parameters (x_{img}, y_{img}), to deal with sudden changes in detection output, resulting from wind disturbances and target occlusion. The relative velocity on the image frame is calculated by the slope of the image positions. We can express the target velocity as given below:

$$\begin{bmatrix} v_{(tx)} \\ v_{(ty)} \end{bmatrix} = \begin{bmatrix} v_{qx_{t-\delta(t)}} \\ v_{qy_{t-\delta(t)}} \end{bmatrix} + \begin{bmatrix} K_x & 0 \\ 0 & K_y \end{bmatrix} \begin{bmatrix} \cos(\psi) & -\sin(\psi) \\ \sin(\psi) & \cos(\psi) \end{bmatrix} \begin{bmatrix} \dot{x}_{img} \\ \dot{y}_{img} \end{bmatrix}. \quad (5)$$

where (K_x, K_y), represent the constants for transformation between the local NED coordinate and the image plane. The computed target velocity components are used to design the lateral and longitudinal velocity control subsytems independently using the log polynomial velocity profile given by Eq. (2). For altitude control, velocity profile is given as,

$$V_z = V_{z0} \log_n \left(1 + ax + bx^2\right) \tag{6}$$

where x represents the ratio of current to initial altitude, and V_{z0} represents the initial velocity in z direction. Further, a flag was set for the quadrotor to descend such that quadrotor was commanded a z velocity only if this flag was equal to 1 as given in Eqs. (8) and (9)

$$flag_{descend} = 1, \text{if } R_{ut} \leq \frac{fp}{1.5}, \tag{7}$$

$$V_z = V_{z0} \log_n \left(1 + ax + bx^2\right), \text{if } flag_{descend} = 1, \tag{8}$$

$$V_z = 0, \text{if } flag_{descend} = 0 \tag{9}$$

The flag was set to 1, if the distance between UAV and the target (R_{ut}) was less than $\frac{fp}{1.5}$, where fp is the camera footprint in meters at the respective altitude. In case of the distance being greater than $\frac{fp}{1.5}$, descend flag was set to 0 and thus the commanded V_z was also equal to zero.

3 Simulation Setup

In this section, we present the details of implementation and results of our proposed framework. The guidance principle was tested in a simulation environment with Microsoft Airsim as the simulator used for quadrotor flight and vision based landing target detection. AirSim is a MAV simulator that uses Unreal Engine as its base, thus being able to simulate realistic scenes and environment. The landing target was in the form of a moving car simulated in unreal engine. Figure 3 shows a screenshot of the simulation environment consisting of the flying quadrotor vehicle modeled in airsim and a ground vehicle which served as the moving landing target. In order to validate our algorithm, we have performed simulations with the setup described above. We have performed multiple experiments with different scenarios for target motion characteristics. In each of the scenarios, the quadrotor is initialized with the command to takeoff to an altitude of 20 m. The landing target is a red colored moving ground vehicle. The target is assumed to be of known color and detection is performed using the standard color based blob detection approach using the downward camera feed of the airsim aerial vehicle. The different landing scenarios evaluated in this work are described below:

- Target moving in an approximate straight line motion with a constant speed. Multiple simulations are performed in this scenario with different fixed target speeds in each case, ranging from 1 m/s to 3 m/s.
- The second scenario introduces a perturbation into the path followed by vehicle in the form of a slope to validate the robustness of proposed algorithm against variations in surface inclination.

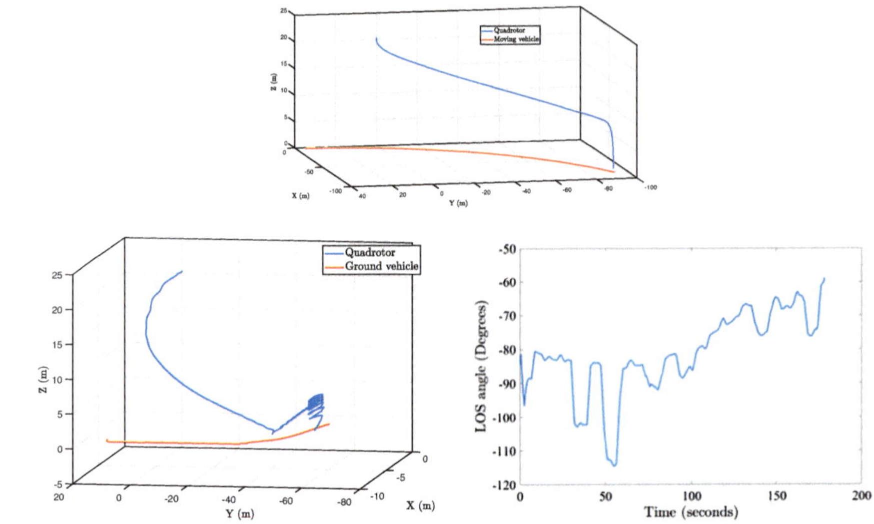

Fig. 2. (a) Landing trajectory followed by UAV when ground vehicle moves in a near straight line trajectory (b) Trajectory followed by the quadrotor while trying to land on a non uniform surface (c) LOS angle profile (Terminal desired LOS angle: -60°)

Figure 2(a) shows the trajectory followed by quadrotor with vision based input for one of the cases where the ground vehicle is moving with a constant speed of 2 m/s. With vision in the loop, the quadrotor follows an efficient trajectory towards the target. Similar trend was observed for target moving at 1 m/s and 3 m/s. Figures 2(b) and (c) shows the performance of the guidance algorithm, when attempting to land on a non uniform surface. Though slight irregularity can be seen at the time of landing, quadrotor is able to land successfully without any knowledge of surface inclination.

4 Experimental Results and Discussion

We have performed a series of tests that involved outdoor data collection, data processing and closed loop controller validation with onboard processing. We perform two sets of experiments to validate the efficacy of proposed controller; In the first set of experiments, we validate the log polynomial controller for a moving target scenario. In this experiment scenario, the quadcopter was commanded to takeoff to an altitude of 10 m and the target was moved in a approximate straight line motion with an approximate velocity of 0.5 m/s (as given by the autopilot mounted on the target) in longitudinal direction and negligible movement in lateral direction. Figure 4(a) shows the trajectory followed by the quadrotor to land on the target. Figures 4(b) and (c) show the longitudinal and lateral velocity profiles of the quadrotor. It can be seen that the longitudinal velocity profile approaches the target speed towards the end and the lateral velocity profile

Fig. 3. A snapshot of the simulation setup consisting of the airsim flying quadrotor vehicle and a ground vehicle as the landing target

approaches 0 thus resulting in zero closing velocity for a successful landing. Figure 5 shows the trajectory followed by the quadcopter when the target was moved in a near circular trajectory. These experiments are conducted in presence of negligible wind conditions or any external disturbances. The log of polynomial controller provides a fine grained control of UAV velocity during the landing process. Using this, an initial accelerating profile ($a = 30, b = -21, n = 10$) is designed to track the target in a time efficient manner, while a continuously decelerating profile ($a = 6, b = 3, n = 10$) is chosen as UAV descends onto the target.

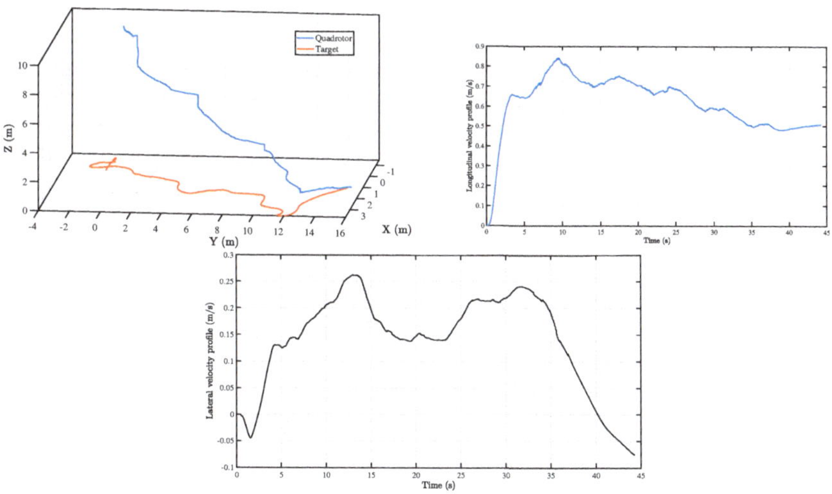

Fig. 4. (a) Trajectory followed by the Quadcopter to land on a moving target. Quadrotor velocity profile in case of moving target landing (a) Longitudinal velocity profile (b) Lateral velocity profile

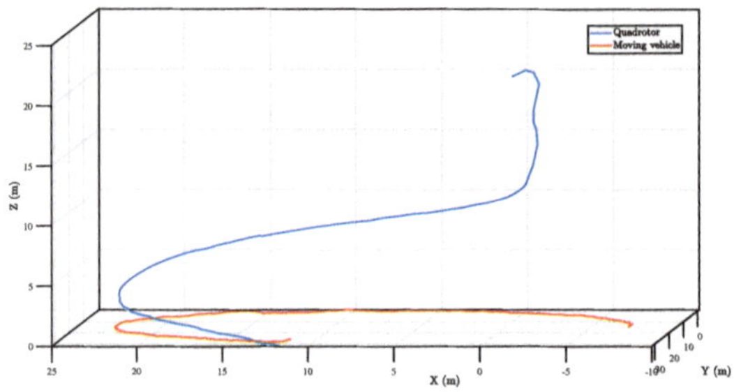

Fig. 5. Trajectory followed by the Quadcopter to land on a moving target in circular motion

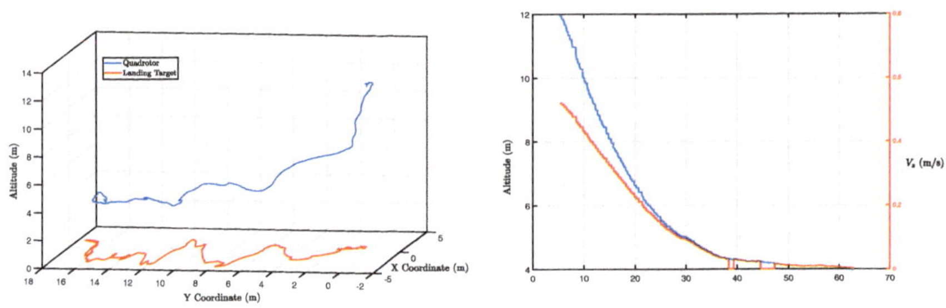

Fig. 6. UAV trajectory and descent velocity while tracking the target and descending till 4 m from the ground

Second set of experiments were aimed at testing the robustness of the landing process. These set of experiments are comprised of different scenarios pertaining to target trajectory and presence of wind. In the first experiment, landing target was made to follow a random trajectory with a near constant speed and the quadrotor was made to manually follow the target at a constant altitude of 12 m. This experiment was done in presence of strong wind conditions with gusts as high as 6 m/s [9]. We have first processed the image data offline. A Kalman filter is used to obtain a robust estimate of trajectory from noisy measurements. Figure 7(a) shows the ground truth, estimated and measured trajectories followed by the landing target. We have then performed a landing experiment in similar high velocity wind conditions, where the loop between vision and control is closed and the processing is done onboard to test the tracking. In this case, both UAV and target start at the same position in XY plane. UAV is initially at an altitude of 12 m above the target. The target is maneuvered manually in random trajectory and UAV tracks the target. Figure 7(b) shows the trajectory followed by the quadrotor and the landing target. Although, an offset can be observed between the two trajectories due to high winds, it can be seen that even in the presence of disturbance, UAV follows a similar trend as the target,

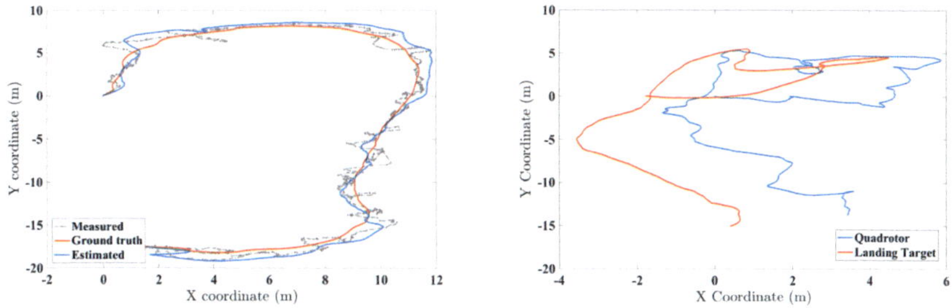

Fig. 7. (a) Trajectory of the ground rover as estimated by the quadcopter (b) Trajectory of the quadrotor tracking the landing target in the presence of high winds (5 m/s steady)

highlighting the controller's efficiency in tracking the target. In the final experiment, target is maneuvered manually in a random sinusoidal trajectory. UAV was commanded to follow the target based on vision measurements and descent till an altitude of 33% of its initial altitude. All other experimental conditions are maintained the same as previous scenario. Figures 6 and b show the trajectory followed by the UAV and the descend velocity profile. In this case for descent velocity, hovering velocity profile parameters ($a = -4.43, b = 9.43$) were chosen in the log polynomial controller. It can be observed that the UAV tracks the target while descending to an altitude of 4 m. At this point V_z becomes zero and UAV continues to track the target at a constant altitude. Through these preliminary experiments, we validate and present the efficacy of our landing strategy and controller in varying scenarios, ranging from straight line movement, tracking and descend in presence of wind disturbances and random trajectory perturbations.

5 Main Experimental Insights

From our experiments, we observed that environmental disturbances such as wind and abrupt changes in target speed or trajectory result in similar effects from the perspective of a camera image. For example, high winds or target

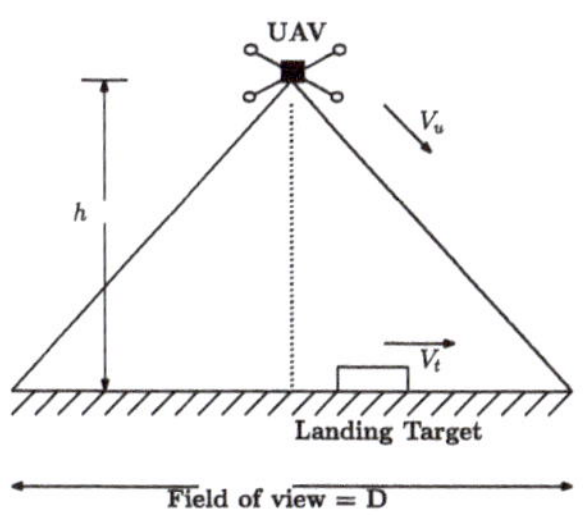

Fig. 8. A depiction of a quadrotor trying to land on a moving target with velocity V_t

changing its speed both result in estimation errors or target going out of view of the camera. Thus, studying the effect of perturbations from computer vision perspective aids in evaluating the conditions to abort and/or reinitialize the landing process, reducing the dependence on knowledge of source of external disturbance. This helps towards developing robust landing strategies, agnostic to external disturbance. We present our approach here with a brief example of mid-landing scenario. Consider an aerial scenario as shown in Fig. 8, where a quadrotor (with a velocity V_u), and with a downward facing camera, is attempting to land on a target moving in straight line with velocity V_t. At time t, quadrotor is at a distance of R from the landing target and an altitude h meters above the ground. Let D be the field of view of the camera in meters at a height h above the ground at time t. The condition for the landing target going out of field of view can be expressed as, $R \geq D \pm \delta, -\dot{R} > 0$. Given such a scenario, the target remains within the field of view of the as long as the velocity of UAV is significant in magnitude in comparison with target velocity and a near zero relative velocity is maintained in the final stages of landing. Given the knowledge of camera parameters and quadrotor velocity profile, the condition for the target to go out of field of view can be estimated.

As explained in the previous section, whenever the target goes out of FOV, we command a zero z velocity. Let h_{OF} be the altitude at which the target went out of field of view. At this instant if V_t was the estimated target velocity and V_u was the calculated commanded velocity, then time required by the target to cover D_{OF} meters (which is the field of fiew of camera when the target was lost) is given by,

$$T = \frac{D_{OF}}{V_t} \tag{10}$$

UAV is commanded a velocity V_u based on the estimated target velocity, such that $V_u \geq V_t$. If the target doesn't come in the view even after t seconds, such that $t \geq T$, it can be safely said that the target has been lost and landing must be aborted or reinitialized using some strategy, for example increasing the altitude.

6 Conclusion and Future Work

Robustness in landing procedure is imperative for any autonomous mission to complete successfully. Our work uses a log polynomial velocity controller for robust autonomous landing on stationary and moving targets. Efficacy of the controller for varying speeds and inclined surface was demonstrated in airsim simulation environment. Controller was validated for straight line and near circular landing scenarios through outdoor hardware experiments, where camera vision was used to sense the landing target. Additionally, robustness of the tracking was demonstrated against wind disturbances and variations in target trajectory. Terminal stage of the landing process, due to ground effect and blindness of the vision system, especially in present of external disturbance is often challenging. Thus, in future we plan to do more landing experiments for different target trajectories involving a concentric smaller target to be used for detection below 1m altitude.

Acknowledgements. – Alvika Gautam is a recipient of TCS PhD research fellowship. – This work was in part supported by UK GCRF EPSRC grant number EP/P02839X/1.

References

1. Geng, L., Zhang, Y., Wang, J., Fuh, J., Teo, S.: Mission planning of autonomous UAVs for urban surveillance with evolutionary algorithms. In: Proceedings of the International Conference on Control and Automation (ICCA), Hangzhou, China, pp. 828–833. IEEE (2013)
2. Waharte, S., Trigoni, N.: Supporting search and rescue operations with UAVs. In: Proceedings of the International Conference on Emerging Security Technologies, pp. 142–147 (2010)
3. Borowczyk, A., Nguyen, D.T., Phu-Van Nguyen, A., Nguyen, D.Q., Saussié, D., Le Ny, J.: Autonomous landing of a quadcopter on a high-speed ground vehicle. J. Guidance Control Dyn. **40**(9), 2373–2380 (2017)
4. Lee, H., Jung, S., Shim, D.H.: Vision-based UAV landing on the moving vehicle. In: Proceedings of the 2016 International Conference on Unmanned Aircraft Systems (ICUAS), Arlington, VA, USA, pp. 1–7 (2016)
5. Ling, K.: Precision landing of a quadrotor UAV on a moving target using low-cost sensors. Master's Thesis, Department of Mechanical and Mechatronics Engineering, University of Waterloo, Ontario, Canada (2014). http://hdl.handle.net/10012/8803
6. Kong, W., Zhang, D., Wang, X., Xian, Z., Zhang, J.: Autonomous landing of an UAV with a ground-based actuated infrared stereo vision system. In: Proceedings of the 2013 IEEE/RSJ International Conference on Intelligent Robots and Systems (IROS), Tokyo, Japan, pp. 2963–2970 (2013)
7. Gautam, A., Sujit, P., Saripalli, S.: Autonomous quadrotor landing using vision and pursuit guidance. IFAC-PapersOnLine **50**(1), 10501–10506 (2017)
8. Gautam, A., Ratnoo, A., Sujit, P.: Log polynomial velocity profile for vertical landing. J. Guidance Control Dyn. **41**, 1–7 (2018)
9. Wind data (2017). https://darksky.net

UAV-Based Automated Labeling of Training Data for Online Water and Land Differentiation

Curtis Klein, Trevor Speckman, Thomas Medeiros, Derek Eells, and Elizabeth Basha(✉)

University of the Pacific, 3601 Pacific Avenue, Stockton, CA 95211, USA
ebasha@pacific.edu

1 Introduction

Wetlands scientists require detailed maps of the land and water contours to correctly understand wetlands water systems. Current methods rely on expensive, manually collected data that are rarely updated, providing obsolete data for modeling. Automating the process would improve data collection and enable scientists to improve wetlands models. The automated solution needs to minimally interact with the fragile environment, to function autonomously to avoid long trips to remote wetland locations, and to provide modularity to support the different measurement techniques for topographic and bathymetric mapping. Unmanned aerial vehicles with varying payloads and autonomous behavior provide a reasonable and effective solution.

Our system focuses on the initial problem of differentiating water and land on a UAV while flying in order to direct other UAVs to perform mapping. The system utilizes an offline, fully automated machine learning training system that automates image labeling and classifier selection. The learned classifier is then automatically programmed on the UAV, which uses it while flying to autonomously differentiate land from water. The UAV then communicates the different geographic regions via radio to the other UAVs to begin mapping operations.

Prior work has explored remote sensing or visual surface classification via aerial robots. Remote sensing of wetlands approaches [1–3] typically post-process data using manual classification and that data may not be current due to the difficulty of obtaining satellite and plane imagery. Most aerial robot prior research has used UAVs to collect data that is then post-processed for use in classification algorithms [4–7]. Some research has focused on online image processing for the purposes of tracking coastlines or similar edges [8]. To the best of our knowledge, no research has focused on online terrain classification by the UAV while flying.

We develop and test our system on the Calaveras River in Stockton, CA. Using data gathered under varying conditions, our prior metaclassifier system [9],

Electronic supplementary material The online version of this chapter (https://doi.org/10.1007/978-3-030-33950-0_10) contains supplementary material, which is available to authorized users.

J. Xiao et al. (Eds.): ISER 2018, SPAR 11, pp. 106–116, 2020.
https://doi.org/10.1007/978-3-030-33950-0_10

expanded to support image data, determines the best classifier model, selecting one with a 97.5% accuracy using two features. We use this classifier model in our online classifier, which we develop and begin to use in field experiments. We also develop a hardware-based automated classification system that the UAV uses to label regions in the image as land or water to develop a training set for the metaclassifier. Field experiments verify that the system provides 100% accuracy in labeling regions. Overall, our system provides an end-to-end automated approach to labeling, classifier training, region classification, and region coordinate communication.

2 Technical Approach

We developed an online system to differentiate water from land based on visual images. The system begins with the automated labeling of images to create a training set, feeds the training set into a metaclassifier that selects the best classification model based on a large set of features and classifiers, uses the model to define regions of interest while flying, and communicates the coordinates of these regions to the specialized UAVs that will map the regions. Figure 1 shows an overview of the system. We will briefly describe each component of the system.

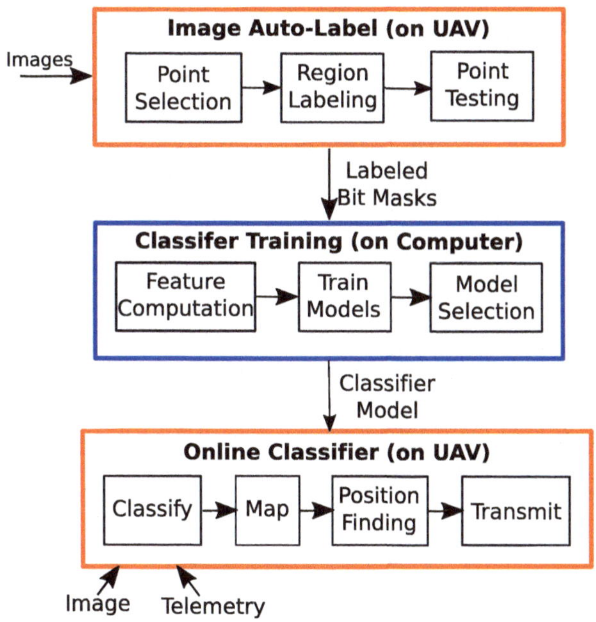

Fig. 1. Online automated differentiation system

2.1 Image Auto-Labeling

The image auto-labeler measurement system is a specialized UAV payload combining a measurement system with a point selection algorithm. The UAV, with this payload, collects an initial photograph of a region, clusters the image, finds coordinates of centroid points of the clusters, and tests those points to determine if it is water or land. This process repeats until a sufficient training set of labeled images exists.

Measurement System. The hardware measurement system is a custom-built, buoy-like device that tests whether a surface is land or water using conductivity through two electrical contacts on opposite sides. The UAV carries the system suspended from a winch and uses the winch to lower the device onto the surface (see Fig. 2). The device also has a mechanical spring-loaded switch attached between the buoy and the winch. While in the air, the switch is closed due to the tension in the line. Once the buoy device reaches a surface, the tension releases and the switch opens, indicating to the UAV to stop lowering. The buoy then determines that the surface is water if an electrical connection exists or land if no connection exists between the contacts.

Fig. 2. Complete system on UAV

Point Selection and Labeling Algorithm. The point selection algorithm consists of three parts: (1) the selection of points to measure, (2) the testing of each point, and (3) the post-processing to create a bit mask.

Algorithm 1 outlines the selection of points. It first reduces the image to a set of clusters based on HSV values. A cluster may exist in multiple spatial locations in the image that share similar HSV values; to provide spatial context, we create contours separating the cluster (note that we use cluster to indicate pixels grouped by HSV values and contour to indicate pixels grouped by both HSV value and spatial location). We then examine the contours to find those with areas large enough for the UAV to successfully test, identify the center point of the contours, and select the top 10 points to measure. By selecting only 10 points, we reduce the time needed to measure each image.

Algorithm 1. Point Selection

```
Convert image to HSV
Cluster image using K-Means
Open morphology
Close morphology

for Each Cluster do
    Create a bit mask of just that cluster
    Find contours of cluster
end for
for Each Contour do
    Determine the contour center point
    if center > 15 pixels from edge then
        Use the selected center point
    else
        Test all points to find point furthest from any hole or edge
    end if
    Store selected point with distance and contour area
end for

Sort contours based on distance from selected point to edge
Select top 10 points
```

Once the points are selected, we order the points from left to right and determine the GPS coordinates of each point. The UAV then flies to each point, measures the conductivity, and labels the point. After measuring all points, the algorithm labels the remaining contours that were not directly measured. It examines the neighboring contours and assigns the label held by the majority of neighbors if more than 25% of the neighbors have a label. If there is no label held by more than 25% of its neighbors, the algorithm skips that contour and moves to the next. The algorithm will iterate through all contours until the entire image is labeled. This results in a bit mask, identifying the water and land regions.

2.2 Metaclassifier Training

We expanded our existing metaclassifer system to support image-based classification [9].

The process starts with a set of images that have been converted to a binary image indicating water or land through the labeling algorithm. The system uses this binary image and the original image to compute features. When computing the features, the algorithm recursively divides the image into quadrants, computes features for each quadrant, and groups similar adjacent cells for expanded features per image. These features consist of the mean and standard deviation of the red, green, blue, hue, saturation, and value of the pixels in the water and land regions (as defined by the bit mask). Each image's feature set consists of the cells' features and the top-level image features.

Once the features are extracted, the system computes all possible combinations up to a maximum feature set size. It will then create a different data set for each feature combination using all of the training images.

Using each set of feature data, our meta-classification system trains 12 different types of classification models such that each model type is trained on all sets of feature data. For each trained model, the system evaluates its success rates based on classifying the verification set. Our system then chooses the model that provides within 2% of the best accuracy and the smallest number of features, which then feeds into the online classification algorithm.

2.3 Online Classification

The online classification algorithm begins with the UAV taking a picture and recording telemetry data including GPS, altitude, pose (rotational quaternion), and other valuable data. The UAV then runs the chosen classifier model to produce a classified binary image. It then maps the water regions of this image through open and close morphology, finding enclosing contours, filtering the regions based on size (calculated from height), and smoothing the contours. Once the UAV has a set of contour points, it calculates the real world position of the points using UAV telemetry (based on [10]).

3 Experiments

We verified the image auto-labeler using manual point selection and automated point selection; we trained the online classification system and confirmed accurate classifications; and we used the automatically labeled images to train the classifier thereby connecting the parts together.

3.1 Overall Setup

Our experiments all use a DJI Matrice 100 system with a custom payload that contains a Raspberry Pi 3, a Garmin LIDAR-Lite v3, a Digi ZigBee radio, and a

Logictech C310 webcam. The Raspberry Pi controls all systems, communicating with both the Matrice and any other payloads (such as the winch) to coordinate the experiments.

To create a data set of images for development, training, and verification, we flew the system over the Calaveras River on the University of the Pacific campus in Stockton, California. The river is narrow with significant vegetation, making it similar to wetlands environments. It has dirt trails along the banks and roads crossing the river, which we can use to ensure that the system handles man-made features that may be along the sides of wetlands (in California most are near active agriculture regions). We collected over 500 images at different times of days and different weather conditions to ensure a wide variance in lighting and that our results hold for a range of scenarios. A small set of images were used for development; we kept the remainder separate for training and validation purposes.

3.2 Image Auto-Labeler

In verifying the image auto-labeler, we manually tested the buoy hardware, simulated the point selection and labeling algorithm, and tested the full system in an automated field experiment.

Manual Point Selection
Before testing the fully autonomous version, we tested the image auto-labeler by manually selecting the test points, ensuring a set of known water and land points. The UAV flies to the provided coordinates, hovers, lowers the buoy using the winch system until the buoy reaches the surface, collects 25 measurements, raises the buoy, and returns. If any measurement is labeled water, the system labels that point as water. We measured 58 points, 30 water and 28 land. All points were labeled accurately with a 100% success rate, showing that our hardware system can correctly label points. Our next step was to verify the point selection algorithm to choose the points to measure.

Automated Point Selection
We first verified the point selection algorithm through simulation to provide a larger number of tests and then confirmed through field experiments.

The simulation used 160 images to verify the operation of the point selection algorithm. We simulate the measurement phase through manually labeled bit

Table 1. Automated point selection simulation results

	Water classification accuracy	False positive percentage	Overall classification accuracy
Mean	96.6	2.9	97.3
Min	47.5	0.0	90.5
Max	100.0	13.4	100.0

masks that identify water and land regions; we also compare the final result of the algorithm to the labeled bit mask to determine the effectiveness of our system.

Table 1 outlines the results of our experiments. On average, the system correctly labels 97.3% of the region, incorrectly labels 3.4% of the water as land, and incorrectly labels 2.9% of the land as water. Of the images, 21 contained no water and were perfectly labeled, providing the best accuracy results.

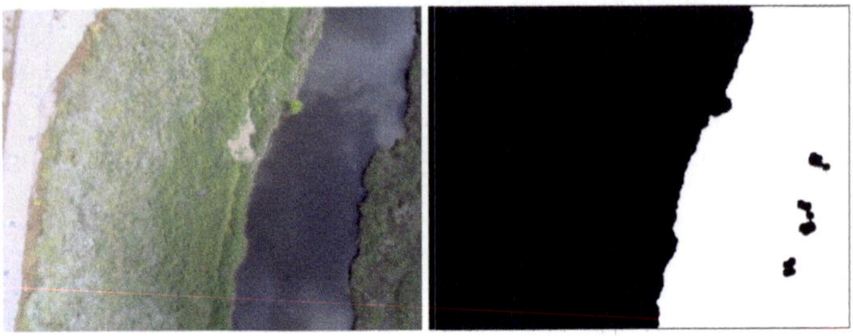

(a) Original Image (b) Labeled Bit Mask

Fig. 3. Worst overall classification accuracy simulation result

Figure 3 shows the worst accuracy result. As the original image shows, there is a large contrast between the land on the left of the image and that on the right; the system incorrectly labeled most of the land on the right as water. This result, while not ideal, would train the system to be cautious while measuring as it would ensure the system sees more water than land, which is safer for our mapping application.

Figure 4 shows the worst result in terms of false negatives, where the system incorrectly labels water as land. The image consists of limited water and the water exists right around bridges; this scenario would require a closer, more detailed image to ensure the UAV does not run into the bridge. We plan to explore hierarchical imaging and overlapping approaches in the future to deal with these specialized scenarios.

We next implemented the algorithm on the UAV and tested automated point labeling on the Calaveras River. The system flew to a defined GPS location, took a picture, returned to land, and computed the points to measure. Once it had the points, it flew to each point in order, deploying the buoy and measuring at each, until all points were measured. After landing, it completed the process by computing the bit mask.

Figure 5 shows the process and result of the experiment. First the UAV took the original image (Fig. 5(a)); then it clustered, smoothed, and contoured the image (Fig. 5(b)); it selected 10 points to measure (Fig. 5(c)); and it constructed the final labeled bit mask of the image (Fig. 5(d)). This process required two flights for one image. The winch appears to be the limiting factor currently; we

(a) Original Image (b) Labeled Bit Mask

Fig. 4. Worst false negative simulation result

plan to better analyze the complete system to verify this and improve it in the future.

This proved the overall operation of the system. We plan to speed up the measurements through winch improvements and examine the different alternatives to labeling the small unmeasured regions. With labeled images, we next focused on the classifier.

3.3 Classifier

Classifier Training
In order to determine the best classifier, we used 500 images collected with our system. We manually labeled our image set (while developing the automatic labeling system) and ran the metaclassifier over all possible combinations of 12 models and 12 features. Table 2 outlines the best five models based on classification accuracy on the training data set. The metaclassifier system selected the J48 model using two features (the red mean and the blue standard deviation). This model is automatically sent to the UAV for the online classifier.

Online Classifier
We successfully implemented the online classifier algorithm on a Raspberry Pi 3. Using the results of the classifier training, we verified basic functionality through in-lab testing by providing the same images and confirming the system had the same classification accuracy.

Our initial field experiment calibrated the camera scaling factor used in the calculation of real world position of the map results (based on [10]). We started with images of a nearby bridge with a known width of 5.79 m (19 feet) and collected three sets of points per image. These points were converted to real world coordinates using our algorithm and provided a distance; we then calibrated the scaling factor to return distances as close to 5.79 m as possible for our typical flight altitude of 60–80 m. After calibrating the scaling factor, our points now have a mean value of 5.76 m with a standard deviation of 0.24.

With that scaling factor, we performed experiments of the online classifier, gathering a data set of 195 images. Figure 6 shows an example result where

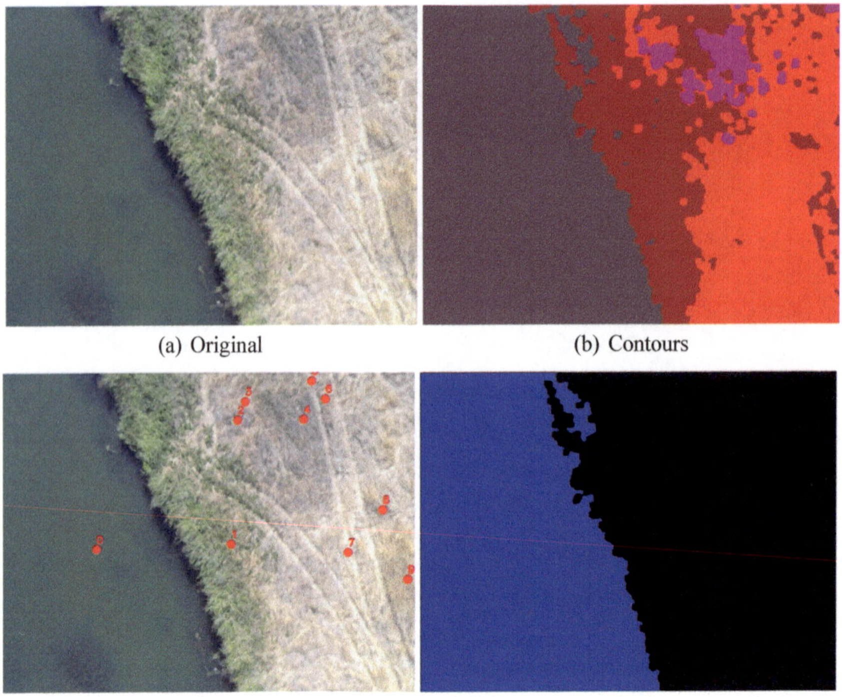

(a) Original (b) Contours

(c) Selected Points (d) Final Bit Mask

Fig. 5. Field experiment results for automated point labeling

Table 2. Classifier training top 5 results

Model	Classification accuracy	Number features	False positives
J48	97.5	3	1.87
J48	97.5	2	2.09
J48	97.3	2	2.13
Random SubSpace	97.1	2	2.65
J48	97.1	1	2.4

the red outlines indicate the identified and mapped regions to communicate. We examined the amount of data to send for each image and, on average, the system communicates 22.1 map locations. In the worst case, the system communicated 78 map locations to define the water regions identified in that image.

We also compared these to manually labeled bit masks to gather accuracy results similar to training. On average, after classification, the results are 90.9% accurate with the best images achieving 100% and the worst achieving 57.1%. Three images are below 70% accurate and 10 images within 70%–79% out of the

Fig. 6. Example field experiment result of online classifier

complete set; we plan to explore these few outliers and add redundant computations to average out lighting impacts in the future.

3.4 Connecting the Pieces

With our automatically labeled set of images, we returned to the metaclassifier system and trained using that set on our previously identified best classifier to confirm the utility of our images. We divided the images into 75% training images and 25% verification images, and trained the J48 model using the same features. Our system provides a 96.3% classification accuracy on the verification set, reasonably close to the 97.5% seen originally. This confirms that we can automatically label images, train classifiers on those images, and use the classifier to differentiate land and water regions.

4 Conclusion

We have designed and implemented our system, verifying functionality through field experiments completed and ongoing.

The image auto-labeler provides an accurate and fully automated approach to labeling images in order to create a training set. This removes the need for a scientist or other person to manually label images, increasing the flexibility of the system to gathering new data and transferring to new locations. The speed of the system is still slower than we would like, providing future work in improving the process.

Once we have a set of labeled images, our classifier training system works well, providing a classification accuracy of 97.5% on manually labeled images and 96.3% on automatically labeled images. Transferring this to the online classifier system also works well although with a reduced classification accuracy of 90.9%.

We plan to explore adding redundancy to the approach to remove outliers and improve the results further.

Overall, our results indicate a complete system that can automatically deal with defining regions to map, removing the need for scientists to interact with data collection and, once flight rules allow, removing the need for people entirely, which helps preserve these fragile ecosystems.

Acknowledgements. We are grateful to USDA-NIFA 2017-67021-25924, and NSF CNS CSR-1217428, which partially supported this work. We also would like to thank Najeeb Najeeb, Zachary Shields, Nicholas Vaughn, and Tristan Watts-Willis for their assistance.

References

1. Baker, C., Lawrence, R., Montagne, C., Patten, D.: Mapping wetlands and riparian areas using landsat ETM+ imagery and decision-tree-based models. Soc. Wetlands **26**, 465 (2006)
2. Ramachandra, T.V., Kumar, U.: Wetlands of greater Bangalore, India: automatic delineation through pattern classifiers. Electron. Green J. **1** (2008)
3. Martyn, R.D., Nobel, R.L., Bettoli, P.W., Maggio, R.C.: Mapping aquatic weeds with aerial color infrared photography and evaluating their control by grass carp. J. Aquat. Plant Manage. **24**, 46–56 (1986)
4. Papakonstantinou, A., Topouzelis, K., Pavlogeorgatos, G.: Coastline zones identification and 3D coastal mapping using UAV spatial data. ISPRS Int. J. Geo-Inf. **5**, 75 (2016). Unmanned Aerial Vehicles in Geomatics
5. DeBell, L., Anderson, K., Brazier, R.E., King, N., Jones, L.: Water resource management at catchment scales using lightweight UAVs: current capabilities and future perspectives. J. Unmanned Veh. Syst. **4**, 7–30 (2016)
6. Casado, M., Gonzalez, R., Kriechbaumer, T., Veal, A.: Automated identification of river hydromophological features using UAV high resolution aerial imagery. Sensors **15**, 27969–27989 (2015)
7. Hung, C., Xu, Z., Sukkarieh, S.: Feature learning based approach for weed classification using high resolution aerial images from a digital camera mounted on a UAV. Remote Sens. **6**(12), 12037–12054 (2014). https://doi.org/10.3390/rs61212037
8. Xu, A., Dudek, G.: A vision-based boundary following framework for aerial vehicles. In: Proceedings of the IEEE/RSJ International Conference on Intelligent Robots and Systems (IROS) (2010)
9. Basha, E., Watts-Willis, T., Detweiler, C.: Autonomous meta-classifier for surface hardness classification from UAV landings. In: Proceedings of the IEEE/RSJ International Conference on Intelligent Robots and Systems (IROS), Vancouver, Canada (2017)
10. Barber, D.B., Redding, J.D., McLain, T.W., Beard, R.W., Taylor, C.N.: Vision-based target geo-location using a fixed-wing miniature air vehicle. J. Intell. Robot. Syst. **47**(4), 361–382 (2006)

Unmanned Aerial Vehicles II

Unmanned Aerial Vehicles II

Shreyansh Daftry

The papers presented in this session and the follow-up panel discussion exemplify important applications of UAVs. Bonatti et al. talked about novel commercial applications in UAV-based aerial filming and developed trajectory optimization algorithms to enable autonomous aerial cinematography. At the other end of the spectrum, there is emerging use of UAVs in industrial and scientific applications. Plowcha et al. designed an autonomous digging UAV for sensor emplacement and developed decision-making algorithms for predicting task-success. Furthermore, there has been considerable progress in enabling long-term autonomy with minimal human intervention. Tian et al. presented multi-UAV systems for search-and-rescue in complex GPS-denied environments such as forests and developed perception and planning algorithms to enable multi-session long-term autonomy. Similarly, Pluckter et al. developed perception algorithms for precision landing in unstructured environments – a key component in enabling extended long-duration autonomous UAV missions. As the autonomous capability of UAVs matures, it becomes increasingly important for the academic community to think harder regarding standards, metrics and benchmarks. The Black-Bird Dataset by Antonini et al. take a small but important step in this direction by releasing a large-scale dataset of aggressive UAV flight.

Autonomous Drone Cinematographer: Using Artistic Principles to Create Smooth, Safe, Occlusion-Free Trajectories for Aerial Filming

Rogerio Bonatti[1(✉)], Yanfu Zhang[2], Sanjiban Choudhury[1], Wenshan Wang[1], and Sebastian Scherer[1]

[1] Robotics Institute, Carnegie Mellon University, Pittsburgh, PA, USA
{rbonatti,sanjibac,basti}@cs.cmu.edu, wenshanw@andrew.cmu.edu
[2] Advanced Technology Research Division, Yamaha Motor Co., Ltd., Shizuoka, Japan
zhangya@yamaha-motor.co.jp

Abstract. Autonomous aerial cinematography has the potential to enable automatic capture of aesthetically pleasing videos without requiring human intervention, empowering individuals with the capability of high-end film studios. Current approaches either only handle off-line trajectory generation, or offer strategies that reason over short time horizons and simplistic representations for obstacles, which result in jerky movement and low real-life applicability. In this work we develop a method for aerial filming that is able to trade off shot smoothness, occlusion, and cinematography guidelines in a principled manner, even under noisy actor predictions. We present a novel algorithm for real-time covariant gradient descent that we use to efficiently find the desired trajectories by optimizing a set of cost functions. Experimental results show that our approach creates attractive shots, avoiding obstacles and occlusion 65 times over 1.25 h of flight time, re-planning at 5 Hz with a 10 s time horizon. We robustly film human actors, cars and bicycles performing different motion among obstacles, using various shot types.

1 Introduction

Aerial vehicles are revolutionizing the way both professional and amateur film makers capture shots of actors and landscapes, increasing the flexibility of narrative elements and allowing the composition of aerial viewpoints which are not feasible using traditional devices such as hand-held cameras and dollies. However, the use of drones for filming today is still extremely difficult due to several motion planning and human-computer interaction challenges.

Electronic supplementary material The online version of this chapter (https://doi.org/10.1007/978-3-030-33950-0_11) contains supplementary material, which is available to authorized users.

J. Xiao et al. (Eds.): ISER 2018, SPAR 11, pp. 119–129, 2020.
https://doi.org/10.1007/978-3-030-33950-0_11

Aerial cinematography incorporates objectives from different robotics areas. Similarly to high-speed flight [2,21], the drone requires smooth and safe trajectories. Following literature in inspection and exploration, the vehicle reasons about viewpoints. In addition, temporal reasoning plays a major role when following a target, analogous to formation flight [32]. Cinematography also needs to consider artistic intent [1], using guidelines such as the rule of thirds, scale and relative angles [3].

Previous approaches in aerial filming do not address the complete problem in a sufficiently generalizable manner to be used in real-life scenarios. Off-line trajectory generation [8,13,28] cannot be used for most practical situations, and the on-line trajectory generation methods that have been proposed have limitations such as ignoring artistic intent or only dealing with limited obstacle representations [22] (ignoring obstacles altogether in many cases).

Our key insight in this work is that this problem can be efficiently solved in real-time as a smooth trajectory optimization. Our contributions in this paper are threefold: (1) we formalize the aerial filming problem following cinematography guidelines for arbitrary types of shots and obstacle shapes, (2) we present an efficient optimization method that exploits covariant gradients of the objective function for fast convergence, and (3) for over 1.25 h of flight time while re-planning, we experimentally show robustness in real-world conditions with different types of shots and shot transitions, actor motions, and obstacle shapes. The supplementary video shows examples of trajectories: https://youtu.be/QX73nBBwd28 (Fig. 1).

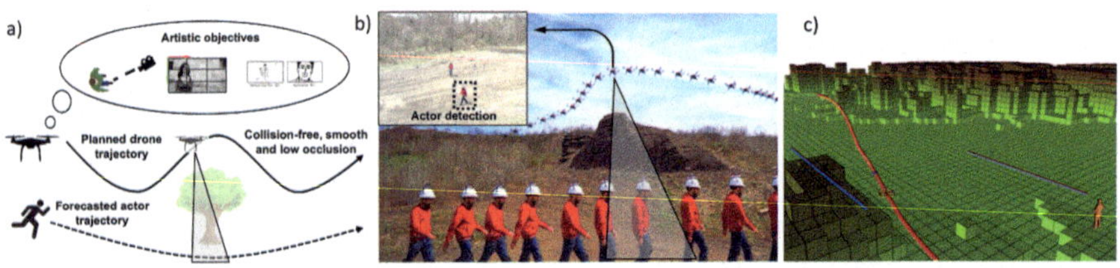

Fig. 1. Aerial cinematographer: (a) The drone reasons about artistic guidelines, forecasts the actor's motion and plans a smooth, collision-free trajectory while avoiding occlusion. (b) Field results produce visually appealing images. The camera detects and tracks the actor, keeping him in the desired screen position. (c) Motion planner view, with actor's motion forecast (purple), desired cinematography guidelines (blue), and optimized trajectory (red).

2 Problem Formulation

The act of filming encompasses a broad spectrum of behaviors and definitions. For example, a single or multiple cameras can be filming a scene, which can be focused on a landscape, and/or contain one or multiple actors in the frame. In this work we focus on single-camera, single-actor scenarios, which are omnipresent in real-life application, and difficult to execute. Therefore, we have a quadrotor with trajectory ξ_q, coupled with a camera, that films one actor with trajectory ξ_a, where $\xi(t)$ maps time $t \in [0, t_f]$ to a configuration. At the end of the paper we also discuss how our definitions and algorithms could be extended to other filming scenarios.

Following literature in cinematography [1,3], we identified a small set of camera positioning parameters that can define a large span of shot types (Fig. 2). We define *static shots* as shots whose parameters remain static over time, independently of the motion of the actor in the environment, and *dynamic shots* as having time-dependent parameters.

While filming, a human finds appropriate camera movements based on a set of implicit and explicit costs. We define a generic framework for aerial filming, where the motion planner's objective is to find a quadrotor's path ξ_q^* that minimizes a surrogate cost function J, that results in a smooth (J_{smooth}), collision-free (J_{obs}), occlusion-free (J_{occ}) trajectory that follows our artistic shot guidelines (J_{shot}) as closely as possible (Eq. 1). The optimal trajectory minimizes the total cost J within a finite time horizon t_f, while belonging to the set of trajectories Ξ that obey the problem's boundary constraints (Eq. 2).

$$J\left(\xi_q\right) = J_{\text{smooth}}\left(\xi_q\right) + \lambda_1 J_{\text{obs}}\left(\xi_q\right) + \lambda_2 J_{\text{occ}}\left(\xi_q, \xi_a\right) + \lambda_3 J_{\text{shot}}\left(\xi_q, \xi_a\right) \quad (1)$$

$$\xi_q^* = \underset{\xi_q \in \Xi}{\text{argmin}} \quad J\left(\xi_q\right), \quad \forall t \in [0, t_f] \quad (2)$$

3 Related Work

Camera control in virtual cinematography is typically on through-the-lens control [4,5,10,18] but disregards real-world limitations such as robot physics constraints and noisy motion predictions. When dealing with arbitrary real-life environments, voxel occupancy maps and truncated signed distance field (TSDF) [23] are common representations, and can supply distance and gradient of a point to nearest object surface. Aerial trajectory generation methods [20,31] are typically designed for aggressive flight and rely on easy to evaluate objective or constraint functions. In general domains, we find techniques with more relaxed requirements [24,29]. We build on CHOMP [25] due to its simple update rule that is amenable to new cost functions.

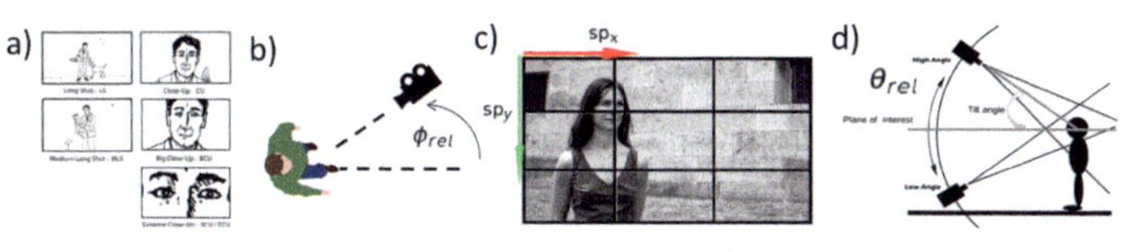

Fig. 2. Shot parameters, adapted from Bowen and Thompson [3]: (a) shot scale ss corresponds to the size of the projection of the actor on the screen; (b) line of action angle $\phi_{rel} \in [0, 2\pi]$; (c) screen position of the actor projection $sp_x, sp_y \in [1]$; (d) tilt angle $\theta_{rel} \in [-\pi, \pi]$

In aerial filming, we first expose works related to navigation using keyframes. Roberts and Hanrahan [28] generate off-line trajectories given potentially infeasible human-defined key-frames and Joubert et al. [13] provide a tool for interactive off-line design of camera trajectories. Similarly, Gebhardt et al. [9] improve a user's time-parametrized trajectory, optimizing it for smooth motions. From user studies, Gebhardt et al. [9] show that smoothness is key to producing visually-appealing videos. Gebhardt et al. [8] also show a smooth trajectory optimization method for user-defined keyframes, including an intuitive user interface. Lino and Christie [17] analytically interpolate between viewpoints while maintaining shot quality. Galvane et al. [6] used this method to control quadrotors, but only in obstacle-free environments. Similarly, Joubert et al. [14] transition between shots for static actors while not colliding with them, but offer no solution to obstacle and occlusion avoidance. Lan et al. [15] also position the drone using keyframes, which defined by a user on the image itself, not on a global coordinate frame.

In a different line of work, Xie et al. [33] generates a set of candidate local camera movements to film landmarks using drones using visual composition rules, and combine the local trajectories into a global smooth camera sequence. Closest to our work, we find Galvane et al. [7], Nägeli et al. [22]. Both methods use high-accuracy indoors motion-capture systems. Galvane et al. [7] use a search-based planner to find feasible trajectories; however, their system adapts to dynamic targets in a purely reactive manner, resulting in a non-smooth behavior. Nägeli et al. [22] apply MPC considering occlusion and safety. However, they plan for short time horizons, use simplistic elliptical representations for all obstacles, and use a high-accuracy motion-capture system. It is not clear if the black-box MPC solver that is used is amenable to other obstacle representations and noise in localization. In contrast, our method works for long time horizons, has a simple unconstrained update rule, operates on TSDFs, has small onboard runtime and can deal with noise in actor motion predictions.

4 Approach

Unlike previous works that operate either with high-accuracy indoor motion capture systems or precision RTK GPS outdoors, we use only conventional GPS, resulting in high noise for both drone localization and actor motion prediction. Therefore we decided to decouple the motion of the drone and the camera. The camera is mounted on a 3-axis independent gimbal and can place the actor on the correct screen position by visual detection, despite errors in the drone's position. By decoupling camera movement, the trajectory to be optimized becomes $\xi_q(t) = [x_q(t)\ y_q(t)\ z_q(t)]^T$, assuming that the drone's orientation ψ_q points towards the actor at all times.

Designing Differentiable Cost Functions for Cinematography
We want trajectories which are smooth, safe, occlusion-free and that follow our artistic guidelines as closely as possible. Following the derivation seen in Sect. 3 of Zucker et al. [34], we define a parametrization-invariant smoothness cost that can

be expressed as a quadratic function, and an obstacle avoidance cost based on a penalization c of the TSDF. We use obstacle avoidance for both the environment and for the dynamic actor, who for this purpose we represent as a moving sphere. In addition, we define two more cost functions specifically for cinematography:

Shot Quality:

$$J_{\text{shot}}\left(\xi_q, \xi_{shot}\right) = \frac{1}{t_f}\frac{1}{2}\int_0^{t_f} ||\xi_q(t) - \xi_{shot}(t)||^2 dt \approx \frac{1}{2(n-1)} Tr(\xi_q^T A_{shot}\xi_q + 2\xi_q^T b_{shot} + c_{shot}) \tag{3}$$

$$\nabla J_{\text{shot}}\left(\xi_q\right) = \frac{1}{n-1}(A_{shot}\xi_q + b_{shot}) \tag{4}$$

Written in a quadratic form, it measures the average squared distance between ξ_q and an ideal trajectory ξ_{shot} that only considers positioning via cinematography parameters. ξ_{shot} can be computed analytically: for each point $\xi_a(t)$ in the actor motion prediction, the drone position lies on a sphere centered at the actor with radius calculated via the shot scale, and angles given by ϕ_{rel} and θ_{rel}, as in Fig. 2.

Occlusion Avoidance:

$$J_{\text{occ}}\left(\xi_q, \xi_a\right) = \int_{t=0}^{t_f}\int_{\tau=0}^{1} c(p(\tau)) \left|\left|\frac{d}{d\tau}p(\tau)\right|\right| d\tau \left|\left|\frac{d}{dt}\xi_q(t)\right|\right| dt, \tag{5}$$

Even though the concept of occlusion for a point is binary, *i.e*, we either have or don't have visibility of the actor, a major contribution of our work is to define a differentiable cost that expresses a viewpoint's occlusion intensity among arbitrary obstacle shapes. Mathematically, we define occlusion as the integral of the TSDF cost c over a 2D manifold connecting both trajectories ξ_q and ξ_a. The manifold is built by connecting each drone-actor position pair in time using the path $p(\tau)$. We then derive the functional gradient:

$$\nabla J_{\text{occ}}\left(\xi_q, \xi_a\right)(t) =$$
$$\int_{\tau=0}^{1} \nabla c(p(\tau))|L||\dot{q}|\left[I - (\hat{\dot{q}} + \tau(\frac{\dot{a}}{|\dot{q}|} - \hat{\dot{q}}))\hat{\dot{q}}^T\right] - c(p(\tau))|\dot{q}|\left[\hat{L}^T + \frac{\hat{L}^T\dot{L}\hat{\dot{q}}^T}{|\dot{q}|} + |L|\kappa^T\right] d\tau, \tag{6}$$

where:

$$q = \xi_q(t), \quad a = \xi_a(t), \quad p(\tau) = (1-\tau)q + \tau a, \quad \hat{v} = \tfrac{v}{|v|},$$
$$\kappa = \tfrac{1}{|\dot{q}|^2}(I - \hat{\dot{q}}\hat{\dot{q}}^T)\ddot{q}, \quad L = a - q.$$

Intuitively, the term $\nabla c(p(\tau))$ is related to variations of the gradient in space, and the term τ acts as a lever for the gradient. The term $c(p(\tau))$ is linked to changes in path length between camera and actor.

Covariant Gradient and Steepest Descent Method

Our objective is to minimize the functional $J\left(\xi_q\right)$ (Eq. 2). Following a first-order Taylor expansion around the current iteration i using gradient $\nabla J(\xi_{qi})$, we build

Algorithm 1, adapted from [25]. We follow a direction of steepest descent on the functional cost, using metric M, which compounds the quadratic terms coming form both J_{smooth} and J_{shot}. M can be seen as an approximation of the Hessian, and only needs to be inverted once, outside of the algorithm's main loop, making the optimization computationally efficient. We follow conventional stopping criteria for descent algorithms, and limit the maximum number of iterations. We use the solution to the previous planning problem concatenated with a straight line segment for future time as the initialization for the next optmization.

Algorithm 1. Optimize (ξ_q)

$M_{inv} \leftarrow (A_{smooth} + \lambda_3 A_{shot})^{-1}$;
for $i = 0, 1, ..., i_{max}$ **do**
  if $(\nabla J(\xi_{qi})^T M_{inv} \nabla J(\xi_{qi}))^2/2 < \epsilon_0$ *or* $(J(\xi_{qi}) - J(\xi_{qi-1})) < \epsilon_1$ **then**
    return ξ_{qi};
  end
  $\xi_{qi+1} = \xi_{qi} - \frac{1}{\eta} M_{inv} \nabla J(\xi_{qi})$;
end
return ξ_{qi};

Gimbal Control Using the Actor's Image Detection
Our system does not use the actor's state estimation coming from a noisy GPS to control the gimbal to keep the actor within the desired screen position. Instead, we only rely on image feedback. Using the monocular images from the camera, the detection module outputs a bounding box to initialize the tracking process at a higher frame rate. We then use a PD controller to keep the center of the bounding box in the desired screen position.

We selected state of the art algorithms with a good speed-accuracy tradeoff for the vision pipeline. We tested three algorithms: single shot detector (SSD) [19], Faster R-CNN [27], and YOLO2 [26]. Faster R-CNN performs the best in terms of precision-recall metrics while operating at a reasonable frame rate. We use MobileNet [12] for feature extraction due to its low memory usage and fast inference speed. The per-frame inference is around 300 ms, running on the a Nvidia TX2 computer. We train all models using a mixture of COCO [16] dataset and our own custom dataset, with a 1:10 ratio, and about 70,000 images in total. We limited the detection categories only to *person*, *car*, *bicycle*, and *motorcycle*, which commonly appear as actors in aerial filming. Finally for actor tracking, we use KCF [11] due to its real-time performance.

5 Experiments

Systems and Experimental Setup
Our drone is the DJI M210 model, coupled with a NVIDIA TX2 computer for both vision and planning pipelines (Fig. 3). The actor wears a Pixhawk PX4 module on a hat that sends his pose to the onboard computer via radio communication. A linear Kalman filter uses these pose estimates to infer his velocity,

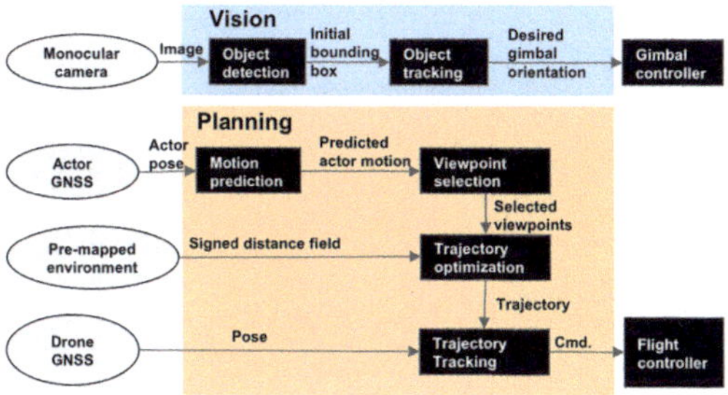

Fig. 3. System architecture. The vision subsystem controls the camera orientation using only the monocular image, independently of the planning subsystem. Planning uses the drone's and actor's current location, and the environment to generate trajectories for the flight controller.

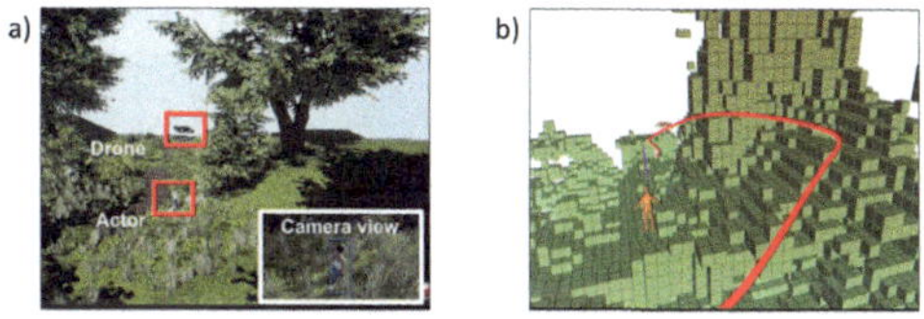

Fig. 4. Photo-realistic simulator used to test the system. Third and first-person renderings shown on the left, and occupancy map with drone trajectory shown on the right.

which is used to predict her trajectory (ξ_a) for the next 10 s. Using a point cloud map of the test site we pre-compute a TSDF map of the region of interest. Re-planning happens at 5 Hz with a 10 s horizon. To simulate the full pipeline and to decide on the relative weights between each cost function (Eq. 1), we built a ROS wrapper to test our software in a photo-realistic environment [30] (Fig. 4).

Experiments

(a) Algorithm robustness: We evaluated our algorithm performing different types of static and dynamic shots, following different types of actors: humans, cars and bicycles at different speeds and motion types. In total, we collected over 1.25 h of flight time while re-planning and avoided obstacles and/or occlusion 65 times. The maximum velocity achieved during the tests was of 7.5 m/s. Figure 5 summarizes the most representative shots, which are also shown in the supplementary video.

(b) Using occlusion cost function: For the same shot type, we planned paths with and without the occlusion cost function (Fig. 6). Our proposed occlusion cost significantly improves the aesthetics of the resulting image, keeping the actor on frame while avoiding obstacles.

(c) Statistical analysis: We evaluate our algorithm on randomized environments (Fig. 7), and display results in Table 1.

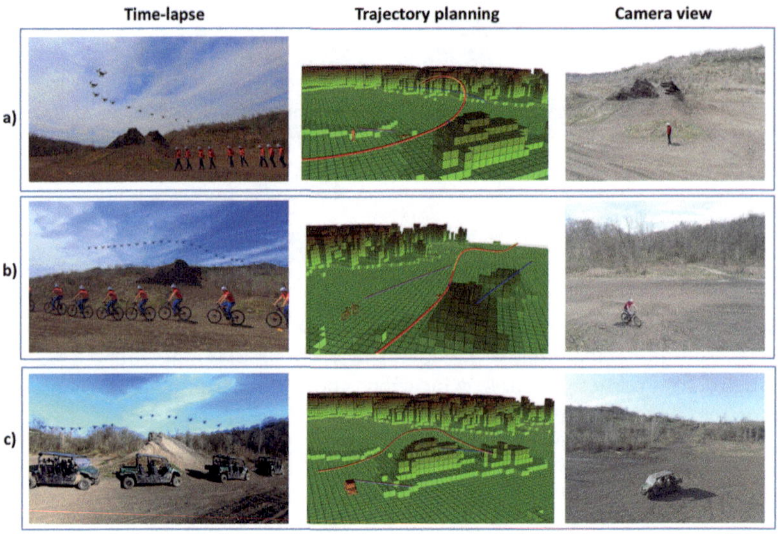

Fig. 5. Results: (a) Circling shot around person, (b) Side shot following biker, (c) Side shot following vehicle. The planned trajectory (red) avoids colliding with and being occluded by the mountain, while remaining smooth even under high actor motion prediction noise. The actor's motion forecast is in purple, and the desired artistic shot is in blue.

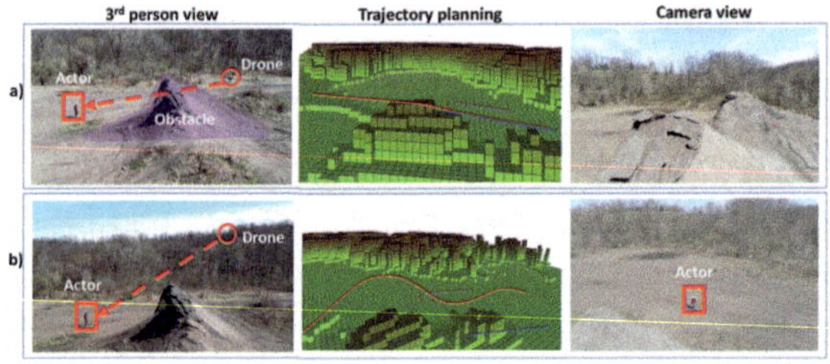

Fig. 6. Comparison of planning (a) without and (b) with occlusion cost function. The occlusion cost function significantly improves the quality of the camera image in comparison with pure obstacle avoidance, for same shot type.

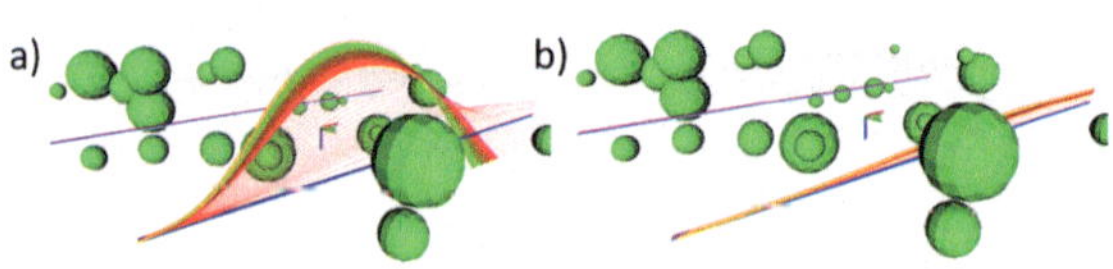

Fig. 7. Randomized environment with obstacles to evaluate planner robustness. (a) Solution including occlusion cost function, and (b) Pure obstacle avoidance.

Table 1. Evaluation of motion planner performance in the randomized environment from Fig. 7. By adding the occlusion cost function we improve actor visibility over 10% in comparison with pure obstacle avoidance in environments with 40 spheres. However, by avoiding occlusion the planner also increases the average distance to the desired artistic trajectory. We generated 100 random configurations for each environment complexity level.

Success metric	Cost functions	Num. spheres in environment		
		1	20	40
Actor visibility along trajectory	$J_{occ} + J_{obs}$	$99.4 \pm 2.2\%$	$94.2 \pm 7.3\%$	$86.9 \pm 9.3\%$
	J_{obs}	$98.8 \pm 3.0\%$	$87.1 \pm 8.5\%$	$75.3 \pm 11.8\%$
Avg. dist. to ξ_{shot}, in m	$J_{occ} + J_{obs}$	0.4 ± 0.4	6.2 ± 11.2	10.7 ± 13.2
	J_{obs}	0.05 ± 0.1	0.3 ± 0.2	0.5 ± 0.3

6 Conclusion and Discussion

In this work we propose and validate a system for autonomous aerial cinematography. Our system is able to execute aesthetically pleasing shots, calculating trajectories that balance motion smoothness, occlusion, and cinematography guidelines. Our experimental results show the algorithm's robustness and speed in a real-life scenarios outdoors, even under noisy actor predictions. In addition, we show that the occlusion cost function we introduced significantly improves the quality of the resulting image, and works for arbitrary obstacle shapes.

There are many key aspects that still need to be solved for us to create a fully autonomous filming system. In terms of actor localization, we are currently working on using only visual inputs to identify her position, with no need for GPS. In addition, one can incorporate an online-mapping module to the vehicle, which could be integrated with minimal changes to our algorithm, but was out of the scope of this work. We are also investigating techniques to automatically select the best artistic intent, *i.e.* type of shot, for a particular scenario, taking into account motion cues of the actor and obstacle configurations. In the longer term, we also plan to adapt our algorithm to multi-drone and multi-actor configurations by adding a cost function to penalize inter-drone sight.

Acknowledgements. We thank Lentin Joseph, Aayush Ahuja, Delong Zhu, and Greg Armstrong for the assistance in field experiments and robot construction. Research presented in this paper was funded by Yamaha Motor Co., Ltd.

References

1. Arijon, D.: Grammar of the Film Language (1976)
2. Barry, A.J., Florence, P.R., Tedrake, R.: High-speed autonomous obstacle avoidance with pushbroom stereo. J. Field Robot. **35**(1), 52–68 (2018)
3. Bowen, C.J., Thompson, R.: Grammar of the Shot. Taylor & Francis (2013)

4. Christie, M., Olivier, P., Normand, J.-M.: Camera control in computer graphics. In: Computer Graphics Forum, vol. 27, pp. 2197–2218. Wiley (2008)
5. Drucker, S.M., Zeltzer, D.: Intelligent camera control in a virtual environment. In: Graphics Interface, pp. 190–190. Citeseer (1994)
6. Galvane, Q., Fleureau, J., Tariolle, F.-L., Guillotel, P.: Automated cinematography with unmanned aerial vehicles. arXiv preprint arXiv:1712.04353 (2017)
7. Galvane, Q., Lino, C., Christie, M., Fleureau, J., Servant, F., Guillotel, P.: Directing cinematographic drones. arXiv preprint arXiv:1712.04216 (2017)
8. Gebhardt, C., Hepp, B., Nägeli, T., Stevšić, S., Hilliges, O.: Airways: optimization-based planning of quadrotor trajectories according to high-level user goals. In: Proceedings of the 2016 CHI Conference on Human Factors in Computing Systems, pp. 2508–2519. ACM (2016)
9. Gebhardt, C., Stevsic, S., Hilliges, O.: Optimizing for aesthetically pleasing quadrotor camera motion (2018)
10. Gleicher, M., Witkin, A.: Through-the-lens camera control. In: ACM SIGGRAPH Computer Graphics, vol. 26, pp. 331–340. ACM (1992)
11. Henriques, J.F., Caseiro, R., Martins, P., Batista, J.: High-speed tracking with kernelized correlation filters. IEEE Trans. Pattern Anal. Mach. Intell. **37**(3), 583–596 (2015)
12. Howard, A.G., Zhu, M., Chen, B., Kalenichenko, D., Wang, W., Weyand, T., Andreetto, M., Adam, H.: Mobilenets: efficient convolutional neural networks for mobile vision applications. arXiv preprint arXiv:1704.04861 (2017)
13. Joubert, N., Roberts, M., Truong, A., Berthouzoz, F., Hanrahan, P.: An interactive tool for designing quadrotor camera shots. ACM Trans. Graph. (TOG) **34**(6), 238 (2015)
14. Joubert, N., Goldman, D.B., Berthouzoz, F., Roberts, M., Landay, J.A., Hanrahan, P., et al.: Towards a drone cinematographer: guiding quadrotor cameras using visual composition principles. arXiv preprint arXiv:1610.01691 (2016)
15. Lan, Z., Shridhar, M., Hsu, D., Zhao, S.: Xpose: reinventing user interaction with flying cameras. In: Robotics Science and Systems (2017)
16. Lin, T.-Y., Maire, M., Belongie, S., Hays, J., Perona, P., Ramanan, D., Dollár, P., Zitnick, C.L.: Microsoft coco: common objects in context. In: European Conference on Computer Vision, pp. 740–755. Springer, Heidelberg (2014)
17. Lino, C., Christie, M.: Intuitive and efficient camera control with the toric space. ACM Trans. Graph. (TOG) **34**(4), 82 (2015)
18. Lino, C., Christie, M., Ranon, R., Bares, W.: The director's lens: an intelligent assistant for virtual cinematography. In: Proceedings of the 19th ACM International Conference on Multimedia, pp. 323–332. ACM (2011)
19. Liu, W., Anguelov, D., Erhan, D., Szegedy, C., Reed, S., Fu, C.Y. and Berg, A.C.: SSD: single shot multibox detector. In: European Conference on Computer Vision, pp. 21–37. Springer, Heidelberg (2016)
20. Mellinger, D., Kumar, V.: Minimum snap trajectory generation and control for quadrotors. In: IEEE International Conference on Robotics and Automation (ICRA), pp. 2520–2525. IEEE (2011)
21. Mohta, K., Watterson, M., Mulgaonkar, Y., Liu, S., Chao, Q., Makineni, A., Saulnier, K., Sun, K., Zhu, A., Delmerico, J., et al.: Fast, autonomous flight in GPS-denied and cluttered environments. J. Field Robot. **35**(1), 101–120 (2018)
22. Nägeli, T., Meier, L., Domahidi, A., Alonso-Mora, J., Hilliges, O.: Real-time planning for automated multi-view drone cinematography. ACM Trans. Graph. (TOG) **36**(4), 132 (2017)

23. Newcombe, R.A., Izadi, S., Hilliges, O., Molyneaux, D., Kim, D., Davison, A.J., Kohli, P., Shotton, J., Hodges, S., Fitzgibbon, A.W.: Kinectfusion: real-time dense surface mapping and tracking. In: 10th IEEE International Symposium on Mixed and Augmented Reality (ISMAR), pp. 127–136. IEEE (2011)
24. Ratliff, N., Zucker, M., Bagnell, J.A., Srinivasa, S.: CHOMP: gradient optimization techniques for efficient motion planning. In: IEEE International Conference on Robotics and Automation, ICRA 2009, pp. 489–494. IEEE (2009)
25. Ratliff, N.D., Silver, D., Bagnell, J.A.: Learning to search: functional gradient techniques for imitation learning. Auton. Robot. **27**(1), 25–53 (2009)
26. Redmon, J., Farhadi, A.: YOLO9000: better, faster, stronger. arXiv preprint (2017)
27. Ren, S., He, K., Girshick, R., Sun, J.: Faster R-CNN: towards real-time object detection with region proposal networks. In: Advances in Neural Information Processing Systems, pp. 91–99 (2015)
28. Roberts, M., Hanrahan, P.: Generating dynamically feasible trajectories for quadrotor cameras. ACM Trans. Graph. (TOG) **35**(4), 61 (2016)
29. Schulman, J., Ho, J., Lee, A.X., Awwal, I., Bradlow, H., Abbeel, P.: Finding locally optimal, collision-free trajectories with sequential convex optimization. In: Robotics: Science and Systems, vol. 9, pp. 1–10. Citeseer (2013)
30. Shah, S., Dey, D., Lovett, C., Kapoor, A.: AirSim: high-fidelity visual and physical simulation for autonomous vehicles (2017). https://arxiv.org/abs/1705.05065
31. Shim, D.H., Kim, H.J., Sastry, S.: Decentralized nonlinear model predictive control of multiple flying robots. In: Proceedings of 42nd IEEE Conference on Decision and control, vol. 4, pp. 3621–3626. IEEE (2003)
32. Turpin, M., Michael, N., Kumar, V.: Trajectory design and control for aggressive formation flight with quadrotors. Auton. Robot. **33**(1–2), 143–156 (2012)
33. Xie, K., Yang, H., Huang, S., Lischinski, D., Christie, M., Kai, X., Gong, M., Cohen-Or, D., Huang, H.: Creating and chaining camera moves for quadrotor videography. ACM Trans. Graph. **37**, 14 (2018)
34. Zucker, M., Ratliff, N., Dragan, A.D., Pivtoraiko, M., Klingensmith, M., Dellin, C.M., Bagnell, J.A., Srinivasa, S.S.: CHOMP: covariant Hamiltonian optimization for motion planning. Int. J. Robot. Res. **32**(9–10), 1164–1193 (2013)

The Blackbird Dataset: A Large-Scale Dataset for UAV Perception in Aggressive Flight

Amado Antonini, Winter Guerra(✉), Varun Murali, Thomas Sayre-McCord, and Sertac Karaman

Laboratory for Information and Decision Systems,
Massachusetts Institute of Technology, Cambridge, USA
{amadoa,winterg,mvarun,rtsm,sertac}@mit.edu

Abstract. The Blackbird unmanned aerial vehicle (UAV) dataset is a large-scale, aggressive indoor flight dataset collected using a custom-built quadrotor platform for use in evaluation of agile perception. Inspired by the potential of future high-speed fully-autonomous drone racing, the Blackbird dataset contains over 10 h of flight data from 168 flights over 17 flight trajectories and 5 environments at velocities up to 7.0 m s^{-1}. Each flight includes sensor data from 120 Hz stereo and downward-facing photorealistic virtual cameras, 100 Hz IMU, ~190 Hz motor speed sensors, and 360 Hz millimeter-accurate motion capture ground truth. Camera images for each flight were photorealistically rendered using FlightGoggles [1] across a variety of environments to facilitate easy experimentation of high performance perception algorithms. The dataset is available for download at http://blackbird-dataset.mit.edu/.

1 Introduction

Aggressive Unmanned Aerial Vehicle (UAV) flight using visual inertial simultaneous localization and mapping (VI-SLAM) has received increasing attention over recent years [1,2]. With the availability of better hardware, aggressive indoor flight maneuvers that were previously only possible using motion capture systems are now becoming achievable using on-board visual inertial state estimation algorithms. In the near future, it is conceivable that complex high-speed tasks, such as fully-autonomous drone racing, will be possible in realtime. To aid in the development of these high-performance algorithms, we provide a large scale, high rate, and high accuracy dataset for the improvement and evaluation of VI-SLAM for agile indoor flight.

Related Work. As can be seen from Table 1, existing UAV datasets focus on either slow speed indoor flight [3] where high accuracy ground truth is achievable, or outdoor high speed flight [4–6] with lower quality ground truth using

Electronic supplementary material The online version of this chapter (https://doi.org/10.1007/978-3-030-33950-0_12) contains supplementary material, which is available to authorized users.

J. Xiao et al. (Eds.): ISER 2018, SPAR 11, pp. 130–139, 2020.
https://doi.org/10.1007/978-3-030-33950-0_12

Table 1. UAV visual inertial datasets comparison

	EuRoC MAV [3]	UPenn fast flight [4]	Zurich Urban MAV [5]	Ours
Environments	2	1	3	**5**[a]
Sequences	11	4	1	**186**
Camera	20 Hz	40 Hz	20 Hz	**120 Hz**
IMU	**200 Hz**	**200 Hz**	10 Hz	100 Hz
Motor encoders	n/a	n/a	n/a	**∼ 190 Hz**
Max distance	130.9 m	700 m	**2 km**	860.8 m
Top speed	2.3 m s^{-1}	**17.5 m s^{-1}**	3.9 m s^{-1}[b]	7.0 m s^{-1}
mm Ground truth	(100 Hz)[c]	n/a	n/a	**3600 Hz**

[a] Additional environments may be rendered using FlightGoggles
[b] Instantaneous velocity from GPS
[c] High accuracy only available for half the sequences

GPS systems. Burri et al. [3] present the EuRoC MAV datasets, a collection of 11 trajectories with an average speed of 1 m s^{-1} and highly accurate ground truth for one half of the sequences. However, the scenes are not representative of typical indoor environments and are captured using comparatively lower rate cameras. Sun et al. [4] present a fast outdoor flight dataset with the same trajectory at 4 different speeds and GPS ground truth. Although this does allow the evaluation of online algorithms in outdoor settings, it does not provide high quality ground truth or different environments. The Zurich Urban MAV Dataset is presented by Majdik, Till, and Scaramuzza. It contains 2 km of visual and inertial data recorded from a tethered UAV flying in an urban setting, but it lacks high-precision ground truth pose. Wang et al. [6] present TorontoCity, a very large UAV dataset with data from multiple perspectives of the city of Toronto captured from different cameras and a LiDAR. TorontoCity focuses on tasks such as segmentation and classification of the environment. It, however does not contain inertial information and cannot be used in the context of visual inertial navigation.

2 Data Collection Setup

UAV Platform. Data was collected using a custom built quadrotor UAV designed for agile autonomous flight, which we call Blackbird (Fig. 1). The UAV caries an Xsens MTi-3 IMU, custom made optical motor encoders for accurate motor speed measurements, a DJI Snail propulsion system, and a NVIDIA Jetson TX2. The body of the vehicle is constructed from 3D printed MarkForged Onyx continuous carbon fiber composite. Rubber dampeners are used to mechanically isolate vibrations from the propulsion system from flight sensors. The physical properties of the quadrotor as well as sensor statistics are shown in Table 2.

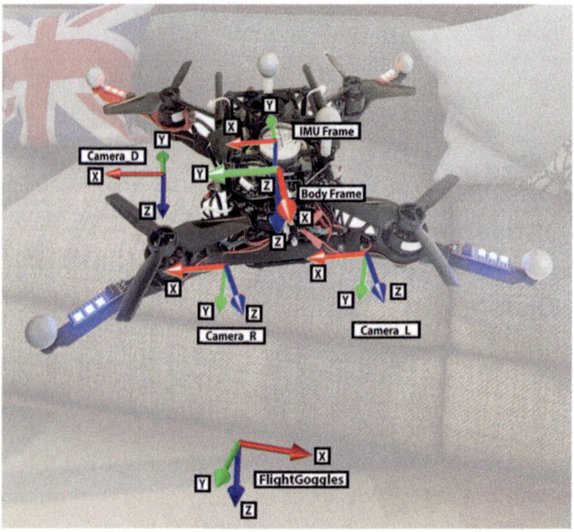

Fig. 1. Coordinate frames in use for this dataset. Note that `Camera_D` and `Body Frame` are coincident, but are translated in the figure for visualization.

Table 2. Quadrotor characteristics

Property	Value	Description
Mass	0.915 kg	Mass with battery
I_{xx}	4.9×10^{-2} kg m^{-2}	X moment of inertia
I_{yy}	4.9×10^{-2} kg m^{-2}	Y moment of inertia
I_{zz}	6.9×10^{-2} kg m^{-2}	Z moment of inertia
Arm length	0.13 m	Center to end of arm
f_x, f_y	665.108 mm	Cameras' focal length
FOV	60.0°	Cameras' vertical FOV
σ^{gyro}	1.2×10^{-4} rad s$^{-1}\sqrt{\text{Hz}}$	Gyroscope noise density
σ_b^{gyro}	4.7×10^{-6} rad s$^{-2}\sqrt{\text{Hz}}$	Gyroscope random walk bias
σ^{accel}	2.0×10^{-3} m s$^{-2}\sqrt{\text{Hz}}$	Accelerometer noise density
σ_b^{accel}	4.4×10^{-5} m s$^{-3}\sqrt{\text{Hz}}$	Accelerometer random walk bias
C_T	2.27×10^{-8} N/rpm^2	RPM to thrust coefficient
Image size	1024 px × 768 px	Image width and height

Experimental Setup. Flights were performed in an 11 m × 11 m × 5.5 m motion capture room, with 24 OptiTrack Prime 17 W cameras providing the 6D pose of the drone at 360 Hz. Each flight in the dataset is between 3–4 min long as the drone traces out a pre-defined periodic trajectory using a non-linear dynamic inversion controller [7]. The drone is controlled and data is recorded by a custom software framework [1] using the Lightweight Communications and Marshalling (LCM) protocol [8].

Visual Data Generation. Visual data was generated in post process using the FlightGoggles photo-realistic image generation system [1]. FlightGoggles uses the ground truth 6D pose of the drone from motion capture to generate images from the viewpoint of each camera on the drone in a virtual environment. The system allows for complete control over the visual appearance of the environment, the rate of camera images (up to the 360 Hz motion capture rate), the number of cameras, and each camera's location and intrinsic and extrinsic properties. The visual data generated by FlightGoggles has been previously validated for use in visual inertial state estimation in [1].

As part of the rendering process, a number of transforms are used to transform NED ground truth data from world frame into FlightGoggles' environment frame. To ensure that all recorded flights in each trajectory takeoff from a common altitude and overlap in the XY plane, T_{mocap}^{norm} is introduced as a per-flight translational offset applied to the ground truth data to correct for offsets introduced during dataset collection. $T_{norm}^{FG_{env}}$ is a per-trajectory common transform that positions flights into the FlightGoggles environment in a collision-free manner. The full transform chain from ground-truth coordinates to render coordinates is shown in Eq. 1.

$$T^{FG_{env}} = T_{norm}^{FG_{env}} \cdot T_{mocap}^{norm} \cdot T^{mocap} \tag{1}$$

where $T^{FG_{env}}$ is the render pose in FlightGoggles' virtual environment.

Sensor Calibration and Temporal Synchronization. The Kalibr package [9] was used to find the noise characteristics of the IMU and the IMU-to-camera transform. A 3 s period at rest is included in every flight to allow for initialization of the time varying IMU bias. Force and torque coefficients of the drone were found experimentally through measurements in a wind tunnel to obtain the relationship between motor speeds and vehicle dynamics. Clock synchronization between motion capture data and on-board sensors was performed using a combination of clock estimation over gigabit ethernet and chrony [10] over the wireless network, with an upper bounded offset of ±5 ms.

3 Dataset Format

Each flight within the dataset contains timestamped values for the following: ground truth 6D pose of the UAV at 360 Hz, IMU measurements at 100 Hz, RPM measurements for each motor at ∼190 Hz, and three camera streams (forward facing stereo pair and downward facing) at 120 Hz. The data is provided as grayscale images, LCM logs, and ROS bags for easy use in typical pipelines. Scripts and binaries necessary to re-render images using FlightGoggles at other rates (up to 360 Hz) or camera parameters and configurations are available at http://blackbird-dataset.mit.edu/.

In addition to the raw data streams, the full calibration information of the UAV system is included in the dataset i.e, IMU noise characteristics, IMU-camera transform, camera intrinsic and extrinsic parameters (as currently rendered), and torque and thrust curves. The dataset's file structure is specified in Fig. 2.

```
BlackbirdDataset
├─ renderConfigs.yaml
├─ fileIndex.csv
├─ downloaderUtility.py
├─ renderUtilities
│  ├─ blackbirdDatasetUtils.py
│  └─ flightGogglesUtils.py
└─ BlackbirdDatasetData
   └─ <trajectoryName>
      └─ <yawType> in {yawConstant, yawForward}
         └─ maxSpeed<V> in {0p5, 1p0, 2p0, ..., 7p0}
            ├─ <trajectoryName>_maxSpeed<V>.{bag, log}
            ├─ groundTruthFlightNormalizationOffset.csv
            ├─ groundTruthPoses.csv
            ├─ images
            │  └─ Camera_<k>_<renderName>.tar, k in {L, R, D}
            │     └─ <utime>_Camera_<k>.png
            └─ videos
               └─ <renderName>.mp4
```

Fig. 2. Dataset file hierarchy.

4 Data Validation

Validation of ground truth data and inertial measurements in both quality and temporal synchronization was performed by comparing raw inertial measurements with derivatives of the ground truth pose using a Savitzky-Golay filter [11]. Figure 3 shows a comparison of the IMU angular rate and acceleration with respect to ground truth. The accuracy of the drone's motor speed sensors were verified through the use of an external tachometer.

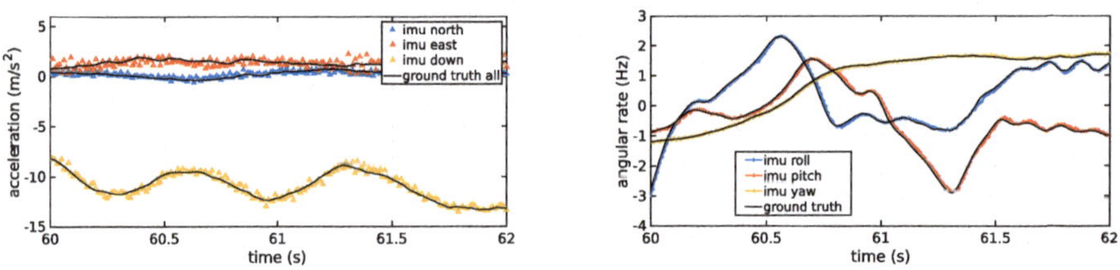

(a) IMU accelerometer vs ground truth. (b) IMU gyroscope vs ground truth.

Fig. 3. Derivative of position and rotational ground truth data compared with accelerometer and gyroscope data for a flight at $4\,\mathrm{m\,s^{-1}}$.

Trajectory Tracking. A feature of the provided dataset is the ability to repeatably run perception algorithms on a nominal trajectory pattern while flying at different speeds with new inertial, dynamical, and visual data. A comparison of the ground truth pose of the same trajectory (Sphinx, see Fig. 6) flown four times at speeds between $1\,\mathrm{m\,s^{-1}}$ and $4\,\mathrm{m\,s^{-1}}$ is shown in Fig. 4, with sufficient tracking accuracy for a user to isolate the speed of flight from other parameters that may affect VI-SLAM algorithms.

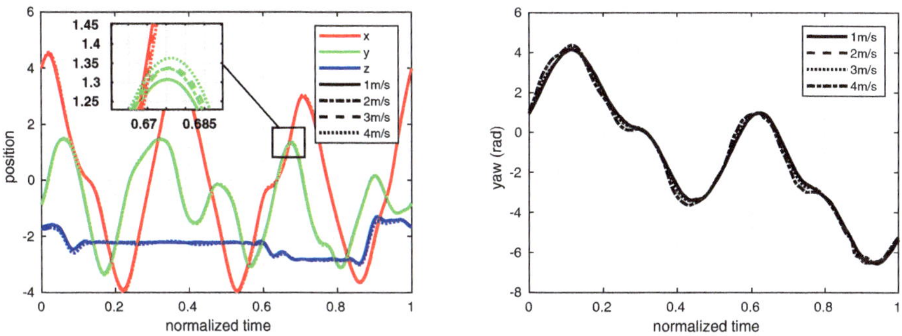

(a) Position tracking

(b) Yaw tracking

Fig. 4. Tracking precision while flying the same trajectory at speeds of 1 to $4\,\mathrm{m\,s^{-1}}$

Table 3. Dataset flights

Trajectory	Constant yaw								Forward-facing							
Top speed ($\mathrm{m\,s^{-1}}$)	0.5	1.0	2.0	3.0	4.0	5.0	6.0	7.0	0.5	1.0	2.0	3.0	4.0	5.0	6.0	7.0
3D figure 8	✓	✓	✓	✓	✓	✓	–	–	–	–	–	–	–	–	–	–
Ampersand	–	M	M	H	–	–	–	–	–	H	H	–	–	–	–	–
Bent dice	–	E	E	M	M	–	–	–	E	E	E	E	–	–	-	–
Clover	–	H	H	H	H	H	H	–	H	H	H	H	H	H	–	–
Dice	–	–	E	E	M	–	–	–	–	E	E	E	–	–	–	–
Flat figure 8	✓	✓	✓	✓	–	✓	–	–	–	–	–	–	–	–	–	–
Half-moon	–	E	E	E	M	–	–	–	–	M	M	M	M	–	–	–
Mouse	–	M	M	M	M	M	M	M	M	M	M	M	M	M	M	M
Oval	–	–	M	M	H	–	–	–	–	M	M	H	H	–	–	–
Patrick	–	E	E	E	E	M	–	–	E	E	E	E	E	–	–	–
Picasso	M	M	M	M	M	M	M	–	M	M	–	H	H	H	–	–
Sid	–	E	E	E	E	E	E	E	M	M	M	M	M	M	–	–
Sphinx	–	H	H	H	H	–	–	–	–	M	M	M	M	–	–	–
Star	–	M	M	M	H	H	–	–	M	M	M	M	M	H	–	–
Thrice	–	E	E	E	E	E	M	M	E	E	E	E	E	E	M	–
Tilted thrice	–	E	E	E	E	E	M	M	E	E	E	E	E	E	E	–
Winter	–	M	M	M	M	M	–	–	M	–	H	H	H	–	–	–

Click flight for grayscale video preview of flight in all rendered environments

Collision Checking. To verify that the trajectories being rendered do not collide with any obstacles in the virtual environment, bullet3 [12] based simulation was used to verify that the trajectories were collision free prior to rendering the camera streams.

5 Datasets

The trajectories were designed to allow for independent variation in the following flight characteristics: speed, yaw, trajectory complexity, and period. The included trajectories are classified by difficulty ([E]asy, [M]edium, [H]ard) according to mean features tracked using the visual inertial state estimation pipeline outlined in [1] and are shown in Table 3. These trajectories range in complexity from an oval with constant yaw and altitude to trajectories with varying speed, altitude, and yaw as they weave through visual obstacles (e.g. the Sphinx) as shown in Fig. 6. For flights that are rendered in multiple environments (see Fig. 5), some environments are harder for state estimation than others.

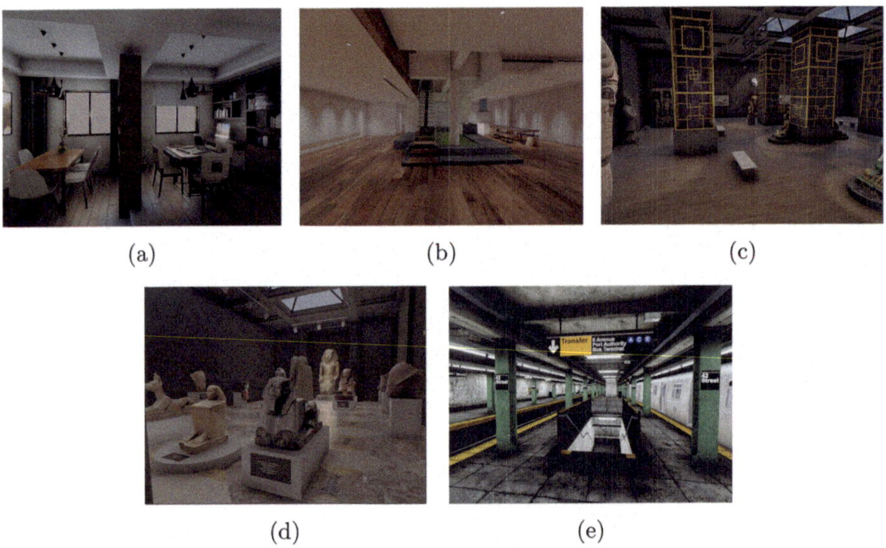

Fig. 5. Five rendering environments for visual data: (a) Butterfly Apartment, (b) Hazelwood Loft, (c) Museum Pillars, (d) Museum Sphinx, and (e) NYC Subway

To generate smoothly trackable trajectories, minimum snap optimization was performed over a set of requested waypoint positions using the non-linear optimization technique from [13], with boundary conditions to make the trajectory periodic. Table 3 shows all the sequences included. There are 163 unique flights of approximately three minutes each for a total of over 10 h and 60 km of ground

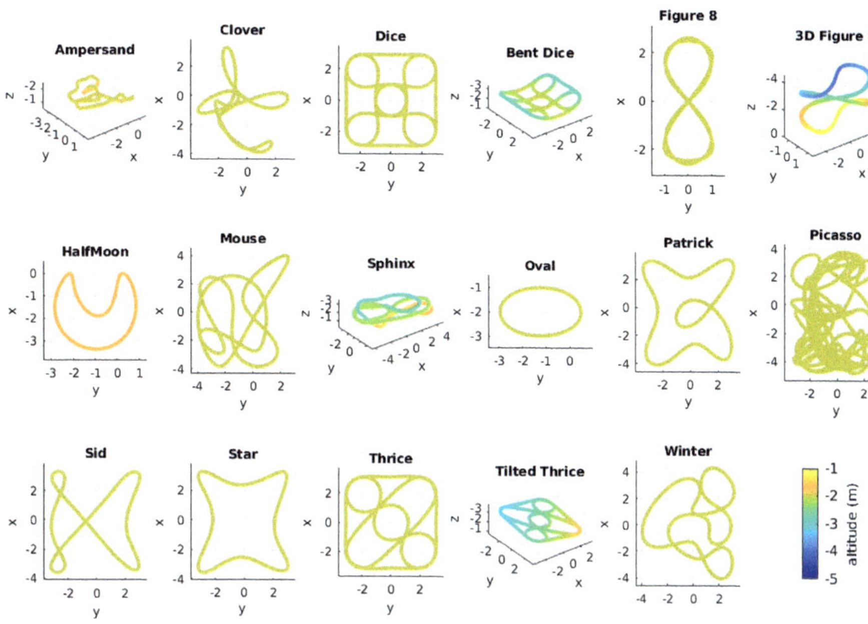

Fig. 6. Top down diagrams of trajectory paths included in this dataset.

truth, inertial, and dynamical sensor data, as well as rendered imagery in multiple environments. Trajectories were designed for specific flight environments (e.g. Sphinx for Museum Sphinx) and are therefore particularly well suited to those environments, however, where it does not result in virtual collisions with objects, the same trajectory can be re-rendered in multiple environments. Flights marked in Table 3 by ✓ do not have visual data associated with them and are intended for calibration use.

6 Known Issues

In this dataset, we synchronized motion capture ground truth data with onboard IMU measurements using camera exposure timestamps provided by OptiTrack, IMU measurement timestamps provided on arrival by our UAV's TX2, and clock sync and drift correction provided by Chrony [10]. However, due to the complexity and stochastic nature of the systems involved, we are only able to guarantee IMU and ground truth temporal alignment to within ± 5 ms across all flights in this dataset. This upperbound was verified in post process by cross correlation of IMU and ground truth measurements for each flight.

Over the course of the various data recording sessions required to create this dataset, many recalibrations of the motion capture groundtruth setup were required due to thermal expansion and contraction of the motion capture support beams. Each recalibration's overall tracking error and groundplane alignment was verified using post process analysis, but are not guaranteed to be exact. Most flights were flown with ≤1.5 mm of mean tracking error and ≤ 1° of ground plane alignment error.

7 Conclusion

The Blackbird dataset allows for the evaluation of the robustness and performance of VI-SLAM algorithms under varying conditions and degrees of agility. It has been used in [14] to evaluate dynamics factors in factor graph VIO. As an additional example, we show the results of running a tracker-based visual inertial odometry method [1] on the Ampersand trajectory for increasing speed of execution. The performance is summarized in Table 4, where drift rate (the percentage error per meter traveled) serves as a metric of the overall state estimation accuracy. The mean features tracked, the average number of features tracked between every pair of consecutive frames, serves as a metric of success for the visual front-end. The structured nature of the dataset allows for clear evaluations of the strengths and failures of particular algorithms (e.g. tracking vs. matching visual front-ends) depending on the type of trajectory and visual environment being flown in.

Table 4. Performance of visual inertial odometry [1] for the same nominal trajectory with varying speed

Speed	Drift rate	Mean features tracked
$1.12\,\mathrm{m\ s^{-1}}$	0.55	118.98
$1.96\,\mathrm{m\ s^{-1}}$	0.95	115.74
$3.08\,\mathrm{m\ s^{-1}}$	1.94	106.83

Due to the large number of aggressive and varied flight patterns available, the Blackbird dataset is well suited to fill a void in the current landscape of available robotics datasets. The inertial, dynamical, and visual sensors provide a complete sensor package for many UAV VI-SLAM algorithms. The photorealistic visual simulation system, FlightGoggles, allows a user to expand beyond the three camera streams provided and re-render images to match the characteristics of their current system, or to evaluate design choices on camera placement, type, and framerate. Variations in visual parameters may all be performed while maintaining the true dynamics and inertial measurements of the UAV, and without sacrificing the high resolution ground truth accuracy provided by an expensive motion capture system.

The ability to repeat nominal trajectories with varying conditions allows for methodical evaluation of perception algorithm performance during high speed flight and enables progressively building capabilities towards more challenging scenarios.

References

1. Sayre-McCord, T., Guerra, W., Antonini, A., Arneberg, J., Brown, A., Cavalheiro, G., Fang, Y., Gorodetsky, A., McCoy, D., Quilter, S., Riether, F., Tal, E., Terzioglu, Y., Carlone, L., Karaman, S.: Visual-inertial navigation algorithm development using photorealistic camera simulation in the loop. In: 2018 IEEE International Conference on Robotics and Automation (ICRA) (2018)
2. Falanga, D., Mueggler, E., Faessler, M., Scaramuzza, D.: Aggressive quadrotor flight through narrow gaps with onboard sensing and computing using active vision. In: 2017 IEEE International Conference on Robotics and Automation (ICRA), pp. 5774–5781. IEEE (2017)
3. Burri, M., Nikolic, J., Gohl, P., Schneider, T., Rehder, J., Omari, S., Achtelik, M.W., Siegwart, R.: The EuRoC micro aerial vehicle datasets. Int. J. Robot. Res. **35**(10), 1157–1163 (2016)
4. Sun, K., Mohta, K., Pfrommer, B., Watterson, M., Liu, S., Mulgaonkar, Y., Taylor, C.J., Kumar, V.: Robust stereo visual inertial odometry for fast autonomous flight. IEEE Robot. Autom. Lett. **3**(2), 965–972 (2018)
5. Majdik, A.L., Till, C., Scaramuzza, D.: The Zurich urban micro aerial vehicle dataset. Int. J. Robot. Res. **36**(3), 269–273 (2017)
6. Wang, S., Bai, M., Mattyus, G., Chu, H., Luo, W., Yang, B., Liang, J., Cheverie, J., Fidler, S., Urtasun, R.: Torontocity: seeing the world with a million eyes. In: 2017 IEEE International Conference on Computer Vision (ICCV), pp. 3028–3036. IEEE (2017)
7. Tal, E., Karaman, S.: Accurate tracking of aggressive quadrotor trajectories using incremental nonlinear dynamic inversion and differential flatness. In: 2018 Proceedings of the 57th IEEE Conference on Decision and Control. IEEE (2018)
8. Huang, A.S., Olson, E., Moore, D.C.: LCM: lightweight communications and marshalling. In: 2010 IEEE/RSJ International Conference on Intelligent Robots and Systems (IROS), pp. 4057–4062 (2010)
9. Furgale, P., Rehder, J., Siegwart, R.: Unified temporal and spatial calibration for multi-sensor systems. In: 2013 IEEE/RSJ International Conference on Intelligent Robots and Systems (IROS), pp. 1280–1286, November 2013
10. Lichvar, M.: Chrony. https://chrony.tuxfamily.org/
11. Savitzky, A., Golay, M.J.: Smoothing and differentiation of data by simplified least squares procedures. Anal Chem. **36**(8), 1627–1639 (1964)
12. Coumans, E.: Bullet physics simulation. In: ACM SIGGRAPH 2015 Courses, p. 7. ACM (2015)
13. Burri, M., Oleynikova, H., Achtelik, M.W., Siegwart, R.: Real-time visual-inertial mapping, re-localization and planning onboard MAVs in unknown environments. In: 2015 IEEE/RSJ International Conference on Intelligent Robots and Systems (IROS), September 2015
14. Antonini, A., Leonard, J., Karaman, S.: Pre-integrated dynamics factors and a dynamical agile visual-inertial dataset for UAV perception. Master's thesis, Massachusetts Institute of Technology (2018)

Search and Rescue Under the Forest Canopy Using Multiple UAS

Yulun Tian[1(✉)], Katherine Liu[1], Kyel Ok[1], Loc Tran[2], Danette Allen[2], Nicholas Roy[1], and Jonathan P. How[1]

[1] Massachusetts Institute of Technology, Cambridge, MA 02139, USA
{yulun,katliu,kyelok,nickroy,jhow}@mit.edu
[2] NASA Langley Research Center, Hampton, VA, USA
{loc.d.tran,danette.allen}@nasa.gov

Abstract. We present an autonomous system consisting of multiple unmanned aerial vehicles (UAVs) for search and rescue under the forest canopy. Our vehicles can be rapidly deployed, can collaboratively explore expanses of terrain efficiently, and are agile enough to operate in reasonably thick forests. We demonstrate the ability to carry out GPS-denied exploration with on-board pose estimation, map inference, and path planning. In addition, we utilize a place recognition system that is able to handle perceptual aliasing unique to a forest environment, and fuse individual areas explored by each vehicle into a globally consistent map. We perform extensive evaluations in both simulation and real-world to demonstrate the effectiveness of our proposed system.

1 Motivation

Lost hikers are often within a mile or two of the last point of detection for extended periods of time, but are undetected for hours at a time because manned aircraft cannot see through the overhead forest canopy. Small autonomous unmanned aircraft systems (UAS), or drones, have been frequently proposed for search and rescue missions under the forest canopy, such as shown in Hampton, VA in Fig. 1. These vehicles can be rapidly deployed, can cover expanses of terrain quickly and are small enough to operate in reasonably thick forests.

2 Problem Statement

Using drones under the forest canopy presents several challenges. Firstly, GPS typically cannot penetrate the overhead foliage, which requires the vehicles to use on-board sensors and a mapping and localization system for state estimation. A second problem is that these vehicles are typically limited by battery life, which limits the search speed. Any search activity is accelerated by using multiple vehicles in parallel, but a multi-vehicle search requires co-ordination between the vehicles in both data fusion and planning. A third problem is that the search can

J. Xiao et al. (Eds.): ISER 2018, SPAR 11, pp. 140–152, 2020.
https://doi.org/10.1007/978-3-030-33950-0_13

be made arbitrarily slow with a poor strategy in terms of where to search next. Using more intelligent search strategies can dramatically accelerate the process.

In this paper we describe an experimental evaluation of a multi-UAS system for search and rescue under the forest canopy. We examine the ability of multiple vehicles to carry out GPS-denied flight using laser-range finders for position estimation and map inference, plan trajectories, and fuse individual maps into a globally consistent model for exploration and search. We evaluate both a map construction process designed for search and exploration in the forest, and the corresponding search process itself.

3 Related Work

The problem of place recognition and map fusion under the forest canopy has received increased attention in recent years. Place recognition is a key component of any SLAM system and has a rich literature for single robot systems equipped with stereo or monocular camera [7,17,18], but these approaches typically require access to multiple images which would be costly to communicate over a network with limited bandwidth, and the high variability of illumination conditions in a forest environment creates challenges for purely image-based solutions.

Using laser-range data, Bosse and Zlot [4] developed a keypoint-based nearest neighbor algorithm for identifying loop closures in forest-type environments, but primarily using an underactuated 3D laser that would be a substantial challenge on a flight vehicle. Tipaldi et al. [22] developed fast laser interest region transform (FLIRT) features for both indoor and outdoor localization. Purely geometric approaches have been examined, including the Geometric LAndmark RElations (GLARE) signatures by Himstedt et al. [10] and the rotationally-invariant extension GLAROT by Kallasi and Rizzini [13]. None of these previous results take advantage of the strong prior structure present in forest environment.

The problems of search and exploration has been extremely well-studied and a comprehensive overview of all such work is outside the scope of this paper. The earliest approaches operated greedily on the nearest "frontier" in the map [23] or maximized expected coverage [20]. Frontier-based approaches have also been demonstrated for multi-UAV exploration in an indoor environment [6]. Later approaches to the exploration problem focused on either greedy [8] or receding horizon [5] information gain. Most recently, there have been results on exploring to find hidden objects [1], but primarily in indoor environments. Surprisingly, the majority of this prior work has not taken into account the dynamics of the vehicle in planning efficient search strategies.

4 Technical Approach

Our experimental investigation focuses on two primary areas, feature extraction for place recognition and map merging, and exploration-based search. We follow the approach described by Giamou et al. [9], extending to multiple vehicles that are communicating with a central ground station[1]. When a new laser scan or scans are received, features are extracted for place recognition and loop closure. Once a loop is closed, the merged map can be inspected by the first responders to plan a route for the rescue personnel.

4.1 Tree Feature Extraction and Map Merging

A forest is a challenging environment for cooperative SLAM due to strong perceptual aliasing and complex occlusions caused by a dense assembly of visually similar trees. Before place recognition can occur, we need to extract stable features from the laser scans. We first cluster the laser return; each cluster of points is then regressed to the parameters of a tree trunk circle $[x, y, r]$ using least squares. We accept a cluster of points as the observation of a tree, if the residual is less than 0.015, the tree has a radius greater than 0.1m, and the observed laser returns cover more than 30% of the arc of the tree trunk circle. Following [14], we correspond a detected tree at time t with a previously detected tree at time t' if the centers are within 1m, and the radii of the trees differ less than 10%.

Once the trees are extracted as features $\{p_i^t\}$ from a scan S^t, a GLARE signature [10] is computed, which is essentially a distribution over the pairwise Euclidean distance $\rho_{i,j}^t$ and angular distance $\theta_{i,j}^+$ between pairs of known trees. For every pair of features (p_i^t, p_j^t) their distance and relative angle in the scan frame $(\theta_{i,j}^t, \rho_{i,j}^t)$ are computed. These are then assigned to bins in a 2D histogram $(\theta_{i,j}^+, \rho_{i,j}^t) \in \text{bin}(n_\theta, n_\rho)$, where n_θ and n_ρ are a quantization of $(\theta^+, \rho_{i,j}^t)$. A Gaussian blur is applied to the histogram of each feature pair, and the GLARE descriptor is a sum over the 2D histograms for each pair. This histogram construction procedure is not rotationally invariant as it depends on an absolute orientation. Kallasi and Rizzini [13] extended the GLARE signature to be rotationally invariant in GLAROT. We use this extension for comparison of geometric signatures between scans.

After receiving laser scans from the flight vehicles, the ground station searches for loop closures in the entire history of scans. The geometric signatures described above are used to prune the search space by retaining only the top $k = 20$ candidate loop closures for each scan S^t. For each candidate loop closure, we perform correspondence graph (CG) matching [3] to verify the loop closure and identify the largest inlier set of correspondences. If the loop closure is accepted, a relative transformation constraint is computed and added to the global pose-graph for trajectory estimation. We use iSAM2 [12] as the back-end SLAM solver. Map merging happens once all UAVs establish a common reference frame through loop closure.

[1] Our experiments assume reliable communication to the ground station, which is generally not the case for operation in a forest environment. However, this is not a limiting assumption in that all of the local inference and map construction happen on-board the vehicle. The loss of the C2 link to the ground station could only delay potential map merges until the link is recovered.

4.2 Exploration-Based Search

Following [19], we formulate the planning problem as a trajectory optimization problem, where a motion planner selects a trajectory of some length that maximizes the probability of finding the target while avoiding observed obstacles. This approach can be represented formally as

$$a_t^* = \text{argmin}_{a_t \in \mathcal{A}} j(b_t, a_t) + h(b_t, a_t), \text{s.t. } g(b_t, a_t) = 0,$$

which is repeatedly re-solved to select the optimal action a_t^* as the robot explores the environment. Here, a_t is an action to be executed at time t, chosen from a set $\mathcal{A}$ of possible actions. The robot's belief b_t is a partial map built from sensor measurements and its current position in that map. To ensure collision-free motion, we use a collision constraint $g(b_t, a_t)$, which returns 1 if the action intersects obstacles or unknown regions in the map, and returns 0 otherwise.

We parameterize the action space A as frontiers of the known map (i.e., the boundaries between known and unknown space). The total estimated cost of choosing action a_t, given b_t, is the sum of the action cost $j(b_t, a_t)$ and the heuristic estimate $h(b_t, a_t)$ of the cost remaining beyond the end of a_t. For exploration-based search, there is rarely a reason to bias one frontier over another in terms of future cost beyond a_t; as a result, we set the heuristic cost $h(\cdot)$ to be the same for all frontiers and choose the action with the minimum instant cost described by $j(\cdot)$. Classical approaches design $j(\cdot)$ such that closer frontiers have lower costs than farther ones, e.g., $j(\cdot)$ returns the distance of the frontier. However, in practice it is often the case that the closest frontier lies behind the flight vehicle. In these cases, traditional cost functions lead to rapidly changing velocity and aggressive turning, which negatively impacts the overall search progress and map quality. On the other hand, purely information-theoretic models of $j(\cdot)$ use submodularity to bound the loss that comes from greedily choosing the next-best frontier; the cost of physically redirecting the vehicle to a new frontier is *not* submodular. In this paper, we propose a new search objective that better respects the vehicle dynamics,

$$j(b_t, a_t) = j_\theta(b_t, a_t) + \lambda j_t(b_t, a_t).$$

Above, $j_\theta(\cdot)$ denotes the orientation change from the vehicle's current heading of the frontier a_t. Incorporating orientation change in the cost function discourages the vehicle from excessive turning and hence produces a smoother overall trajectory. In addition, we include a separate, weighted cost term $j_t(\cdot)$ based on the classical Euclidean distance to the frontier. Intuitively, the user-specified weight λ encodes a *trade-off* between the smoothness and overall length of the trajectory. As we shall see in the experimental evaluations, the proposed objective function, with a suitably chosen weight, results in a longer but smoother and more time-efficient trajectory. Lastly, using an objective function based on orientation change enables *online* selection of the next-best frontier. Specifically, the currently selected frontier can be updated, whenever a new candidate with a significantly higher score is detected. In contrast, a dynamic-agnostic planner

Fig. 1. Two quadrotors exploring the forest at NASA Langley Research Center. A video of the flight available at https://groups.csail.mit.edu/rrg/videos/iser2018

(e.g., with a purely distance-based objective) is less compatible with this approach, as it might continuously command large change in heading, leading to potentially unstable and dangerous flight behavior.

5 Results and Experiments

We evaluated the performance of the proposed system in both simulation and real-world flight tests. Both experiments used the same set of parameters for frontier exploration (Sect. 4.2), but a few vehicle-related parameters differed to compensate for the differences in the simulation and the real world. For example, the limit on maximum allowed acceleration was set to be much higher in real-world experiments, in order to compensate for external forces such as the wind that are not modeled in simulation.

5.1 Simulation Setup

We benchmarked our proposed system with the baseline in simulation. Vehicle dynamics and inertial measurements unit (IMU) readings for a simulated vehicle were generated using the Drake toolbox [21] and a quadrotor model as described in [16]. To test the integration with the full stack, we also utilized the Pixhawk [15] Software-In-The-Loop (SITL) to simulate the motor commands from the Pixhawk, which were fed back into the Drake dynamics model. We used the Unity game engine to simulate 2D laser scans in a randomly generated forest environment at roughly 30 Hz with a 270° field of view and 30 m range. All sensor measurements were simulated with low noise. Laserscans and inertial measurements were passed to a laser-based Extended Kalman Filter (EKF) that incorporated incremental odometry measurements from iterative closest point (ICP) on two laser scans with inertial measurements. The vehicle state from the estimator was then used to build a map, which was used by our proposed planner. Finally, the simulated setpoints generated by the motion planner were passed to the Pixhawk SITL.

Table 1. Comparison with baseline planner in three different search areas in simulation. Each planner was evaluated multiple times inside each area and average completion time and flight speed are recorded. In all three cases, the proposed planner was able to fly faster and cover the area using much shorter amount of time compared to the baseline planner.

Area	Frontier planner	Duration (s)	Average speed (m/s)
1	Proposed	**313.50**	**0.84**
1	Baseline	504.87	0.69
2	Proposed	**340.52**	**0.82**
2	Baseline	448.91	0.71
3	Proposed	**302.41**	**0.80**
3	Baseline	477.96	0.72

5.2 Frontier-Based Planning Results in Simulation

We evaluated the proposed frontier-based exploration planner (Sect. 4.2) in simulation. A single UAV was tasked with covering a $20\,\text{m} \times 20\,\text{m}$ search area in a randomly generated forest. To benchmark our proposed planner, we also implemented a baseline planner that greedily selected the closest frontier [23] without online update. For a comprehensive evaluation, we assigned different search areas with varying search difficulty (e.g., density of obstacles) to the vehicle. For each search area, multiple exploration missions were carried out and average performances are recorded in Table 1. The proposed planner clearly outperformed the baseline planner in terms of both the total time to complete the mission and the average speed during flight. Figure 2 shows the trajectory of the proposed planner (a)–(d) and the baseline planner (e)–(h) in an example search area. Note that using the proposed planner resulted in a much smoother overall trajectory compared to the baseline. In addition, the proposed planner is clearly more time-efficient. At $t = 250\,\text{s}$, the proposed planner nearly completed the mission, while there was still a significant portion of the map uncovered by the baseline planner.

5.3 Outdoor Flight Setup

Real flight tests were performed in the forest at NASA LaRC, shown in Fig. 1. Each vehicle was a modified DJI F450 carrying a horizontally mounted Hokuyo UTM-30LX laser rangefinder, a Pixhawk PX4 unit providing inertial measurements and motor commands, a downward-facing LidarLite for altitude measurements, and an Intel NUC computer for on-board computation. The Hokuyo produced laser measurements at a rate of 40 Hz over an angular field of view of 270° with 0.25° angular resolution. The inertial measurements and LidarLite measurements were processed at 100 Hz. Our software stack included a 2D scan-matcher and an EKF [2] for state estimation, 3D occupancy mapping using the Octomap library [11], path planning using A* search in the projected 2D map,

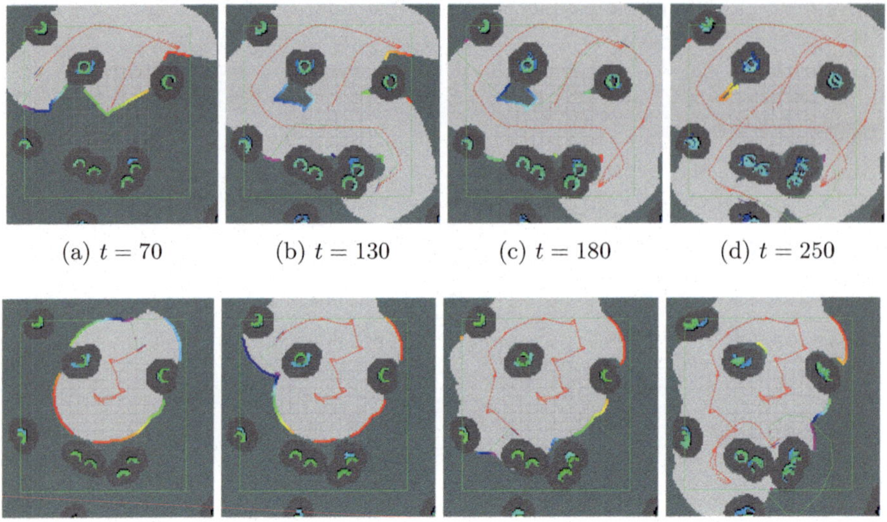

(a) $t = 70$ (b) $t = 130$ (c) $t = 180$ (d) $t = 250$

(e) $t = 70$ (f) $t = 130$ (g) $t = 180$ (h) $t = 250$

Fig. 2. Vehicle trajectories and partial maps in simulation at different time steps (seconds). (a)–(d) show the proposed planner; (e)–(h) show the baseline planner. Green rectangle denotes the search area assigned to the vehicle. Colored point clouds shows the set of all frontiers with colors representing different scores (magenta to red in decreasing score). Red path shows the trajectory taken by the vehicle. The trajectory in (a)–(d) is much smoother and more time-efficient compared to the trajectory in (e)–(h).

as well as the exploration planner presented in Sect. 4.2. All of state estimation, mapping, and planning happened online and on-board the vehicles, and only map merging occurred online on an off-board ground station. Raw laser scans from each vehicle were communicated via 5.8 GHz WiFi to the ground station for map merging. In addition, the vehicles were commanded to fly at 1.8 m altitude with a maximum allowed velocity of 2.0 m/s, and a maximum acceleration of 0.4 m/s^2.

5.4 Tree Feature Detection and Place Recognition Results

We evaluated the proposed tree feature detection and place recognition algorithm (Sect. 4.1) in a forest dataset collected at Middlesex Fells Reservation, MA [9]. This dataset contained traversals of the same forest trail in opposite directions, hence including a large number of loop closures. To detect ground truth loop closures, we estimated the sensor rig's trajectory using the same 2D scan-matcher and EKF described in Sect. 5.3. The 281 m path was divided into two trajectories to simulate a cooperative SLAM scenario; see Fig. 6(a). Figure 6(b) shows the resulting precision-recall curves of both the proposed approach and previous approach that directly clustered FLIRT point clouds without shape fitting

(a) Vehicle 1 at t=30

(b) Vehicle 1 at t=45

(c) Vehicle 1 at t=90

(d) Vehicle 1 at t=120

(e) Vehicle 2 at t=30

(f) Vehicle 2 at t=45

(g) Vehicle 2 at t=90

(h) Vehicle 2 at t=120

Fig. 3. Camera images from on-board GoPro camera at different time steps (seconds); First two rows show camera images from the first vehicle and the last two rows show images from the second vehicle. Map merging happens at $t = 45$. Frames are approximately aligned with the sensor data shown in Fig. 4.

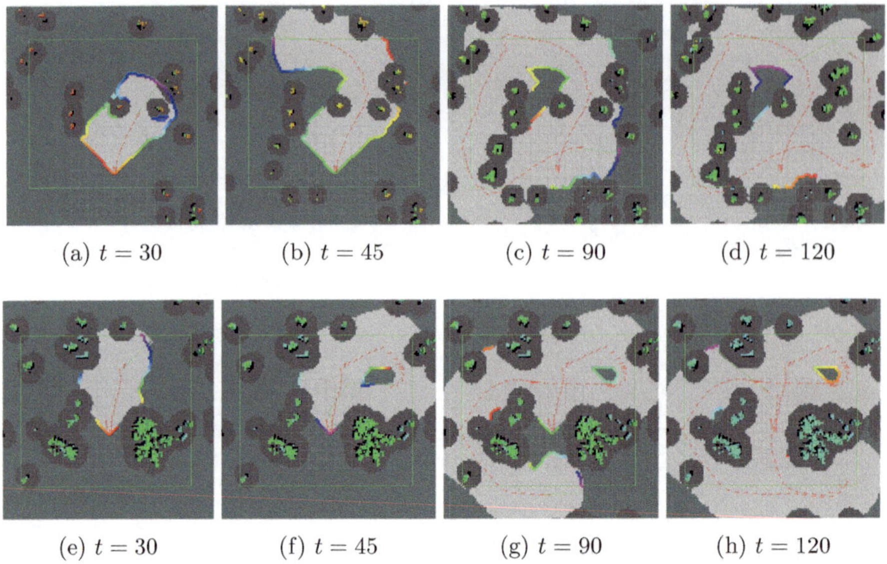

Fig. 4. Vehicle trajectories in real flight tests at different time steps (seconds). (a)–(d) shows the trajectory of the first vehicle and (e)–(h) shows the second vehicle. Map merging happens off-board on the ground station at $t = 45$.

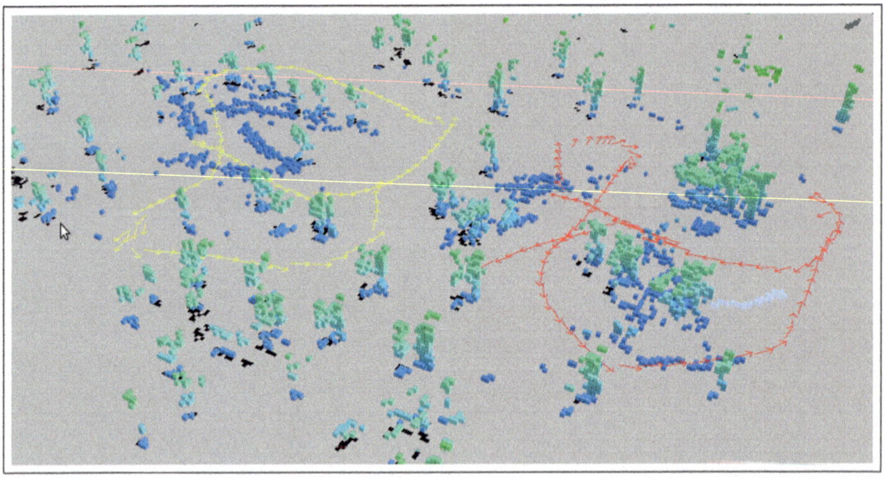

Fig. 5. The merged map resulting from collaborative exploration of the NASA LaRC forest by two vehicles. As the vehicles (paths shown in red and yellow) explore the center area of the forest, they establish inter-trajectory loop closures which enables the local maps to be merged. The resulting joint map is represented as a voxel grid with an altitude-based colormap for occupied cells (blue to green in increasing altitude), grey for unoccupied cells, and dark grey for unvisited cells (none in this figure).

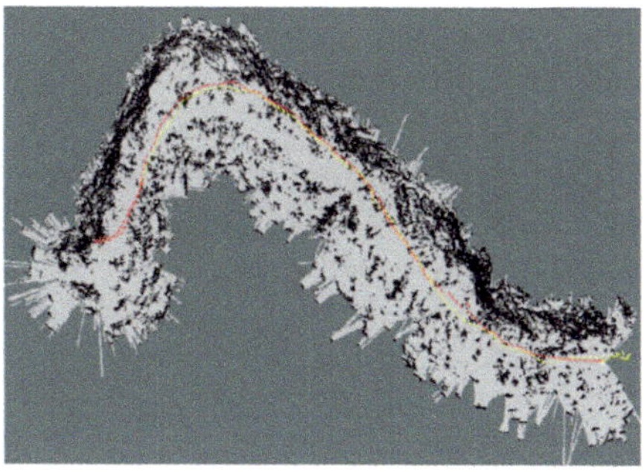

(a) Two trajectories after loop closure

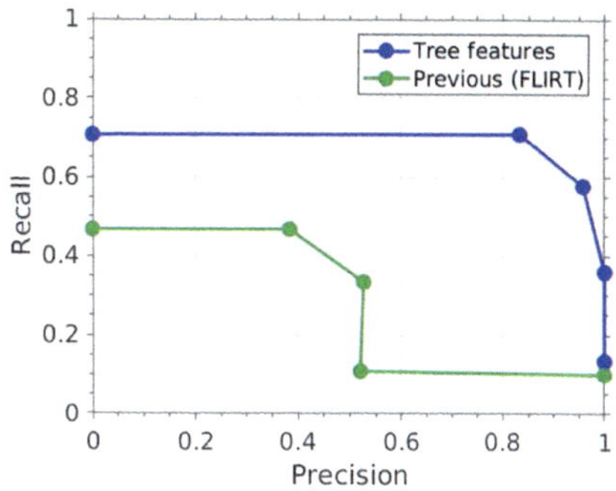

(b) Precision-Recall curve

Fig. 6. Performance of the Tree features vs. clustered FLIRT features on the Middlesex forest dataset.

Table 2. Flight durations and average speeds for flight tests at NASA LaRC.

	Duration (s)	Average speed (m/s)
Vehicle 1	122.89	2.66
Vehicle 2	135.25	2.65

to the semantic class of trees. In the forest environment, the performance of the proposed approach unsurprisingly outperformed the previous work. In our evaluation, we imposed a stricter criterion when detecting ground truth loop closures. Specifically, previous work [9] classified a proposed loop closure as a true positive if the corresponding two poses estimated by EKF are sufficiently close to each other. This criterion was sufficient for evaluation of place recognition performance which is the main focus of [9]. However, in our application, the discovered loop closures were eventually used by the back-end SLAM solver to refine trajectory estimate through pose-graph optimization. Hence, it is important to ensure that the relative transformations (factors in the pose-graph) extracted from the loop closures are accurate and do not contain outliers. To check this, we compared each extracted transformation with the transformation estimated by EKF. A proposed loop closure was declared a true positive if and only if the difference between the extracted transformation and the EKF estimate is sufficiently small in terms of both translation and rotation.

5.5 Real World Planning and Map Merging Results

The vehicles were started at different locations, with unknown relative positions. The mission was specified by non-overlapping search bounds of size $17\,\text{m} \times 20\,\text{m}$ for each vehicle. Each vehicle was tasked with observing the entire search region. The map was initially unknown; as the vehicles individually completed the coverage task, they established inter-trajectory loop closures based on similar tree features observed during flight. The individual maps from both vehicles were then merged realtime on the ground station. Figure 5 shows the final map that resulted from the example mission. The progress of planners throughout the flight is shown in Fig. 4 and roughly aligned on-board camera images are shown in Fig. 3; mission duration and average speeds are reported in Table 2. Note that due to measurement noise, the calculated average speeds were higher than the maximum allowed speed (2.0 m/s) even after applying a low-pass filter to the raw velocities. Since the proposed planner preferred frontiers with smaller orientation change, we observed that the flight behavior exhibited bias towards a natural spiral pattern.

6 Main Experimental Insights

While one-step greedy frontier selection is computationally convenient, one of the goals in the search and rescue domain is to efficiently search the forest environment. We observed that choosing the nearest frontier in heavily occluded environments such as forests led to myopic and inefficient planning, especially if a new frontier can only be selected once the current frontier has been explored. Often times, the nearest frontier was located close to the previous frontier, which led to the vehicle flying a short distance, then stopping to select the next frontier to visit. An example of such inefficiencies is shown in Fig. 2(e)–(h), where the dynamics-agnostic planner frequently stopped when performing search in the simulated forest, and covered the search window in a longer time than necessary.

A naive improvement to the frequent stopping problem could be solved by constantly re-planning, instead of requiring that the vehicle reach the intended frontier. However, constantly re-planning using only distance in the cost function may cause the vehicle to continuously command large changes in heading, which generally leads to flights that are less smooth. For slower quadrotor flight, this may be acceptable, but for time sensitive tasks such as search and rescue, we would like the vehicle to keep up search speed when possible. Additionally, limits on vehicle acceleration, coupled with the myopic planner results in the vehicle flying short, slower trajectories, as the trajectory to the next best (i.e. closest) frontier is short and does not allow the vehicle to pick up speed.

While we keep the computational tractability of one-step greedy frontier selection, our approach encodes the behavior of a longer-horizon, dynamics-aware planner by putting a heuristic cost on the deviation of a frontier from the vehicle's current heading vector. For example, in the real forest flights shown in Fig. 4, the first vehicle first visited the top-right corner of the search window first, favoring the direction of the vehicle velocity, before heading to the bottom section in a naturally occurring spiral pattern.

While the new planner demonstrates a desirable emergent search behavior, we observe that the bias to follow the current velocity vector sometimes results in the vehicle flying past a small patch of unobserved space, and having to return later in the mission (as in Fig. 4, where the second vehicle misses the tear-dropped shaped section of unknown space). A further sophisticated approach may involve a clustering approach to encourage the planner to visit small patches of unobserved space. We leave this for future work.

References

1. Aydemir, A., Sjöö, K., Folkesson, J., Pronobis, A., Jensfelt, P.: Search in the real world: active visual object search based on spatial relations. In: Proceedings of the International Conference on Robotics and Automation (ICRA) (2011)
2. Bachrach, A., Prentice, S., He, R., Roy, N.: RANGE-robust autonomous navigation in GPS-denied environments. J. Field Rob. **28**(5), 644–666 (2011)
3. Bailey, T., Nebot, E.M., Rosenblatt, J., Durrant-Whyte, H.F.: Data association for mobile robot navigation: a graph theoretic approach. In: Proceedings of the International Conference on Robotics and Automation (ICRA) (2000)
4. Bosse, M., Zlot, R.: Place recognition using keypoint voting in large 3D lidar datasets. In: Proceedings of the International Conference on Robotics and Automation (ICRA) (2013)
5. Bourgault, F., Makarenko, A.A., Williams, S.B., Grocholsky, B., Durrant-Whyte, H.F.: Information based adaptive robotic exploration. In: Proceedings of the International Conference on Intelligent Robots and Systems (IROS) (2002)
6. Cesare, K., Skeele, R., Yoo, S.-H., Zhang, Y., Hollinger, G.: Multi-UAV exploration with limited communication and battery. In: Proceedings of the International Conference on Robotics and Automation (ICRA) (2015)
7. Cummins, M., Newman, P.: Appearance-only SLAM at large scale with FAB-MAP 2.0. Int. J. Rob. Res. **30**(9), 1100–1123 (2011)
8. Feder, H.J.S., Leonard, J.J., Smith, C.M.: Adaptive mobile robot navigation and mapping. Int. J. Rob. Res. Spec. Issue Field Serv. Rob. **18**(7), 650–668 (1999)
9. Giamou, M., Babich, Y., Habibi, G., How, J.: Stable laser interest point selection for place recognition in a forest. In: Proceedings of the International Conference on Intelligent Robots and Systems (IROS) (2017)
10. Himstedt, M., Frost, J., Hellbach, S., Böhme, H.-J., Maehle, E.: Large scale place recognition in 2D LIDAR scans using geometrical landmark relations. In: Proceedings of the International Conference on Intelligent Robots and Systems (IROS) (2014)
11. Hornung, A., Wurm, K.M., Bennewitz, M., Stachniss, C., Burgard, W.: OctoMap: an efficient probabilistic 3D mapping framework based on octrees. Auton. Rob. **34**(3), 189–206 (2013)
12. Kaess, M., Johannsson, H., Roberts, R., Ila, V., Leonard, J., Dellaert, F.: iSAM2: incremental smoothing and mapping with fluid relinearization and incremental variable reordering. In: Proceedings of the International Conference on Robotics and Automation (ICRA) (2011)
13. Kallasi, F., Rizzini, D.L.: Efficient loop closure based on FALKO LIDAR features for online robot localization and mapping. In: Proceedings of the International Conference on Intelligent Robots and Systems (IROS) (2016)
14. Kukko, A., Kaijaluoto, R., Kaartinen, H., Lehtola, V.V., Jaakkola, A., Hyyppä, J.: Graph SLAM correction for single scanner MLS forest data under boreal forest canopy. ISPRS J. Photogrammetry Remote Sens. **132**, 199–209 (2017)
15. Meier, L., Tanskanen, P., Heng, L., Lee, G.H., Fraundorfer, F., Pollefeys, M.: PIXHAWK: a micro aerial vehicle design for autonomous flight using onboard computer vision. Auton. Rob. **33**(1–2), 21–39 (2012)

16. Mellinger, D., Michael, N., Kumar, V.: Trajectory generation and control for precise aggressive maneuvers with quadrotors. Int. J. Rob. Res. **31**(5), 664–674 (2012)
17. Milford, M.: Visual route recognition with a handful of bits. In: Proceedings of the Robotics: Science and Systems (RSS) (2012)
18. Paul, R., Newman, P.: FAB-MAP 3D: topological mapping with spatial and visual appearance. In: Proceedings of the International Conference on Robotics and Automation (ICRA) (2010)
19. Richter, C., Roy, N.: Learning to plan for visibility in navigation of unknown environments. In: Proceedings of the International Symposium on Experimental Robotics (ISER) (2016)
20. Stachniss, C., Burgard, W.: Exploring unknown environments with mobile robots using coverage maps. In: Proceedings of the International Joint Conferences on Artificial Intelligence (IJCAI) (2003)
21. Tedrake, R.: Drake: a planning, control, and analysis toolbox for nonlinear dynamical systems (2014)
22. Tipaldi, G.D., Spinello, L., Burgard, W.: Geometrical flirt phrases for large scale place recognition in 2D range data. In: Proceedings of the International Conference on Robotics and Automation (ICRA) (2013)
23. Yamauchi, B., Schultz, A., Adams, W.: Mobile robot exploration and map building with continuous localization. In: Proceedings of the International Conference on Robotics and Automation (ICRA) (1998)

Predicting Digging Success for Unmanned Aircraft System Sensor Emplacement

Adam Plowcha[1(✉)], Yue Sun[2], Carrick Detweiler[1], and Justin Bradley[1]

[1] Computer Science and Engineering Department, University of Nebraska, Lincoln, USA
pplowcha@cse.unl.edu

[2] Mechanical Engineering Department, University of Nebraska, Lincoln, USA

Abstract. We have developed an autonomous, digging, Unmanned Aircraft System (UAS) for sensor emplacement. A key challenge is quickly determining whether or not a particular digging activity will lead to successful emplacement, thereby allowing the system to potentially try another location. We have designed a first-of-its-kind decision-making algorithm using a Markov Decision Process to autonomously monitor the activity of a digging UAS activity to quickly decide if success is likely. Further, we demonstrate through many experimental trials that our method outperforms other decision-making methods with an overall success rate of 82.5%.

Keywords: Sensor emplacement · UAS · Field robotics · Markov Decision Process

1 Motivation and Problem Statement

We have developed a small Unmanned Aircraft System (UAS) that can autonomously fly to a remote location to emplace a sensor underground (shown in Fig. 1) [13]. Unlike prior approaches for UAS sensor deployment [1,5,8], our approach *digs* into the ground to place the sensor underground for explicit measurements or to conceal the sensor. An aircraft-mounted digging system provides greater flexibility in locations that can be reached and can minimize the time to deploy a sensor. To minimize the weight

Fig. 1. UAS with digging auger.

This material is based upon work supported by USSTRATCOM through the 55th Contracting Squadron under Contract No. FA4600-12-D-9000 and was also funded, in part, by NSF-1638099 and USDA-2017-67021-25924.

Electronic supplementary material The online version of this chapter (https://doi.org/10.1007/978-3-030-33950-0_14) contains supplementary material, which is available to authorized users.

J. Xiao et al. (Eds.): ISER 2018, SPAR 11, pp. 153–164, 2020.
https://doi.org/10.1007/978-3-030-33950-0_14

of the digging system and maximize the UAS range, the system is optimized for a particular type of soil. This optimization may lead to emplacement failure if the system attempts to dig in soil that varies locally from the target type, and since the system is operating far from the operator, autonomy is required to determine if emplacement is likely to succeed.

The key contribution of this work is the development and experimental validation of algorithms to quickly determine whether or not a particular digging activity will lead to successful emplacement. An emplacement may fail, for example, if the soil is too compact or has extraneous obstructions (e.g. roots, rocks), or the sensor may get stuck or partially emplaced, which we consider the worst case failure. If these situations can be accurately detected early, the UAS can extract the sensor and try an alternate location. Setting a time limit on a digging attempt provides an easy litmus test but can lead to either wasting resources on what will be an unsuccessful dig, or aborting a potentially successful dig and subsequently wasting energy flying to a new location. In either case, faster, more reliable prediction will yield improved emplacement results. To illustrate, in our trials we find that 29% of our attempts succeed within 20 s, but of those that go longer there is a significant variance in the time it takes to succeed, and only 48% fail to reach the target depth.

Success in emplacement is best characterized as a stochastic event. Although knowing the soil type can reduce uncertainty, the likelihood of unforeseen obstructions slowing or stopping the progress is high and difficult to predict. With this uncertainty in mind, we have designed a first-of-its-kind decision-making algorithm for a digging UAS that enables autonomous monitoring of digging to quickly decide if success is likely or if another digging location should be selected. Markov Decision Processes (MDP) provide optimal decision-making under uncertainty [10], and as such, we develop an MDP to predict the outcome of a single digging event. We also provide a comparison of the performance of our MDP against binary decision trees and a support vector machine, which are commonly used classification techniques. Finally, we compare these methods to the performance of an expert human operator in predicting the outcome during operation of *153 experimental trials involving emplacement of sensors in six different soil types.*

2 Technical Approach

Figure 2 shows a detailed diagram of the UAS auger system, which we describe briefly here for context[1]. The auger is mounted to a DJI Matrice 100 with a custom weight-distributing chassis, a micro-controller, a depth sensor, and a DC motor which turns the auger. The auger is hollow, allowing a sensor of up 25 mm in diameter and 76 mm length to be placed inside. Upon reaching target depth, the auger mechanism (with sensor inside) can be released. Alternately, the auger can be reversed, removing the sensor to try an alternate location.

[1] Additional details can be found in [13].

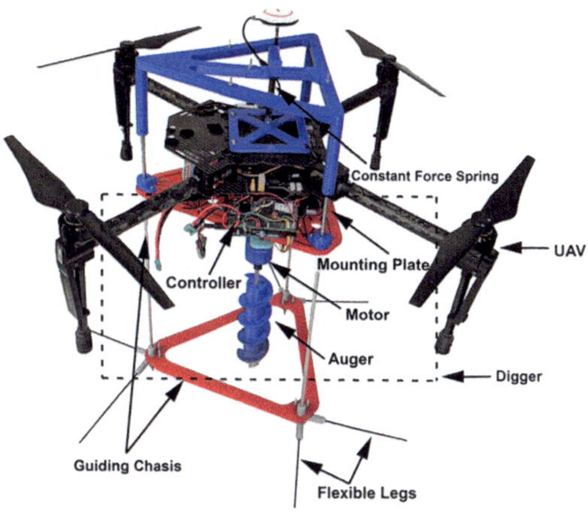

Fig. 2. Details of UAS auger system

2.1 Digging System State Characterization

To characterize the digging process we leverage the following key performance metrics that, together, provide insight into the status of the digging operation: motor RPM, motor current, auger depth, and elapsed digging time. For instance, decreasing motor rpm and increasing motor current draw coupled with increasing auger depth is often indicative of a potentially successful dig. Conversely, high motor rpm with low motor current draw might indicate that the auger has been impeded by something solid and will not be able to achieve a greater depth. In many scenarios the system may initially achieve a certain auger depth, and subsequently fail to advance, indicating a potential failure. This is not captured well by the above metrics, so we add the metric *rate of digging progress* to help identify this scenario.

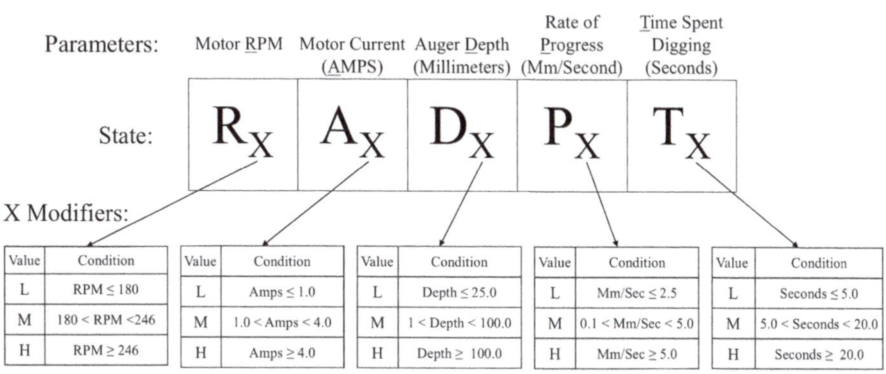

Fig. 3. State description - a state with high motor RPM, low amps, low depth, low rate of progress, and low time would be $R_H A_L D_L P_L T_L$.

We impose three discrete measures, "Low", "Medium", and "High", on each metric to create a state space with sufficient fidelity to represent the digging process via an automaton. This results in $3^5 = 243$ system states, a description of which can be seen in Fig. 3.

3 Experimental Setup

An understanding of the nuances of digging helps develop the correct methods and models we use to predict digging success. We analyzed the results of 153 digging trials in a testbed, shown in Fig. 4. Soil types were provided to us by our collaborators as described in [13]. To meet these specifications, a testbed having six different soil compositions with a range of compactness and moisture levels representative of the specified target environment was created. During each trial we logged the system state described in Fig. 3 at a rate of 10 Hz. We label a dig "successful" when the auger reaches its target depth of 100 mm. If the auger does not reach its target depth, then it is manually stopped by the operator when system parameters indicate that success is unlikely or approximately 120 s have passed, and the trial is considered a digging failure.

From the 153 trials, we use eight randomly chosen successful trials and eight randomly chosen failed trials as the training data for our various decision making models. By using approximately 10% of our data for training, we avoid over fitting our data when running those models against the data from the remaining 137 trials. We have chosen four decision-making methods for evaluation and describe each one below.

Fig. 4. Test site with different soil types.

3.1 Markov Decision Process

A MDP is a tuple, $S = \{S, A, P(s, a, s'), R(s, a, s'), \gamma\}$, representing states, actions, transition probabilities, rewards, and a discount factor, respectively.

The MDP can be solved to generate an optimal policy, π, representing which action should be taken in any state. MDPs in UAS have predominantly been used for path planning [2,6]. In this case, the stochasticity represented by the MDP captures the dynamic uncertainty from the assumed noise and disturbances potentially preventing the vehicle from attaining the commanded position. MDPs have also been used in UAS for other purposes such as image classification [11], search and rescue decisions [14], and target detection [4]. In each of these cases the stochasticity (i.e. randomness) of their data was suited to the application of a MDP. Similarly, for our digging success prediction, the transitions between states are highly uncertain, sometimes being equi-probabilistically distributed among 3–4 outgoing transitions from a state. Moreover, we have numerous ending states as seen in Table 1, with numerous paths to those states. This uncertainty stems from the soil, what may be in the soil, the failure conditions of the motor and auger, as well as how well a particular auger performs in a particular soil type. This uncertainty is well represented by a MDP.

Our MDP consists of 243 states, as described in Sect. 2. The actions are either: (1) continue digging or (2) stop digging. We generated transition probabilities from our 16 randomly chosen training trials. We created our reward table after examining our data using the following rationale. Rewards were increased to encourage stopping when a successful outcome seemed unlikely. This includes situations where very little digging depth is achieved over a fixed period of time, or when the system was in a state where high motor current draw and low motor RPM indicated a potential stall situation. States in which the system had achieved its target depth were considered termination states and were assigned the largest reward for the stopping action. To encourage continuation, we assigned higher rewards if a positive rate of digging progress was observed, or the system had been digging only a short time. To further tune our MDP reward table, we ran the initial MDP-generated policy against random successful and unsuccessful digs observing the MDP's "decision" at each time interval within the files. This allowed us to find areas within our reward table that could be adjusted to increase the chance of the policy leading to a successful outcome. For instance, we discovered that we needed to increase the reward for continuing while in a state that had the attributes of medium time with medium progress, i.e. $R_X A_X D_X P_M T_M$. The discount factor, $\gamma = 0.96$, was tuned by comparing success rates at γ values near 1.0 and then decreasing γ until success rates became maximal. We then used version 1.6 of the MDPToolbox [3] for Matlab® to generate our policy which was then run against our remaining data sets.

3.2 Decision Tree

Decision trees are supervised learning methods that use a training set of data to compute a relationship between a set of input attributes and a target attribute. This relationship is a model in the form of a binary decision tree that can be used to predict the target attribute of new data [9]. In our case, the input attributes consist of the digging system's measured parameters (time, motor current draw, motor amps, and depth of auger), and the target attribute is the decision to continue or stop digging. We choose this method as the relatively small number of attributes should generate a model that will be fast and efficient to implement on the hardware of our digging system. Additionally, it has been successfully employed to classify soil conditions from UASs in the past [1].

Using the same 16 randomly chosen digging trials we used as training sets for our MDP, we create a binary classification tree based on the four attributes our system sensors provided using Matlab®'s "fitctree()" method [7].

3.3 Support Vector Machine

A support vector machine (SVM) is a supervised learning method that classifies data from a training set by defining a hyperplane with maximal distance from the respective data points within the set. New data can then be classified by its position relative to that hyperplane. We choose to evaluate this method in our situation for the sake of completeness as there are cases when SVMs are reported to outperform decision trees [12] as well as vice-versa [1].

We create our SVM-based predictive model using Matlab®'s "fitcsvm()" method with the default linear kernel and the same training sets as our previous methods.

3.4 Human Operator Decision

The human operator made predictions about the overall outcome of a digging evolution based on the experience gained from the numerous observations made while creating and testing the digging system. The digger control software used to log the raw sensor data also allowed the user to indicate a prediction of success or failure at any given moment during a dig. Clicking a button labeled "Succeed" or "Fail" would insert a marker at that particular time into the log file (see the upper right corner of Fig. 5). On average, the human operator would indicate a prediction after twenty seconds of digging based on observations of the visual and audible clues from the digging system as well as observing the sensor data received from the system on the digging control software. In this way, the human operator had access to the same data used by the machine-based methods of decision making.

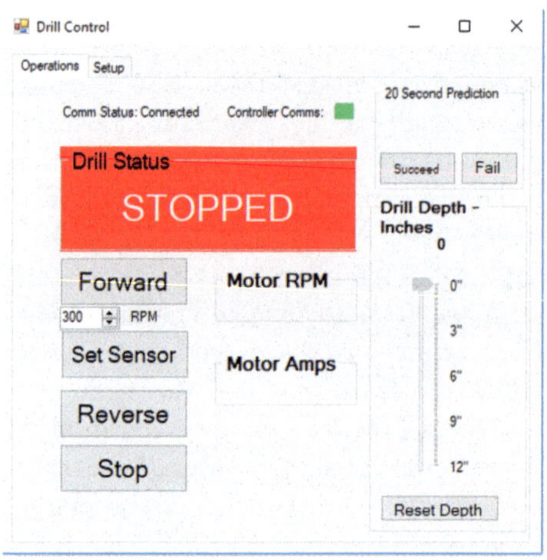

Fig. 5. Digger control software.

4 Experimental Results

Figure 6 shows an example profile of a successful digging attempt. It shows an increase in depth and current usage (with a corresponding drop in RPM) as the dig progresses. Figures 7 and 8 show example profiles of failed digging attempts. Figure 7 shows a consistently high motor RPM with low current usage and depth stopping at about 30 mm. We call this a failure due to spinning. Figure 8 shows the opposite type of failure where something in the soil causes the auger bit to drastically slow down or stop with a commensurate rise in motor current draw. We call this a failure due to stalling.

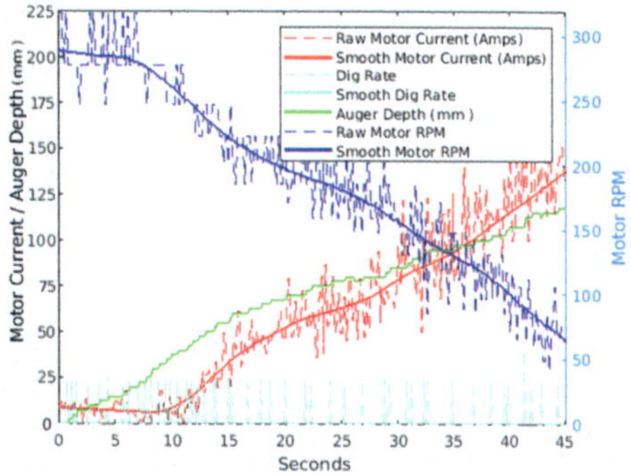

Fig. 6. Raw data from a successful dig.

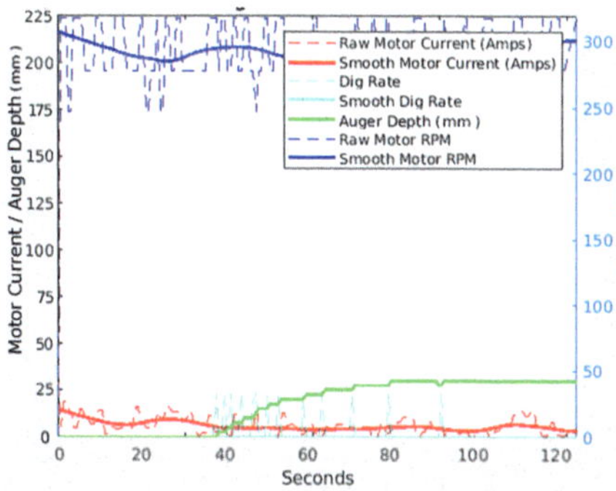

Fig. 7. Raw data from a dig failure due to spinning.

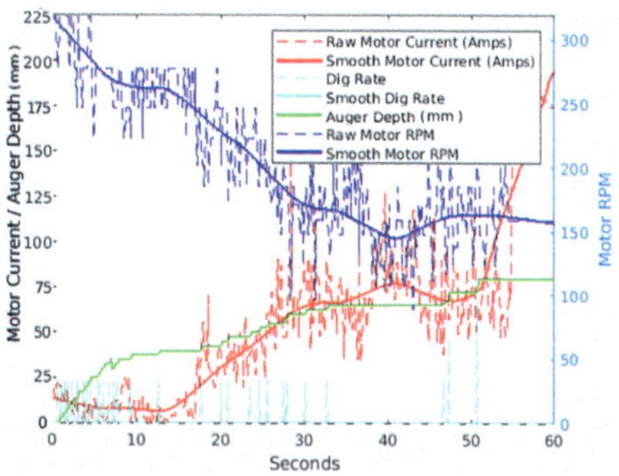

Fig. 8. Raw data from a dig failure due to stalling.

Table 1 shows the final states which define both successful and unsuccessful digs. Of the trials, 80 were successful with the digger reaching its target depth. Those successes were distributed among 19 different end states. There were 73 trials that failed where the digger did not reach its target depth. Those failures were distributed among 19 different end states. The large number of final end states shows the stochasticity of the digging and that a single metric cannot be used to detect failure.

The majority of our successful digs (45 out of 80) occurred within the first 20 s as evidenced by the states ending with T_L (low time) and T_M (medium time). We can also separate the majority of the failed digging evolutions into two subgroups based on their RPM and current draw. Trials either failed due to the auger stalling, as indicated by states starting with R_L (low RPM) or R_M

(medium RPM) and then A_M (medium amps) or A_H (high amps), or the auger was freely spinning, as indicated by states starting with R_H (high RPM) and then A_L (low amps) or A_M (medium amps).

In some instances there is little difference between the end states of a successful dig and an unsuccessful one. The first row in Table 1 shows that the successful dig and unsuccessful dig differ only in a single separation of modifier for depth (i.e., D_H vs. D_M). This is due to the range of values used to specify medium depth (D_M) vs. high depth (D_H) and medium time (T_M) vs. high time (T_H). D_M spans a range of 75 mm ending just below the D_H mark of 100 mm. T_H is any value greater than 20 s. Examination of the data logs for the specific trials in question show that two of the unsuccessful digs were at the lower end of the D_M range, one had been stalled at 90 mm for over 30 s, and all three trials had been digging for over 90 s. The four successful digs all achieved the target depth of 100 mm in 55 to 61 s. A human operator in control of the dig used judgment as to when to stop or to allow the digging to continue based on visual and audible clues. Again, this highlights the stochasticity of a digging event as failure is difficult to predict.

Table 1. Trial end states labeled with **R**PM, **A**mps, **D**epth, **P**rogress, and **T**ime, in terms of **L**ow, **M**edium, and **H**igh as defined in Fig. 3

Successful attempts		Failed attempts	
End state	Quantity	End state	Quantity
$R_L A_M D_H P_L T_H$	5	$R_L A_M D_M P_L T_H$	3
$R_L A_M D_H P_M T_H$	2	$R_L A_H D_L P_L T_H$	1
$R_L A_M D_H P_H T_M$	2	$R_L A_H D_M P_L T_H$	11
$R_L A_H D_H P_L T_H$	4	$R_L A_H D_M P_M T_H$	1
$R_L A_H D_H P_M T_H$	2	$R_M A_L D_M P_L T_H$	2
$R_M A_L D_H P_H T_M$	1	$R_M A_M D_L P_L T_H$	1
$R_M A_M D_H P_L T_H$	3	$R_M A_M D_M P_L T_H$	1
$R_M A_M D_H P_M T_H$	1	$R_M A_H D_M P_L T_H$	4
$R_M A_M D_H P_H T_L$	1	$R_H A_L D_L P_L T_M$	2
$R_M A_M D_H P_H T_M$	4	$R_H A_L D_L P_L T_H$	7
$R_M A_H D_H P_M T_H$	4	$R_H A_L D_L P_M T_M$	1
$R_H A_L D_H P_L T_H$	2	$R_H A_L D_M P_L T_H$	12
$R_H A_L D_H P_M T_H$	4	$R_H A_L D_M P_M T_M$	1
$R_H A_L D_H P_H T_L$	3	$R_H A_L D_M P_H T_L$	1
$R_H A_L D_H P_H T_M$	12	$R_H A_L D_M P_H T_M$	1
$R_H A_M D_H P_L T_H$	7	$R_H A_M D_L P_L T_H$	2
$R_H A_M D_H P_M T_H$	2	$R_H A_M D_M P_L T_H$	20
$R_H A_M D_H P_H T_L$	2	$R_H A_M D_M P_M T_H$	1
$R_H A_M D_H P_H T_M$	19	$R_H A_H D_M P_L T_H$	1

4.1 MDP

We tested our MDP-derived policy against the 137 remaining data sets not used for training by iterating over the timed sequence of data and using the policy to decide whether to continue to the next time step or stop. Our MDP policy had an overall success rate of 82.5% in predicting the outcome of a dig. While overall success is an important metric, we are more concerned with instances when our MDP predicts success when success is not possible (a false positive) and how quickly it correctly decides to abort a dig when success is not achievable (a true negative). Recall that our worst case is when energy is wasted by a system that continues to dig when it will not succeed as this energy could be used to fly to a new location for another attempt at sensor emplacement. Figure 9 shows the true/false positive/negative rates of our four decision making methods. All "stop" decisions based on the true negatives recognized by our MDP occurred within 20 s.

There were nine instances where the MDP policy correctly made predictions counter to the human operator. We highlight the following two cases as examples.

In our 40th trial, the human operator observed what appeared to be a stalled state where the auger had stopped advancing at about midway to the target depth and had remained there for approximately 13 s. In the two seconds prior to the operator making a prediction, the auger began a slightly perceptible advance downward, but the human operator determined this was not enough to allow it to achieve the desired depth and recorded a prediction of "failure." However, that slight advance in depth was enough to put the state of progress into the medium category (P_M) which the MDP policy recognized as a "continue" condition, and the auger eventually reached the desired depth.

Conversely, in our 85th trial, the human operator noted a very reasonable advance in depth that had reached three quarters of the way to the target depth by the time a prediction was required. The human operator recorded a prediction of "success" but did not realize the commensurate increase in motor current draw and decrease in motor RPM were indicating a situation where the auger bit might stop turning completely. However, as the system was in a state with high amperage (A_H) and low RPM (R_L), the MDP policy recognized a "stop" condition and correctly predicted failure.

4.2 Decision Tree

The decision tree model was similarly tested against the 137 remaining data sets not used for training by iterating through each time-step in the trial data and applying the model to determine if digging should continue or stop. Our decision tree-based model had an overall success rate of 61.3%. Its false positive rate was 65.6%.

4.3 Support Vector Machine

The SVM model was also tested against the 137 trials not used for its training using the same method as described above. It had an overall success rate of

		Condition	
		Successful Dig	Unsuccessful Dig
MDP Predicts	Success	TPR: 81%	FPR: 08%
	Failure	FNR: 19%	TNR: 92%

		Condition	
		Successful Dig	Unsuccessful Dig
Dec Tree Predicts	Success	TPR: 85%	FPR: 66%
	Failure	FNR: 15%	TNR: 34%

		Condition	
		Successful Dig	Unsuccessful Dig
SVM Predicts	Success	TPR: 76%	FPR: 100%
	Failure	FNR: 24%	TNR: 0%

		Condition	
		Successful Dig	Unsuccessful Dig
Human Predicts	Success	TPR: 44%	FPR: 17%
	Failure	FNR: 56%	TNR: 83%

Fig. 9. Decision method True/False Positive/Negative rates out of 137 trials

23.3% with a false positive rate of 100% and a true negative rate of 0%, which means that the SVM did not successfully predict a failure in any of the 137 trials it was simulated against.

4.4 Human Operator

The goal of human operator was to make a prediction at approximately the 20-s mark of each digging evolution. As 61 of the 153 digging trials reached their target depth within 20 s, the human operator made 91 predictions with an overall success rate of 69.5% and a false positive rate of 16.7%.

5 Main Experimental Insights

Initially we believed the human operator was the benchmark for correct prediction due to the significant experience with the system and additional knowledge of the environmental conditions (e.g. the type of soil the system was digging in). Our expert human operator was only able to successfully predict the outcome of a digging evolution 69.5% of the time, but also with a 16.7% false positive rate. The decision tree and SVM both underperform the human operator (although the decision tree only by a few percentage points). While both methods can handle noise within their training data, the amount of noise within the particular training sets may have impacted their overall effectiveness. The MDP-generated policy performs better than the other three methods with an overall success rate of 82.5% and only an 8% false positive rate, and this low false positive rate is critical in our application as premature stopping is preferred to continued digging when success is not possible.

A key contribution of this paper is the insight gained due to the nature of the problem. It was initially perceived that a simple timer to determine failure would be sufficient. However, in practice, many surprising cases emerged in which little auger progress was observed for over a minute just prior to rapid success. This showcases the variability in digging operations and how algorithmic decision makers can enhance performance.

6 Conclusion and Future Work

A digging UAS provides an excellent means of reaching and placing sensors in a variety of locations. This mobility comes with costs associated with the digging mechanism including the need to predict success/failure. Since failure of a single digging activity is likely, predicting the event and relocating for another attempt as quickly as possible is critical for mission success.

Our work shows that a MDP can be developed, trained, and successfully used to predict the outcome of a digging evolution with 82.5% accuracy which exceeds those of other decision making methods, including expert human operators. Our next goal is to deploy our algorithm on an improved version of our UAS digging system for real-time testing. Furthermore, we would like to extend the capability of algorithm to not only predict success or failure, but to also select appropriate courses of action based on the specific type of failure detected.

Acknowledgments. Thanks to Dr. Sebastian Elbaum, Dr. Brittany Duncan, Andrew Rasmussen, Ajay Shankur, Jacob Hogberg, and Aaron Clare and the staff at Horning State Farm for their assistance.

References

1. Anthony, D., Basha, E., Ostdiek, J., Ore, J.P., Detweiler, C.: Surface classification for sensor deployment from UAV landings. In: 2015 IEEE International Conference on Robotics and Automation (ICRA), pp. 3464–3470, May 2015
2. Bethke, B., Bertuccelli, L., How, J.: Experimental demonstration of adaptive MDP-based planning with model uncertainty. In: AIAA Guidance, Navigation and Control Conference and Exhibit, p. 6322 (2008)
3. Chadès, I., Chapron, G., Cros, M.J., Garcia, F., Sabbadin, R.: MDPtoolbox: a multi-platform toolbox to solve stochastic dynamic programming problems. Ecography **37**(9), 916–920 (2014)
4. Chanel, C.P.C., Teichteil-Königsbuch, F., Lesire, C.: Multi-target detection and recognition by UAVs using online pomdps. In: AAAI, pp. 1381–1387 (2013)
5. Corke, P., Hrabar, S., Peterson, R., Rus, D., Saripalli, S., Sukhatme, G.: Autonomous deployment and repair of a sensor network using an unmanned aerial vehicle. In: 2004 IEEE International Conference on Robotics and Automation, Proceedings, ICRA 2004, vol. 4, pp. 3602–3608. IEEE (2004)
6. Dadkhah, N., Mettler, B.: Survey of motion planning literature in the presence of uncertainty: considerations for UAV guidance. J. Intell. Rob. Syst. **65**(1–4), 233–246 (2012)

7. The Mathworks, Inc., Natick, Massachusetts: MATLAB version 9.3.0.713579 (R2017b) (2017)
8. Pister, K.S.: Tracking vehicles with a UAV-delivered sensor network (2001)
9. Rokach, L., Maimon, O.: Data Mining with Decision Trees: Theory and Application, vol. 2. World Scientific (2014)
10. Russell, S.J., Norvig, P.: Artificial Intelligence: A Modern Approach, 3rd edn., vol. 3. Prentice Hall, Upper Saddle River (2009)
11. Scott, K.K.: Occlusion-aware sensing and coverage in unmanned aerial vehicle (UAV) networks. Ph.D. thesis, University of Cincinnati (2016)
12. Shafri, H., Ramle, F.: A comparison of support vector machine and decision tree classifications using satellite data of Langkawi Island. Inf. Technol. J. **8**(1), 64–70 (2009)
13. Sun, Y., Plowcha, A., Nail, M., Elbaum, S., Terry, B., Detweiler, C.: Unmanned aerial auger for underground sensor installation. In: Accepted to appear 2018 IEEE/RSJ International Conference on Intelligent Robots and Systems (IROS 2018) (2018)
14. Waharte, S., Trigoni, N.: Supporting search and rescue operations with UAVs. In: 2010 International Conference on Emerging Security Technologies (EST), pp. 142–147. IEEE (2010)

Geometric Control and Trajectory Optimization for Bidirectional Thrust Quadrotors

Michael Watterson(✉), Ahmed Zahra, and Vijay Kumar

University of Pennsylvania, Philadelphia, PA 19146, USA
{wami,ahzahra,kumar}@seas.upenn.edu

Abstract. Geometric control of quadrotors provides a way to control the 3D position and a yaw angle of the robot with a larger stability region than linearized controllers, but is still singular for some orientations. This singularity exists even if $SO(3)$ is parameterized in a non-singular way such as with unit quaternions. Recent advances using the Hopf Fibration have eliminated these singularities using multiple coordinate charts and switching between them to avoid singularities. We further extend the envelope of geometric control for quadrotors to a system where the propellers can reverse direction and produce negative forces using additional charts. We show that our proposed controller is stable in hovering a robot upright and upside down. We then show how to extend trajectory generation using differentially flat coordinates to optimize trajectories using multiple charts to produce trajectories which can transition between upright and upside down.

Keywords: Aerial robotics · Geometric control · Trajectory optimization · Planning

1 Introduction

With the advent of bidirectional thrust propellers and motors, hobbyists have been demonstrating flying upside down with manual control. There are many reasons why doing the same in autonomous operation would be beneficial. Agile maneuvers such as going through vertical windows [11] are more efficient if the robot does not need to reverse roll direction in mid air. Some sensor suites can also benefit from being reconfigured from pointing upward to pointing downward by flying a robot upside down instead of needing to add additional actuated gimbals. For these reasons, we seek to develop quadrotor control algorithms to fly with forward and reverse thrust, upright and inverted.

Electronic supplementary material The online version of this chapter (https://doi.org/10.1007/978-3-030-33950-0_15) contains supplementary material, which is available to authorized users.

J. Xiao et al. (Eds.): ISER 2018, SPAR 11, pp. 165–176, 2020.
https://doi.org/10.1007/978-3-030-33950-0_15

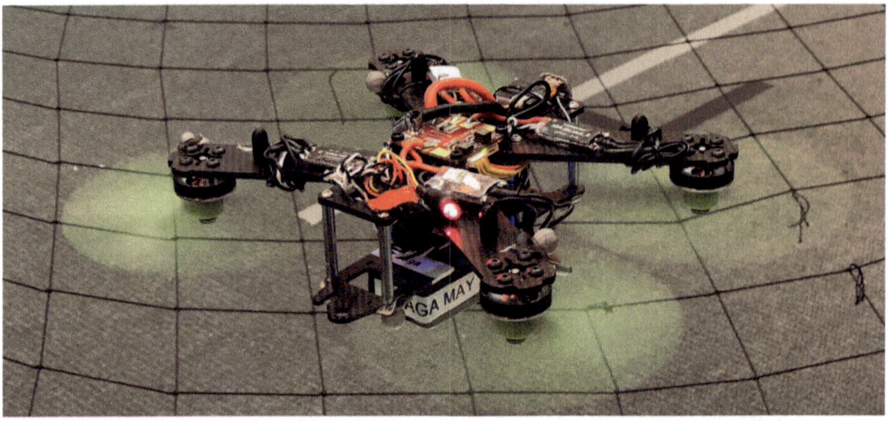

Fig. 1. We propose a new geometric controller for quadrotors which works with bi-directional motors to operate both forward and reversed. Here our platform is hovering upside down with its motors reverse.

1.1 Related Work

Many different approaches have been developed over the years to implement nonlinear controllers for multirotor aerial vehicles [1,3,5,7,8,14]. Traditionally, these control schemes implement attitude and position control separately. There are however some works that have combined them both into one control block [5].

Although several of these previous works have used geometrical control methods [1,3,7,8,14] to implement aggressive maneuvers on multirotor aerial vehicles, some [1,6,15] have explored inverted flight. In one of these approaches [1], rotor pitch actuation was used on a quadrotor to allow it to change the direction of the thrust vector with respect to the body frame, which would be necessary for stabilization purposes for very large quadrotors, but adds a lot more complexity to the control scheme through the additional degrees of freedom. In addition, the added mechanical complexity is too heavy for small agile quadrotors. In the other approach [15], reinforcement learning was used on a helicopter to allow it to perform inverted hovering and simple, low-speed acrobatic maneuvers by learning a non-linear model of the helicopter's dynamics. It does not discuss how this approach can be extended to run aggressive maneuvers, or as to how to extend this approach to multirotor vehicles. This paper builds on the works exploring geometric control methods on $SE(3)$, and shows how they can be applied on a quadrotor to perform aggressive maneuvers in inverted and upright flight. [6] uses reachability sets to recover from brief periods of open loop rotation during the inverted phase.

Generating aggressive trajectories with inversions classically has two limitations. In [12], they use a quadrotor state which does not include the yaw of the robot. This controls the robot in a subset of it real state space, and their method cannot be easily extended to work when yaw matter. In addition, their method requires optimizing a set of acceleration set-points and interpolating between them. We instead would like to work in continuous time, so we use a minimal number of variables to represent our trajectory. The methods of [14,17] are singular during certain orientations due to using the yaw. They can only generate aggressive trajectories which roll by 90° or pitch by 90°, but not both. The Hopf Fibration controller [18] can solve both of these issues by using multiple charts to represent the quadrotor state. The manifold trajectory generation technique presented in [19] is a good start for this problem but needs some modification.

1.2 Contributions of This Work

In this work, we extend the geometric controller in [18] to work with normal and inverted flight. We extend the work of [10,14] to generate trajectories through multiple charts using a spline-based trajectory optimization. Finally, we verify our controller and planner on a hardware testbed.

2 Technical Approach

To achieve our goal of forward and inverted thrust quadrotor control we require an extension of the geometric control model which can work with the propellers in both directions. We would also like our controller to achieve feedback control for arbitrary states on $SE(3)$, which we extend from our previous work [18] to work with both directions of thrust. Finally, to demonstrate trajectories, we propose a trajectory optimization algorithm to generate dynamically feasible trajectories for arbitrary oriented waypoints based off the work of [14].

2.1 Geometric Control with Bidirectional Thrust

As explored in [9,14,18], geometric controllers produce a feedback control law under the under-actuated constraint that a quadrotor's thrust is restricted to be parallel to the quadrotor's b_3 body direction.

As in [9,14,18], we can relate individual thrust speeds s_i to a force f along b_3 and moments m_1, m_2 and m_3 along b_1, b_2 and b_3 using an invertible transform A. A depends on the characteristic radius of the robot L, and thrust and moment coefficients of the props (k_f and k_m).

$$\begin{bmatrix} m_1 \\ m_2 \\ m_3 \\ f \end{bmatrix} = \begin{bmatrix} 0 & k_f L & 0 & k_f L \\ -k_f L & 0 & k_f L & 0 \\ k_m & -k_m & k_m & -k_m \\ k_f & k_f & k_f & k_f \end{bmatrix} \begin{bmatrix} s_1^2 \text{sign}(s_1) \\ s_2^2 \text{sign}(s_2) \\ s_3^2 \text{sign}(s_3) \\ s_4^2 \text{sign}(s_4) \end{bmatrix} = AS \tag{1}$$

$$S = A^{-1} \begin{bmatrix} m_1 \\ m_2 \\ m_3 \\ f \end{bmatrix} \tag{2}$$

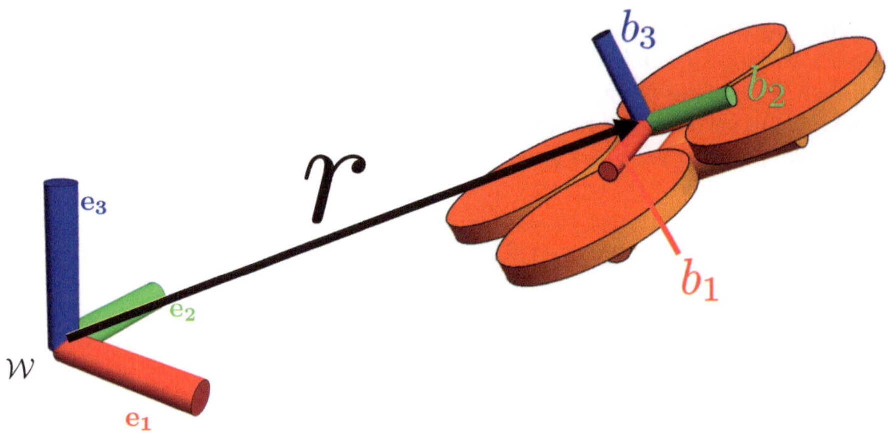

Fig. 2. A quadrotor ($\mathcal{B}$ frame) position (r) and orientation (R) are measured with respect to the inertial frame $\mathcal{W}$.

Unlike previous geometric control works, we define s_i to be positive for normal thrust direction and negative for reverse. In this work, we only consider the case where the robot's position and yaw is controllable, thus we assume $s_i > 0$ for all i or $s_i < 0$ for all i. For simplicity we write $\text{sign}(s)$ instead of $\text{sign}(s_i)$ (Fig. 2).

The equations of motion are then expressed with respect to the inertia tensor H and the mass of the robot m:

$$\begin{aligned} m\ddot{r} &= m\begin{bmatrix} 0 \\ 0 \\ -g \end{bmatrix} + R \cdot fe_3 \\ H\dot{\omega} &= \begin{bmatrix} m_1 \\ m_2 \\ m_3 \end{bmatrix} - \omega \times (H\omega) \end{aligned} \tag{3}$$

To derive a feedback control law we define ξ and a, b, c as in [18] to be:

$$\xi = m\left(\ddot{r} + \begin{bmatrix} 0 \\ 0 \\ g \end{bmatrix}\right) \tag{4}$$

Thus

$$\xi = Rfe_3 \tag{5}$$

$$\frac{\text{sign}(s)\xi}{||\xi||} = Re_3 = \begin{bmatrix} a \\ b \\ c \end{bmatrix} \tag{6}$$

Like in [18], we use the Hopf Fibration to derive a desired orientation q_{des} for the rotation based of a, b, c and a desired body-frame yaw ψ.

$$q_{des} = \frac{1}{\sqrt{2(1+c)}} \begin{bmatrix} 1+c \\ -b \\ a \\ 0 \end{bmatrix} \otimes \begin{bmatrix} \cos\frac{\psi}{2} \\ 0 \\ 0 \\ \sin\frac{\psi}{2} \end{bmatrix} \tag{7}$$

Taking the derivative of Eq. 6 can be plugged into the angular velocity equations in [18] to find angular velocities for the attitude controller using the modification:

$$\text{sign}(s)\left(\frac{\xi^T\xi \cdot I - \xi\xi^T}{||\xi||^3} \cdot \dot{\xi}\right) = \begin{bmatrix} \dot{a} \\ \dot{b} \\ \dot{c} \end{bmatrix} \tag{8}$$

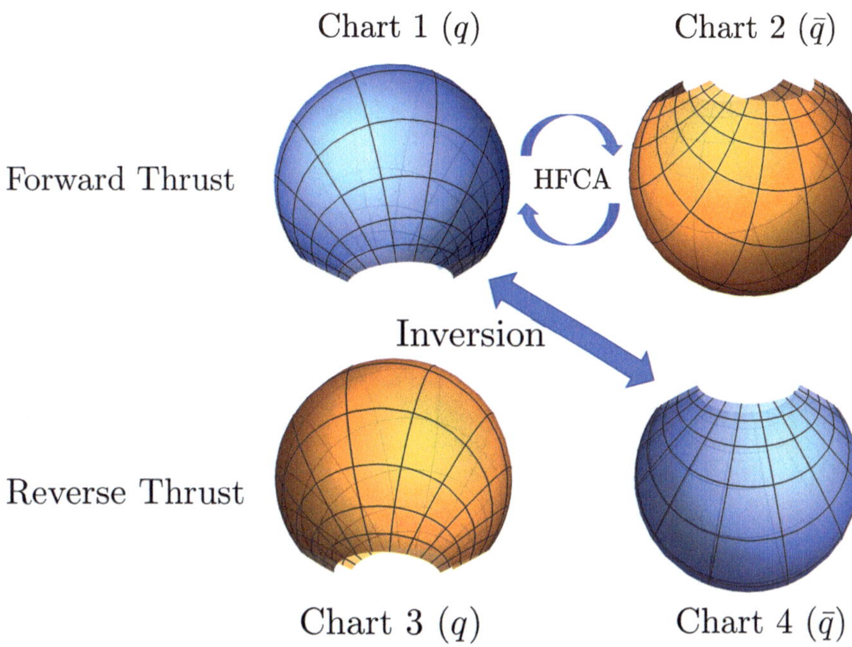

Fig. 3. With the addition of reversing the motor direction, we require two additional charts in addition to those explained in [18].

2.2 Extra Hopf Fibration Charts

Switching between a hover and inverted position requires an additional two charts not used in [18]. Since the body frame b_3 direction is downwards during hover with reversed motors, the blue charts in Fig. 3 both represent hover states. The second chart used in both hover and inverted mode is:

$$\bar{q}_{des} = \frac{1}{\sqrt{2(1-c)}} \begin{bmatrix} -b \\ 1-c \\ 0 \\ a \end{bmatrix} \otimes \begin{bmatrix} \cos\frac{\gamma}{2} \\ 0 \\ 0 \\ \sin\frac{\gamma}{2} \end{bmatrix} \tag{9}$$

During chart transition, we can equate the two charts to find a new yaw angle γ to make the rotation and angular velocities continuous.

2.3 Planning

As in [2,4,10,14,16], we can represent the trajectory of the robot as a 4-dimensional polynomial spline which represents the 3-D position of the robot and its yaw angle ψ. This 4-dimensional coordinates are called flat-coordinates which can be mapped to the full 6-dimensional state and control inputs for the quadrotor [14], given that the trajectory is continuous up to jerk.

We can express angular velocity constrains on ω_1 and ω_2 as in [18] with the property:

$$\omega_1^2 + \omega_2^2 = \dot{a}^2 + \dot{b}^2 + \dot{c}^2 \tag{10}$$

We then can optimize the trajectory using the nonlinear optimizer in [19] using a minimum jerk cost functional.

With a nominal trajectory in $\mathbb{R}^3$, which obeys the dynamic limits of the robot, we can seek to find a full trajectory in $SO(3)$ using the Hopf Fibration. Along the trajectory we calculate $\ddot{r} + g$ and check to see if it is in the upper half plane or the lower half plane. Assuming that the switch is instantaneous, this gives a set of charts for the Hopf Fibration controller. The non-instantaneous case would happen in a trajectory where the robot is accelerating sideways and free-falling in z. With using splines, this would only happen if two adjacent waypoints are both representing free fall and the optimal time and location put both of them, so the resultant trajectory is constant in acceleration. In normal operation this would never happen, so we can safely ignore this case. With respect to spline coefficients α this optimization becomes:

$$\begin{aligned} &\min f(\alpha) \\ &s.t.\ g(\alpha) < 0 \end{aligned} \tag{11}$$

To generate a trajectory which switches between normal and inverted states, we need to enforce that the thrust of the quadrotor $\ddot{r} + g$ is sideways during the trajectory at an intermediate waypoint halfway between the two charts. As shown in Fig. 4, this is a non-convex constraint to with the additional constraint that $\ddot{r} + g \neq 0$ to avoid the singularity in the geometric controller by a threshold ϵ. Instead, we enforce a conservative convex approximation of this constraint by linearizing the convex constraint. This implicitly enforces the planned trajectory to be in a restricted homotopy class of solutions.

From the set of charts, we can calculate a nominal trajectory ψ_n such that $\omega_3 = 0$. Note that ψ is discontinuous for all of the chart transitions. If we do

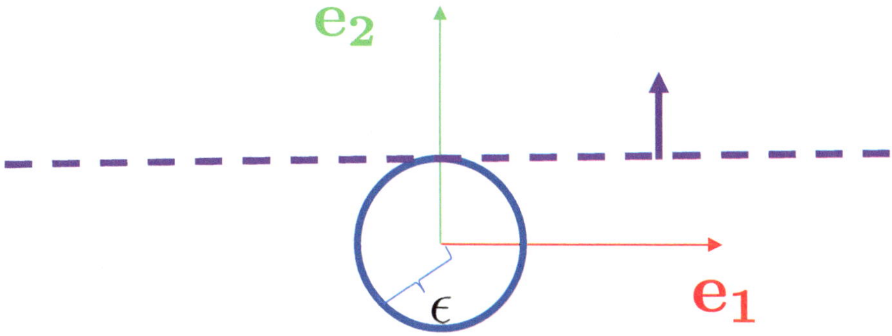

Fig. 4. Constraining the quadrotor to be horizontal requires a non-convex constraint to be outside of a ball ϵ, with $\ddot{r}_3 + 3 = 0$. We linearize this constraint by choosing a tangent line to this ball and constrain to the half-plane outside the ball.

not care about the yaw of the robot, this will be the optimal trajectory. However, if we want to incorporate the initial and final yaw of the robot, we need to account for the yaw. In addition, letting the yaw drift arbitrarily is counter-productive for any quadrotor system with onboard sensors, which encompasses most commercially available platforms.

The goal of the optimization algorithm is to optimize a deviation trajectory ψ_d such that the final trajectory $\psi = \psi_n + \psi_d$ obeys end conditions and minimizes angular velocity or acceleration. We match the cost function of this optimization to that of the original one, such that the same derivative of the physical system is being minimized. For example, if we minimize jerk on the first optimization, we minimize angular velocity on the second. For snap, we minimize the angular acceleration, and so on. To be able to work with coefficients instead of a continuous function with infinite dimensions, we first fit a spline to the nominal trajectory ψ_n. This can be done by using Chebyshev polynomial interpolation with one polynomial per yaw chart. This allows us to formulate the jerk version of the optimization problem as:

$$\begin{aligned} &\min \ ||\dot{\psi}_d||^2 \\ &s.t. \ (\psi_n + \psi_d)(0) = \psi_0 \\ &\qquad (\psi_n + \psi_d)(T) = \psi_f \end{aligned} \tag{12}$$

If we want to add explicit limits to $\dot{\psi}_d$ we can, but since the cost functional already penalizes this term, this is equivalent to using a barrier function to enforce constraints.

3 Experiments

3.1 Testbed

We test our algorithms on an off-the-shelf racing quadrotor platform shown in Figs. 1 and 5 hovering with forward and reverse thrust modes. The on-board flight controller is a Pixracer which is running a modified version of the PX4 stack [13] to support the reverse thrust and using both orientation and angular velocities simultaneously in the attitude controller [14].

The quadrotors position is controlled using a motion capture system running at 100 Hz. A base station computer handles the trajectory generation and user interface. These components are not computational intensive and could be run on-board the computer on a slightly larger platform [10]. The base station and the robot communicate over a standard Wifi network.

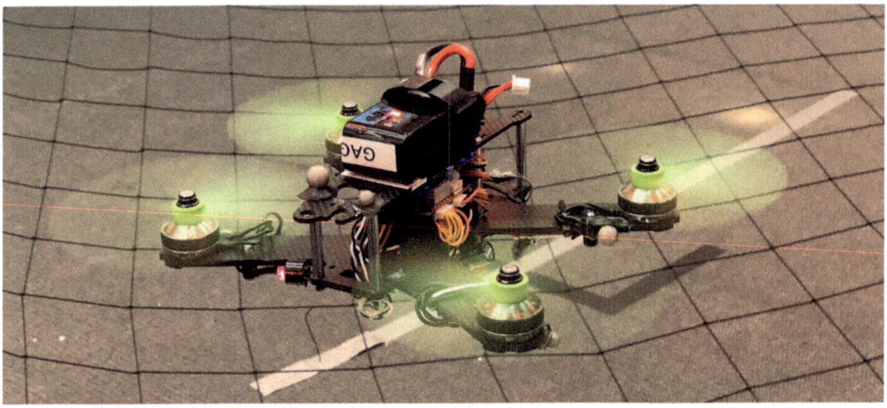

Fig. 5. Our quadrotor testbed hovering with forward thrust. The reverse thrust mode can be seen in Fig. 1

We used bi-directional propellers for our quadrotor which are symmetric when spun forward and backwards except with opposite thrust direction. Regular racing quadrotor propellers would not work for this application, and we found that they did not have enough thrust to lift the platform when reversed. In Fig. 6, we show our calibrated thrust-throttle data and linear fit. The throttle is a $PX4$ control input which represents a 0 to 1 range of minimum to maximum PWM sent to the motor. The linear fit is good enough for our control experiments, although a piecewise linear model with a cutoff at 5N would have lower residual.

3.2 Results

We ran a step input where we switched the chart from forward hover (1) to reverse hover (3). This results in an error in roll of π radians, which the quadrotor briefly looses altitude while recovering. Figure 7 shows a plot of the chart change during this flipping maneuver, as well as the angle error experienced during the flip.

From the plot, it is easy to see that the angle error goes up to π radians when the flip is initiated and converges to around 0 once the maneuver is completed. Figure 8 shows the path taken by the quadrotor while executing this maneuver.

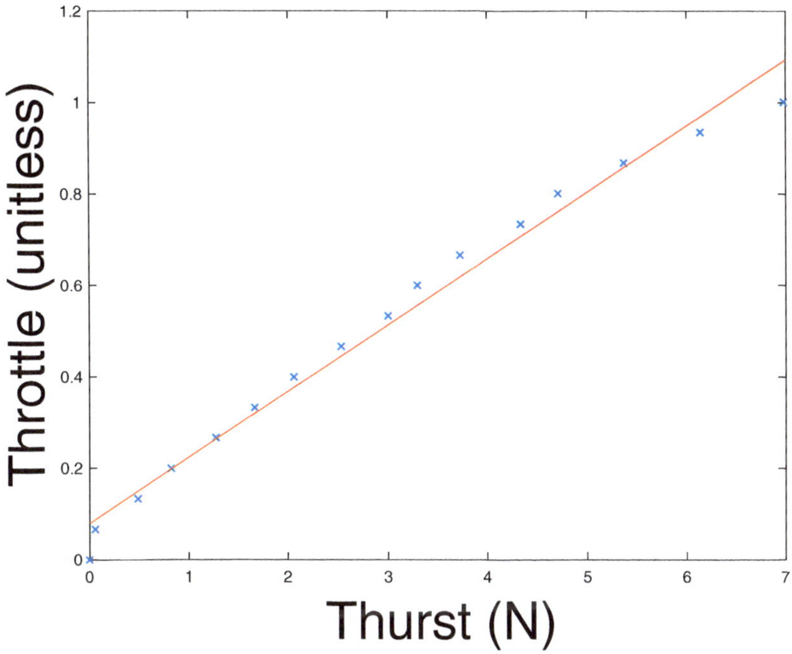

Fig. 6. We calibrated the throttle vs thrust curve for our platform using a thrust stand. Here we have chosen a linear fit model, which has a good fit for the data.

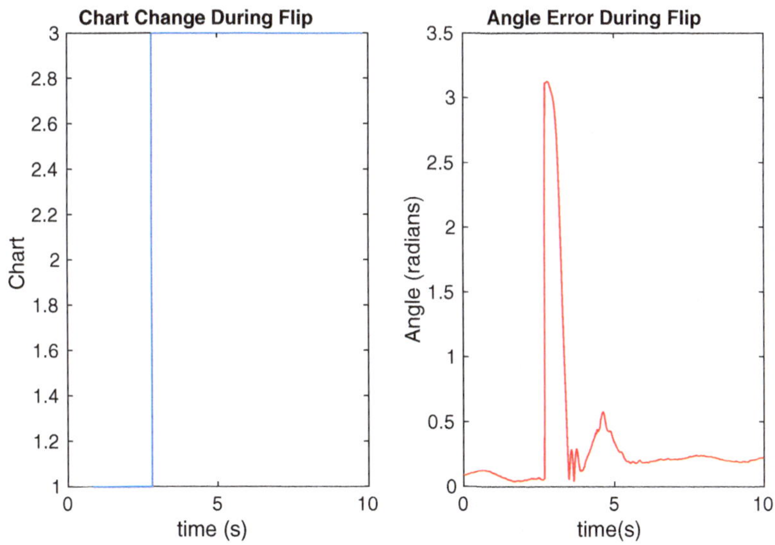

Fig. 7. Plot of the chart and the angle error during the flipping maneuver

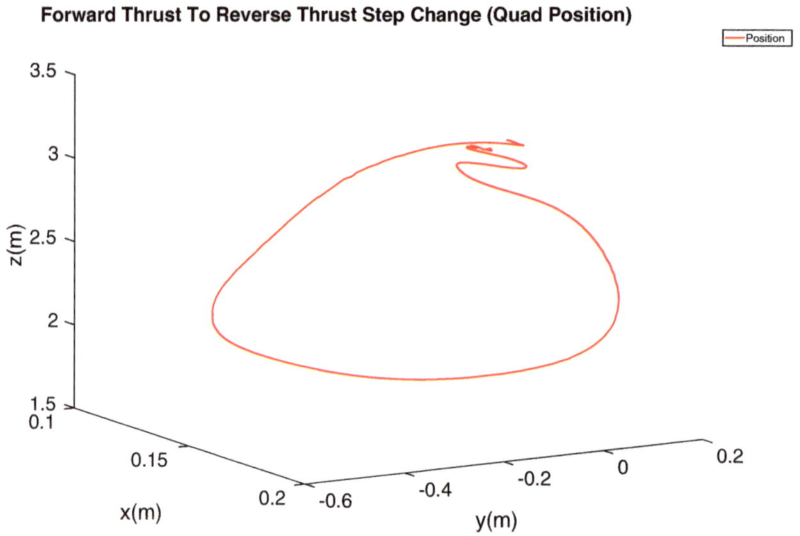

Fig. 8. Position of the quadrotor with 180 flip step input. The robot temporarily looses a meter in altitude, but is able to successfully recover from a large error.

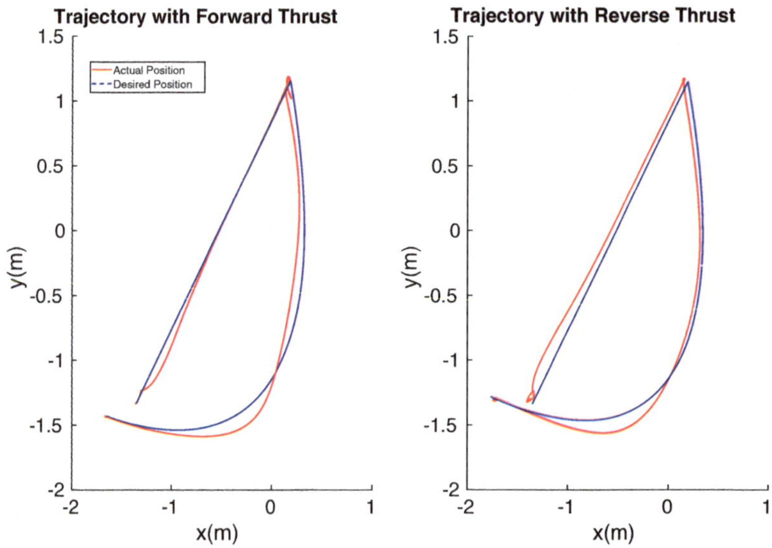

Fig. 9. Trajectory performance of our quadrotor executing trajectories entirely in the forward direction and entirely upside down.

As can be seen from the plot, the quadrotor drops in height quite a bit while executing this maneuver, and so to avoid any crashes, this maneuver should only be executed if there's enough distance between the quadrotor and the ground.

Figure 9 shows a plot of a simple 3-waypoint minimum jerk trajectory being executed once in upright flight, and once under inverted flight. It is interesting to note that these plots are not identical as we would expect them to be, which is most likely due to the quadrotor having some asymmetrical difference when flying upright and upside down.

In addition to the controller results shown in Sect. 3.2, we demonstrate the results of our trajectory optimization algorithm using multiple charts to invert the quadrotor. With the convexification of the constraint in Fig. 4, we make the robot do a backflip while translating. We set the location of the start and end waypoint and constrain the trajectory to be inside our volume. The resulting optimized trajectory implicitly goes upwards in order to achieve the correct acceleration to perform the flip.

4 Conclusion

We have presented a new geometric controller for quadrotors with bidirectional motors using multiple charts. Our controller is able to stabilize the robot in hover and inverted modes which we demonstrate with a hardware testbed platform. Finally we show how to optimize trajectories using an extension of prior work in trajectory optimization for quadrotors and propose experiments to demonstrate this.

References

1. Cutler, M., How, J.P.: Analysis and control of a variable-pitch quadrotor for agile flight. J. Dyn. Syst. Meas. Contr. **137**(10), 101002–101002-14 (2015)
2. Deits, R., Tedrake, R.: Efficient mixed-integer planning for UAVs in cluttered environments. In: 2015 IEEE International Conference on Robotics and Automation (ICRA), pp. 42–49 (2015)
3. El-Badawy, A., Bakr, M.: Quadcopter aggressive maneuvers along singular configurations: an energy-quaternion based approach. J. Control Sci. Eng. **2016**, 10 (2016)
4. Flores, M.E.: Real-time trajectory generation for constrained nonlinear dynamical systems using non-uniform rational B-spline basis functions. Ph.D. thesis (2007)
5. Foehn, P., Scaramuzza, D.: Onboard state dependent LQR for agile quadrotors. In: International Conference on Robotics and Automation. IEEE (2018)
6. Gillula, J.H., Huang, H., Vitus, M.P., Tomlin, C.J.: Design of guaranteed safe maneuvers using reachable sets: autonomous quadrotor aerobatics in theory and practice. In: ICRA. IEEE (2010)
7. Goodarzi, F., Lee, D., Lee, T.: Geometric nonlinear PID control of a quadrotor UAV on SE(3). In: 2013 European Control Conference (ECC), pp. 3845–3850, July 2013
8. Lee, T., Leok, M., McClamroch, N.H.: Nonlinear robust tracking control of a quadrotor UAV on SE(3). Asian J. Control **15**(2), 391–408 (2013)
9. Lee, T., Leoky, M., McClamroch, N.: Geometric tracking control of a quadrotor UAV on SE(3). In: CDC (2010)

10. Liu, S., Watterson, M., Mohta, K., Sun, K., Bhattacharya, S., Taylor, C.J., Kumar, V.: Planning dynamically feasible trajectories for quadrotors using safe flight corridors in 3-D complex environments. IEEE Rob. Autom. Lett. **2**(3), 1688–1695 (2017)
11. Loianno, G., Brunner, C., McGrath, G., Kumar, V.: Estimation, control, and planning for aggressive flight with a small quadrotor with a single camera and IMU. IEEE Rob. Autom. Lett. **2**(2), 404–411 (2017)
12. Lupashin, S., D'Andrea, R.: Adaptive fast open-loop maneuvers for quadrocopters. Auton. Rob. **33**, 89–102 (2012)
13. Meier, L., Honegger, D., Pollefeys, M.: PX4: a node-based multithreaded open source robotics framework for deeply embedded platforms. In: International Conference on Robotics and Automation. IEEE (2015)
14. Mellinger, D., Kumar, V.: Minimum snap trajectory generation and control for quadrotors. In: International Conference on Robotics and Automation. IEEE (2011)
15. Ng, A.Y., Coates, A., Diel, M., Ganapathi, V., Schulte, J., Tse, B., Berger, E., Liang, E.: Autonomous inverted helicopter flight via reinforcement learning. In: Ang, M.H., Khatib, O. (eds.) Experimental Robotics IX, pp. 363–372. Springer, Heidelberg (2006)
16. Preiss, J.A., Hausman, K., Sukhatme, G.S., Weiss, S.: Trajectory optimization for self-calibration and navigation. In: Robotics: Science and Systems (RSS) (2017)
17. Thomas, J., Loianno, G., Pope, M., Hawkes, E.W., Estrada, M.A., Jiang, H., Cutkosky, M.R., Kumar, V.: Planning and control of aggressive maneuvers for perching on inclined and vertical surfaces. In: ASME, August 2015
18. Watterson, M., Kumar, V.: Control of quadrotors using the hopf fibration on SO(3). In: International Symposium on Robotics Research (ISRR) (2017)
19. Watterson, M., Liu, S., Sun, K., Smith, T., Kumar, V.: Trajectory optimization on manifolds with applications to SO(3) and R3xS2. In: Robotics Science and Systems (RSS) (2018)

Precision UAV Landing in Unstructured Environments

Kevin Pluckter(✉) and Sebastian Scherer

Robotics Institute, Carnegie Mellon University, Pittsburgh, PA, USA
kpluckte@cs.cmu.edu, basti@cmu.edu

Abstract. Autonomous landing of a drone is a necessary part of autonomous flight. One way to have high certainty of safety in landing is to return to the same location the drone took-off from. Implementations of return-to-home functionality fall short when relying solely on GPS or odometry as inaccuracies in the measurements and drift in the state estimate guides the drone to a position with a large offset from the initial position. This can be particularly dangerous if the drone took-off next to something like a body of water. Current work on precision landing relies on localizing to a known landing pattern, which requires the pilot to carry a landing pattern with them. We propose a method using a downward facing fisheye lens camera to accurately land a UAV from where it took off on an unstructured surface, without a landing pattern. Specifically, this approach uses a position estimate relative to the take-off path of the drone to guide the drone back. With the large Field-of-View provided by the fisheye lens, our algorithm can provide visual feedback starting with a large position error at the beginning of the landing, until 25 cm above the ground at the end of the landing. This algorithm empirically shows it can correct the drift error in the state estimation and land with an accuracy of 40 cm.

Keywords: Precision landing · Fisheye lens · UAV

1 Introduction

Landing a quadrotor autonomously is essential for the ubiquitous presence of drones. Majority of UAVs currently rely on GPS based state-estimation for landing, but drift builds up over time and these systems are unreliable if in a GPS denied environment. Given this, a drone cannot reliably land in the same spot and could potentially attempt to land in an unforgiving location, such as a nearby tree. Our proposed solutions prevents these issues by re-traversing the proven take-off path back to its starting position. We are able to precisely land the drone with this method via a monocular fisheye camera. Majority of current

Electronic supplementary material The online version of this chapter (https://doi.org/10.1007/978-3-030-33950-0_16) contains supplementary material, which is available to authorized users.

J. Xiao et al. (Eds.): ISER 2018, SPAR 11, pp. 177–187, 2020.
https://doi.org/10.1007/978-3-030-33950-0_16

work on precision landing focuses on landing in a known structured environment (helipad or runway) [11]. Given a known landing pattern, these algorithms are able to identify the target and use its geometry to robustly estimate the relative state of the drone. Much work has gone into different landing pattern designs to enable accurate pose estimation [16,18,19]. While these methods result in accurate landings, they require the drone to land on a specific type of landing pattern. This would require drone operators to carry a landing pattern with them in order to ensure the safe return and landing of their drone. Our work focuses on a new area, landing at the UAVs starting position in an unstructured and unknown environment. This enables drones to be autonomously deployed in the field and return to their starting position while following the take-off path in reverse.

In order to accurately land a quadrotor in an unstructured environment, without prior knowledge of the take-off location several problems must be addressed: the drone must be able to localize relative to where it has taken off from and be able to guide itself to ensure a safe landing. This work addresses these problems with a method inspired by [6]. During the take-off a set of images are recorded, and during landing, the drone localizes to these images and descends along a similar path back to its initial position, as seen in Fig. 1. The approach improves the safety of landing twofold, by landing in the same starting position, there is a high likelihood of the location still being safe, and by taking a path similar to the one from take-off enables the drone to avoid obstacles that were avoided during take-off. To the best of our knowledge, we present the first algorithm for precision landing of a drone in unstructured environments, such as those in Fig. 2, with an average accuracy of 40 cm. In summary, the contributions of this paper are:

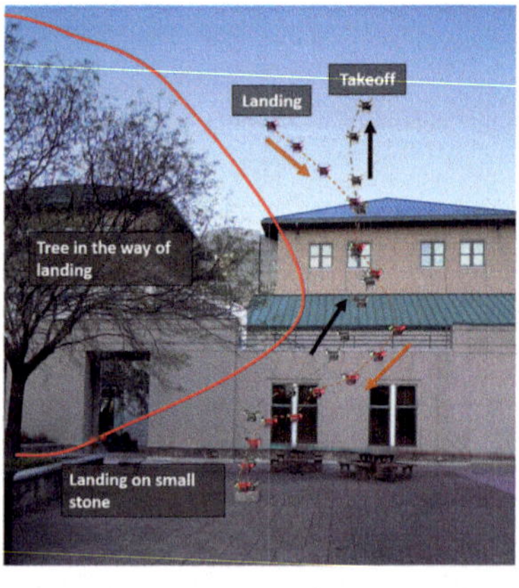

Fig. 1. Precision Landing Demonstration: the take-off path (black), and the landing path (orange). The key challenges are for the drone to safely and precisely land. This equates to the drone avoiding the tree that was avoided during take-off, and landing on the stone the drone started on. Our algorithm successfully completes both tasks.

- An algorithm for safe precision landing of quadrotors in unstructured environments.
- Experimental results on precision landing in various environments with baseline comparisons.

2 Related Work

This section will go over work done in the area of precision landing and adjacent fields that have influenced the proposed method.

Precision Landing
Within the area of precision landing for vertical takeoff and landing (VTOL) vehicles, much work has focused on using helipad design. Specifically, helipads with concentric circles and with a H in the center are common along with other custom designed helipads [11]. Approaches using helipads with either an H or T on them use pretrained neural networks to identify the letter and then using the known geometry of either the letter or the surrounding circle to estimate 6DOF pose relative to the landing pattern [16,18,19]. A shortcoming of these methods is losing information on the landing pattern during the approach. Based on the size of the landing pattern and the field-of-view (FOV) of the camera, the quad-rotor will no longer be able to see the landing pattern and be unable to estimate it's pose. Approaches using more custom landing patterns use unique circular patterning with varying size to ensure good recognition and pose estimation close up and far away [2,3,12]. Additionally, these methods rely solely on circle detection and identifying the circles that form the correct pattern, neural networks are not used. Another approach to tackle the problem of FOV is by using two different cameras, one for approaching from a distance and another once close up [1]. As mentioned before though, these methods require there to be a landing pattern of known size to be able to successfully perform an accurate landing. Without a landing pattern, other techniques and assumptions must be applied to recognize the area and land.

Fig. 2. Landing locations: depicted are two of the landing locations used during testing.

Visual Teach and Repeat

One particular method that has been used in similar scenarios to relatively localize and traverse long distances using a monocular camera is Visual Teach and Repeat (VTR). Primarily focusing on ground robots, this technique will "teach" a path to a robot via piloted traversal of the path, and then be able to robustly and accurately traverse the path using a monocular camera and other base level sensors for odometry [6,7]. In order to estimate a 6DOF pose estimate, the algorithm must first recognize the image from the teach pass that it is nearest to and then estimate the pose. The pose estimation is enabled with a monocular camera with the assumption of local ground planarity. After the robot re-localizes itself, it relies upon visual odometry to take it to the next keyframe in the path.

Recent work has begun to expand this concept to quad-rotors [13,14]. The first of these papers shows a proof of concept for VT&R with a drone equipped with a downward facing camera and laser range finder. This approach builds a local 3D map is during the teach pass that the drone then localizes to during the repeat pass. The approach shows promising results, but did not look into how the technique would be affected by varying altitudes. The second paper proposes a more fully developed VTR for drones, but uses a forward facing camera and a qualitative position estimation. From these various approaches to VTR, we have developed the proposed method for precision landing, using a similar architecture, but a different method for generation of motion command and transitioning between keyframes.

Visual Servoing

The premise of visual servoing is to to directly control the robot using vision [4,5]. One type of visual servoing is Position Based Visual Servoing (PBVS), where features are extracted in an image from an object of known size to command an input towards a goal pose. PBVS is similar in nature to the methods used landing patterns where a model is known. In the proposed work, there is no model of the position from which the drone took off, but a planarity assumption for the ground from which the quad-rotor took off can be made. From this assumption, we can estimate pose from one camera frame to another if the height at each position is known. With scale abmiguity is solved, a method similar to PBVS can be used for the controller input.

3 Approach

The approach is inspired by Visual Teach and Repeat [6], where the take-off is the teach pass and the landing is the repeat pass. In this section, we will discuss the details of this method: the teach pass (take-off), the repeat pass (landing), position estimation, and control will be explained. The entire pipeline can be seen in Fig. 3, which runs at 15 Hz onboard on an NVIDIA Jetson TX2.

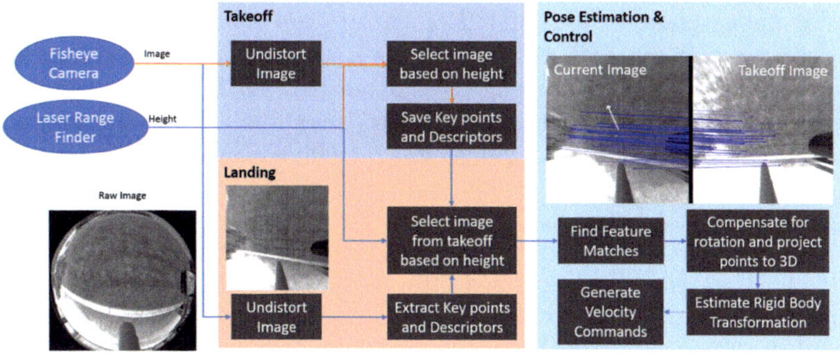

Fig. 3. Pipeline of algorithm with steps during take-off having orange arrows and steps during landing having blue arrows. Additionally, a sample of a raw image, undistorted image, and matched images are shown. In the matched image sample above the Pose Estimation and Control section shows how matches are found on the image and a resultant white arrow displays the motion required to minimize the position error.

Take-off & Landing

During take-off, a set of images is recorded in regular intervals with a fisheye lens camera. The wide FOV of the fisheye provides key information at the start of the take-off when the camera is very close to the ground and good information at higher altitudes. Images are recorded at a higher rate while the drone is close to the ground and at a slower rate as the drone ascends higher. Details on the image recording rate and other parameters used for the method can be found here. Once all of the images are recorded, key points are found and ORB descriptors are extracted and saved to be used during the landing phase. This process only takes seconds and allows for faster run-time during landing by not performing repetitive calculations.

For landing, the drone comes back to the final position from take-off at 8 m height. Then the current image is compared with the image taken at an altitude just below the current altitude of the drone. A relative position estimate to the image from take-off is estimated in the form of a 2-D rigid body transform. This estimate is then used to command the drone towards that position. While the drone approaches the position of the image from take-off, it constantly descends. Once the drone goes below the height of the current take-off image being compared to, the next closest image is used. This process commands the drone along a similar path from take-off back to it's original position on the ground.

To initialize the landing process, the drone uses its state-estimate to fly above where it started. Once the drone is in position it needs to be able to localize. In order to accomplish this task we used a downward facing fisheye camera and a set of images collected during take-off. A 180° FOV fisheye lens was selected as it provides much more information about the ground below the drone than a pinhole lens. This is particularly beneficial at the start and end of the landing, allowing for a larger offset at the start, 8m above ground, and for more reliable motion commands until it is 0.25 m above the ground. The images are projected to a flat image plane [9,10] and ORB features are extracted from the drone's current image and matched the descriptors from the corresponding image from take-off in order to localize during the descent [15]. ORB features were selected over learned neural network features and features made for spherical images as literature shows that they are much slower [17,20], which would hinder the speed of the control loop.

Once features are matched, the pose is estimated relative to an image from take-off. First, the feature points' position is corrected using the estimated roll and pitch. Then they are projected into 3-D space using a planar assumption, the measured height, and intrinsic camera parameters. These two steps can be accomplished by rearranging the rotational flow correction equation from [8] and use it for correcting the rotation of an individual point and project it into 3D space:

$$\mathbf{P}_z = \mathbf{Z}\cos(\theta_x)\cos(\theta_y) \tag{1}$$

$$\mathbf{P}_x = \frac{(-p_x - \sin(\theta_y)f)\mathbf{P}_z}{f} \tag{2}$$

$$\mathbf{P}_y = \frac{(-p_y + sin(\theta_x)f)\mathbf{P}_z}{f} \tag{3}$$

Where $\mathbf{Z}$ is the measured height from the laser range finder, θ is the current estimated roll and pitch of the drone, $\mathbf{P}$ is a feature point's metric coordinates in 3D with respect to the drone's current position, p is the feature point's position in the camera frame, and f is the estimated focal length of the image. Once both sets of points are projected onto the ground plane, a 2D rigid body transformation can be calculated and give a metric relative position error. Random sample consensus (RANSAC) is used to reject outliers while finding the rigid body transform. If there is a low consensus during RANSAC, the drone ascends to increase its FOV until it can again match with the image from take-off. This pipeline can be seen in the Pose Estimation & Control section in Fig. 3.

Once the position error is estimated, it is used as an input to a PI position controller using the drone's onboard state-estimate. The drone receives velocity commands from the position controller during the descent at 15 Hz until it is 0.25 m above the ground when it completes the landing. At 1.5 m the drone aligns with the concurring image from take-off before finally descending.

4 Experiments

In this section, we discuss the experiments and results. We perform a drone landing experiment to demonstrate the performance and robustness of our precision landing algorithm, as well as present a baseline comparison to landing using only state estimate. Additionally, data is collected and presented for comparison of the efficacy of both fisheye and pinhole camera lens.

A second experiment was performed to test the algorithms robustness to scene change. Landing tests were performed with a van next to the drone for takeoff, and before landing, the van was moved to a set further distance as shown

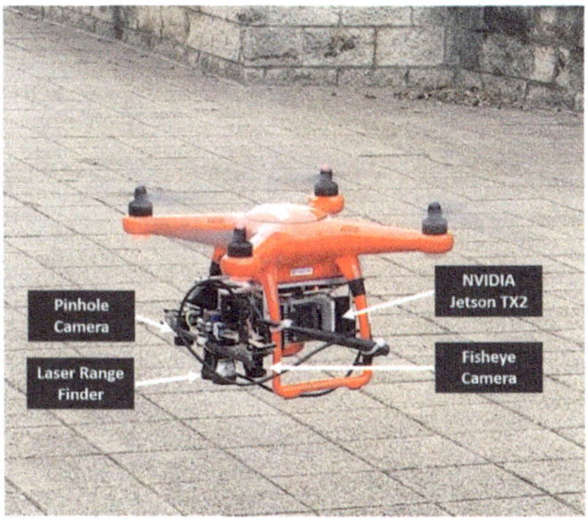

Fig. 4. Experimental Setup: experiments were conducted with the X-Star Premium Quadcopter by Autel Robotics. A NVIDA Jetson TX2, ueye camera equipped with fisheye camera lens, and SF30 laser range finder are equipped. The Jetson TX2 is connected to the X-Star Premium's embedded computing device. A second ueye camera is attached with a pinhole lens for offline comparison.

in Fig. 6. The setup for our experiments can be seen in Fig. 4 and all the data collected can be found here.

Landing Experiments and Comparisons

The experimental procedure is as follows. First, an environment is selected and the drone is placed in its initial position. For measurement purposes, the drone is placed next to either a natural landmark, such as a corner of a tile, or an artificial landmark, such as a golf tee in the grass. Second, the drone is manually piloted with varying paths taken (i.e. straight, diagonally, curved, etc.). Once the takeoff is completed, it is flown in a random direction in varying amounts to simulate drift from the drone's state-estimate. Finally, the landing procedure is initiated and the drone autonomously lands. After the drone has landed, measurements are taken for the drone's displacement from the initial position. These will be compared with results using a GPS & Intertial Measurement Unit (IMU) based Extended Kalman Filter (EKF) state-estimation. This test has been performed in a wide range of environments, grass, turf, and stone tiling. Wind speeds are recorded from METAR report from the Allegheny County Airport (KAGC) to observe the algorithms robustness to high wind speeds and gusty weather. The performance for the proposed method and the GPS-IMU based EKF can be seen in Table 1.

Data is recorded from both the fisheye and pinhole cameras and analyzed offline to determine the reliability of position estimation. The algorithm is ran on the data and if RANSAC cannot determine a consensus on a rigid body transform, the image instance is considered a failed localization. The percentage of failed localizations across multiple trials can be seen in Fig. 5. This shows that the fisheye lens can localize well even when close to the ground and has a lower failure rate throughout the landing than the pinhole.

Table 1. Precision-landing and state-estimate landing results

Method	Environment	Number of Trials	Accuracy (cm)	Standard Deviation (cm)	Wind Speed (m/s)	Max Gust (m/s)
Precision Landing	Patio	18	38	23	3	5
Precision Landing	Grass 1	17	44	42	3	5
Precision Landing	Turf	10	34	17	7	12
Precision Landing	Grass 2	10	45	24	8	15
State Estimate	Patio	10	183	74	2	0
State Estimate	Grass 1	10	359	205	2	0

Scene Change Experiments

The experimental procedures for the scene change experiments are the same as the landing with the addition that a van moves from a set initial position to a new set position further away. Three sets of five experiments were ran with each set having a different starting and ending offset of the van from the drone's initial position. The results can be seen in Fig. 7.

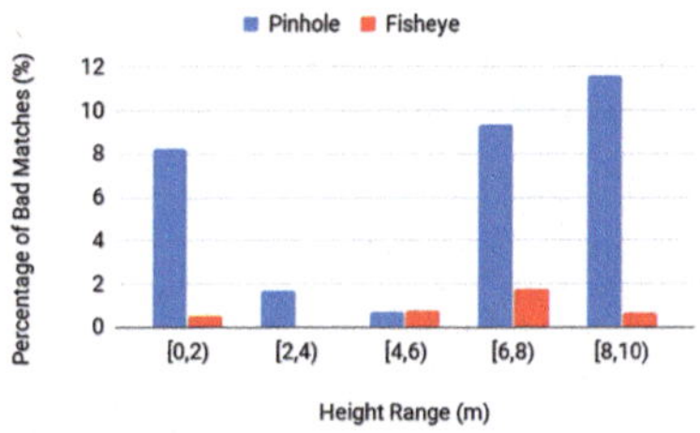

Fig. 5. Comparison of Fisheye and Pinhole Camera Lens for the task of localization. Each bar displays the percentage of bad position estimations during landing across various trials in the different environments. The fisheye camera has fewer bad matches than pinhole camera.

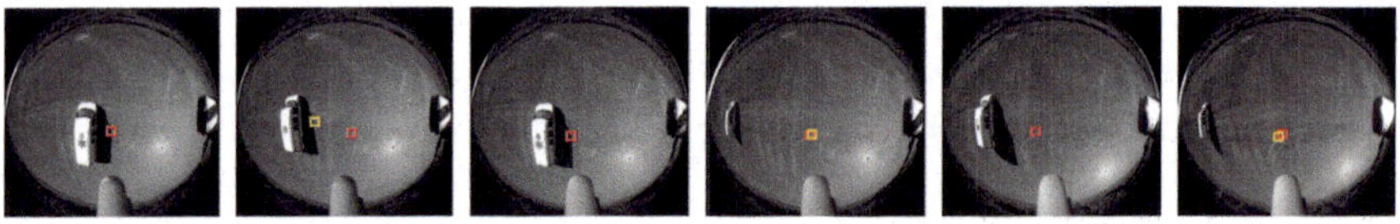

Fig. 6. Displays the three scene change experiments. From left to right is each experiment: Short-Med, Short-Long, and Med-Long, with the take-off location in red and landing in yellow.

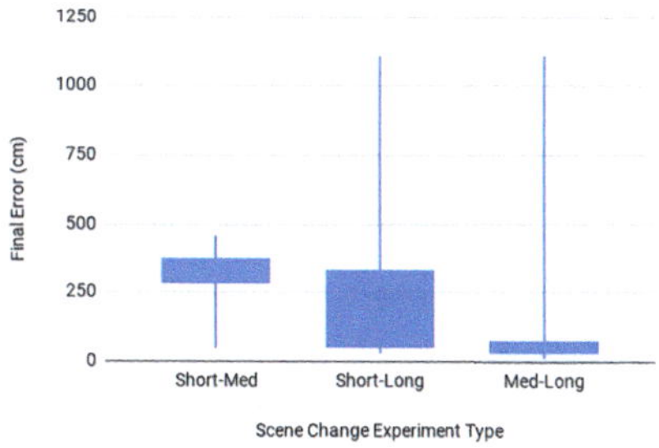

Fig. 7. Results from scene change experiments. Each box-and-whisker plot shows the resulting accuracy across the three scene change tests. Short-Med means the van moved from 4 ft from the drone to 16 ft from the starting point. For Short-Long, the van moves from 4 ft away to 41 ft. For Med-Long the van moves from 16 ft to 41 ft.

5 Conclusions and Future Work

The results from the experiments show that our algorithm can safely and accurately land a quadrotor to its original takeoff position in various environments, and that it outperforms the IMU-GPS EKF state-estimate by a statistically significant margin. Additionally, our algorithm proves to be robust to wind as the results do not vary with the wind conditions. There have been two observed failure case, if the ground plane has a sudden large change in height, in which case the planar assumption is no longer valid and an accurate rigid body transform cannot be calculated, and when the scene has a large amount of change.

From the fisheye and pinhole camera lens comparisons, we conclude that the fisheye lens, although having high radial distortion, provides more information for the task of autonomous precision landing. This is particularly useful in two scenarios. Firstly, when a drone initially starts its descent. This can be seen in the results when there is a large offset from the end of the takeoff to the beginning of the landing. Overall, this results in a more robust landing capability as the drone relies less on the state-estimate. Secondly, when the drone is closer to the ground and there are higher levels of wind. In essence, due to the fisheye lens' greater FOV, even if pushed off-course by wind, it maintains visual of the image from takeoff and can recover.

The scene change experiment demonstrated the effects of a change in the environment for the landing algorithm with three different scenarios. The first where a large object that provides many of the features in the images moves to a new position relatively close to where it started. This causes the landing algorithm to be drawn off course and land close to the area with the highest number of the original features. This short coming gives interesting insight on potential other uses for the proposed method. One, landing on a moving landing pattern. If the landing pattern is highly textured, even if the landing pad were moving in the environment, the drone would be drawn towards it. Second, if a highly textured landing pattern was used, the drone could change where is lands relatively easily. The other two scenarios where the van moved far from its initial position resulted in accurate landings. Each of these scenarios had one edge case of the drone following the van, likely because the van had not moved far enough away. From these two results it can be seen that if strong features are removed, the drone can still land accurately.

There are several interesting areas to be explored for future work. One topic to further explore is landing on a moving landing site with a highly textured surface. Another area for future work would be the final control piece for the landing when the camera can no longer match features.

Acknowledgements. This work was supported by Autel Robotics under award number A018532.

References

1. Arora, S., Jain, S., Scherer, S., Nuske, S., Chamberlain, L., Singh, S.: Infrastructure-free shipdeck tracking for autonomous landing. In: 2013 IEEE International Conference on Robotics and Automation (ICRA), pp. 323–330. IEEE (2013)
2. Bai, W., Pan, F., Xing, B.Y., Pan, C., Pei, M.X.: Visual landing system of UAV based on ADRC. In: 2017 29th Chinese Control And Decision Conference (CCDC), pp. 7509–7514. IEEE (2017)
3. Benini, A., Rutherford, M.J., Valavanis, K.P.: Real-time, GPU-based pose estimation of a UAV for autonomous takeoff and landing. In: 2016 IEEE International Conference on Robotics and Automation (ICRA), pp. 3463–3470. IEEE (2016)
4. Chaumette, F., Hutchinson, S.: Visual servo control. i. basic approaches. IEEE Robot. Autom. Mag. **13**(4), 82–90 (2006)
5. Chaumette, F., Hutchinson, S.: Visual servo control ii advanced approaches [tutorial]. IEEE Robot. Autom. Mag. **14**(1), 109–118 (2007)
6. Clement, L., Kelly, J., Barfoot, T.D.: Monocular visual teach and repeat aided by local ground planarity. In: Field and Service Robotics, pp. 547–561. Springer, Cham (2016)
7. Clement, L., Kelly, J., Barfoot, T.D.: Robust monocular visual teach and repeat aided by local ground planarity and color-constant imagery. J. Field Robot. **34**(1), 74–97 (2017)
8. Honegger, D., Meier, L., Tanskanen, P., Pollefeys, M.: An open source and open hardware embedded metric optical flow CMOS camera for indoor and outdoor applications. In: 2013 IEEE International Conference on Robotics and Automation (ICRA), pp. 1736–1741. IEEE (2013)
9. Kannala, J., Brandt, S.S.: A generic camera model and calibration method for conventional, wide-angle, and fish-eye lenses. IEEE Trans. Pattern Anal. Mach. Intell. **28**(8), 1335–1340 (2006)
10. Kannala, J., Heikkilä, J., Brandt, S.S.: Geometric camera calibration. Wiley Encyclopedia of Computer Science and Engineering (2008)
11. Kong, W., Zhou, D., Zhang, D., Zhang, J.: Vision-based autonomous landing system for unmanned aerial vehicle: a survey. In: 2014 International Conference on Multisensor Fusion and Information Integration for Intelligent Systems (MFI), pp. 1–8. IEEE (2014)

12. Merz, T., Duranti, S., Conte, G.: Autonomous landing of an unmanned helicopter based on vision and inertial sensing. In: Experimental Robotics IX, pp. 343–352. Springer, Heidelberg (2006)
13. Nguyen, T., Mann, G.K., Gosine, R.G., Vardy, A.: Appearance-based visual-teach-and-repeat navigation technique for micro aerial vehicle. J. Intell. Robot. Syst. **84**(1–4), 217–240 (2016)
14. Pfrunder, A., Schoellig, A.P., Barfoot, T.D.: A proof-of-concept demonstration of visual teach and repeat on a quadrocopter using an altitude sensor and a monocular camera. In: 2014 Canadian Conference on Computer and Robot Vision (CRV), pp. 238–245. IEEE (2014)
15. Rublee, E., Rabaud, V., Konolige, K., Bradski, G.: ORB: an efficient alternative to sift or surf. In: 2011 IEEE international conference on Computer Vision (ICCV), pp. 2564–2571. IEEE (2011)
16. Saripalli, S., Montgomery, J.F., Sukhatme, G.S.: Visually guided landing of an unmanned aerial vehicle. IEEE Trans. Robot. Autom. **19**(3), 371–380 (2003)
17. Simo-Serra, E., Trulls, E., Ferraz, L., Kokkinos, I., Fua, P., Moreno-Noguer, F.: Discriminative learning of deep convolutional feature point descriptors. In: Proceedings of the IEEE International Conference on Computer Vision, pp. 118–126 (2015)
18. Guili, X., Zhang, Y., Ji, S., Cheng, Y., Tian, Y.: Research on computer vision-based for uav autonomous landing on a ship. Pattern Recogn. Lett. **30**(6), 600–605 (2009)
19. Yang, S., Scherer, S.A., Zell, A.: An onboard monocular vision system for autonomous takeoff, hovering and landing of a micro aerial vehicle. J. Intell. Robot. Syst. **69**(1–4), 499–515 (2013)
20. Zhao, Q., Feng, W., Wan, L., Zhang, J.: SPHORB: A fast and robust binary feature on the sphere. Int. J. Comput. Vision **113**(2), 143–159 (2015). https://doi.org/10.1007/s11263-014-0787-4

Micro Aerial Vehicles

Micro Aerial Vehicles

Greg Dudek

This session featured several papers dealing with data collection and measurement from a distance, in particular using micro-air vehicles, but also from other platforms. In the first paper, the authors exploited the non-contact nature of an aerial vehicle, using on board state estimation, to develop a prototype that might be suitable for deployment to inspect a nuclear facility. In the second paper, the authors considered methods that exploited the versatile mobility of a flying vehicle to inspect all parts of a structure where conventional vehicle would have difficulty. A pair of LIDAR sensors, along with a conventional camera, were used to autonomously plan a trajectory achieved inspection and coverage of a large bridge. In the third paper, the authors considered how to minimize the travel time while collecting data over a terrain of interest, an example of hierarchical reinforcement learning. While that work was deployed in a marine environment, the approach was also suitable for use from the air. In the fourth paper, the authors considered a very unusual sensing paradigm to perform localization, and demonstrated some promising experimental results. Finally, in the last paper in this section, the authors addressed the challenging problem of night-time landing of autonomous air vehicles. The approach using thermal infrared (FLIR) sensing and was validated and demonstrated to be effective operating in the Mars simulation environment at the Jet Propulsion Laboratory. Overall the ensemble of papers have a good sense of the versatility of aerial mapping, navigation, inspection and localization in challenging scenarios. The authors also touched on the significant logistic and validation challenges posed by this kind of work.

Nuclear Environments Inspection with Micro Aerial Vehicles: Algorithms and Experiments

Dinesh Thakur[1(✉)], Giuseppe Loianno[2], Wenxin Liu[1], and Vijay Kumar[1]

[1] GRASP Lab, University of Pennsylvania, 3330 Walnut Street, Philadelphia 19104, USA
{tdinesh,wenxinl,kumar}@seas.upenn.edu
[2] Tandon School of Engineering, New York University, 6 MetroTech Center, Brooklyn, NY 11201, USA
loiannog@nyu.edu

Abstract. In this work, we address the estimation, planning, control and mapping problems to allow a small quadrotor to autonomously inspect the interior of hazardous damaged nuclear sites. These algorithms run onboard on a computationally limited CPU. We investigate the effect of varying illumination on the system performance. To the best of our knowledge, this is the first fully autonomous system of this size and scale applied to inspect the interior of a full scale mock-up of a Primary Containment Vessel (PCV). The proposed solution opens up new ways to inspect nuclear reactors and to support nuclear decommissioning, which is well known to be a dangerous, long and tedious process. Experimental results with varying illumination conditions show the ability to navigate a full scale mock-up PCV pedestal and create a map of the environment, while concurrently avoiding obstacles.

Keywords: Inspection · Aerial robotics

1 Introduction

Nuclear site decommissioning is a complex and tedious process, which includes clean-up of radioactive materials and progressive dismantling of the site as shown in Fig. 1. Because of the finite life of a reactor, the decommissioning is an essential step in a nuclear power plant. This process is also critical if there is an accident, as was the case in the Fukushima Daiichi Nuclear Power Plant in 2011 [2]. For planning the decontamination and decommission, surveys of the inside of the

D. Thakur, G. Loianno—These authors contributed equally.

Electronic supplementary material The online version of this chapter (https://doi.org/10.1007/978-3-030-33950-0_17) contains supplementary material, which is available to authorized users.

J. Xiao et al. (Eds.): ISER 2018, SPAR 11, pp. 191–200, 2020.
https://doi.org/10.1007/978-3-030-33950-0_17

containment vessel are crucial. Figure 2 shows a simplified cross-section of a multi-story containment for a boiling water reactor (BWR). The primary goal of this work is to develop a fully autonomous system that is capable of inspecting inside damaged sites such as ones seen in Fig. 1. We consider an example problem scenario of visually inspecting inside the Primary Containment Vessel (PCV) of a damaged nuclear power plant unit.

Fig. 1. Left - the exterior of one of the damaged units at Fukushima Daiichi [1], Right - processed image of the pedestal area inside of Unit 2 collected after the meltdown [1].

Micro Aerial Vehicles (MAVs), equipped with on-board sensors, are ideal platforms for autonomous inspection of cluttered, confined, and hazardous environments. The use of aerial platforms in nuclear settings poses several challenges such as the need to navigate in damaged/unknown environments, without GPS, under low illumination, without communication link to human-pilot or user interface, etc. Furthermore, a damaged nuclear site generally poses constraints on access (entry hatches can be less than 0.1–0.3 m in diameter) and the operating conditions can be adversely affected by condensation and fog.

A survey of various autonomous systems deployed in inspection of damaged areas is provided in [3,4]. A combination of ground and air vehicle is used in [5] to inspect a building after an earthquake. Autonomous inspection of penstocks and tunnels using aerial vehicles is done in [6].

Focusing on nuclear environment, wall climbing robots have been popular for inspecting undamaged steam generators at nuclear sites [7]. The authors in [8] provide an overview of aerial robots for radiation mapping, while [9] develops a drone equipped with sensors for nuclear risk characterization. Inspection of outdoor environments is primarily done with the aid of GPS. In the case of GPS-denied environments, remote teleoperation is used [10]. In [11] a robot is used in nuclear sites for damage assessment. Unmanned construction equipment and robots were used for surveillance work and cleaning up rubble outside buildings [12]. Finally, [13] provides various robots that have been deployed in nuclear settings.

All the previous solutions were remotely teleoperated without considering the autonomy component required by the complex and remote operations in a

Fig. 2. Cross-sectional view of a BWR containment (left) [14]. Mock-up of the PCV pedestal and control rod drive (CRD) ramp (right).

nuclear (eventually damaged) scenario. Moreover, the physical size of the robots range from 1 m to 10 m. These systems are big with weights usually ranging from 2 kg to more than 1000 kg and can carry bulky sensors. Our problem of navigating inside a PCV is very different. Previous attempts to visually inspect the reactor Unit 1 with ground robots have been unable to complete the missions as they got stuck in confined environments of the damaged sites.

This work presents multiple contributions. First, we develop a fully autonomous system with state estimation, control and mapping modules for PCV inspection task concurrently running onboard a custom designed 0.16 m platform. Second, using the onboard map created during the navigation, the vehicle is able to automatically replan its path to avoid obstacles. Finally, we conduct a set of preliminary studies and experiments in different conditions in a mock-up representing a PCV pedestal at Southwest Research Institute (SwRI). To the best of our knowledge, this is the first fully autonomous system of this size and scale applied to inspect the interior of a full scale mockup PCV for supporting nuclear decommissioning. Although we motivate by specific example of inspecting inside a PCV where the entry point is of order of 0.1 to 0.2 m, this system can be used in autonomous inspection of any damaged/tight regions.

2 Technical Approach

The platform is a 0.16 m diameter, 236 g quadrotor (Fig. 3) using Qualcomm®, Snapdragon™Flight™ with a 2 cell Lipo battery 7.4 V. The proposed solution is based on our previous work [15], where we focused on the ability to perform aggressive maneuvers while tracking a simple trajectory. In the proposed scenario, there are several additional challenges that need to be addressed. The vehicle needs to navigate considering different illumination conditions, concurrently creating a map of the environment and replanning its path to avoid obstacles, while reaching the final mission goal. In the following, we present a brief overview of the key onboard approaches for estimation, mapping and planning.

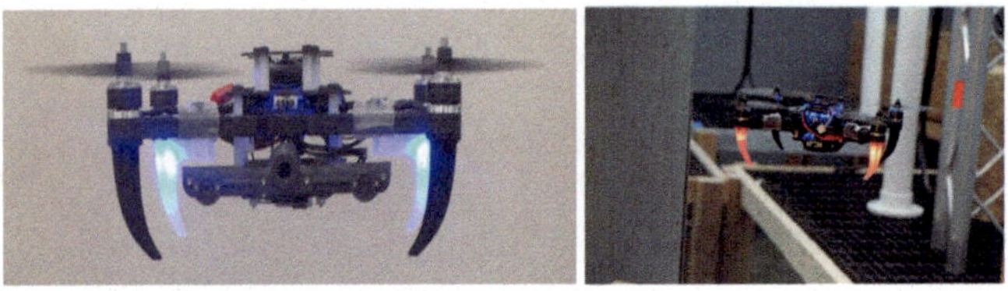

Fig. 3. The small 236 g platform used for the nuclear inspection task.

2.1 State Estimation and Control

As shown in [15], the Visual Inertial Odometry (VIO) system localizes the rigid body with respect to the inertial frame combining Inertial Measurement Unit (IMU) data and downward facing Video Graphics Array (VGA) resolution camera with 170° field of view. The prediction step is based on the IMU integration. The measurement update step is given by the standard 3D landmark perspective projection onto the image plane leading to the Extended Kalman Filter (EKF) updates. An additional Unscented Kalman Filter (UKF) is able to estimate the state of the vehicle for control purpose at 500 Hz. From a control point of view, we use a proportional and derivate nonlinear controller both for the position and attitude loops [15–17]. The control inputs $\tau, \mathbf{M}$ are chosen as

$$\begin{aligned} \mathbf{M} &= -k_R \mathbf{e}_R - k_\Omega \mathbf{e}_\Omega + \Omega \times J\Omega - \mathbf{J}\left(\hat{\Omega} R^\top R_C \Omega_C - R^\top R_C \dot{\Omega}_C\right) \\ \tau &= \left(-k_x \mathbf{e}_x - k_v \mathbf{e}_v + mg\mathbf{e}_3 + m\ddot{\mathbf{x}}_d\right) \cdot R\mathbf{e}_3 \end{aligned} \tag{1}$$

with $\ddot{\mathbf{x}}_d$ the desired acceleration, k_x, k_v, k_R, k_Ω positive definite terms. The subscript C denotes a commanded value. The quantities $\mathbf{e}_R, \mathbf{e}_\Omega, \mathbf{e}_x, \mathbf{e}_v$ are the orientation, angular rate, and translation errors respectively, defined in [16,17] and R the orientation, Ω the angular rate in the body frame. The attitude loop runs on the Digital Signal Processor (DSP) available on the board at 1 kHz, whereas the position loop runs on the Advanced RISC Machines (ARM) processing unit concurrently with the estimation and planning pipelines.

2.2 Mapping

The frontward-facing stereo camera is employed to create a dense map of the environment (see Fig. 3). The stereo camera mapping algorithm uses two rectified images of 640×480 resolution at 15 Hz with 90° horizontal field of view to determine the location of obstacles in the environment. To generate a high frame rate map for real-time obstacle avoidance, the rectification process of the two images is split into two separate threads. Both images are concurrently rectified within 3 ms on the ARM processor. A block matcher algorithm produces a disparity map by searching for matches along the epipolar line and provides distance information. The tradeoff between speed and accuracy can be explored by changing the block matcher filter size. Larger filter window sizes generate

Fig. 4. Obstacle course setup (left), Generated map with the path avoiding obstacles.

less noise in disparity maps at a higher computational cost. The disparity information is then used to generate a 3D pointcloud of the environment from the pinhole camera projection model. We use 3D voxel grids to discretize the space and the 3D points are used as votes to obtain a voxel grid occupancy map for planning.

2.3 Planning

Given a pre-generated global polynomial trajectory and depth map of the environment we use the work in [18] for on-line reactive planning. The generated trajectory deviates from the global path according to encountered obstacles. Quintic B-splines are used to ensure the required smoothness of the trajectory, which are continuous up to the forth derivative of position (snap). Locality of trajectory changes due to changes in the control points, this means that is a change in one control point affects only a few segments in the entire trajectory. Closed-form solutions are available for position at a given time, derivatives with respect to time (velocity, acceleration, jerk, snap) and integrals over squared time derivatives, thus allowing real-time performance.

3 Experimental Results

An important step prior to deployment is the evaluation of developed technologies in mock-up facilities. In this section, we report on the experiments carried out in a mock-up of the PCV pedestal (5 m diameter) and CRD ramp as shown in Fig. 2. Three different tests were conducted to validate the proposed setup and strategy. During all the tests, the nominal velocity of the vehicle was limited to 0.5 m/s.

3.1 Obstacle Avoidance Course

For these experiments the vehicle is supposed to navigate along an obstacle course that consists of multiple vertical 102 mm PVC pipes, 25.4 mm bundled vertical tubes, and aluminum trusses. During the test, the UAS started from a

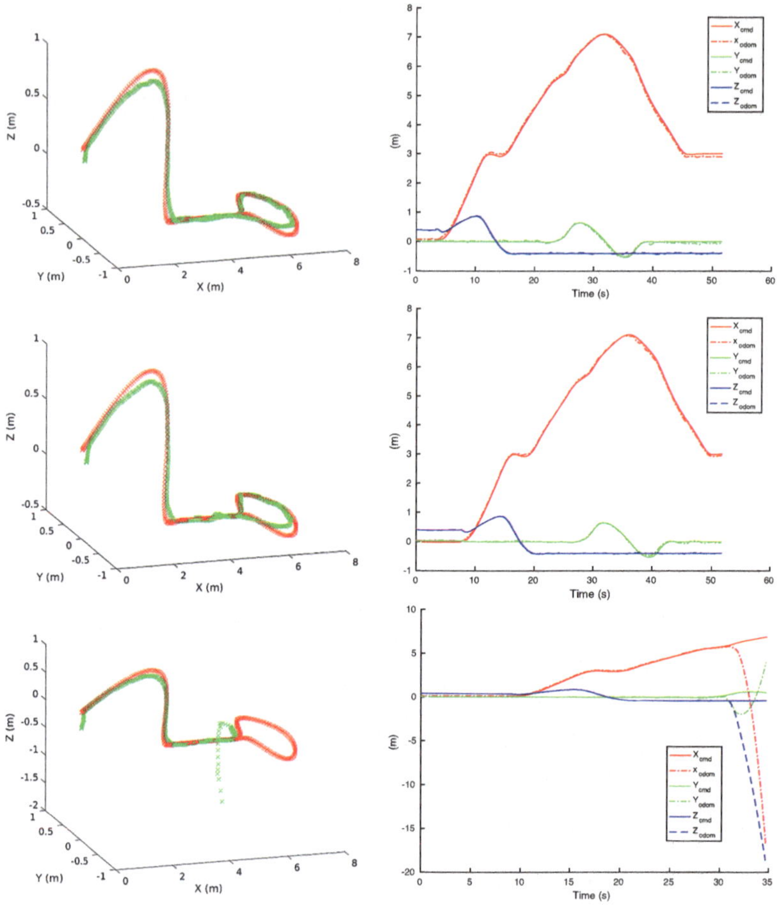

Fig. 5. Experiments in different light conditions showing the global trajectory tracking (left column) and tracking error by components (right column) for normal 12 V (up), medium 8 V (center), low 7.5 V (bottom) with crash/drift.

predefined location on the floor and was given a single waypoint located at the far end of the obstacle field. As there were essentially no overhead obstacles, the altitude of the UAS was manually restricted to force it to travel through the obstacle field as opposed to the lower-cost flight path located above the obstacles. Figure 4 shows the obstacle course and the generated map of one of the multiple obstacle avoidance test trials.

3.2 Luminance Tests

The current platform is not yet equipped with a lighting source. However, to compensate for variations in light, we can exploit the automatic camera exposure algorithm. To determine an estimate of the illuminance required for the

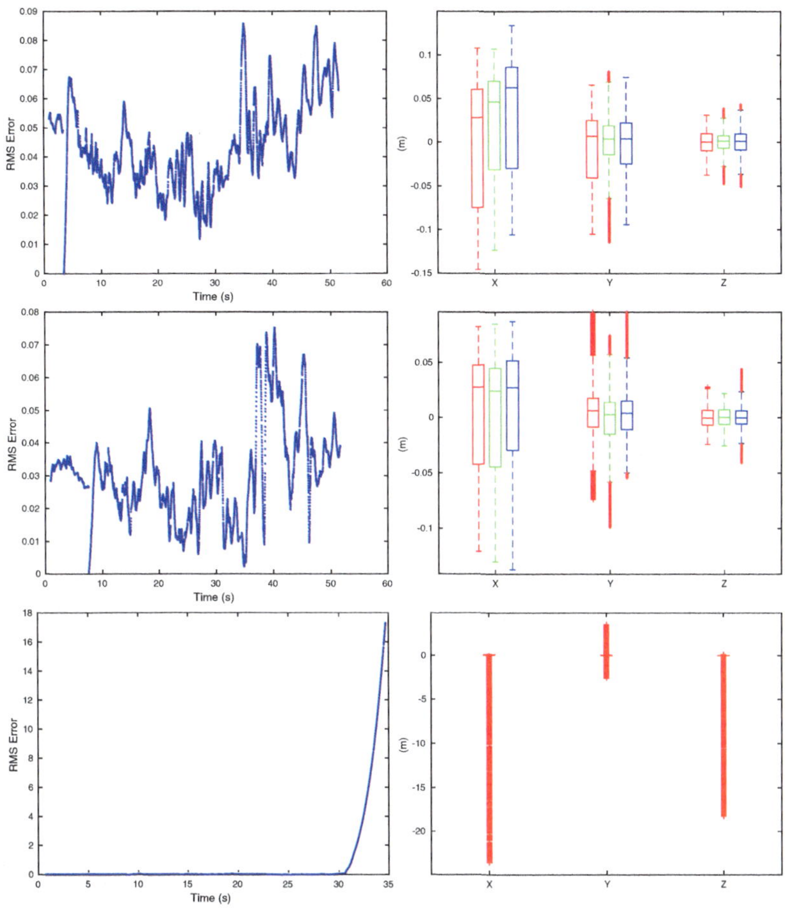

Fig. 6. Experiments in different light conditions showing RMSE in X, Y and Z (left column) and maximum errors (right column), for normal 12 V (up), medium 8 V (center), low 7.5 V (bottom) with crash/drift.

vehicle to accurately extract visual features to maintain adequate localization, experiments with varying luminance were conducted. The PCV pedestal was fitted with multiple LED light strips whose luminous emittance was controlled directly by varying the drive voltage of the lights. Experiments were conducted at three different voltage levels (with all external sources turned off), to find the lowest illuminance to operate the localization algorithm. The corresponding lux values were measured at 7 different locations at constant height above the grates inside the pedestal. Data was collected with an LX1330B digital light meter. The average values corresponding to the voltage levels are 41.8 lx @12V, 9.5 lx @8V, 0.5 lx @7.5V. Figure 5 (left column) shows the trajectories followed by the UAS in different luminance, whereas in Fig. 5 (right column) we report the control errors between the commanded position and tracked positions. Figure 6 (left column) shows the RMSE (Root Mean Squared Errors), whereas Fig. 6 (right column) presents distribution of the tracking error across three runs. Finally, the reader should notice that the 7.5 V experiment was performed only once to avoid additional damage to the vehicle due drift the in VIO and subsequent crash.

3.3 PCV Inspection

The third, and most complex, test scenario consisted of multiple stationary and suspended obstacles (of varying sizes) at unknown locations inside of the pedestal. The vehicle starts from the external point, moves into the cluttered pedestal area and performs the site inspection and environment mapping while avoiding obstacles (Fig. 7). The planning and obstacle avoidance operate in the 3D space and the vehicle safely maneuvers from the pedestal mock-up. Multiple runs with varying obstacle locations were conducted.

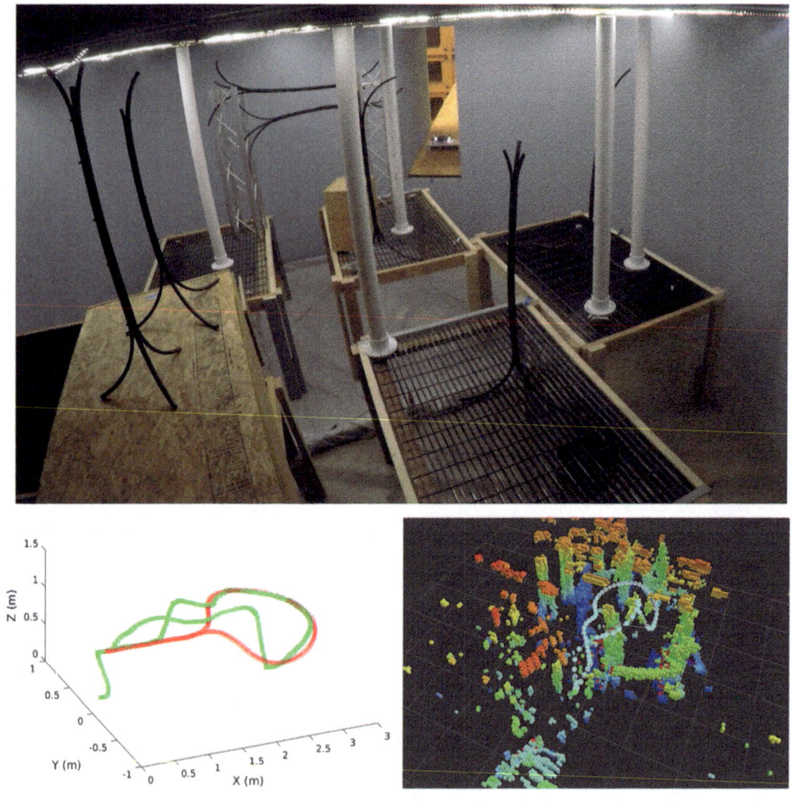

Fig. 7. PCV pedestal setup with unknown obstacles for experiments (top). The red is the original trajectory as if obstacles were not in the path, whereas the green is the state estimate including replanning phases (bottom left). The generated obstacle map (bottom right).

4 Main Experimental Insights

There are several insights that can be obtained from our experiments. First, we showed the possibility to concurrently run, onboard an aerial platform with limited computational capability, the state estimation, control, mapping and planning algorithms with obstacle avoidance to solve a complex nuclear inspection task. Second, we were able to successfully detect and avoid obstacles 0.25 m in diameter with a stereo camera even with a relatively small stereo baseline. Finally, experimental evidence shows that an onboard LED payload will be necessary to allow navigation in very dark conditions with illuminance lower than roughly 8 lx.

The replanning strategy can be improved to avoid local minima by utilizing a global planner. To achieve higher speeds, faster sensors and algorithms are needed to locate the obstacles in shorter amount of time. This is essential in operations like the presented one, where there are mission time constraints in addition to spatial ones. We believe these experimental insights will allow the development of smaller scale platforms up to 0.1 m in diameter and will aid in on-board LED payload design to enable the vehicle to perform autonomous navigation and obstacle avoidance in reduced lighting conditions.

Acknowledgments. We would like to acknowledge the Richard Garcia and Monica Garcia from SwRI who enabled us to conduct these experiments in PCV mock-up at the San Antonio, TX facility. This work was supported by the TEPCO L99048MEC grant, Qualcomm Research, ARL grants W911NF-08-2-0004, W911NF-17-2-0181, ONR grants N00014-07-1-0829, N00014-14-1-0510, ARO grant W911NF-13-1-0350, NSF grants IIS-1426840, IIS-1138847, DARPA grants HR001151626, HR0011516850. This work was supported in part by C-BRIC, one of six centers in JUMP, a Semiconductor Research Corporation (SRC) program sponsored by DARPA.

References

1. Tokyo Electric Power Company: Fukushima daiichi nuclear power plant photo collection. http://photo.tepco.co.jp/en/
2. World Nuclear Association: Decommissioning nuclear facilities. http://www.world-nuclear.org/information-library/nuclear-fuel-cycle/nuclear-wastes/decommissioning-nuclear-facilities.aspx
3. Murphy, R.R.: A decade of rescue robots. In: 2012 IEEE/RSJ International Conference on Intelligent Robots and Systems, pp. 5448–5449, October 2012
4. Murphy, R.R., Tadokoro, S., Kleiner, A.: Disaster Robotics, pp. 1577–1604. Springer, Cham (2016)
5. Michael, N., Shen, S., Mohta, K., Kumar, V., Nagatani, K., Okada, Y., Kiribayashi, S., Otake, K., Yoshida, K., Ohno, K., Takeuchi, E., Tadokoro, S.: Collaborative mapping of an earthquake-damaged building via ground and aerial robots. J. Field Robot. **29**(5), 832–841 (2012)
6. Ozaslan, T., Loianno, G., Keller, J., Taylor, C.J., Kumar, V., Wozencraft, J.M., Hood, T.: Autonomous navigation and mapping for inspection of penstocks and tunnels with MAVS. IEEE Robot. Autom. Lett. **2**(3), 1740–1747 (2017)

7. Li, J., Wu, X., Xu, T., Guo, H., Sun, J., Gao, Q.: A novel inspection robot for nuclear station steam generator secondary side with self-localization. Robot. Biomimetics **4**(1), 26 (2017)
8. Connor, D., Martin, P.G., Scott, T.B.: Airborne radiation mapping: overview and application of current and future aerial systems. Int. J. Remote Sens. **37**(24), 5953–5987 (2016)
9. Boudergui, K., Carrel, F., Domenech, T., Guénard, N., Poli, J.P., Ravet, A., Schoepff, V., Woo, R.: Development of a drone equipped with optimized sensors for nuclear and radiological risk characterization. In: 2011 2nd International Conference on Advancements in Nuclear Instrumentation, Measurement Methods and their Applications, pp. 1–9, June 2011
10. Aleotti, J., Micconi, G., Caselli, S., Benassi, G., Zambelli, N., Bettelli, M., Zappettini, A.: Detection of nuclear sources by UAV teleoperation using a visuo-haptic augmented reality interface. Sensors **17**(10) (2017). http://www.mdpi.com/1424-8220/17/10/2234
11. Kawatsuma, S., Fukushima, M., Okada, T.: Emergency response by robots to Fukushima-Daiichi accident: summary and lessons learned. Industr. Robot Int. J. Robot. Res. Appl. **39**(5), 428–435 (2012)
12. Ohno, K., Kawatsuma, S., Okada, T., Takeuchi, E., Higashi, K., Tadokoro, S.: Robotic control vehicle for measuring radiation in Fukushima Daiichi nuclear power plant. In: 2011 IEEE International Symposium on Safety, Security, and Rescue Robotics, pp. 38–43, November 2011
13. Tokyo Electric Power Company: Application of robot technology. http://www.tepco.co.jp/en/decommision/principles/robot/index-e.html
14. Wikipedia: Boiling water reactor. https://en.wikipedia.org/wiki/Containment_building/
15. Loianno, G., Brunner, C., McGrath, G., Kumar, V.: Estimation, control, and planning for aggressive flight with a small quadrotor with a single camera and IMU. IEEE Robot. Autom. Lett. **2**(2), 404–411 (2017)
16. Lee, T., Leoky, M., McClamroch, N.H.: Geometric tracking control of a quadrotor UAV on SE(3). In: 49th IEEE Conference on Decision and Control (CDC), pp. 5420–5425, December 2010
17. Mellinger, D., Kumar, V.: Minimum snap trajectory generation and control for quadrotors. In: IEEE International Conference on Robotics and Automation, Shangai, China, pp. 2520–2525 (2011)
18. Usenko, V., von Stumberg, L., Pangercic, A., Cremers, D.: Real-time trajectory replanning for MAVs using uniform B-splines and a 3D circular buffer. In: IEEE/RSJ International Conference on Intelligent Robots and Systems, pp. 215–222, September 2017

View Planning and Navigation Algorithms for Autonomous Bridge Inspection with UAVs

Prajwal Shanthakumar[1(✉)], Kevin Yu[1], Mandeep Singh[2], Jonah Orevillo[1], Eric Bianchi[1], Matthew Hebdon[1], and Pratap Tokekar[1]

[1] Virginia Tech, Blacksburg, USA
{prajwal,klyu,j0n4h0,beric7,mhebdon,tokekar}@vt.edu
[2] Indian Institute of Technology Kanpur, Kanpur, India
mandeeps@iitk.ac.in

1 Introduction

Current methods for bridge inspection involve significant manual effort, with humans in harnesses and operators controlling cranes. It is time consuming, often requires closure of roads, expensive and potentially dangerous. With drone technology maturing and several commercial solutions entering the market, using UAVs for the task of bridge inspection is a great alternative to make bridge inspection faster, safer and less expensive [1].

Manual flight of UAVs, however, requires skillful piloting, and can be challenging in windy conditions or when there is no line-of-sight (long bridges, flight under the bridge, etc.). GPS based approaches for autonomous or assisted manual flights may not be possible because GPS can be unreliable around the bridge structure. Thus, GPS denied navigation for safe and accurate flight around bridges is the main motivation for our work [2].

Further, even if the GPS problem is not present, planning way-point based missions in the global frame of reference is challenging. Choosing waypoints can be tedious and would require knowledge of the bridge structure in the global frame. Planning for altitude is difficult because the bridge can slope up or down and the terrain around the bridge is often uneven. The bridge could also be over water. Hence, we explore the approach of navigating the UAV in a frame relative to the bridge structure.

In this paper, we describe our work on designing a fully autonomous UAV system to inspect bridges. Section 2 describes the UAV system and our strategy for autonomous bridge inspection. Section 3 discusses experiments and results. We conclude in Sect. 4 and indicate the direction of future work.

Electronic supplementary material The online version of this chapter (https://doi.org/10.1007/978-3-030-33950-0_18) contains supplementary material, which is available to authorized users.

J. Xiao et al. (Eds.): ISER 2018, SPAR 11, pp. 201–210, 2020.
https://doi.org/10.1007/978-3-030-33950-0_18

2 Technical Approach

The UAV system and the algorithm for autonomous bridge inspection are described below.

2.1 System Description

The UAV built for bridge inspection weighs 3.5 kg and is shown in Fig. 1. The following are the components onboard:

- Pixhawk Autopilot with built-in IMU and Compass, running PX4 firmware [3]: for low-level flight control.
- NVIDIA Jetson TX2 running ROS (Robot Operating System): for processing sensor data and publishing the desired velocity to the Pixhawk.
- Scanse Sweep 2D LIDARs [4]: to estimate bridge structure and enable autonomous navigation.
- PointGrey Flea3 Monocular Camera (global shutter, 60 FPS, 2.0MP): for visual odometry.
- GoPro Camera: to collect images of the bridge structure for defect identification.

Fig. 1. UAV for bridge inspection.

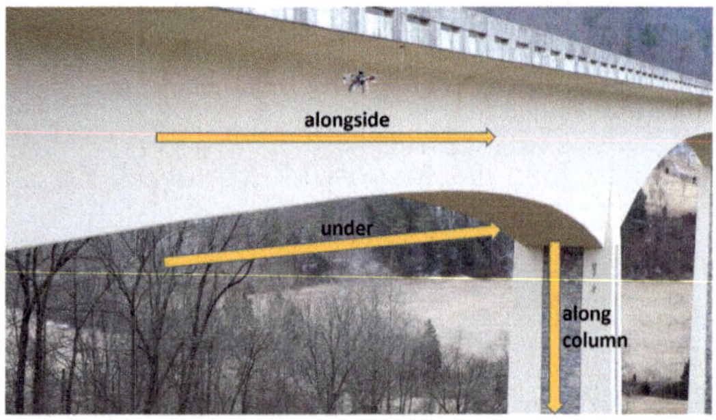

Fig. 2. Illustration of some useful navigation routines.

2.2 Algorithm Description

Our current approach for complete inspection coverage of all bridge surfaces is to autonomously execute a series of maneuvers from a library of navigation routines (Fig. 2). The algorithm consists of three parts: (1) Local Controller for execution of each of the routines through real-time LIDAR based navigation. (2) Supervisor to determine if we have completed a navigation routine and can progress to the next (based on special cases in LIDAR data and/or Visual Odometry) (3) Coverage Planner to find the sequence of navigation routines to employ for complete visual coverage of the bridge. While the UAV autonomously navigates the bridge, a camera onboard records images of the bridge that can be examined for defects and/or used for reconstruction.

2.2.1 Local Controller

For safe and accurate autonomous traversal of bridge surfaces, the employed algorithm must be fast enough to not only run real-time but also robust to factors such as wind disturbances, unreliable GPS and Compass.

We employ 2D LIDAR (a rotating 1D LIDAR) based local control for this purpose. Using two 2D LIDARs (one horizontal and one vertical), we implement routines for traversing beside the bridge, along columns and under the deck. We exploit geometry inherent in bridges and treat bridge surfaces as planes. A 2D LIDAR scan of a plane shows up as a line as shown in Fig. 3.

A Hough Transform based approach is used to find the specific line of interest that approximates a bridge surface. PID control is employed to drive the UAV parallel to the bridge surface, maintaining a fixed distance to it. The software architecture is illustrated in Fig. 4. A more detailed description of Estimation and Control is provided below.

Fig. 3. White dots: LIDAR data; Red line: best fit using Hough Transform.

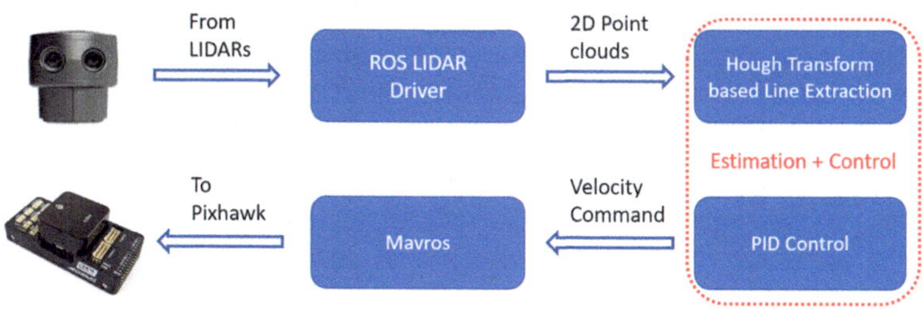

Fig. 4. Software architecture for autonomous navigation.

Estimation

To understand how bridge surfaces are estimated, consider, for instance, the problem of estimating the *vertical span* of the bridge girder surface (highlighted in red in Fig. 5). The data used for this purpose is the 2D point cloud from a vertically scanning LIDAR. The following are the steps involved (explained with ref. to Figs. 5, 6 and 7).

- Points very close (on the UAV itself) and very far (ground plane, distant trees, etc.) are first filtered out (circled in blue in Fig. 7).
- The Hough Transform is then used to find lines in the filtered data. Let's assume the three lines as shown in Figs. 5 and 7 are found with enough confidence.
- We are interested in extracting only the red line which approximates the vertical span of the girder surface. This is a line with an expected slope of $\approx 90°$ and width 3–5 m. Therefore, the yellow line with slope $\approx 10°$ and the green line with width ≈ 1 m can be filtered out (Figs. 5 and 7).

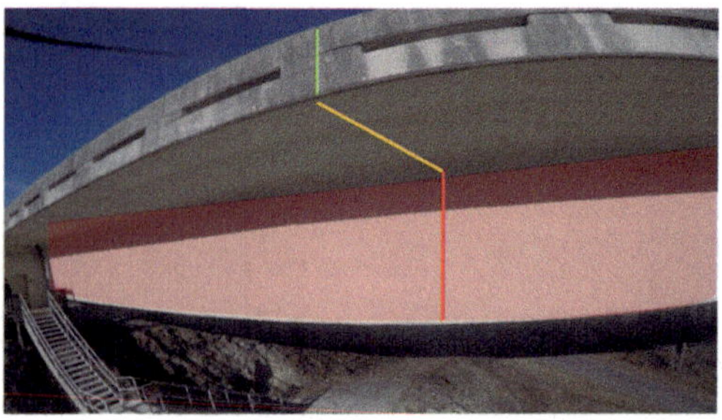

Fig. 5. The bridge girder surface highlighted in red.

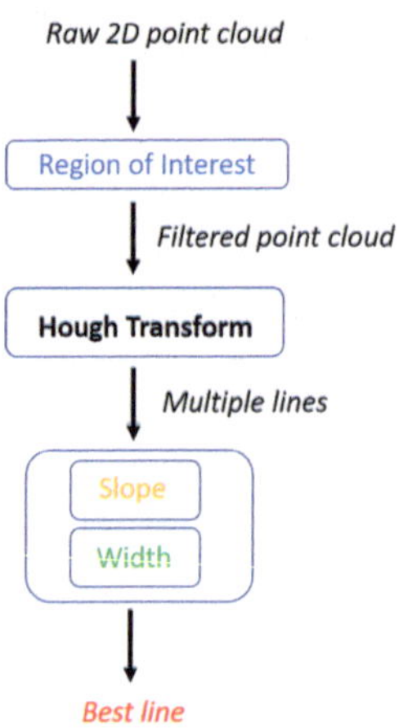

Fig. 6. Steps involved in the Hough Transform based Line Extraction process

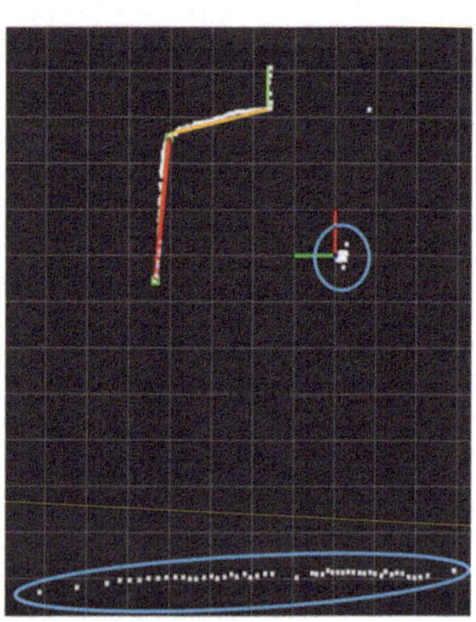

Fig. 7. Visualization (rviz) of filtered out points and extracted lines

Control

The goal of the control algorithm is to drive the UAV parallel to the (estimated) bridge surfaces. This involves two independent PID loops along two perpendicular axes to maintain position with respect to the surface. A constant nominal velocity along the third axis drives the UAV parallel to the bridge surface.

As an example, for flight beside the bridge girder (Fig. 8):

- One PID loop maintains the desired distance along the axis perpendicular to the girder surface (green axis).
- Another PID loop maintains altitude w.r.t the bridge girder (red axis).
- A nominal velocity drives the UAV along the axis parallel to the girder surface (blue axis).

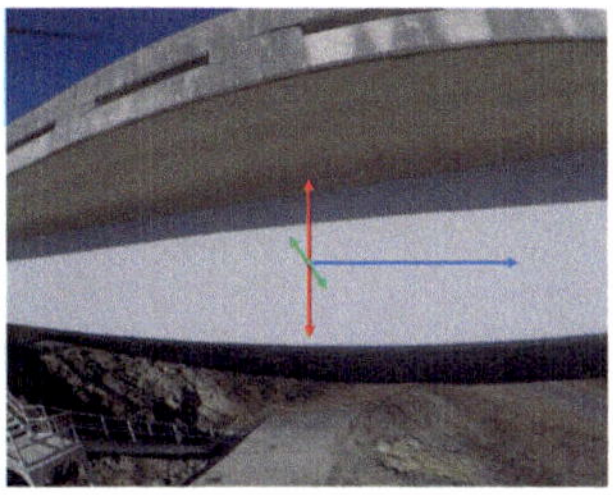

Fig. 8. Independent PID loops control position along two axes; constant velocity along the third.

2.2.2 Supervisor

The role of the Supervisor is to switch between navigation routines that the Local Controller executes, in order to execute the tour for complete coverage planned by the Coverage Planner.

Special cases in the data from the two 2D LIDARs have been identified to determine when to switch from one routine to another. For instance, consider the scenario where the UAV is traversing up a bridge column as shown in Figs. 9(a)(b). On encountering the bridge girder, the number of points in the horizontal cut of the LIDAR increases dramatically as shown in Figs. 9(c)(d). At this juncture, the Supervisor can switch from the Column Following routine to the Girder Following routine.

However, even a coarse estimate of global position can be used to better inform the Supervisor. For this purpose, we are exploring Visual Odometry. Semi-direct approaches such as SVO [5] and Visual-Inertial Odometry algorithms such as VINS-Mono [6] were found to be prone to error when subject to rotational motion. To adapt existing VO for bridge inspection, removal of outliers (clouds and dynamic objects) from features detected by the indirect methods such as ORB SLAM [7] is underway. Integration of feedback from LIDAR to correct the scale drift of ORB SLAM 2 is also in progress.

It is to be noted that the Supervisor need not necessarily be fully automated. Once the UAV has completed coverage of one bridge surface using the Local Controller, we could have a human operator switch the UAV to another routine (bridge surface). This can be done in tricky scenarios such as when switching from flight under the bridge to flight beside the bridge.

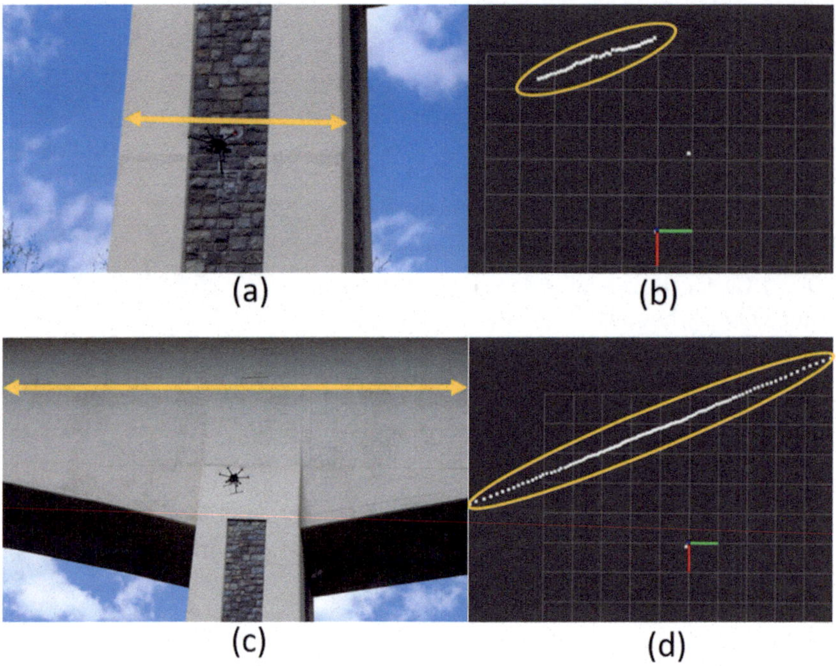

Fig. 9. Switching from column follow to girder follow.

2.2.3 Coverage Planner

To plan a tour for full visual coverage of the bridge structure, we partition the bridge into a set of planar surfaces as shown in Fig. 10. Each surface can be approximated by a polygon. Each polygon has two nodes, one of which will be chosen as the entry node and the other as the exit node. The local controller will ensure navigation between the entry and exit nodes parallel to the bridge surface. The Supervisor will identify that the exit node of the current polygon/entry node of the next polygon has been reached and pick the appropriate routine to continue the tour.

The visual coverage problem is that of finding the sequence in which the bridge surfaces must be traversed, as well as determining the entry node for each of the polygons representing the surfaces.

We formulate visual coverage as a Generalized Traveling Salesperson Problem (GTSP). A GTSP solver takes as input a graph whose nodes have been partitioned into disjoint clusters. The output of the solver is a tour in which each cluster is visited exactly once, while minimizing the total cost to visit all clusters.

In our GTSP formulation, each cluster is a polygon containing an entry and exit node as shown in Fig. 11. There exists an edge between every pair of nodes lying in separate clusters. For any cluster X, let X_1 and X_2 respectively represent its entry and exit nodes. Then, the cost on an edge from node A_1 in cluster A to node B_1 in cluster B is given by,

$$C(A_1, B_1) = D(A_1, A_2) + D(A_2, B_1)$$

where $D(M, N)$ represents the Euclidean distance between nodes M and N in 3D space, i.e. the cost is the sum of the distance to cover the current polygon (surface) and the distance to reach the entry node of the next polygon from the exit node of the current polygon.

We can form a full cost matrix of our graph by using estimates of distance between every pair of nodes. This is the input to our GTSP solver, Generalized Large Neighborhood Search (GLNS) [8]. Once we obtain a solution from the GLNS solver we are able to get the tour, which visits every cluster exactly once. By using a GTSP formulation, we can guarantee that we visit every polygon, which represents sections of the bridge, and solve for an optimal tour.

Fig. 10. Bridge partitioned into surfaces with entry/exit nodes (blue dots).

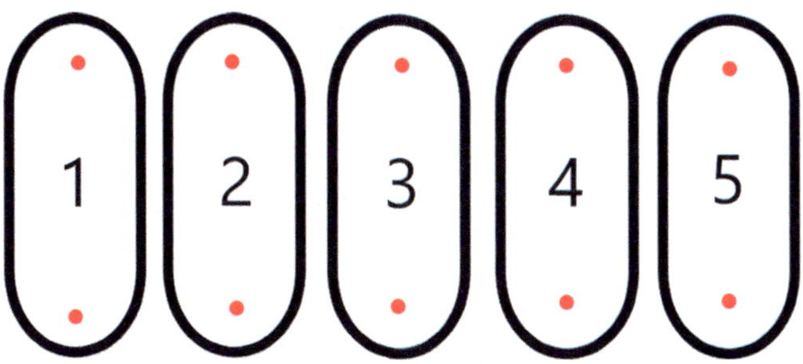

Fig. 11. GTSP clusters with entry/exit nodes (red dots) corresponding to Fig. 10.

3 Experiments and Results

We conducted experimental flights at the Virginia Tech Transportation Institute (VTTI) Smart Road Bridge, which is 53 m tall, 609 m long, 5 span, variable height concrete box girder bridge (Fig. 12). We tested LIDAR based autonomous flight (no GPS) with a GoPro camera onboard collecting images of the bridge structure.

3.1 Flight Beside Bridge Girder

The objective of the experiment was to have the UAV fly alongside the bridge with a nominal velocity of 0.5 m/s maintaining a horizontal separation of 4.5 m to the girder. The LIDAR scan (rotation) rate was 2 Hz, and hence the control signal (velocity in the direction perpendicular to the bridge) was also issued at 2 Hz. (The LIDAR can scan at upto 10 Hz, and a higher control rate can be employed, for instance, in windy conditions. The algorithmic computations only take approximately 1 ms.). Figure 13 shows the control signal driving the UAV to the desired separation of 4.5 m from the structure. Hence, we could experimentally verify that accurate flight alongside the bridge at a fixed horizontal distance from it is feasible, which is helpful for consistent data (image) collection during flight.

Fig. 12. Experiment at the VTTI Smart Road Bridge.

Maintaining the correct altitude with respect to the bridge structure proved to be more challenging. Since the terrain around the bridge is not even, using a downward facing LIDAR is not an option for maintaining altitude. We cannot rely on the barometer alone, not only because of barometric drifts and inaccuracies, but also because the bridge may slope up or down. Hence, we need some method

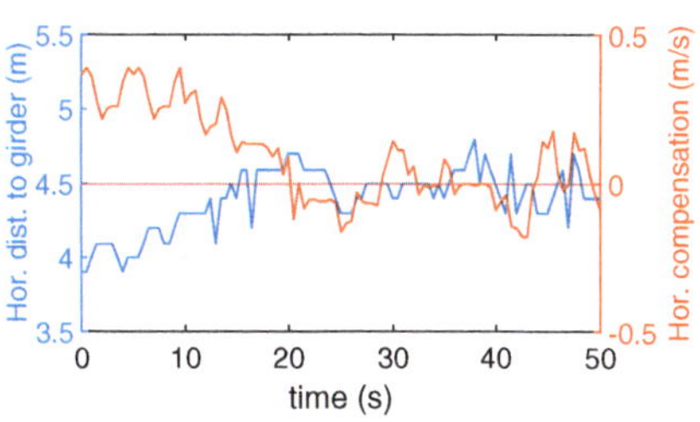

Fig. 13. Correction velocity in the direction perpendicular to the bridge structure maintaining the desired distance of 4.5 m from it.

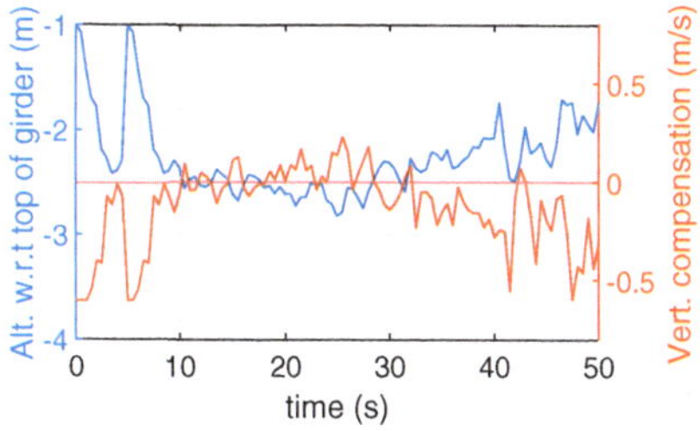

Fig. 14. Correction velocity in the vertical direction maintaining the desired altitude of 2.5 m below the top of the girder

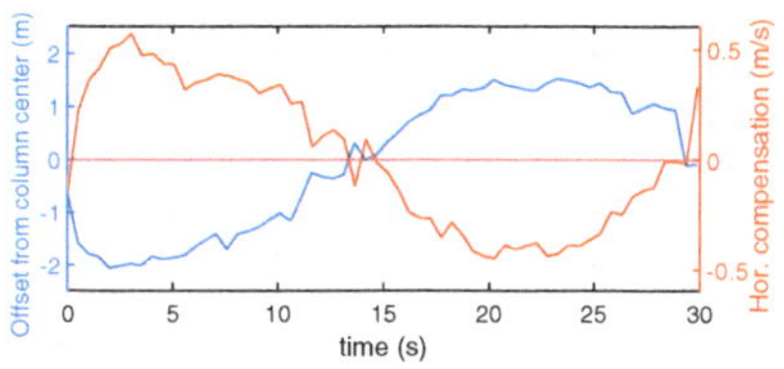

Fig. 15. Correction velocity in the direction tangential to the bridge column centering the UAV w.r.t the column

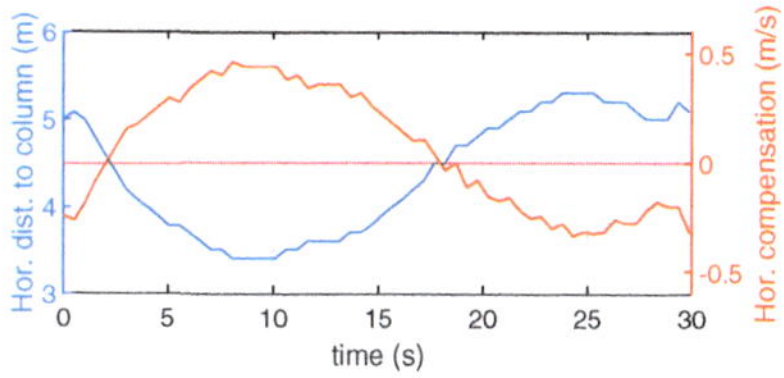

Fig. 16. Correction velocity in the direction perpendicular to the bridge structure maintaining the desired distance of 4.5 m from it

to hold position w.r.t. the bridge structure itself. For this purpose, we use the vertical cut from the 2D LIDAR rotating in the vertical plane that gives us a cross-section of the bridge. The vertical cut was used to maintain an altitude of 2.5 m below the top of the girder as shown in Fig. 14.

3.2 Flight Along Bridge Column

The objective of this experiment was to have the UAV traverse up and down a bridge column with a vertical nominal velocity of 0.5 m/s, while compensating in the horizontal direction to center and maintain a separation of 4.5 m w.r.t the column. The experimental results in Figs. 15 and 16 show the control algorithm compensating to maintain center and the desired distance to the column.

4 Conclusion and Future Work

We presented a fast (real-time) LIDAR based approach for autonomous navigation using two 2D LIDARS scanning in horizontal and vertical planes. We described strategies to switch between navigation routines to cover various bridge surfaces. We envision full coverage of the bridge using three routines for UAV flight: (1) beside the girder, (2) along bridge columns, and, (3) under the bridge deck. (1) and (2) have already been tested and (3) is conceptually similar to (1) and (2). The GTSP algorithm is implemented similar to our prior work [9]. Future work will involve experiments to test the integration of routines and execution of the full GTSP tour. The LIDAR based approach can also be used to provide safety guarantees in assisted manual flights. Operator (pilot) input can be rejected if it will result in the UAV getting too close to the bridge structure. Adapting ideas from our present approach to different types of bridges (which could require working with other sensors such as Stereo camera for navigation) is also a future direction.

Acknowledgements. This work has been funded by the Center for Unmanned Aircraft Systems (C-UAS), a National Science Foundation-sponsored industry/university cooperative research center (I/UCRC) under NSF Award No. IIP-1161036 along with significant contributions from C-UAS industry members.

References

1. Zink, J., Lovelace, B.: Unmanned aerial vehicle bridge inspection demonstration project. Technical report (2015)
2. Gillins, M.N., Gillins, D.T., Parrish, C.: Cost-effective bridge safety inspections using unmanned aircraft systems (UAS). In: Geotechnical and Structural Engineering Congress 2016, pp. 1931–1940 (2016)
3. Meier, L., Honegger, D., Pollefeys, M.: PX4: a node-based multithreaded open source robotics framework for deeply embedded platforms. In: 2015 IEEE International Conference on Robotics and Automation (ICRA), May 2015
4. SCANSE. http://scanse.io/
5. Forster, C., Pizzoli, M., Scaramuzza, D.: SVO: fast semi-direct monocular visual odometry. In: 2014 IEEE International Conference on Robotics and Automation (ICRA), pp. 15–22, May 2014
6. Qin, T., Li, P., Shen, S.: VINS-Mono: a robust and versatile monocular visual-inertial state estimator. arXiv preprint arXiv:1708.03852 (2017)
7. Mur-Artal, R., Montiel, J.M.M., Tardós, J.D.: ORB-SLAM: a versatile and accurate monocular slam system. IEEE Trans. Robot. **31**(5), 1147–1163 (2015)
8. Smith, S.L., Imeson, F.: GLNS: an effective large neighborhood search heuristic for the generalized traveling salesman problem. Comput. Oper. Res. **87**, 1–19 (2017)
9. Yu, K., Budhiraja, A.K., Tokekar, P.: Algorithms and experiments on routing of unmanned aerial vehicles with mobile recharging stations. IEEE (2018, to appear)

Policy Search on Aggregated State Space for Active Sampling

Sandeep Manjanna[1(✉)], Herke Van Hoof[2], and Gregory Dudek[1]

[1] Mobile Robotics Lab (MRL), Center for Intelligent Machines, McGill University, Montreal, QC, Canada
{msandeep,dudek}@cim.mcgill.ca
[2] Amsterdam Machine Learning Lab (AMLAB), University of Amsterdam, Amsterdam, The Netherlands
h.c.vanhoof@uva.nl

1 Motivation, Problem Statement, and Related Work

We present an anytime [1] adaptive sampling technique that generates paths to efficiently measure and then mathematically model a scalar field by performing non-uniform measurements in a given region of interest. In particular, the class of scalar field we are interested is some physical or virtual parameter that varies with location, such as depth of the sea floor or the probability of finding a lost object. As the measurements are collected at each sampling location, we can compute an estimate of the large-scale variation of the phenomenon of interest. We compute a sampling path that minimizes the expected time to accurately model the phenomenon of interest by visiting high information regions using non-myopic path generation based on reinforcement learning.

As an example application, we consider monitoring the health of coral reefs by sampling visual data from the surface using an autonomous surface vehicle (ASV) shown in Fig. 1a. Increase in the sea surface temperatures has resulted in widespread coral bleaching at an ever-increasing rate [2] (Fig. 1b). Improved monitoring would enhance the currently poor understanding of the spatial and temporal dynamics of coral bleaching. Since we are sampling from the surface, higher information gain is provided in shallower regions where visibility is better.

Novel contributions of this work include (1) the computation of a sampling technique based on policy search, (2) a non-uniform (multi-resolution) representation of the state space to aid better spatial sampling and (3) a statistically significant evaluation through rigorous experiments with real satellite image data and real robots in the field.

Active sampling refers to the act of strategically planning paths based on the observations made until the current time-step. Exhaustively sampling each point of an unknown survey region [3,4] can be tedious and impractical if the

Electronic supplementary material The online version of this chapter (https://doi.org/10.1007/978-3-030-33950-0_19) contains supplementary material, which is available to authorized users.

J. Xiao et al. (Eds.): ISER 2018, SPAR 11, pp. 211–221, 2020.
https://doi.org/10.1007/978-3-030-33950-0_19

(a) (b) (c)

Fig. 1. (a) Custom made differential drive ASV used for coral reef surveys. (b) A coral with bleached spots [2]. (c) Aerial image of the reef surveyed in our field experiments.

survey space is large and/or the phenomenon of interest has only a few regions with important information (*hotspots*) [5,6]. Also it has been observed that for low-pass multi-band signals, uniform sampling can be inefficient and sampling rates far below the Nyquist rate can still be information preserving [7]. This is the key guiding principle behind active and non-uniform sampling [6,8,9].

In our approach, a continuous two-dimensional sampling region is discretized into uniform grid-cells, such that the robot's position $\mathbf{x}$ can be represented by a pair of integers $\mathbf{x} \in \mathbb{Z}^2$. Each grid-cell (i, j) is assigned a score $\mathrm{q}(i, j)$ indicating the expected goodness of the visual data in that cell. The goal is to maximize the total accumulated score J over a trajectory τ within a fixed amount of time T. To specify the robot's behavior we use a parametrized policy $\pi_{\boldsymbol{\theta}}(\mathbf{s}, \mathbf{a}) = p(\mathbf{a}|\mathbf{s}; \boldsymbol{\theta})$ that maps the current state $\mathbf{s}$ of sampling to a distribution over possible *actions* $\mathbf{a}$. Our aim will be to automatically find good parameters $\boldsymbol{\theta}$, after which the policy can be deployed without additional training on new problems.

2 Technical Approach

Our algorithm gets trained with a generic score-map (q) generated by the satellite data from areas that exemplify the target environments, for example images of coral reefs. The system is trained to achieve paths that preferentially cover *hotspots* at the earlier stages of exploration. These learned parameters then define a *policy* π (in the sense of reinforcement learning) that is then used on the satellite image or any other sensor map of the target coral reef (Fig. 1c) to generate an explicit action plan. During the test phase, an action with maximum probability is chosen at a given state. Thus, the policy does not need to be re-trained for each new reef map. This property is a key feature of our approach.

In our approach, we formalize the sampling problem as a Markov Decision Process (MDP). We take the state $\mathbf{s}$ to include the position of the robot $\mathbf{x}$ as well as the map q containing the per-location score for the visual data, $\mathbf{s} = (\mathbf{x}, \mathrm{q})$. The action space A consists of four actions (move North, East, South, or West). Transitions deterministically move the agent in the desired direction. Once the visual data at the current cell (i, j) is sampled, the score $\mathrm{q}(i, j)$ is reduced to 0.

The discounted reward function is defined as $\gamma^t q(\mathbf{x})$, with the discount factor $0 \leq \gamma \leq 1$ encouraging the robot to sample cells with high scores in early time steps t.

2.1 Policy Gradient Method

Policy gradient methods use gradient ascent for maximizing the expected return $J_\theta = \mathbb{E}_{\tau_\theta}[\sum_{t=1}^{|\tau_\theta|} \mathrm{q}(\mathbf{x}_t)\gamma^t]$. The gradient of the expected return ($\nabla_\theta J_\theta$) guides the direction of the parameter update ($\theta_{k+1} = \theta_k + \eta \nabla_\theta J_\theta$, where η is the learning rate). The likelihood ratio policy gradient [10] is given by,

$$\nabla_\theta J_\theta = \int_\tau \nabla_\theta p_\theta(\tau) R(\tau) d\tau \tag{1}$$

This expression depends on the correlation between actions and previous rewards, which are 0 in expectation and cause additional variance. We use the Policy Gradient Theorem (PGT) algorithm and the GPOMDP algorithm [11–14] for computing the policy gradient as it yields relatively low-variance updates. Accordingly, the policy gradient is given by,

$$\nabla_\theta J_\theta = \frac{1}{m} \sum_{i=1}^{m} \sum_{t=0}^{H-1} \nabla_\theta \log \pi_\theta(a_t^{(i)} | s_t^{(i)}) \left(\sum_{j=t}^{H-1} r(s_j^{(i)}, a_j^{(i)}) - b(s_t^{(i)}) \right). \tag{2}$$

In this equation, the gradient is based on m sampled trajectories from the system, with $\mathbf{s}_j^{(i)}$ the state at the jth time-step of the ith sampled roll-outs. Furthermore, b is a variance-reducing baseline. In our experiments, we set the baseline to the observed average reward.

2.2 Feature Aggregation

A popular method to define stochastic policies over a set of deterministic actions is the use of the Gibbs distribution as policy (also referred to as Boltzman exploration of softmax policy). We consider a commonly used linear Gibbs softmax policy parameterization [12,15] given by,

$$\pi(\mathbf{s}, \mathbf{a}) = \frac{e^{\theta^T \phi_{\mathbf{s},\mathbf{a}}}}{\sum_{\mathbf{b}} e^{\theta^T \phi_{\mathbf{s},\mathbf{b}}}}, \quad \forall \mathbf{s} \in S; \mathbf{a}, \mathbf{b} \in A, \tag{3}$$

where $\boldsymbol{\phi}_{\mathbf{s},\mathbf{a}}$ is an l-dimensional feature vector characterizing state-action pair $(\mathbf{s}, \mathbf{a})$ and $\boldsymbol{\theta}$ is an l-dimensional parameter vector.

The final feature vector $\phi_{\mathbf{s},\mathbf{a}}$ is formed by concatenating a vector $\phi'_s \delta_{aa'}$ for every action $a' \in \{\text{North}, \text{East}, \text{South}, \text{West}\}$, where $\phi'_s \subset \mathbb{R}^k$ is a feature representation of the state space, and $\delta_{aa'}$ is the Kronecker delta. Thus, the final feature vector has $4 \times k$ entries, 75% of which corresponding to non-chosen actions will be 0 at any one time step. We consider five different types of robot-centric

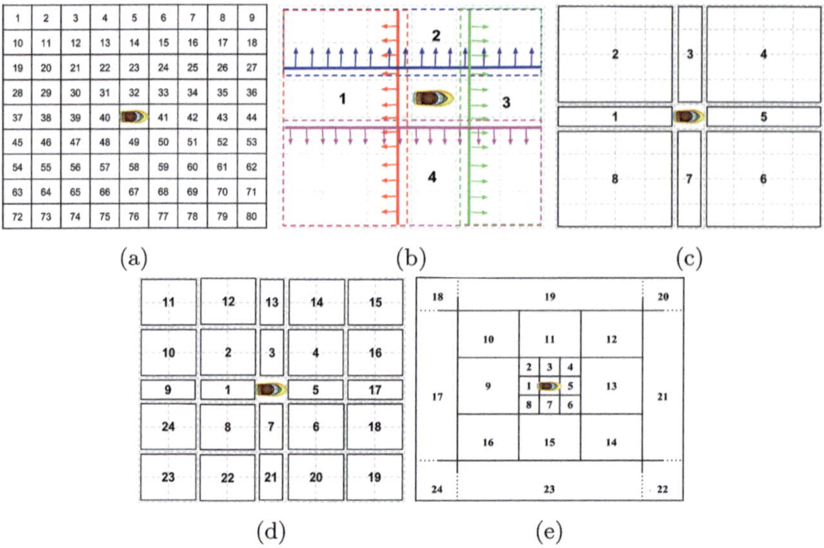

Fig. 2. Robot-centric feature space aggregations. (a) Uniform-grid feature aggregation. (b) 4-feature aggregation. (c) 8-feature aggregation. (d) 24-feature aggregation. (e) Multi-resolution feature aggregation.

feature designs ($\phi'_{\mathbf{s}}$). The first one is to consider a vector with all the scores in the score-map q as presented in Fig. 2a. This feature vector grows in length as the size of the sampling region increases resulting in higher computation times for bigger regions. The four other kinds of feature aggregations are illustrated in Fig. 2b–e. These aggregations have a fixed number of features, corresponding to the average scores in the feature map in each of the indicated areas, irrespective of the size of the sampling region.

Figure 2e depicts a multi-resolution aggregation where the feature cells grow in size along with the distance from the robot. This results in high resolution features close to the robot and lower resolution features further from the robot's current position. The aggregated feature design is only used to achieve better policy search [16], but the robot action is still defined at the grid-cell level.

3 Experiments

We use a custom-made (Fig. 1a) ASV to survey the coral reefs from the surface to collect visual data. We train our policy based sampling algorithm with the scoremap (Fig. 3b) computed as a multispectral function of the satellite image of a reef (Fig. 3a). The ASV uses the trained parameters to generate sampling paths over the map of interest. We present two test scenarios: (1) A densely populated reef shown in Fig. 4a, and (2) A reef with scattered coral heads shown in Fig. 4c. In the examples in this paper we use aerial images of the reef in Holetown, Barbados.

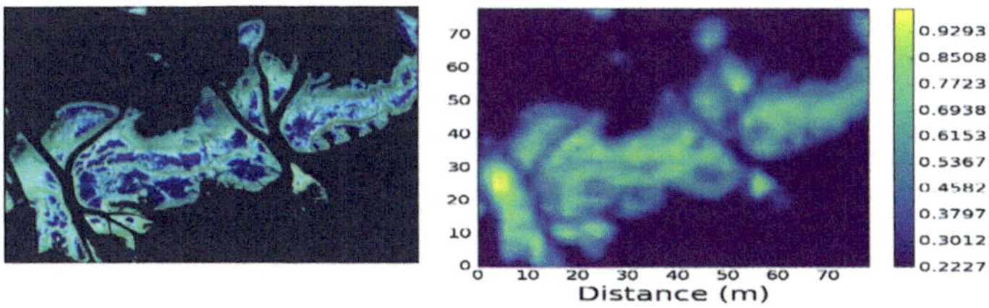

(a) (b)

Fig. 3. Experimental setup and scenarios. (a) Satellite image of a reef used for training. (b) Scoremap used for training. The colorbar indicates the interestingness or a score for the presence of corals.

We compare our sampling algorithm with a traditional exhaustive sampling technique using boustrophedonic path [4].

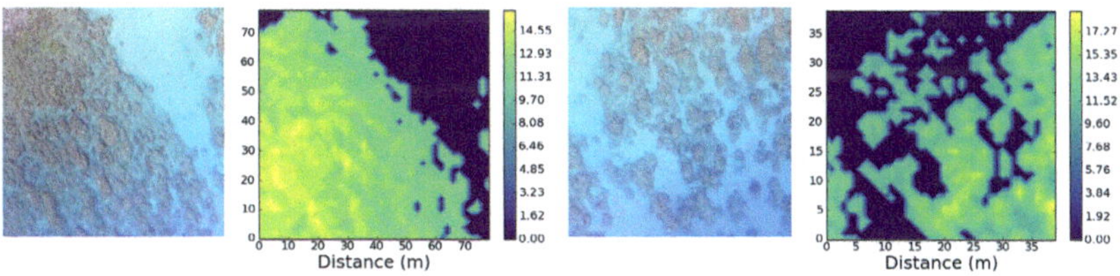

(a) (b) (c) (d)

Fig. 4. Experimental scenarios. (a) and (c) Reef images for test scenarios 1 and 2 respectively. (b) and (d) Scoremaps of the test scenarios 1 and 2 respectively. Colorbars of the scoremaps indicate a score for the presence of corals.

4 Results

Comparing different feature aggregations presented in Sect. 2.2 shows that multi-resolution aggregated features achieve the highest discounted total rewards (Fig. 5a). Also for the uniform grid aggregation, the computation increases quadratically with the size of the area map (Fig. 5b). These results further strengthen our observation (which follows from the nature of reward discounting under gentle assumptions) that immediate actions are influenced by nearby rewards and the farther low-resolution features enhance non-myopic planning of the complete trajectory. Hence, we used multi-resolution representation for further experiments.

Figure 6a–d present the paths generated by our sampling technique and the boustrophedonic sampler on both the test scenarios for 800 time-steps. The path generated by our technique clearly minimizes the sampling over sand and

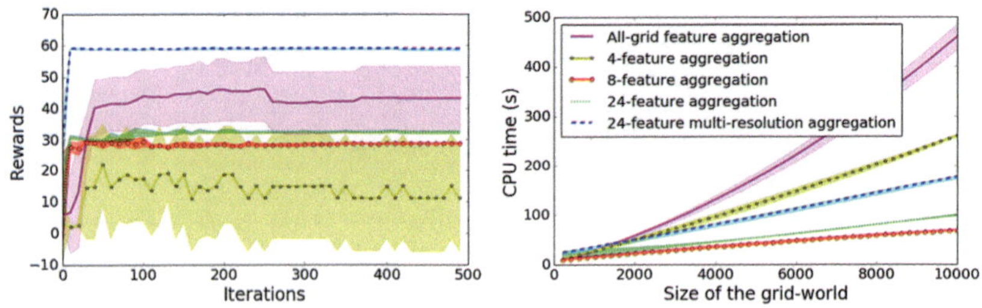

(a) Discounted total rewards (b) CPU-time vs. world size

Fig. 5. Results from evaluation of different feature aggregations. Shaded region indicates the standard deviation over five trials on three different sized maps.

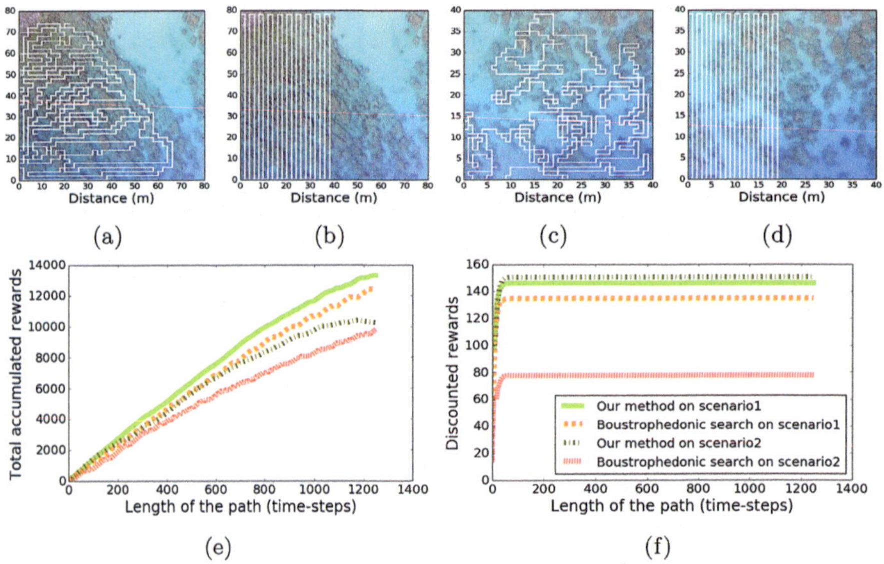

Fig. 6. Results comparing the policy search based sampling algorithm with boustrophedonic sampler. (a) and (c) Policy based sampling path for scenario1 and scenario2 respectively. (b) and (d) Boustrophedonic path for scenario1 and scenario2 respectively. (e) and (f) present the plots for total accumulated rewards and discounted rewards against the length of different sampling paths as indicated in the legends. We used $\gamma = 0.9$ for discounting over time.

maximizes the visual sampling of corals. The total score collected by both the approaches is comparable (Fig. 6e); however, the discounted reward achieved by our method is significantly higher when running on the scattered coral-head scenario (Fig. 6f). This indicates that the proposed approach tends to gather most score early on in the trajectory, resulting in better anytime performance.

5 Field Experiments

The example application considered in this paper is to collect the visual data of corals from the surface of water. Hence, the shallower the regions visited, the better is the quality of the coral images. In field experiments, we collect visual data of the reef with our sampling method and evaluate our technique for this specific application using the bathymetric data as a measure for shallowness of the region covered. Figure 7 presents the images captured at different locations of the reef region with varying depths. These images strengthen our hypothesis of covering shallower reefs to achieve high quality visual data of the corals.

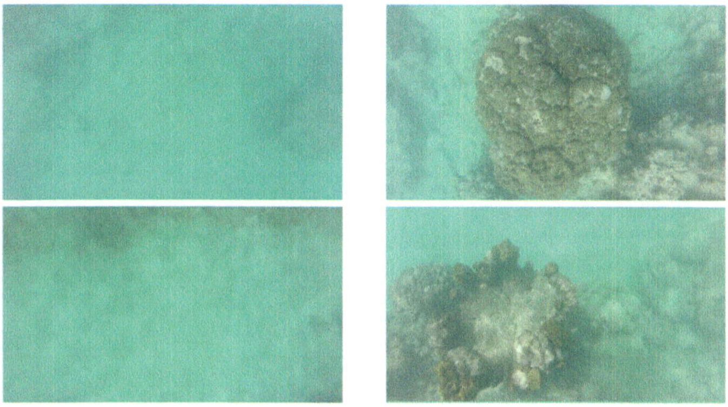

(a) (b)

Fig. 7. Quality of the visual data. Column (a) presents two images with no useful information. The reef is either too deep (>20 ft) for good visual samples from the surface or there are no coral-heads. Column (b) presents good quality visual samples of the coral-heads from shallower regions.

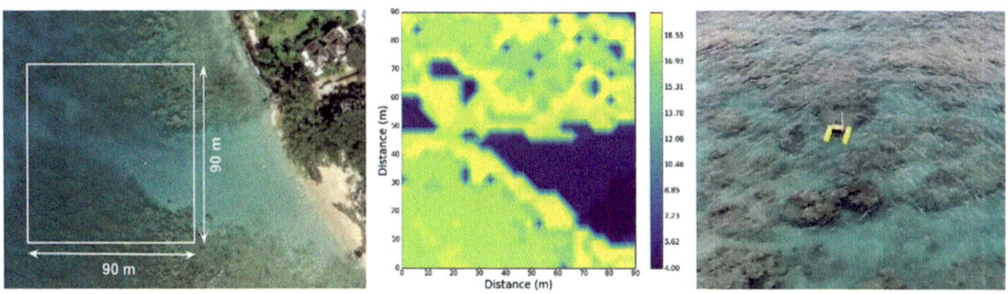

(a) (b) (c)

Fig. 8. Setup for field experiments. (a) Folkestone Marine Reserve area in Barbados with several coral outcrops. The region of interest is marked with a rectangular box. (b) Scoremap generated by processing the satellite images of the region of interest. Colorbar of the scoremap indicates the interestingness or score for the presence of corals. (c) Aerial image of the ASV performing visual data sampling.

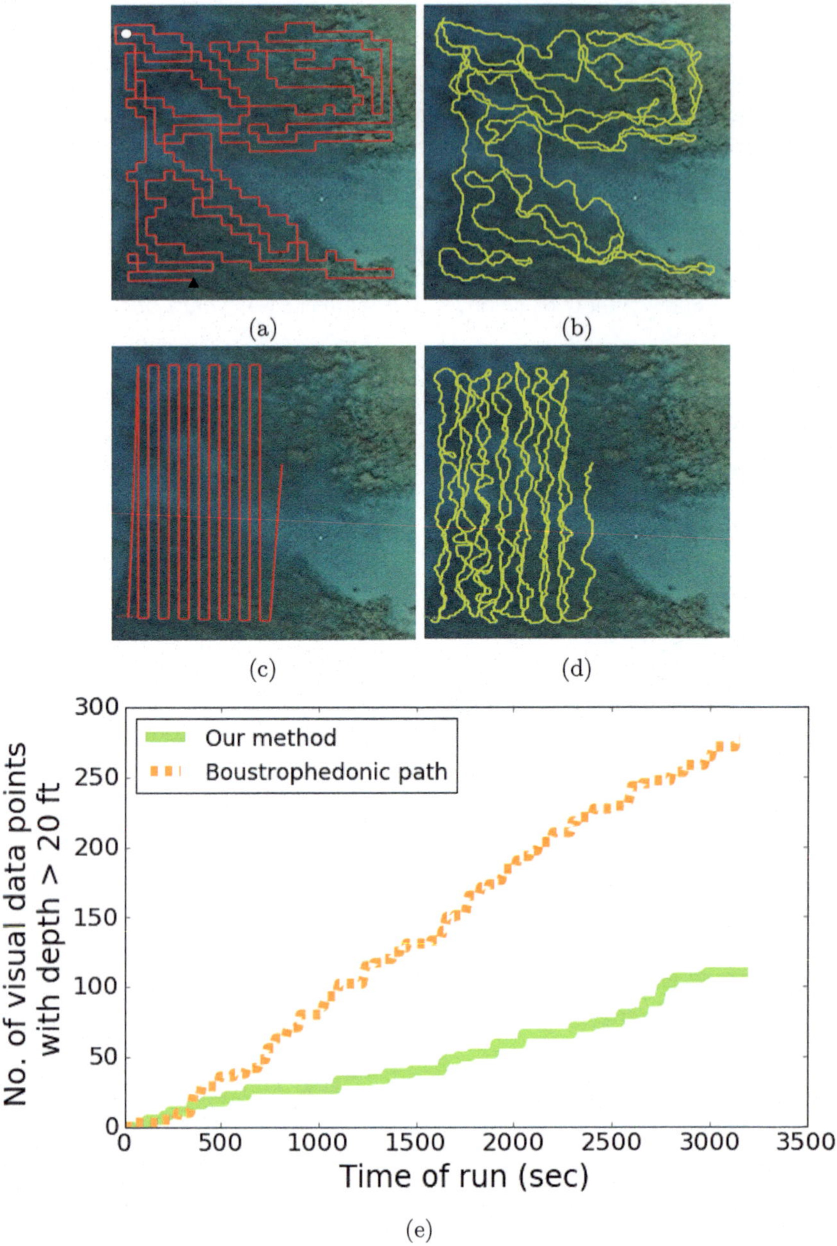

Fig. 9. Results from field experiments. (a) Planned path for visual data sampling using our non-uniform policy based sampling technique. White and black dots represent the start and end points of the path respectively. (b) The trajectory of the boat in field. The discrepancy between the planned and executed path is due to the sea-surface conditions during the trials. (c) and (d) are the planned and executed boustrophedonic paths. (e) Plot illustrating that the number of non-informative visual samples collected by a boustrophedonic sampler is almost three times the ones collected by our method.

We conducted field experiments at Folkestone Marine Reserve in Barbados (Fig. 8a), over a shallow region known to have several coral outcrops. The size of the region of interest considered in these experiments is 90 m × 90 m. The scoremap (Fig. 8b) is computed as a multispectral function of the satellite image of the reef in this region. The reef region is discretized into a grid-world to fit the scoremap of size 30×30. We deployed a custom made differential drive ASV (Figs. 1a and 8c) equipped with a sensor suite consisting of: a downward facing camera, a sonar pinger (1 Hz), a GPS receiver for localization, and a water-quality sensor-pack.

Our policy based sampling system was trained on the same scoremap used in Sect. 3 (Fig. 3b). The path generated by our method is presented in Fig. 9a. The generated path covers most of the high-scoring regions according to the scoremap used (Fig. 8b). This path is limited to a run of 40 min. The path in Fig. 9b illustrates the actual path executed by the autonomous boat and it is observed that the sea surface conditions have a considerable impact on the smoothness of the trajectory executed. It is possible for the policy from our approach to flexibly adapt to such distortions as it can be re-evaluated at each time step. Thus visiting spots that it missed now on a later pass, or skip parts that it accidentally visited too early.

We compare the coverage performed by our policy search method with a traditional exhaustive coverage technique using boustrophedonic path. It took 75 min for boustrophedonic path to completely cover the region of interest. Figure 9c illustrates the boustrophedonic path to perform sampling for the first 40 min. Figure 9e presents the total number of visual data points collected from regions which are deeper than 20 feet (i.e. visual data samples that are not useful to monitor the health of the corals) plotted against the time spent surveying the region. The total number of visual data points considered in this plot is constant for both the techniques. The comparison plot illustrates that the number of non-informative visual samples collected by a boustrophedonic sampler is more than twice the non-informative visual samples collected by our non-uniform policy based sampling method.

6 Conclusions and Experimental Insights

One of the novel contributions of this paper is to explore non-uniform state aggregation in policy search in the context of robotic path planning. The results suggest that such aggregation can have a major impact on the efficiency of state exploration and modeling as demonstrated by exhaustive experiments using real-but-stored data from real field deployments. We further validated this expectation in our field deployments. The incremental sampling-and-modeling paradigm we use, can be applied to many different domains where the benefits of efficient sample acquisition should accrue, but in the marine measurement domain in particular it is irrefutable that increased sampling efficiency has a major impact on the scale and feasibility of modeling efforts. For example, on the North and South Bellairs reefs where our experiments were conducted, the impact of weather and sea conditions (including tidal variations) place a significant premium on efficient sampling and have often curtained a measurement session prematurely. Likewise, this makes the use of anytime algorithms (like ours) especially important since an experiment may have to be terminated without much prior notice [1].

The direct application of our approach presupposes ongoing localization. For the near-shore ocean at moderate scales, this can be reliably achieved using traditional GPS, differential GPS, and related methods [17]. This need for ongoing localization applies to most methods, but by using non-uniform spatial sampling, it may be possible to better account for some types of pose estimation errors.

References

1. Zilberstein, S., Russell, S.J.: Anytime sensing, planning and action: a practical model for robot control. IJCAI **93**, 1402–1407 (1993)
2. Hoegh-Guldberg, O.: Climate change, coral bleaching and the future of the world's coral reefs. Marine Freshw. Res. **50**(8), 839–866 (1999)
3. Xu, A., Viriyasuthee, C., Rekleitis, I.: Optimal complete terrain coverage using an unmanned aerial vehicle. In: IEEE International Conference on Robotics and Automation (ICRA), pp. 2513–2519 (2011)
4. Choset, H., Pignon, P.: Coverage path planning: The boustrophedon cellular decomposition. In: Field and Service Robotics, pp. 203–209. Springer, London (1998)
5. Manjanna, S., Dudek, G.: Data-driven selective sampling for marine vehicles using multi-scale paths. In: IEEE/RSJ International Conference on Intelligent Robots and Systems (IROS), pp. 6111–6117 (2017)
6. Low, K.H., Dolan, J.M., Khosla, P.: Adaptive multi-robot wide-area exploration and mapping. In: Proceedings of the International Joint Conference on Autonomous Agents and Multiagent Systems, pp. 23–30 (2008)
7. Venkataramani, R., Bresler, Y.: Perfect reconstruction formulas and bounds on aliasing error in sub-Nyquist nonuniform sampling of multiband signals. IEEE Trans. Inf. Theory **46**(6), 2173–2183 (2000)
8. Rahimi, M., Hansen, M., Kaiser, W.J., Sukhatme, G.S., Estrin, D.: Adaptive sampling for environmental field estimation using robotic sensors. In: International Conference on Intelligent Robots and Systems (IROS), pp. 3692–3698. IEEE (2005)
9. Sadat, S.A., Wawerla, J., Vaughan, R.: Fractal trajectories for online non-uniform aerial coverage. In: IEEE International Conference on Robotics and Automation (ICRA), pp. 2971–2976 (2015)
10. Williams, R.J.: Simple statistical gradient-following algorithms for connectionist reinforcement learning. Mach. Learn. **8**(3–4), 229–256 (1992)
11. Baxter, J., Bartlett, P.L.: Infinite-horizon policy-gradient estimation. J. Artif. Intell. Res. **15**, 319–350 (2001)
12. Sutton, R.S., McAllester, D.A., Singh, S.P., Mansour, Y.: Policy gradient methods for reinforcement learning with function approximation. In: Advances in Neural Information Processing Systems, pp. 1057–1063 (2000)
13. Deisenroth, M.P., Neumann, G., Peters, J.: A survey on policy search for robotics. Found. Trends® Robot. **2**(1–2), 1–142 (2013)
14. Kober, J., Bagnell, J.A., Peters, J.: Reinforcement learning in robotics: a survey. Int. J. Robot. Res. **32**(11), 1238–1274 (2013)

15. Barto, A.G., Bradtke, S.J., Singh, S.P.: Real-time learning and control using asynchronous dynamic programming. University of Massachusetts at Amherst, Department of Computer and Information Science (1991)
16. Singh, S.P., Jaakkola, T., Jordan, M.I.: Reinforcement learning with soft state aggregation. In: Advances in Neural Information Processing Systems, pp. 361–368 (1995)
17. Dudek, G., Jenkin, M.: Inertial sensing, GPS, and odometry. In: Siciliano, B., Khatib, O. (eds.) Springer Handbook of Robotics, pp. 477–490. Springer, Cham (2016)

4-DOF Magnetic Field Based Localization for UAV Navigation

Harald Gietler[1(✉)], Christoph Böhm[2], Tobias Mitterer[1], Lisa-Marie Faller[1], Stephan Weiss[2], and Hubert Zangl[1]

[1] Sensors and Actuator Group, Alpen-Adria University Klagenfurt, Klagenfurt, Austria
Harald.Gietler@aau.at

[2] Control of Networked Systems Group, Alpen-Adria University Klagenfurt, Klagenfurt, Austria

Abstract. This paper presents a magnetic field based 4-degrees of freedom (DOF) localization approach for unmanned aerial vehicles (UAVs). The sensor system is comprised of two elements, a 3-axis magnetic field source attached to the mobile object, i.e. on the UAV, and a 3-axis magnetic field sensor being on a fixed position, e.g. ground. The transmitters are driven by signals in the low radio frequency (RF) range which ensures robustness against reflections and absorbing objects, making it ideally suitable for indoor usage. Additionally, its accuracy increases with decreasing distance between transmitter and receiver. Our system is thus complementary to vision based systems which experience difficulties when navigating close to ground level. This is mainly caused by their focal length and the loss of features. The proposed concept is validated in simulation and additionally with a rapid-prototyping-based experimental setup.

1 Introduction

Along with the growing number of autonomous UAVs, the demand of accurate localization approaches for navigational tasks increases. Recent research focuses on indoor navigation, a global positioning system (GPS)-denied environment, using visual and inertial sensors for pose estimation [1]. However, their accuracy decreases when approaching objects. Reasons are, e.g. motion blur and low overlap between consecutive images. Alternatively, ultra wide band (UWB) modules have been used for indoor localization enabling high accuracy [4] and the capability of data transmission at the same time. However, a direct line of sight is necessary between transmitter and receiver to maintain accuracy [5,11]. This issue is addressed by placing redundant UWB modules, which increases the chance to cover the targeted object. Quasi-static magnetic field based

The research leading to these results has received funding from the Austrian Ministry for Transport, Innovation and Technology (BMVIT) under the grant agreement 855777 (MODULES).

J. Xiao et al. (Eds.): ISER 2018, SPAR 11, pp. 222–231, 2020.
https://doi.org/10.1007/978-3-030-33950-0_20

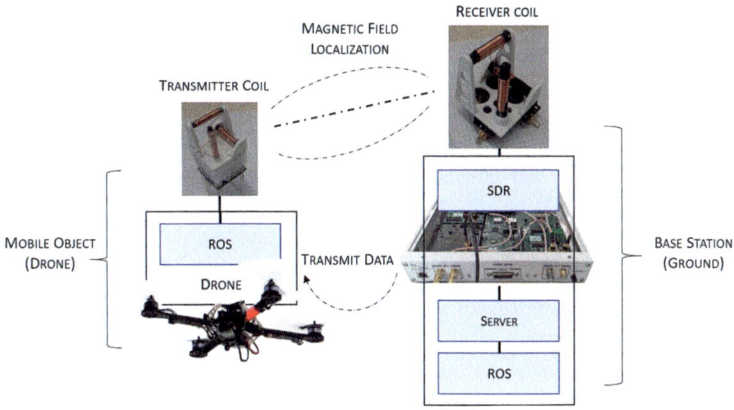

Fig. 1. Schematic overview of the system as developed for magnetic navigation.

sensors can be used to estimate the position of the UAV and its rotation around the z-axis.

In this paper, we introduce a tracking system (see Fig. 1), which is comprised of a 3-axis magnetic field source (transmitter) and a 3-axis magnetic field sensor (receiver). The quasi-static magnetic field source is made of a stack of three orthogonal coils, which are driven in time-interleaved operation. The sensor measures the generated magnetic field strength and calculates the relative position and yaw angle of the source with respect to the sensor. Each sensor coil measures one component of the generated magnetic field vector, using the fact that the local magnetic flux is proportional to the induced coil voltage. Since both, the transmitter and the receiver consist of 3 orthogonally placed coils, 9 unique measurements can be done.

2 Proposed System Concept

The proposed concept consists of a freely movable transmitting coil stack, e.g. attached to a UAV, and a receiving coil arrangement with a fixed position and attitude, e.g. on the ground (compare Fig. 1). The three coils are placed orthogonally to each other to cover the three spatial axis. Assuming near hovering of the UAV during landing, it is legitimate to assume that the z-axis of the transmitter and receiver are collinearly aligned and therefore 4-DOF remain. The rectangular driven transmitting coils are embedded in parallel oscillator circuitries and L-C impedance matching networks in order to boost the coil current i_L. Consequently, i_L is sinusoidally shaped and has a higher amplitude compared to the input current.

2.1 Magnetics Modelling

The current i_L generates a quasi-static magnetic field when flowing through the transmitter. For the landing scenario, the transmitting coil can be approximated by a magnetic dipole, i.e. if the position of the sensor is within the near field region and at least several times the coil radius away [6]. Then, the corresponding magnetic flux density $\mathbf{B}$ in air is given by

$$\mathbf{B}(\mathbf{r}, \mathbf{m}) = \begin{bmatrix} B_x \\ B_y \\ B_z \end{bmatrix} = \frac{\mu_0}{4\pi} \left(\frac{3\mathbf{r}(\mathbf{m}\mathbf{r})}{|\mathbf{r}|^5} - \frac{\mathbf{m}}{|\mathbf{r}|^3} \right). \tag{1}$$

The quantities $\mathbf{r} = \begin{bmatrix} x \ y \ z \end{bmatrix}^T$, $\mathbf{m}$ and μ_0 denote the relative position with respect to the source, the magnetic moment of the transmitting coil and the permeability of free space, respectively. The magnetic moment can be expressed as $\mathbf{m} = -N_{tx} i_L \mathbf{A}_T$, where N_{tx} and $\mathbf{A}_T$ represent the number of coil turns of the transmitting coil and its cross-sectional area, respectively. As the magnetic moment indirectly models the attitude [3] of the transmitting coil, it can be rewritten as

$$\mathbf{m} = -N_{tx} i_L |\mathbf{A}_T| \mathbf{R}, \tag{2}$$

where $\mathbf{R} \in \mathbb{R}^{3\times 3}$ is a rotational matrix. $\mathbf{R}$ can be modeled as rotation around the z-axis $\mathbf{R_z}$ with α being the rotation angle.

$$\mathbf{R_z} = \begin{bmatrix} \cos\alpha & -\sin\alpha & 0 \\ \sin\alpha & \cos\alpha & 0 \\ 0 & 0 & 1 \end{bmatrix} \tag{3}$$

For small receiving coils, a constant magnetic flux density over the cross sectional area can be assumed and the quasi-static magnetic field induces a voltage in the receiving coils, following the equation

$$V = -N_{rx} \mathbf{A}_R \frac{\delta \mathbf{B}}{\delta t}. \tag{4}$$

N_{rx} and $\mathbf{A}_R$ represent the number of turns of the receiving coil and its cross-sectional area, respectively. The induced voltage depends on the derivative of the magnetic flux density $\mathbf{B}$, which links the magnetic field source and the inductive sensor. Without any further synchronization between the transmitter and the receiver only the absolute value of the induced voltage's amplitude can be measured. The missing sign of the introduced voltage leads to additional ambiguities. In fact, all the possible sign permutations of the 9 unique voltage measurements need to be taken into account. Each of them would correspond to an underlying pose.

2.2 Estimation and Calibration Approach

To determine the transmitter's position and yaw angle from the measured magnetic field, an optimization algorithm is used. Three transmitter as well receiver coils result in 9 unique equations. Compared to 4-DOF, this results in an over-determined set of equations, which can be solved by non-linear least squares (NLS) based optimization approaches. The so-called "trust-region reflective" algorithm is employed to solve the NLS problem [10]. This method is very robust and can be used for ill-conditioned problems. Compared to other non-linear optimization approaches such as the Levenberg-Marquardt algorithm, it converges faster. The NLS problem can be represented by

$$\min \sum_{\mathbf{r},\mathbf{m} \in \mathbf{S}} (V - V_{IND})^2, \tag{5}$$

where V_{IND} represents the measured induced voltage and $\mathbf{S}$ represents the space of feasible poses. The introduced concept is suggested to be used for supporting landing manoeuvers. Thus, the receiver is mounted at ground level. Note, that e.g. negative z-positions are restricted for the landing scenario. In order to increase the measured signals, magnetic cores with high permeability are used. Therefore, gain factors c_t and c_r are added to the model of the transmitter and receiver coils leading to

$$\mathbf{m} = -c_t N_{tx} i_L |\mathbf{A}_T| \mathbf{R}, \tag{6}$$

$$V = -c_r N_{rx} \mathbf{A}_R \frac{\delta \mathbf{B}}{\delta t}. \tag{7}$$

The same optimization algorithm as described above is applied to (5) with respect to the gain factors in order to find the best calibration values for c_t and c_r based on as many known transmitter position and poses as possible.

3 Implementation

As shown in Fig. 1, the transmitter as well as the receiver coils are mounted in a 3D-printed skeleton. The dimensions of the coil stacks and its parameters are summarized in Table 1. While the dimensions are given for the prototype, the realization in a 3D-printed setup further enables future optimization regarding efficiency (weight, size, positioning, etc.) of the transmitter coils.

Table 1. Coil stack dimensions

TX coil stack	Value	RX coil stack	Value
x, y	26 mm	x, y	52 mm
z	45 mm	z	65 mm
Weight	11.2 g	Weight	65.5 g
Core radius	3 mm	Core radius	8 mm
Core length	25 mm	Core length	50 mm
Coil turns N_{tx}	55	Coil turns N_{rx}	64

On the transmitter side, the coils are driven by 3.3 V pulse-width modulation (PWM) outputs of a micro controller with a switching frequency of

$f_{SW} = 457$ kHz. The sending slots of the three transmitting coils are time-interleaved. A full sequence lasts 60 ms, which further also defines the maximum update rate of the system. The consumed current of the transmitting circuitry is approximately 4 mA. Note, that the PWM signal causes a sinusoidal current in the transmitting coil, due to the oscillation circuitry connected to the coil. On the receiver side, the induced voltage V_{IND} of each coil is amplified using a chain of low-noise amplifiers [2]. Subsequently, the signal is sampled with a high-resolution analog to digital converter (ADC). The digital signal is further low-pass filtered and an envelope detection is performed to retrieve the amplitude of the induced voltage. The analog to digital conversion as well as the signal processing are performed on a universal software radio peripheral (USRP). It is worth noting, that the amplified environmental noise floor is approximately 0.4 mV. This leads to noise in the reconstructed pose which, opposed to optical systems, decreases with decreasing distance. In order to retrieve the sign of the induced voltage with respect to the transmitter, transmitter and receiver are time-synchronized. The required synchronization accuracy depends on the used switching frequency and has to be at least lower than half the switching period. As introduced in (4), the induced voltage is proportional to the derivative of the magnetic flux density **B**. Since the current i_L is the only time depended quantity of **B**, it defines the contribution to V_{IND}. The derivative of the sinusoidally shaped current i_L introduces a 90° phase shift and the angular frequency ω as linear amplification factor. Consequently, higher switching frequencies lead to higher induced voltages and subsequently, increased range. Contrary, higher switching frequencies require more precise time-synchronization strategies. Figure 2 shows the relationship between switching frequency, required time-synchronization accuracy and the achievable range. The achievable range corresponds to a signal-to-noise ratio (SNR) of 20 dB with respect to 1 V, which results in useful pose estimation results. In general, it is not easy to achieve wireless time-synchronization in the µs range. However, recent approaches achieve accuracies in the ns region [7,9]. Note, that the achievable range also depends on various other parameters such as coil current i_L or coil geometries. These

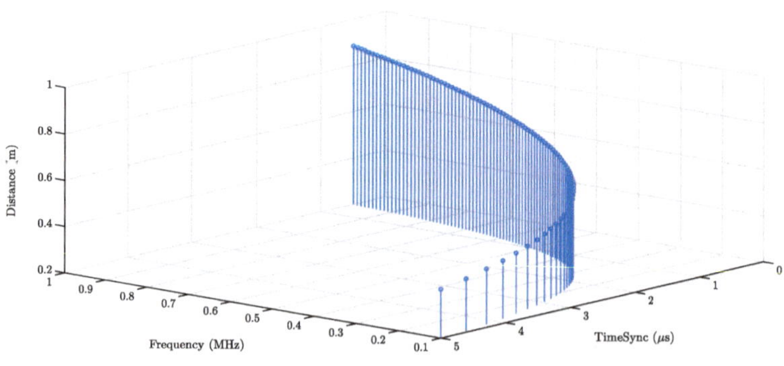

Fig. 2. The non-linear relationship between switching frequency, required time-synchronization accuracy and the achievable range is reported.

parameters have been kept fixed for the results shown in Fig. 2 at the previously described values.

4 Simulation Results

A MATLAB based simulation environment was used to proof the validity of the concept. The model parameters are taken from Table 1 and measurements using the experimental prototype. An arbitrarily chosen trajectory demonstrates that the accuracy of the introduced approach increases with decreasing distance. The receiver coil is located at the origin and is aligned with the z-axis. The transmitter is moving with a velocity of $0.2\,\mathrm{ms}^{-1}$. Figure 3 showcases the performance of the proposed localization system. To stabilize the convergence behavior of the solver, the previous estimation result is used as initial condition for consecutive optimization steps. However, if the relative noise level exceeds a certain threshold, convergence of the solver is not guaranteed. The achieved range, for the measured noise-level and given coil currents on the prototyping setup, is approximately 1 m. The maximum achievable distance depends on several factors. Note that some of them are design parameters. On one hand, the magnetic field decreases with distance to the power of 3. On the other hand, it is proportional to the generating coil current. Using stronger currents would therefore increase the maximum achievable distance. Further aspects are sensitivity and noise level, which heavily depend on the used amplification and processing hardware as well as environmental influences. Another design parameter is the used switching frequency. Higher frequencies lead to higher induced voltages, but reduces object penetration depth. Keeping these factors in mind, it can be expected that reasonable estimation of distances up to meters-range can be achieved.

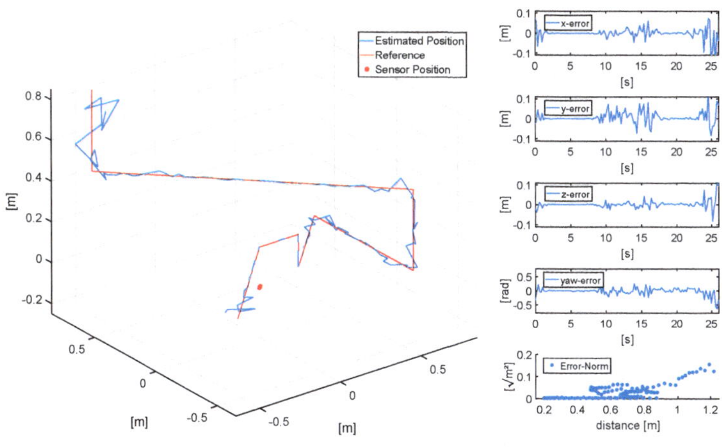

Fig. 3. A comparison between the reference trajectory shown as red waveform, and the estimated trajectory shown as blue waveform. Additionally, the errors in each dimension as well as the Euclidean error norm over the distance are presented.

5 Experimental Results

For experimental measurements, the OptiTrack [8] tracking system was used to get ground truth values of the transmitters and receivers pose, i.e. position and orientation. The mean error of the tracking system is approximately 0.7 mm for both, transmitter and receiver. Based on the measurements of the induced voltage in the receiver, the position and yaw angle of the transmitter are estimated. These estimates are compared to the ground truth values of the tracking system. Figure 4 reports a first experiment where mainly x- and y-movements are performed, while the z-position and yaw-angle remain almost constant. It can be seen, that the experiment includes positive and negative x- and y-positions. The estimation error is located in the low cm range.

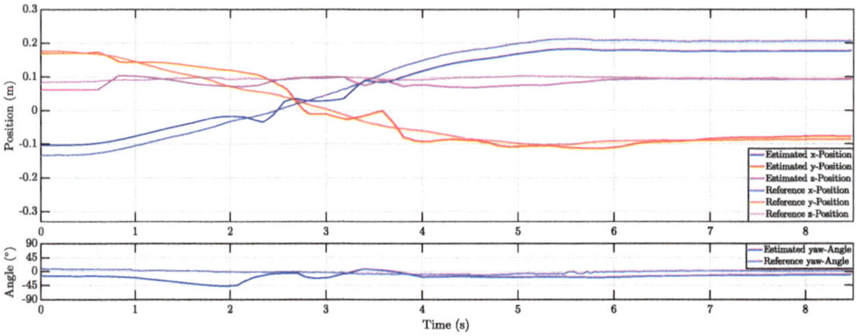

Fig. 4. An experiment where mainly movements in x- and y-direction are shown. The dotted lines represent the ground truth reference values, whereas the solid waveforms represent the estimates of the proposed sensor system.

A second experiment, including movements in each spatial axis, is shown in Fig. 5. The resulting estimation accuracy is comparable with the first experiment.

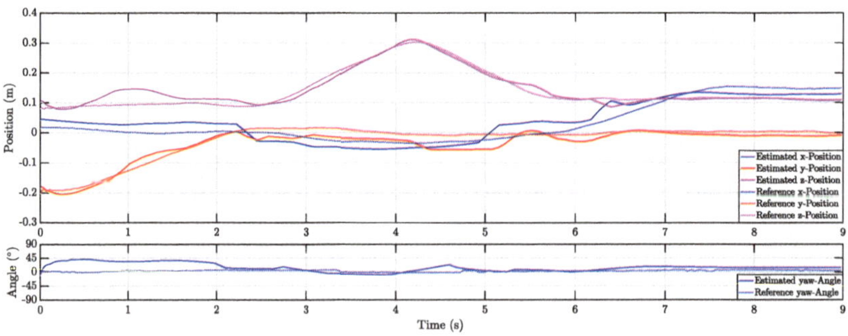

Fig. 5. Movements in all three spatial axis is shown. The dotted lines represent the ground truth reference values, whereas the solid waveforms represent the estimates of the proposed sensor system.

In the third experiment the localization behavior when altering the yaw-angle is investigated. The position components of the pose are kept almost constant. Again, the position error is located in the cm range, while the average yaw-angle estimation error is 9°, which is reported in Fig. 6.

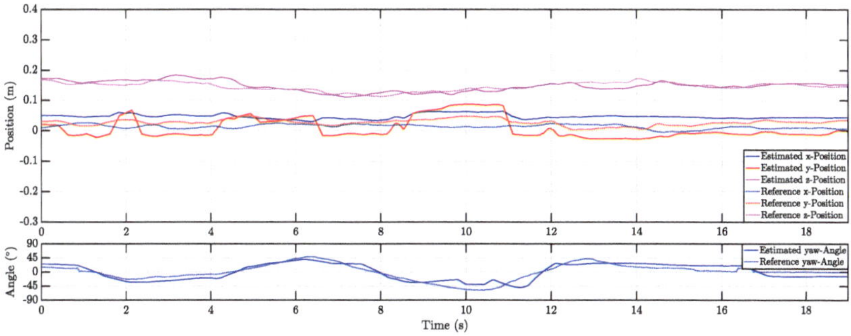

Fig. 6. Angular motion around the z-axis is investigated. The dotted lines represent the ground truth reference values, whereas the solid waveforms represent the estimates of the proposed sensor system.

The fourth experiment investigates the maximum achievable range for the current configuration. When the induced voltage falls below a certain threshold the sensor is deactivated and the last pose is kept, which can be seen in Fig. 7. The achieved range in this experiment was approximately 70 cm.

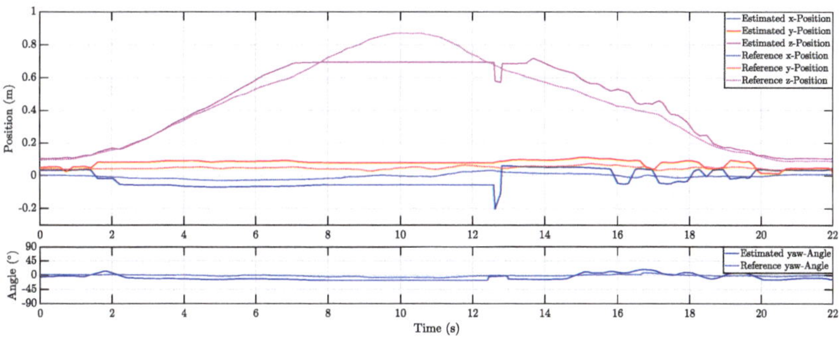

Fig. 7. The maximum range of the used hardware is investigated.

These experiments demonstrate the accuracy of the suggested approach. The noise influence, which becomes mainly obvious in the reconstruction of the pose, is due to the spatial arrangement of the transmitter and receiver coils. As described previously, orthogonality of the used coils is an important factor to avoid cross-talk. In order to eliminate these influences, we further suggest to use a more advanced coil arrangement. Separating the receiving coils can improve the estimation accuracy, due to the reduction of cross-talk. However, the achieved resolution of the employed localization exceeds that of the commonly used camera-based systems for landing scenarios. Additionally, it can be expected that the approach is robust with respect to rain, fog, dust and snow.

6 Conclusion

We introduced a quasi-static magnetic field based localization system for mobile platforms which is capable of estimating 4-DOF. The functionality of the introduced concept has been verified based on simulations and experimental results. It was shown, that the system's accuracy increases with decreasing distance. Additionally, also no direct line of sight is necessary. Concluding, we can state that the system we propose can well be used as complementary to camera-based localization approaches. Also because the proposed magnetic principle does not require a direct line of sight, but is unaffected by obstruction or fog.

References

1. Bloesch, M., Burri, M., Omari, S., Hutter, M., Siegwart, R.: Iterated extended Kalman filter based visual-inertial odometry using direct photometric feedback. Int. J. Robot. Res. **36**(10), 1053–1072 (2017). https://doi.org/10.1177/0278364917728574. https://doi.org/10.3929/ethz-b-000187364
2. Faller, L.M., Mitterer, T., Leitzke, J.P., Zangl, H.: Design and evaluation of a fast, high-resolution sensor evaluation platform applied to MEMS position sensing. IEEE Trans. Instrum. Meas. **67**(5), 1014–1027 (2018). https://doi.org/10.1109/TIM.2017.2771955
3. Ge, X., Wang, Y., Ding, N., Wu, X., Wang, Y., Fang, Z.: An electromagnetic tracking method using rotating orthogonal coils. IEEE Trans. Magn. **48**(12), 4802–4810 (2012). https://doi.org/10.1109/TMAG.2012.2203917
4. Hausman, K., Weiss, S., Brockers, R., Matthies, L., Sukhatme, G.S.: Self-calibrating multi-sensor fusion with probabilistic measurement validation for seamless sensor switching on a UAV. In: 2016 IEEE International Conference on Robotics and Automation (ICRA), pp. 4289–4296 (2016). https://doi.org/10.1109/ICRA.2016.7487626
5. Heydariaan, M., Mohammadmoradi, H., Gnawali, O.: Toward standard non-line-of-sight benchmarking of ultra-wideband radio-based localization. In: 2018 IEEE Workshop on Benchmarking Cyber-Physical Networks and Systems (CPSBench), pp. 19–24 (2018). https://doi.org/10.1109/CPSBench.2018.00010
6. Jackson, J.D.: Classical Electrodynamics, 3rd edn. Wiley, New York (1999). http://cdsweb.cern.ch/record/490457
7. Lim, R., Maag, B., Thiele, L.: Time-of-flight aware time synchronization for wireless embedded systems. In: EWSN, pp. 149–158 (2016)
8. NaturalPoint, Inc. DBA OptiTrack: Optitrack website (2018). https://optitrack.com/

9. Pérez-Solano, J.J., Felici-Castell, S.: Improving time synchronization in wireless sensor networks using Bayesian inference. J. Netw. Comput. Appl. **82**, 47–55 (2017). https://doi.org/10.1016/j.jnca.2017.01.007. http://www.sciencedirect.com/science/article/pii/S1084804517300164
10. Yuan, Y.X.: Recent advances in trust region algorithms. Math. Program. **151**(1), 249–281 (2015). https://doi.org/10.1007/s10107-015-0893-2
11. Zhang, C., Kuhn, M.J., Merkl, B.C., Fathy, A.E., Mahfouz, M.R.: Real-time noncoherent UWB positioning radar with millimeter range accuracy: theory and experiment. IEEE Trans. Microw. Theory Tech. **58**(1), 9–20 (2010)

Robust Vision-Based Autonomous Navigation, Mapping and Landing for MAVs at Night

Shreyansh Daftry[1(✉)], Manash Das[2], Jeff Delaune[1], Cristina Sorice[1], Robert Hewitt[1], Shreetej Reddy[3], Daniel Lytle[4], Elvin Gu[5], and Larry Matthies[1]

[1] Jet Propulsion Laboratory, California Institute of Technology, Pasadena, USA
Shreyansh.Daftry@jpl.nasa.gov
[2] Indian Institute of Technology, Kharagpur, India
[3] University of Pennsylvania, Philadelphia, USA
[4] University of Southern California, Los Angeles, USA
[5] Carnegie Mellon University, Pittsburgh, USA

Abstract. This paper is about vision-based autonomous flight of MAVs at night. Despite it being dark almost half of the time, most of the work to date has addressed only daytime operations. Enabling autonomous night-time operation of MAVs with low SWaP on-board sensing capabilities is still an open problem in current robotics research. In this paper, we take a step in this direction and introduce a robust vision-based perception system using thermal-infrared cameras. We present this in the context of safe autonomous landing on rooftop-like structures, and demonstrate the efficacy of our proposed system through extensive real-world flight experiments in outdoor environments at night.

Keywords: Micro Aerial Vehicles · Thermal-infrared · Night operation · Vision-based navigation · 3D mapping · Autonomous landing

1 Introduction

Micro Aerial Vehicles (MAVs) have built a formidable résumé by making themselves useful in a number of important applications, from disaster scene surveillance and package delivery to robots used in aerial imaging, architecture and construction. As these aerial robots aspire for long-term autonomous operation, the ability to navigate safely becomes critical. While vision-based methods for autonomous navigation of MAVs have been well studied in general [7,9,24], most of these approaches have addressed only daytime operation under normal illumination conditions using standard visible-spectrum cameras. Enabling operations at night would significantly enhance the tactical value of such a system.

In this paper, we propose to use thermal-infrared (TIR) cameras to enable vision-based autonomous operations for MAVs at night. The nature of TIR

J. Xiao et al. (Eds.): ISER 2018, SPAR 11, pp. 232–242, 2020.
https://doi.org/10.1007/978-3-030-33950-0_21

modality makes them highly robust to low-illumination conditions (including total darkness, as shown in Fig. 1) and other environmental effects such as the presence of fog, smoke and dust. Furthermore, COTS, uncooled, TIR cameras based on micro-bolometer focal planes are now small enough to be practical payloads on resource-constrained MAVs.

Despite the aforementioned advantages, these cameras also have several characteristics that are challenging for vision algorithms - limited resolution, low SNR, rolling shutter and motion-blur distortion, etc. Hence, the performance of traditional computer vision techniques developed for electro-optical imagery may not directly translate to the thermal domain. However, to date there have been very few attempts to study the use of thermal-infrared modality for visual navigation and mapping. Even fewer have validated them experimentally in the context of autonomous robotics through real-world experiments. Our goal and primary contribution in this work was to develop a robust end-to-end visual navigation, mapping and autonomous landing system, using a passive monocular thermal-infrared camera as its primary sensing modality.

Fig. 1. An instance of a scene captured with visible-spectrum camera during the day (left), at night (right) and with a TIR camera at night (right). It can be seen that the nature of TIR modality enables the MAV to have more robust night vision.

2 Related Work

Vision in Low-Illumination Conditions. The problem of autonomous navigation and mapping in low-illumination conditions has been widely discussed in the robotics community, in the context of unmanned ground vehicles or automated driving. Most of these approaches have either focused on the use of active sensors such as lidar [1,16], or the use of active illuminators [10,17]. Other approaches [18,20] have used consumer cameras at night, but rely on strong priors from the urban environment. These sensors and methods thus do not scale to MAVs that are constrained to low SWaP and operate in unstructured environments.

Vision with Thermal-Infrared. In the last few years, there have been several applications of thermal cameras in machine vision - ranging from object recognition and human activity detection [15] to visual odometry and SLAM [2,25].

See [14] for a comprehensive review. In particular, thermal-infrared modality has proven to be very useful in challenging environments, such as in the presence of obscurants [5,21] and darkness [6], where vision-based systems would otherwise fail [4]. The work of [19] is most closely related to ours, where a thermal stereo-pair is used onboard a MAV for visual navigation.

3 Technical Approach

In this section, we describe our proposed approach for robust autonomous flight at night. The primary focus of this paper is going to be the perception system, which is responsible for estimating the 3D scene structure, using a stream of thermal images and the associated pose of the vehicle. Figure 2 provides a schematic overview of the framework. The core components includes state-estimation and monocular dense 3D mapping. In addition, we also discuss the technical approach for our application of safe autonomous landing on rooftop-like structures.

3.1 State Estimation

Our state-estimation pipeline is based on VINS-MONO [23] - a tightly coupled visual-inertial odometry algorithm. It is an optimization-based SLAM framework that takes as input calibrated thermal images and inertial data, and produces high-accuracy pose estimates of the MAV in metric scale. There were various reasons to use this particular algorithm - it runs in real-time onboard the MAV, achieves state-of-the-art performance across several benchmark datasets in the visible domain [8] and has a formulation to compensate for rolling shutter distortions. In our experience, large rolling shutter distortions can be detrimental to VIO performance during high-speed flights. It is to be noted that while we evaluate the performance of our state-estimation pipeline, as discussed in Sect. 4.2, a comprehensive benchmarking of different VIO algorithms in the thermal-infrared domain is beyond the scope of this paper.

3.2 Monocular Dense Reconstruction

Structure from Motion (SfM) approaches [26] can be used to reconstruct 3D scene geometry from a moving monocular camera. However, such SfM approaches produce sparse maps that are not suited for collision-free navigation: typically the resulting point cloud of 3D features is highly noisy, and more crucially, the absence of visual features in regions with low texture does not necessarily imply free space. In contrast, dense approaches to 3D reconstruction use camera motion and variable-baseline stereo for depth map estimation. Here, we use a dense approach based on the REMODE (REgularized MOnocular Depth Estimation) algorithm [22]. The motivation behind using it is two fold: First, it performs the depth estimation with direct pixel intensities instead of indirect feature descriptors, making it more robust for low-contrast TIR images. Second, the highly parallelized formulation allows fast onboard computation on a GPU.

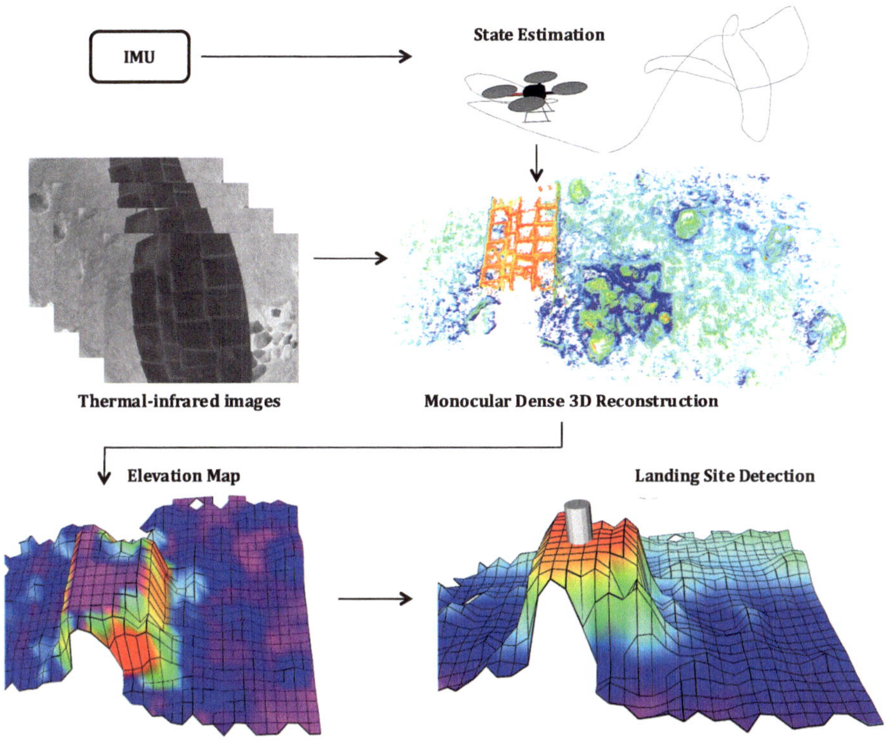

Fig. 2. System overview. The key components of the perception pipeline shown above include visual-inertial state estimation, 3D mapping and safe landing site detection

In the following we give a brief overview of the REMODE algorithm. The algorithm computes a dense depth map for selected reference views where the depth computation for a single pixel is formalized as a Bayesian estimation problem. A depth-filter is initialized for all pixels in every newly selected reference image $\mathbf{I}_r$ and every subsequent image $\mathbf{I}_k$ is then used to perform a recursive Bayesian update step of the depth estimates. Given a new observation $\{\mathbf{I}_k, \mathbf{T}_{wk}\}$, where $\mathbf{T}_{wr} \in SE(3)$ describe the pose of reference frame $\mathbf{r}$ relative to world frame $\mathbf{w}$, the 95% depth-confidence interval $[d_i^{\min}, d_i^{\max}]$ of the depth filter corresponding to pixel i into the image $\mathbf{I}_k$ and find a segment of the epipolar line l. Using a zero-mean sum of squared differences (ZMSSD) score on a 8×8 patch, we then search the pixel $\mathbf{u}_i'$ on the epipolar line l that has the highest correlation with the reference pixel $\mathbf{u}_i$. A depth measurement d_i^k corresponding to a pixel i in the image $\mathbf{I}_k$ is computed by triangulating the corresponding points $\mathbf{u}_i$ and $\mathbf{u}_i'$ from the views r and k respectively. After the depth map converged, we enforce its smoothness by applying a total-variation based image filter [3].

In practise, both the camera intrinsic matrix and the pose estimates from visual-inertial state estimation are not accurate enough. This results in inaccurate estimation of the fundamental matrix $\mathbf{F}$, and as a result depth-filters do not converge. In contrast, **F** derived from only the image correspondences are more accurate as they inherently capture the epipolar geometry. Thus, we use a direct optimization framework based on DSO [11], that computes the relative camera pose between two observations $\mathbf{T}_{rk}$, such that the intensity difference of the corresponding points $\mathbf{u}_i$ and $\mathbf{u}'_i$ is minimum using the cost function E_i

$$E_i = \sum_{i \in \mathcal{N}_p} \left\| \mathbf{I}_r[\mathbf{u}_i] - \mathbf{I}_k[\mathbf{u}'_i] \right\|_\gamma \tag{1}$$

To compensate for automatic gain changes between frames, an affine brightness transfer function given by $e^{-a_k}(\mathbf{I}_k - b_k)$ is added to Eq. 1. Thus, resulting in the following cost function:

$$E_i = \sum_{i \in \mathcal{N}_p} w_{\mathbf{u}_i} \left\| (\mathbf{I}_r[\mathbf{u}_i] - b_r) - \frac{e^{a_r}}{e^{a_r}} (\mathbf{I}_k[\mathbf{u}'_i] - b_k) \right\|_\gamma \tag{2}$$

where $w_{\mathbf{u}_i}$ is a gradient-dependent weighting given by:

$$w_{\mathbf{u}_i} = \frac{c^2}{c^2 + \left\| \nabla \mathbf{I}(\mathbf{u}) \right\|_2^2} \tag{3}$$

The relative camera pose $\mathbf{T}_{rk}$ here is initialized using the pose from the state-estimation pipeline. Given a depth measurement per pixel, we simply use the camera model to obtain the corresponding 3D point.

3.3 Vision-Based Safe Autonomous Landing

While the above described perception pipeline is applicable for vision-based navigation and obstacle avoidance in general, we study this specifically in the context of autonomous landing. To land safely and autonomously on different structures is an essential capability for many scenarios - perch-and-stare missions, to conserve or recharge batteries, etc. At minimum, this requires an ability to find safe landing spots through a semantic interpretation of the environment.

The first step towards this is to generate a 2.5D interpretation of the terrain. We use a recently developed robot-centric elevation mapping framework [12]. The map is stored as a 2-dimensional grid, where the height in each cell is modeled as a normal distribution. As the robot moves and generates instantaneous point clouds, the height estimates for each cell is updated using a recursive Bayesian update step. The goal of the original work was to develop a local map representation that serves foot-step planning for walking robots over and around obstacles. However, we find that the local two-dimensional elevation map is an efficient on-board map representation for MAVs [13] that are flying outdoors - it allows us to keep a safe distance to the surface and to detect and approach suitable landing spots. By tightly coupling the local map to the robot's pose, the framework can efficiently deal with drift in the pose estimate.

Once we have the elevation map, we are now in a position to determine the criteria for safe landing spots. We define a suitable landing site as: (1) approximately planar and level, (2) sufficiently large to permit MAV ingress, landing, and egress; and (3) free of obstacles to ensure vehicle safety. A cost function is defined using the above, and used to reason about a suitable landing site; then, we use motion planning and control to navigate to the desired location.

Fig. 3. (a) Quadrotor platform, (b) test setup during the day, and (c) at night.

4 Experiments and Results

In this section we analyze the qualitative and quantitative performance of our proposed system on publicly available benchmark datasets, and through real-world flight experiments on a MAV at night.

4.1 Experimental Setup

Hardware Platform. We use a Luminier QAV400 quad-rotor platform equipped with a down-ward facing thermal-infrared camera (We do experiments with both FLIR Boson and A65) and a MPU-9250 IMU, as shown in Fig. 3(a). Our flight computer is a Nvidia Jetson TX2 board, which incorporates a quad-core ARM processor and an embedded GPU with 256 CUDA cores.

Mission Scenario. In all our tests, we consider the perch-and-stare scenario, where the quad-rotor is tasked to explore a given area using a set of pre-defined waypoints, find a suitable elevated platform for surveillance and autonomously land on it. We created an analog test setup at JPL's Marsyard with artificial obstacles and a landing platform, as shown in Fig. 3.

System Overview. The software perception pipeline runs onboard the MAV. We use the ground-station only for communication and mission-level task planning. On average, the quadrotor was flying 2–3 m above the surface and used an elevation map of size 5×5 m, with a resolution of 10 cm per cell. During each flight, the quadrotor was restrained using a light-weight tether. Its to be noted that the tether is only for compliance to federal regulations and does not limit the feasibility of a free flight.

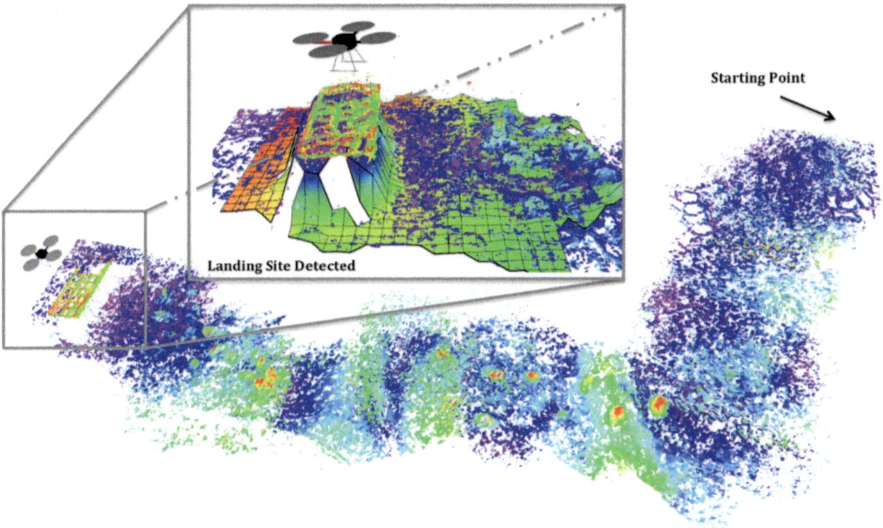

Fig. 4. Qualitative results from a flight test showing the 3D reconstruction of the scene (textured with thermal intensity), elevation map and the detected landing site.

4.2 Performance Evaluation

Several flight tests were conducted at different times of the night - ranging from dusk to midnight, to evaluate the robustness of the proposed method across varying illumination, thermal conditions and flight speed. Figure 4 shows the system performance during one of the flight sequence. From takeoff to landing, the state-estimation module was able to successfully track the pose of the vehicle. Furthermore, we visualize the 3D reconstruction of the scene generated during the flight. Note: The point cloud is textured using the thermal intensity images. It can be seen that even though the quadrotor was flying in total darkness, over a terrain with low texture, it is still able to accurately reconstruct the scene. This reconstruction is both dense and semantically rich; several important properties regarding the terrain can easily be inferred - the shape and size of rocks, thermal properties of the different terrains such as bedrock vs. sand pits, etc.

Overall, the MAV was able to complete each mission successfully - navigate, map the environment and find a suitable landing site. Furthermore, to quantitatively evaluate the performance of our system, we setup an outdoor motion-capture arena to obtain ground-truth for all our flight experiments. The average position drift of the pose estimation w.r.t ground truth in x, y and z axis was 0.1112 m, 0.02556 m and 0.04978 m respectively, which is around 1% of the distance traveled. Similarly, an average orientation drift of 3.171° was observed across all the flight tests.

Computationally, the state estimation pipeline runs at 30 Hz and consumes 2 cores on CPU, while the monocular dense reconstruction runs at 1 Hz and uses about 35% of the GPU, on average. The elevation mapping and landing package uses another core, and the fourth core is reserved for the camera driver, communication, and control. It was observed that an update rate of 1 Hz is sufficient to always maintain a dense elevation map below the MAV. However, when moving in a straight line, the local elevation map behind the MAV is more populated than in the front. In the future, we will modify the MAV to have a slightly forward facing camera or consider using a multi-camera setup in order to have better distribution.

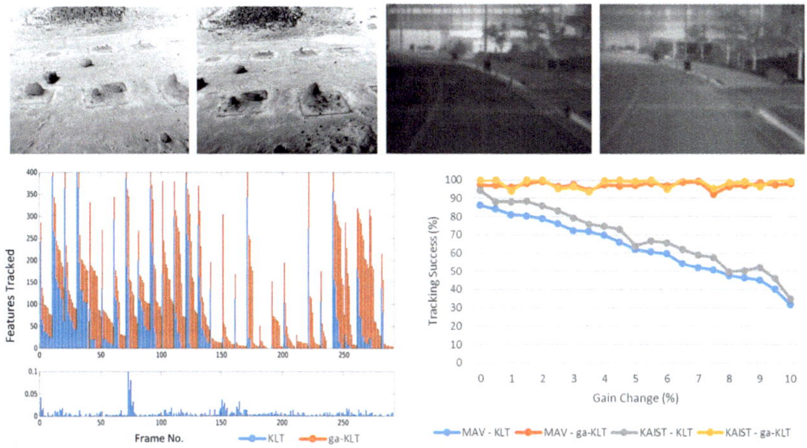

Fig. 5. Analysis of degraded performance due to AGC. (Top) Nearby frames from an image sequence illustrating the significant change in appearance as a result of AGC. (Bottom-Left) Features tracked and the corresponding gain change for one of the day-time flight sequence, (Bottom-Right) Tracking success as a function of gain-change.

4.3 Experimental Insights

A key insight obtained in this work is that applying vision-based techniques developed for visible-spectrum directly to thermal-infrared domain requires careful design considerations to account for the nuances of thermal-imaging. In particular, accounting for photometric effects like gain-change and rolling-shutter distortion can improve the robustness of perception tasks.

Automatic Gain Change. In the context of vision-based methods with thermal-images, one of the most important challenge we identified was with respect to rapid automatic gain change (AGC). As hot objects move in and out of the TIR camera's field of view, large regions of pixels saturate, and the sensor adjusts gain to darken the image in an attempt to avoid saturation. The resulting change in intensity can be dramatic from one frame to the next (See Fig. 5), invalidating the constant brightness assumption enforced in most vision-front end solutions. To quantify the effect of such rapid gain, we compare the feature tracking performance of standard KLT tracker to a gain-adaptive version (using the open-source code from [27]). Note: These experiments were conducted with thermal images from the FLIR A65 camera; the effects of AGC in FLIR Boson camera was minimal and hence we did not see similar improvements in tracking performance by compensating for AGC change.

Rolling Shutter Distortion. We qualitatively evaluated our state-estimation pipeline with- and without rolling shutter compensation. It was observed that while this can be neglected for MAV flights in slow dynamics, rolling distortions can be significant during aggressive motions; leading to state-estimation failure. In future work, we will also extend our dense reconstruction pipeline to compensate for rolling shutter effects.

5 Conclusion

In this paper, we take a step towards enabling autonomous MAV flight at night. We present a real-time perception system that employs thermal-infrared cameras for sensing, models its nuances during algorithm design and selection, and is able to perform robust state-estimation and monocular dense 3D mapping. Quantitatively and qualitatively, we show improvements in performance over simply re-using visible-spectrum algorithms, through extensive real-world experiments. We have demonstrated our system in the context of safe autonomous landing on rooftop-like structures. However, our method and findings apply to any aspect of experimental robotics where an autonomous system has to operate at night.

Acknowledgement. The research was carried out at the Jet Propulsion Laboratory, California Institute of Technology, under a contract with the National Aeronautics and Space Administration.

References

1. Barfoot, T.D., McManus, C., Anderson, S., Dong, H., Beerepoot, E., Tong, C.H., Furgale, P., Gammell, J.D., Enright, J.: Into darkness: visual navigation based on a lidar-intensity-image pipeline. In: Robotics Research, pp. 487–504. Springer (2016)
2. Borges, P.V.K., Vidas, S.: Practical infrared visual odometry. IEEE Trans. Intell. Transp. Syst. **17**(8), 2205–2213 (2016)
3. Bresson, X., Esedoglu, S., Vandergheynst, P., Thiran, J.P., Osher, S.: Fast global minimization of the active contour/snake model. J. Math. Imaging Vis. **28**(2), 151–167 (2007)
4. Brunner, C., Peynot, T., Vidal-Calleja, T., Underwood, J.: Selective combination of visual and thermal imaging for resilient localization in adverse conditions: day and night, smoke and fire. J. Field Robot. **30**(4), 641–666 (2013)
5. Burian, F., Kocmanova, P., Zalud, L.: Robot mapping with range camera, CCD cameras and thermal imagers. In: 2014 19th International Conference on Methods and Models in Automation and Robotics (MMAR), pp. 200–205 (2014)

6. Chen, L., Sun, L., Yang, T., Fan, L., Huang, K., Xuanyuan, Z.: RGB-T SLAM: a flexible slam framework by combining appearance and thermal information. In: 2017 IEEE International Conference on Robotics and Automation (ICRA), pp. 5682–5687 (2017)
7. Daftry, S., Zeng, S., Khan, A., Dey, D., Melik-Barkhudarov, N., Bagnell, J.A., Hebert, M.: Robust monocular flight in cluttered outdoor environments. arXiv preprint arXiv:1604.04779 (2016)
8. Delmerico, J., Scaramuzza, D.: A benchmark comparison of monocular visual-inertial odometry algorithms for flying robots. Memory **10**, 20 (2018)
9. Dey, D., Shankar, K.S., Zeng, S., Mehta, R., Agcayazi, M.T., Eriksen, C., Daftry, S., Hebert, M., Bagnell, J.A.: Vision and learning for deliberative monocular cluttered flight. In: Proceedings of the International Conference on Field and Service Robotics (FSR) (2015)
10. Dubbelman, G., van der Mark, W., van den Heuvel, J.C., Groen, F.C.: Obstacle detection during day and night conditions using stereo vision. In: IEEE/RSJ International Conference on Intelligent Robots and Systems, IROS 2007, pp. 109–116 (2007)
11. Engel, J., Koltun, V., Cremers, D.: Direct sparse odometry. IEEE Trans. Pattern Anal. Mach. Intell. **40**(3), 611–625 (2018)
12. Fankhauser, P., Bloesch, M., Gehring, C., Hutter, M., Siegwart, R.: Robot-centric elevation mapping with uncertainty estimates. In: Mobile Service Robotics, pp. 433–440. World Scientific (2014)
13. Forster, C., Faessler, M., Fontana, F., Werlberger, M., Scaramuzza, D.: Continuous on-board monocular-vision-based elevation mapping applied to autonomous landing of micro aerial vehicles. In: 2015 IEEE International Conference on Robotics and Automation (ICRA), pp. 111–118 (2015)
14. Gade, R., Moeslund, T.B.: Thermal cameras and applications: a survey. Mach. Vis. Appl. **25**(1), 245–262 (2014)
15. Han, J., Bhanu, B.: Human activity recognition in thermal infrared imagery. In: IEEE Computer Society Conference on Computer Vision and Pattern Recognition-Workshops 2005 CVPR Workshops, p. 17 (2005)
16. Husain, A., Jones, H., Kannan, B., Wong, U., Pimentel, T., Tang, S., Daftry, S., Huber, S., Whittaker, W.L.: Mapping planetary caves with an autonomous, heterogeneous robot team. In: Aerospace Conference, pp. 1–13. IEEE (2013)
17. Mascarich, F., Khattak, S., Papachristos, C., Alexis, K.: A multi-modal mapping unit for autonomous exploration and mapping of underground tunnels. In: 2018 IEEE Aerospace Conference, pp. 1–7 (2018)
18. Milford, M.J., Wyeth, G.F.: SeqSLAM: visual route-based navigation for sunny summer days and stormy winter nights. In: 2012 IEEE International Conference on Robotics and Automation (ICRA), pp. 1643–1649. IEEE (2012)
19. Mouats, T., Aouf, N., Chermak, L., Richardson, M.A.: Thermal stereo odometry for UAVs. IEEE Sens. J. **15**(11), 6335–6347 (2015)
20. Nelson, P., Churchill, W., Posner, I., Newman, P.: From dusk till dawn: localisation at night using artificial light sources. In: 2015 IEEE International Conference on Robotics and Automation (ICRA), pp. 5245–5252 (2015)
21. Papachristos, C., Mascarich, F., Alexis, K.: Thermal-inertial localization for autonomous navigation of aerial robots through obscurants. In: 2018 International Conference on Unmanned Aircraft Systems (ICUAS), pp. 394–399 (2018)
22. Pizzoli, M., Forster, C., Scaramuzza, D.: Remode: probabilistic, monocular dense reconstruction in real time. In: 2014 IEEE International Conference on Robotics and Automation (ICRA), pp. 2609–2616. IEEE (2014)

23. Qin, T., Li, P., Shen, S.: VINS-Mono: a robust and versatile monocular visual-inertial state estimator. arXiv preprint arXiv:1708.03852 (2017)
24. Shen, S., Michael, N., Kumar, V.: Autonomous indoor 3D exploration with a micro-aerial vehicle. In: 2012 IEEE International Conference on Robotics and Automation (ICRA), pp. 9–15. IEEE (2012)
25. Vidas, S., Sridharan, S.: Hand-held monocular slam in thermal-infrared. In: 2012 12th International Conference on Control Automation Robotics & Vision (ICARCV), pp. 859–864 (2012)
26. Wu, C.: Towards linear-time incremental structure from motion. In: 2013 International Conference on 3D Vision-3DV 2013, pp. 127–134. IEEE (2013)
27. Zach, C., Gallup, D., Frahm, J.M.: Fast gain-adaptive KLT tracking on the GPU. In: IEEE Computer Society Conference on Computer Vision and Pattern Recognition Workshops, CVPRW 2008, pp. 1–7. IEEE (2008)

Robot Learning I

Robot Learning I

Kostas Bekris

The first session on Learning covered aspects on how perception or control processes trained on simulated data can be transferred to real robots. This included a paper by Li et al. on an automated rendering pipeline for rapidly generating labeled datasets for instance-based image segmentation for robotic manipulation. The paper by Viereck et al. focused on transferring control policies learned in simulation onto a real robot. This was performed in the context of a "category level" manipulation task where a control policy was learned to enable a robot to perform a mating task for novel objects. The contribution by Julian et al. first used simulation to jointly learn a set of diverse policies that were parameterized in a way to make them easily reusable and composable. It was then the responsibility of the composition process to achieve transferability to the real system. Finally, the work by Mao et al. investigated object perception for classification or recognition based on touch sensing, aiming at enabling robots to operate even in darkness. It was shown that a classifier trained over simulated wraps using a continuum manipulator was transferable to identifying real world objects with the real system.

Learning Accurate Objectness Instance Segmentation from Photorealistic Rendering for Robotic Manipulation

Siyi Li[1(✉)], Jiaji Zhou[2], Zhenzhong Jia[2], Dit-Yan Yeung[1], and Matthew T. Mason[2]

[1] The Hong Kong University of Science and Technology, Clear Water Bay, Hong Kong
{sliay,dyyeung}@cse.ust.hk
[2] Carnegie Mellon University, Pittsburgh, USA
{jiajiz,zhenzjia,matt.mason}@cs.cmu.edu

Abstract. Recent progress in computer vision has been driven by high-capacity deep convolutional neural network (CNN) models trained on generic large datasets. However, creating large datasets with dense pixel-level labels is extremely costly. In this paper, we focus on the problem of instance segmentation for robotic manipulation using rich image and depth features. To avoid intensive human labeling, we develop an automated rendering pipeline for rapidly generating labeled datasets. Given 3D object models as input, the rendering pipeline produces photorealistic images with pixel-accurate semantic label maps and depth maps. The synthetic dataset is then used to train an RGB-D segmentation model by extending the Mask R-CNN framework for depth input fusion. Our results open up new possibilities for advancing robotic perception using cheap and large-scale synthetic data.

Keywords: Instance segmentation · Robotic manipulation · Synthetic data

1 Introduction

Over the past two decades, robotic solutions have been increasingly deployed in various domains, including households, industrial manufacturing, warehouse automation, and so on. However, it is still challenging for robots to perform complex tasks autonomously in unstructured and unknown environments. Grasping and manipulating [1] individual items in cluttered and constrained space is such a problem remaining largely unsolved. Consider the process of picking up all the objects from a cluttered bin, the objects could be arranged with partial or

Electronic supplementary material The online version of this chapter (https://doi.org/10.1007/978-3-030-33950-0_22) contains supplementary material, which is available to authorized users.

J. Xiao et al. (Eds.): ISER 2018, SPAR 11, pp. 245–255, 2020.
https://doi.org/10.1007/978-3-030-33950-0_22

even full occlusion. Heavy stacking of several homogeneous objects also poses non-trivial challenges for the robot to determine the correct number of those instances. These issues must be addressed from the perception side.

Remarkable progress has been made by utilizing deep learning to perception problems in recent years. Those promising results are typically obtained by training Convolutional Neural Networks (CNNs) using an extensive amount of labeled data. Though there exist unsupervised methods such as convolutional autoencoders to reduce the required labeled training samples, the performance of these approaches is less satisfying. Therefore a key question is how to create such large datasets with minimal human labor.

Recently an increasingly popular approach is to avoid manual annotations by using synthetic data generated by simulators. While labeled data are difficult to obtain in the real world, they are very easy to generate in the simulator. It has been shown that CNNs trained in synthetic data exhibit reasonable generalization performance on several perception tasks such as object detection and pose estimation [2,3]. This opens up new promises for transferring deep perception models trained in simulators to real-world robotic scenarios.

In this paper we study perception for robotic manipulation in cluttered environments. The perception capability to segment each object instance is crucial for successful manipulation. We focus on the instance segmentation problem, which requires the detection of all objects in an image while also precisely segmenting each instance. We develop an automated rendering pipeline to generate labeled datasets for training CNNs. Our pipeline takes 3D CAD models as input and produces photorealistic labeled images with enough variety. We train an RGB-D objectness instance segmentation system by extending Mask R-CNN [4] to leverage depth sensor information. Through extensive experiments, we show that models trained on photorealistic synthetic data allow for accurate object instance segmentation in real-world manipulation scenarios.

2 Related Work

Object detection and segmentation has been dramatically improved in recent years. For object detection, one successful line of work considers the task as region proposal followed by classification and scoring. The so-called Region-based CNN (R-CNN) framework [5,6] has evolved much both in terms of accuracy and efficiency. The success of semantic segmentation dates back to the advent of Fully Convolutional Networks (FCNs) [7]. The recent Mask R-CNN [4] extends R-CNN framework by adding a small FCN branch for predicting segmentation masks and is currently the state of the art for object instance segmentation. To utilize deep learning in robotic vision, most existing work attempts to make use of the real-world data in an efficient manner. These methods either set up a self-supervised training pipeline in real world [8] or apply transfer learning by using a small amount of real-world samples [9].

Synthetic data has been extensively used in computer vision. Modern open-world computer games have been used to create perception benchmark [10]. 3D object models are also increasingly used for training neural networks for object detection [11], human pose estimation [12] and viewpoint estimation [13].

Synthetic data has also been applied to robotic vision. Sadeghi *et al.* [14] learn a policy for real-world quadrotor obstacle avoidance by using a simulator with varied 3D scenes and textures. [15] demonstrates that the behavior of a high-level control policy trained in a physics simulator can be transferred to real-world quadrotor for autonomous target following. In [2], the authors propose a self-supervised learning system for object detection, where the initial model is pre-trained using synthetic images and fine-tuned over real-world images. [3] explores the domain randomization technique in pose estimation by training only in simulated images with non-realistic random textures.

3 Synthetic Data Generation

Figure 1 shows the pipeline for synthetic data generation. Our rendering system is based on Blender [16]. Our 3D CAD model database is a subset of YCB object dataset [17] containing 20 objects. For each scene, we first choose a subset of objects from the model database with random initial poses above the bin. Then the Bullet physics [18] engine simulates the objects falling into the bin by gravity before settling down into a stable equilibrium state. The final poses will then be rendered with the specific setting of lighting condition and camera view to generate the final RGB output.

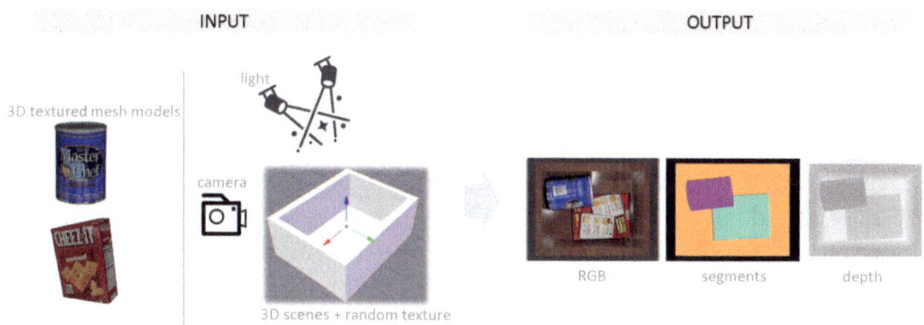

Fig. 1. Our pipeline for generating synthetic data.

Fig. 2. Sample RGB images from our generated dataset.

Object Poses: We maintain a set of dropping positions relative to the bin to make sure that object will mostly drop into the bin. The orientations are sampled from $\{(0,0,0),(\pi/2,0,0),(0,\pi/2,0),(0,0,\pi/2)\}$ to cover all possible views.

Lighting and Camera: We use point lighting source, which is varied with respect to location and intensity to create shadows and over exposures. The projective camera has a resolution of 640×480, focal length of 35 mm. The camera position is sampled from a pyramid on top of the bin center.

Domain Randomization: We also domain randomize the background bin. In particular, the random textures are chosen from random RGB values or sampled from online material texture resources. Figure 2 shows some samples.

4 Learning Objectness Instance Segmentation

Our instance segmentation system is based on Mask R-CNN [4]. We develop several techniques to leverage Mask R-CNN with depth input for instance segmentation in the manipulation scenario.

Objectness Training: Instead of performing multi-class instance segmentation, here we only care about objects versus non-objects. Therefore we utilize *objectness* Mask R-CNN to distinguish objects from non-objects (background). Intuitively, learning the notion of objectness instance makes it easier for the model to generalize to novel objects unseen in the training set.

Model Architecture: Our model uses the ResNet-50 [19] with Feature Pyramid Network (FPN) [20] as the convolutional backbone due to its accuracy and efficiency. The original Mask R-CNN only takes an RGB image as input and produces a set of detection bounding boxes and mask maps. To make full use of RGB-D cameras, we extend the Mask R-CNN framework to leverage depth information. Figure 3 shows the proposed multi-modal network architecture, where both the RGB input and the depth input (replicated across channels to match the RGB) are fed into the shared ResNet-50 tower. Then the feature maps are concatenated across channels, followed by an additional convolutional layer to merge the features. The rest of the network remains the same. One advantage of this design is that both the RGB input and depth input can utilize available ResNet-50 backbone pre-trained in ImageNet [21] dataset. The whole network is then fine-tuned for the segmentation task.

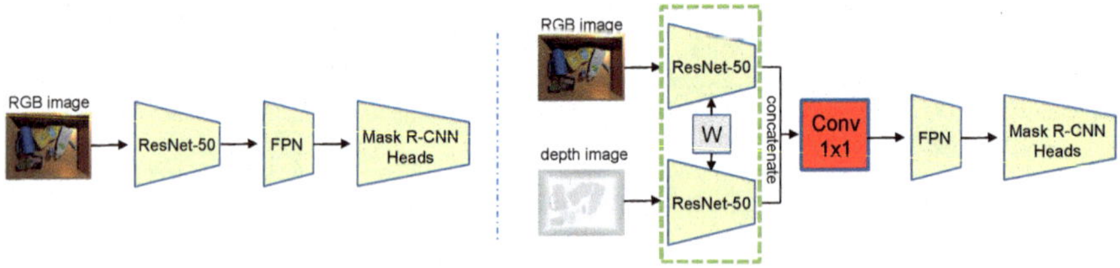

Fig. 3. An illustration of the original Mask R-CNN architecture (left) and the proposed RGB-D Mask R-CNN architecture (right).

5 Multi-view Real-World Data Generation

We also develop an automated approach to efficiently generate real-world labeled data by leveraging robots with wrist mounted cameras. We program the robot to take color and depth images from multiple views by generating camera poses in a pyramid on top of the table center with a gazing constraint. For each scene, the robot takes more than fifty views of images, five of which are manually labeled. We then use a correspondence algorithm described below to label each point with consistent instance identity (ID) from the separate three views in the fused point cloud. Finally, we perform projection of the labeled point cloud onto each unlabeled view and generate segmentation masks for each of them. Figure 4 shows the pipeline of the fusing algorithm.

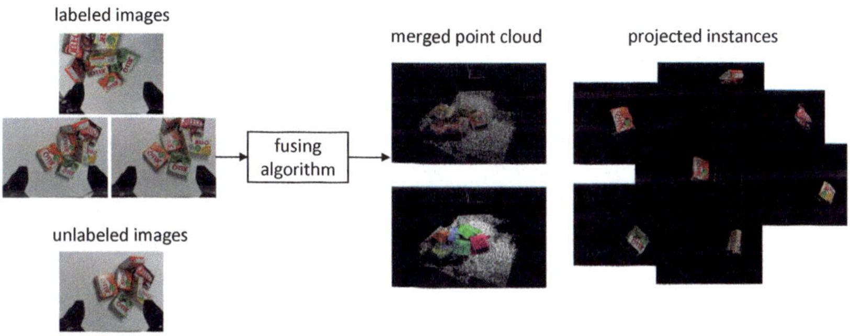

Fig. 4. Multi-view label fusing pipeline.

Multi-view Instance Matching: For two different labeled views, the two point clouds corresponding to the same instance usually do not share the same label ID. We assume that for the labeled views, the corresponding point clouds of the same instance viewed from different camera poses still share a certain amount of overlapping area. This can be then used to decide instance correspondence and merging, as described as follows.

1. We construct an occupancy grid out of the existing merged point cloud (initialized from the first point cloud): each grid g_{ijk} stores the points that belong to the grid and the label of the grid l_{ijk} is determined as the majority label among all points in the grid.
2. To merge a new labeled view, for each point p in the point cloud PC with label l that is mapped to a grid g_{ijk} with label l_{ijk}, we then add the overlapping count among label ID l in the new view with l_{ijk} in the existing aggregate point cloud. These counts between labels of the new view and labels of the existing views form a correlation matrix.
3. We use best weighted bipartite graph matching treating the correlation matrix as a weight matrix and merge the points and labels according to the matching result.

6 Experiments

In this section, we present a series of experiments to evaluate performance of the proposed objectness instance segmentation system. We first conduct extensive ablation studies on synthetic data to validate both the data generation design choices and the capability of the Mask R-CNN model. Then we evaluate the full system on the real-world dataset generated by the multi-view instance matching algorithm.

6.1 Implementation Details

We train the model using images of size 640×480 without resizing. We train on 2 GTX 1080Ti GPUs for 100k iterations, with a learning rate of 0.025 which is reduced by 10 times at the 40k iteration. We use Synchronized SGD with a weight decay of 0.0009 and a momentum of 0.9. Other hyper-parameters are set according to the original paper [4]. To leverage features trained from a larger image domain, the convolutional backbone ResNet-50 is pre-trained on ImageNet [21] for 1000-way object classification. The whole model is then fine-tuned on our generated synthetic dataset to learn general objectness instance segmentation.

6.2 Evaluation Metrics

We use the commonly average precision (AP) metric. Given an image, the model outputs a number of mask predictions sorted by confidence values. A mask prediction is considered as positive if its IoU (intersection-over-union) with the ground truth is above a certain threshold. By choosing different number of predictions, it gives a plot of the precision against the recall values. AP can be viewed as finding the area under the precision-recall graph. We report the COCO-style metrics [22] including AP_{50} (AP over IoU 0.5), AP_{75} (AP over IoU 0.75) and AP (average AP over IoUs ranging from 0.5 to 0.95 with a step of 0.05).

6.3 Data Generation Design Choices

With the ability of rapidly generating a large amount of labeled data, one may ask questions of "how much data is needed?" and "which type of data is most valuable?". To that end, we compare the following six different strategies for the data generation process:

(a) **single-object**, where only one object instance is rendered for each image
(b) **multi-object**, where multiple object instances are rendered for each image
(c) **multi-object with varying texture**, where multiple object instances are rendered with varying background textures for each image
(d) **multi-object with varying lighting**, where multiple object instances are rendered with varying lighting conditions for each image

(e) **multi-object with varying camera views**, where multiple object instances are rendered with varying camera perspective views for each image
(f) **multi-object with varying all**, where multiple object instances are rendered with varying all the above axes for each image.

For each of the strategy, we render a training set of 500 images and train a model with RGB input only. For strategy (a), we render 2500 images to roughly match the number of instance annotations. Those models are then evaluated on a common test set with 500 images generated with all variations.

Table 1 left compares the performance of different models trained with different training sets on the common test set. We first observe that the most crucial aspect of the data generation process is to have multiple objects in the scene. If the model only sees single-object scenes in the training process, it can hardly generalize to heavily occluded and cluttered multi-object cases. Furthermore, we see that domain randomization on the proposed three axes all consistently improves the generalization performance of the trained model and hence all of them are necessary to generate a dataset with enough variety.

Table 1. Left: instance segmentation mask AP results on the common test set of models trained using different data generation strategies. Right: instance segmentation mask AP on the synthetic test set of models with different inputs. Units are percentage numbers.

Training data	a	b	c	d	e	f	Input	RGB	Depth	Fusion
AP	0.89	21.26	34.80	33.03	36.57	**46.78**	AP	74.13	69.63	**81.35**
AP_{50}	0.99	36.97	57.98	51.47	58.02	**70.66**	AP_{50}	89.49	87.54	**92.59**
AP_{75}	0.99	21.62	36.10	35.96	39.25	**50.49**	AP_{75}	80.12	74.95	**85.38**

6.4 Evaluation on Synthetic Data

To evaluate the capability of the objectness Mask R-CNN model on synthetic images, we formally create a larger training set consisting of 8000 images with varying background texture, lighting and camera views. A test set consisting of 2000 images is rendered separately. We compare Mask R-CNN trained on RGB only, Mask R-CNN trained on depth only and Mask R-CNN trained on both inputs. Table 1 right shows the results. We find that when only using single input, Mask R-CNN performs significantly better when trained on RGB data, indicating that in most cases the color textures provide more discriminative information than the geometric structure. Moreover, RGB and depth fusion with joint training achieves the best performance. Depth signal proves to be very useful for extreme cases such as very dark lighting condition or heavily occluded small objects. For example, we can observe from Fig. 5 that the fusion method usually constructs better-quality masks and reliably detects small object parts even when they are heavily occluded. Also the high AP values in Table 1 indicate that the proposed segmentation system can almost perfectly fit the synthetic data.

Fig. 5. Qualitative results of models with different inputs on two synthetic scenes.

6.5 Evaluation on Real-World Data

To construct the real-world dataset, we collect both the RGB and depth images under 10 different manipulation scenes using the wrist mounted camera. The ground-truth masks are then generated semi-automatically by the multi-view instance matching algorithm in Sect. 5. The final dataset contains about 600 images. Figure 6 shows the object configurations of the different manipulation scenes.

For training, we render a synthetic dataset consisting of 10000 images with random object configurations from the 20 objects of the 3D database. Note that the training object set is only a subset of the real-world objects. For example, scenarios 8–10 in Fig. 6 all contain novel objects not in the training object database, including Red Bull bottles, Colgate toothpaste boxes, real bananas and so on. This allows us to further evaluate the generalization ability of the proposed segmentation system.

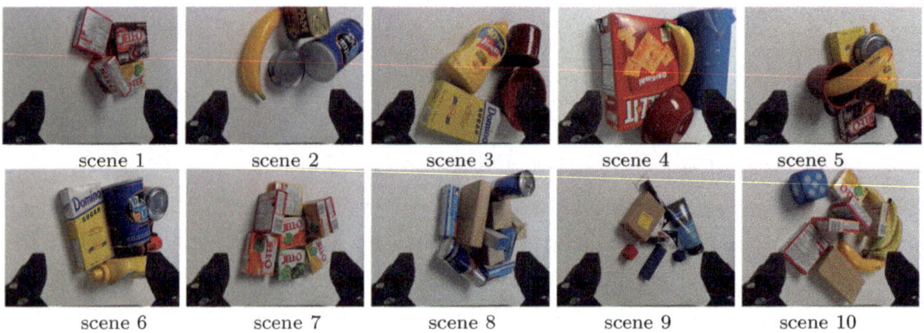

Fig. 6. Object configurations of the multi-view real-world data.

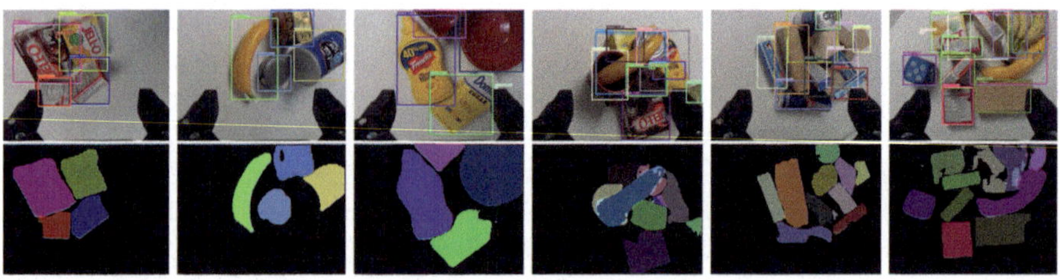

Fig. 7. Qualitative results of RGB-only model on real-world scenes.

To overcome the data bias between synthetic data and real-world data, we take the following data augmentation procedures. For the RGB data, we propose to paste real-world background images to the background part of the original synthetic image. This adapts the model to be resistant to real-world distractors such as the gripper in the image corner. For the depth data, we add random noise to the perfect synthetic depth image since real-world depth sensor observations are usually noisy. Besides the aforementioned data augmentation, we do not use any labeled real-world data in the training phase.

Table 2. Instance segmentation mask AP results of RGB-only (upper row) and fusion (lower row) models on real-world data. Units are percentage numbers.

Scene	1	2	3	4	5	6	7	8	9	10	All
AP	**66.46**	**53.37**	**78.40**	51.50	**36.30**	**56.87**	**30.48**	37.40	**42.26**	**42.31**	**39.11**
	65.73	49.98	77.56	**52.20**	27.62	49.06	24.50	**37.54**	31.19	38.20	34.81
AP_{50}	84.84	**84.17**	99.46	74.63	**65.43**	**79.30**	**56.70**	**63.95**	**67.13**	**69.49**	**65.72**
	88.82	77.42	**100**	**86.15**	53.50	74.09	46.83	61.31	53.45	58.06	59.12
AP_{75}	76.24	**50.22**	**83.54**	**51.84**	**34.18**	61.99	**27.02**	41.95	**43.24**	**42.86**	**39.43**
	76.35	42.42	77.39	46.15	25.54	**62.03**	24.51	**45.16**	25.57	42.17	34.72

To compare the generalization ability of models with different inputs, we evaluate both the RGB-only model and the fusion model on the real-world dataset. Table 2 shows the quantitative results of the two models on the presented 10 manipulation scenes. We can observe that the overall AP scores of the two models both decrease compared to the performance on the synthetic dataset. This is reasonable since the models are directly transferred without any adaptation. Nevertheless, both models can still achieve a AP score over 30, which can be considered to be pretty usable in most manipulation scenarios. From the per-scene AP results, we also find that both models achieve high accuracy for known object cases (scene 1–6), indicating that the proposed photorealistic rendering pipeline minimizes the gap between simulator and reality. Scene 5 and 7 are two challenging highly cluttered cases where different objects are heavily stacked together. Thus our models achieve significantly lower accuracy compared to the simple cases. Note that scene 7 is harder since two adjacent homogeneous objects may be falsely detected as one, making it difficult to determine the correct number of instances. As for the novel objects cases (scene 8–10), our model exhibits very good generalization ability. Most of the instances can be correctly detected and segmented even with occlusion. Figure 7 further shows the representative qualitative results.

For the real-world evaluation, we observe that the RGB-only model performs better than the fusion model on most cases. This is mainly due to the noisiness of the depth information from the depth sensor. The real-world depth images usually contain holes due to some reflective part of the objects. Although we try to add simple random noise in the data augmentation process, the gap between different depth signals still exists. Trying to simulate the real-world depth noise in the rendering process is a possible extension to alleviate this problem.

7 Conclusion and Future Work

In this paper, we have explored the potential of utilizing synthetic data for real-world robotic perception scenarios. We develop an automated rendering pipeline to generate photorealistic images with pixel-level semantic annotations given the scanned 3D CAD models. Large datasets can be easily collected without any human intervention. We propose an objectness instance segmentation system by extending the Mask R-CNN framework for multi-modal input fusion. Extensive experiments show that the proposed system exhibits good generalization performance from synthetic data to real-world scenarios. For future work, we plan to build a larger 3D object database and further improve the capability of the proposed system.

Acknowledgments. This research has been supported by General Research Fund 16207316 from the Research Grants Council of Hong Kong.

References

1. Zhou, J., Paolini, R., Johnson, A.M., Bagnell, J.A., Mason, M.T.: A probabilistic planning framework for planar grasping under uncertainty. IEEE Robot. Autom. Lett. **2**(4), 2111–2118 (2017)
2. Mitash, C., Bekris, K.E., Boularias, A.: A self-supervised learning system for object detection using physics simulation and multi-view pose estimation. In: IROS (2017)
3. Tobin, J., Fong, R., Ray, A., Schneider, J., Zaremba, W., Abbeel, P.: Domain randomization for transferring deep neural networks from simulation to the real world. In: IROS (2017)
4. He, K., Gkioxari, G., Dollár, P., Girshick, R.: Mask R-CNN. In: ICCV (2017)
5. Girshick, R., Donahue, J., Darrell, T., Malik, J.: Rich feature hierarchies for accurate object detection and semantic segmentation. In: CVPR (2014)
6. Ren, S., He, K., Girshick, R., Sun, J.: Faster R-CNN: towards real-time object detection with region proposal networks. In: NIPS (2015)
7. Long, J., Shelhamer, E., Darrell, T.: Fully convolutional networks for semantic segmentation. In: CVPR (2015)
8. Zeng, A., Yu, K.T., Song, S., Suo, D., Walker, E., Rodriguez, A., Xiao, J.: Multi-view self-supervised deep learning for 6D pose estimation in the amazon picking challenge. In: ICRA (2017)
9. Schwarz, M., Milan, A., Periyasamy, A.S., Behnke, S.: Rgb-d object detection and semantic segmentation for autonomous manipulation in clutter. Int. J. Robot. Res. **37**(4–5), 437–451 (2018)
10. Richter, S.R., Hayder, Z., Koltun, V.: Playing for benchmarks. In: ICCV (2017)
11. Pepik, B., Stark, M., Gehler, P., Schiele, B.: Multi-view and 3D deformable part models. IEEE Trans. Pattern Anal. Mach. Intell. **37**(11), 2232–2245 (2015)
12. Varol, G., Romero, J., Martin, X., Mahmood, N., Black, M.J., Laptev, I., Schmid, C.: Learning from synthetic humans. In: CVPR (2017)
13. Su, H., Qi, C.R., Li, Y., Guibas, L.J.: Render for CNN: viewpoint estimation in images using CNNs trained with rendered 3D model views. In: ICCV (2015)
14. Sadeghi, F., Levine, S.: $(CAD)^2RL$: real single-image flight without a single real image. In: RSS (2017)

15. Li, S., Liu, T., Zhang, C., Yeung, D.Y., Shen, S.: Learning unmanned aerial vehicle control for autonomous target following. In: IJCAI (2018)
16. Blender. http://www.blender.org/
17. Calli, B., Singh, A., Walsman, A., Srinivasa, S., Abbeel, P., Dollar, A.M.: The YCB object and model set: towards common benchmarks for manipulation research. In: ICAR (2015)
18. Bullet. http://bulletphysics.org/
19. He, K., Zhang, X., Ren, S., Sun, J.: Deep residual learning for image recognition. In: CVPR (2016)
20. Lin, T.Y., Dollár, P., Girshick, R., He, K., Hariharan, B., Belongie, S.: Feature pyramid networks for object detection. In: CVPR (2017)
21. Russakovsky, O., Deng, J., Su, H., Krause, J., Satheesh, S., Ma, S., Huang, Z., Karpathy, A., Khosla, A., Bernstein, M., et al.: Imagenet large scale visual recognition challenge. Int. J. Comput. Vis. **115**(3), 211–252 (2015)
22. Lin, T.Y., Maire, M., Belongie, S., Hays, J., Perona, P., Ramanan, D., Dollár, P., Zitnick, C.L.: Microsoft COCO: common objects in context. In: ECCV (2014)

Adapting Control Policies from Simulation to Reality Using a Pairwise Loss

Ulrich Viereck[1(✉)], Xingchao Peng[2(✉)], Kate Saenko[2(✉)], and Robert Platt[1(✉)]

[1] College of Computer and Information Science, Northeastern University, Boston, USA
{uliv,rplatt}@ccs.neu.edu

[2] Department of Computer Science, Boston University, Boston, USA
{xpeng,saenko}@bu.edu

Abstract. This paper proposes an approach to domain transfer based on a pairwise loss function that helps transfer control policies learned in simulation onto a real robot. We explore the idea in the context of a "category level" manipulation task where a control policy is learned that enables a robot to perform a mating task involving novel objects. We explore the case where depth images are used as the main form of sensor input. Our experimental results demonstrate that proposed method consistently outperforms baseline methods that train only in simulation or that combine real and simulated data in a naive way.

1 Motivation and Problem Statement

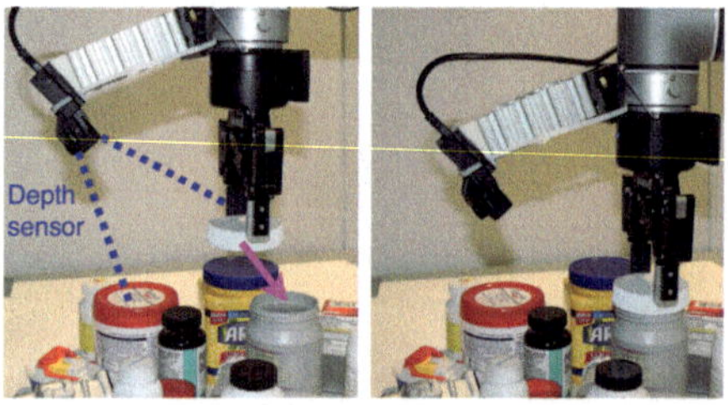

Fig. 1. Our goal is to learn a controller that uses depth image feedback to mate the cap to the bottle in the presence of clutter. Experimental setup on UR5 robot with a Intel RealSense depth sensor mounted as shown.

Recently, there has been a lot of interest in using deep neural networks to learn "pixels-to-torques" visuomotor controllers: robotic controllers that take sequential image data as input and produce low level motor commands as output. Ideally, we would learn these controllers with training data collected using real robotic hardware [5]. However, this approach is rarely feasible because of the large amounts of training experience typically required to train deep neural networks. Instead, it is convenient to learn pixels-to-torques control policies using simulated data.

Unfortunately, this exposes us to the *dataset shift problem* [7]. When the simulation is not sufficiently accurate, then the control policy learned in simulation may not work well in reality. There are two types of simulation errors that

J. Xiao et al. (Eds.): ISER 2018, SPAR 11, pp. 256–266, 2020.
https://doi.org/10.1007/978-3-030-33950-0_23

typically can cause domain shift problems: (1) errors simulating images that the robot would observe, and (2) errors simulating real world contact and frictional dynamics. This paper limits consideration to non-contact tasks and therefore the focus is on domain shift errors caused by image simulation and depth image simulation in particular.

In this paper, we propose an approach to transferring visuomotor control policies learned on simulated depth data to real world observations. The key idea is to reduce the gap between simulation and reality by augmenting the simulated data used to train the system with a small amount of real robot data. Each piece of real robot data is paired with a piece of simulated data that corresponds to the same robot state. We train the neural network using a loss function that has two terms: a *task loss* that encodes the desired robotic behavior and a *pairwise loss* that penalizes networks that do not represent real and simulated data the same way.

We make two contributions relative to prior work: (1) we propose a neural network architecture that combines the pairwise loss approach to domain transfer with a pixels-to-torques controller; (2) we characterize the method for depth image data rather than for RGB data. Unlike [9] which uses the pairwise loss function as part of the state estimator and only explores single task instances, our approach learns controllers that can solve "category level" manipulation tasks. We find that the approach can work well even when the real data is produced in a simplified version of the actual robotic scenario that is experienced at test time.

2 Related Work

This paper complements a variety of recent literature on simulation-to-reality learning for robotics tasks. One example is recent work that uses "domain randomization" of simulated images to affect better transfer to reality [8]. Another example is work using GANs to affect the transfer [1]. In contrast, the approach followed here is simpler than GAN-based approaches and more relevant to depth data than domain randomization methods.

The most related approach to ours is that of [9]. It also uses the pairwise loss function to minimize domain shift between simulated and real data. However, it only explores single task instances (e.g., placing a specific rope object on a specific scale object), whereas we solve "category level" manipulation tasks where object shape and size varies from one instance of the task to the next. Furthermore, [9] uses keypoint prediction on RGB image as an intermediate state representation, whereas we estimate a distance-to-goal function directly. We also use depth data which tends to be less affected by domain shift due to lighting and background than RGB data.

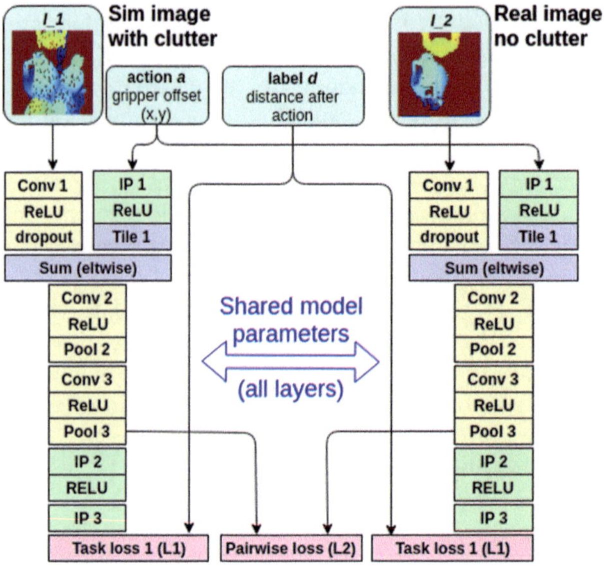

Fig. 2. Architecture of the controller neural network. The pairwise loss at pool3 favors networks that give real and simulated images similar representations.

3 Technical Approach

3.1 Controller Network Architecture

We learn a pixels-to-torques controller that takes depth images as input and outputs manipulator displacements. The controller is based on a method proposed in our prior work [10] where we estimate a distance function with respect to a goal state using a neural network. Given an image and a candidate manipulator displacement, the distance function predicts expected distance-to-goal on the following time step. We select a manipulator displacement by sampling a set of candidate displacements and selecting the one that is predicted to move closest to a goal. Essentially, this method learns a value function over the cross product of observation and action (depth image and manipulator displacement). However, instead of using reinforcement learning, we train the neural network directly by using supervised learning with distance targets produced by our simulator. Specifically, we create a dataset by sampling from a space of possible task scenarios and initial conditions. For each sample, we simulate the depth image that would be observed and calculate distance-to-goal after performing the associated displacement.

3.2 Supervised Domain Adaptation

This paper characterizes a domain transfer technique based on the following loss function (similar to what was originally proposed in [9]):

$$\begin{aligned}\mathscr{L} = \alpha & \sum_{(I,a)\in\mathbb{X}_S} \|g\left(f(I,a;\theta_f),\theta_g\right) - y(I,a)\|_1 \\ + \beta & \sum_{(I,a)\in\mathbb{X}_T} \|g\left(f(I,a;\theta_f),\theta_g\right) - y(I,a)\|_1 \\ + \gamma & \sum_{(I_S,I_T,a)\in\mathbb{X}_{ST}} \|f(I_S,a;\theta_f) - f(I_T,a;\theta_f)\|_2. \end{aligned} \quad (1)$$

Here, we write the neural network as a composition of two functions, f and g. f denotes the "early" part of the network comprised of convolutional layers that learn feature representations for the images and actions. g denotes the "late" fully connected layers that encode the regression task. The output of the end-to-end network for an image I and action (i.e. displacement) a is written $g(f(I,a;\theta_f);\theta_g)$, where θ_f denotes the parameters of f and θ_g denotes parameters of g. The loss is evaluated over a training dataset comprised of two parts: the data from source domain and the data from the target domain. The source domain consists of simulated images and actions $(I,a) \in \mathbb{X}_S$ and the associated labels $y(I,a)$ (these are the distance-to-goal that would result from taking action a from the state that produced image I). The target domain $\mathbb{X}_T$ consists of real images and actions paired with the associated labels. The set $\mathbb{X}_{ST}$ consists of triples (I_S, I_T, a) where I_S and I_T are the simulated-real image pair and a is the action that was taken. We assume that the cardinality of $\mathbb{X}_S$ is much larger than either $\mathbb{X}_T$ or $\mathbb{X}_{ST}$, i.e. that we have many more simulated images than real images. The first two terms of Eq. 1 are called *task losses.* The third is the *pairwise loss.* The first and second task losses are minimized when the neural network fits the simulated and real data well, respectively. The third is minimized when the network assigns both simulated and real depth images the same high-level encoding. The pairwise loss is critical: by "preferring" networks that encode real and simulated data similarly, this term facilitates good transfer from simulation to reality. The approach is implemented by the neural network architecture illustrated in Fig. 2. It takes as input a pair of depth images, I_1 and I_2 (matching paired images from simulation and reality), and an action $a = (x, y) \in \mathbb{R}^2$. It learns a function, $d(I,a) \in \mathbb{R}_{>0}$, that describes the distance between the object in the hand and the target after displacing the manipulator by a.

3.3 Data Simulation

We train using a combination of real and simulated data. The simulated portion of the dataset is generated using OpenRAVE [2] to generate 100k 64×64 pixel depth images. Each depth image is taken for a random robot configuration with clutter objects selected randomly from a set of more than 250 objects and placed randomly in the vicinity of the target bottle. For each of the 100k scenes, we sample 100 actions (i.e. hand displacements) and estimate the distance-to-goal that would result if that action were executed. Finally, we simulate missing pixel noise by setting the value of each pixel to 0 with a 10% probability. We also collect 7260 labeled real training images on the robot (UR5, see Fig. 1) and measure the corresponding ground truth distance-to-goal. Importantly, the real images used for training do *not* have clutter. As a result, it is easier to obtain training data semi-automatically because it is unnecessary to reproduce simulated clutter on the real system. These real images were paired with 7260 simulated images that portray the same robotic state.

3.4 Unsupervised Domain Adaptation

One alternative way to tackle the simulated-to-real knowledge transfer is to leverage unsupervised domain adaptation algorithms [6]. Assume two distributions p and q are sampled from simulated and real domain, respectively, we apply Maximum Mean Discrepancy (MMD) [3] to align p and q. Denoting the reproducing kernel Hilbert space (RKHS) with a characteristic kernel k by $\mathbb{H}_k$. The MMD $d_k(p,q)$ is defined as

$$d_k^2(p,q) \triangleq \|\boldsymbol{E}_p[f(I_S)] - \boldsymbol{E}_q[f(I_T)]\|^2_{\mathbb{H}_k} \tag{2}$$

where I_S and I_T are unpaired images from source (simulated) and target (real) domains. The main limitation of the algorithm is the computation complexity is quadratic. In this paper, following [4,6], we adopt the unbiased estimation of MMD which can be computed with cost $O(n)$. Specifically,

$$\begin{aligned} d_k^2(p,q) \triangleq \frac{2}{n_s} \sum_{i=1}^{n_s/2} [&k(f(I_S^{2i-1}), f(I_S^{2i})) + k(f(I_T^{2i-1}), f(I_T^{2i})) \\ &- k(f(I_S^{2i-1}), f(I_T^{2i})) - k(f(I_S^{2i}), f(I_T^{2i-1}))] \end{aligned} \tag{3}$$

where n_s is the number of instances in source domain. The training loss with unsupervised domain adaptation is:

$$\begin{aligned} \mathscr{L} = \alpha &\sum_{(I,a)\in\mathbb{X}_S} \|g\left(f(I,a;\theta_f),\theta_g\right) - y(I,a)\|_1 \\ &+ \gamma d_k^2(p,q). \end{aligned} \tag{4}$$

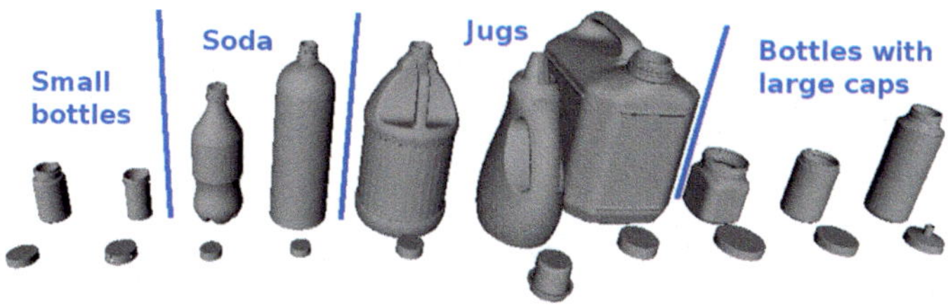

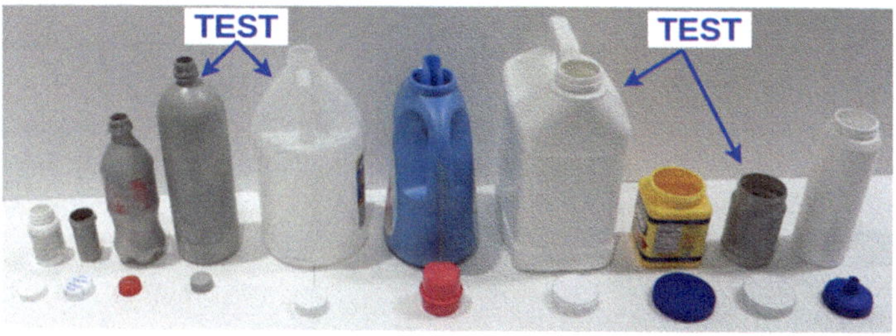

Fig. 3. Simulated and real bottles used in our experiments. From left to right: *vitamin, pills, soda, water, 1-gal, laundry, 2-gal, corn starch, peanut butter, sport bottle.* Test bottles are shown with arrows.

4 Experiments

4.1 Task Description

We evaluate this approach using the *cap-on-bottle* task, where the robot must align a cap with a bottle opening. This is a challenging task because we are learning a "category level" manipulation skill where the system must learn to perform the task for novel object instances in the presence of randomly placed clutter. We test our methods in simulation, as well as on a real robot using a Robotiq 85 two-fingered hand mounted on a UR5 arm. The Intel RealSense SR300 is mounted to the UR5 wrist as shown in Fig. 1. The RealSense creates depth images that are input to our controller and used to estimate desired hand displacements.

Testing on the robot proceeds as follows. A test bottle is placed on a table surrounded by clutter intended to make the task more challenging. Most of the clutter objects are bottles *with* caps, which makes the task even harder. The bottle cap is manually placed into the robotic hand. The position of the bottle is unknown to the algorithm but measured for the purposes of experimental evaluation. The gripper is initialized to a random offset within a 10 cm box centered on the bottle at a height of 5 cm above the table. At each iteration, the controller acquires a depth image, samples 1k manipulator displacements, moves the gripper in the direction predicted to reduce distance-to-goal by the most, and moves toward the table by 1 cm. After executing 5 iterations, we measure how close the cap is to the bottle opening (Fig. 4).

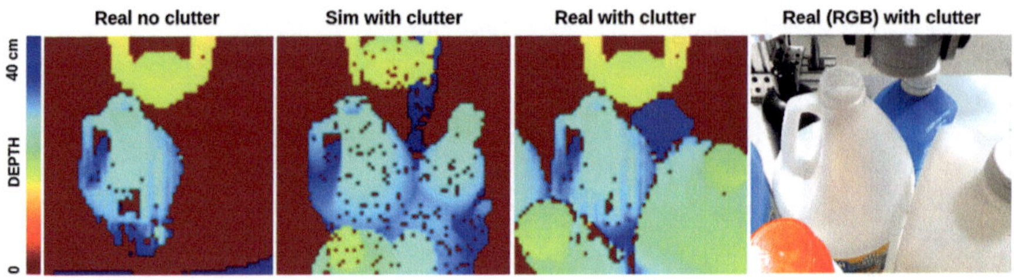

Fig. 4. Depth images generated for training and testing. For training *Real no clutter* are paired and augmented by *Sim with clutter* images. Testing images are *Real with clutter* images. All these images show the same 1-gal bottle for comparison of different image domains, but we do not test on bottles that are in the training set (see Sect. 4 for details). The *Real (RGB) with clutter* images are not used (reference only).

4.2 Experimental Settings

Our proposed approach is to train the neural network using both the task and pairwise loss (supervised adaptation) over both simulated and real images. We compare this approach against five alternative setups: (1) training the network using only real images *without* clutter; (2) training using only real images *with* clutter; (3) training using simulated images with clutter; (4) training using simulated images with clutter with *unsupervised adaptation* on real images *with* clutter (MMD method) (5) training using simulated images with clutter combined with *labeled* real images *without* clutter. Among the five setups, pairwise losses are directly applied to layer *pool3*. For the MMD method, we add a fully connected layer with 512 channels after layer *conv1* and apply MMD loss on the top of the fully connected layer. We choose layer *conv1* as it only contains image information, which makes more sense to align, as opposed to later layers which also contain pose information. We will refer to the combination of the five methods and the proposed approach as the six "domain transfer methods".

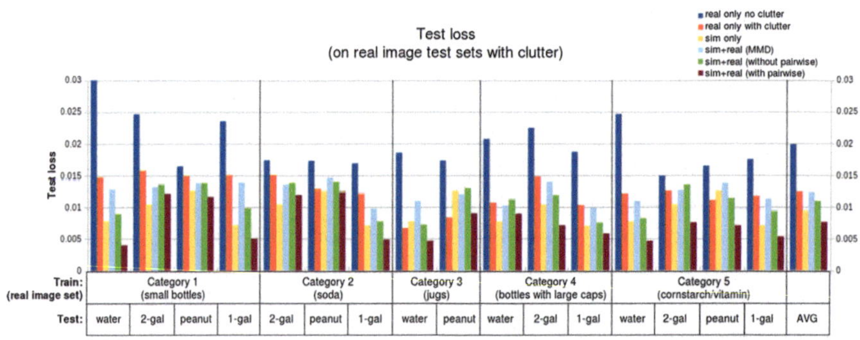

Fig. 5. Test losses on real image test sets with clutter. See text for details.

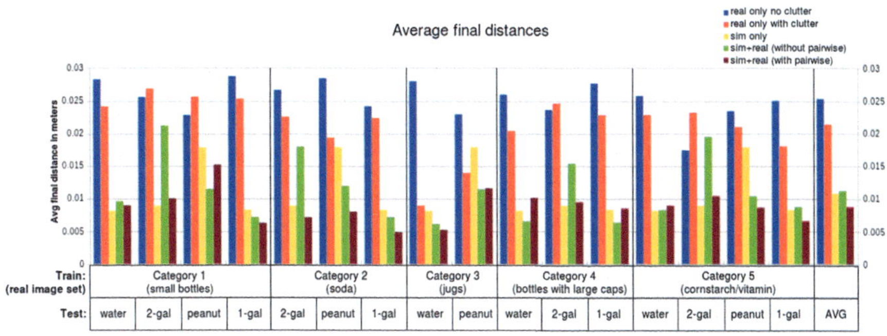

Fig. 6. Average final distances to cap on real image test sets with clutter. See text for details.

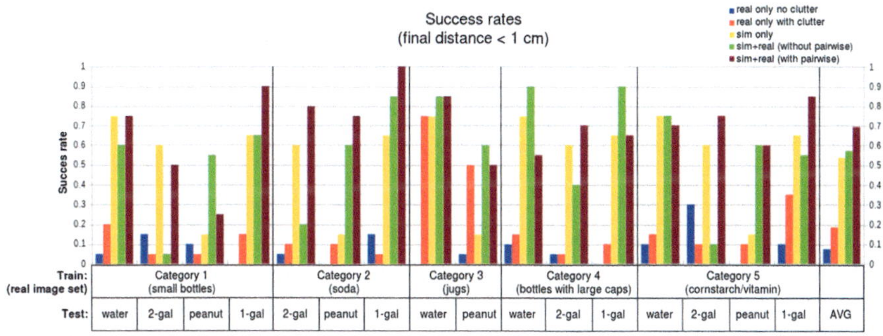

Fig. 7. Average success rates on real image test sets with clutter. See text for details.

We evaluate the six domain transfer methods over a set of five different problem scenarios. In each problem scenario, we train using data derived from a different object category. The five object categories are: (i) small pill bottles; (ii) soda bottles; (iii) jugs; (iv) bottles with large caps; (v) cornstarch and vitamin bottles. These categories are illustrated in Fig. 3. A separate network is trained for each object category and is evaluated on as many as four different test objects (indicated in the figure with arrows). We do not test on objects that happen also to be in the object category used for training.

4.3 Evaluation

Figures 5, 6 and 7 show the results grouped by object category. For example, the "Category 1" results show performance for each of the six domain transfer methods evaluated on four different test objects (water, 2-gal, peanut, and 1-gal). Figure 5 show L1 test loss of the learned network (lower is better). Figure 6 shows the average final distance between cap and bottle opening after the last control iteration. Sometimes the controller diverged or the cap was moved to a distractor object. Therefore the final distance is set to 3 cm for final distances > 3 cm to reduce the effect of diverged runs on the average. Each bar is an average of 20 trials (lower is better). Figure 7 shows the success rate of the end-to-end controller (we define "success" to occur when the cap final position is within 1 cm of the position of the bottle top). Each bar is an average of 20 trials (higher is better). Note that for the unsupervised domain adaptation baseline "sim+real (MMD)" we have only results for the "test loss", not for "final distance" and "success rates". However, comparing the three bar plots for the other domain transfer methods shows that the test loss roughly reflects the performance on the robot.

The baseline "sim only" consists of 100k simulated images as described in Sect. 3. The dataset used for domain transfer method "sim+real (MMD)" contains 100k labeled simulated images with 726 unlabeled real images with clutter per bottle in the category (e.g. 2×726 for category 2 "soda" or 3×726 for category 3 "jugs"). We use real images *with* clutter, since this matches closely the distribution of the target domain (we test with clutter). However, our approach is able to use real images *without* clutter with better performance. The datasets for "sim+real (without pairwise)" and "sim+real (with pairwise)" contain *labeled* real images *without* clutter paired with the corresponding simulated images in the same robot state. The real images in the dataset are repeated to obtain 100k real images to balance the 100k simulated images. To prevent overfitting to the real images, the weight for the task loss of the real images is set to 0.1 (the weight for the task loss of the simulated images remains 1.0). The weight for the pairwise loss for "sim+real (with pairwise)" is set to 0.1 and the weight for MMD loss for MMD baseline is set to 0.05 (weight loss parameters were determined by hyperparameter tuning).

In Fig. 5, we show the experimental results of the unsupervised domain adaptation method. The MMD-based method works better than "real image with clutter," despite the fact that it does not use any supervision for the real images. However, it performs worse than "sim+real without pairwise loss". After some empirical analysis, we think it is mainly caused by three reasons. Firstly, the real data is unlabeled in the unsupervised domain adaptation setting. Secondly, the real images used for training and testing are also biased by themselves. Thirdly, the domain adaptation loss is not directly applied to align the action information while the test loss is highly relevant to that.

We experimented with clutter and no clutter combinations in the sim-real image pairs. Figure 8 shows the test loss when trained on different combinations of having clutter in sim or real part of the image pairs. For this experiment we trained on bottles from category 5 (this category contains cornstarch and vitamin bottle) and tested on all available bottles. There are two averages across test bottles: "AVG" is the average over all bottles (including bottles trained on) and "AVG (holdout only)" excluding the training bottles. Note that the "sim-real" datasets also contains 100k sim images *with* clutter. The main insight of this experiment is that using sim image *with clutter* and real image *without* clutter gives almost as good of a performance as real image *with* clutter in the case of applying pairwise loss. Also, the gap between "with" and "without" pairwise

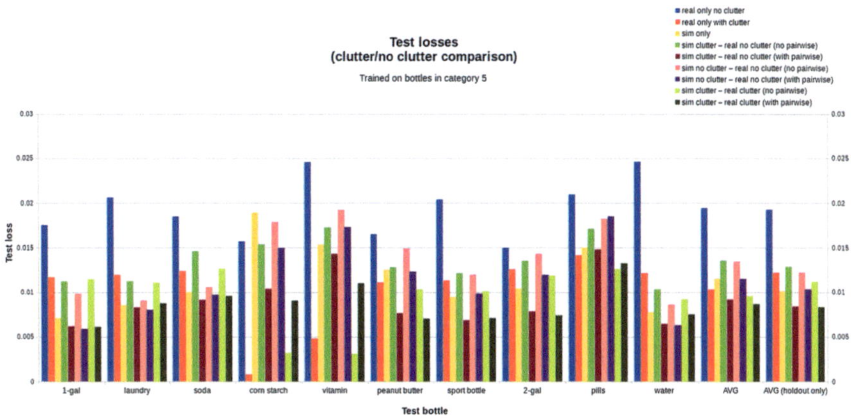

Fig. 8. Effect of different clutter/no clutter combinations in the pairwise sim-real image dataset. This plot also shows the effect adding pairwise loss. Without pairwise loss the network tends to overfit to the real images. In particular the network trained on "real only with clutter" and "sim clutter - real clutter (no pairwise)" overfit to the two training bottles "corn starch" and "vitamin bottle" in the category 5 bottles.

is the largest. Note how the nets containing *real images with clutter* overfits to "corn starch" and "vitamin" if no pairwise loss is applied. This shows that the pairwise loss has a regularization effect and prevents overfitting to the training data.

5 Discussion and Conclusion

The group of bars at the far right of Figs. 5, 6 and 7 labeled "AVG" summarize the comparison. Each bar (i.e. each domain transfer method) is an average of all experiments for that domain transfer type. This illustrates a few key results. First, the domain transfer method using our proposed task-pairwise loss function does best overall (lowest loss, lowest average final distance, highest task success rate). Second, the method using only real images does worst, probably because we do not train on enough different objects to facilitate generalization to novel objects. Third, training using only simulated images does quite well (54% success rate): not quite as well as we can do using our proposed method (70% success rate), but not nearly as badly as training on only real data.

The unsupervised domain adaptation approach performs worse than trained on "sim only" (see Fig. 5). The reason might be that simulated and real depth images look very similar (they are both depth images with similar distributions of gray scale values) and the domain adaptation does not have much of an effect.

Overall, we conclude that training visuomotor policies for category-level tasks in simulation is a promising approach, and that by collecting a small amount of labeled real data in simplified scenarios and using the pairwise loss, we can improve performance on real systems.

References

1. Bousmalis, K., Irpan, A., Wohlhart, P., Bai, Y., Kelcey, M., Kalakrishnan, M., Downs, L., Ibarz, J., Pastor, P., Konolige, K., et al.: Using simulation and domain adaptation to improve efficiency of deep robotic grasping. arXiv preprint arXiv:1709.07857 (2017)
2. Diankov, R.: Openrave. http://openrave.org
3. Gretton, A., Borgwardt, K.M., Rasch, M.J., Schölkopf, B., Smola, A.: A kernel two-sample test. J. Mach. Learn. Res. **13**(Mar), 723–773 (2012)
4. Gretton, A., Sejdinovic, D., Strathmann, H., Balakrishnan, S., Pontil, M., Fukumizu, K., Sriperumbudur, B.K.: Optimal kernel choice for large-scale two-sample tests. In: Advances in Neural Information Processing Systems, pp. 1205–1213 (2012)
5. Levine, S., Pastor, P., Krizhevsky, A., Quillen, D.: Learning hand-eye coordination for robotic grasping with deep learning and large-scale data collection. arXiv preprint arXiv:1603.02199 (2016)
6. Long, M., Cao, Y., Wang, J., Jordan, M.I.: Learning transferable features with deep adaptation networks. arXiv preprint arXiv:1502.02791 (2015)
7. Quionero-Candela, J., Sugiyama, M., Schwaighofer, A., Lawrence, N.D.: Dataset Shift in Machine Learning. The MIT Press, Cambridge (2009)
8. Tobin, J., Fong, R., Ray, A., Schneider, J., Zaremba, W., Abbeel, P.: Domain randomization for transferring deep neural networks from simulation to the real world. In: 2017 IEEE/RSJ International Conference on Intelligent Robots and Systems (IROS), pp. 23–30. IEEE (2017)
9. Tzeng, E., Devin, C., Hoffman, J., Finn, C., Abbeel, P., Levine, S., Saenko, K., Darrell, T.: Adapting deep visuomotor representations with weak pairwise constraints. arXiv preprint arXiv:1511.07111 (2015)
10. Viereck, U., Pas, A., Saenko, K., Platt, R.: Learning a visuomotor controller for real world robotic grasping using simulated depth images. In: Conference on Robot Learning (CoRL) (2017)

Scaling Simulation-to-Real Transfer by Learning Composable Robot Skills

Ryan Julian[1(✉)], Eric Heiden[1], Zhanpeng He[1], Hejia Zhang[1], Stefan Schaal[1], Joseph Lim[1], Gaurav Sukhatme[1], and Karol Hausman[2]

[1] University of Southern California, Los Angeles, CA, USA
{rjulian,heiden,zhanpenh,hejiazha,sschaal,limjj,gaurav}@usc.edu
[2] Google Brain, Mountain View, CA, USA
karolhausman@google.com

Abstract. We present a novel solution to the problem of simulation-to-real transfer, which builds on recent advances in robot skill decomposition. Rather than focusing on minimizing the simulation-reality gap, we learn a set of diverse policies that are parameterized in a way that makes them easily reusable. This diversity and parameterization of low-level skills allows us to find a transferable policy that is able to use combinations and variations of different skills to solve more complex, high-level tasks. In particular, we first use simulation to jointly learn a policy for a set of low-level skills, and a "skill embedding" parameterization which can be used to compose them. Later, we learn high-level policies which actuate the low-level policies via this skill embedding parameterization. The high-level policies encode how and when to reuse the low-level skills together to achieve specific high-level tasks. Importantly, our method learns to control a real robot in joint-space to achieve these high-level tasks with little or no on-robot time, despite the fact that the low-level policies may not be perfectly transferable from simulation to real, and that the low-level skills were not trained on any examples of high-level tasks. We illustrate the principles of our method using informative simulation experiments. We then verify its usefulness for real robotics problems by learning, transferring, and composing free-space and contact motion skills on a Sawyer robot using only joint-space control. We experiment with several techniques for composing pre-learned skills, and find that our method allows us to use both learning-based approaches and efficient search-based planning to achieve high-level tasks using only pre-learned skills.

R. Julian and E. Heiden—Equal Contribution.

Electronic supplementary material The online version of this chapter (https://doi.org/10.1007/978-3-030-33950-0_24) contains supplementary material, which is available to authorized users.

J. Xiao et al. (Eds.): ISER 2018, SPAR 11, pp. 267–279, 2020.
https://doi.org/10.1007/978-3-030-33950-0_24

1 Introduction

Motivation

The Constructivist hypothesis proposes that humans learn to perform new behaviors by using what they already know [1]. To learn new behaviors, it proposes that humans leverage their prior experiences across behaviors, and that they also generalize and compose previously-learned behaviors into new ones, rather than learning them from scratch [2]. Whether we can make robots learn so efficiently is an open question. Much recent work on robot learning has focused on "deep" reinforcement learning (RL), inspired by achievements of deep RL in continuous control [3] and game play domains [4]. While recent attempts in deep RL for robotics are encouraging [5–7], performance and generality on real robots remains challenging.

A major obstacle to widespread deployment of deep RL on real robots is data efficiency: most deep RL algorithms require millions of samples to converge [8]. Learning from scratch using these algorithms on a real robot is therefore a resource-intensive endeavor, *e.g.* by requiring multiple robots to learn in parallel for weeks [9]. One promising approach is to train deep RL algorithms entirely in faster-than-real-time simulation, and transfer the learned policies to a real robot.

Problem Statement

Our contribution is a method for exploiting hierarchy, while retaining the flexibility and expressiveness of end-to-end RL approaches.

Consider the illustrative example of block stacking. One approach is to learn a single monolithic policy which, given any arrangement of blocks on a table, grasps, moves, and stacks each block to form a tower. This formulation is succinct, but requires learning a single sophisticated policy. We observe that block stacking–and many other practical robotics tasks–is easily decomposed into a few reusable primitive skills (e.g. locate and grasp a block, move a grasped block over the stack location, place a grasped block on top of a stack), and divide the problem into two parts: learning to perform and mix the skills in general, and learning to combine these skills into particular policies which achieve high-level tasks.

Related Work

Our approach builds on the work of Hausman *et al.* [10], that learns a latent space which parameterizes a set of motion skills, and shows them to be temporally composable using interpolation between coordinates in the latent space. In addition to learning reusable skills, we present a method which learns to compose them to achieve high-level tasks, and an approach for transferring compositions of those skills from simulation to real robots. Similar latent-space methods have been recently used for better exploration [11,12] and hierarchical RL [13–15].

Our work is related to parameter-space meta-learning methods, which seek to learn a single shared policy which is easily generalized to all skills in a set, but do not address skill sequencing specifically. Similarly, unlike recurrent meta-learning methods, which implicitly address sequencing of a family of sub-skills to achieve goals, our method addresses generalization of single skills while providing an explicit representation of the relationship between skills. We show that explicit representation allows us to combine our method with many algorithms for robot autonomy, such as optimal control, search-based planning, and manual programming, in addition to learning-based methods. Furthermore, our method can be used to augment most existing reinforcement learning algorithms, rather than requiring the formulation of an entirely new family of algorithms to achieve its goals.

Previous works proposed frameworks such as Associative Skill Memories [16] and probabilistic movement primitives [17] to acquire a set of reusable skills. Other approaches introduce particular model architectures for multitask learning, such as Progressive Neural Networks [18] or Attention Networks [19].

Common approaches to simulation-to-real transfer learning include randomizing the dynamic parameters of the simulation [20], and varying the visual appearance of the environment [21]. Another approach is explicit alignment: given a mapping of common features between the source and target domains, domain-invariant state representations [22], or priors on the relevance of input features [23], can further improve transferability.

Our method can leverage these techniques to improve the stability of the transfer learning process in two ways: (1) by training transferable skills which generalize to nearby skills from the start and (2) by intentionally learning *composable* parameterizations of those skills, to allow them to be easily combined before or after transfer.

2 Technical Approach

Our work synthesizes two recent methods in deep RL–pre-training in simulation and learning composable motion policies–to make deep reinforcement learning more practical for real robots. Our strategy is to split the learning process into a two-level hierarchy (Fig. 1), with low-level skill policies learned in simulation,

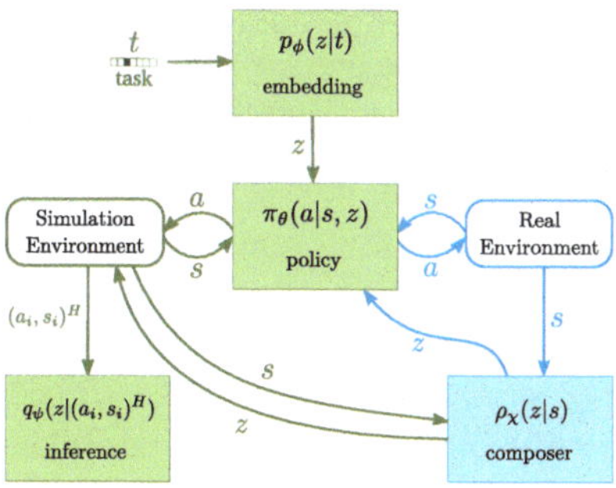

Fig. 1. Block diagram of proposed architecture for transfer learning. Shared low-level skill components are shown in green. The high level task-specific component is shown in blue.

and high-level task policies learned or planned either in simulation or on the real robot, using the imperfectly-transferred low-level skills.

Skill Embedding Learning Algorithm
In our multi-task RL setting, we pre-define a set of low-level skills with IDs $T = \{1, \ldots, N\}$, and accompanying, per-skill reward functions $r_{t\in T}(s, a)$.

In parallel with learning the joint low-level skill policy π_θ as in conventional RL, we learn an embedding function p_ϕ which parameterizes the low-level skill library using a latent variable z. Note that the true skill identity t is hidden from the policy behind the embedding function p_ϕ. Rather than reveal the skill ID to the policy, once per rollout we feed the skill ID t, encoded as s one-hot vector, through the stochastic embedding function p_ϕ to produce a latent vector z. We feed this same value of z to the policy for the entire rollout, so that all steps in a trajectory are correlated with the same value of z.

$$\mathfrak{L}(\theta, \phi, \psi) = \max_{\pi} \mathbb{E}_{\substack{\pi(a,z|s,t)\\ t\in T}} \left[\sum_{i=0}^{\infty} \gamma^i \hat{r}(s_i, a_i, z, t) \middle| s_{i+1} \sim p(s_{i+1}|a_i, s_i) \right] \quad (1)$$

where

$$\hat{r}(s_i, a_i, z, t) = \alpha_1 \mathbb{E}_{t\in T}\left[\mathbb{H}\{p_\phi(z|t)\}\right] + \alpha_2 \log q_\psi(z|s_i^H) + \alpha_3 \mathbb{H}\{\pi_\theta(a_i|s_i, z)\} \\ + r_t(s_i, a_i)$$

To aid in learning the embedding function, we learn an inference function q_ψ which, given a trajectory window s_i^H of length H, predicts the latent vector z which was fed to the low-level skill policy when it produced that trajectory. This allows us to define an augmented reward which encourages the policy to produce distinct trajectories for different latent vectors. We learn q_ψ in parallel with the policy and embedding functions, as shown in Eq. 1.

We also add a policy entropy bonus $\mathbb{H}\{\pi_\theta(a_i|s_i, z)\}$, which ensures that the policy does not collapse to a single solution for each low-level skill, and instead encodes a variety of solutions. All the above reward augmentations arise naturally from applying a variational lower bound to an entropy-regularized, multi-task RL formulation which uses latent variables as the task context input to the policy. For a detailed derivation, refer to [10].

The full robot training and transfer method consists of three stages.

Stage 1: Pre-Training in Simulation while Learning Skill Embeddings
We begin by training in simulation a multi-task policy π_θ for all low-level skills, and a composable parameterization of that library $p_\phi(z|t)$ (i.e. a skill embedding). This stage may be performed using any deep RL algorithm, along with the modified policy architecture and loss function described above. Our implementation uses Proximal Policy Optimization [24] and the MuJoCo physics engine [25].

The intuition behind our pre-training process is as follows. The policy obtains an additional reward if the inference function is able to predict the latent vector which was sampled from the embedding function at the beginning of the rollout.

This is only possible if, for every latent vector z, the policy produces a distinct trajectory of states s_i^H, so that the inference function can easily predict the source latent vector. Adding these criteria to the RL reward encourages the policy to explore and encode a set of diverse policies that can perform each low-level skill in various ways, parameterized by the latent vector.

Stage 2: Learning Hierarchical Policies
In the second stage, we learn a high-level "composer" policy, represented in general by a probability distribution $\rho_\chi(z|s)$ over the latent vector z. The composer actuates the low-level policy $\pi_\theta(a|s,z)$ by choosing z at each time step to compose the previously-learned skills. This hierarchical organization admits our novel approach to transfer learning: by freezing the low-level skill policy and embedding functions, and exploring only in the pre-learned latent space to acquire new tasks, we can transfer a multitude of high-level task policies derived from the low-level skills.

This stage can be performed directly on the real robot or in simulation. As we show in Sect. 3, composer policies may treat the latent space as either a discrete or continuous space, and may be found using learning, search-based planning, or even manual sequencing and interpolation. To succeed, the composer policy must explore the latent space of pre-learned skills, and learn to exploit the behaviors the low-level policy exhibits when stimulated with different latent vectors. We hypothesize that this is possible because of the richness and diversity of low-level skill variations learned in simulation, which the composer policy can exploit by actuating the skill embedding.

Stage 3: Transfer and Execution
Lastly, we transfer the low-level skill policy, embedding and high-level composer policies to a real robot and execute the entire system to perform high-level tasks.

3 Experiments

Point Environment
Before experimenting on complex robotics problems, we evaluate our approach in a point mass environment. Its low-dimensional state and action spaces, and high interpretability, make this environment our most basic test bed. We use it for verifying the principles of our method and tuning its hyperparameters before we deploy it to more complex experiments. Portrayed in Fig. 2 is a multi-task instantiation of this environment with four goals (skills).

At each time step, the policy receives as state the point's position and chooses a two-dimensional velocity vector as its action. The policy receives a negative reward equal to the distance between the point and the goal position.

After 15,000 time steps, the embedding network learns a multimodal embedding distribution to represent the four tasks (Fig. 2). Introducing entropy regularization [10] to the policy alters the trajectories significantly: instead of steering to the goal position in a straight line, the entropy-regularized policy encodes a distribution over possible solutions. Each latent vector produces a different solution. This illustrates that our approach is able to learn multiple distinct solutions for the same skill, and that those solutions are addressable using the latent vector input.

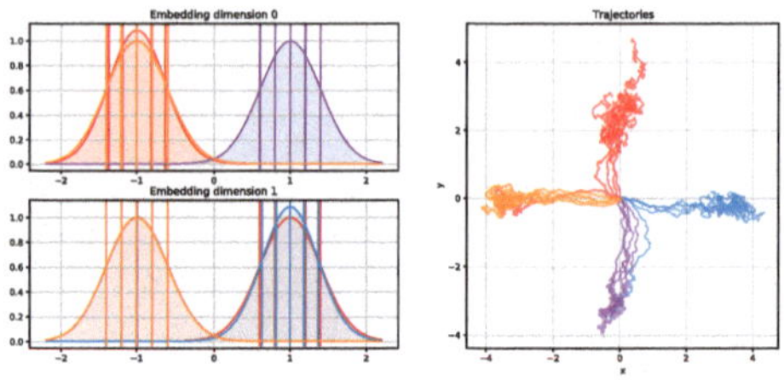

Fig. 2. Skill embedding distribution which successfully disentangles four different tasks using the embedding function.

Sawyer Experiment: Reaching

We ask the Sawyer robot to move its gripper to within 5 cm of a goal point in 3D space. The policy receives a state observation with the robot's seven joint angles, plus the cartesian position of the robot's gripper, and chooses incremental joint movements (up to 0.04 rad) as actions (Fig. 3).

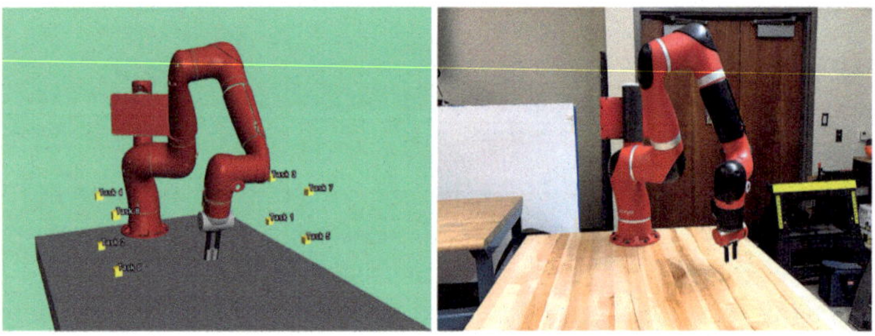

Fig. 3. Multitask environment in simulation (left) and reality (right) for the reaching experiment. In pre-training, the robot learns a low-level skill for servoing its gripper to each of eight goal positions.

We trained the low-level policy on eight goal positions in simulation, forming a 3D cuboid enclosing a volume in front of the robot (Fig. 4). The composer policies feed latent vectors to the pre-trained low-level skill policy to achieve high-level tasks such as reaching previously-unvisited points (Fig. 5).

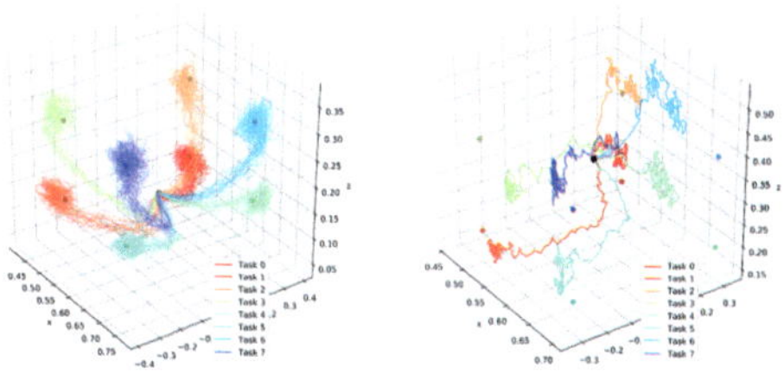

Fig. 4. Gripper position trajectories for each skill from the low-level pre-trained reaching policy, after 100 training iterations. Left: simulation, right: real robot.

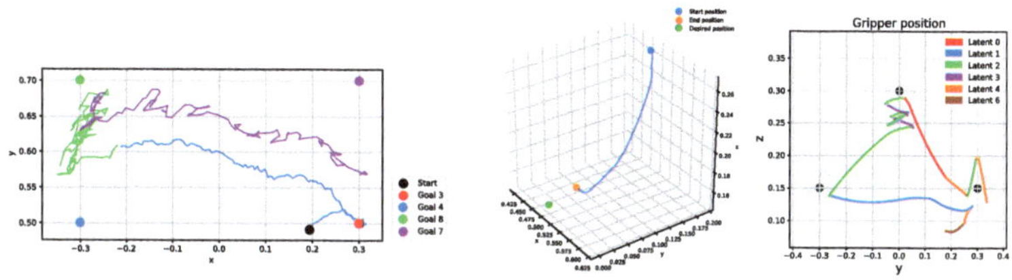

Fig. 5. 1. Gripper position trajectory while interpolating latents between task 3, 4, 8 and 7 for the embedded reaching skill policy on the real robot. 2. Gripper position for composed reaching policy, trained with DDPG to reach an unseen point. 3. Gripper position trajectory reaching points (black) in a triangle within a radius of 5 cm composed by search-based planning in the latent space.

Composition Experiments

All Sawyer composition experiments use the same low-level skill policy, pre-trained in simulation. We experimented both with composition methods which directly transfer the low-level skills to the robot (direct), and with methods which use the low-level policy for a second stage of pre-training in simulation before transfer (sim2real).

Task Interpolation in the Latent Space (Direct)

We evaluate the embedding function to obtain the mean latent vector for each of the 8 pre-training tasks, then feed linear interpolations of these means to the latent input of the low-level skill policy, transferred directly to the real robot. For a latent pair (z_a, z_b), our experiment feeds z_a for 1 s, then $z_i = \lambda_i z_a + (1 - \lambda_i) z_b$ for $\lambda_i \in [0, 1]$ for 1 s, and finally z_b for 1 s. We observe that the linear interpolation in latent space induces an implicit motion plan between the two points, despite the fact that pre-training never experienced this state trajectory. In one experiment, we used this method iteratively to control the Sawyer robot to draw a U-shaped path around the workspace (Fig. 5.1).

End-to-End Learning in the Latent Space (sim2real)

Using DDPG [3], an off-policy reinforcement learning algorithm, we trained a composer policy to modulate the latent vector to reach a previously-unseen point. We then transferred the composer and low-level policies to the real robot. The policy achieved a gripper distance error of 5 cm, the threshold for task completion as defined by our reward function (Fig. 5.2).

Search-Based Planning in the Latent Space (sim2real and Direct)
We used Uniform Cost Search in the latent space to find a motion plan (i.e. sequence of latent vectors) for moving the robot's gripper along a triangular trajectory. Our search space treats the latent vector corresponding to each skill ID as a discrete option. We execute a plan by feeding each latent in the sequence to the low-level policy for ∼1 s, during which the low-level policy executes in closed-loop.

In simulation, this strategy found a plan for tracing a triangle in less than 1 min, and that plan successfully transferred to the real robot (Fig. 5.3). We replicated this experiment directly on the real robot, with no intermediate simulation stage. It took 24 min of real robot execution time to find a motion plan for the triangle tracing task (Fig. 6).

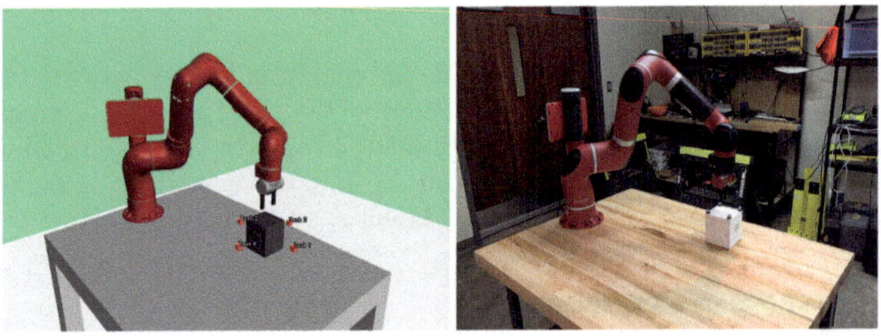

Fig. 6. Multitask environment in simulation (left) and reality (right) for the box pushing experiment. Four goals are located on the table at a distance of 15 cm from the block's initial position.

Sawyer Experiment: Box Pushing
We ask the Sawyer robot to push a box to a goal location relative to its starting position, as defined by a 2D displacement vector in the table plane. The policy receives a state observation with the robot's seven joint angles, plus a relative cartesian position vector between the robot's gripper and the box's centroid. The policy chooses incremental joint movements (up to ±0.04 rad) as actions. In the real experimental environment, we track the position of the box using motion capture and merge this observation with proprioceptive joint angles from the robot.

We trained the low-level pushing policy on four goal positions in simulation: up, down, left, and right of the box's starting point (Fig. 7). The composer policies feed latent vectors to the pre-trained skill policy to push the box to positions which were never seen during training (Figs. 8 and 9).

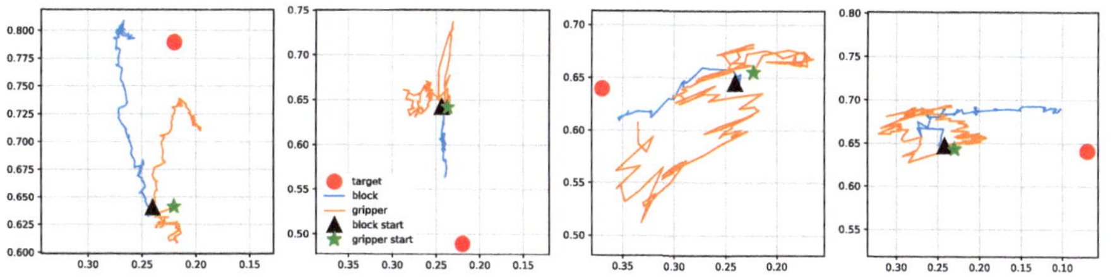

Fig. 7. Trajectories of the block (blue) and gripper position (orange) created by execution of the embedded skill policy for each of the four pre-trained skills (red dots).

Composition Experiments

Task Interpolation in the Latent Space (Direct)

We evaluated the embedding function to obtain the mean latent vector for each of the four pre-trained pushing skills (i.e. up, down, left, and right of start position). We then fed the mean letant of adjacent skills (e.g. $z_{\text{up-left}} \sim mean\{z_{\text{up}}, z_{\text{left}}\}$) while executing the pre-trained policy directly on the robot (Fig. 8).

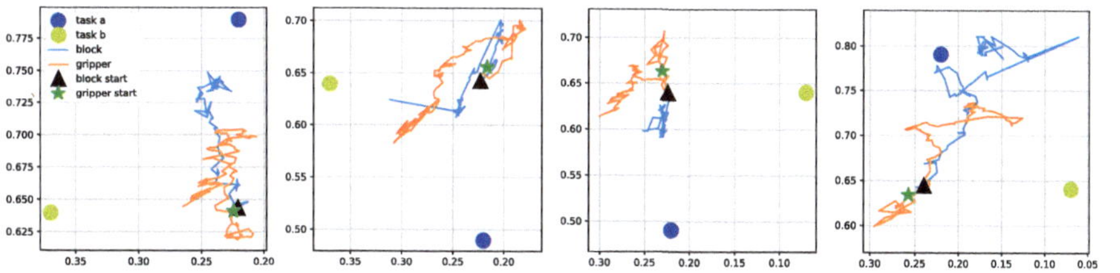

Fig. 8. Trajectories of the block (blue) and gripper position (orange) created by executing the pre-trained embedded skill policy while feeding the mean latent vector of two neighboring skills. The robot pushes the block to positions between the goal locations of the two pre-trained skills.

We find that in general this strategy induces the policy to move the block to a position between the two pre-trained skill goals. However, magnitude and direction of block movement was not easily predictable from the pre-trained goal locations, and this behavior was not reliable for half of the synthetic goals we tried.

Search-Based Planning in the Latent Space (sim2real)

Similar to the search-based composer on the reaching experiment, we used Uniform Cost Search in the latent space to find a motion plan (sequence of latent vectors) for pushing the block to unseen goal locations (Fig. 9). We found that search-based planning was able to find a latent-space plan to push the block to any location within the convex hull formed by the four pre-trained goal locations. Additionally, our planner was able to push blocks to some targets significantly

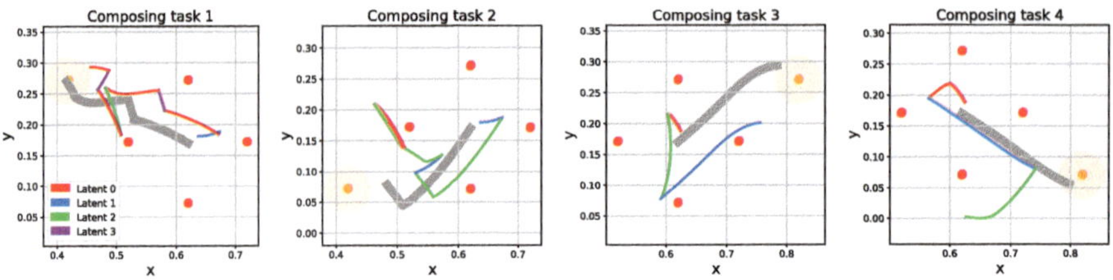

Fig. 9. Search-based planning in the latent space achieves plans for pushing tasks where the goal position (orange dot) for the block is outside the region of tasks (red dots) on which the low-level skill was trained. Colored line segments trace the path of the robot gripper, and indicate which skill latent was used for that trajectory segment. The block's position is traced by the gray lines.

outside this area (up to 20 cm). Unfortunately, we were not able to reliably transfer these composed policies to the robot.

We attribute these failures to transfer partially to the non-uniform geometry of the embedding space, and partially to the difficulty of transferring contact-based motion policies learned in simulation, and discuss these results further in Sect. 4.

4 Main Experimental Insights

The point environment experiments verify the principles of our method, and the single-skill Sawyer experiments demonstrate its applicability to real robotics tasks. Recall that all Sawyer skill policies used only joint space control to actuate the robot, meaning that the skill policies and composer needed to learn how using the robot to achieve task-space goals without colliding the robot with the world or itself.

The Sawyer composition experiments provide the most insight into the potential of latent skill decomposition methods for scaling simulation-to-real transfer in robotics. The method allows us to reduce a complex control problem–joint-space control to achieve task-space objectives–into a simpler one: control in latent skill-space to achieve task-space objectives.

We found that the method performs best on new skills which are interpolations of existing skills. We pre-trained on just eight reaching skills with full end-to-end learning in simulation, and all skills were always trained starting from the same initial position. Despite this narrow initialization, our method learned a latent representation which allowed later algorithms to quickly find policies which reach to virtually any goal inside the manifold of the pre-training goals. Composed policies were also able to induce non-colliding joint-space motion plans between pre-trained goals (Fig. 5).

Secondly, a major strength of the method is its ability to combine with a variety of existing, well-characterized algorithms for robotic autonomy. In addition to model-free reinforcement learning, we successfully used manual programming (interpolation) and search-based planning on the latent space to quickly reach both goals and sequences of goals that were unseen during pre-training (Figs. 5, 8 and 9). Interestingly, we found that the latent space is useful for control not only in its continuous form, but also via a discrete approximation formed by the mean latent vectors of the pre-training skills. This opens the method to combination with large array of efficient discrete planning and optimization algorithms, for sequencing low-level skills to achieve long-horizon, high-level goals.

Conversely, algorithms which operate on full continuous spaces can exploit the continuous latent space. We find that a DDPG-based composer with access only to a discrete latent space (formed from the latent means of eight pre-trained reaching skills and interpolations of those skills) is significantly outperformed by a DDPG composer that leverages the entire embedding space as its action space (Fig. 10). This implies that the embedding function contains information on how to achieve skills beyond the instantiations the skill policy was pre-trained on.

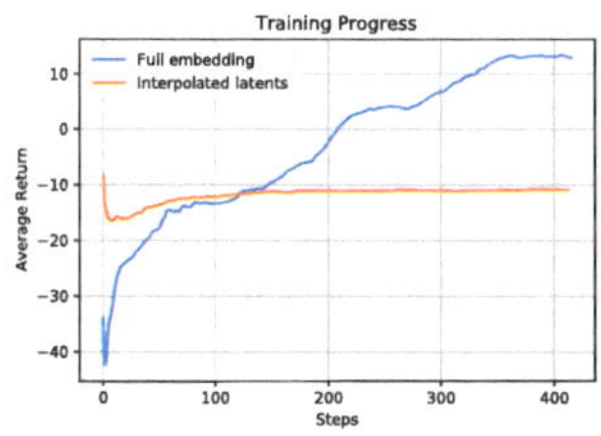

Fig. 10. Training returns for a composing policy using an embedding trained on eight-goal reacher to reach a new point between the trained goals.

The method in its current form has two major challenges.

First is the difficulty of the simulation-to-real transfer problem even in the single-skill domain. We found in the Sawyer box-pushing experiment (Fig. 7) that our ability to train transferable policies was limited by our simulation environment's ability to accurately model friction. This is a well-known weakness in physics simulators for robotics. A more subtle challenge is evident in Fig. 4, which shows that our reaching policy did not transfer with complete accuracy to the real robot despite it being free-space motion task. We speculate that this is a consequence of the policy overfitting to the latent input during pre-training in simulation. If the skill latent vector provides all the information the policy needs to execute an open-loop trajectory to reach the goal, it is unlikely to learn closed-loop behavior.

The second major challenge is constraining the properties of the latent space, and reliably training good embedding functions, which we found somewhat unstable and hard to tune. The present algorithm formulation places few constraints on the algebraic and geometric relationships between different skill embeddings. This leads to counterintuitive results, such as the mean of two pushing policies pushing in the expected direction but with unpredictable magnitude (Fig. 8), or the latent vector which induces a reach directly between two goals (e.g A and B) actually residing much closer to the latent vector for goal A than for goal B (Fig. 5). This lack of constraints also makes it harder for composing algorithms to plan or learn in the latent space.

5 Conclusion

Our experiments illustrate the promise and challenges of applying of state-of-the-art deep reinforcement learning to real robotics problems. For instance, our policies were able to learn and generalize task-space control and even motion planning skills, starting from joint-space control, with no hand-engineering for those use cases. Simultaneously, the training and transfer process requires careful engineering and some hand-tuning. In the case of simulation-to-real techniques, our performance is also bounded by our ability to build and execute reasonable simulations of our robot and its environment, which is not a trivial task.

In future work, we plan to study how to learn skill embedding functions more reliably and with constraints which make them even more amenable to algorithmic composition, and further exploring how to learn and plan in latent space effectively. We also plan to combine our method with other transfer learning techniques, such as dynamics randomization [20], to improve the transfer quality of embedded skill policies in future experiments. Our hope is to refine the method into a easily-applicable method for skill learning and reuse. We also look forward to further exploring the relationship between our method and meta-learning techniques, and the combination of our method with techniques for learning representations of the observation space.

Acknowledgements. The authors would like to thank Angel Gonzalez Garcia, Jonathon Shen, and Chang Su for their work on the garage[2] reinforcement learning for robotics framework, on which the software for this work was based. This research was supported in part by National Science Foundation grants IIS-1205249, IIS-1017134, EECS-0926052, the Office of Naval Research, the Okawa Foundation, and the Max-Planck-Society. Any opinions, findings, and conclusions or recommendations expressed in this material are those of the author(s) and do not necessarily reflect the views of the funding organizations (https://github.com/rlworkgroup/garage).

References

1. Cooper, P.A.: Paradigm shifts in designed instruction: from behaviorism to cognitivism to constructivism. Educ. Technol. **33**(5), 12–19 (1993)
2. Drescher, G.L.: Made-Up Minds: A Constructivist Approach to Artificial Intelligence. MIT Press, Cambridge (1991)
3. Lillicrap, T.P., Hunt, J.J., Pritzel, A., Heess, N., Erez, T., Tassa, Y., Silver, D., Wierstra, D.: Continuous control with deep reinforcement learning. CoRR, abs/1509.02971 (2015)
4. Mnih, V., Kavukcuoglu, K., Silver, D., Rusu, A.A., Veness, J., Bellemare, M.G., Graves, A., Riedmiller, M., Fidjeland, A.K., Ostrovski, G., Petersen, S., Beattie, C., Sadik, A., Antonoglou, I., King, H., Kumaran, D., Wierstra, D., Legg, S., Hassabis, D.: Human-level control through deep reinforcement learning. Nature **518**(7540), 529–533 (2015)

5. Levine, S., Finn, C., Darrell, T., Abbeel, P.: End-to-end training of deep visuomotor policies. JMLR **17**(1), 1334–1373 (2016)
6. Chebotar, Y., Hausman, K., Zhang, M., Sukhatme, G., Schaal, S., Levine, S.: Combining model-based and model-free updates for trajectory-centric reinforcement learning (2017)
7. Gu, S., Holly, E., Lillicrap, T., Levine, S.: Deep reinforcement learning for robotic manipulation with asynchronous off-policy updates. In: ICRA. IEEE (2017)
8. Duan, Y., Chen, X., Schulman, J., Abbeel, P.: Benchmarking deep reinforcement learning for continuous control. arXiv, 48:14 (2016)
9. Levine, S., Pastor, P., Krizhevsky, A., Ibarz, J., Quillen, D.: Learning hand-eye coordination for robotic grasping with deep learning and large-scale data collection. IJRR **37**(4–5), 421–436 (2018)
10. Hausman, K., Springenberg, J.T., Wang, Z., Heess, N., Riedmiller, M.: Learning an embedding space for transferable robot skills. In: ICLR (2018)
11. Gupta, A., Mendonca, R., Liu, Y., Abbeel, P., Levine, S.: Meta-reinforcement learning of structured exploration strategies. CoRR, abs/1802.07245 (2018)
12. Eysenbach, B., Gupta, A., Ibarz, J., Levine, S.: Diversity is all you need: learning skills without a reward function. CoRR, abs/1802.06070 (2018)
13. Heess, N., Wayne, G., Tassa, Y., Lillicrap, T., Riedmiller, M., Silver, D.: Learning and transfer of modulated locomotor controllers. CoRR, abs/1610.05182 (2016)
14. Haarnoja, T., Hartikainen, K., Abbeel, P., Levine, S.: Latent space policies for hierarchical reinforcement learning. CoRR, abs/1804.02808 (2018)
15. Co-Reyes, J.D., Liu, Y., Gupta, A., Eysenbach, B., Abbeel, P., Levine, S.: Self-consistent trajectory autoencoder: hierarchical reinforcement learning with trajectory embeddings. ArXiv e-prints, June 2018
16. Pastor, P., Kalakrishnan, M., Righetti, L., Schaal, S.: Towards associative skill memories. In: Humanoids, November 2012
17. Rueckert, E., Mundo, J., Paraschos, A., Peters, J., Neumann, G.: Extracting low-dimensional control variables for movement primitives. In: ICRA, May 2015
18. Rusu, A.A., Rabinowitz, N.C., Desjardins, G., Soyer, H., Kirkpatrick, J., Kavukcuoglu, K., Pascanu, R., Hadsell, R.: Progressive neural networks. CoRR, abs/1606.04671 (2016)
19. Rajendran, J., Prasanna, P., Ravindran, B., Khapra, M.M.: ADAAPT: a deep architecture for adaptive policy transfer from multiple sources. CoRR, abs/1510.02879 (2015)
20. Peng, X.B., Andrychowicz, M., Zaremba, W., Abbeel, P.: Sim-to-real transfer of robotic control with dynamics randomization. CoRR, abs/1710.06537 (2017)
21. Sadeghi, F., Levine, S.: CAD2RL: real single-image flight without a single real image. In: RSS (2017)
22. Tzeng, E., Devin, C., Hoffman, J., Finn, C., Peng, X., Levine, S., Saenko, K., Darrell, T.: Towards adapting deep visuomotor representations from simulated to real environments. CoRR, abs/1511.07111 (2015)
23. Kroemer, O., Sukhatme, G.S.: Learning relevant features for manipulation skills using meta-level priors. CoRR, abs/1605.04439 (2016)
24. Schulman, J., Wolski, F., Dhariwal, P., Radford, A., Klimov, O.: Proximal policy optimization algorithms. CoRR, abs/1707.06347 (2017)
25. Todorov, E., Erez, T., Tassa, Y.: MuJoCo: a physics engine for model-based control. In: IROS (2012)

Sim-to-Real Transferable Object Classification Through Touch-Based Continuum Manipulation

Huitan Mao[1(✉)], Junius Santoso[2], Cagdas Onal[2], and Jing Xiao[2]

[1] Department of Computer Science, UNC Charlotte, Charlotte, USA
hmao4@uncc.edu
[2] Robotics Engineering, WPI, Worcester, USA
jsantoso@wpi.edu

Abstract. It is important to investigate object perception for classification or recognition based on touch sensing, especially when robots are operating in darkness or the objects are difficult to capture by vision sensors. In this work, we present a new form of continuum manipulator equipped with sparse touch sensing, validate the effectiveness of automatic generation of the touch-based continuum wraps, and the effectiveness of object classification based on the continuum wraps. Using the indirect object shape information encoded in the robot shape, we demonstrate that a classifier trained from the simulated continuum wraps is transferable to identify the real world objects with real continuum wraps.

Keywords: Continuum manipulation · Tactile sensing · Object perception

1 Introduction

In order to broaden the real-world application of robotics, the importance of advancing the capability of robotic manipulation is unquestionable. Most of the existing research in robotic manipulation assumes that the target object is known or visible, or can be made visible by adding a light source. However, in many real-world scenarios, such as underwater or in a smoke-filled room (in a building on fire), a target object may not be made visible. Therefore, it is necessary to investigate robotic manipulation based on other sensing modalities.

Tactile or force sensing can be very useful in providing more information about surrounding objects. Indeed, there is recent research on using tactile sensing for object recognition [1], exploration and manipulation [2–4], and shape

H. Mao and J. Santoso—The authors equally contributed to this paper.

Electronic supplementary material The online version of this chapter (https://doi.org/10.1007/978-3-030-33950-0_25) contains supplementary material, which is available to authorized users.

J. Xiao et al. (Eds.): ISER 2018, SPAR 11, pp. 280–289, 2020.
https://doi.org/10.1007/978-3-030-33950-0_25

estimation [5,6]. Touch sensors are usually put on the finger tips of the grippers, which are used to touch a target object to collect contact points. Object recognition is achieved solely based on the collected contact points in [1] or in combination with visual perception [7]. Such touch-based perception is especially useful for perceiving transparent objects, which can be missed by visual sensing.

Continuum manipulators are more flexible for manipulating objects of a wide range of shapes and sizes in very cluttered environments due to their compliant and soft nature. In particular, a continuum manipulator can get into a small hole to fetch an object that a conventional, articulated manipulator cannot [8]. Since it is often dark in such a hole and difficult to bring in an extra light source, touch-based continuum manipulation is most desirable. If the target object is not visible, then object perception (i.e., classification and/or recognition) is also needed from touch-based continuum manipulation.

In our recent work [9], we introduced a shape-based approach for object classification and recognition through continuum manipulation. The main idea is to use the continuum robot as a tool to indirectly "measure" the object shape. That is, the shape of a continuum arm during whole-arm wraps of a target object, which can be transparent (and thus not visible), is used to indirectly characterize the shape of the object and this information is used for classification and recognition. For an object of any shape, a continuum wrap is generated automatically by a touch-based approach. However, the work is tested only in simulation with the assumption that the continuum arm is covered by tactile sensors. In reality, existing continuum manipulators are often not equipped with tactile sensors. One difficulty is that the body of the robot is deformable. A kinematics-based contact detection and localization approach for continuum robots is presented in [10], for which an external tracking system is required but may not always be available in real-world scenarios.

In this paper, we study touch-based identification of object categories using a new form of continuum manipulator consisting of origami-based modules [11] and tactile sensors attached at each section. Using these new manipulators with sparse tactile sensing, we aim to achieve touch-based wrapping of objects and experimentally validate the following conjecture: the shape-based classifier introduced in [9] and trained in simulation can be readily transferred to classifying real objects with real touch-based continuum manipulation. We envision the robot to be used in a search and rescue scenario where it could be exposed to low-light environment hence could benefit from the touch-based manipulation.

2 Technical Approach

We next explain the manipulators we built for this study, sensors, and the touch-based motion planning strategy for generating continuum wraps, and the classification method of unknown objects.

2.1 Manipulators and Sensors

The continuum manipulators we built for this study consist of multiple origami continuum modules connected in series. Each continuum module contains a foldable origami body, three brushed DC micro-motors with pulley systems, and a controller board that offers on-board sensory measurements, feedback control, and module-to-module communication. The foldable body is made out of Polyethylene terephthalate (PET) films and constructed based on the Yoshimura crease pattern. This unique tubular structure with a diameter of 7 cm is capable of bending in two directions and extending/retracting, while maintaining its structure and resisting torsion. The foldable structure is connected to an acrylic plate on the top and to the PCB on the bottom where the motors are secured. Three nylon cables secured to the motor shafts and spanning the length of the structure along the edges are used to drive the segment. Each motor is equipped with a magnetic encoder for position control.

We built two continuum manipulators for this study. Figure 1(a) shows the 3-section manipulator used in the experiments with planar wraps, and each section has an active white module with motors on it and a passive green foldable body to expand the robot workspace. Improved from the 3-section manipulator, Fig. 1(b) shows the 4-section manipulator used for generating spatial wraps benefiting to its more compact and lightweight modules and larger touch sensor contact areas.

Each section is characterized using three parameters (s, κ, ϕ), where s is the section length, κ is the curvature, and ϕ is the orientation angle [8,11]. Using the inverse kinematics of a continuum section developed in [11], we can then find the required cable lengths (l_1, l_2, l_3), that will shape the module into the desired configuration (s_d, θ_d, ϕ_d). The cable lengths are converted into encoder positions, which are then sent to the low-level controller as reference signals.

We constructed the touch sensors acting as bumper switches using two copper sheets adhered onto a parallel plate structure made out of the same material used for the origami collapsible body. One copper sheet is connected to the control board's digital I/O pin while the other is connected to ground reference. When the copper sheets touch each other due to depression of the structure, an electrical circuit is completed hence signaling a touch on the continuum section. For each continuum section we place the touch sensor at the module and sandwiched between consecutive modules as shown in Fig. 1.

2.2 Touch-Based Continuum Wrapping

A general touch-based motion planning strategy is introduced in [9] for a continuum manipulator to progressively generate wraps around an object under the guidance of the contacts made along the way without knowing the object model. Starting from the initial configuration, the robot motion alternates between the *enclosing motion step* to make contacts with the object, and the *advancing motion step* to move forward towards wrapping around the object, until a wrap is formed or no further motion is feasible due to the physical limits of the manipulator.

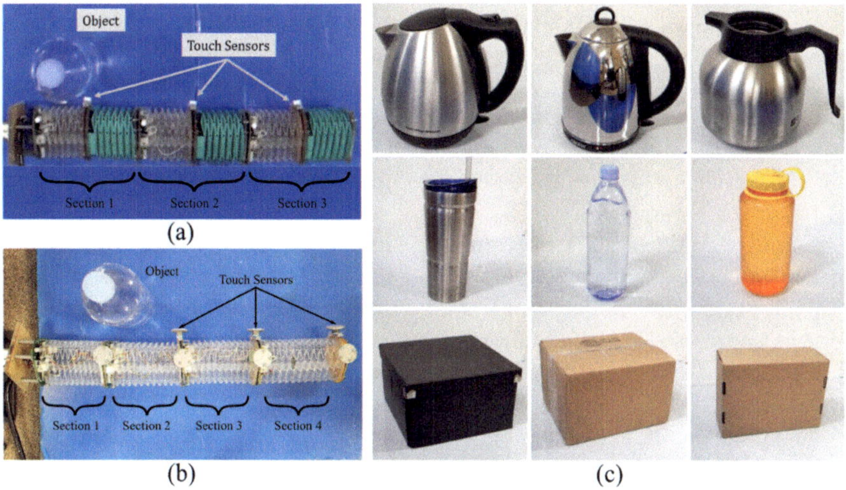

Fig. 1. An object (a transparent water bottle) was placed near the base of arms for manipulation: (a) a touch sensor was mounted in the middle of each origami module in a 3-section arm, (b) a touch sensor was mounted at the distal end of each module in a 4-section arm. The diameter of the continuum section is 7 cm. (c) A list of objects used in the experiments.

Such continuum wraps can be efficiently generated within hundreds of milliseconds (time of planning and collision checking combined) in simulation [9].

To achieve an *advancing motion step*, as described in [9], contact localization and estimation of the tangential and normal directions of the local contact patches are required; whereas, we relax this requirement in this paper through extrapolating the robot section endpoints based on their local frames. Therefore, our planner only needs to know whether each manipulator section is in contact or not from the tactile sensing to plan the next move, which makes it more effective to guide the manipulator hardware to achieve touch-based continuum wraps.

2.3 Classification of Unknown Objects

As introduced in Sect. 1, we aim to experimentally validate a shape-based classifier using continuum wraps [9]. The shape of the continuum robot wrapping around an object is described by a chord histogram descriptor, which approximates the robot shape using many 3D chords and statistically captures its shape based on the chord parameters. We first trained a linear SVM classifier in simulation using wraps generated around the simulated objects and next applied the trained model to classify the real-world objects using real-world wraps. The objects in simulation were scaled to roughly match the dimensions of the real-world objects.

3 Experimental Results with Planar Wraps

In our experiments with planar wraps, we used the 3-section manipulator as shown in Fig. 1(a). The robot manipulator is initially set at a straight-line configuration with full contraction, and the testing object is placed near the arm base. For each section, $s \in [0.085, 0.145](m)$, $\kappa \in [-10.69, 10.69](1/m)$, and $\phi \in [-\pi, \pi]$.

A linear SVM classifier is trained in simulation using wraps from 10 water bottles, 10 boxes, and 6 teapots. For each object one wrap was generated. Figure 2 shows a few planar continuum wraps around the objects in simulation. We next conducted classification of three different real-world objects: a teapot, a water bottle, and a card box, through real-world continuum wraps.

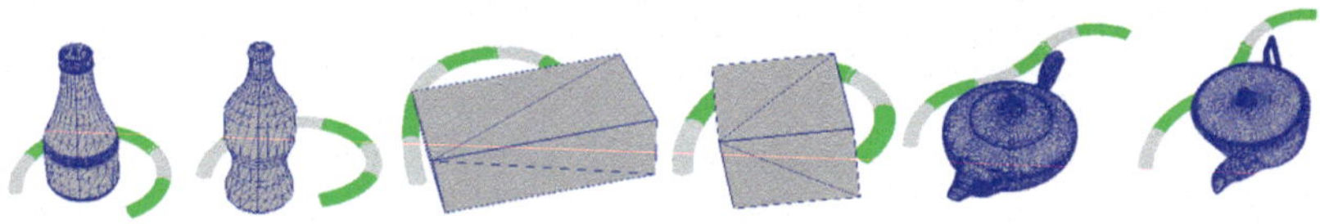

Fig. 2. A few examples of the planar continuum wraps generated on different objects in simulation. The simulated arm has 3 sections and each section is colored in white for the first half and in green for the second half.

The attached video shows the wrapping process of the real-world objects, and Fig. 3 shows a few motion snapshots. Note that the goal of such wraps is to encode the shape of an unknown object into the shape of the manipulator, as opposed to achieving tight grasps of objects with known models [8]. Therefore, the wraps do not need to be enclosing.

The final robot configurations commanded by the planner were used for classification. Table 1 summarizes object and wrap information and the classification results. It can be seen that the classifier learned solely from simulation is already effective in classifying the real-world objects, as the probabilities of correct classification are more than two times higher than that of a random guess (about 0.33). However, the box was mis-classified to be a bottle because the contact on Sect. 1, which was closest to the base, was missed by the sensor, and therefore the planner kept commanding Sect. 1 to bend more while it was actually stopped by the contact. Since the curvature of Sect. 1 is a distinctive feature for classification as shown in Fig. 3, the classifier with the inaccurate data of a larger curvature resulted in the misclassification.

4 Experimental Results with Spatial Wraps

We use the 4-section robot arm (Fig. 1(b)) to conduct spatial continuum wraps around the objects to collect spatial shape information. The arm sections of this robot are more compact (rather than having two modules connected as one

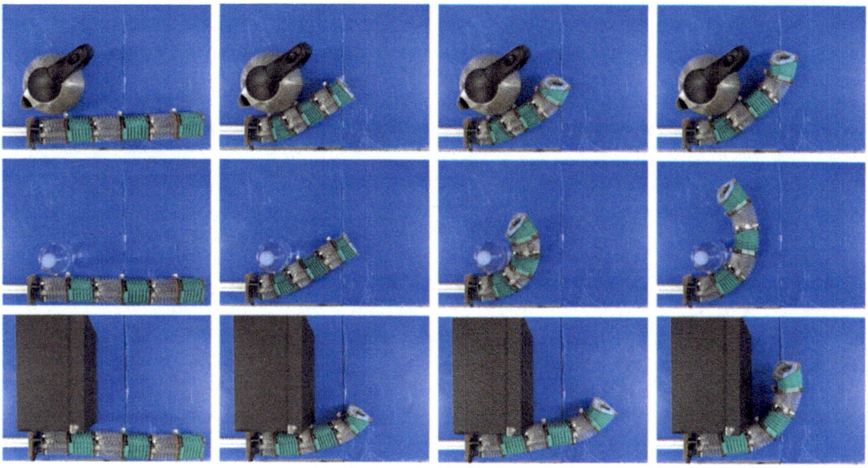

Fig. 3. The motion snapshots of the continuum wraps generated on the real objects. The rightmost sub-figure in each row shows the final wrapping configuration.

Table 1. Object dimension, number of intermediate configurations to generate the wraps, SVM prediction and its probability using 1 planar wrap for each object.

Object	Dimension (cm)	# of configurations	SVM prediction	Probability
Bottle	$8 \times 8 \times 24$	59	Bottle	0.81
Box	$27 \times 16 \times 26$	61	Bottle	0.83
Teapot	$24 \times 18 \times 21$	42	Teapot	0.71

robot section) and lightweight, which makes it more suitable for spatial wraps. It also overcame the problems of missed contacts with increased contact areas of the touch sensors. Different wraps covering different areas on the objects were conducted to collect object spatial shape information. The 3D chords generated from different wraps are accumulated into one histogram as an overall representation of the object shape. We next explain how the robot arm is lifted to conduct spatial wraps and present the object classification results.

4.1 Robot Arm Lifting

The arm is lifted up by keeping the first module of the robot at a certain configuration during the experiment and then changed when switching to other wrapping planes. See Fig. 5(b) for such examples when the robot arm is lifted up. The benefit of having this additional module is that the other modules (Sects. 2, 3 and 4) can be fully used for generating the wraps, and they only need to undergo less strain and stress when wrapping around the objects. Alternatively, the modules could be mounted vertically to minimize the effect of gravity; however this configuration cannot always be possible in a real life scenario, hence we decided to proceed with the former.

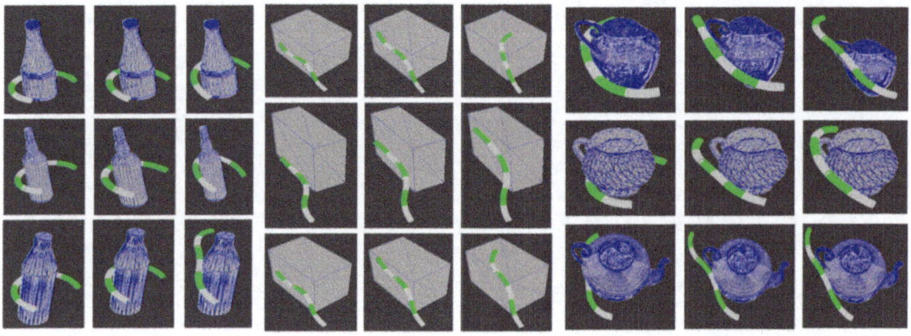

(a) bottles (b) boxes (c) teapots

Fig. 4. The spatial wraps in simulation. Each row in each subfigure is the three wraps around the same object. The simulated arm has 3 sections and each section is colored in white for the first half and in green for the second half.

4.2 Classification Results

We trained a linear SVM classifier using the same set of training objects used in Sect. 3, and the shape of each object was captured using 3 wraps. Figure 4 shows the wraps around the training objects generated in simulation. Figure 5 shows examples of such planar and spatial wraps around the real world objects. For testing dataset, we considered 3 object categories and 3 objects from each category (Fig. 1(c)).

The attached video shows the motion process of the continuum wraps. As mentioned earlier, such wraps are the result of local motion generation, and they are just used to encode the object shape onto the robot arm shape. Therefore, some (for instance the wraps around the boxes) may only locally conform the robot shape to the object shape and achieve partial wrapping of the object.

We noticed that the robot final wrapping configurations deviate from the robot motion commands sent by the motion planner (major cause for the misclassification of the box in Sect. 3). There are two main reasons. First, our robot arms currently do not have the proprioceptive sensors to achieve precise closed-loop control. Second, some contacts may be missed by the current sparse touch sensing on our robot arms. Therefore, in order to more precisely identify the final robot shape, we used an external vision tracking system to identify the robot final configuration when the wrapping process is terminated. This wrapping configuration is next used for generating the 3D chords and conducting the final object category classification.

Table 2 summarizes the objects used, the average number of robot configurations to generate the wraps, and the SVM prediction results. Overall, the classifier was able to correctly recognize 2 bottles, 3 boxes, and 2 teapots. The

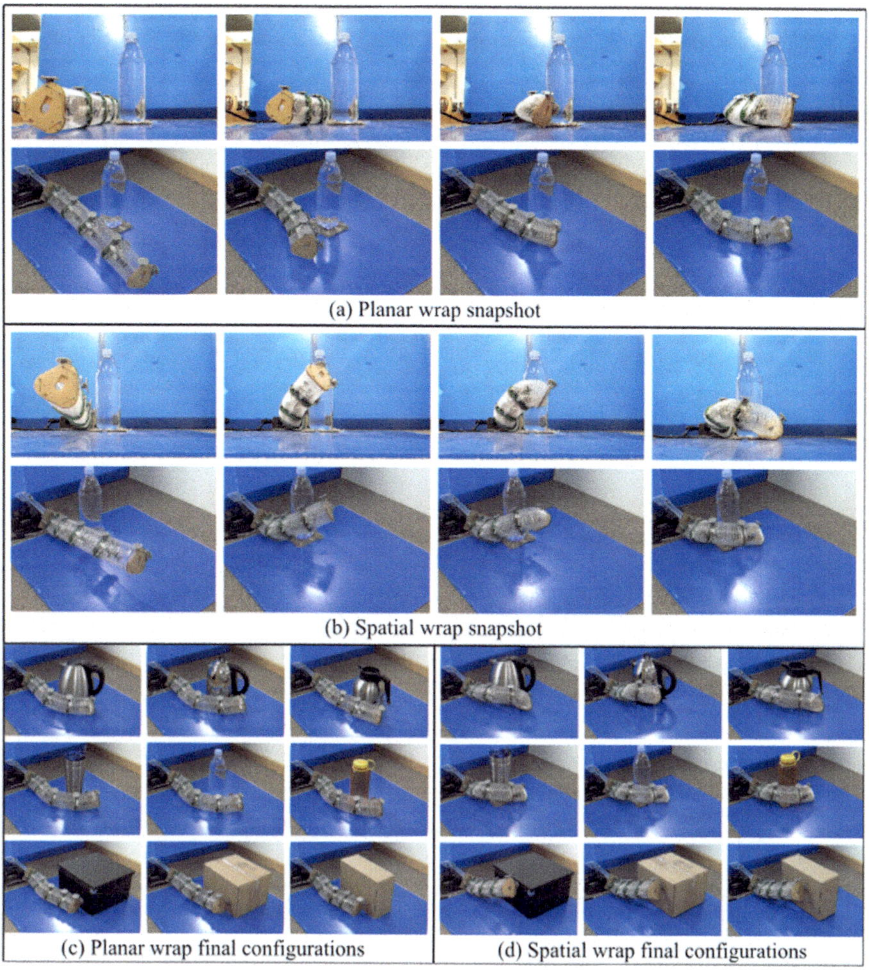

Fig. 5. The motion snapshots of (a) planar and (b) spatial continuum wraps around the bottle, where the top and bottom rows show the motion of the same wrap from two view angles respectively, (c) final configurations for the remaining objects with planar wraps, (d) spatial wrap final configurations.

boxes are easier to be correctly identified since the robot arm conforms to the side surfaces of the boxes and therefore has distinctive straight sections in contact. The wraps of the bottles and the teapots typically have more curved robot sections but differ in lengths due to their dimensions. The confusion of classifying the bottles and the teapots is because sometimes the handles of the teapots (more distinctive features) are not captured. This can be improved by using longer robot sections and more dense touch sensing.

Table 2. Object dimension, average number of intermediate configurations to generate one wrap, SVM prediction and its probability using 3 spatial wraps for each object.

Object	Dimension (cm)	Avg. # of config.	SVM prediction	Probability
Bottle1	8 × 8 × 24	45	Bottle	0.52
Bottle2	9 × 9 × 21	43	Teapot	0.56
Bottle3	8 × 8 × 21	40	Bottle	0.51
Box1	27 × 16 × 26	12	Box	0.74
Box2	25 × 21 × 17	10	Box	0.71
Box3	27 × 10 × 18	15	Box	0.78
Teapot1	24 × 18 × 21	36	Teapot	0.46
Teapot2	21 × 15 × 21	40	Teapot	0.51
Teapot3	22 × 16 × 16	38	Bottle	0.51

5 Experimental Insights

Our results have demonstrated that the shape-based classifier trained solely from simulation is able to generalize to real-world objects. This confirms our two key insights. First, because object classification is based on the shapes of the continuum arm wrapping around the objects and not the shapes of the objects directly, the classifier has the advantage of avoiding direct sensing and perception of the shape of an unknown target object as well as the associated limitations (such as low object visibility) and all the sensing uncertainties involved that can negatively affect classification accuracy. Second, the continuum wraps generated on objects in the same category have similar shapes, which are captured by the intrinsic parameters of the continuum arm, no matter if the objects wrapped are virtual or real.

Since conducting many real-world continuum wraps can be time-consuming, it is significant that the classifier trained purely in simulation showed considerable effectiveness in classifying real objects. This could make classifier training more efficient and feasible for classifying a large number of categories of many real objects from touch-based continuum wrapping.

6 Future Improvement

The overall system can be improved in multiple ways for better results and robustness. First, soft modules with longer length and more dense touch sensing can help better capture the unique features of the object shapes, for instance, the handles of the teapots. Moreover, more dense touch sensing can also help reduce the cases where the motion planner kept curving the robot sections while the actual contacts are blocking the robot motion. Second, more sophisticated gravity compensation should be considered in order to better lift up the robot arm and form spatial wraps around different areas on the objects. Third, since the final robot configuration is the result of both desired motion commands and the contacts with the object, the final shape of the robot can be better identified by equipping robots with proprioceptive sensors and feedback control rather than relying on an external tracking system.

The manipulators we built for this study are for object perception purposes and hence have limited payloads. Once the unknown objects are recognized via touch wraps, force-closure continuum graspings can be generated for further manipulation of the objects using the algorithms in [8]. We would also like to enable handling heavy objects by further improving the robot payloads.

Acknowledgements. This work is supported by NSF grant IIP-1439695, CMMI-1752195.

References

1. Zhang, M.M., Atanasov, N., Daniilidis, K.: Active end-effector pose selection for tactile object recognition through Monte Carlo tree search. In: IEEE/RSJ International Conference on Intelligent Robots and Systems (IROS) (2017)
2. Sommer, N., Li, M., Billard, A.: Bimanual compliant tactile exploration for grasping unknown objects. In: International Conference on Robotics and Automation (ICRA). IEEE (2014)
3. Li, Q., Schürmann, C., Haschke, R., Ritter, H.J.: A control framework for tactile servoing. In: Robotics: Science and Systems (2013)
4. Yi, Z., Calandra, R., Veiga, F., van Hoof, H., Hermans, T., Zhang, Y., Peters, J.: Active tactile object exploration using Gaussian processes. In: IEEE/RSJ IROS (2016)
5. Jia, Y.-B., Tian, J.: Surface patch reconstruction from one-dimensional tactile data. IEEE Trans. Autom. Sci. Eng. **7**(2), 400–407 (2010)
6. Mao, H., Xiao, J.: Object shape estimation through touch-based continuum manipulation. In: International Symposium of Robotics Research (ISRR) (2017)
7. Falco, P., Lu, S., Cirillo, A., Natale, C., Pirozzi, S., Lee, D.: Cross-modal visuo-tactile object recognition using robotic active exploration. In: IEEE ICRA (2017)
8. Li, J., Xiao, J.: Progressive planning of continuum grasping in cluttered space. IEEE Trans. Robot. (TRO) **32**(3), 707–716 (2016)
9. Mao, H., Zhang, M.M., Xiao, J., Daniilidis, K.: Shape-based object classification and recognition through continuum manipulation. In: IEEE/RSJ IROS (2017)
10. Bajo, A., Simaan, N.: Kinematics-based detection and localization of contacts along multisegment continuum robots. IEEE Trans. Robot. **28**(2), 291–302 (2012)
11. Santoso, J., Skorina, E.H., Luo, M., Yan, R., Onal, C.D.: Design and analysis of an origami continuum manipulation module with torsional strength. In: IEEE/RSJ IROS (2017)

Robot Navigation

Robot Navigation

Thierry Peynot

The Navigation session focused on methods to enable safe and effective navigation in a variety of challenging environments, including pliable vegetation, brittle surfaces, very large buildings, rivers with varying winds and currents, or busy urban roads. A key common requirement of the methods presented in this session was the ability to predict the behavior of the robot(s), the environment (especially upon interacting with it) and/or other agents in the scene. In the first paper, Ordonez et al. presented a method to characterise pliable navigation, allowing an autonomous ground vehicle to drive through it safely, whenever possible. The multi-legged robot in the paper by Tennakoon et al. safely and effectively probed the terrain with its foot, before stepping, to determine if it would collapse. The paper by Nieto et al. presented an on-line task coordination strategy for multi-robot systems. The paper by Moulton et al. demonstrated a method to estimate environmental force maps that can help an autonomous surface vehicle to navigate in the presence of winds and currents. Finally, in the paper by Meghjani et al., predicting the intention and trajectory of other vehicles allowed a self-driving car to navigate in a typical urban environment. The discussion concluding this session raised the question of how well an autonomous mobile robot can, and needs to, predict the behaviour of other agents and of the environment. For example, it was noted that human behavior is considered quite unpredictable, yet, humans have been able to drive in traffic in practice.

Characterization and Traversal of Pliable Vegetation for Robot Navigation

Camilo Ordonez[1(✉)], Ryan Alicea[1], Brandon Rothrock[2], Kyle Ladyko[1], Jeremy Nash[2], Rohan Thakker[2], Shreyansh Daftry[2], Mario Harper[1], Emmanuel Collins[1], and Larry Matthies[2]

[1] Florida State University, Tallahassee, Fl 32310, USA
cordonez@fsu.edu
[2] Jet Propulsion Laboratory, Pasadena, USA

Abstract. Outdoor mobile robots currently treat vegetation as obstacles that need to be avoided. In order to have less conservative robots that fully exploit their motion capabilities, it is required to obtain models of the interaction of vegetation with the vehicle. This work proposes and experimentally verifies models of interaction of wheeled and tracked vehicles with pliable vegetation. In addition, a methodology to map perceptual features of the environment to resistive forces experienced by the robots is presented.

1 Introduction

Vegetated terrains are pervasive in the real world and thus it is crucial for autonomous systems to be able to treat them in a non-binary fashion. That is, robots should be able to reason about the vegetation in their surroundings and decide if and how to traverse such challenges. Of particular interest are terrains that are taller than or of comparable height to the vehicle hull (hereafter referred to as tall vegetation).

Current state of the art utilizes a conservative approach that classifies such vegetated terrains as obstacles and then plans routes that navigate around them. For example, dense point clouds obtained using lidar data have been used to segregate the environment into classes representing whether the terrain was traversable or non-traversable [1]. Also, the spaces between foliage have been used as the workspace within which paths could be generated [2].

In relevant work on identification of vegetated terrain, a multispectral sensor was created out of a near-infrared camera and a visible light video camera [3], leading to a robust chlorophyll detector in natural environments. Additional work focused on segmentation of lidar data into three classes using local three-dimensional point cloud statistics [4,5]. A perception methodology more geared towards vegetation traversal has also been developed [6]. In that research a

Electronic supplementary material The online version of this chapter (https://doi.org/10.1007/978-3-030-33950-0_26) contains supplementary material, which is available to authorized users.

J. Xiao et al. (Eds.): ISER 2018, SPAR 11, pp. 293–304, 2020.
https://doi.org/10.1007/978-3-030-33950-0_26

combination of lidar and radar was employed to detect large tree trunks that can be occluded behind thick foliage.

A 3D algorithm to detect and classify obstacles has also been proposed [7]. The approach classifies the obstacles as belonging to different obstacle types (rocks, vegetation, etc). It also included a terrain load bearing model, which was coupled with a suspension model of the vehicle to generate estimates of safe traversal speeds for the different terrains. Vegetation traversal has been studied [8], where uniform patches of vegetation were characterized using a lumped frictional model. The developed terrain models were then used to plan trajectories that exploited the vehicle momentum to traverse the vegetation patches while avoiding vehicle immobilization. More detailed stem models of vegetation have been proposed [9]. However, that work was limited to simulations and did not incorporate perception.

In this paper, we propose two different yet complementary approaches to characterize pliable terrains. One approach relies on traversals of the terrain and measurements of proprioceptive data to estimate a lumped resistance force experienced by the vehicle. The second approach relies on probing the terrain via a manipulator equipped with a torque-force sensor. Using this data, a physics-based model of the local vegetation is developed. In addition, this work presents a regression model which learns the mapping between perceptual features from exteroceptive sensors to interaction forces between the robot and the vegetation. The remainder of the paper is structured as follows: Sect. 2 describes the experimental platforms, Sect. 3 details the technical approach, Sect. 4 summarizes the experimental results, and Sect. 5 contains concluding remarks and future work.

2 Experimental Platforms

As shown in Fig. 1, this work utilizes two robotic platforms. The Roman tracked vehicle equipped with a Robosimian 7 degree of freedom manipulator developed at JPL, a Velodyne VPL-16 lidar, and 3 Intel RealSense ZR300 cameras. The second robot is a Husky skid steered vehicle from ClearPath equipped with a Velodyne VPL-16 lidar, 2 Intel RealSense ZR300's, current sensors per motor, a Microstrain 3DM-GX3-25 IMU, and a force bumper developed in-house containing two front-facing load cells.

3 Technical Approach

The overall approach is decomposed into two stages. A terrain modeling and characterization module and a regression system to map from perception data into resistance forces experienced by the robot.

3.1 Vegetation Characterization

Due to the different capabilities of the robots employed, the terrain modeling is platform dependent and is conducted as detailed below.

Fig. 1. Experimental platforms used in this work. (a) The tracked Roman platform equipped with a manipulator to probe the terrain. (b) The skid steered Husky robot.

Terrain Modeling with the Roman: Given the manipulator available on the Roman, it is possible to conduct a bottom up approach to terrain characterization. As illustrated in Fig. 2, the approach starts by modeling stems of vegetation as thin rods anchored on the ground, with rotational stiffness k, and rotational damping b. The second order linear differential equation describing the unforced dynamics of the stem can be expressed by

$$J\ddot{\theta} + b\dot{\theta} + k\theta = 0, \tag{1}$$

where J is the stem's moment of inertia, θ is the angle of the stem, and $\dot{\theta}$ and $\ddot{\theta}$ are its angular velocity and acceleration.

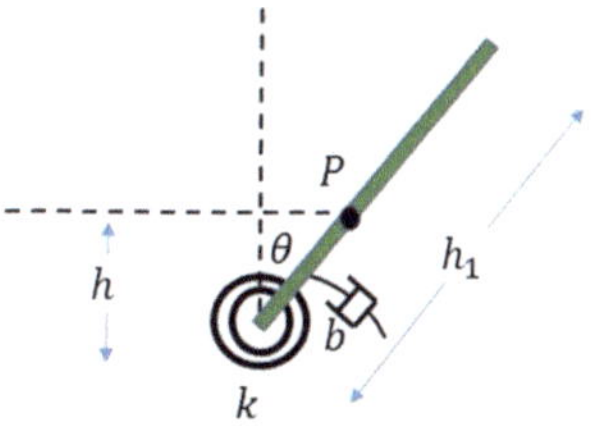

Fig. 2. Model of a vegetation stem of length h_1, rotational stiffness k and rotational damping b. Here, h represents the location at which an external force F interacts with the stem.

The robot longitudinal dynamics when interacting with one or more stems can be expressed as

$$M\ddot{x} + D(x, \dot{x}) + R_{res} = F, \tag{2}$$

where M is the robot's mass and F is the tractive force exerted by the robot tracks on the terrain. This force is required to accelerate the vehicle and to overcome the rolling resistance R_{res} and the drag D generated by the tall vegetation. Here, we assume that the terrain underneath the vegetation has the same properties as non-vegetated terrain in its proximity.

Assuming a very small stem inertia J, and given that the robot bumper makes contact with the stems at a height h, one can model the drag on the robot resulting from n stems using

$$D = \frac{1}{h}\sum_{i=1}^{n} \alpha_i k_i \theta_i \cos^2\theta_i + \frac{1}{h}\sum_{i=1}^{n} \alpha_i b_i \dot{\theta}_i \cos^2\theta_i, \tag{3}$$

where $\alpha_i = 1$ if stem i is in contact with the vehicle, and 0 otherwise.

Terrain Modeling with the Husky: In the absence of a manipulator, it is complicated to capture the interaction of individual stems. In addition, although it is expected that the overall resistance force varies with robot velocity, it was found that in practice is very complicated to traverse the same region twice at different speeds because during the first traversal the robot permanently destroys some of the vegetation. For these reasons, in the case of the Husky robot, the resistance of the vegetation is modeled via a lumped force f. Assuming an applied tractive force F, vehicle mass M, and acceleration a, the robot forward dynamics are governed by

$$F - f = Ma. \tag{4}$$

As the robot moves through the terrain, F is estimated from the current sensors using $F = \frac{\eta k_t G_r}{R}(i_l + i_r)$, where k_t is the torque constant, G_r is the gear ratio of the motor, R is the wheel radius, i_l and i_r are the left and right motor currents, and η is the efficiency of the motor and gearbox.

Employing a grid based map of the terrain as shown in Fig. 3, the net resistance f^k experienced by the vehicle at time k is given by

$$f^k = \sum_{i=1}^{n} \frac{\alpha_i^k f_i}{l}, \tag{5}$$

where n is the number of cells in the map, α_i^k is the overlap-length of the vehicle with cell i at time k, f_i is the resistance force of cell i and l is the length of the robot.

In the field, the resistance values f_i for each cell are estimated as follows: the robot is commanded to move in straight line at constant velocity. Then, provided the robot localization, the instantaneous robot-cell overlap values α_i^k are computed. To account for small accelerations imposed by the controller and ground interaction, the robot acceleration a^k is measured with an on board IMU. Combining (4) and (5),

$$F^k - Ma^k = \sum_{i=1}^{n} \frac{\alpha_i^k f_i}{l} \tag{6}$$

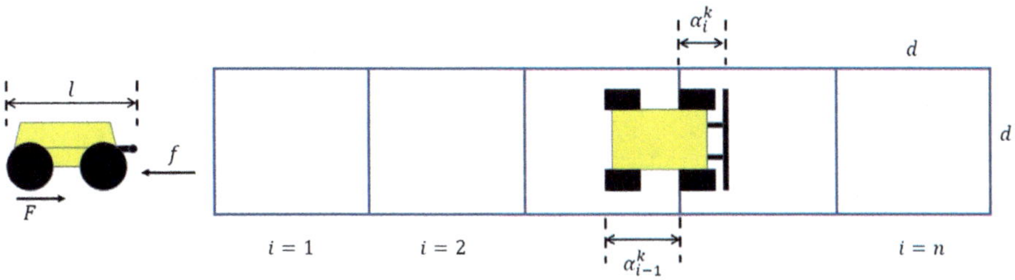

Fig. 3. Vehicle terrain interaction model for the Husky. A lumped resistance force f_i is estimated for each grid cell in the robot map. Each cell is assumed of size $d \times d$. α_i^k represents the overlap length of the vehicle with cell i at time k.

As the robot moves through the terrain, it is possible to obtain m observations of (6) and estimate the unknown cell resistance forces using linear least squares as follows:

$$Y = Hq + \epsilon, \tag{7}$$

where

$$Y = \begin{bmatrix} F^1 - Ma^1 \\ F^2 - Ma^2 \\ \vdots \\ F^m - Ma^m \end{bmatrix}, \; H = \begin{bmatrix} \frac{\alpha_1^1}{l} & \frac{\alpha_2^1}{l} & \dots & \frac{\alpha_n^1}{l} \\ \frac{\alpha_1^2}{l} & \frac{\alpha_2^2}{l} & \dots & \frac{\alpha_n^2}{l} \\ \vdots & \vdots & & \vdots \\ \frac{\alpha_1^m}{l} & \frac{\alpha_2^m}{l} & \dots & \frac{\alpha_n^m}{l} \end{bmatrix}, \text{ and } q = \begin{bmatrix} f_1 \\ f_2 \\ \vdots \\ f_n \end{bmatrix},$$

ϵ represents the error in the resistance force prediction. Since (7) is an overdetermined system, q is estimated as the least squares solution, $\hat{q} = (H^T H)^{-1} H^T Y$.

3.2 Mapping from Perception to Resistance Force

The objectives of the perception system is to predict terrain properties from vision and lidar sensors. The current and force measurements are used to directly estimate the lumped resistance directly underneath the robot as described above. These estimated parameters are accumulated into a 3D voxel grid, and used to train a prediction model to infer these parameters on a local region of the map surrounding the robot. The voxel resolution we are using is $5\,\text{cm}^3$.

The vision sensors used on the robot consist of Intel RealSense ZR300 assisted stereo cameras and a VLP-16 lidar. The stereo cameras are primarily used for localization using ORB-SLAM2 [11]. The 3D voxel map is based on the JMS architecture described in [12], and is populated using lidar data. We use the multi-return mode of the Velodyne to record both the strongest and last return, illustrated in Fig. 5, to predict vegetation density. The mean difference between these two returns is stored for each voxel. The resulting voxel map contains a single scaler for each cell, and is processed in $5 \times 5 \times 5$ voxel subvolumes by the prediction model.

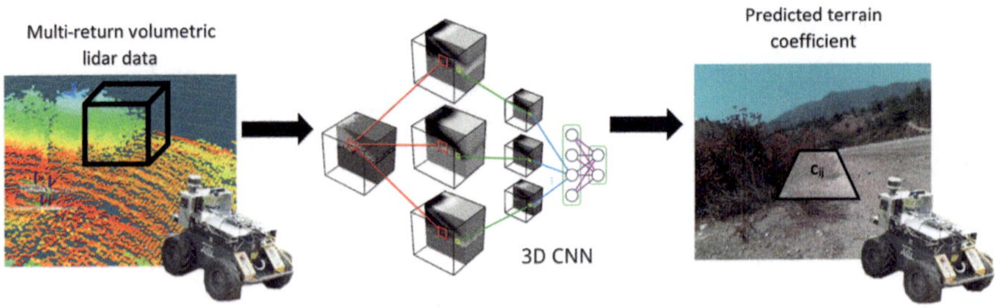

Fig. 4. Overview of the terrain prediction system. The terrain is divided into 25 cm^3 subvolumes indicated by the box outline. Multi-return statistics are collected from the lidar to populate each 5 cm^3 voxel in the map, which is then processed by a 3D CNN to predict resistance values for each subvolume. See text for details.

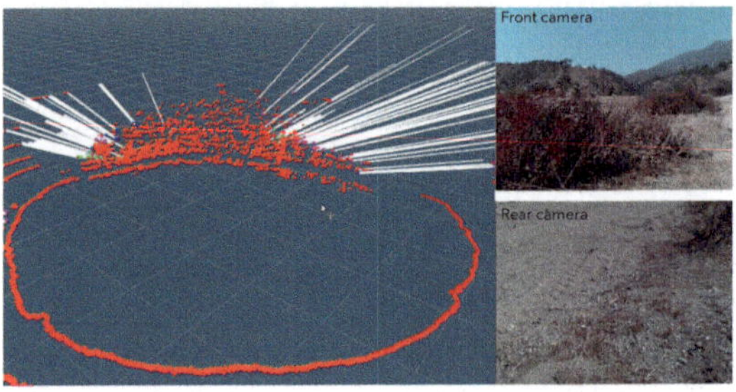

Fig. 5. Velodyne multi-return mode consists of the strongest (red dot) and last return (white ray), which measures how well the laser can penetrate through the vegetation, and helps predict vegetation density.

The prediction model is based on a volumetric convolutional neural network [13]. The CNN model utilizes 3D convolution to reason about local spatial context, followed by a dense fully-connected layer and a regression output layer, as shown in Fig. 4. The loss function is a regularized negative log-likelihood, optimized using stochastic gradient descent. The input to the network is a 3D tensor containing a subvolume of the map with dimensions 25 m × 25 m × 25 m. The output is the lumped resistance, as described above, and is run as a sliding window within the region of interest.

4 Results

This section presents results of vegetation characterization with the both platforms and perception results on the Husky.

4.1 Roman Platform

The unforced dynamics of the stem were experimentally verified on real vegetation with high speed video in previous work [9]. Here, field experiments were conducted on scenarios like the one of Fig. 6b. A rigid bar with a width equivalent to that of the robot was held by the gripper to capture the full resistance which would be experienced by the robot during traversal. The end-effector position trajectory of Fig. 6a was commanded. During each probe, the manipulator pushed the bar forward in order to deflect an area of vegetation. While the vegetation was held at a constant deflection, the forces seen by the force plate at the wrist were used to estimate the stiffness, k, of the vegetation. Once the stiffness was estimated, the damping component, b, was obtained by analysis of the constant velocity periods of the probe.

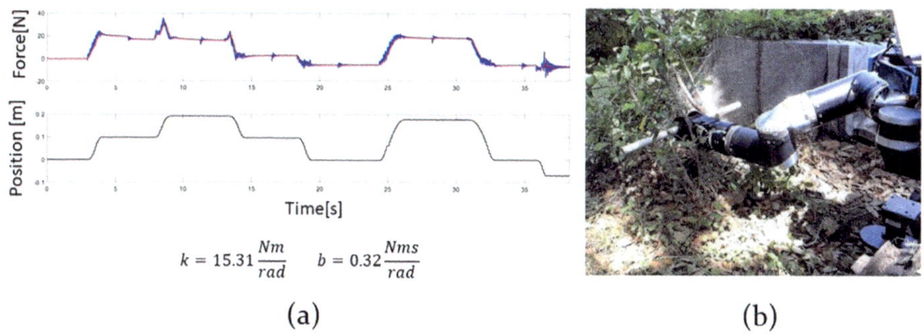

Fig. 6. Estimation of stem model parameters (k, b) through manipulation. The manipulator pushes the stem by following the shown displacement trajectory. The force plot corresponds to the normal force sensed by the torque-force sensor installed in the wrist of the arm. The regions of the trajectory where the arm maintains a constant position are used to estimate k and regions of constant velocity are used to estimate b.

The estimated stem models can then be used to quantify the difficulty of traversal through a patch of vegetation. A traversal experiment was conducted in the scenario shown in Fig. 7a. First, the manipulator was used to obtain a lumped stiffness and damping value for each grouping of stems. This was done by probing each patch from different orientation angles. The rolling resistance on the terrain was estimated by driving the vehicle at constant speed, 0.55 m/s, on flat ground adjacent to the vegetation clusters under observation. From forward simulations of the vehicle dynamics using (2), the energetic cost of traversal was computed as $E = \int_0^t Fv dt$, where F is the applied tractive force and v is the robot velocity. Energetic costs only include the period of time which the robot was in contact with the stem.

The energetic cost required to traverse the area in Fig. 7a was measured to be 767 J while the predicted value was 620 J, resulting in a percent error of 19.14%.

In Fig. 8, a more practical situation is presented in which the robot is required to traverse a patch of randomly distributed stems with known stiffness and damping properties. A Particle Swarm Optimization method [10] was implemented in order to find the optimal entry point (vehicle position along the Y-axis and orientation with respect to the X-axis) to the patch to yield the minimum energetic

(a) (b)

Fig. 7. The vegetated areas used to validate the stem model with the Roman. (a) Leafy and dense underbrush (k: 96.34 $\frac{Nm}{rad}$; b: 70.9$\frac{Nms}{rad}$ (b) Cluster of thin saplings (k: 159.0 $\frac{Nm}{rad}$; b: 80.2 $\frac{Nms}{rad}$).

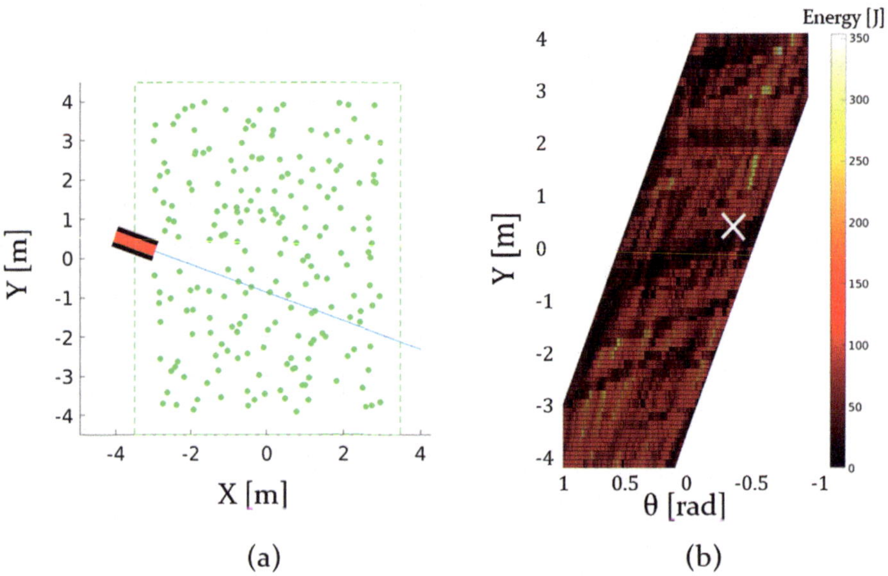

Fig. 8. (a) Vegetation patch with randomly distributed stems. (b) Energetic cost of traversal of the vegetation patch from different entry positions and orientations. The lowest cost entry configuration is highlighted with an X. Rolling resistance was ignored in these simulations.

cost of traversal. In the optimization, the vehicle was constrained to move in rectilinear motion but with an arbitrary heading such that the vehicle traversed the full width of the patch.

4.2 Husky Platform

Estimation of Lumped Resistance Force: The Husky robot was commanded to traverse the vegetated terrain shown in Fig. 9 from four different starting locations. In all experiments, the robot moved in a straight line and low level data including currents, position, and IMU was logged. One run was performed at a speed of 0.25 m/s and was employed to estimate the resistance force from each cell via the method described in Sect. 3.1. The estimation results are summarized in Fig. 10. Runs 2 and 3 were performed at of 0.5 m/s and run 4 at 1 m/s. In all these runs, the vehicle started on short grass, transitioned to medium grass, and finished on dense brush, which correlates with the estimated resistance values of Fig. 10.

Fig. 9. Husky traversing a vegetation patch from different starting positions.

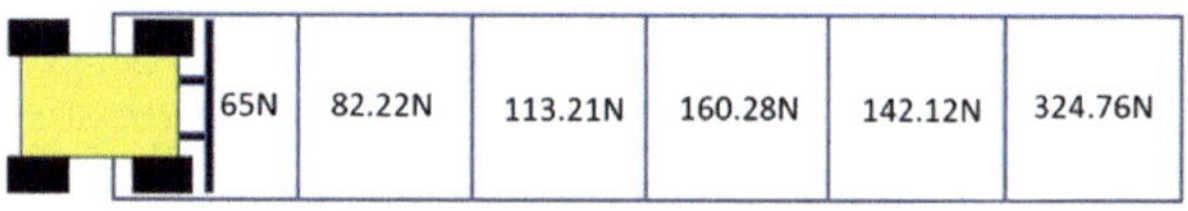

Fig. 10. Estimated lumped resistance force per cell. Each cell has a size of $1\,\mathrm{m} \times 1\,\mathrm{m}$

Once the lumped resistance forces per cell are estimated, a forward simulation of the vehicle over the terrain can be performed and used to compare the predicted vs actual energy mechanical energy required to traverse the vegetation patch. Figure 11 shows a typical energy comparison in the scenario of Fig. 9. The errors in energy predictions for the four experimental runs are summarized in Table 1.

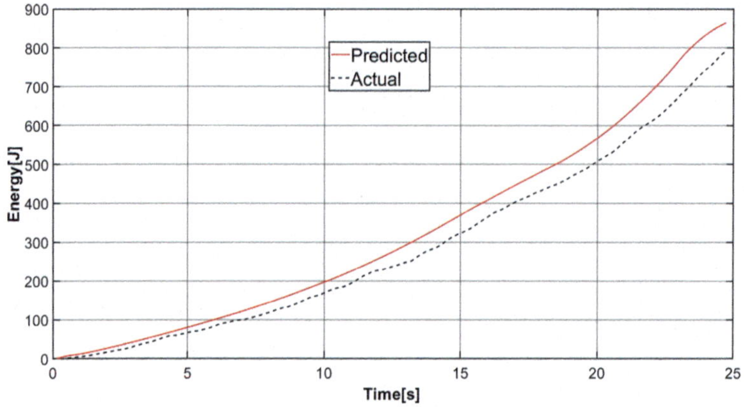

Fig. 11. Predicted vs. experimental mechanical energy.

Table 1. Energy prediction errors

Run	Speed [m/s]	Energy prediction error [%]
1	0.25	9.01
2	0.50	11.97
3	0.50	14.22
4	1.00	23.25

Perception: Our experiments are conducted on a region of terrain approximately 100 m in length that is traversed with the Husky robot. This environment contains a diversity of vegetation that includes dirt, light traversable vegetation, and heavy brush that is mostly untraversable. Lumped resistance is recorded along the traverse path of the robot, and is used to train the prediction model in a self-supervised manner. For these experiments, some of the terrain is dense enough to completely halt the robot and violate the constant velocity assumption. These discrepancies were manually removed and reannotated to provide a cleaner map for training the prediction algorithm.

Figure 12 illustrates some example results from our prediction model. The left image contains an elevation map to illustrate the locations of the vegetation. The black line through the center of the map illustrates the traverse path of the robot. The right image shows the output of the terrain predictions overlaid on the map, with red being the highest resistance, and green the lowest. Note that the bands on the bottom of the image are artifacts from humans walking along side the robot, which are correctly filtered out by the model. For evaluation, we discretize the lumped resistance into three classes of zero, light, and heavy resistance, and measure accuracy simply as classification error. Using a 4-fold cross-validation, our model has a prediction accuracy of 74% in this scenario.

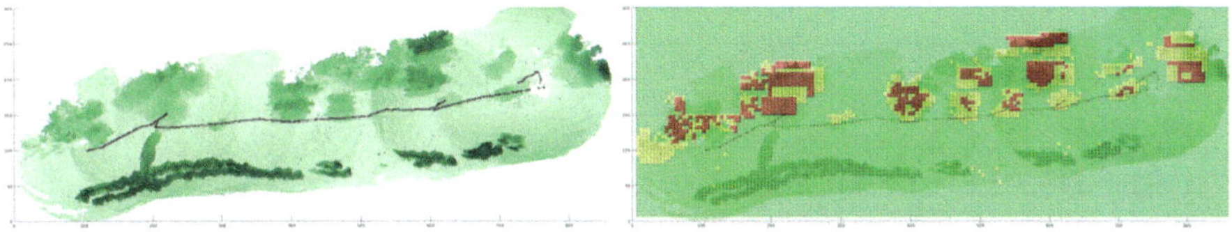

Fig. 12. Terrain classification results. The left image shows an elevation map of the terrain, with the traverse path of the robot illustrated by the black line. The right image shows the predicted lumped resistance overlayed over the map.

5 Conclusions and Future Work

This work developed two different methodologies to characterize the interaction of a vehicle with pliable vegetation. One approach is suitable for robots equipped with a manipulator and characterizes the local terrain by using a stem model with experimentally estimated rotational stiffness and damping properties. The second approach relies on motor current sensing to characterize the local terrain via a lumped resistance force. Both approaches were experimentally validated on short vegetation patches $\approx$6 m and were shown to provide energy prediction errors of less than 20%.

Additionally, an approach based on particle swarm optimization was developed in simulation to determine the optimal entry pose to a vegetation patch of randomly distributed stems.

The approach to perception shows promise in its ability to predict the resistance to traversal in previously unexplored regions. Though as the length of the trajectory increases and the terrain becomes more varied, the quality with which the existing sensors and terrain model are able capture the terrain interaction begins to degrade. This produces a noisy ground-truth signal for the perception system leading to a reduction in prediction performance. Future designs can likely be improved by better sensor design and placement as well as higher fidelity terrain models.

Several important challenges remain open such as autonomous selection of probing regions on the vegetation, dealing with occlusions, and estimation of overall patch geometry prior to traversal. It is expected that some of these challenges can be alleviated via cooperation with an aerial vehicle. Future work will explore the coupling of the proposed methodology with a sampling based motion planner to autonomously generate energy-efficient trajectories. In addition, this work will be applied to legged platforms.

Acknowledgements. This work was supported by the collaborative participation in the Robotics Consortium sponsored by the U.S. Army Research Laboratory under the Collaborative Technology Alliance Program, Cooperative Agreement DAAD 19-01-2-0012. The U.S. Government is authorized to reproduce and distribute reprints for Government purposes not withstanding any copyright notation there on.

References

1. Krusi, P., Furgake, P., Bosse, M., Siegwart, R.: Driving on point clouds: motion planning, trajectory optimization, and terrain assessment in generic nonplanar environments. J. Field Robot. **34**(5), 940–984 (2017)
2. Vandapel, N., Kuffner, J., Amidi, O.: Planning 3-D path networks in unstructured environments. In: IEEE International Conference on Robotics and Automation, Barcelona, pp. 4624–4629 (2005)
3. Bradley, D., Thayer, S., Stentz, A., Rander, P.: Vegetation detection for mobile robot navigation. Technical report (2003)
4. Hebert, M., Vandapel, N.: Terrain classication techniques from ladar data for autonomous navigation. In: Collaborative Technology Alliances Conference (2003)
5. Lalonde, J., Vandapel, N., Huber, D., Hebert, M.: Natural terrain classication using three dimensional ladar data for ground robot mobility. J. Field Robot. **23**, 839–861 (2006)
6. Matthies, L., Bergh, C., Castano, A., MAcedo, J., Manduchi, R.: Obstacle detection in foliage with ladar and radar. In: International Symposium of Robotics Research (2003)
7. Talukder, A., Manduchi, R., Owens, K., Matthies, L., Castano, A., Hogg, R.: Autonomous terrain characterization and modeling for dynamic control of unmanned vehicles. In: IEEE/RSJ Conference on Intelligent Robots and Systems (IROS) (2002)
8. Ordonez, C., Gupta, N., Chuy, O., Collins, E.: Momentum based traversal of mobility challenges for autonomous ground vehicles. In: IEEE International Conference on Robotics and Automation (2013)
9. Ordonez, C., Alicea, R., Rothrock, B., Ladyko, K., Harper, M. Karumanchi, S., Matthies, L., Collins, E.: Modeling and traversal of pliable materials for tracked robot navigation. In: SPIE Defense, Security, and Sensing Conference Unmanned Systems Technology XIX (2018)
10. Trelea, I.C.: The particle swarm optimization algorithm: convergence analysis and parameter selection. Inf. Process. Lett. **85**(6), 317–325 (2003)
11. Mur-Artal, R., Tardós, J.D.: ORB-SLAM2. an open-source SLAM system for monocular, stereo, and RGB-D cameras. IEEE Trans. Robot. **33**, 1255–1262 (2017)
12. Bajracharya, M., Ma, J., Howard, A., Matthies, L.: Real-time 3D stereo mapping in complex dynamic environments. In: International Conference on Robotics and Automation-Semantic Mapping, Perception, and Exploration (SPME) Workshop (2012)
13. Maturana, D., Scherer, S.: Voxnet: A 3D convolutional neural network for real-time object recognition. In: 2015 IEEE/RSJ International Conference on Intelligent Robots and Systems (IROS) (2015)

Safe Terrain Probing Method for Multi-legged Robots Operating on Brittle Surfaces

Eranda Tennakoon[1,2(✉)], Navinda Kottege[2], Thierry Peynot[1], and Jonathan Roberts[1]

[1] Science and Engineering Faculty, Queensland University of Technology, Brisbane, Australia
kulatunga.tennakoon@hdr.qut.edu.au
[2] Robotics and Autonomous Systems Group, CSIRO, Pullenvale, Australia
navinda.kottege@csiro.au

Abstract. Multi-legged robots working in challenging environments are often required to walk on fragile terrain that may collapse upon traversal. We propose a method to test the robot's next foothold for collapses without endangering it. In this method, the robot's body is posed in a favourable manner and the terrain is probed to test for collapses using its foot. We also present a study in identifying collapses by using the force profile and the foot tip displacement of the probing foot during probing.

Keywords: Terrain probing · Legged robots · Brittle terrain · Terrain collapse

1 Introduction

One of the key use cases for autonomous robots is aiding in rescue missions [1]. Operating on such missions often require robots to traverse dangerous and difficult terrain [2]. This type of terrain often deforms upon traversal and puts the robot and its mission at risk. Therefore, the effectiveness of robots aiding in search and rescue operations largely depends on their ability to successfully traverse such terrain.

By design, legged robots offer a unique advantage on difficult terrain since they only require a series of small footholds as opposed to pathways or tracks. There have been some significant advances in legged robot locomotion on difficult terrain as demonstrated with BigDog [3] using whole body control to react to external disturbances, ANYmal [4] using pose optimisation to adapt to changing environments and Weaver [5] using adaptive methods to reduce the cost of transport on rough terrain. However, most of these methods are reactive methods that require the robot to enter potentially dangerous terrain before they can

Electronic supplementary material The online version of this chapter (https://doi.org/10.1007/978-3-030-33950-0_27) contains supplementary material, which is available to authorized users.

J. Xiao et al. (Eds.): ISER 2018, SPAR 11, pp. 305–317, 2020.
https://doi.org/10.1007/978-3-030-33950-0_27

respond. This increases the overall risk as the robot may find itself in situations where it would be too late to recover and continue the mission or its actions would lead to damage. Therefore, it is useful to test the terrain before moving the robot into an area that can collapse.

The normal force exerted by a robot leg and the vertical foot displacement was related to the terrain stiffness by Kroktov in early work in this area [6]. Hoepflinger et al. [7] used robot terrain probing to determine terrain friction and then predict terrain friction in un-traversed terrain using unsupervised machine learning methods. However, the use of terrain probing to determine terrain collapses has only received limited coverage in the literature. Tokuda et al. [8] using a pressure sensor on the sole of the robot's foot to show a change in the sensor voltage when the robot probes and incites a collapse at the foothold is one such example. Their study was limited to a laboratory model of collapsible foothold which they termed as a *breakable foothold.* Ambe and Matsuno [9] used the robot leg to probe the terrain to test if the foothold collapses under the exerted force. However, the method they use require them to set a threshold force to test on the robot leg to create the necessary constraints. This is not robust for real terrain situations as it is often not possible for the robot to test for a predefined threshold force. Furthermore, their method is not generalisable to robots with a greater number of legs as the problem of calculating foot contact forces becomes indeterminate when the robot has more than 3 legs in contact with the terrain [10].

In this paper, we introduce a novel terrain probing strategy for multi-legged robots operating on brittle terrain. Our approach is similar to [9] but is generalisable for any multi-legged robot and does not require the robot to meet a predefined threshold force. The proposed method moves the robot's centre of mass to a favourable location within it's support polygon to allow the robot to probe with more force while keeping the force applied on the terrain by the support feet below safe, pre-determined limits. We also show that we can classify whether terrain collapses occur on the brittle terrain by using the force profile and the foot tip displacement of the probing foot.

2 Method

We define that a terrain collapse has occurred when the robot's probing foot pierces through the brittle terrain layer and the foothold is no longer able to support the robot's weight component. If a collapse occurs during probing, this fact could be recorded and sent to a foothold planner so that it is taken into consideration in the next foothold planning process. If no collapse occurs during the probing, the maximum exerted force on the terrain is recorded and the robot can safely step on to the tested foothold. By keeping a record of these exerted maximum forces, a force restriction can be assigned to the particular foothold to keep the robot safe. In subsequent steps, the robot will need to make sure not to exceed this force restriction during motion since the foothold has not been tested to withstand a higher force.

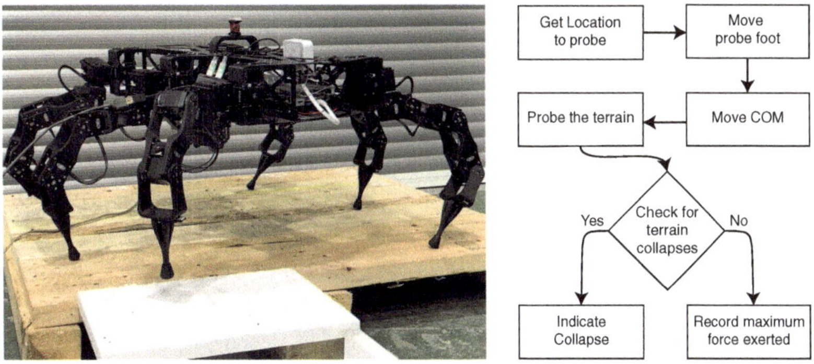

(a) (b)

Fig. 1. Hexapod robot Bullet probing a terrain sample (a) and the overview of the proposed method (b).

We propose a method to effectively pose the robot's body (move the body without moving the feet) for one probing motion on the terrain. An overview of this method is shown in Fig. 1(b). First, the probing foot is moved on top of the foothold that needs to be tested, without making contact with it. Then the Center of Mass (COM) of the robot is posed strategically as proposed by our method. Finally, the robot is made to probe the given foothold and identify any collapses.

In this work, we assume that the robot is provided with a set of force restrictions on the robot's support feet which must not be exceeded during motion and an external footstep planner provides the robot with a feasible foothold that needs to be tested. Only the forces acting along the gravitational axis are considered since collapses are mainly inflicted by the robot's weight on the terrain. For simplicity, we assume that the terrain is locally flat (at the foothold) and does not break apart parallel to the surface and that it can withstand sufficient frictional force. The COM is assumed to be fixed at the body center of the robot and moves with the body center. The body of the robot is sufficiently heavy which restricts any large changes in the COM position.

2.1 Strategy for Moving the COM and Probing the Terrain

The COM of the robot is posed iteratively towards a goal location using a gradient descent approach. A definition of the parameters and variables used are provided in Table 1. The outcome of the algorithm is a vector that determines the direction and the distance of movement of the COM for each iteration.

The potential to pose the COM towards a goal location is defined as $U_{goal} = \frac{1}{2}\epsilon\|O_{(c)} - O_{(g)}\|^2$. This drives the COM closer to the given goal at each iteration. The support polygon made up of the support feet (excluding the probing foot) will be referred to as the support polygon (SP) of the robot throughout this paper. The potential to keep the COM within SP is defined as $U_{polygon,i}$ for the ith polygon edge where $U_{polygon,i} = \frac{1}{2}\gamma(\frac{1}{\|O_{(c)}-O_{(p,i)}\|} - \frac{1}{D})^2$ for $\|O_{(c)} - O_{(p,i)}\| \leq D$, else $U_{polygon,i} = 0$. In the event of a terrain collapse during probing, this helps to keep the robot from stumbling or falling over. Any errors in the modelling of the COM position is also mitigated due to setting the threshold D.

To keep the robot from exceeding the force restrictions set on it, two potentials are defined. $U_{foot\,tip,j}$ pushes the COM away from the given foot tip (indexed as j) where $U_{foot\,tip,j} = \frac{1}{2}\eta(\frac{1}{F_{diff1,j}} - \frac{1}{F_{s1}})^2$ for $F_{diff1,j} \leq F_{s1}$, else $U_{foot\,tip,j} = 0$. In some cases, the repulsion from two support feet can get mutually cancelled and the COM can move between the two support feet towards an SP edge (e.g. when the goal location for the COM is set outside the SP). When the COM moves towards an SP edge, the weight component on the two support feet making up that edge increases. Therefore, a virtual foot is defined on each SP edge at a point closest to the COM and a potential is defined similar to $U_{foot\,tip,j}$ using the force restrictions and the force values of the two support feet that makes each SP edge. $U_{polygonF,k}$ pushes the COM away from the SP edge (in terms of forces on the feet) where $U_{polygonF,k} = \frac{1}{2}\beta(\frac{1}{F_{diff2,k}} - \frac{1}{F_{s2}})^2$ for $F_{diff2,k} \leq F_{s2}$, else $U_{polygonF,k} = 0$.

Table 1. Variables and parameters used

Variable	Definition
$O_{(c)}$	Current COM location
$O_{(g)}$	COM Goal location
$O_{(p,i)}$	Nearest point from the COM to the *ith* SP edge
$\|O_{(c)} - O_{(g)}\|$	Distance from the goal location to the COM
ϵ	Parameter imposing rate of potential change to goal
γ	Parameter imposing rate of potential change to SP edge
η	Parameter imposing rate of potential change to foot tip
β	Parameter imposing rate of potential change to SP edge force
D	Parameter for the safety distance to SP edge
$F_{current,j}$	Current force at foot tip j
$F_{rstrc,j}$	Tested force restriction value for foot tip j
F_{s1}	A safety force threshold for the foot tips
F_{s2}	A safety force threshold for each SP edge
$F_{diff1,j}$	$F_{rstrc,j} - F_{current,j}$
$F_{diff2,k}$	$(F_{rstrc,l1} + F_{rstrc,l2}) - (F_{current,l1} + F_{current,l2})$ where $l1$ and $l2$ are the two support feet that makes the *kth* SP edge

The COM can be driven to a goal position by moving it in the direction of the greatest descent of the resulting potential field which can be calculated as $G = -\Delta U$ where U is the potential (specified above) and G is the gradient.

Using definitions for G and U_{goal} the gradient to the goal location can be calculated as $G_{goal} = -\epsilon(O_{(c)} - O_{(g)})$.

Similarly, the gradient from the ith polygon edge $g_{polygon,i}$, is calculated using $U_{polygon,i}$ (Eq. 1). The total rejection from the SP edge $G_{polygon}$ is then calculated using

$$g_{polygon,i} = \begin{cases} \gamma(\frac{1}{\|O_{(c)}-O_{(p,i)}\|} - \frac{1}{D})\frac{(O_{(c)}-O_{(p,i)})}{\|O_{(c)}-O_{(p,i)}\|^3}, & \text{if} \|O_{(c)} - O_{(p,i)}\| \leq D. \\ 0, & \|O_{(c)} - O_{(p,i)}\| > D. \end{cases} \quad (1)$$

and

$$G_{polygon} = \sum_{i}^{n} g_{polygon,i} \quad (2)$$

where n is the number of SP edges of the robot.

Finally, the gradient due to the force restrictions on the support feet can be calculated using the potential definitions $U_{foot\,tip,j}$ and $U_{polygonF,k}$ as follows:

$$g_{foot\,tip,j} = \begin{cases} \beta(\frac{1}{F_{diff1,j}} - \frac{1}{F_{s1}})(\frac{1}{F_{diff1,j}})^2\frac{(O_{(c)}-O_{(f,j)})}{\|O_{(c)}-O_{(f,j)}\|}, & \text{if } F_{diff1,j} \leq F_{s1}. \\ 0, & F_{diff1,j} > F_{s1}. \end{cases} \quad (3)$$

$$g_{polygonF,k} = \begin{cases} \beta(\frac{1}{F_{diff2,k}} - \frac{1}{F_{s2}})(\frac{1}{F_{diff2,k}})^2\frac{(O_{(c)}-O_{(p,k)})}{\|O_{(c)}-O_{(p,k)}\|}, & \text{if} F_{diff2,k} \leq F_{s2}. \\ 0, & F_{diff2,k} > F_{s2}. \end{cases} \quad (4)$$

The sum of their effect is then calculated using

$$G_{force\,restrictions} = \sum_{j}^{N} g_{foot\,tip,j} + \sum_{k}^{M} g_{polygonF,k} \quad (5)$$

where the number of support feet is given by N and the number of SP edges is given by M. The sum of the calculated gradient values given in Eq. 6 is a vector that determines the direction and the distance of movement of the COM for each iteration. This value is then mapped directly to the Cartesian plane.

$$G_{COM} = G_{goal} + G_{polygon} + G_{force\,restrictions} \quad (6)$$

However, at very large G_{COM} values the move distance for the COM will be very large. Therefore, the displacement vector (S) for the COM is defined as

$$S = \begin{cases} C\frac{G_{COM}}{\|G_{COM}\|}, & \text{if} G_{COM} > C \\ 1 & G_{COM} \leq C \end{cases} \quad (7)$$

Here, C is a constant that is decided experimentally considering the minimum move distance of the COM allowed by the *robot controller* (controller responsible for moving the COM of the robot to the desired location) and the number of move iterations typically required to move the COM to a given goal.

After the COM is moved to a favourable location for the probing action, the probing foot is lowered in a direction normal to the foothold surface along the gravitational axis. The probing motion is ended when either of the following 3 conditions are reached: (1) the probing foot reaches the torque limit of any of its motors, (2) the workspace limit of the probing leg is reached, or (3) one of the support feet lifts off the ground (e.g. when the robot tilts back when pushing on the probe point).

2.2 Identifying Terrain Collapses

Thanks to the probing action, we want to classify whether the terrain at the given foothold belongs to the class *collapse* and *non-collapse*, i.e. we want to determine whether the terrain at a given foothold collapses or not. A set of features were chosen experimentally from the force profile of the probing foot and the probing foot position relative to the robot body during probing. These selected features are: (1) the maximum force recorded during probing (F_{max}), (2) the foot tip travel distance from the start of probing motion at F_{max}, (3) the maximum foot tip travel distance from the start of probing motion to the end, (4) the force at the foot tip at the end of probing, and (5) the torque at each motor of the probing leg at F_{max}. A classification is performed to determine whether the chosen feature vector allows the robot to classify the occurrence of terrain collapses effectively. The reference labelling of terrain samples used for training and for testing is based on the observation whether a collapse occurred or not during probing, made by an expert.

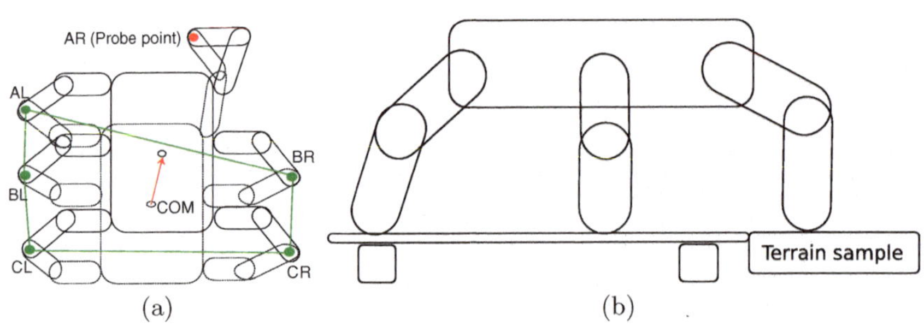

Fig. 2. COM moved by posing the robot within the SP (top view). Legs were named AR,BR,CR,CL,BL and AL clockwise (a). Illustration of robot probing (side view) (b).

3 Experiments and Setup

The experiments were done using the hexapod *Bullet* (Fig. 1(a)). Bullet has 5 degrees of freedom in each leg and has the configuration *yaw-pitch-roll-roll-roll.* The links and corresponding joints are named Coxa (with a yaw and a pitch joint), Femur, Tibia, and Tarsus. The specifications for the robot and computer used are given in Table 2. For the experiments in this section, we used the following values for the parameters of our proposed method: $\epsilon = 0.02\,Jm^{-2}$, $D = 0.12m$, $\gamma = 1.3925 \times 10^5\,Jm^2$, $\eta = 2.67 \times 10^7\,JN^2$, $\beta = 4.25 \times 10^6\,JN^2$, $F_{s1} = 0.05N$ and $F_{s2} = 0.1N$.

Table 2. Hardware Specifications of Bullet hexapod and the computer used.

Type	Description
General	Mass: 9.51 kg
	Dimensions (Body): length = 500 mm × width = 280 mm
Servomotors	30 × Dynamixel MX-106
Computer	Intel core i5 PC (16 GB RAM) running ROS on Ubuntu 16.04

3.1 Experiment 1: Moving the COM

We first test how the proposed method moves the COM to a given goal location. Ideally, the robot is able to exert the most force when the COM is moved right on top of the probe point (i.e. weight vector going through the probe point). However, the COM movement is restricted to within the SP and therefore, the COM needs to be moved to a favourable location within the SP. One such favourable location is the closest point to the probe point from within the SP as it gives a higher weight component of the robot on to the probe point [11]. We set the goal location as the probe point itself. This moves the COM to the desired favourable location and at the same time allows to test how the method responds to goal locations outside the SP. For simplicity, the COM is moved assuming that the probing leg is connected to the body center of the robot at a point with no offset and the probing point was chosen such that the effects of this offset on the robot probing are minimized. No force restrictions were set on the support feet for these experiments. Figure 2(a) shows the initial position of the COM of the robot with the COM moved to a final position by posing the robot body.

Next, two sets of tests were performed to study the system's ability to handle force restrictions on the support feet. We are aiming to use the maximum probing force exerted on the foothold as the force restriction (Sect. 2). However, for the purpose of testing the method, simulated force restrictions were set on the support feet for these experiments. In the first test, a force restriction value was set to only one of the legs to see how the method behaves. The leg that was closest to the probing point (arbitrarily selected) was chosen to set the force restriction to highlight the effects.

In the second test, a random set of force restrictions were set on all the support feet. The random force values were generated ensuring that the robot was not starting off with invalid initial conditions. The robot was then made to move with the given force restrictions using the proposed method. The robot was also made to do a probing on a known (by an expert) collapsible terrain sample. This was done when the robot was at the final COM position after moving with the aforementioned random force restrictions.

3.2 Experiment 2: Identifying Terrain Collapses

In this experiment, we performed a simple study of identifying terrain collapses. The robot was made to probe a point within the workspace of the probing leg where we place different terrain samples. To simplify the experiment and to allow the robot to probe with the same initial conditions, the robot's COM was not moved in these experiments and the probing was done with the robot at its initial pose. Based on the previous experiment we chose the probe point such that the robot's COM was close enough to exert a considerable force, yet not too close to lose generality. However, in the actual implementation, this step can be done after the robot has moved the COM to a favourable location.

The samples in Table 3 were chosen for their common occurrence in urban environments as well as easy repeatability of experiments. Figure 2(b) shows how the robot is made to probe a collapsible terrain sample. After the robot probes each sample, the force profile at the foot tip and the foot tip position (relative to the robot center) were recorded. The features were extracted from the data and then labeled as *collapse* or *non collapse* depending on the result of the probing. A simple binary classification was done using a Support Vector Machine (SVM). A comparison of classification accuracy was also done using discriminant analysis and k-nearest neighbours to evaluate the discrimination power of the chosen feature vector with different classifiers.

Table 3. Terrain samples

Non-collapsible	Collapsible
Brick	Damp peat slab
Hardwood	Thin ice
Hard ice	Styrofoam (5 mm)
Styrofoam (15 mm)	Styrofoam (10 mm)
Gravel	

4 Results and Analysis

4.1 Results and Analysis of Experiment 1

The robot was made to probe the chosen point in the terrain after each movement of the COM (for each iteration of the gradient descent algorithm). Figure 3 shows

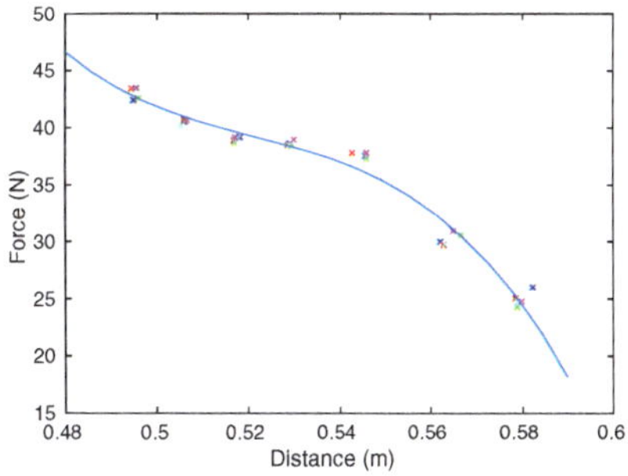

Fig. 3. Maximum force calculated at each probe for each movement iteration of the COM vs. distance from the probe point to the COM for multiple runs of the experiment (shown in different colours). A curve fitting was done to show the variation.

the force at the foot vs. the distance between the COM and probe point for five trials along with a least squares fitted curve (Polynomial of degree 3). In the first iteration of the gradient descent (in each run of the experiment), the probing motion was ended because the probing leg reached the workspace limit. In the remaining iterations, the probing motion was ended by the robot's feet (AL and BR which are closest to the probing foot) lifting off the platform. The force required to lift the support feet increased at each movement of the COM indicating that the selected goal location for the COM position is favourable as each iteration transferred more weight of the robot onto the probing foot.

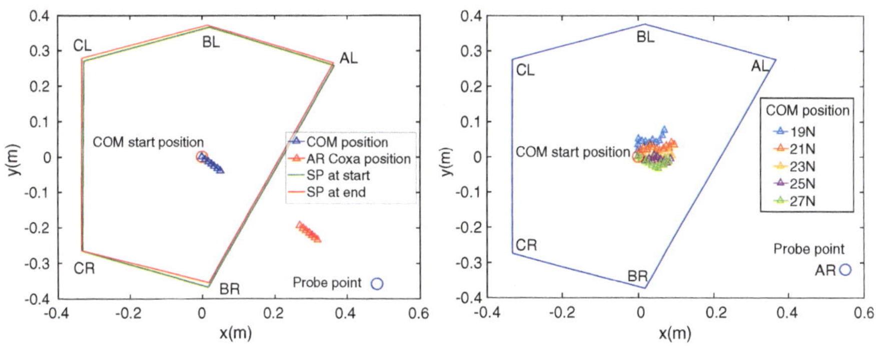

(a) (b)

Fig. 4. COM movement during probing relative to the initial SP, with (a) showing COM movement with no force restriction on any foot and (b) showing COM movement for five trials, each with a unique force restriction set on the BR foot.

In Fig. 4(a) the movement of the COM (relative to the initial SP) for one of the experiment runs in Fig. 3 is presented. Here, the movement of the Coxa joint with the COM indicates that both points came closer to the probe point at each iteration and thereby minimizing any effects that could be introduced by the offset. The amount of weight the robot transfers to the probing leg is decided by the position of the COM, but the amount of force it can exert on the terrain can be limited by the distance from the probe point (probe leg's point of contact with the terrain) to a given joint in the probing leg as well as the orientation of each joint. Examining the effects of those aspects on the probing force is beyond the scope of this study. However, a new goal location for the COM can easily be set by considering the effects of the aforementioned aspects. It was also observed that the initial and final shape of the SP was different due to position errors of the legs. The proposed gradient method approach was robust to these errors as the new SP was calculated and considered for the algorithm in each iteration.

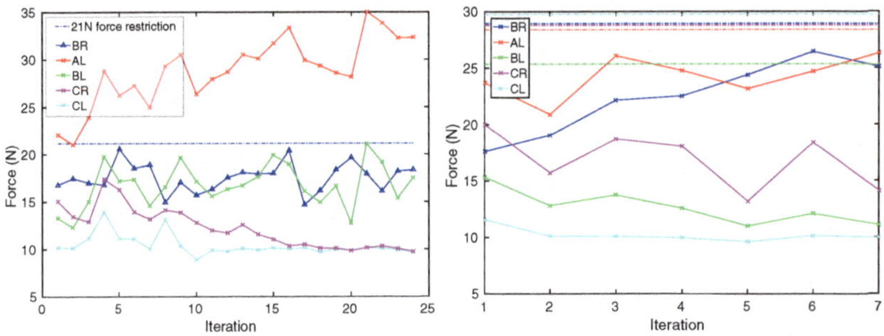

(a) (b)

Fig. 5. Force profile with each iteration with force restriction set to 21 N is shown in (a) and the force profile of support feet with each iteration of COM movement with random force restriction on support feet is shown in (b). The dashed lines represent the force restriction for each respective foot.

Next, moving the COM with force restrictions was tested by first imposing a force restriction only on one foot. The Fig. 4(b) shows the COM movement when different force restrictions were set on just one foot. Here, the feet of the robot is marked as AR, BR and CR on the right-hand side and AL, BL, and CL on the left-hand side. In this case, AR is the probing foot marked as *probe point* in the Fig. 4(b) and BR is the support foot with the force restriction imposed. As the force restriction on the BR foot gets more strict, the COM was pushed further away from the foot. Figure 5(a) shows the force profile of the BR foot for the force restriction of 21 N set on it compared with the force profiles of all the support feet. It can be observed that the BR foot always stayed under the imposed force restriction which was the goal of the proposed method.

Figure 5(b) shows the force profile of each support foot at each iteration when random force restrictions were set to all of the support feet. The proposed strategy was successful in moving the COM of the robot within the SP while not exceeding the imposed force restrictions. The force profile of the closest feet to the probe location kept increasing gradually as expected when the COM came close to the SP edge. To demonstrate the intended final outcome of the proposed method a simple probing was done on a Styrofoam piece once the robot reached the final COM position with the given force restrictions. Figures 6(a) and (b) show the force profile of the probing foot and support feet respectively while probing the Styrofoam piece (collapsible foothold). Clear changes were observed at the collapse point in the force profiles of the probing foot and the support feet. The probing was done ensuring that the force restrictions on the support feet were not exceeded (Fig. 6(b)).

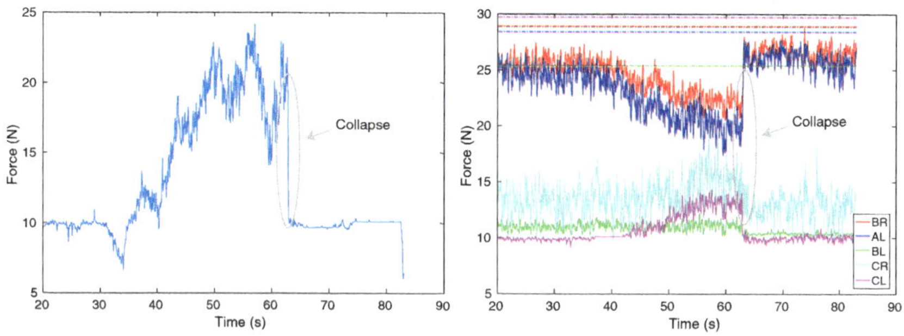

(a) Force at the probing foot (b) Forces at the support feet

Fig. 6. Forces at the feet when the robot was probing a Styrofoam piece.

4.2 Results and Analysis of Experiment 2

Based on the terrain probing, a study of the terrain collapses was done using simple classification methods. A total of 47 terrain samples were tested and the features were labeled as *collapse* or *non-collapse* depending on the outcome of probing. The trials were separated 50/50 for the training set and the test set by random sampling of the data. An F1 score of greater than 0.88 was obtained for multiple experimental runs evaluated with a support vector machine (SVM) classifier (with a Gaussian kernel), indicating good classification results.

Next, the chosen set of features was evaluated for its ability to classify the occurrence of terrain collapses on previously unobserved terrain data. For this analysis, one of the terrain materials was left out of the training data and the classifier was made to classify this as a collapse or a non-collapse based on the training done on the remaining data. This binary classification was done using the 3 classifiers: SVM, discriminant analysis (DA) and k-nearest neighbours (KNN). Table 4 shows the number of samples each classifier correctly labeled against the number of samples tested. From the results, we see that all the classifiers managed to correctly classify the majority of the samples. This indicates that the chosen feature vector performs well for identifying terrain collapses.

Table 4. Classification for a material not in the training data using different classifiers.

Label	Material	SVM	DA	KNN
Collapse	Thin ice	4/5	5/5	5/5
	Styrofoam (5mm)	5/5	5/5	5/5
	Styrofoam (10 mm)	4/6	6/6	6/6
	Damp peat slab	5/5	5/5	5/5
Non Collapse	Brick	5/5	5/5	5/5
	Gravel	4/5	4/5	4/5
	Hard wood	5/5	5/5	5/5
	Styrofoam (15 mm)	4/6	6/6	6/6
	Hard ice	5/5	5/5	5/5

5 Conclusion

We presented a terrain probing method to aid legged robots to safely identify terrain collapses on brittle terrain. This method was shown to be robust to position errors of the leg joints as the COM was moved within the support polygon of the robot's support feet to allow more force to be exerted on the terrain. The COM movement is executed without exceeding the predetermined force restrictions at the robot's support feet. This indicates that the proposed method is applicable in real terrain conditions where the robot's ability to effectively probe the terrain is compromised. It is especially relevant when the robot is already on brittle terrain with the risk of collapse if additional force is exerted on support feet while probing terrain ahead.

A simple binary classification of terrain collapses was also performed using the force profile and the foot displacement of the probing foot. The chosen feature vector comprising the force profile and the foot tip displacement of the probing leg was shown to be effective in determining whether terrain collapses occurred during probing.

In future work, we plan to extend the proposed method to a continuous gait (probe before step) that also takes lateral forces on the probing leg into consideration to gain further information about the terrain.

Acknowledgements. The authors would like to thank Fletcher Talbot, James Brett, Pubudu Aravinda and Benjamin Tam for their support during this work.

References

1. Murphy, R.R., Tadokoro, S., Nardi, D., Jacoff, A., Fiorini, P., Choset, H., Erkmen, A.M.: Search and rescue robotics. In: Siciliano, B., Khatib, O. (eds.) Springer Handbook of Robotics, pp. 1151–1173. Springer (2008)

2. Messina, E., Jacoff, A., Scholtz, J., Schlenoff, C., Huang, H., Lytle, A., Blitch, J.: Statement of requirements for urban search and rescue robot performance standards. Draft report, DHSSTD and NIST, USA (2005)
3. Raibert, M., Blankespoor, K., Nelson, G., Playter, R.: Bigdog, the rough-terrain quadruped robot. IFAC World Congr. **41**(2), 10822–10825 (2008)
4. Hutter, M., Gehring, C., Jud, D., Lauber, A., Bellicoso, C.D., Tsounis, V., Hwangbo, J., Bodie, K., Fankhauser, P., Bloesch, M., Diethelm, R., Bachmann, S., Melzer, A., Hoepflinger, M.: ANYmal - a highly mobile and dynamic quadrupedal robot. In: IEEE/RSJ International Conference on Intelligent Robots and Systems (IROS), pp. 38–44 (2016)
5. Bjelonic, M., Kottege, N., Homberger, T., Borges, P., Beckerle, P., Chli, M.: Weaver: hexapod robot for autonomous navigation on unstructured terrain. J. Field Robot. **35**(7), 1063–1079 (2018)
6. Krotkov, E.: Active perception for legged locomotion: every step is an experiment. In: IEEE International Symposium on Intelligent Control (1990)
7. Hoepflinger, M.A., Hutter, M., Gehring, C., Blösch, M., Siegwart, R.: Unsupervised identification and prediction of foothold robustness. In: IEEE International Conference on Robotics and Automation (ICRA), pp. 3293–3298 (2013)
8. Tokuda, K., Toda, T., Koji, Y., Konyo, M., Tadokoro, S., Alain, P.: Estimation of fragile ground by foot pressure sensor of legged robot. In: IEEE/ASME International Conference on Advanced Intelligent Mechatronics (AIM), pp. 447–453 (2003)
9. Ambe, Y., Matsuno, F.: "Leg-grope walk": strategy for walking on fragile irregular slopes as a quadruped robot by force distribution. ROBOMECH **3**(1), 7 (2016)
10. Schmucker, U., Schneider, A., Rusin, V., Zavgorodniy, Y.: Force sensing for walking robots. In: International Symposium on Adaptive Motion of Animals and Machines (AMAM), pp. 25–30 (2005)
11. Garcia, J., Wood, G., Barrera-Mora, F.: Reactions on rigid legs of rectangular tables. Appl. Math. **6**(03), 599 (2015)

On-Line Coordination Tasks for Multi-robot Systems Using Adaptive Informative Sampling

Carlos Nieto-Granda[1(✉)], John G. Rogers III[2], Nicholas Fung[2], Stephanie Kemna[3], Henrik I. Christensen[1], and Gaurav Sukhatme[3]

[1] University of California San Diego, San Diego, USA
cnietogr@ucsd.edu
[2] US Army Research Lab, Adelphi, USA
[3] University of Southern California, Los Angeles, USA

Abstract. Robotic platforms have continued to advance in autonomous capabilities and as such, have been deployed in increasingly more complex scenarios. In this work, we look to address the utilization of multiple robots to take Received Signal Strength Indication (RSSI) measurements within an environment that has been mapped a priori and contains many radio transmitters. The sensor measurements are used to build a Gaussian Process Model of the maximum RSSI across all transmitters in the environment. The algorithm evaluates points of interest within the environment and directs robots based on the highest expected information gain per meter of travel. To coordinate multiple robots, the environment is divided into partitions based on the expected travel cost to visit each evaluation point within the partition. By assigning each robot to a different partition, path disruption is reduced, and exploration is more efficient than exploring random areas. The algorithm was tested and evaluated in both in simulation and on real-world platforms.

Keywords: Distributed multi-robot exploration · Environment dynamic-partitioning · Cooperative multi-robot teams

1 Introduction

Multiple cooperative robots have the potential to enable fast information gathering and more efficient coverage and monitoring of large areas. Potential search applications include search and rescue missions under disaster scenarios. Mapping applications include the measurement of algae abundance in a lake environment. For military applications, multi-robot systems promise more efficient operation under both searching and mapping applications. Besides, such systems are more robust to adversarial attacks. In civil applications (e.g., pollution monitoring, surveillance, search and rescue), the use of several inexpensive, heterogeneous, agile platforms is an appealing alternative to monolithic single robot

J. Xiao et al. (Eds.): ISER 2018, SPAR 11, pp. 318–327, 2020.
https://doi.org/10.1007/978-3-030-33950-0_28

systems. In many of these applications, time or scale are critical components in gathering the information. Collaborative multi-robot teams can exhibit significant advantage over individual robots working by themselves when time or scale is essential.

We address a scenario in which the goal is to map the received signal strength indication (RSSI) within an environment with one or more radio signals. For this experiment, the map of the area and points of interest therein are known a priori. Multiple robots are tasked with measuring the RSSI at these points of interest to build an RSSI model of the environment. The algorithm tasks each robot according to the expected information gain per meter of travel. Further, the algorithm partitions the map according to the effort required to measure every point of interest within each partition. By assigning each robot to a different partition, the system reduces the disruptions that result from robots crossing into each other's paths.

1.1 Related Work

Informative sampling in robotics was pioneered by Krause et al. [1], Singh et al. [2] and Low et al. [3]. The goal is to task robotic platforms to collect information within some resource constraint efficiently. Kemna et al. [4] implement a decentralized online approach to explore an environment. Voronoi partitions are used to divide the work and coordinate the efforts between the multiple robotic systems. Duvallet and Tews [5] also used GP regression in a structured environment, by mapping WiFi strength in an industrial setting. The focus of their work was on demonstrating a particle-filter based approach for localization, after the construction of a GP map. In contrast, we focus on the construction of the GP model through online modeling. Ouyang et al. [6] developed a decentralized multi-robot active sensing approach where robots coordinate their actions whenever their sensing would affect the same GP component. Atanasov et al. [7] developed an efficient decentralized and optimal solution for a multi-sensor active information acquisition problem that is nonmyopic and memory-efficient. They show they can reduce the complexity in the planning horizon and the number of sensors. Their work was applied to multi-robot active SLAM.

1.2 Contributions

In this paper, we present an approach for our coordinating multiple robots to model the RSSI of an area. Each robot makes individual decisions based on an information adaptive rate sampling approach. The robots and coordinated using a partitioning algorithm that divides the environmental space into equal partitions. This work builds on the work of Kemna, et al. [4] and is meant to serve as the analogous Unmanned Ground Vehicle (UGV) implementation.

Specifically, our contributions include:

- evaluation of points within an environment according to an information rate adaptive metric and tasking robots to these points in an efficient order

- efficient partitioning of an environment for multi-robot cooperative missions
- implementation and validation of the algorithm within a real-world environment on robotic platforms

2 Technical Approach

We look to utilize a group of robots to autonomously and efficiently build a model of the RSSI within an indoor environment. To this end, we developed an algorithm to select a good location on the map to measure efficiently. The algorithm estimates the information gain that would be attained by traveling to each evaluation point and selects the best based according to an information-gain rate metric. That is, the point that would yield the most information, in the form of entropy reduction, per meter of travel when the robot moves to this location and takes a measurement. To coordinate multiple robots, the central control unit divides the environment into partitions that require an equal amount of work to evaluate thoroughly.

We evaluated the algorithm in simulation under several different maps using teams of one, two, three, and four robots. The simulation experiments used a ground truth model of the environment obtained by taking extensive measurements in a real-world environment using teleoperation. We further this work by implementing the algorithm on real-world robot platforms and execute the experiment online. We run experiments using two and three robots with one, two, and three radio sources in many different indoor environments.

2.1 Information Gain Rate

The expected information gain for each sensor measurement is measured in entropy reduction. We calculate the differential entropy of a normal distribution:

$$h(\mathbf{x}) = \ln(\sigma\sqrt{2\pi\epsilon}) \tag{1}$$

We observe that the entropy is calculated according to the variance of the distribution and as such, repeated measurements will lower the entropy of the model.

Equation 1 gives us the expected information gain for a sensor measurement at a specific point, and previous efforts have sought to minimize the overall entropy of the model by tasking robots to the point of interest with the highest entropy value. Instead, we look to evaluate each evaluation point based on the path that the robot will take and the sensor measurements it will take along this path. We calculate the utility of a point of interest ($\mathbf{x}_{poi}$) from a point of origin ($\mathbf{x}_0$) as:

$$U(\mathbf{x}_0, \mathbf{x}_{poi}) = \frac{\sum h(\mathbf{x}_1), h(\mathbf{x}_2), ..., h(\mathbf{x}_{poi})}{(dist[\mathbf{x}_0, \mathbf{x}_{poi}] + \alpha)} \tag{2}$$

where $\mathbf{x}_1, \mathbf{x}_2, ..., \mathbf{x}_{poi}$ are the series of points that are traversed to achieve the shortest path and $dist[\mathbf{x}_0, \mathbf{x}_{poi}]$ is the distance of this path. α is a constant value parameter used to weight the utility according to the computing demands of generating new evaluations from the robot's new location. This evaluation will penalize the selection of very short paths, as these decisions will necessitate frequent calculations of new paths and utilities. This experimentation used a value of $\alpha = 1$. Further experimentation under different systems may require different values of α according to computing power to evaluate paths and entropy based on the GP model.

2.2 Coordination and Task Allocation

In practical application of this algorithm, a significant problem arises with multiple robots. Path planning for multiple agents can prove to be complex and are computationally expensive. Each computed path must include not only traditional two-dimensional obstacle and kinematic constraints but also any potential interference from dynamic obstacles; including the interference caused by other robots in the team. Alternatively, paths can be computed under the assumption that the robots will not cross paths and instead rely on local path planning to maneuver around dynamic obstacles as they appear while traversing the global path. This approach can result in long delays or even robot collisions depending upon the robustness of the obstacle avoidance capabilities of the robots.

This algorithm takes the latter approach to minimize computation time. We also take an additional step to reduce the occurrences of robots crossing into each other's paths. We divide the environment into partitions based on the expected travel costs required to visit and measure each point of interest within the partition. In this manner, we endeavor to create partitions that require similar amounts of time to completely measure. It would, perhaps, be more useful to divide the space into partitions in which each partition would instead yield equal utility. However, the challenge with this approach is that the utility is based on the model of the environment that is continually updated through sensor measurements. As such, the initial partitioning would be performed under complete ignorance of the signal model. Partitioning during execution could be undertaken, but would require additional computation time and would continually change based on new measurements altering the signal model.

To divide the space, we represent the environment in a graph structure and perform a normalized graph cut. The graph structure is generated through the representation of each point of interest as a node. An edge is established between nodes if there is a line of sight between the points and the edge is weighted according to the distance. The Floyd-Warshall Algorithm [8] is used to iteratively search through the weighted graph and establish shortest paths between each pair of points and store these values in a distance matrix C. We use the result to approximate the shortest distance from any point of interest to every other point of interest. These distances are used as an approximation of the travel cost between the points. These approximations can be computed before the mission start, saving computation during run time and used with Eq. 2 to calculate the utility of traveling to each point of interest.

Fig. 1. Example of partitions selected by normalized graph cuts on inverse path cost affinity matrix for between two and four robots, shown on the three environments used in this paper.

We define a "spatial affinity" matrix W such that each value $W_{i,j} = \frac{1}{C_{i,j}}$. A high spatial affinity value $W_{i,j}$ indicates that two points i and j are spatial closes. We also establish a diagonal matrix D such that each value $D_{i,i}$ is the sum of edges adjacent to vertex i. We can then use these matrices to perform a normalized graph cut, described in the work of Shi and Mallik [9]. We find the eigenvector corresponding to the second smallest eigenvalue of the matrix:

$$D^{-1}(D - W)D^{-1} \tag{3}$$

Elements of the second smallest eigenvector are grouped according to the positive or negative value. This process creates a binary partition of the points of interest. These groupings can be further partitioned through the recursive application of the process to create more segments to accommodate more than two robots. Each subsequent cut is performed on the partition that has the lowest normalized cut cost metric. Figure 1 shows an example of the partitioning performed on a sample environment derived from a map constructed from a real-world testing environment.

After the environment has been partitioned, the robots are spatially separated by tasking them to separate partitions. If there are more partitions than robots, each robot can be tasked to a different partition prioritizing the partitions that have the highest entropy. As the robots explore their own partition and the overall entropy is reduced, a robot can be tasked to a different partition if its assigned partition is reduced in entropy by a significant amount.

2.3 Real World Implementation

The algorithm was implemented utilizing two and three robot teams utilizing the Jackal platform, created by Clearpath Robotics as seen in Fig. 2. Each Jackal

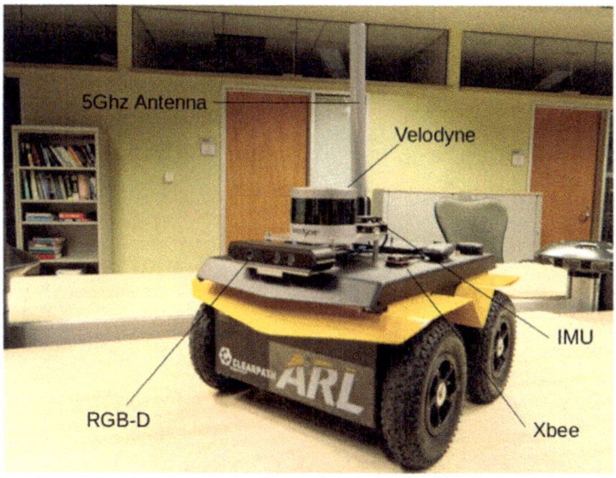

Fig. 2. The Jackal robot used to gather RSSI data, indicating its sensors. The RSSI data was used to build the ground truth GP models.

is equipped with an X-Bee radio to take RSSI measurements, a PicoStation for communications, and a Velodyne VLP-16 LiDAR to detect walls and obstacles. A separate computer is designated as the central control unit. This unit performs the environment partitioning and tasks the robots during run time. Additional X-Bee radios were stationed throughout the environment to serve as radio signal transmitters.

The experiment was performed on the University of California, San Diego campus. Four different locations within Atkinson and Jacobs Halls were used. Each environment was mapped a priori using the same robots while under teleoperation. The robots were tasked from a central control unit within the same environment.

3 Results

3.1 Simulation Results

The system was evaluated through several different simulation experiments. Ground truth data were collected using a robot teleoperated through a real-world environment. A thorough search of the environment was performed while collecting sensor data to train a ground truth GP model. In the simulation, sensor data was generated according to this ground truth data along with added random noise.

We ran simulation experiments with between one and four robots in three different environments. A sample output comparing the results under a varied number of robots for a specific environment can be seen in Fig. 3. The coordination through partitioning can be seen in Fig. 1. We do not include the complete simulation results, but we generally see an improvement in performance with more robots, as expected. We also see improvement in efficiency using the partitioning approach to provide coordination for the robots.

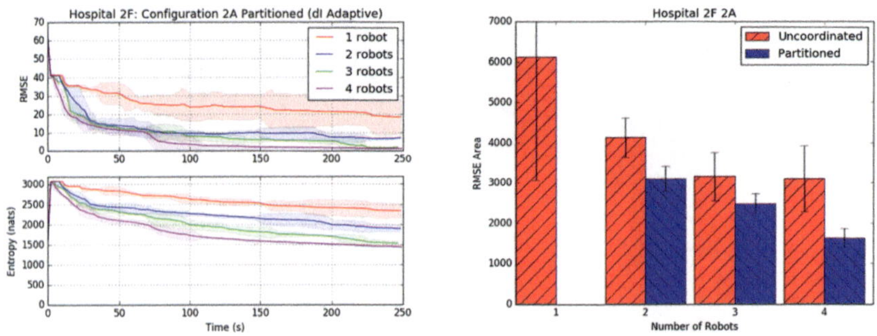

Fig. 3. Simulated results from a sample run using informative adaptive sampling and normalized graph cut-based partitioning for coordination; Left is RMSE and entropy curves for one to four robots; Right is the area under the RMSE curve

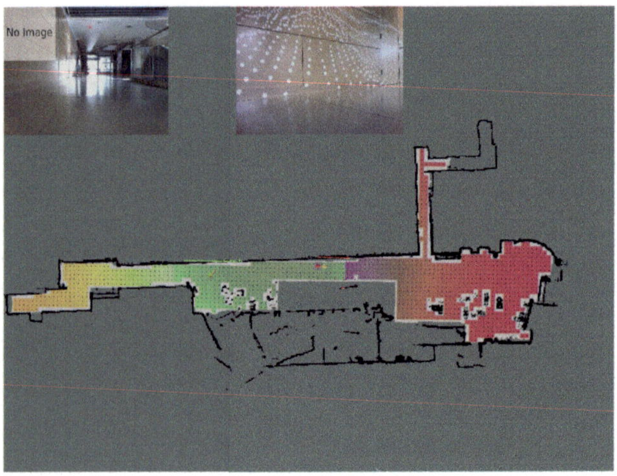

Fig. 4. An example of a map partition: 3 robots, Jacobs Hall UCSD

3.2 Field Experiments

The experiment incorporated real-world operation of the robots within environments over Atkinson and Jacobs Halls on the campus of the University of California, San Diego. We endeavored to include large environments with obstacles that are typical of office buildings. An example of a partitioned test environment can be seen in Fig. 4.

The primary metric we use to evaluate the systems are the entropy decay over time. As the robots explore the area and take sensor measurements, the overall entropy of the sensor model is reduced. Efficient exploration and measurements will reduce the entropy at a faster pace than a straightforward greedy, closest waypoint algorithm. We test the system under four conditions: two or three robots under uncoordinated or partitioned environments.

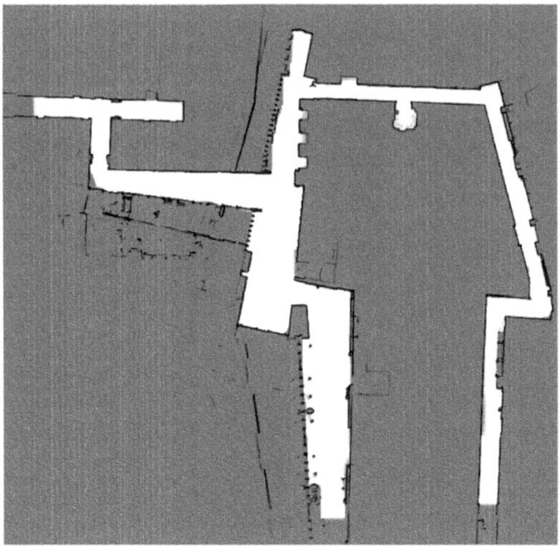

Fig. 5. Map of the test environment: a portion of Atkinson Hall on the campus of the University of California, San Diego

A map of the area incorporated from the experiment on the first floor of Atkinson Hall can be seen in Fig. 5. The experimental results from this environment can be seen in Fig. 6 alongside results for another environment in the Computer Science & Engineering (CSE) building. The overall summary of entropy values for these environments can be seen in Table 1. We expected that the partitioning step would task each robot to different portions of the map to reduce the amount of interference they impact on each other's paths. From this graph, we note that there is a modest improvement between the partitioned and uncoordinated approaches for Atkinson Hall. We see a more pronounced improvement for the experimentation in the CSE building. For this environment, we see a more considerable improvement when coordinating the robots through map partitioning. Although we do not include the graphs from the other environments here, we see similar improvements across all maps.

Table 1. The area under entropy curves for information rate adaptive sampling, uncoordinated and partitioned environments for the Atkinson First Floor and Computer Science & Engineering Building Environments

Atkinson, 1st Floor

	Uncoordinated	Partitioned
2 Robots	89.8505	88.4639
3 Robots	90.6349	87.9373

Computer Science & Engineering Building

	Uncoordinated	Partitioned
1 Robot	—	209.8928
2 Robots	185.6054	177.1364
3 Robots	176.7638	155.7035

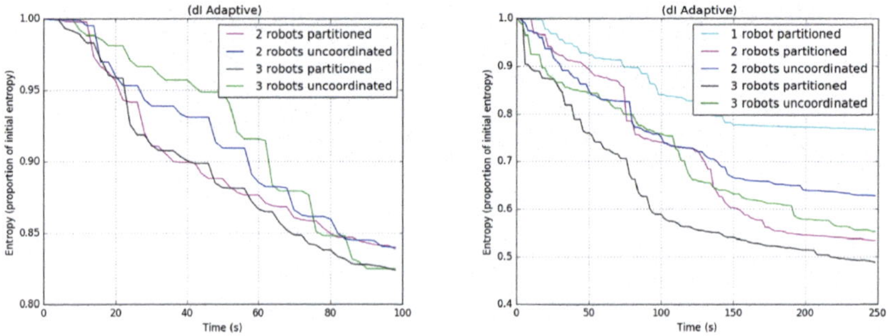

Fig. 6. Comparison between 2 or 3 robots under partitioned and uncoordinated environments using information rate adaptive sampling for the first floor of Atkinson Hall (Fig. 5) and the Computer Science & Engineering Building

4 Main Experimental Insights

The improvements we see under this experiment are not as extensive as we predicted. We theorize that there are some reasons for this result. Of primary consideration are the restrictive environments that were tested. Each test environment incorporates only a portion of the UCSD building. Because of the architecture of the environment and our testing limitations, we could not establish sufficient communications throughout the entirety of the building. This necessitated that we artificially limit the area. As such, the size of the map and environment partitions were limited in size. We hypothesize that larger environments would result in more pronounced improvements through partitioning.

We also note that each environment consists mostly of long, narrow hallways. As with many academic and office buildings, the architecture of the environment contains long hallways with numerous side rooms. Because the building was in regular use during our experimentation, we could not open the offices and classrooms to our mapping and exploration. This limits not only the overall size of the maps but also the significant decisions that the algorithm must make when tasking robots to specific locations. Therefore the algorithm will have a lesser impact on the overall efficiency of the exploration.

5 Conclusions and Future Work

There are two main weaknesses in the algorithm that we target for improvement in our future work. First is that the system centers around a central controller. This controller executes the partitioning and robot tasking and as such, must maintain communications with the robot throughout execution. In the future, we look to evolve the algorithm into a distributed system so that the robots can operate independently and collect information before returning into the communication range to share the data.

The second weakness is that the system operates only under known environments. There are several applications such as disaster relief where the environment could be known or semi-known if, for example, a hallways collapses and becomes impossible. Our future work looks into furthering the system so that it can operate under unknown locations. It would build a map while simultaneously taking sensor data for the signal model. This modification necessitates that the partitioning of the environment changes dynamically as more of the map is created.

Acknowledgements. This work was supported by the U.S. Army Combat Capabilities Development Command Army Research Lab (ARL) through the Distributed and Collaborative Intelligent Systems and Technology (DCIST) Collaborative Research Alliance (CRA).

References

1. Krause, A., Singh, A., Guestrin, C.: Near-optimal sensor placements in Gaussian processes-theory, efficient algorithms and empirical studies. J. Mach. Learn. Res. **9**(May), 235–284 (2008)
2. Singh, A., Krause, A., Guestrin, C., Kaiser, W., Batalin, M.: Efficient planning of informative paths for multiple robots. In: International Joint Conference on Artificial Intelligence (IJCAI) (2007)
3. Low, K.H., Dolan, J.M., Khosla, P.: Adaptive multi-robot wide-area exploration and mapping. In: Proceedings of the 7th International Conference on Autonomous Agents and Multiagent Systems (AAMAS), pp. 23–30 (2008)
4. Kemna, S., Rogers III, J.G., Nieto-Granda, C., Young, S., Sukhatme, G.S.: Multi-robot coordination through dynamic voronoi partitioning for informative adaptive sampling in communication-constrained environments. In: International Conference on Robotics and Automation (ICRA), May 2017
5. Duvallet, F., Tews, A.D.: WiFi position estimation in industrial environments using Gaussian processes. In: IEEE/RSJ International Conference on Intelligent Robots and Systems (IROS), pp. 2216–2221. IEEE (2008)
6. Ouyang, R., Low, K.H., Chen, J., Jaillet, P.: Multi-robot active sensing of non-stationary gaussian process-based environmental phenomena. In: International Conference on Antonomous Agents and Multiagent Sytems (AAMAS) (2014)
7. Atanasov, N., Le Ny, J., Daniilidis, K., Pappas, G.J.: Decentralized active information acquisition: theory and application to multi-robot SLAM. In: ICRA, pp. 4775–4782 (2015)
8. Floyd, R.W.: Algorithm 97: shortest path. Commun. ACM **5**(6), 345 (1962)
9. Shi, J., Malik, J.: Normalized cuts and image segmentation. IEEE Trans. Pattern Anal. Mach. Intell. **22**(8), 888–905 (2000)

External Force Field Modeling for Autonomous Surface Vehicles

Jason Moulton[1], Nare Karapetyan[1], Alberto Quattrini Li[2(✉)], and Ioannis Rekleitis[1]

[1] Computer Science and Engineering Department, University of South Carolina, Columbia, SC, USA
{moulton,nare}@email.sc.edu, yiannisr@cse.sc.edu

[2] Computer Science Department, Dartmouth College, Hanover, USA
alberto.quattrini.li@dartmouth.edu

Abstract. Operating in the presence of strong adverse forces is a particularly challenging problem in field robotics. In most robotic operations where the robot is not firmly grounded, such as aerial, surface, and underwater, minimal external forces are assumed as the standard operating procedures. The first action for operating in the presence of non-trivial forces is modeling the forces and their effect on the robots motion. In this work an Autonomous Surface Vehicle (ASV), operating on lakes and rivers with varying winds and currents, collects wind and current measurements with an inexpensive custom-made sensor suite setup, and generates a model of the force field. The modeling process takes into account depth, wind, and current measurements along with the ASVs trajectory from GPS. In this work, we propose a method for an ASV to build an environmental force map by integrating in a Gaussian Process the wind, depth, and current measurements gathered at the surface. We run extensive experimental field trials for our approach on real Jetyak ASVs. Experimental results from different locations validate the proposed modeling approach.

Electronic supplementary material The online version of this chapter (https://doi.org/10.1007/978-3-030-33950-0_29) contains supplementary material, which is available to authorized users.

J. Xiao et al. (Eds.): ISER 2018, SPAR 11, pp. 328–338, 2020.
https://doi.org/10.1007/978-3-030-33950-0_29

1 Introduction

While robots become increasingly common, from the robotic vacuum cleaner and warehouse product-moving robots, to the prospect of autonomous cars, robots are assumed to operate largely undisturbed. Most fielded robots operate on fairly flat grounds, with minimal wind, wave, and current forces. Contrary to these scenarios, we consider robots that are best suited to operate in environments restrictive to humans. As such, capabilities to operate in unknown/dynamic environments, and in the presence of adverse external forces, are required to ensure that robots become ubiquitous and safe in many applications – such as safe inspection of infrastructure [16], search and rescue [13], environmental sampling [12], monitoring of water quality [10], and mapping inaccessible regions in more efficient, less cost prohibitive means.

Fig. 1. UofSC autonomous surface vehicle outfitted with anemometer, depth sonar, and current sensors.

In this work, using an Autonomous Surface Vehicle (ASV) – see Fig. 1 – we provide the following contributions: a reliable inexpensive platform for collecting depth, wind, and current data in different environments and conditions; and a data processing and model derivation approach for spatially varying environments.

1.1 Motivation

Currently, as observed in a variety of experiments – see Figs. 2 and 3 for some examples – an ASV relying on a conventional robotic actuator controllers for way-point navigation is unable to maintain its course when faced with non trivial external forces such as heavy wind or current. Due to the PID controller being tuned for conditions which are stable, there is no simple method to provide tuned coefficients for dynamic environments where wind and currents are always changing. This drives our motivation to pro-actively model, plan for, and adapt to these dynamics so the robot can maintain its course and not miss large swaths of its planned trajectory.

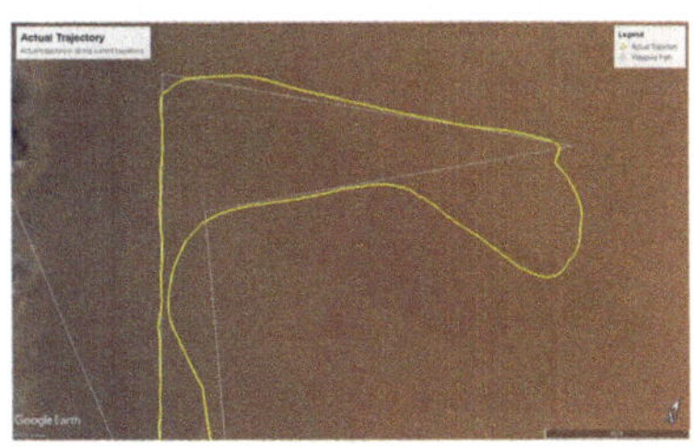

Fig. 2. ASV unable to maintain course in heavy current when turning in downstream directions. The white line represents ideal path and the yellow line the actual (GPS) trajectory of the ASV.

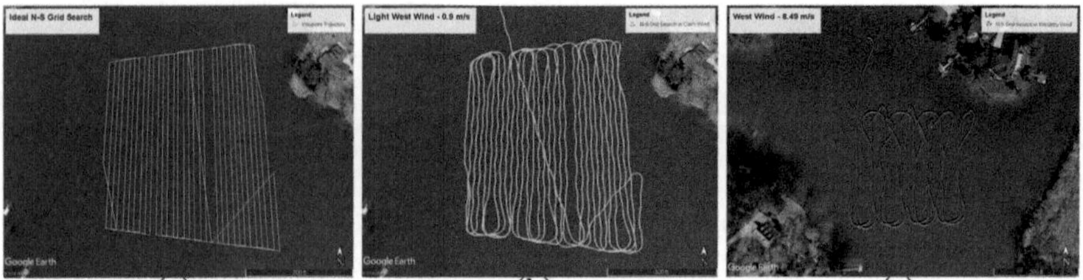

(a) (b) (c)

Fig. 3. Ideal North-South (a) grid search missions are used to provide a baseline for measuring tracking performance. Trajectories for grid searches conducted in (b) calm conditions and (c) a subset in 8.49 m/s wind illustrating the effects of the wind on the ASV.

1.2 Related Work

Current research modeling wind and currents mainly focuses on large scales, with applications to oceanic navigation and power generation. In the problem addressed in this work, the effects of environmental forces impacting the platform are greater considering the size of the ASV. For instance the work of Soman *et al.* [19] reviewed existing wind prediction strategies for optimizing efficiency and profits in power generation applications. Their work focuses both on long term and short term forecasting, but mostly as it pertains to direction, since the turbines they are designing for are stationary. One more closely related study is by Al-Sabban *et al.* [2], which focuses on the effect of wind on Unmanned Aerial System (UAS). They implement a hybrid Gaussian distribution of a wind field and a slightly modified Markov Decision Process (MDP) to identify the optimal path and optimal power consumption trajectory for a UAS. This work may prove valuable in our future work, but at the current time, we are concerned with an environment with leeward and windward effects caused by surrounding landscape. Another work, by Encarnaçao and Pascoal [4], studied the problem of developing control systems for marine crafts that are able to follow trajectories to track another boat, under the effects of water currents. Their models center on inertial tracking and compensation of roll, pitch, and yaw rates after the force is sensed to provide course corrections. In our work, we aim at being more proactive, in that we actively measure and model environmental variables with the end-state intent on taking corrective control measures prior to the robot coming under the effects of the wind or currents. Finally, Hsieh *et al.* [5] have provided many contributions related to the problem of mapping the effects current phenomena with their design of a control strategy for collaborative underwater robots to track coherent structures and manifolds on generally static conservative flows [6,9,14]. Huynh *et al.* presented a path planning method for minimizing the energy consumption of an autonomous underwater vehicle (AUV) [7]. Their work addresses varying ocean disturbances that are assumed to not exceed the capabilities of the AUV. These works, while addressing aspects of collecting and modeling dynamic environmental characteristics, do not address fully the same time and space considerations that impact our lightweight, small-scale ASV. To establish models that can drive future adaptive controls and online planning strategies for our ASV, we adopt the following technical approach.

2 System and Methodology

In this section, we present the proposed hardware setup, calibration and verification of inexpensive sensors in a controlled environment, experimental environments, and data collection and processing required to develop comprehensive models of external forces. We break this into two components, the physical and technical characteristics associated with the design and build of our ASV, and the data gathering/processing approach we take to model depth, current, and wind. The modeling process takes into account wind and current measurements, the ASVs trajectory from GPS, compass, and IMU data, together with the morphology and bathymetry of the environment.

We deployed our methods on a Mokai Es-Kape platform[1] termed *Jetyak*, based on the WHOI Jetyak [8] ASV and conducted extensive calibration of the operational parameters in benign conditions. Furthermore, we performed several experiments to collect data under a variety of conditions. Such data has been used to perform predictions to evaluate different machine learning techniques. In the following a brief overview of the platform is presented – see our work [15] for more details on our version of the Jetyak. The baseline system, capable of way-point navigation and teleoperated control, consists of an ArduPilot [11] PixHawk microcontroller utilizing IMU, GPS, and telemetry data. The sensors employed are intentionally inexpensive to enable their adoption by a large number of end-users. For instance, our anemometer[2] is readily available for building and programming as a kit. It provides wind speed and direction updates at 4 Hz costing about 100 US dollars, whereas a state-of-the-art ultrasonic anemometer with user accessible serial output costs over 1,500 US dollars. We employ the same constraint for the depth sonar and water current sensors, with none of the sensors costing more than 200 US dollars. Selected water current sensors[3] are simple hall effect sensors. The depth sensor[4] mounted on the ASV is National Marine Electronics Association (NMEA) 0183 compliant[5] and outputs serial text to allow Robot Operating System (ROS) [17] based software direct access to the measurements. In addition to 2.4 GHz remote control and 900 MHz remote frequency communications, we employ a Raspberry Pi to host the ROS framework recording all telemetry, depth, current, and wind data collected.

[1] http://www.mokai.com/mokai-es-kape/.
[2] https://learn.sparkfun.com/tutorials/weather-meter-hookup-guide.
[3] http://www.raymarine.com/view/index.cfm?id=555.
[4] http://www.cruzpro.co.nz/active.html.
[5] https://www.nmea.org/content/nmea_standards/nmea_0183_v_410.asp.

Mounting inexpensive, off-the-shelf, sensors requires experimentation both on the mounting position and orientation, as well as reverse engineering to build and calibrate the micro-controller servicing the sensor. Being a 1-d sensor, the depth sonar requires the smallest effort, since it returns depth readings directly to the USB port of the Raspberry Pi. The only challenge with mounting a depth sensor is ensuring it remains submerged and avoids cavitation interference resulting from the physical properties of the ASV's hull. The same challenge applies to selecting the location and orientation of the current sensors. Cavitation, air pockets, and water deflection from the hull and the other sensors have an adverse impact on the water flow through the sensor. The least challenging sensor to mount is the anemometer, simply requiring a stand-off bracket to attach it to the mast. Prior to installation, calibration of the current, wind, and depth sensors is performed to ensure the accuracy of the recorded data. A full tutorial on building the discussed platform with all diagrams and sensor driver code is available on the UofSC Autonomous Field Robotics Lab (AFRL) website [1], including lessons learned for programming inexpensive micro-controllers and placement/alignment of sensors.

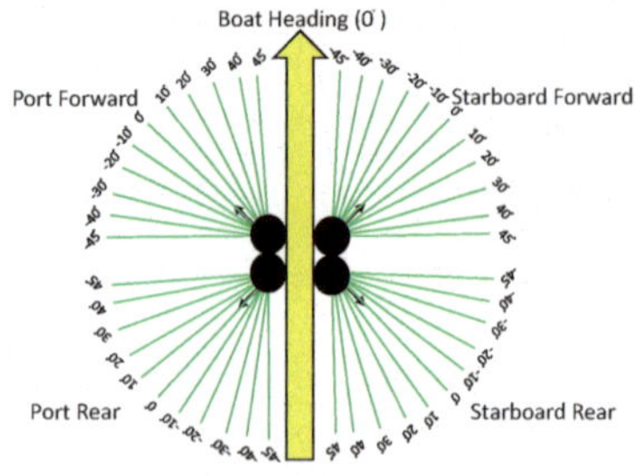

Fig. 4. Mounting configuration for surface current sensors allows selection of the highest sensor reading for F1 and its next highest neighbor for F2. Calculations for speeds and directions are explained below.

The facilities of the Hydraulics Laboratory at Civil and Environmental Engineering department were used to calibrate the sensors and establish ground truth estimates. Throughout all calibrations, the PixHawk with ArduPilot code, ROS node, Arduino microcontroller programs are run to ensure proper operation when added to the ASV. Laboratory experiments using controlled flow demonstrated that the sensors are effective for water current component measurement at a $\pm 45^\circ$ orientation to the sensor. As such, different setups have been tested to achieve the best coverage. After several attempts to provide accurate current readings covering all directions, we arrived at the proposed solution of four current sensors offset from each other at 90° and offset from the ASV's reference frame by 45°; see Fig. 4 for the proposed configuration. Using the highest measurement and highest neighbor measurement, we are able to accurately calculate the current in the boat's reference frame and transform it to the world reference frame by subtracting the boat's x and y velocity components. The high-level calculations are described in the following section.

2.1 Data Gathering and Processing

The wind and current sensors record data are influenced by the motion of the ASV. More specifically, the sensor measurement ($\overrightarrow{R}$) is the vector sum between the motion of the ASV ($\overrightarrow{A}$) and the real value of the physical phenomenon ($\overrightarrow{W}$). The motion of the ASV can be inferred by the GPS velocity and compass sensors. Therefore, the true value can be estimated as $\overrightarrow{W} = \overrightarrow{R} - \overrightarrow{A}$.

The current sensors are mounted in fixed locations and measure scalar current velocities (f). We select the component forces for ($\overrightarrow{F}$) by assigning the highest measured force to $F1$ and the highest of $F1's$ two neighboring sensor readings to $F2$. The forces are then offset by 45° to account for the angular velocities read by the sensor as the ASV ($\overrightarrow{A}$) traverses the current ($\overrightarrow{C}$). In our optimal setup, we have aligned four sensors, one offset at 45° on each quadrant of the boat, illustrated in Fig. 4. This provides the equation $\overrightarrow{C} = \overrightarrow{F} - \overrightarrow{A}$ which we will solve using the trigonometric properties resulting from our sensor array alignment.

Once accurately aligned, the collected data were combined using a Gaussian Process (GP) mapping technique [18] to build a model of forces and depth. Initial modeling and correlation uses boat heading, GPS velocity of the boat, depth measurements, wind speed and direction measurements, and four current measurements as input. More formally, given a phenomenon $f(\mathbf{x})$, a GP can be used to estimate $f(\mathbf{W})$ at locations $\mathbf{W} = [\mathbf{w}^1, \mathbf{w}^2, \dots \mathbf{w}^k]$ with a posterior distribution fitted over noisy measurements $\mathbf{Y} = [y^1, y^2, \dots, y^n]$ collected by the robots at the corresponding GPS locations $\mathbf{X} = [\mathbf{x}^1, \mathbf{x}^2, \dots, \mathbf{x}^n]$:

$$p(f(\mathbf{W}) \mid \mathbf{WX}, \mathbf{Y}) \sim \mathcal{N}(\mu_{\mathbf{W}}, \mathbf{\Sigma_W}). \quad (1)$$

As typically done in the mainstream approach, assuming a zero-mean GP, the estimate of the phenomenon is given by the mean vector $\mu_{\mathbf{W}} = K(\mathbf{W}, \mathbf{X})\text{cov}(\mathbf{Y})^{-1}\mathbf{Y}$, where $\text{cov}(\mathbf{Y}) = K(\mathbf{X}, \mathbf{X}) + \sigma_n^2 I_q$ is the correlation between observed values and σ_n^2 is the noise affecting the measurements $\mathbf{Y}$. The covariance matrix is calculated as $\mathbf{\Sigma_W} = K(\mathbf{W}, \mathbf{W}) - K(\mathbf{W}, \mathbf{X})\text{cov}(\mathbf{Y})^{-1}K(\mathbf{W}, \mathbf{X})^T$. For accuracy evaluation, we tested different $\boldsymbol{K}()$ kernels. Kernels used for this initial comparison include linear, ExpQuad, Matern 3/2, and radial basis function (RBF), among which best performance was achieved by Matern 3/2, expressed by the following equation:

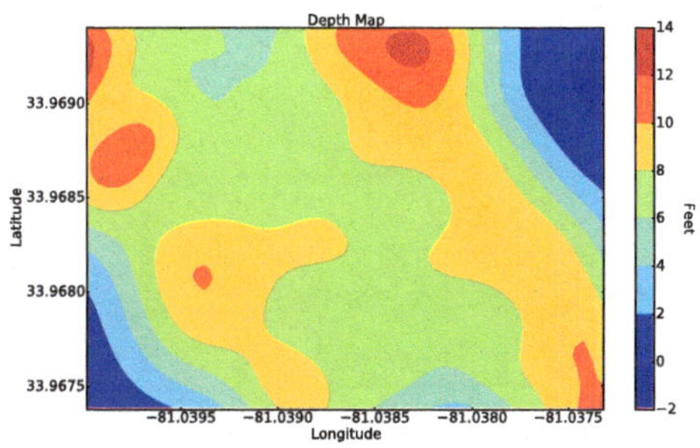

Fig. 5. Depth map of Congaree River bottom resulting from GP.

$$k(\mathbf{x}^i, \mathbf{x}^j | \Theta) = \sigma_n^2 (1 + \frac{\sqrt{3}r}{\sigma_l}) exp(-\frac{\sqrt{3}r}{\sigma_l}), \quad (2)$$

where r is the Euclidean distance between $\mathbf{x}^i$ and $\mathbf{x}^j$, and σ_l is a positive parameter. Using the observations $\mathbf{X}$ and $\mathbf{Y}$ through the optimization of hyperparameters of the GP, predictions can be obtained.

3 Experiments

Experiments were carried out with a Jetyak equipped with the sensor suite described above in two different environments: a lake (Fig. 3) in a 100 m × 100 m region, relatively calm; and a river (Fig. 6), where the conditions are changing over time depending on rain and planned discharges by the local hydro-electric company. Our methodology uses standard grid-search patterns to establish a baseline for performance comparison.

We then process the collected data in two stages. First, we verify all time stamp and world orientation data by converting the PixHawk data logs to keyhole markup language (KML) for visual inspection of the missions, sequences and trajectories of the robots during the field trial. Once time stamps are verified to contain no gaps, we then process the ROS bag file by a Python script to align time stamps from each sensor, based on an approximate time synchronizer scheme[6]. It is at this point where we can verify the integrity of the data collected. Once complete, we are able to deem our system, scripts, calculations, and data as sound for further processing. The second stage processes the data for transformations, visualizations, and GP predictions as described in the previous section.

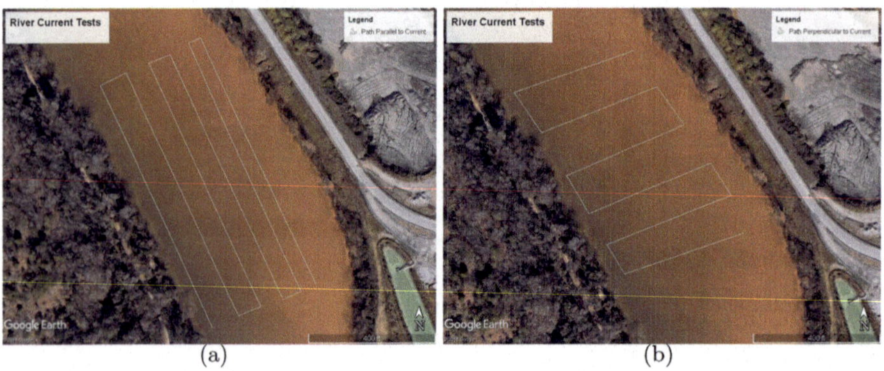

Fig. 6. Baseline testing pattern parallel (a) to the predominant current and perpendicular (b) to the predominant current on the Congaree River, SC.

Some preliminary experiments were completed to fine tune the PID controller parameters and to assess the effects of taking the ASV, whose controller was tuned in the lake, to the river and evaluating its performance. As shown in Fig. 3(c), clearly the ASV was not able to maintain the planned trajectory.

The main set of experiments included data collection under different conditions. In particular, first, we collected wind and current measurements by tying the ASV so that is stationary. In this way, a baseline is available to compare

[6] http://wiki.ros.org/message_filters/ApproximateTime.

with, when data is collected as the ASV is moving. Second, we planned different waypoint missions characterized by different patterns – i.e., parallel and, perpendicular to current patterns – to collect data from different orientations.

Initial collected data in Fig. 5 reflects the resulting topographical map of the riverbed after data collection using the coverage path in Fig. 6(b). Wind and current data collected from the Congaree River clearly characterizes highly dynamic currents, changing winds, and a highly variant depth – see Fig. 9 for recorded wind and current measurements; and Figs. 7 and 8 for the predicted wind speed and direction, and current speed and direction, respectively, over the region. Figure 9(a) illustrates successful data collection and modeling of wind forces experienced in moderately windy conditions. Our system recorded sustained winds of 8.5 m/s and gusts reaching 12.5 m/s, accurate measurements according to the local weather recording station, which recorded sustained winds of 9 m/s and gusts at 14 m/s during the 3 h trial. Given the mostly open terrain and the effects of the shoreline on wind patterns, our measurements are in-line with the recording station.

As illustrated in Fig. 9, the average actual current measured while the ASV was navigating the cross sections of the river was 5.0 m/s, compared to the docked stationary current measurement of 3.0 m/s. The challenge with surface

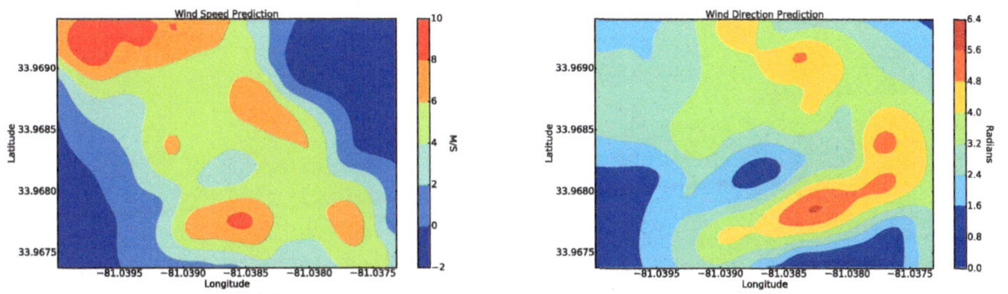

(a) (b)

Fig. 7. (a) Wind Speed Prediction Map for Congaree River, SC. (b) Wind Direction Prediction Map for Congaree River, SC.

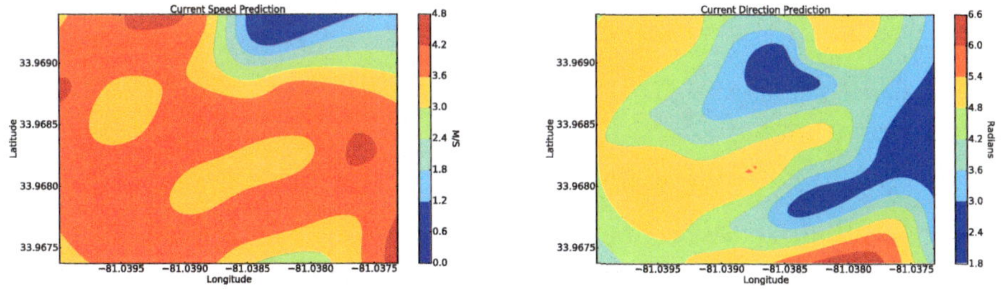

(a) (b)

Fig. 8. (a) Current Speed Prediction Map for Congaree River, SC. (b) Current Direction Prediction Map for Congaree River, SC.

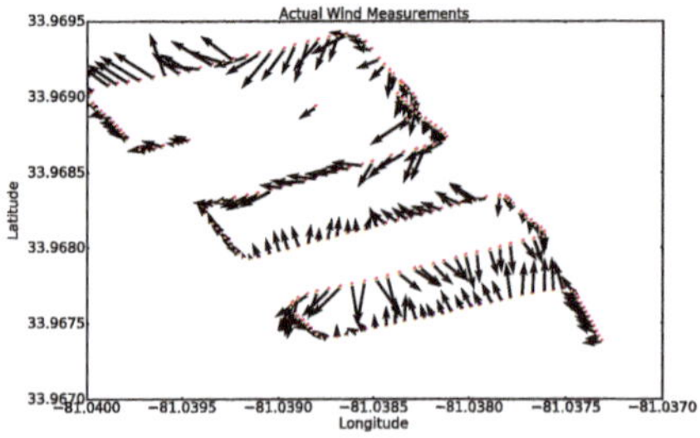

(a)

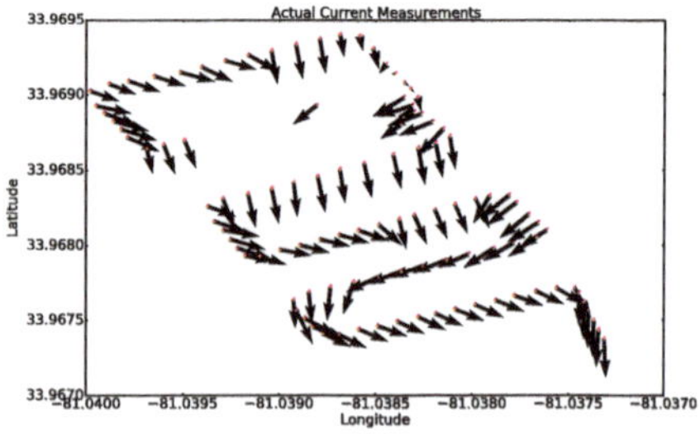

(b)

Fig. 9. Field trial on the Congaree River, SC on the trajectory in Fig. 6(b). (a) Recorded wind measurements. (b) Recorded current measurements.

currents is establishing the ground truth due to the high degree of fluctuation with them. In addition to the previously mentioned lab calibrations, we ensure measurements are in the correct order of magnitude at the beginning of each field trial by recording the free floating GPS velocity. In this particular field trial, we approximated the ground truth current speed to be 2.5 m/s. Due to the hull displacement, as expected, we recorded a slightly lower velocity using this free float method.

4 Discussion

Our technical and experimental contributions in this paper have produced a baseline of inexpensive tools and methods for recording the external forces – wind and current – acting on an ASV. GP regression modeling allowed us to predict for a large area with sparse measurements collected during the experiments, as shown in Figs. 7 and 8. From the beginning of this endeavor, calibration of the analog current sensors to accurately reflect surface current required continuous refinement. Lab configuration and testing provided us an initial guideline on the best sensor configuration for a static vessel. Furthermore, on-free-float and static ASV testing confirmed our measurements to be reasonable for the conditions. Currently, this system is stable for recording and observing nature's phenomena in action.

Of particular interest in these experiments is the confirmation of the close correlation between the depth (Fig. 5) and the affected current patterns in Fig. 8. Verifying that our inexpensive system is able to collect and confirm US Geological Survey studies [3], encourages extending our research. As such, to improve the precision of predictions, our current work involves studying and analyzing different GP kernels to better emulate the short temporal livelihood of the predictions. The long-term goal is to combine all three models – i.e., current, depth, wind – into a comprehensive impact model that can be used to improve coverage and search algorithms and develop proactive controls to enable ASV operation in highly dynamic environments.

Acknowledgment. The authors would like to thank the National Science Foundation for its support (NSF 1513203). The authors are grateful to Professor Enrica Viparelli from the Civil and Environmental Engineering Department for her assistance in ground truth measurements at the Hydraulics Laboratory.

References

1. AFRL: Jetyak documentation (2018). https://afrl.cse.sc.edu/afrl/resources/JetyakWiki/index.html. Accessed 19 July 2018
2. Al-Sabban, W.H., Gonzalez, L.F., Smith, R.N.: Wind-energy based path planning for unmanned aerial vehicles using Markov decision processes. In: IEEE International Conference on Robotics and automation (ICRA), pp. 784–789 (2013)
3. Buchanan, T., Somers, W.: Discharge Measurements at Gaging Stations: U.S. Geological Survey Techniques of Water-Resources Investigations, vol. 3. USGS Online Library, Washington, DC (1969)
4. Encarnaçao, P., Pascoal, A.: Combined trajectory tracking and path following: an application to the coordinated control of autonomous marine craft. In: IEEE Conference on Decision and Control (CDC), vol. 1., pp. 964–969. IEEE (2001)
5. Hsieh, M.A., Forgoston, E., Mather, T.W., Schwartz, I.B.: Robotic manifold tracking of coherent structures in flows. In: IEEE International Conference on Robotics and automation (ICRA), pp. 4242–4247 (2012)
6. Hsieh, M.A., Hajieghrary, H., Kularatne, D., Heckman, C.R., Forgoston, E., Schwartz, I.B., Yecko, P.A.: Small and adrift with self-control: using the environment to improve autonomy. In: International Symposium on Robotics Research (ISRR). Springer, pp. 387–402 (2018)
7. Huynh, V.T., Dunbabin, M., Smith, R.N.: Predictive motion planning for AUVs subject to strong time-varying currents and forecasting uncertainties. In: IEEE International Conference on Robotics and Automation (ICRA), pp. 1144–1151 (2015)
8. Kimball, P., et al.: The WHOI Jetyak: an autonomous surface vehicle for oceanographic research in shallow or dangerous waters. In: IEEE/OES Autonomous Underwater Vehicles (AUV), vol. 10, pp. 1–7 (2014)
9. Kularatne, D., Bhattacharya, S., Hsieh, M.A.: Optimal path planning in time-varying flows using adaptive discretization. IEEE Robot. Autom. Lett. **3**(1), 458–465 (2018)
10. Quattrini Li, A., Rekleitis, I., Manjanna, S., Kakodkar, N., Hansen, J., Dudek, G., Bobadilla, L., Anderson, J., Smith, R.N.: Data correlation and comparison from multiple sensors over a coral reef with a team of heterogeneous aquatic robots. In: International Symposium of Experimental Robotics (ISER), Tokyo, Japan (2016)
11. Mackay, R.: Ardupilot (2016). https://github.com/ArduPilot/ardupilot_wiki/blob/master/rover/source/docs/gettit.rst
12. Manjanna, S., Quattrini Li, A., Smith, R.N., Rekleitis, I., Dudek, G.: Heterogeneous multirobot system for exploration and strategic water sampling. In: IEEE International Conference on Robotics and Automation (ICRA), Brisbane, Australia, pp. 4873–4880 (2018)
13. Matos, A., Martins, A., Dias, A., Ferreira, B., Almeida, J.M., Ferreira, H., Amaral, G., Figueiredo, A., Almeida, R., Silva, F.: Multiple robot operations for maritime search and rescue in euRathlon 2015 competition. In: MTS/IEEE OCEANS, Shanghai, China, pp. 1–7 (2016)

14. Michini, M., Hsieh, M.A., Forgoston, E., Schwartz, I.B.: Robotic tracking of coherent structures in flows. IEEE Trans. Robot. **30**(3), 593–603 (2014)
15. Moulton, J., Karapetyan, N., Bukhsbaum, S., McKinney, C., Malebary, S., Sophocleous, G., Quattrini Li, A., Rekleitis, I.: An autonomous surface vehicle for long term operations. In: MTS/IEEE Oceans, Charleston, SC, USA (2018)
16. Murphy, R.R., Dreger, K.L., Newsome, S., Rodocker, J., Steimle, E., Kimura, T., Makabe, K., Matsuno, F., Tadokoro, S., Kon, K.: Use of remotely operated marine vehicles at Minamisanriku and Rikuzentakata Japan for disaster recovery. In: IEEE International Symposium on Safety, Security, and Rescue Robotics (SSRR), pp. 19–25 (2011)
17. Quigley, M., et al.: ROS: an open-source robot operating system. In: ICRA workshop on Open Source Software, Kobe, Japan, vol. 3, p. 5 (2009)
18. Rasmussen, C.E., Williams, C.K.I.: Gaussian Processes for Machine Learning. MIT Press, Cambridge (2006)
19. Soman, S.S., Zareipour, H., Malik, O., Mandal, P.: A review of wind power and wind speed forecasting methods with different time horizons. In: North American Power Symposium, pp. 1–8 (2010)

Context-Aware Intention and Trajectory Prediction for Urban Driving Environment

Malika Meghjani[1,2(✉)], Shashwat Verma[2], You Hong Eng[2], Qi Heng Ho[2,3], Daniela Rus[4], and Marcelo H. Ang Jr.[3]

[1] Singapore University of Technology and Design, Singapore, Singapore
malika_meghjani@sutd.edu.sg
[2] Singapore-MIT Alliance for Research and Technology, Singapore, Singapore
{shashwat,youhong,qiheng}@smart.mit.edu
[3] National University of Singapore, Singapore, Singapore
mpeangh@nus.edu.sg
[4] Massachusetts Institute of Technology, Cambridge, USA
rus@csail.mit.edu

Abstract. This paper addresses intention and trajectory prediction of exo-vehicles in an urban driving environment. Urban environments pose challenging scenarios for self-driving cars, specifically pertaining to traffic light detection, negotiating paths at the intersections and sometimes even overtaking illegally parked cars in narrow streets. This complex task of autonomously driving while considering anomalous situations make urban driving conditions unique when compared to highway driving. In order to overcome these challenges, we propose to use road contextual information to predict driving intentions and trajectories of surrounding vehicles. The intention prediction is obtained using a recurrent neural network and the trajectory is predicted using a polynomial model fitting of the past lateral and longitudinal components of the vehicle poses and road contextual information. The integrated process of intention and trajectory prediction is performed in real-time by deploying and testing on a self-driving car in a real urban environment.

1 Introduction

Urban driving scenarios involve unique challenges when compared to highway driving due to complexity of the environment and corresponding road-rules. Given the vast range of road-rules that may or may not be applicable to anomalous urban driving situations, it becomes crucial to use road contextual information for autonomous and intelligent driving. An example of the anomalous situation is illegally parked car in a single lane on a bi-directional street which

Electronic supplementary material The online version of this chapter (https://doi.org/10.1007/978-3-030-33950-0_30) contains supplementary material, which is available to authorized users.

J. Xiao et al. (Eds.): ISER 2018, SPAR 11, pp. 339–349, 2020.
https://doi.org/10.1007/978-3-030-33950-0_30

requires the ego vehicle to overtake from the opposite side of the road. In such scenarios, using road contextual information such as number and direction of lanes, lane width and distance to the upcoming intersection, can help in human-like decision making. Hence, we propose to use contextual information for intention and trajectory prediction of surrounding vehicles to plan a path for the ego-vehicle.

Our system comprises of five main modules as illustrated in Fig. 1. Specifically, these include:

- **Perception** module comprising of vision-based obstacle detection and classification system along with LiDAR-based point cloud clustering for identifying the road region.
- **Sensor fusion and tracking** module performs data association between the vision and LiDAR systems and tracks the vehicles in the global map frame.
- **Behavior analysis and prediction** module converts the vehicle tracks from map frame to lane coordinate system and uses road contextual information for behavior prediction.
- **Decision making and planning** module uses the predicted behavior of exo-vehicles for decision making of the ego-vehicle. This decision is then translated into a path using the planner.
- **Speed and steering control** module is a low-level controller used to execute the planned path.

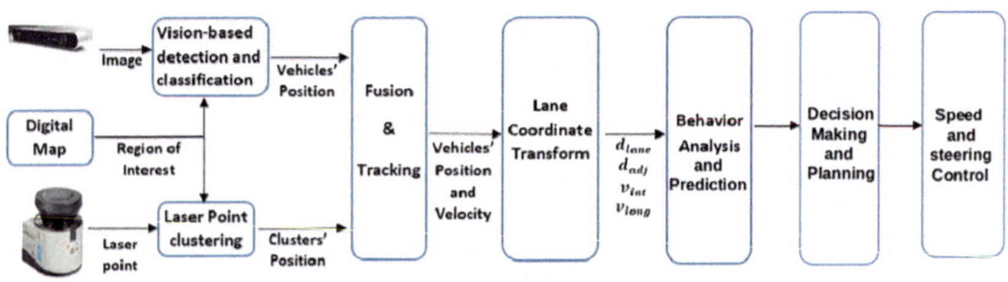

Fig. 1. System overview comprising of five modules: perception, sensor fusion and tracking, behavior analysis and prediction, decision making and planning, speed and steering control.

In our initial work [1], we focused on the perception, sensor fusion and tracking modules along with a naive behavior analysis and prediction method. The behavior prediction system was rule-based and could only predict binary outcome of changing or keeping lanes for a single leading vehicle.

In this paper, we focus on predicting intentions as well as trajectories of multiple exo-vehicles. Specifically, the predicted intentions help us distinguish between left and right lane changes for surrounding vehicles based on the learned driving behaviors. The driving intent is predicted using a recurrent neural network (RNN) instead of a rule-based method since, RNNs can capture subtle urban driving behaviors which otherwise cannot be independently represented by road-rules. On the contrary, RNNs are not very promising for trajectory prediction as they require a huge range of data to model various driving patterns. So, we predict the trajectory of the target vehicle based on its short-term tracked motion, the predicted intention and the road-contextual information. This allows us to predict trajectories without explicitly learning a range of driving patterns a-priori. Hence, our proposed context-aware prediction method can provide an early intention and trajectory prediction with reasonably acceptable time horizon in the future.

2 Related Work

There is an extensive amount of literature related to intention and trajectory prediction for motion planning of ego-vehicle [2]. Specifically, in [3] the authors propose a method adapted based on structural RNN to predict lane change intention of exo-vehicles. Unlike our approach, their method is specifically applicable for highway driving and does not predict the trajectories of the vehicles. Another method based on RNN is proposed in [4] for trajectory prediction, also for highway driving. Though the reported results are encouraging, this method requires a large amount of labeled data to model a wide-range of possible trajectories. Ding et al. [5] also propose a neural network based model for trajectory prediction. They address the data dependency issue by using a simulator and recording implicit driving behaviors of human subjects while they controlled the simulated vehicles. In our work, we use a neural network as well but to only predict the intention of the surrounding vehicles. The trajectory is predicted based on past vehicle states. We also use road contextual information which reduces the need of huge amount of data for learning intentions.

A real-time trajectory planning approach is also proposed by Li et al. in [6]. They present a hierarchical motion planning framework comprising of a high-level behavior planner to generate a coarse reference path and a low-level trajectory planner providing locally feasible candidate paths. The candidate paths are then evaluated using an objective function to select the optimal collision-free, smooth and dynamically feasible path. Their proposed approach, unlike our method, does not account for the intentions of the surrounding vehicles and instead passively plans paths in the available collision-free space. In [7,8] the human driving intentions and its uncertainties are taken into account for motion planning of the ego-vehicle. However, this work particularly considers the decision making task of navigating at the intersections which does not require any trajectory prediction.

In this paper, we aim to combine intention and trajectory prediction methods for driving behavior analysis in an urban environment. Similar to our goal, Galceran et al. [9] propose an integrated approach of inference and decision making for autonomous driving. They use hand engineered driving behaviors for model fitting trajectories and detecting change point for most likely behavior prediction. In contrast, our proposed approach does not use predefined trajectories but instead uses RNN for intention prediction and an online polynomial model fitting algorithm for trajectory prediction which is motivated by [10].

3 System Overview

The goal of this paper is to provide real-time prediction of intentions and trajectories of surrounding vehicles such that we can accordingly plan path for the ego-vehicle. Given the individual trajectory of each vehicle i, from the sensor fusion and tracking module, for recent time steps $\tau_i = t_1, ..., t_n$, we use it as input to intention and trajectory prediction algorithms. The intention prediction is obtained using a RNN. Specifically, we propose to use a RNN comprising of a single long-short term memory (LSTM) unit to represent and predict the driving intention I_i of the target vehicle for one-time step in the future. The LSTM outputs a probability distribution over the set of intentions which in our case is lane keeping, right lane change and left lane change. The trajectory of the target vehicle is then predicted using a polynomial curve which is obtained using the vehicle's initial and expected final states. The expected final state is inferred based on the intention predicted by LSTM.

4 Technical Approach

In order to validate our proposed approach we train and test our system based on the data collected from the ego-vehicle. This allows us to have access to the ground truth information which is independent of any perception errors. Our proposed approach has much less dependency on the range of training data as we enhance it with the road contextual information.

4.1 Road Context Information

We obtained the road contextual information from OpenStreetMap [11] and the lane center information was later modified according to our autonomous vehicle dynamics. An overlay of the road geometry information illustrating the lane center and direction is presented in Fig. 2 (left) along with an instance of ego-vehicle's path while it is changing lanes (right). The lane change path indicated in red is referenced against the lane center marking represented in yellow.

4.2 Intention Prediction

We use urban driving data collected from our ego-vehicle for training and testing the intention prediction module of our system. The data comprises of vehicle's state and road contextual information which are the input features for the LSTM network. Specifically, these features include:

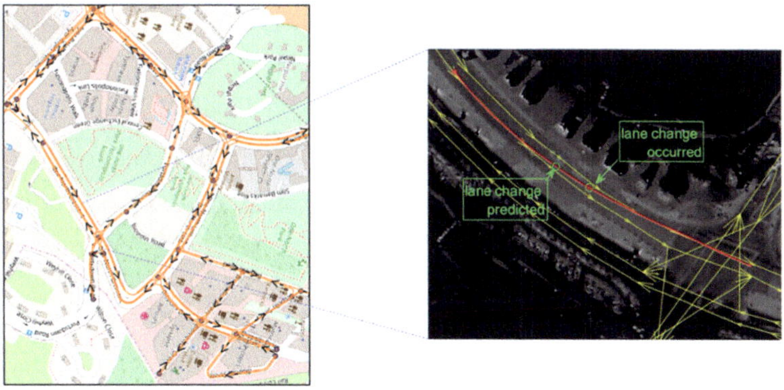

Fig. 2. The digital map overlaid with the lane center in yellow (left). A close-up of the digital map illustrating lane-change instance in red (right).

- δx: Change in lateral pose compared to previous time step
- δy: Change in longitudinal pose compared to previous time step
- ll: If left lane exists (Boolean 0/1)
- rl: If right lane exists (Boolean 0/1)
- d_{center}: Distance from the center of the lane. This value ranges from -1.5 m to 1.5 m for an average lane width of 3 m.

The LSTM network is trained on sample sequences of 1 second comprising of 4 samples. The network outputs predictions in real-time for 1 time step i.e., 0.25 s in future. The LSTM output is converted into a probability distribution over three classes: lane keeping, right lane change and left lane change.

4.3 Trajectory Prediction

Once the intention of the vehicle is known, the next critical step is to predict its trajectory in order to plan path for the ego-vehicle. The trajectory that a vehicle will take in a particular instant is stochastic in nature. First of all, different drivers might have different driving behaviors. For example, some drivers are more aggressive while others are more conservative. This affects how fast the lane change is performed and how the vehicle slows down when approaching an intersection. Second, even for the same driver, those behaviors might change depending on the situation of the road. However, given the road structure, intention of the driver and the past poses of the target vehicle, it is possible to reasonably predict its most likely trajectory.

Given the road contextual information, we know the adjacent lanes that the exo-vehicle can potentially transit to and thus we can generate possible hypotheses accordingly. For example, if the exo-vehicle travels on the left lane of a two lane road, it either continues on its current lane or makes a right lane change. In other words, left lane change is not possible and there are only two hypotheses (lane keeping and right lane change). For each hypothesis, we will model the lateral component by a 5th-order polynomial and longitudinal component by a 4th-order polynomial as suggested in [10]. The polynomial is solved by providing the initial and final states of the target vehicle. The lateral displacement $d(t)$ is of the form:

$$d(t) = c_5 t^5 + c_4 t^4 + c_3 t^3 + c_2 t^2 + c_1 t + c_0 \tag{1}$$

where $c_{i,i=\{0,1,2,3,4,5\}}$ are coefficients. Given a start time, $t_0 = 0$, a predefined end time $t_1 = 5$, initial and final states, the coefficients are solved using (2).

$$\begin{bmatrix} t_0^5 & t_0^4 & t_0^3 & t_0^2 & t_0 & 1 \\ t_1^5 & t_1^4 & t_1^3 & t_1^2 & t_1 & 1 \\ 5t_0^4 & 4t_0^3 & 3t_0^2 & 2t_0 & 1 & 0 \\ 5t_1^4 & 4t_1^3 & 3t_1^2 & 2t_1 & 1 & 0 \\ 20t_0^3 & 12t_0^2 & 6t_0 & 2 & 0 & 0 \\ 20t_1^3 & 12t_1^2 & 6t_1 & 2 & 0 & 0 \end{bmatrix} \cdot \begin{bmatrix} c_5 \\ c_4 \\ c_3 \\ c_2 \\ c_1 \\ c_0 \end{bmatrix} = \begin{bmatrix} d_0 \\ d_1 \\ \dot{d}_0 \\ \dot{d}_1 \\ \ddot{d}_0 \\ \ddot{d}_1 \end{bmatrix} \tag{2}$$

The initial state is used to determine d_0, $\dot{d}_0$ and $\ddot{d}_0$. The final displacement, d_1 is chosen depending upon the hypothesis whereas, vehicle's final lateral speed and acceleration $\dot{d}_1$ and $\ddot{d}_1$ are both zero. The longitudinal displacement $s(t)$ is of the form:

$$s(t) = c_4 t^4 + c_3 t^3 + c_2 t^2 + c_1 t + c_0 \tag{3}$$

Similar to (1), the coefficients c_i are solved using (4), given start time, $t_0 = 0$, a predefined end time $t_1 = 5$, initial and final states.

$$\begin{bmatrix} t_0^4 & t_0^3 & t_0^2 & t_0 & 1 \\ 4t_0^3 & 3t_0^2 & 2t_0 & 1 & 0 \\ 4t_1^3 & 3t_1^2 & 2t_1 & 1 & 0 \\ 12t_0^2 & 6t_0 & 2 & 0 & 0 \\ 12t_1^2 & 6t_1 & 2 & 0 & 0 \end{bmatrix} \cdot \begin{bmatrix} c_4 \\ c_3 \\ c_2 \\ c_1 \\ c_0 \end{bmatrix} = \begin{bmatrix} s_0 \\ \dot{s}_0 \\ \dot{s}_1 \\ \ddot{s}_0 \\ \ddot{s}_1 \end{bmatrix} \tag{4}$$

The initial state is used to determine s_0, $\dot{s}_0$ and $\ddot{s}_0$. The final displacement, s_1 is unknown and is not required in the equation. The vehicle's longitudinal speed and acceleration are chosen to be $\dot{s}_1 = \dot{s}_0 + \ddot{s}_0 * t$ and $\ddot{s}_1 = \ddot{s}_0$, respectively.

5 Experimental Results

We train and test our prediction algorithms using our self-driving car in real urban environment. Specifically, we use the circuit represented in Fig. 2 which is situated in One-North region of Singapore city. The training set was collected for ≈8 km and tested for 2.3 km over the same path. The perception, sensor fusion and tracking modules are used from our previous work [1]. In this section, we exclusively discuss the intention and trajectory prediction results for the ego-vehicle from the behavior analysis and prediction module of our system as illustrated in Fig. 1. The test on ego-vehicle was specifically chosen to validate our proposed prediction algorithms without any influence of the perception and ground-truth data labeling errors.

5.1 Intention Prediction

We compare intention prediction results from LSTM with our rule-based method from [1]. For the rule-based method, we considered the distance to lane center and lateral velocity of the vehicle to decide on the probability of lane change. It is important to note that in our previous work we did not distinguish between left and right lane changes. However, LSTM based intention prediction can uniquely classify the two lane changes, but we combined their probabilities for a fair comparison to the rule-based method. We compare and report the precision-recall values for the two methods in Table 1. The threshold values for the two methods were independently selected based on the ROC curve shown in Fig. 3.

Table 1. Comparison of rule-based method with our proposed LSTM based approach for intent prediction.

	Class	Precision	Recall
LSTM	Lane keep	0.97	0.80
	Lane change	0.34	0.80
Rule-based	Lane keep	0.93	0.75
	Lane change	0.24	0.57

It is evident from Table 1, that LSTM based method provides significantly better precision recall rate. In addition to these performance metrics, we also calculated the average predicted time to lane change (TTLC) for the test data. The TTLC is calculated as the time difference between when a prediction is made and when the vehicle actually crossed the dividing line between the two lanes. The average TTLC provided by rule-based method is 3.25 s and using LSTM it is 3.69 s (≈14% better).

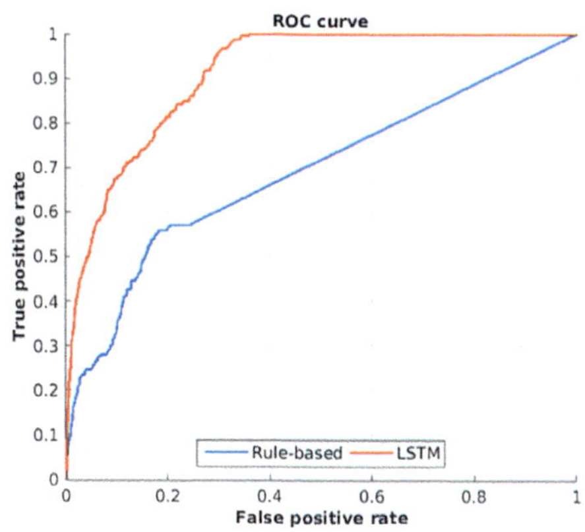

Fig. 3. ROC curve to compare rule-based method and LSTM.

The intention prediction of the ego-vehicle using LSTM for three classes (lane keeping, right lane change and left lane change) is presented in Fig. 4. The ground-truth information is represented in Fig. 4a which is obtained from the indicator values of the ego-vehicle. The corresponding predicted values from LSTM are illustrated in Fig. 4b.

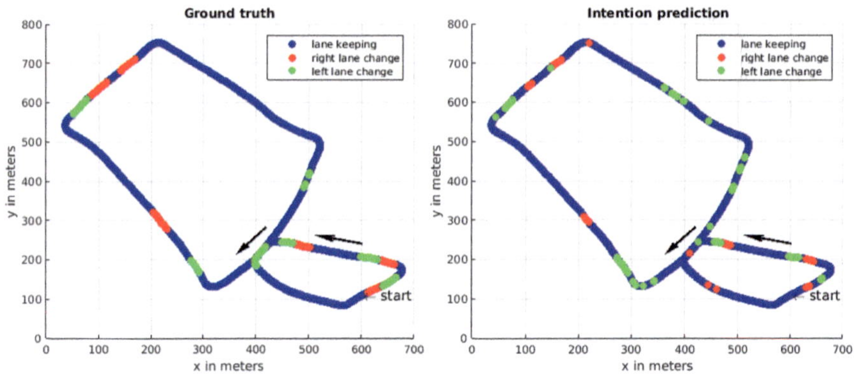

(a) Ground truth (b) Predicted intention

Fig. 4. Intention prediction of ego-vehicle using LSTM.

Note that there are some delays in prediction due to the fact that the indicator signals are given even before the intended maneuver is started. We also observed that slight off-centering of the vehicle from the lane center often triggers a lane change prediction, resulting in a false positive. This particular outcome can be tuned by varying the threshold for the probability of lane change based on the ROC from Fig. 3. Similar results were observed for exo-vehicle and an instance is presented at the following link: https://youtu.be/1GZMFLk5bk4.

5.2 Trajectory Prediction

An example of the ego-vehicle performing lane change on a two lane road is shown in Fig. 5. The road boundaries and the road separator are indicated using grey solid and dashed line, respectively. The positions marked as stars are the known vehicle positions in the last 5 sampling times. The goal is to predict the likely trajectories from the last sampling point. Given the road structure, we know that the vehicle could either keep lane or lane change. These two trajectories are indicated as dotted purple line and solid orange line for lane keeping and lane changing, respectively. The duration of prediction is 5 s into the future at sampling interval of 0.25 s. A time varying instance of the predicted vehicle trajectories for the two intents is presented at the following link: https://youtu.be/9-8kmSBtMEo.

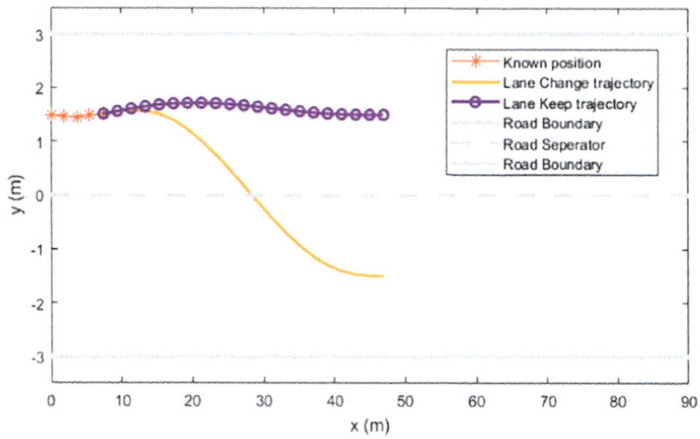

Fig. 5. Two hypothesis trajectories: lane keep and lane change for a vehicle traveling on a two lanes road.

Figure 6 shows the accuracy of the trajectory prediction algorithm for three different look-ahead times: 0.25 s, 1.0 s and 2.0 s. It is important to explain what it means by look-ahead time. If the look-ahead time is 1.0s, for example, we are trying to predict the vehicle's position at $t + 1.0$ s given only the vehicle's position until time t. Since there are two potential trajectories to choose from as shown in Fig. 5, we choose the trajectory based on the driver's intention given by the LSTM network.

For 0.25 s look-ahead, the trajectory prediction is perfectly in-line with the predicted path, indicating that the intention of the driver is inferred correctly for this short look-ahead time. The motion model used to generate the lane keeping and lane changing hypotheses, also managed to predict the position of the vehicle accurately, as vehicles generally maintain their momentum at this short interval.

Similarly for 1.0 s look-ahead, the LSTM network is able to accurately predict the instance of lane change intention of the driver. However, there are some position errors while predicting the trajectory as the motion model needs to predict longer into the future. Lastly, for 2.0 s look ahead, there is a noticeable error in trajectory prediction at the start of the lane change maneuver which is caused due to the delay in predicting the lane change intention. In addition to that, there are errors due to the motion model as there is more uncertainty now to predict position 2.0 s into the future.

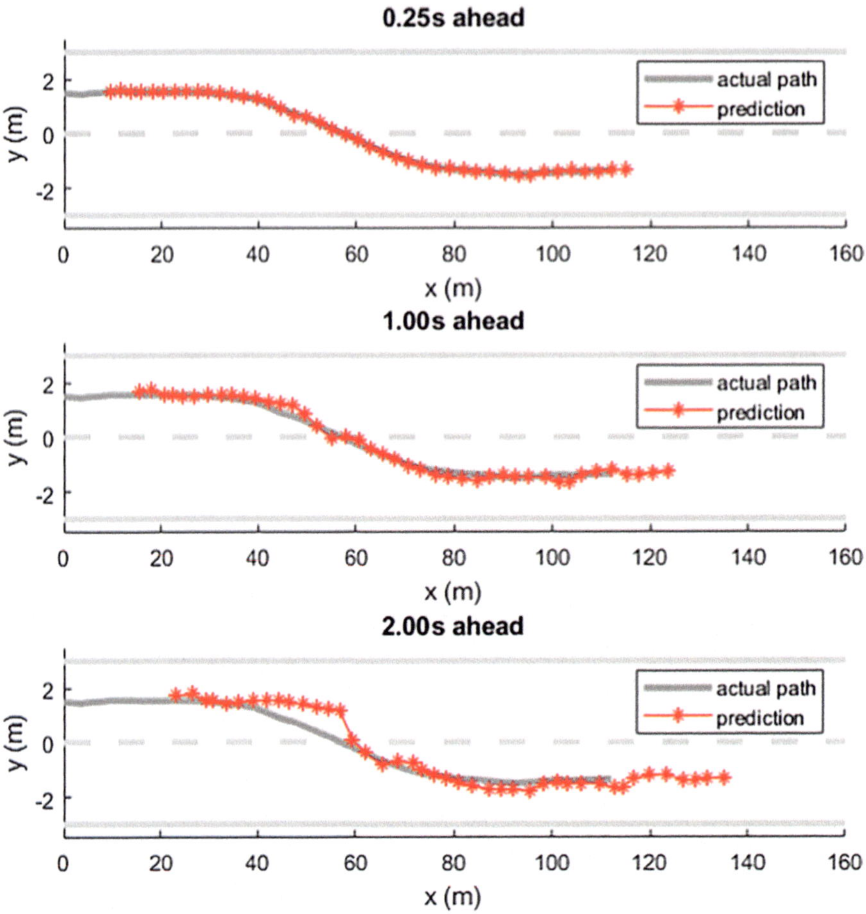

Fig. 6. Predicted trajectory for different look ahead times: 0.25 s, 1 s and 2 s. The thick grey line shows the actual path.

6 Conclusion

The results from our intention and trajectory prediction algorithms seem promising without requiring big datasets that are generally needed by the existing deep learning prediction methods. Instead, we use road contextual information which helps us to significantly improve our prediction outcome. We validated our proposed LSTM based intention prediction method which outperforms the rule-based method without explicitly accounting for anomalous scenarios. Our trajectory prediction method has proven to provide real-time results with very low error rates. Lastly, the integrated output of our intention and trajectory prediction algorithms is shown to be applicable for decision making and path planning by the autonomous vehicles in urban driving environments [12].

Acknowledgment. This research was supported by the National Research Foundation, Prime Minister's Office, Singapore, under its CREATE programme, Singapore-MIT Alliance for Research and Technology (SMART) Future Urban Mobility (FM) IRG. We would also like to acknowledge the support of NVIDIA Corporation through its NVAIL program and NUS-NVIDIA MoU.

References

1. Verma, S., Eng, Y.H., Kong, H.X., Anderson, H., Meghjani, M., Leong, W.K., Shen, X., Zhang, C., Ang, M.H., Rus, D.: Vehicle detection, tracking and behavior analysis in urban driving environments using road context. In: 2018 IEEE International Conference on Robotics and Automation (ICRA). IEEE (2018)
2. Paden, B., Čáp, M., Yong, S.Z., Yershov, D., Frazzoli, E.: A survey of motion planning and control techniques for self-driving urban vehicles. IEEE Trans. Intell. Veh. **1**(1), 33–55 (2016)
3. Patel, S., Griffin, B., Kusano, K., Corso, J.J.: Predicting future lane changes of other highway vehicles using RNN-based deep models. arXiv preprint arXiv:1801.04340 (2018)
4. Altché, F., De La Fortelle, A.: An LSTM network for highway trajectory prediction. In: 2017 IEEE 20th International Conference on Intelligent Transportation Systems (ITSC), pp. 353–359. IEEE (2017)
5. Ding, C., Wang, W., Wang, X., Baumann, M.: A neural network model for driver's lane-changing trajectory prediction in urban traffic flow. Math. Probl. Eng. **2013**, 8 (2013)
6. Li, X., Sun, Z., Cao, D., He, Z., Zhu, Q.: Real-time trajectory planning for autonomous urban driving: framework, algorithms, and verifications. IEEE/ASME Trans. Mechatron. **21**(2), 740–753 (2016)
7. Bandyopadhyay, T., Won, K.S., Frazzoli, E., Hsu, D., Lee, W.S., Rus, D.: Intention-aware motion planning. In: Algorithmic Foundations of Robotics X, pp. 475–491. Springer (2013)
8. Sezer, V., Bandyopadhyay, T., Rus, D., Frazzoli, E., Hsu, D.: Towards autonomous navigation of unsignalized intersections under uncertainty of human driver intent. In: 2015 IEEE/RSJ International Conference on Intelligent Robots and Systems (IROS), pp. 3578–3585. IEEE (2015)
9. Galceran, E., Cunningham, A.G., Eustice, R.M., Olson, E.: Multipolicy decision-making for autonomous driving via changepoint-based behavior prediction: theory and experiment. Auton. Robots **41**(6), 1367–1382 (2017)
10. Houenou, A., Bonnifait, P., Cherfaoui, V., Yao, W.: Vehicle trajectory prediction based on motion model and maneuver recognition. In: 2013 IEEE/RSJ International Conference on Intelligent Robots and Systems (IROS), pp. 4363–4369. IEEE (2013)
11. OpenStreetMap contributors. Planet dump. https://planet.osm.org, https://www.openstreetmap.org (2017)
12. Meghjani, M., Luo, Y., Ho, Q.H., Cai, P., Verma, S., Rus, D., Hsu, D.: Context and intention aware planning for urban driving. In: IEEE/RSJ International Conference on Intelligent Robots and Systems (IROS) (2019)

Robot Learning II

Robot Learning II

Robert Platt

This session contained six papers related to applications of deep learning to various robotics control problems. The first paper focused on interactive machine learning where the agent learns a control policy while a human supervisor provides corrective feedback. The paper developed a novel deep learning version of the COACH method known as Deep-COACH. The second paper developed an approach to using neural networks to learn a forward model that could be used to compensate for unmodelled robot dynamics. The method was tested in a robotic manipulation context. The third paper developed an approach to learning a controller for re-grasping based only on tactile feedback. The method is based on a variational auto encoder that used LSTM modules to handle sequence data. The fourth paper performed a comparison of a variety of approaches to modelling and control of a wheeled inverted pendulum. The fifth paper used deep learning to estimate desired offsets between hand, tool, and environment during manipulation interactions. The sixth explored approaches to modelling and policy learning for a 24-actuator tensegrity robot known as SUPERball. Discussions were followed on questions about when deep learning was most suited to robotics and when more traditional model based approaches might work better. The general consensus is that both model-based and model-free methods are needed but that it is presently unclear how to integrate these types of methods in general.

Interactive Learning with Corrective Feedback for Policies Based on Deep Neural Networks

Rodrigo Pérez-Dattari[1(✉)], Carlos Celemin[2], Javier Ruiz-del-Solar[3], and Jens Kober[2]

[1] Universidad de Chile, Av. Tupper, 2007 Santiago, Chile
rodrigo.perez.d@ing.uchile.cl
[2] Delft University of Technology, Mekelweg 2, Delft, Netherlands
{c.e.celeminpaez,j.kober}@tudelft.nl
[3] AMTC Center, Universidad de Chile, Av. Tupper, 2007 Santiago, Chile
jruizd@ing.uchile.cl

Abstract. Deep Reinforcement Learning (DRL) has become a powerful strategy to solve complex decision making problems based on Deep Neural Networks (DNNs). However, it is highly data demanding, so unfeasible in physical systems for most applications. In this work, we approach an alternative Interactive Machine Learning (IML) strategy for training DNN policies based on human corrective feedback, with a method called Deep COACH (D-COACH). This approach not only takes advantage of the knowledge and insights of human teachers as well as the power of DNNs, but also has no need of a reward function (which sometimes implies the need of external perception for computing rewards). We combine Deep Learning with the COrrective Advice Communicated by Humans (COACH) framework, in which non-expert humans shape policies by correcting the agent's actions during execution. The D-COACH framework has the potential to solve complex problems without much data or time required. Experimental results validated the efficiency of the framework in three different problems (two simulated, one with a real robot), with state spaces of low and high dimensions, showing the capacity to successfully learn policies for continuous action spaces like in the Car Racing and Cart-Pole problems faster than with DRL.

Keywords: Reinforcement Learning · Deep Learning · Interactive Machine Learning · Learning from Demonstration

1 Introduction

Deep Reinforcement Learning (DRL) has obtained unprecedented results in decision-making problems, such as playing Atari games [1], or beating the world

Electronic supplementary material The online version of this chapter (https://doi.org/10.1007/978-3-030-33950-0_31) contains supplementary material, which is available to authorized users.

J. Xiao et al. (Eds.): ISER 2018, SPAR 11, pp. 353–363, 2020.
https://doi.org/10.1007/978-3-030-33950-0_31

champion in GO [2]. Nevertheless, in robotic problems, DRL is still limited in applications with real-world systems [3]. Most of the tasks that have been successfully addressed with DRL have two common characteristics: (1) they have well-specified reward functions, and (2) they require large amounts of trials, which means long training periods (or powerful computers) to obtain a satisfying behavior. These two characteristics can be problematic in cases where (1) the goals of the tasks are poorly defined or hard to specify/model (reward function does not exist), (2) the execution of many trials is not feasible (real systems case) and/or not much computational power or time is available, and (3) sometimes additional external perception is necessary for computing the reward/cost function.

On the other hand, Machine Learning methods that rely on transfer of human knowledge, Interactive Machine Learning (IML) methods, have shown to be time efficient for obtaining good performance policies and may not require a well-specified reward function; moreover, some methods do not need expert human teachers for training high performance agents [4–6]. In previous years, IML techniques were limited to work with low-dimensional state spaces problems and to the use of function approximation such as linear models of basis functions (choosing a right basis function set was crucial for successful learning), in the same way as RL. But, As DRL has shown, by approximating policies with Deep Neural Networks (DNNs) it is possible to solve problems with high-dimensional state spaces, without the need of feature engineering for preprocessing the states. If the same approach is used in IML, the DRL shortcomings mentioned before can be addressed with the support of human users who participate in the learning process of the agent.

This work proposes to extend the use of human corrective feedback during task execution to learn policies with state spaces of low and high dimensionality in continuous action problems (which is the case for most of the problems in robotics) using deep neural networks.

We combine Deep Learning (DL) with the corrective advice based learning framework called COrrective Advice Communicated by Humans (COACH) [6], thus creating the Deep COACH (D-COACH) framework. In this approach, no reward functions are needed and the amount of learning episodes is significantly reduced in comparison to alternative approaches. D-COACH is validated in three different tasks, two in simulations and one in the real-world.

2 Related Work

This paper proposes a novel alternative to adapt policies, combining IML and DL. Specifically, we focus on techniques which transfer the teacher's knowledge based on occasional human feedback that may be either evaluative or corrective. Evaluative feedback has been used similarly to RL in methods wherein a human teacher communicates the desirability of the executed action or policy, with validations in problems of state spaces of either low dimensionality [4,5] or high dimensionality [7,8]. In contrast, corrective feedback is given by the teacher directly in the action's domain in order to modify the magnitude computed by the policy. To the best of our knowledge, corrective feedback has been only validated in problems with state spaces of low dimensionality [6,9].

3 Deep COACH

With COACH, a human teacher can advise a correction signal to the actions that the agent executes. If the agent executes an action a that the human considers to be erroneous, then s/he would indicate the direction in which the action should be corrected (increment or decrement); thus, COACH was proposed for problems with continuous actions. Each dimension of the action would have a corresponding correction signal h with values 0, -1 or 1 which produces an error signal with arbitrary magnitude e that is used to directly shape the policy in a supervised manner. Thus, $error = h \cdot e$, where $h = 0$ indicates that no correction has been advised. $h = \pm 1$ indicates the direction of the advised correction.

In this framework, we use two types of neural network architectures: feed forward fully-connected (FNN) for low-dimensional state problems, and convolutional neural networks (CNN) for high-dimensional state problems, e.g., raw image state spaces. In both cases the policies are updated every time feedback is received and also by sampling from a memory buffer B with a fixed frequency every b time steps. Every time the user advises a correction, the buffer B is fed with the current state and a label generated by adding the action taken with the error correction $y_{label} = a + error$. In the case of the CNN architecture, the convolutional layers are trained offline before the interactive process for learning a low-dimensional representation of the state. The state is embedded in the latent space of an autoencoder trained with a database of the agent exploring the environment. In Algorithm 1, the pseudocode of D-COACH is presented.

In the original COACH, it is proposed that each dimension should be trained independently [10], which has the advantage of creating a working framework that does not need any prior information about the problem in order to give corrections. We call this type of policy updating *decoupled* training, so a correction in an specific action dimension does not modify the magnitude of the actions in other axes for the same corresponding state. However, in this work we consider that for some problems it may be advantageous to exploit prior user knowledge about relations between the different dimensions of the actions. In this way, a correction in one of the action axes may be used to update more than one dimension. We call this case *coupled* training.

4 Experiments and Results

Our proposed algorithm is validated experimentally in three different problems: (i) Cart-Pole (continuous action), which is a simulated task with low-dimensional state space; (ii) Car Racing, a simulated task with high-dimensional state space (raw pixels of an image); and (iii) Duckie Racing, a task with a real robot featuring a high-dimensional state space (raw pixels of an image).

Algorithm 1. D-COACH

1: **Require:** error magnitude e, buffer update interval b, buffer sampling size N, buffer size K, pre-trained encoder parameters (if convolutional)
2: **Init:** $B = []$ # *initialize memory buffer*
3: **for** t = 1,2,... **do**
4:     **observe** state s_t
5:     **execute** action $a_t = \pi_t(s)$
6:     **feedback** human corrective advice h_t
7:     **if** h_t is not **0** **then**
8:         $error_t = h_t \cdot e$
9:         $y_{label(t)} = a_t + error_t$
10:         **update** $\pi(s)$ using SGD with pair $(s_t, y_{label(t)})$
11:         **update** $\pi(s)$ using SGD with a mini-batch sampled from B
12:         **append** $(s_t, y_{label(t)})$ to B
13:         **if** length(B) $> K$ **then**
14:             $B = B[2 : K + 1]$
15:     **if** mod(t, b) is 0 and B is not $\emptyset$ **then**
16:         **compute** $\pi(s)$ using SGD with a mini-batch sampled from B

The experiments with the simulated agents are intended to compare the complete D-COACH presented in Algorithm 1, along with a version of it without buffer (ignoring lines 2 and 11–16), and with a well known DRL agent (Deep Deterministic Policy Gradient DDPG [11] implemented by OpenAI [12]). The comparison is carried out by plotting the cumulative reward obtained at each episode by the agent as a function of time. In the case of D-COACH, the obtained reward is only used as a performance metric. Also, the results are presented as a function of time instead of episodes (except in the Duckie Racing experiment), because episodes can have variable duration depending on the policy. Hence, the episode scale would not properly show the time taken by the learning process, which is an important characteristic, since D-COACH is meant to work with real robots. The simulated environments, Cart-Pole and Car Racing, were ran at 22.5 and 20.5 FPS, respectively. These experiments were carried out using human teachers and simulated teachers. Humans had approximately 5 min to practice teaching in each environment. The learning curves of agents trained by 10 human teachers were obtained and averaged; the learning curves of agents trained by a simulated teacher were repeated 30 times and averaged. Along with the algebraic mean, the confidence intervals that represent the 60th percentile of the data were plotted. In the case of the Car Racing problem, it was observed that coupled training was advantageous when the teachers were humans. The designed coupled signals are shown in Table 1.

The hyper-parameters of the neural networks used in these experiments were tuned with preliminary experiments. Different combinations of them were tested by a human teacher and the ones that made the training easiest were selected (see Fig. 1). The D-COACH error magnitude constant e was set to **1** in this paper.

Table 1. Values of h in the Car Racing problem for human teachers. When feedback is given, the generated correction acts over more than one dimension of the action. For instance, the feedback signal *forward* means that the agent should simultaneously increase its acceleration and decrease its brake.

Feedback	h (direction, acceleration, brake)
Forward	(0, 1, −1)
Back	(0, −1, 1)
Left	(−1, −1, 0)
Right	(1, −1, 0)

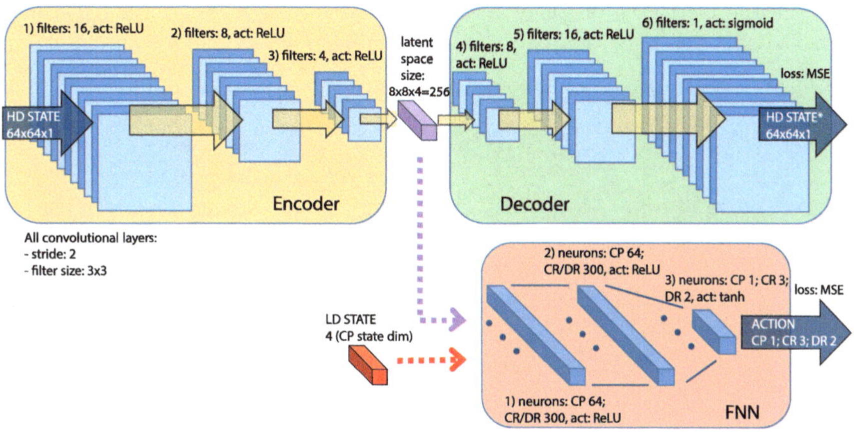

Fig. 1. D-COACH neural networks architecture. Variations between environments are specified with the acronyms CP (Cart-Pole), CR (Car Racing) and DR (Duckie Racing). HD STATE: high-dimensional state space. LD STATE: low-dimensional state space.

4.1 Validation of Replay Buffer with Simulated Teachers

The use of experience replay has been extensively validated in DRL; however, in this approach, we still consider it necessary to test its impact. Unlike DRL, where the policy is updated with information collected from every time step, in COACH-like methods there only is new data to update the policy when feedback is given by the teacher, so the amount of data used to update the policy may be lower than in the RL case. Since the original COACH has been widely validated with real human teachers in several tasks, we carried out most of the comparisons using a simulated teacher (a high performance policy standing-in as teacher, which was actually trained with D-COACH and a real human teacher) in this work, like in some of the experiments presented in [6], in order to compare the methods under more controlled conditions.

The simulated teacher generates feedback using $h = \text{sign}(a_{teacher} - a_{agent})$, whereas the decision of advising feedback at each time step is given by the probability $P_h = \alpha \cdot \exp(-\tau \cdot timestep)$, where $\{\alpha \in \mathbb{R}\ |0 \leq \alpha \leq 1\}$ and $\{\tau \in \mathbb{R}\ |0 \leq \tau\}$.

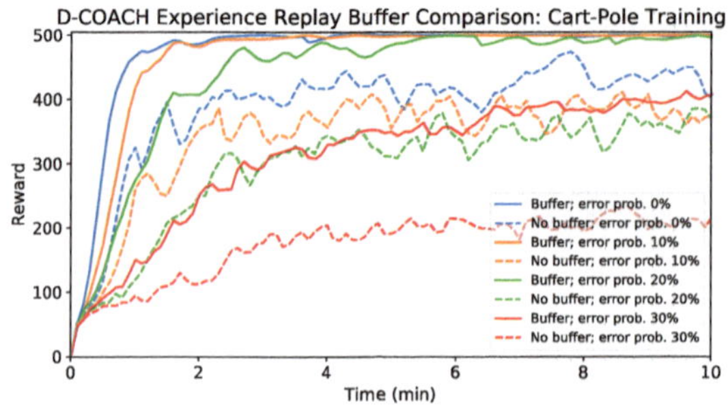

Fig. 2. Comparison between using or not experience replay buffer for different values of P_{err} in the Cart-Pole problem. Buffer: $K = 200$; $b = 10$; $N = 50$. P_h: $\alpha = 0.6$; $\tau = 0.0003$. Simulated teacher network learning rate: 0.0003.

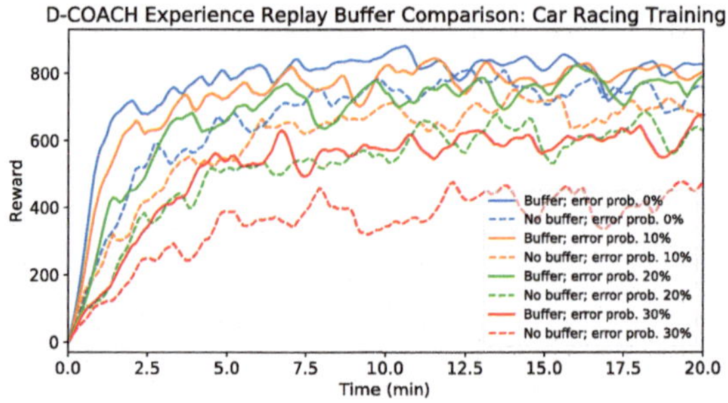

Fig. 3. Comparison between using or not experience replay buffer for different values of P_{err} in the CarRacing problem. Buffer: $K = 1000$; $b = 10$; $N = 100$. P_h: $\alpha = 0.6$; $\tau = 0.000015$. Simulated teacher network learning rate: 0.0003.

Additionally, since human teachers occasionally advise wrong corrections, a probability of giving erroneous feedback P_{err} is added to the model. The variable P_{err} indicates the probability that at least one dimension of h is multiplied by -1 when feedback is given.

A comparison of D-COACH with and without the use of an experience replay buffer is carried out by means of the simulated teacher. To test the behavior of these scenarios when erroneous feedback is added, different values of P_{err} are selected. These results can be seen in Figs. 2 and 3 (for better readability, no confidence intervals were added).

In Figs. 2 and 3 the learning curves show a large difference between the processes of learning that use experience replay buffer with respect to the cases without the buffer. In the case without the buffer, which is more similar to the original COACH, it is possible to see that the learning agent is not benefiting from the advised corrections as much as it can do when the pieces of advice are kept in the memory. For instance, we can see that D-COACH learns more from corrections with 20% of mistakes when using the buffer than in the case of perfect corrections, but without any buffering. This means the buffer is necessary

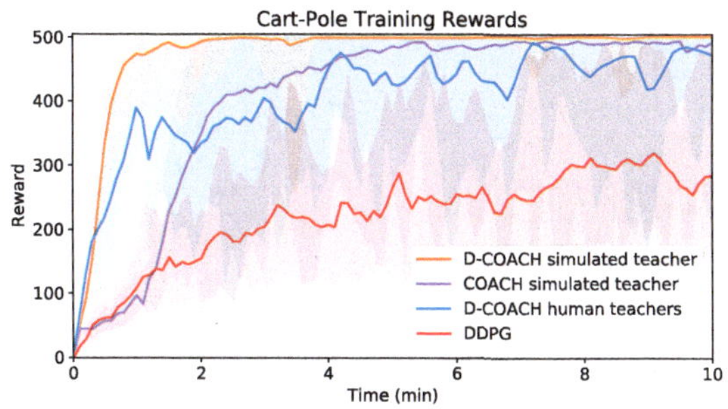

Fig. 4. Cart-Pole training. Buffer: $K = 200$; $b = 10$; $N = 50$. P_h: $\alpha = 0.6$; $\tau = 0.0003$. Human teacher network learning rate: 0.003; Simulated teacher network learning rate: 0.0003.

for increasing the use of the information available, even when this information is corrupted and not clean.

4.2 Comparison of DRL and D-COACH Using Real Human Teachers

These experiments are intended to compare the learning process of D-COACH (simulated teacher and human teacher) with the DRL algorithm DDPG. Taking into account that the Cart-Pole problem has a low dimensional state space, the original COACH, based on basis functions, is also included in the comparison. In this case, $P_{err} = 0\%$ was used for the simulated teachers. The results of this problem are shown in Fig. 4, wherein it is possible to see that COACH-like methods outperform the DRL agent with a large difference. When using the simulated teacher, D-COACH learns faster than the original COACH. The performance of D-COACH with human teachers decreases with respect to the simulated teacher. This is because human teachers are not perfect and make mistakes, but they are being compared with a simulated teacher with $P_{err} = 0\%$, which means that it makes no mistakes. Also because the simulated teacher model is quite simple to represent the complexity of the human behavior, then, although it is not very realistic, it is still useful for comparisons of interactive learning strategies under similar conditions.

In Fig. 5 the learning curves of the Car Racing problem are presented. Again, D-COACH results in a fast convergence. Unlike reported results of DRL algorithms for this problem, in the very early minutes D-COACH reaches high performance policies that have not been obtained by most of the DRL approaches, to the best of our knowledge. If we compare a policy trained with D-COACH for approximately 75 min by an experienced teacher against several state-of-the-art DRL approaches, it can be seen that it outperforms most of them (see Table 2). The problem is considered to be solved if the agent gets an average score of 900 or more over 100 random tracks. However, we observed that this value can substantially vary between

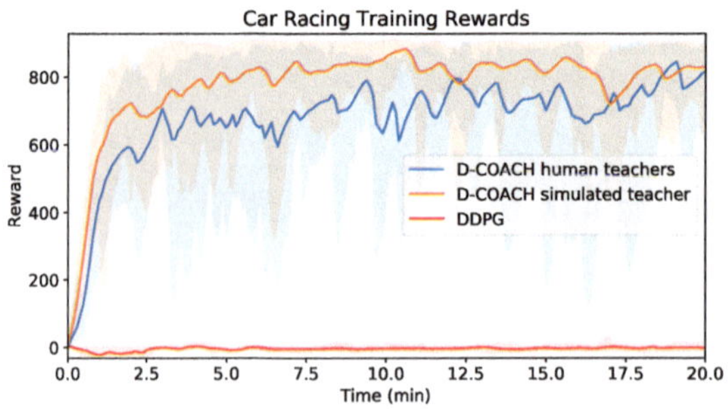

Fig. 5. Racing Car training. Buffer: $K = 1000$; $b = 10$; $N = 100$. P_h: $\alpha = 0.6$; $\tau = 0.000015$. Human teacher network learning rate: 0.001; Simulated teacher network learning rate: 0.0003.

Table 2. Car Racing state-of-the-art learning algorithms comparison. DRL results taken from [13].

Method	Average score over 100 random tracks
DQN	343 ± 18
A3C (continuous)	591 ± 45
A3C (discrete)	652 ± 10
Ceobillionaire's algorithm (unpublished)	838 ± 11
Full World Model	906 ± 21
D-COACH (experienced teacher)	**895–909 ± 18–80**
	Average over 20 evaluations: 903 ± 46

different evaluations, so in Table 2, the obtained range of values over 20 evaluations is presented for D-COACH.

4.3 Validation in a Real System

In the third problem that we called Duckie Racing, an agent has to learn to drive a Duckiebot (from the project Duckietown [14] with modifications from the Chile Duckietown Team[1]) autonomously through a track based on raw visual information of an onboard camera. The actions in this problem are the forward velocity and the steering angle of the Duckiebot. Two tasks are set for this environment: (i) driving the Duckiebot freely through the track, with permission to drive in both lanes, and (ii) driving the Duckiebot only in the right lane, which demands more accuracy in driving. In this problem, an episode stops if the robot leaves the track/right lane, or after 30 s. The performance index in this task is the percentage of the total track length traveled during the episode. Hence the faster and more accurate the Duckiebot drives, the more distance it will travel.

[1] https://github.com/Duckietown-Chile/.

This problem is not used for comparisons of the methods, but only as a validation of D-COACH using experience replay, which showed to be the best alternative in the previous problems. Figure 6 shows the learning curve for each of the tasks explored in this environment with a real robot and a real human teacher. The curves and the video[2] attached to this paper show that the system quickly learns to drive properly through the road based only on the human corrections. As expected, the policy is faster when the robot has the freedom to drive over both lanes. Learning this task with RL would definitely take more training time, and might need an external perception system to compute the reward function, whereas with D-COACH this performance index does not have any influence on the learning process, rather it is used for descriptive and comparative purposes. Figure 7 shows a comparison between an observation and its reconstruction generated by the autoencoder.

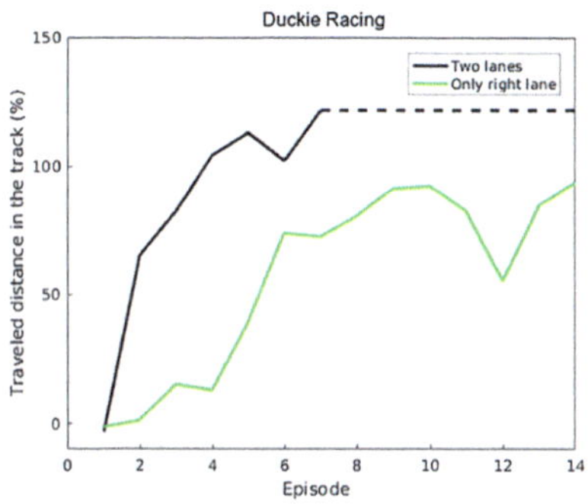

Fig. 6. Duckie Racing training.

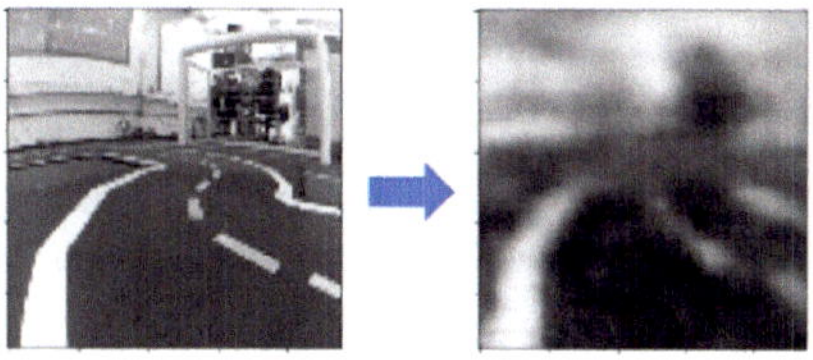

Fig. 7. Duckie Racing autoencoder input (left) vs output (right).

5 Conclusions

This work presented D-COACH, an algorithm for training policies modeled with DNNs interactively with corrective advice. The method was validated in a problem of low-dimensionality, along with problems of high-dimensional state spaces like raw pixel observations, with a simulated and a real robot environment, and also using both simulated and real human teachers.

The use of the experience replay buffer (which has been well tested for DRL) was re-validated for this different kind of learning approach, since this is a feature not included in the original COACH. The comparisons showed that the use of memory resulted in an important boost in the learning speed of the agents, which were able to converge with less feedback, and to perform better even in cases with a significant amount of erroneous signals.

[2] https://youtu.be/vcEtuRrRIe4.

The results of the experiments show that teachers advising corrections can train policies in fewer time steps than a DRL method like DDPG. So it was possible to train real robot tasks based on human corrections during the task execution, in an environment with a raw pixel level state space. The comparison of D-COACH with respect to DDPG, shows how this interactive method makes it more feasible to learn policies represented with DNNs, within the constraints of physical systems. DDPG needs to accumulate millions of time steps of experience in order to obtain good performances as shown in [11]. However, this is not always possible with real systems.

Acknowledgements. This work was partially funded by FONDECYT Project 1161500 and CONICYT/PIA Project AFB180004. A portion of it has taken place in the University of Chile Duckietown's headquarters, *FabLab U. de Chile* (http://www.fablab.uchile.cl/). Special thanks to Matias Mattamala, who provided the necessary tools to do the tests with the Duckiebots.

References

1. Mnih, V., Kavukcuoglu, K., Silver, D., Graves, A., Antonoglou, I., Wierstra, D., Riedmiller, M.: Playing Atari with deep reinforcement learning. In: NIPS Deep Learning Workshop (2013)
2. Silver, D., Huang, A., Maddison, C.J., Guez, A., Sifre, L., van den Driessche, G., Schrittwieser, J., Antonoglou, I., Panneershelvam, V., Lanctot, M., Dieleman, S., Grewe, D., Nham, J., Kalchbrenner, N., Sutskever, I., Lillicrap, T., Leach, M., Kavukcuoglu, K., Graepel, T., Hassabis, D.: Mastering the game of go with deep neural networks and tree search. Nature **529**, 484–489 (2016)
3. Gu, S., Holly, E., Lillicrap, T., Levine, S.: Deep reinforcement learning for robotic manipulation with asynchronous off-policy updates. In: IEEE International Conference on Robotics and Automation (ICRA) (2017)
4. Akrour, R., Schoenauer, M., Sebag, M.: Preference-based policy learning. In: Joint European Conference on Machine Learning and Knowledge Discovery in Databases (2011)
5. Knox, W.B., Stone, P.: Interactively shaping agents via human reinforcement: the TAMER framework. In: Fifth International Conference on Knowledge Capture (2009)
6. Celemin, C., Ruiz-del Solar, J.: An interactive framework for learning continuous actions policies based on corrective feedback. J. Intell. Robotic Syst. **95**, 1–20 (2018)
7. Christiano, P., Leike, J., Brown, T.B., Martic, M., Legg, S., Amodei, D.: Deep reinforcement learning from human preferences (2017). arXiv preprint arXiv:1706.03741
8. Warnell, G., Waytowich, N., Lawhern, V., Stone, P.: Deep TAMER: Interactive agent shaping in high-dimensional state spaces (2017). arXiv preprint arXiv:1709.10163
9. Argall, B.D., Browning, B., Veloso, M.: Learning robot motion control with demonstration and advice-operators. In: IEEE/RSJ International Conference on Intelligent Robots and Systems (IROS) (2008)

10. Celemin, C., Ruiz-Del-Solar, J.: Teaching agents with corrective human feedback for challenging problems. In: IEEE Latin American Conference on Computational Intelligence (LA-CCI) (2017)
11. Lillicrap, T.P., Hunt, J.J., Pritzel, A., Heess, N., Erez, T., Tassa, Y., Silver, D., Wierstra, D.: Continuous control with deep reinforcement learning. In: International Conference on Learning Representations (ICLR) (2016)
12. Dhariwal, P., Hesse, C., Klimov, O., Nichol, A., Plappert, M., Radford, A., Schulman, J., Sidor, S., Wu, Y.: OpenAI baselines (2017). https://github.com/openai/baselines
13. Ha, D., Schmidhuber, J.: World models (2018). arXiv preprint arXiv:1803.10122
14. Paull, L., Tani, J., Ahn, H., Alonso-Mora, J., Carlone, L., Cap, M., Chen, Y.F., Choi, C., Dusek, J., Fang, Y., Hoehener, D., Liu, S.Y., Novitzky, M., Okuyama, I.F., Pazis, J., Rosman, G., Varricchio, V., Wang, H.C., Yershov, D., Zhao, H., Benjamin, M., Carr, C., Zuber, M., Karaman, S., Frazzoli, E., Del Vecchio, D., Rus, D., How, J., Leonard, J., Censi, A.: Duckietown: an open, inexpensive and flexible platform for autonomy education and research. In: IEEE International Conference on Robotics and Automation (ICRA) (2017)

Force-Torque Sensor Disturbance Observer Using Deep Learning

Kamal Mohy el Dine(✉), Jose Sanchez, Juan Antonio Corrales, Youcef Mezouar, and Jean-Christophe Fauroux

Université Clermont Auvergne, SIGMA Clermont, Institut Pascal, BP 10448, 63000 Clermont-Ferrand, France
kamal.mohy_el_dine@sigma-clermont.fr

Abstract. Robots executing force controlled tasks require accurate perception of the applied force in order to guarantee precision. However, dynamic motions generate non-contact forces due to the inertia. These non-contact forces can be regarded as disturbances to be removed such that only the forces generated by contacts with the environment remain. This paper presents an observer based on a recurrent neural network that estimates the non-contact forces measured by a force-torque sensor attached at the end-effector of a robotic arm. The approach is proven to also work with an external load attached to the robotic arm. The recurrent neural network observer uses signals from the joint encoders of the robotic arm and a low-cost inertial measurement unit to estimate the wrenches (i.e. forces and torques) generated due to gravity, inertia, centrifugal and Coriolis forces. The accuracy of the proposed observer is experimentally evaluated by comparing the measurements of the attached force-torque sensor to the observer's non-contact forces estimation. Additionally, the pure contact force estimation is evaluated against an external force-torque sensor.

1 Introduction

In order for robots to perform tasks involving physical interactions in unknown and dynamic environments, such as physical human-robot collaboration, haptic control (e.g. bilateral teleoperation) and locomotion; it is necessary that they accurately and timely perceive contacts with the environment. Thus, estimation of external contact forces is required to guarantee the success of robotic tasks where contacts are present. In order to estimate contact forces, force-torque sensors are usually attached to the robot's end-effector to measure wrenches generated while interacting with the environment. However, as the force-torque sensor

K.M. el Dine and J. Sanchez—These authors contributed equally to the work.

Electronic supplementary material The online version of this chapter (https://doi.org/10.1007/978-3-030-33950-0_32) contains supplementary material, which is available to authorized users.

J. Xiao et al. (Eds.): ISER 2018, SPAR 11, pp. 364–374, 2020.
https://doi.org/10.1007/978-3-030-33950-0_32

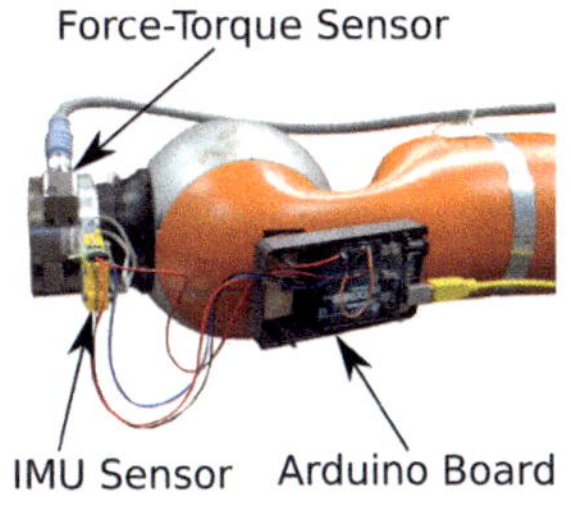

Experimental setup showing the force-torque sensor and the IMU mounted on a KUKA robot.

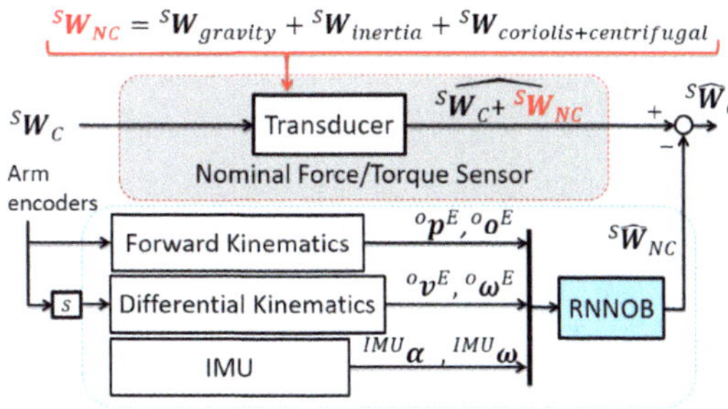

Recurrent neural network observer to estimate the non-contact forces (o: robot frame, E: end-effector frame, S: sensor frame, IMU: IMU frame).

Fig. 1. Experimental setup and our proposed force observer.

measures both internal and external forces, it is necessary to first estimate the internal (*non-contact*) forces caused by gravity, inertia, Coriolis and centrifugal forces. Once these non-contact forces have been estimated, they can be then subtracted from the force-torque sensor output to obtain the pure external (*contact*) forces, as shown in the block diagram in Fig. 1b.

Several works have focused on estimating non-contact forces on a force-torque sensor by incorporating acceleration signals. For instance, García et al. used an observer based on a state-space system that included the dynamics of the robot [1]. Equations based on the inertia matrix of a known load attached to the force-torque sensor were used in [2] to estimate non-contact forces. Similarly, Kubus and Wahl estimated the inertial parameters of the attached load by identification, then the corresponding internal forces were calculated using the Newton-Euler formulation [3].

Instead of requiring accurate estimation of parameters such as the inertia matrix, mass and the center of mass position, other approaches have relied on machine learning methods to directly estimate forces. One of the first works in robotics to use neural networks for estimating forces, was described in [4], where a feedforward neural network approximated two-dimensional forces based on the robot's joint positions, velocities and accelerations. More recently, Kollmitz et al. estimated a six-dimensional contact force on a robot platform using a time-delay neural network [5], where the network delayed the inputs (wrench and acceleration measurements) in order to include temporal information. Another network architecture that makes use of sequential data are recurrent neural networks (RNNs). This type of networks has been successfully applied to estimate forces in robotic tasks. For instance, a force distribution map on a person's limb generated by contacts with a hospital gown was estimated in [6] and contact transients during a snap-fit assembly task were detected in [7]. Based on the recent success in applying RNNs to estimate forces in robotic tasks, we propose to train a recurrent neural network using the robot's proprioceptive information and a low-cost accelerometer to estimate directly the non-contact forces.

2 Technical Approach

The wrench output of the force-torque sensor can be expressed as:

$$\boldsymbol{W} = \boldsymbol{W}_{NC} + \boldsymbol{W}_C = \left[\boldsymbol{F}, \boldsymbol{\Gamma}\right]^T \tag{1}$$

where $\boldsymbol{W}_{NC} = [\boldsymbol{W}_{gravity}, \boldsymbol{W}_{inertia}, \boldsymbol{W}_{coriolis+centrifugal}]^T$ is the disturbance wrench affecting the sensor due to the non-contact forces and torques $\boldsymbol{f}_{NC}$ and $\boldsymbol{\tau}_{NC}$ respectively. $\boldsymbol{W}_C$ is the pure contact wrench due to contact forces and torques $\boldsymbol{f}_C$ and $\boldsymbol{\tau}_C$ respectively. $\boldsymbol{F}$ and $\boldsymbol{\Gamma}$ are the force and torque values expressed in the sensor frame S respectively as:

$$\boldsymbol{F} = \boldsymbol{f}_{NC} + \boldsymbol{f}_C \tag{2}$$

$$\boldsymbol{\Gamma} = \boldsymbol{\tau}_{NC} + \boldsymbol{\tau}_C \tag{3}$$

Using the Newton-Euler approach, $\boldsymbol{f}_{NC}$ and $\boldsymbol{\tau}_{NC}$ can be expanded and Eqs. (2) and (3) can be rewritten as:

$$\boldsymbol{F} = m\boldsymbol{\alpha} - m\boldsymbol{g} + \dot{\boldsymbol{\omega}} \times m\boldsymbol{c} + \boldsymbol{\omega} \times (\boldsymbol{\omega} \times m\boldsymbol{c}) + \boldsymbol{f}_C \tag{4}$$

$$\boldsymbol{\Gamma} = \boldsymbol{I}\dot{\boldsymbol{\omega}} + \boldsymbol{\omega} \times (\boldsymbol{I}\boldsymbol{\omega}) + m\boldsymbol{c} \times \boldsymbol{\alpha} - m\boldsymbol{c} \times \boldsymbol{g} + \boldsymbol{\tau}_C \tag{5}$$

where $\boldsymbol{\omega}$ is the angular velocity vector of the sensor with respect to its frame, $\boldsymbol{\alpha}$ and $\dot{\boldsymbol{\omega}}$ are the linear and angular acceleration vectors respectively, $\boldsymbol{g}$ is the vector corresponding for gravity, m is the mass of the load, $\boldsymbol{c}$ is its center of mass coordinates vector and $\boldsymbol{I}$ is a 3×3 symmetric matrix representing the inertia matrix in the sensor frame.

In standard control approaches, when accuracy is not critical, the sensor measurements in Eqs. (2) and (3) are used in their default form taking both contact and noncontact forces as the feedback signal to the controller. However, when accurate force control is required, non-contact forces need to be removed as they can lead to wrong reference force values fed back to the controller. In order to estimate these forces precisely, the ten inertial parameters of the load should be known, namely: m, $\boldsymbol{c}$ and the values of $\boldsymbol{I}$. In the literature, researchers usually use identification methods to obtain these values and apply Eqs. (4) and (5) to calculate the non-contact forces and torques. However, the accuracy of estimating non-contact forces and torques based on identification is dependent on the accuracy of the center of mass position of the load $\boldsymbol{c}$ and the calculation of the kinematic vectors $\boldsymbol{\alpha}$, $\boldsymbol{\omega}$ and $\dot{\boldsymbol{\omega}}$ in the same frame.

To overcome these inaccuracies, we propose an observer based on a recurrent neural network (RNNOB) to estimate directly non-contact forces independently of the twist and acceleration transformations. Since the involved signals are sequential, an RNN architecture using Long Short-term Memory (LSTM) units, as described in [8], is ideal to correlate the sensor's kinematics to its wrench output. Figure 1 shows the sensors mounted on the arm along with block diagram of the RNNOB showing how the non-contact forces are estimated and then canceled from the force-torque sensor measurements.

3 Experiments: Data Collection and Testing

As shown in Eqs. (4) and (5), the non-contact forces are directly related to angular velocities and accelerations, linear accelerations and the sensor orientation (due to gravity effects). Thus, in order to train the observer it is necessary to generate a broad spectrum of states for these variables. To this end, two datasets were collected using the joint encoders of a KUKA LWR-4 arm with the ATI Gamma[1] force-torque sensor and an Adafruit (L3GD20H + LSM303)[2] inertial measurement unit (IMU) mounted on its wrist. The first dataset was collected without an external load (see Fig. 1), while the second had the Shadow Dexterous Hand[3] attached as an external load, as shown in Fig. 2. The measurement and average rates for the sensors used, as depicted in Fig. 1, are described below:

1. **ATI Gamma:** produces a six-dimensional wrench expressed in the sensor's frame S at 1,000 Hz.
2. **Adafruit (L3GD20H + LSM303):** generates linear accelerations and angular velocities expressed in the IMU frame at 300 Hz.
3. **Joint encoders:** Provide, through forward and differential kinematics, the end-effector orientation (in quaternion representation) plus linear and angular velocities expressed in the robot frame Σ_O at 500 Hz.

Since the sensors operate at different rates, the data was recorded at 500 Hz to have a uniform sampling rate. Thus, the force-torque sensor is effectively downsampled and the last output of the IMU is kept until a new sample is published.

3.1 Data Collection Without Load

For this dataset, the data collected was generated by *manual* and *automatic* trajectories. The manual data was collected by setting the robot controller to gravity compensation mode and then moving the wrist manually to various poses in the workspace with random velocities and accelerations. The automatic data (e.g. without human intervention) was generated by moving the robot between random points in its workspace using various trapezoidal velocity profiles. The manual data was collected in six trials, each about four minutes long and the automatic data was collected in ten trials with an average time of two minutes each. These datasets were then combined into one training dataset, where one trial of each of the manual and automatic data were separated to create a test dataset. The validation of the RNN during training used 20% of the training dataset.

[1] http://www.ati-ia.com/products/ft/ft_models.aspx?id=Gamma.
[2] https://www.adafruit.com/product/1714.
[3] https://www.shadowrobot.com/products/dexterous-hand/.

3.2 Data Collection with External Load

Besides the manual and automatic trajectories, described in Subsect. 3.1, an additional *sinusoidal* trajectory on the xy-plane of the robot (see Fig. 2) frame was generated for this dataset. During this sinusoidal trajectory the end-effector was rotated around each axis sequentially (i.e. roll, pitch and then yaw).

The manual data was collected in five trials, each about five minutes long; the automatic data was collected in ten trials with an average time of two minutes each, and the sinusoidal data was collected in one trial about three minutes long. These datasets were then combined into one training dataset, where one trial of each of the manual and automatic data were separated to create a test dataset. The sinusoidal data was used entirely for training. The validation of the RNN during training used 20% of the training dataset. Additionally to the testing set described above, a pure *rotational* motion test was used to validate the proposed observer against gravitational forces. This rotational test can be seen in Fig. 7, where the angular rotations are shown at the bottom of the figure.

3.3 Collision Test

To show the accuracy of the proposed approach in estimating pure contact forces, an ATI Mini45[4] force-torque sensor was used as reference. The arm was set to gravity compensation mode to move the external load and collide it ten times with the reference force-torque sensor fixed on the table as shown in Fig. 2.

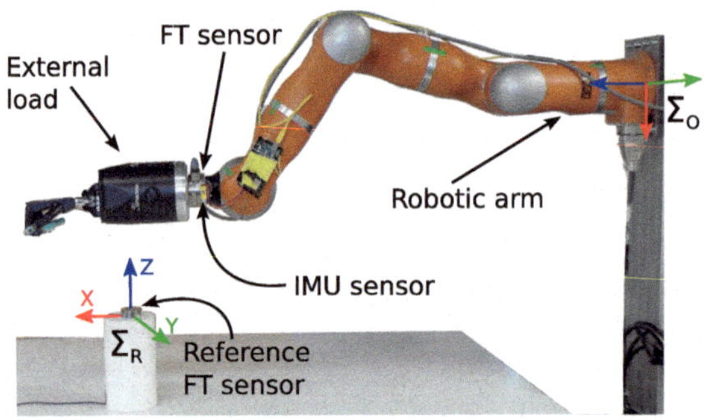

Fig. 2. Collision test setup.

4 Results

4.1 RNN Model

To validate our RNN observer, we trained different models using the position (p), orientation (o) and the twist (v, ω) derived from the joint encoders[5]; and

[4] https://www.ati-ia.com/products/ft/ft_models.aspx?id=Mini45.

[5] Note that for simplicity, when the subscripts are not indicated the quantities are assumed to be of the end-effector expressed on the robot frame.

the linear acceleration obtained from the IMU (${}^{IMU}\alpha$). Besides different input features, models with different hyperparameters such as the number of epochs, sequence length and learning rate, were also tested. However, the difference in performance between these networks was not statistically noticeable. Thus, the three different models based on the *pose* (p, o), *orientation and twist* (o, v, ω), and *linear acceleration, orientation and twist* $({}^{IMU}a, o, v, \omega)$ were evaluated. All these models were trained for 20 epochs using an architecture with two hidden layers, with 15 and 10 LSTM units respectively, and a sequence length for the input layer of 20 time steps (0.2 s). In the output layer, we applied Stochastic Gradient Descent with a learning rate of 0.01 to minimize the mean square error of the regression problem. A hyperbolic tangent sigmoid function was used as the activation function between the hidden layers and for the output layer a linear activation function was used.

Figure 3 shows the performance achieved by the models based on different input features on the dataset with an external load as described in Sect. 3.2. It can be clearly seen that the model incorporating the measurements from the IMU $({}^{IMU}a, o, v, \omega)$ achieves the best overall performance. However, for the rotation test, as the accelerations are small and only the orientation changes, this model is slightly outperformed by the one using only pose information (p, o).

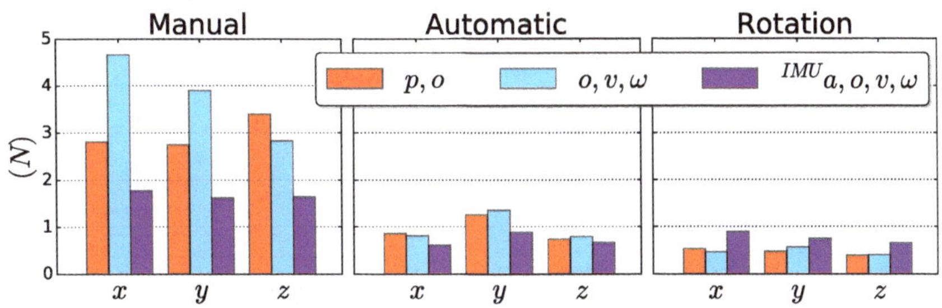

(a) Linear errors for the three types of test motions.

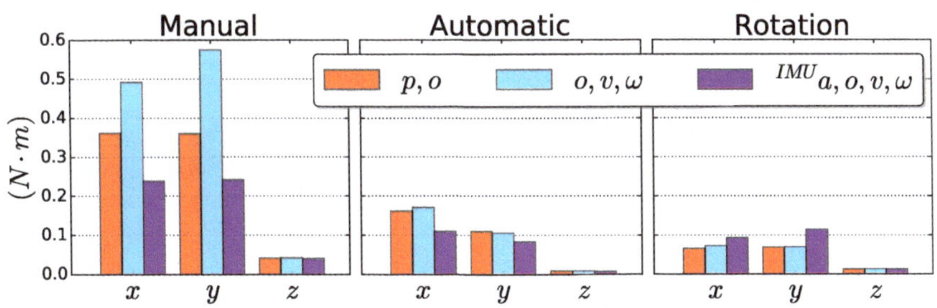

(b) Angular errors for the three types of test motions.

Fig. 3. RMS errors for the RNN models based on: (1) pose (p, o), (2) orientation and twist (o, v, ω) and (3) linear acceleration, orientation and twist $({}^{IMU}a, o, v, \omega)$.

4.2 Force Estimation Without Load

The root mean square (RMS) errors on the test datasets of the automatic and manual motions without load, as described in Sect. 3.1, are summarized in Table 1. This table shows the force estimation of the RNN observer for previously unseen automatic and manual trajectories compared to the measurement of the force-torque sensor. Since no load was attached to the sensor, except for the top plate which has a negligible mass, the generated wrenches remained very low. The force values were under $\pm 1.2\,\mathrm{N}$ while the torques did not exceed $\pm 0.6\,\mathrm{N \cdot m}$ (see Fig. 4).

Table 1. RMS errors on the automatic and manual test datasets without an external load.

Automatic		N	$N \cdot m$
	X	0.0432	0.0010
	Y	0.0380	0.0007
	Z	0.0268	0.0009

Manual		N	$N \cdot m$
	X	0.1943	0.0022
	Y	0.1814	0.0022
	Z	0.1100	0.0012

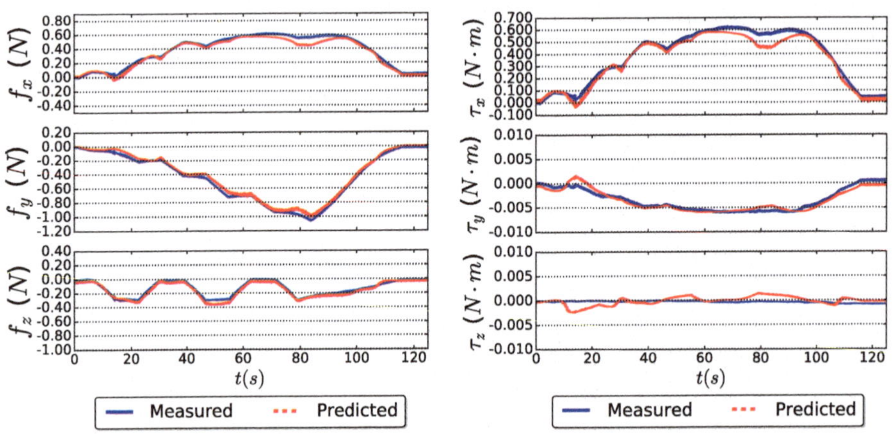

Fig. 4. Non-contact forces (${}^{S}\widehat{\boldsymbol{W}}_{NC}$) estimated by the RNNOB for an unseen automatic trajectory with no load attached to the force-torque sensor. The y-axis shows the forces and torques as measured in the force-torque sensor frame.

4.3 Force Estimation with External Load

To compare the performance of the proposed observer, we implemented an *analytical* observer using the method described in [3]. As this method requires knowledge of the inertial parameters, we estimated these parameters using least squares on a trial of the manual dataset (described in Sect. 3.2). Both methods were compared against the wrenches measured by the force-torque sensor

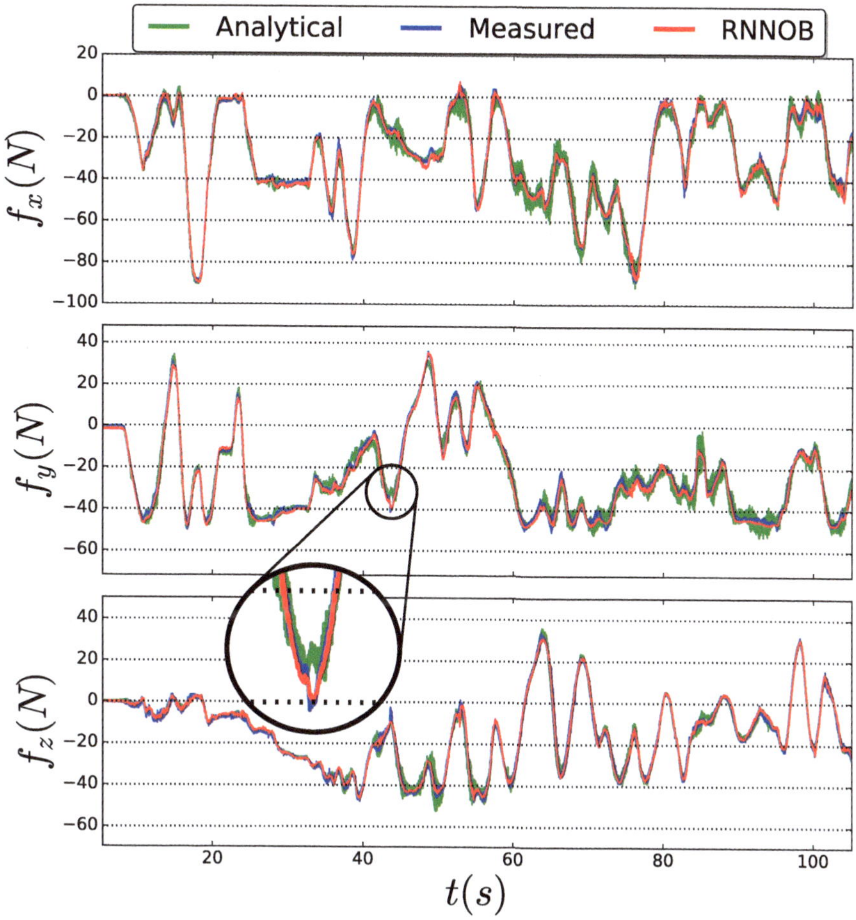

Fig. 5. Non-contact wrench estimation of the proposed RNNOB compared with an analytical-based approach and as measured by the force-torque sensor for an unseen manual trajectory with an external load. The y-axis shows the forces along x, y and z as measured in the force-torque sensor frame respectively.

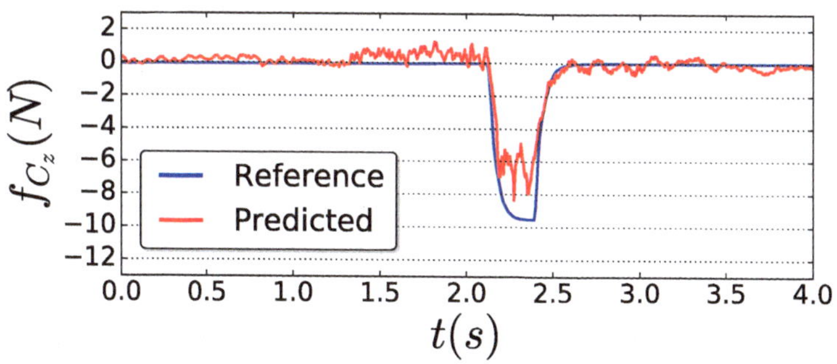

Fig. 6. Contact force estimation of the proposed approach compared to the reference force measurement.

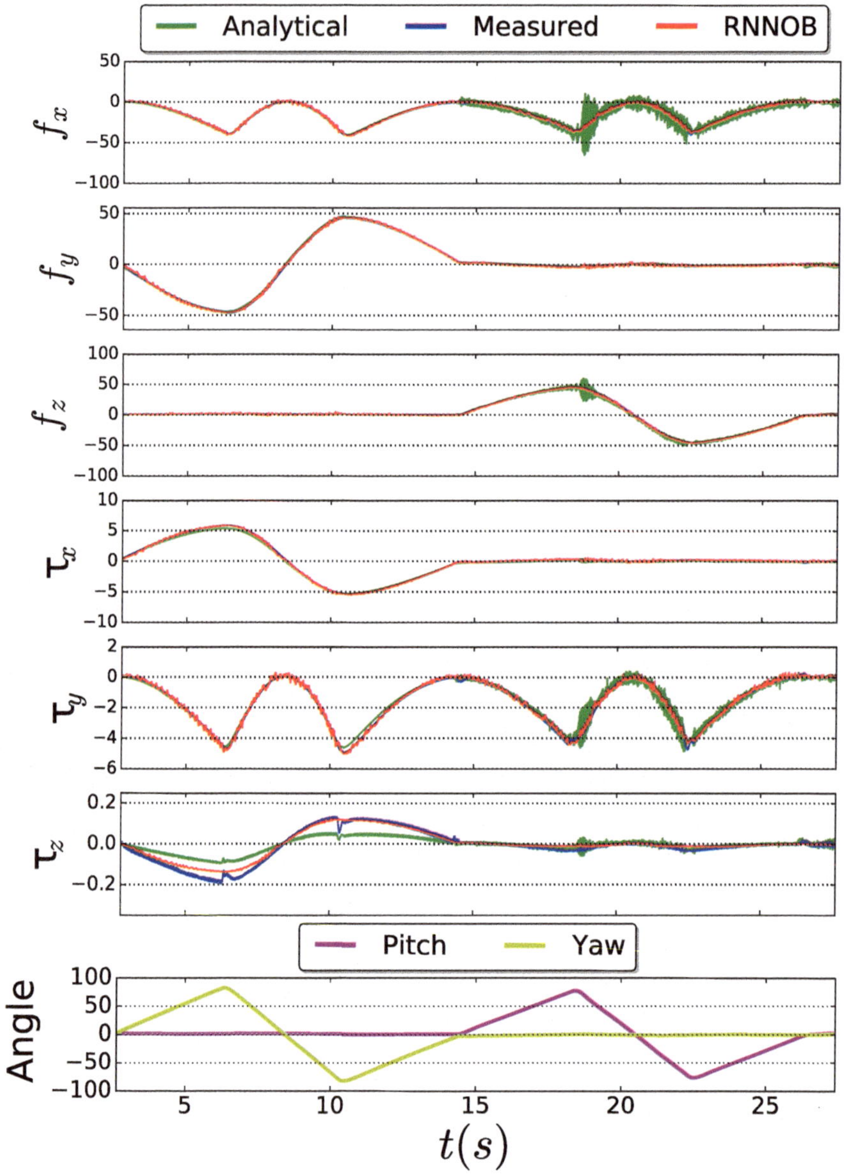

Fig. 7. Non-contact wrench estimation of the proposed RNNOB compared with an analytical-based approach and as measured by the force-torque sensor. The wrench was generated by rotating the end-effector, with an attached load, around the roll, pitch and yaw axes to experience gravitational forces along the three axis of the sensor. The first three rows show the forces in the x, y and z axes of the force-torque sensor frame and are expressed in N. The next three rows show the torques and are expressed in $N \cdot m$. The last row shows the rotations around the sensor's y-axis (pitch) and z-axis (yaw) expressed in degrees. The roll angle is not shown since it has no significant effect as the sensor's x-axis is along the gravity vector $\boldsymbol{g}$.

and their corresponding RMS errors are shown in Table 2. These results were obtained for the three motion tests described in Sect. 3.2, namely, manual, automatic and rotational. A visual example of the force estimation for both, RNN and analytical, observers against the measured forced for a previously unseen manual trajectory can be seen in Fig. 5.

Additionally, the efficiency of these methods in estimating gravitational forces was evaluated by the rotational test as it can be seen in Fig. 7. Table 2 clearly shows that the proposed RNNOB outperforms the analytical observer in all motion tests.

Table 2. The root mean square error on the dataset with an external load for the proposed RNNOB and the analytical method based on identification.

		Manual		Automatic		Rotational	
		RNNOB	Analytical	RNNOB	Analytical	RNNOB	Analytical
Error (N)	f_x	1.6578	2.8934	0.8075	2.4436	0.8405	2.6900
	f_y	1.7235	2.6943	1.1334	2.8558	0.7189	1.2193
	f_z	1.5420	2.2989	0.7829	1.5147	0.7082	1.2204
Error ($N \cdot m$)	τ_x	0.2538	0.3358	0.1384	0.2835	0.0911	0.1408
	τ_y	0.2332	0.3420	0.0995	0.2020	0.1053	0.1601
	τ_z	0.0403	0.0513	0.0089	0.0132	0.0118	0.0301

4.4 Collision Test

The mean error for the ten collisions was of 1.197 ± 0.338 N. One example of the contact tests is shown in Fig. 6, the contact force is expressed in the reference sensor frame on the table. It can be seen from the figure that the proposed observer can estimate the contacts that are larger than 2 N.

5 Conclusions

The article presented a recurrent neural network observer that estimates the non-contact forces acting on the force-torque sensor, namely: inertial, gravitational, centrifugal and Coriolis forces. The proposed observer is model-free and was able to accurately estimate the non-contact forces when the force-torque sensor is subject to high dynamic motions without the need of an identification process. Additionally, in a variety of highly dynamic motions, the RNNOB outperformed the analytical method based on the identification of the inertial parameters. Furthermore, the input signals used by the proposed observer were unfiltered and thus, no additional delay is introduced. Besides, pure contact forces can be obtained by subtracting the output of observer from the output of the force-torque sensor such that external contacts can be detected.

The observer is limited by its inability to estimate zero forces when the robot is motionless . This is a known issue of RNNs and some work has been proposed to alleviate this problem, as suggested in [6]. Another interesting future improvement would be to combine the analytical observer with the RNNOB to address the shortcomings of our proposed observer, as well as addressing safety issues by relying on a model-based approach.

Acknowledgements. Kamal Mohy El Dine works for the European project Robots to Re-Construction (Bots2ReC). His research leading to these results has received funding from the European Union's Horizon 2020 research and innovation programmes under grant agreement No. 687593.

Jose Sanchez is sponsored by the French government research program Investissements d'Avenir through the RobotEx Equipment of Excellence (ANR-10-EQPX-44) and the IMobS3 Laboratory of Excellence (ANR-10-LABX-16-01), by the European Union through the program Regional competitiveness and employment 2007–2013 (ERDF - Auvergne region) and by the Auvergne region.

References

1. García, J.G., Robertsson, A., Ortega, J.G., Johansson, R.: Force and acceleration sensor fusion for compliant robot motion control. In: IEEE International Conference on Robotics and Automation, pp. 2709–2714, Barcelona, Spain (2005)
2. Kröger, T., Kubus, D., Wahl, F.: 6D force and acceleration sensor fusion for compliant manipulation control. In: IEEE/RSJ International Conference on Intelligent Robots and Systems, pp. 2626–2631, Beijing, China (2006)
3. Kubus, D., Wahl, F.M.: Scaling and eliminating non-contact forces and torques to improve bilateral teleoperation. In: IEEE/RSJ International Conference on Intelligent Robots and Systems, pp. 5133–5139, St. Louis, USA (2009)
4. Smith, A.C., Mobasser, F., Hashtrudi-Zaad, K.: Neural-network-based contact force observers for haptic applications. IEEE Trans. Robot. **22**(6), 1163–1175 (2006)
5. Kollmitz, M., Büscher, D., Schubert, T., Burgard, W.: Whole-body sensory concept for compliant mobile robots. In: IEEE International Conference on Robotics and Automation, Brisbane, Australia (2018). (To appear in)
6. Erickson, Z., Clegg, A., Yu, W., Turk, G., Liu, C.K., Kemp, C.C.: What does the person feel? learning to infer applied forces during robot-assisted dressing. In: IEEE International Conference on Robotics and Automation, pp. 6058–6065, Singapore (2017)
7. Karlsson, M., Robertsson, A., Johansson, R.: Detection and control of contact force transients in robotic manipulation without a force sensor. In: IEEE International Conference on Robotics and Automation, pp. 21–25, Brisbane, Australia (2018)
8. Graves, A., Mohamed, A.R., Hinton, G.: Speech recognition with deep recurrent neural networks. In: IEEE International Conference on Acoustics. Speech and Signal Processing, pp. 6645–6649, Vancouver, Canada (2013)

Learning to Grasp Without Seeing

Adithyavairavan Murali(✉), Yin Li, Dhiraj Gandhi, and Abhinav Gupta

The Robotics Institute, Carnegie Mellon University, Pittsburgh, USA
{amurali,yinl2,dgandhi,gabhinav}@cs.cmu.edu

Abstract. Can a robot grasp an unknown object without seeing it? In this paper, we present a tactile-sensing based approach to this challenging problem of grasping novel objects without prior knowledge of their location or physical properties. Our key idea is to combine touch based object localization with tactile based re-grasping. To train our learning models, we created a large-scale grasping dataset, including more than 30K RGB frames and over 2.8 million tactile samples from 7800 grasp interactions of 52 objects. Moreover, we propose an unsupervised auto-encoding scheme to learn a representation of tactile signals. This learned representation shows a significant improvement of 4–9% over prior works on a variety of tactile perception tasks. Our system consists of two steps. First, our touch localization model sequentially "touch-scans" the workspace and uses a particle filter to aggregate beliefs from multiple hits of the target. It outputs an estimate of the object's location, from which an initial grasp is established. Next, our re-grasping model learns to progressively improve grasps with tactile feedback based on the learned features. This network learns to estimate grasp stability and predict adjustment for the next grasp. Re-grasping thus is performed iteratively until our model identifies a stable grasp. Finally, we demonstrate extensive experimental results on grasping a large set of novel objects using tactile sensing alone. Furthermore, when applied on top of a vision-based policy, our re-grasping model significantly boosts the overall accuracy by 10.6%. We believe this is the first attempt at learning to grasp with only tactile sensing and without any prior object knowledge. For supplementary video and dataset see: cs.cmu.edu/GraspingWithoutSeeing.

1 Introduction

Consider the task of grasping a slippery glass bottle. We use vision to determine the location and object properties such as shape. Based on these estimates, we plan how to approach and make contact with the bottle. However, not until we get tactile feedback by touching, can we adjust our hands for a reliable grasp. In many cases, the hand completely occludes the object after contact, severely diminishing the use of hand-eye coordination; yet in all these cases we humans

Electronic supplementary material The online version of this chapter (https://doi.org/10.1007/978-3-030-33950-0_33) contains supplementary material, which is available to authorized users.

J. Xiao et al. (Eds.): ISER 2018, SPAR 11, pp. 375–386, 2020.
https://doi.org/10.1007/978-3-030-33950-0_33

are invariably successful in grasping the objects. In fact, we are even capable of grasping objects solely based on touching. A good example is when we probe around on a nightstand for grabbing a glass of water. The sense of touch plays a vital role in grasping. Yet, most of our current grasping algorithms are primarily dependent on visual sensing. In the recent Amazon Picking Challenge, only one of 26 teams used a tactile sensor [1]. Can a robotic manipulator learn to grasp solely based on touching and without using vision? Moreover, can the robot incorporate both visual inputs and tactile feedback for robust grasping?

In this work, we present the first general framework for learning to grasp with only tactile sensing and without prior object knowledge. Our goal is to scale to a diverse set of unknown objects. To this end, we focus on 2D planar grasps of a single object. To start with, we design a *localization module* to obtain an approximate location of the object. Intuitively, we control the robot to sequentially "touch-scan" the grasp plane until hitting the object and we use a particle filter to aggregate the measurements and track the target location.

With all the uncertainty of object location, tactile sensing and kinematics, how can the robot reliably grasp the object? Our core idea is to treat grasping as a multi-step process with error recovery. Specifically, we propose a *re-grasping module* that refines the initial grasp with multiple re-trials. To extract rich meaningful features for the re-grasping task, we use a recurrent auto-encoder to learn an unsupervised representation from all the unlabelled data. These features are then fed to another neural network that simultaneously estimates grasp stability, and predicts the adjustment for the next grasp. Our framework will iterate on the grasps until our network estimates a high chance of success or the number of trials reaches a predefined limit.

Our high-capacity deep network requires a large-scale tactile dataset for training, which is missing in the community. We have thus created a new dataset of grasping with both tactile and visual sensing. Specifically, we record images, haptic measurements as the robot gripper encloses its fingers on an object, high-level re-grasp actions sent to the motion planner and labels of whether an object has been successfully grasped. Our publicly available dataset includes 7.8K interactions with 52 unique objects with material labels. We hope that it will serve as a major resource for future research on visio-haptic manipulation.

Our method is trained using our dataset, and tested on 20 unseen objects. We systematically vary different components of our framework and benchmark the performance. To begin with, we demonstrate that our unsupervised representation learning produces rich tactile features for a variety of passive (material recognition) and active (re-grasping) tasks. Next, we show that haptic based re-grasping improves a baseline policy, with the ground truth object location provided by vision-based localization. Finally, with touch based localization, our full method achieves a grasping accuracy of 40.0% using tactile sensing alone. We believe this is one of the first results of grasping a large set of unknown objects without seeing. Furthermore, we explore combining haptic and visual sensing for robust grasping. Our results indicate that our multi-step re-grasping with tactile feedback (1) improves the robustness of grasp execution and (2) offers an easy plug-in for existing grasp planning methods.

2 Related Work

Grasping is one of the fundamental problems in robotic manipulation and we refer readers to recent surveys [2–4].

Vision Based Grasping. Visual perception has been the primary modality for sensing, grasp planning and execution. Several work on model-based grasping make use of visual information like point clouds/images to estimate physical properties of objects (e.g., shape [5] or pose [6]), and finally to generate control commands for grasping. Sensing detailed physical properties from visual inputs can be exceedingly challenging, and might not be necessary for finding desired controls. Therefore, recent papers have focused on learning-based approaches [7, 8]. These methods directly map input visual data to the control signals for open-loop grasping. Recently, a lot of progress has been made in this direction by using deep models [9–12]. However, using visual inputs alone leads to errors such as slippage due to low-friction or wrong grasp location due to self-occlusion.

Tactile Exploration. In contrast, humans make great use of tactile signals for grasping and can even grasp unknown objects without using visual sensing [13]. Recent work in robotics has also explored the use of haptics for sensing an object's shape, pose, location or attributes [14–16]. For example, if the location of an object is known, the shape can be estimated by actively touching its parts [17, 18]. Similarly, given the 3D models of objects, several recent work seek to infer the 6DOF pose of the objects with a series of information-gathering actions [19–21]. However, these results have neither been considered for the task of grasping nor can generalize to unknown objects. The most relevant work are from [22] and [23]. Pezzementi et al. [22] built occupancy grid mapping using tactile sensing of unknown 2D objects. Kaboli et al. [23] proposed a pre-touch strategy to localize novel objects in a 3D workspace. These work are similar to our touch localization step, yet they failed to complete the full pipeline of grasping.

Re-grasping with Tactile Sensing. Haptic feedback is widely used for closed-loop control when executing a grasp, also known as re-grasping. Early work [24] focused on analytical solutions for 2D planar grasp given ideal tactile sensing of a known object shape. For real world tactile data, hand crafted rules can be highly effective if object shape is known [25]. Several recent works addressed the task of re-grasping or assessing grasp stability without prior object knowledge [26–31]. However, they all rely on a good initial grasp given by another sensor modality. The most relevant work are from [32,33] and [34]. Based on tactile feedback, Dang et al. [32] learned to predict grasp stability [35], which is further used to guide grasping. Their method can generalize to unknown objects but requires accurate object locations. Moreover, their approach only used simulated data with hand-designed features. Koval et al. [33] utilized haptic feedback to learn both pre and post contact push-grasping policies. Their method accounts for inaccurate sensing of object location and pose, yet is limited to objects with known shapes. Conversely, our method learns tactile based re-grasping policies

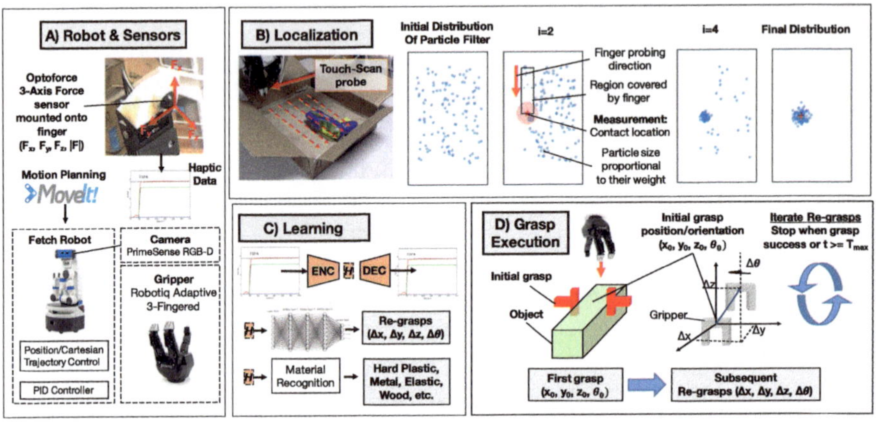

Fig. 1. Overview of our system and approach. (A) Our robot and sensors: Fetch robot, 3-Fingered underactuated Robotiq gripper, 3-Axis OptoForce force sensor, RGB-D head camera (B) Our touch based object localization with particle filtering to aggregate evidences of the object's location. (C) Our unsupervised learning scheme for haptic features (D) Our re-grasping model: Based on haptic features from current grasp, we estimate grasp stability and predict how to adjust the grasp.

with neither prior knowledge of the object (shape/physics) nor necessarily a good initial grasp. In addition, our approach makes use of large-scale real-world visual and haptic data to learn how to grasp. Moreover, Felip et al. [34] presented a full tactile grasping pipeline (exploration and re-grasping) with a wrist force-torque sensor, fingertip tactile sensors and a fully actuated multi-fingered gripper. They used a set of hand-crafted rules/features and demonstrated success on a small set of novel objects. Conversely, our tactile perception modules are learned from data and only uses the fingertip tactile sensors. We show that our learned model can be applied to successfully grasp a larger set of novel objects, including deformable and elastic ones.

Grasping Datasets. Large-scale datasets have fueled the success of visual learning methods for grasping [12]. However, there have been only few attempts in terms of tactile datasets and they focus on passive tasks e.g. material recognition [36], or are limited to a small set of objects [37,38]. Our work is also intertwined with the unsupervised learning of tactile feature representations. Previous work has primarily used hand-crafted features [25], bag-of-words [39] and sparse coding [40].

3 Dataset

In this section, we present the effort on creating our visio-haptic dataset for grasping. Large-scale haptic dataset for grasping is important for learning high capacity deep models. Unfortunately, this kind of dataset is missing in the community. We seek to bridge this gap by collecting a new grasping dataset that includes both visual and haptic sensor data. Specifically, our dataset consists of 7800 grasp interactions with 52 different objects. Each grasp interaction lasts for 3.5–4 s and is recorded with:

• *RGB Frames.* We capture images of four specific events of a grasping: for the initial scene, before, during and after grasp execution. These images have a resolution of 1280×960.

• *Haptic Measurements.* Tactile signals are measured by force sensors mounted on each of the three fingers of the gripper. The sensor measures the magnitude (F) and the direction of forces (F_x, F_y, F_z) at 100 Hz.

• *Grasping Actions and Labels.* We record the pose of all 2D planar grasps, including the initial grasp $(x_0, y_0, z_0, \theta_0)$ and subsequent re-grasps $(x_t, y_t, z_t, \theta_t)$. We also record whether the re-grasp succeeded.

• *Material Labels.* We label material categories (7) for each object, including metal, hard plastic, elastic plastic, stuffed fabric, wood, glass and ceramic.

Data Collection. To collect this dataset, we sample and execute a large set of grasps. The robot will lift up objects and automatically detect successful grasps. A major issue with this data collection process is how we can get more successful grasps. It is easy to collect failure cases by applying random grasps but it is difficult to collect successful grasps, which are critical for learning. To address this issue, we used an existing vision based grasping policy to sample an initial grasp from a pre-learned visual grasping policy [41]. We collect two sets of data and combine them to form our final dataset. The first set includes all 52 objects with 50–55 initial grasps. Each initial grasp is followed by a single random re-grasp. The grasps in this set have a higher rate of success. On the other hand, our second set contains a subset of 7 objects covering different types of materials. For each object in this set, we sample 80–100 initial grasps, and allow 2–3 random re-grasps, resulting in a higher failure rate.

Dataset Statistics. Overall, our dataset includes more than 30K RGB frames and over 2.8 million of tactile samples from 7800 grasp interactions of 52 objects. We provide grasping actions and labels for each interaction, as well as material labels for each object. To the best of our knowledge, this is by far the largest dataset for vision-haptic grasping. Our dataset is publicly available at: cs.cmu.edu/GraspingWithoutSeeing.

4 Technical Approach

Consider samples of tactile responses during grasping in Fig. 2. These signals encode important information about the object in contact–the magnitude implies the material, temporal force profile across the fingers indicates the shape and grasp stability. Therefore, our core intuition is that these haptic signals can be used to refine grasping, even if we start with a low quality initial grasp. Thus, our

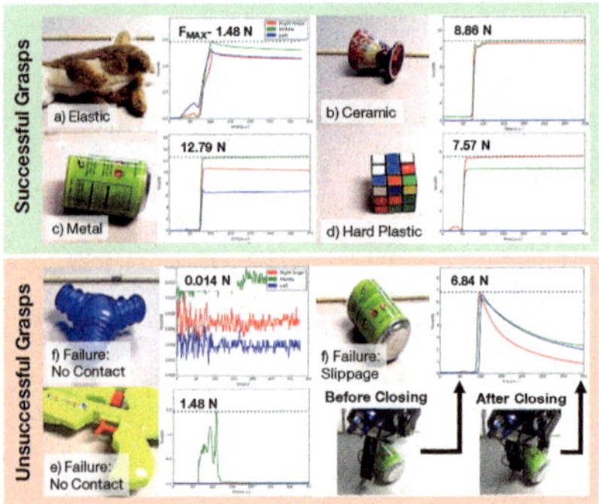

Fig. 2. Tactile responses from successful/failure grasps of different objects. We plot the time series of force magnitude from our sensors on three fingers (red: right, green: middle, blue: left). We record signals before and after the gripper closes. These signals provide critical cues for grasping. And we propose to learn (1) tactile features from these signals and (2) re-grasping polices using the learned features.

method first localizes the object by touching, generate a random initial grasp and further refine the grasp by a few re-trials. Figure 1 presents the overview of our setup and method. Our method has three major components: (1) localization of objects; (2) learning tactile features for re-grasping; and (3) learning re-grasping policies. Algorithm 1 summarizes our complete pipeline.

Object Localization. Our localization module finds the target object with a sequence of touch-scans and tracks the belief (in 2D space) using a particle filter, similar to [33]. Touching the object might change its location. This displacement is usually small, yet is determined by how the robot moves (u_t), and the physical properties of the object and its environment. We simplify the *motion* model by assuming a Gaussian distribution independent of u_t: $p(x_t|x_{t-1}, u_t) = \mathcal{N}(x_{t-1}, \sigma^2 I)$, where σ is a small noise. For the *measurement* model, any location on the 2D plane can be either free space (no contact) or occupied by the object (contact). We either increase (occupied) or decrease (free space) the weights of particles that lies within the vicinity (a sphere of radius 2.5 cm for our experiments) of the location, an example of which is shown in Fig. 1. With a noisy estimate of the object location, we generate an initial grasp by randomly sampling from the parameter space (π_{random}). Executing such a grasp is likely to fail, as this sample can be far away from feasible grasps. We thus rely on iteratively adjustments to reach a successful grasp.

Learning Tactile Features. We used a recurrent auto-encoder to learn a tactile feature representation for re-grasping. The encoder M_{ENC} encodes a time series of tactile data (denoted as $O \in R^{12}$) and robot control signals (f) into a low dimensional latent space H. The decoder M_{DEC} (not used during testing) reconstructs the input data ($\hat{O} = M_{DEC}(H)$) from the latent space H. f specifically refers to the position of the under-actuated gripper and takes on the maximum possible value if no object is grasped/enclosed. M_{ENC} and M_{DEC} had LSTM units and were trained with $L2$ loss using stochastic gradient descent.

Algorithm 1 Grasping Without Seeing

Localize object with vision/touch
Sample g_1 from $\pi_{vision}/\pi_{random}$
Execute g_1 on robot
Collect first haptic measurement O_1
for i = $2{:}T_{max}$ **do**
  Encode $H_{i-1} \leftarrow M_{ENC}(O_{i-1}, f_{i-1})$
  Compute $p_{i-1} = M_{stability}(H_{i-1})$
  if $p_{i-1} > p_{threshold}$ **then**
    break
  else
    Compute $g_i = \pi_{regrasps}(H_{i-1})$
    Execute g_i on robot
    Collect haptic measurement O_i
  end
end

Learning to Re-grasp. We treat grasping as a multi-step process that can be trained with self-supervision as prior work [10]. Formally, given a current grasp g and its tactile measurement (O, f), we encode them as $H = M_{ENC}(O, f)$. When testing, we iteratively apply the predicted $\Delta g = \pi_{re-grasp}(H)$ to current grasp g. We execute $g + \Delta g$ until $p = M_{stability}(H)$ predicts a high rate of success. Both $\pi_{re-grasp}$ and $M_{stability}(H)$ are simple feed-forward networks with five fully connected layers with a final sigmoid function. Our output action Δg is a 4D vector parameterized by the change of the gripper's position and orientation $(-\pi/4 \leq \Delta\theta \leq \pi/4)$. The policy was trained with a multi-label binary classification loss: $\sum_{i=1}^{K}\sum_{j=1}^{D}\delta(u_{i,j})\text{Cross-Entropy}(y_{ij}, \hat{y})$. $\hat{y}$ corresponds to the success/failure label, y_{ij} is the sigmoid output, $D(=5)$ is the number of discretized bins, $K(=4)$ is the number of control parameters and $\delta(u_{i,j})$ is an indicator function.

5 Experimental Evaluation

Our results are summarized into two parts. First, we evaluate the learned haptic features for material recognition and grasp stability estimation. Second, we benchmark our tactile re-grasping module and report results for the full tactile-only grasping pipeline. For grasping and re-grasping, we physically test our method on the objects shown in Fig. 3.

Fig. 3. Our test set of 20 novel objects. Set A contains slightly harder objects to grasp (e.g., toy guns) compared to Set B. This test setting is very challenging as testing objects are not presented in the training set.

Learning Tactile Features. We present results for material recognition and grasp stability estimation using our learned features in Table 1.

Table 1. Tactile features for material recognition and grasp stability estimation.

Feature type	Material recognition		Grasp stability estimation	
	Deep network	SVM	Deep network	SVM
Auto-encoder (ours)	**42.86**	40.68	**85.92**	84.50
Sparse coding [40,42]	36.35	35.93	81.37	80.12
Hand crafted [15]	33.50	33.66	82.54	82.66

• *Material Recognition.* The task is to classify 7 different materials (plastic, wood, metal, etc.) in our dataset using tactile signals during grasping. Our classifiers are trained on a subset of the training set (80%) and tested on the remaining held out set. The features learned from our auto-encoder outperforms sparse coding [42] and hand crafted features [15] by a significant margin (at least 4–7%). The deep network also performs at least comparably or slightly better than the SVM for all features. In particular, our haptic features with deep networks improves the traditional method of sparse coding with SVM by 5–8%. Furthermore, we show the confusion matrix for material recognition in Fig. 4b. The majority of the error comes from hard objects that are composed of wood/metal/glass being mis-classified as hard plastic. This result demonstrates that our haptic features encode physical properties of the object.
• *Grasp stability estimation.* The task is to estimate whether the grasp will be successful given tactile signals during grasping. We train the classifiers on the training set and apply them on our full test set (580 trials on 20 unseen objects), reporting the accuracy for binary classification. Overall, the results of grasp stability estimation follow the same trend of material recognition. These results demonstrate that our haptic features encode physical properties of the object and provide a strong support to our design of the re-grasping model i.e. the combination of our learned haptic features and deep networks. To better understand our tactile features for grasping, we visualize the t-SNE embedding of the learned features and plot example results of our grasp stability estimation in Fig. 4a. We observe that the main failure modes are from that (1) the part of the finger containing the haptic sensor may not come into contact with the object; and (2) the object may slip in the gripper.

Table 2. Evaluation of our re-grasping model (a) and our full grasping pipeline (b).

Initial Grasp	Re-grasp	Acc (%) Set A	Set B	A+B
Random	-	16.3	32.5	24.4
Random	Random	17.5	28.8	23.2
Random	Ours	33.8	41.3	**37.5**

(a) **Re-grasping Model**: Breakdown of grasping accuracy. All methods use oracle object locations (from vision).

Obj Loc	Initial Grasp	Re-grasp	Acc (%) (Set B)
Touch	Random	-	26.0
Touch	Random	Ours	**40.0**
Vision	Vision	-	51.3
Vision	Vision	Ours	**61.9**

(b) **Full Pipeline**: Grasping accuracy of our full pipeline.

Tactile Based Grasping. We further evaluate our tactile based grasping module. We first separately evaluate our touch based re-grasping model, followed by the full pipeline. Finally, we look into whether using our re-grasping model along with a vision-based policy improves overall grasping performance.

• *Re-grasping Model.* We evaluate the re-grasping model and report grasping accuracy in Table 2a, with results averaged over 160 trials (8 canonical poses/object) per experiment. In this study, we assume an oracle object location is given. We compare three different settings: a single random re-grasp, multiple random re-grasps, and our re-grasping model. For fair comparison, we set the number of trials for random re-grasps equal to the maximum number of trials (=4) of our model. Grasping accuracy on Set B is always higher than Set A. For the full set, the baseline accuracy is 24.4%, where the first (and only) grasp is sampled from a random policy without re-grasping. Our re-grasping model gets an accuracy of 37.5%, outperforming the baseline by 13.1%.

• *Grasping without Seeing.* We further test our full pipeline including localization and re-grasping. In this case, we simplify our benchmark by only considering our test set B and use 5 trials per object. This is primarily limited by the run time of our experiments. Our results are shown in Table 2b. Our pipeline improves the baseline of random grasping by 14% and reaches an accuracy of 40% with only tactile sensing. Our method is the first learning based approach for a complete grasping of multiple novel objects using only the sense of touch.

• *Visio-Haptic Grasping.* Our last experiment (see Table 2b) combines the proposed re-grasping model with a vision-based policy from [41]. Combining vision based grasping with our tactile based re-grasping can largely improve the accuracy by 10.6%. These results provide a strong evidence for the need of joint visio-haptic grasping to improve robustness. Our re-grasping model can also be readily plugged onto existing grasp planning methods.

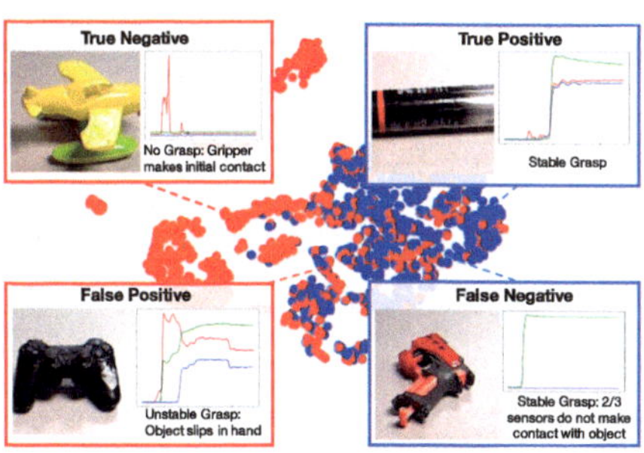

(a) Visualization of learned haptic features using t-SNE Embedding. Red and blue correspond to failed and successful grasps. We also plot four typical examples for grasp stability estimation.

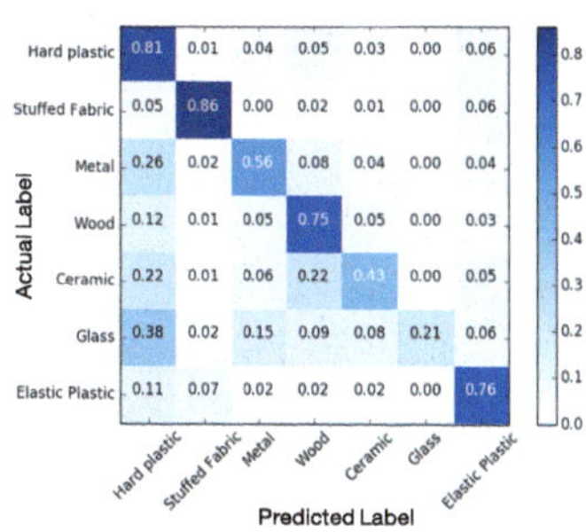

(b) Confusion matrix for material recognition on a held out test set. Using our learned haptic features, we achieve an accuracy of 42.86%.

Fig. 4. Visualization of the tactile features (a) and the confusion matrix (b).

6 Conclusion and Future Work

In this paper, we demonstrate one of the first attempts of learning to grasp novel objects using only tactile sensing and without prior knowledge about the object. The core of our method lies in the combination of (a) a simple method of touch based localization (b) unsupervised learning of rich tactile features and (c) a learning based method for re-grasping using haptic feedback. First, we created a large-scale dataset for visio-haptic grasping to evaluate our method and to facilitate future research. With this dataset, we used a auto-encoder to learn rich features from raw tactile signals. These features proved effective for both passive tasks like material recognition and active tasks like re-grasping, and displayed an improvement of around 4–9% over prior methods. Finally, we show that our novel re-grasping model can progressively improve the grasping, leading to significantly higher success rate even from a noisy initial grasp. Our method achieved a grasping accuracy of 40.0% with just tactile sensing and we demonstrate that this re-grasping model can be combined with existing vision based grasping to further improve the accuracy by about 10%. We hope that our method together with our dataset could provide valuable insights for solving the challenging problem of autonomous grasping.

Looking forward, the object shape could first be reconstructed with tactile exploration followed by grasp planning, instead of a random initial grasp. Another major failure mode with our current setup is partial observability - the regions of the robot's finger not covered by the sensor might come in contact and push the object. This affects all stages of our pipeline and could be mitigated by using novel skin and wrist force-torque sensors with better spatial coverage.

Acknowledgements. This work was supported by Google Focused Award and ONR MURI N000141612007. Abhinav Gupta was supported in part by Sloan Research Fellowship and Adithya Murali was partly supported by a Uber Fellowship. The authors would also like to thank Brian Okorn, Reuben Aronson, Ankit Bhatia, Lerrel Pinto, Tess Hellebrekers and Suren Jayasuriya for discussions.

References

1. Correll, N., Bekris, K., Berenson, D., Brock, O., Causo, A., Hauser, K., Okada, K., Rodriguez, A., Romano, J., Wurman, P.: Analysis and observations from the first amazon picking challenge. IEEE T-ASE **15**(1), 172–188 (2017)
2. Bicchi, A., Kumar, V.: Robotic grasping and contact: a review. In: ICRA (2000)
3. Tegin, J., Wikander, J.: Tactile sensing in intelligent robotic manipulation-a review. Ind. Robot. Int. J. **32**, 64–70 (2005)
4. Bohg, J., Morales, A., Asfour, T., Kragic, D.: Data-driven grasp synthesis-a survey. IEEE Trans. Robot. **30**, 289–309 (2014)
5. Miller, A.T., Knoop, S., Christensen, H.I., Allen, P.K.: Automatic grasp planning using shape primitives. In: ICRA (2003)
6. Collet, A., Berenson, D., Srinivasa, S.S., Ferguson, D.: Object recognition and full pose registration from a single image for robotic manipulation. In: ICRA (2009)

7. Saxena, A., Driemeyer, J., Kearns, J., Osondu, C., Ng, A.Y.: Learning to grasp novel objects using vision. In: ISER (2006)
8. Watter, M., Springenberg, J., Boedecker, J., Riedmiller, M.: Embed to control: a locally linear latent dynamics model for control from raw images. In: NIPS (2015)
9. Levine, S., Finn, C., Darrell, T., Abbeel, P.: End-to-end training of deep visuomotor policies. JMLR **17**, 1334–1373 (2016)
10. Pinto, L., Gupta, A.: Supersizing self-supervision: learning to grasp from 50k tries and 700 robot hours. In: ICRA (2016)
11. Levine, S., Pastor, P., Quillen, D.: Learning hand-eye coordination for grasping with deep learning and large-scale data collection. In: ISER (2016)
12. Mahler, J., Liang, J., Niyaz, S., Laskey, M., Doan, R., Liu, X., Ojea, J.A., Goldberg, K.: Dex-net 2.0: deep learning to plan robust grasps with synthetic point clouds and analytic grasp metrics. RSS (2017)
13. Johansson, R.S., Flanagan, J.R.: Coding and use of tactile signals from the fingertips in object manipulation tasks. Nat. Rev. Neurosci. **10**, 345 (2009)
14. Chu, V., McMahon, I., Riano, L., McDonald, C.G., He, Q., Perez-Tejada, J.M., Arrigo, M., Fitter, N., Nappo, J.C., Darrell, T., et al.: Using robotic exploratory procedures to learn the meaning of haptic adjectives. In: ICRA (2013)
15. Spiers, A., Liarokapis, M., Calli, B., Dollar, A.: Single-grasp object classification and feature extraction with simple robot hands and tactile sensors. IEEE Trans. Haptics **9**, 207–220 (2016)
16. Yuan, W., Zhu, C., Owens, A., Srinivasan, M., Adelson, E.: Shape-independent hardness estimation using deep learning and a gelsight tactile sensor. In: ICRA (2017)
17. Yi, Z., Calandra, R., Veiga, F., van Hoof, H., Hermans, T., Zhang, Y., Peters, J.: Active tactile object exploration with gaussian processes. In: IROS (2016)
18. Meier, M., Schopfer, M., Haschke, R., Ritter, H.: A probabilistic approach to tactile shape reconstruction. IEEE Trans. Robot. **27**, 630–635 (2011)
19. Petrovskaya, A., Khatib, O.: Global localization of objects via touch. IEEE Trans. Robot. **27**, 569–585 (2011)
20. Saund, B., Chen, S., Simmons, R.: Touch based localization of parts for high precision manufacturing. In: ICRA (2017)
21. Javdani, S., Klingensmith, M., Bagnell, J.A., Pollard, N.S., Srinivasa, S.S.: Efficient touch based localization through submodularity. In: ICRA (2013)
22. Pezzementi, Z., Reyda, C., Hager, G.D.: Object mapping, recognition, and localization from tactile geometry. In: ICRA (2011)
23. Kaboli, M., Feng, D., Yao, K., Lanillos, P., Cheng, G.: A tactile-based framework for active object learning and discrimination using multimodal robotic skin. IEEE Robot. Autom. Lett. **2**, 2143–2150 (2017)
24. Fearing, R.: Simplified grasping and manipulation with dextrous robot hands. IEEE J. Robot. Autom. **2**, 188–195 (1986)
25. Hsiao, K., Chitta, S., Ciocarlie, M., Jones, E.G.: Contact-reactive grasping of objects with partial shape information. In: IROS (2010)
26. Romano, J.M., Hsiao, K., Niemeyer, G., Chitta, S., Kuchenbecker, K.J.: Human-inspired robotic grasp with tactile sensing. Trans. Robot. **27**, 1067–1079 (2011)
27. Bekiroglu, Y., Laaksonen, J., Jorgensen, J.A., Kyrki, V., Kragic, D.: Assessing grasp stability based on learning and haptic data. IEEE Trans. Robot. **27**, 616–629 (2011)
28. Chebotar, Y., Hausman, K., Su, Z., Sukhatme, G.S., Schaal, S.: Self-supervised regrasping using spatio-temporal tactile features and reinforcement learning. In: IROS (2016)

29. Dang, H., Allen, P.K.: Grasp adjustment on novel objects using tactile experience from similar local geometry. In: IROS (2013)
30. Dragiev, S., Toussaint, M., Gienger, M.: Uncertainty aware grasping and tactile exploration. In: ICRA (2013)
31. Calandra, R., Owens, A., Upadhyaya, M., Yuan, W., Lin, J., Adelson, E.H., Levine, S.: The feeling of success: does touch sensing help predict grasp outcomes? In: Conference on Robot Learning (2017)
32. Dang, H., Weisz, J., Allen, P.K.: Blind grasping: Stable robotic grasping using tactile feedback and hand kinematics. In: ICRA (2011)
33. Koval, M.C., Pollard, N.S., Srinivasa, S.S.: Pre-and post-contact policy decomposition for planar contact manipulation under uncertainty. IJRR **35**, 244–264 (2016)
34. Felip, J., Bernabe, J., Morales, A.: Contact-based blind grasping of unknown objects. In: IEEE-RAS International Conference on Humanoid Robots (2012)
35. Dang, H., Allen, P.K.: Stable grasping under pose uncertainty using tactile feedback. Auton. Robot. **36**, 309–330 (2013)
36. Erickson, Z., Chernova, S., Kemp, C.C.: Semi-supervised haptic material recognition for robots using generative adversarial networks, arXiv (2017)
37. Chebotar, Y., Hausman, K., Su, Z., Molchanov, A., Kroemer, O., Sukhatme, G., Schaal, S.: BIGS: BioTac grasp stability dataset. In: ICRA Workshop (2016)
38. Chebotar, Y., Hausman, K., Kroemer, O., Sukhatme, G., Schaal, S.: Generalizing regrasping with supervised policy learning. In: ISER (2016)
39. Schneider, A., Sturm, J., Stachniss, C., Reisert, M., Burkhardt, H., Burgard, W.: Object identification with tactile sensors using bag-of-features. In: IROS (2009)
40. Madry, M., Bo, L., Kragic, D., Fox, D.: ST-HMP: unsupervised spatio-temporal feature learning for tactile data. In: ICRA (2014)
41. Murali, A., Pinto, L., Gandhi, D., Gupta, A.: CASSL: curriculum accelerated self-supervised learning. In: ICRA (2018)
42. Roberge, J.-P., Rispal, S., Wong, T., Duchaine, V.: Unsupervised feature learning for classifying dynamic tactile events using sparse coding. In: ICRA (2016)

Semi-parametric Approaches to Learning in Model-Based Hierarchical Control of Complex Systems

Munzir Zafar(✉), Areeb Mehmood, Mouhyemen Khan, Shimin Zhang, Muhammad Murtaza, Victor Aladele, Evangelos A. Theodorou, Seth Hutchinson, and Byron Boots

Georgia Institute of Technology, Atlanta, GA 30332, USA
mzafar7@gatech.edu

Abstract. For systems with complex and unstable dynamics, such as humanoids, the use of model-based control within a hierarchical framework remains the tool of choice. This is due to the challenges associated with applying model-free reinforcement learning on such problems, such as sample inefficiency and limits on exploration of state space in the absence of safety/stability guarantees. However, relying purely on physics-based models comes with its own set of problems. For instance, the necessary limits on expressiveness imposed by committing to fixed basis functions, and consequently, their limited ability to learn from data gathered on-line. This gap between theoretical models and real-world dynamics gives rise to a need to incorporate a learning component at some level within the model-based control framework. In this work, we present a highly redundant wheeled inverted-pendulum humanoid as a testbed for experimental validation of some recent approaches proposed to deal with these fundamental issues in the field of robotics, such as: 1. Semi-parametric Gaussian Process-based approaches to computed-torque control of serial robots [1] 2. Probabilistic Differential Dynamic Programming framework for trajectory planning by high-level controllers [2,3] 3. Barrier Certificate based safe-learning approaches for data collection to learn the dynamics of inherently unstable systems [4]. We discuss how a typical model-based hierarchical control framework can be extended to incorporate approaches for learning at various stages of control design and hierarchy, based on the aforementioned tools.

Keywords: Hierarchical control · Model-based control · Wheeled inverted pendulum humanoids · Semi-parametric model · Safe learning · Probabilistic trajectory optimization

---

Electronic supplementary material The online version of this chapter (https://doi.org/10.1007/978-3-030-33950-0_34) contains supplementary material, which is available to authorized users.

J. Xiao et al. (Eds.): ISER 2018, SPAR 11, pp. 387–397, 2020.
https://doi.org/10.1007/978-3-030-33950-0_34

1 Introduction

Wheeled inverted pendulum (WIP) systems offer fast and efficient locomotion along with the ability to deal with very heavy payloads. This ability allows them to compensate large external forces by readily adjusting their center of mass (CoM). **Golem Krang** [6] is a tree-structured serial robot with two serial manipulators (each with 7-degrees-of-freedom (DoF)) mounted on a 3-DoF serial link (base, spine, torso). The system is mounted on a differential drive able to dynamically balance itself in an inverted pendulum configuration (see Fig. 1).

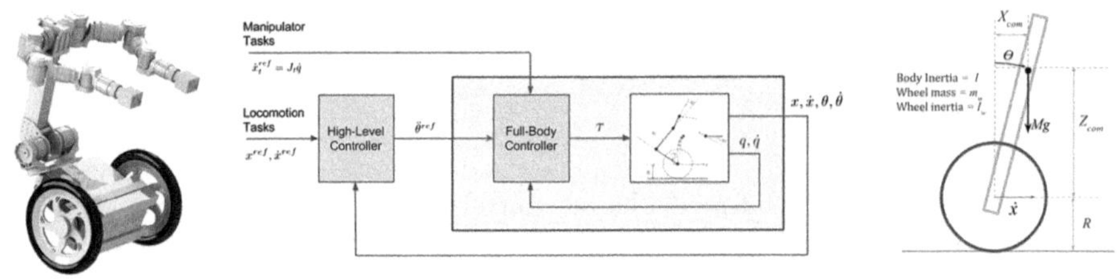

Fig. 1. Krang (left) with hierarchical control (middle) and its 2D model (right)

A large body of literature exists to deal with the nonlinear, unstable, and under-actuated dynamics of WIP system [13–15]. Based on this, efforts are invested on WIP humanoid by attempting to control WIP independently from the control of upper body manipulators [10]. However, this approach comes with limitations in performing tasks that impose constraints on the kinematics of upper body while satisfying locomotion and stability objectives. This motivates the case for a unified approach in handling both locomotion and manipulation.

We utilize a framework similar to [5] for whole-body control of **Krang** through a hierarchical optimization-based approach and relying on physics-based modeling of the system. The framework has two levels of hierarchy: (1) *low-level controller* that considers the complete dynamics and optimizes only for one time-step, and (2) *high-level controller* that plans for longer horizons on a simplified model of the system for computational efficiency. The approach is grounded in a physics-based model of the system, associated non-holonomic constraints, and stability consideration of the zero dynamics of the system [16].

We then introduce learning approaches at various stages of the control design and hierarchy to deal with model uncertainty. The aim is to improve control performance given new observations in an online manner. We propose the use of recently developed semi-parametric approaches using Gaussian Processes (GP) to represent unknown non-linearities. We discuss the use of GP-based approaches at three stages of the design: (1) *safe-learning for data collection* on unstable dynamics, (2) *inverse dynamics (ID) using semi-parametric model* for low-level controller's torque computations, and (3) *trajectory planning and optimization* in the high-level controller.

2 Technical Approach

When a WIP humanoid is moving on the ground under nonholonomic constraints of no-slip/skid on the wheels, the robot has $n_{DoF} = 3+n_{tree}$ DoFs which include: (1) Heading $\dot{x}$ of the robot (2) Spin $\dot{q}_0$ of the robot (3) Pitch $\dot{q}_1$ of the base link and 4) Relative motions of n_{tree} child links in the tree structure mounted on the base link. Choosing the minimum set of n_{DoF} coordinates, equations of motion (EOM) can be derived using Kane's formulation [8] as, unlike Euler-Lagrange, it is able to deal with the quasi-velocity $\dot{x}$. The system is under-actuated since the first 3 DoFs are driven by only 2 actuators *i.e.* the left and right wheel motors.

It is possible to derive the model using Featherstone's approach [9] if we do not insist on the minimum set of coordinates. Generic simulation platforms that provide efficient computation tools for highly articulated structures, like DART, model this system using $n_{full} = 8 + n_{tree}$ coordinates, which include 6 for position/orientation of the floating base, 2 for relative angles of the wheels, and n_{tree} joint motions. The formulation in minimum set of coordinates helps with analysis of zero dynamics of the system, but the use of n_{full} coordinates is indispensable if we are to make use of off-the-shelf algorithms for fast computation of quantities we need for ID control, such as the robot's inertia matrix. For this reason, equations of motion (EOM) using both sets of coordinates are useful. One informs design of the high level controller and analysis of the zero dynamics, while the latter is useful for the low-level controller that needs fast computation of full dynamics of the robot. Given EOM in n_{full} coordinates, we can perform a similarity transformation on them, using a transform Jacobian relating speeds in minimum coordinates to n_{full} speeds, to obtain EOM in minimum coordinates. Similarly, the transform Jacobian can be multiplied with task Jacobians in n_{full} coordinates to obtains Jacobians for minimum set of coordinates.

Figure 1 gives an overview of the hierarchical control approach. It is similar to [5]'s work on a bipedal system that has demonstrated effective performance for walking humanoids. This is the first time that a similar approach is developed for whole-body control of WIP humanoid (see [16] for a detailed account). The low-level controller uses Quadratic Programming (QP) based optimization over joint accelerations to achieve several operational space objectives framed as weighted sums of quadratic costs. The high-level controller uses DDP over the full horizon for trajectory planning of CoM given a locomotion target, and also in a receding horizon fashion online for closed-loop tracking of the optimal trajectory [11,12]. The simplified model is able to capture the essential parts of the zero dynamics of the full system, which explains how the formulation of the high-level controller ensures stability of the zero dynamics.

The above hierarchical approach is model-based. These models, although of immense use, are based on assumptions that may not always hold (like rigid bodies, simplified friction behaviors, and linearity and invariability of identifiable parameters). In Golem Krang, all joints are actuated by harmonic drives with high frictions, and the relationship between current commands and actual torques is not reported to be linear. Thus, it violates key assumptions at play in the hierarchical whole-body control framework discussed above. To address this, we propose the use of semi-parametric representations of our dynamics for predictive control in both levels of the hierarchy.

Semi-parametric representations can be used for predictive control of dynamics in both hierarchical levels. For low-level controller, the dynamics can be represented similar to [1]. GPs incorporate parametric basis functions as priors while augmenting non-parametric kernel functions to learn unknown non-linearities missed out by their parametric counterparts. For high-level controller, semi-parametric approach to trajectory optimization proposed in [2] can be utilized. GPs are used to represent uncertainty in the simplified model for Krang and trajectory optimization is performed in belief space. Finally, to address the problem of collecting data on our inherently unsafe and unstable dynamics, the safe-learning framework of [4] can be used. This framework relies on barrier certificates in the presence of GP-based uncertainties to reason about the safety of exploration in unknown regions of the state space for gathering data for learning.

Fig. 2. Krang in its balanced pose carrying a cup on a tray

3 Experiments and Results

The platform, Golem Krang (Fig. 2), is equipped with an on-board computer, and precise position encoders for all joints. Real-time sensing and control is possible with many dedicated CAN-based physical interfaces, and the real-time operating system installed on the computer. For inter-process communication (IPC) among sensing/control and application processes with real-time guarantees, ACH [18] is used as the underlying framework. The lower body of Krang is very heavy (~100 kg), making it suitable to perform dynamic tasks by manipulating its weight torque intelligently. These characteristics make this platform especially suited for trying out the various ideas for dynamic control of complex systems discussed earlier. This section is divided into three subsections. We first discuss results from the simulation-based investigations of model-based control that are geared specifically to overcome various challenges in implementing model-based hierarchical control on this platform. Secondly, we discuss our result on semi-parametric learning and control of a 7DOF arm of this robot in simulation. Finally, we list the set of follow-up experiments needed to implement the full pipeline discussed in the preceding paragraphs.

3.1 Model-Based Hierarchical Control

A verification of the hierarchical control approach for whole-body control of Krang was done using DART simulation of Krang. Parameters for the model were chosen to reflect properties of the physical system. A simulation video is provided in the supplementary material[1].

Figure 3 shows the reference trajectory determined by DDP on the simplified model, and the state of the simplified model that results by applying the receding horizon control in a closed-loop manner on the simplified model. The control and state trajectories produced by this high-level controller are used as references for the low-level controller, which is also responsible for controlling the position and orientation of the end-effector. We see that the state follows the reference trajectory very closely with disturbances occurring during fast transitions owing to the disturbances caused by unmodeled full body dynamics (i.e. not modeled by the simplified model used by the high-level controller). Also, the last plot to the right is the plot of six joint torques from the full body. It is clear that the entire body is participating in performing the three tasks.

Snapshots of the full body at 4 instances during execution are shown in Fig. 4. Figure 4A shows the perspective and side views of the case when we perform ID based low-level control for the full body. This control assumes perfect knowledge of model parameters. And it shows that the hierarchical controller is able to perform unified control of locomotion and manipulation. This is demonstrated by the cup on the tray carried by the arms. If we turn off whole-body control and perform simple locomotion task on the simplified model, the cup ends up falling. This is due to the changes in pitch of the body affecting the orientation of the

[1] Visit https://vimeo.com/user90167025/review/292912849/f662c39b8f.

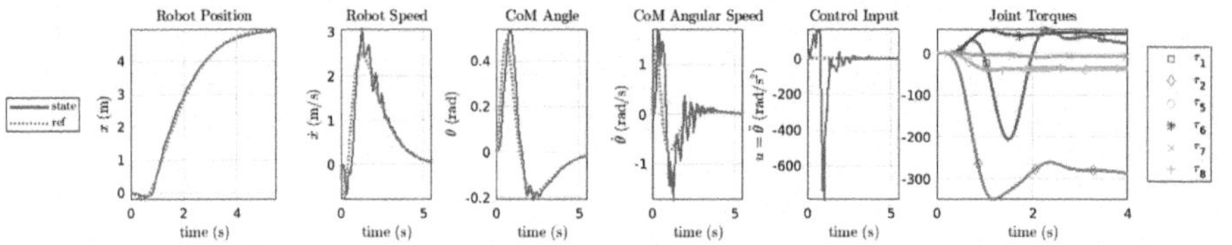

Fig. 3. Simplified model reference and state trajectories (first 5 plots). The last plot shows the resulting joint torques for $t \in (0, 4)$ seconds.

end-effector. This however is prevented when the control is unified as shown in the figure.

In order to perform this experiment on the hardware, we needed to verify two more cases in simulation. The waist joint on the robot supports the bulk of weight of the upper body, and should not be unlocked after reaching the desired height. We investigated in simulation if it is necessary to unlock this joint to perform the tray-carrying task. We achieved control of simulated robot with a locked waist joint, by (a) excluding waist joint acceleration in the vector of decision variables for QP-based optimization in the low-level controller, and (b) deleting the columns and rows in the various Jacobian-based cost functions [16], and EOMs for ID that correspond to the waist joint. Figure 4B shows the results. We see that when the waist is unable to lift the upper body to satisfy unified objectives, the base joint has to tilt the body a lot more to ensure CoM targets are being followed as the arms are busy in maintaining tray orientation.

A third point of investigation before hardware experiments could begin is the use of IK for control of the arms. This allows us to relax the assumption of perfect knowledge of system parameters, as kinematic references are followed by joint level PID controllers. Figure 4C shows the result. Again the waist joint is locked in this formulation. A key point to note for IK is that the robot joints are all harmonic drives with the effect of frictions much more significant than other dynamic effects. We reflected this fact in choosing our simulation frictions. In our experience, IK is a lot easier to perform in the presence of high viscous frictions, as the frictions are helpful in stabilizing the system around any configuration. This however is not the case for base, waist and torso joints. This is due to the inertial and gravitational effects of the articulated structures mounted on these three joints far outweigh the respective frictions on these joints. Therefore the use of ID-based control for these joints is indispensable. In the absence of perfect system knowledge, learning based approaches discussed above have to be utilized.

3.2 Semi-parametric Learning on Krang's Arm

To verify the semi-parametric based approach on low-level controller, we first implement it on one of the serial manipulators on Krang. This is a necessary investigation before deploying the framework on the whole system. Non-linear error terms were included in the dynamics with the objective of GP capturing such dynamics. We successfully implemented it and verified it in simulation.

The experimental pipeline in performing the semi-parametric based approach incorporated the following steps: follow reference trajectory in operation space and collect training/testing datasets, compute hyper-parameters from training set, and perform online control.

The point cloud of end-effector positions are shown in Fig. 5. Hyper-parameters were determined based on minimizing the log-marginal likelihood and online control is performed on test trajectory. The computation time for performing online control was of the order 5∼12 ms per query point. The performance of GP-based torque computation outperformed purely parametric based

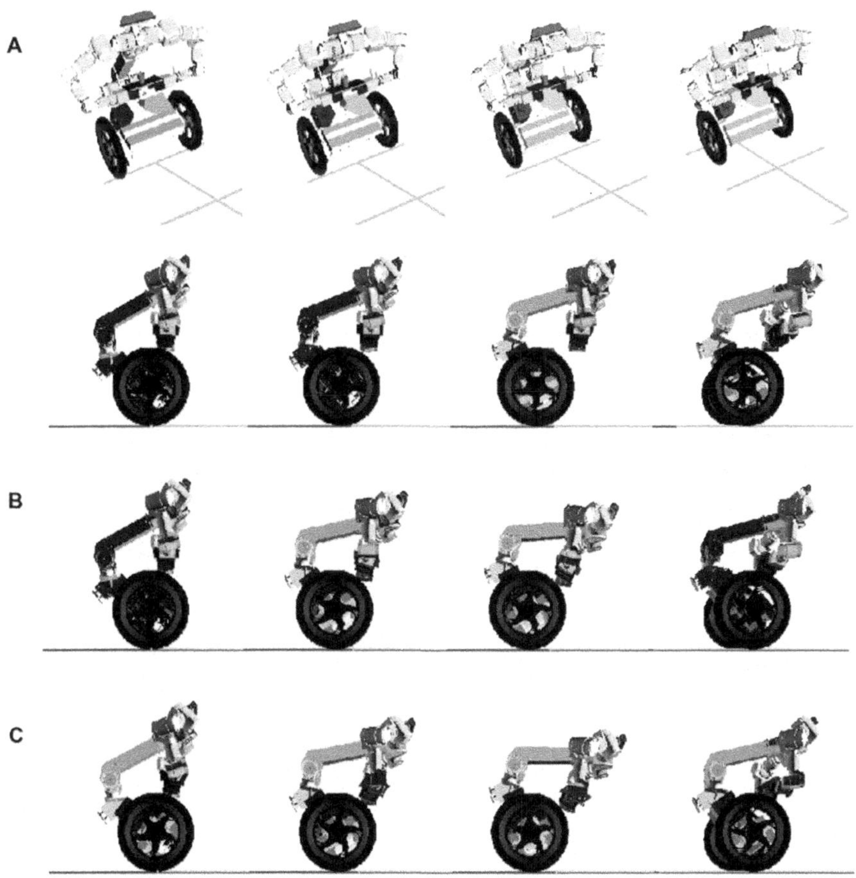

Fig. 4. Whole-body control of Krang unifying locomotion with end-effector orientation control to carry a tray with a cup. MPC based high-level controller in all cases. (A) ID based low-level controller (B) Same as A with locked waist joint. (C) Low-level controller with IK based arm control and ID based lower body control (locked waist joint).

rigid-body dynamics torque computations as shown in Fig. 6. Joint 7 is not included because we are not interested in performing learning on this joint due to its relative ease of control through feedback controller. In the joint-space, the normalized mean-square error is computed and the ratio of improvement is tabulated in Table 1.

3.3 Follow-Up Experiments

After having verified the results in simulations, we have the following set of hardware experiments lined up: (a) perform MPC based control of WIP on the hardware with all body joints locked except the wheels, (b) attempt whole-body control using IK on arms and ID on the lower body (this attempt will have degraded or unstable performance due to erroneous dynamic model estimates), (c) safe learning based data collection to train GPs for WIP dynamics, (d) probabilistic differential dynamic programming [2] for the trajectory generation in the high-level controller using the GPs trained in the previous step, and (e) semi-parametric computed torque control [1] for ID-based low-level control of the full body.

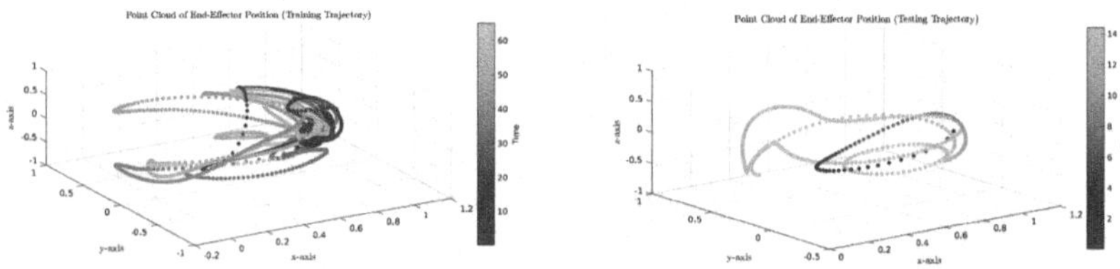

Fig. 5. Point cloud data of end-effector positions for data collection of training and testing sets.

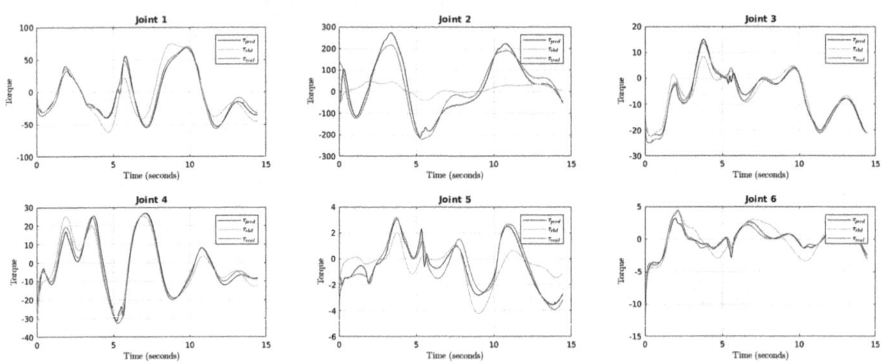

Fig. 6. Red - measured torque, Blue - GP-predicted torque, Green - RBD-based torque. The GP-predicted torque matches the measured torque fares much better than RBD-based torque.

Table 1. Table with normalized mean squared error for each joint. The error is computed between actual torque and predicted torque.

Joint 1	Joint 2	Joint 3	Joint 4	Joint 5	Joint 6
17.4	33.5	5.8	5.1	12.3	2.01

4 Experimental Insights

Some of the key experimental insights from this work are:

1. The hierarchical whole-body control approach performs significantly better (in simulation) than the traditional implementation for WIP humanoid control. For example, in our previous implementation [7], the arms are being manipulated using joint-level PID controllers on trajectories produced by IK. A separate process running an LQR-based balancing algorithm balances the robot by rapidly updating the CoM as the robot changes its pose. In our experience with this implementation, the behavior of the robot is not suited to provide guarantees to higher-level AI algorithms. As the controllers are not unified/coordinated, it is impossible to predict the effects of one on the other, and thus makes it very difficult to plan trajectories for performing complex tasks. As an example, when moving forward, change in pitch of the robot forces the end-effector's position to change if the upper body control is not coordinated with CoM motion. Similarly, during large-force interactions, planning body poses that prevent large shifts in equilibrium positions is only possible with a unified approach. In our simulation-based experiments, we have verified that the hierarchical controller performs better in executing coordinated manipulation and locomotion task.
2. For high-level controllers, the simplification of the robot's body to a single rigid link with an equivalent CoM is sufficient. In the bipedal robots, this simplification is quite common, but this is the first time a similar simplification is being done for whole-body control of a WIP humanoid. Given that WIP dynamics are more unstable, it was important to establish that a similar simplified model succeeds to capture the essential parts of the dynamics.

Furthermore, based on the ongoing experiments, we expect to inquire the following set questions:

1. Does the safe learning approach proposed by [4] for collecting data on unstable systems and demonstrated on a simulated quadrotor example, generalize to other unstable systems like WIP humanoids, and how well does it work in practice? Are they effective in improving the control policy from the policy at work while collecting data?
2. Can fast approximate GP prediction [3] and QP-based ID on semi-parametric models of a 25 DOF system be made fast enough to allow real-time control of the system?

3. Does performing DDP in the belief space [2] significantly outperform the deterministic implementation of DDP for the high-level controller, in terms of robustness and stability?

For future work, adaptive interaction of controllers operating at different time-scales is a promising area of investigation. The data collected on manual fine-tuning of the performance based on adjustment of various tuneable parameters can help develop a framework for adaptive online change of the weights/gains of the controllers according to requirements of the tasks being performed. For example, when taking a sharp turn, we may like to enforce stiffness in high-level controller objectives to make the low-level controller strictly follow its targets, but we may want to relax this stiffness when, say, moving in a straight line.

References

1. Nguyen-Tuong, D., Peters, J.: Using model knowledge for learning inverse dynamics. In: 2010 IEEE International Conference on Robotics and Automation (ICRA). IEEE (2010)
2. Pan, Y., Theodorou, E.: Probabilistic differential dynamic programming. In: Advances in Neural Information Processing Systems (2014)
3. Pan, Y., et al.: Prediction under uncertainty in sparse spectrum Gaussian processes with applications to filtering and control. International Conference on Machine Learning (2017)
4. Wang, L., Theodorou, E.A., Egerstedt, M.: Safe learning of quadrotor dynamics using barrier certificates. arXiv preprint arXiv:1710.05472 (2017)
5. Feng, S.: Online hierarchical optimization for humanoid control (2016)
6. Stilman, M., Olson, J., Gloss, W.: Golem krang: dynamically stable humanoid robot for mobile manipulation. In: 2010 IEEE International Conference on Robotics and Automation (ICRA). IEEE (2010)
7. Canete, L., Takahashi, T.: Disturbance compensation in pushing, pulling, and lifting for load transporting control of a wheeled inverted pendulum type assistant robot using the extended state observer. In: 2012 IEEE/RSJ International Conference on Intelligent Robots and Systems (IROS). IEEE (2012)
8. Kane, T.R.: Dynamics of nonholonomic systems. J. Appl. Mech. **28**(4), 574–578 (1961)
9. Featherstone, R.: Robot dynamics algorithms (1984)
10. Stilman, M., et al.: Robots using environment objects as tools the 'MacGyver' paradigm for mobile manipulation. In: 2014 IEEE International Conference on Robotics and Automation (ICRA). IEEE (2014)
11. Jacobson, D.H., Mayne, D.Q.: Differential dynamic programming (1970)
12. Tassa, Y., Erez, T., Todorov, E.: Synthesis and stabilization of complex behaviors through online trajectory optimization. In: 2012 IEEE/RSJ International Conference on Intelligent Robots and Systems (IROS). IEEE (2012)
13. Chan, R.P.M., Stol, K.A., Halkyard, C.R.: Review of modelling and control of two-wheeled robots. Annu. Rev. Control **37**(1), 89–103 (2013)
14. Nagarajan, U., Kantor, G., Hollis, R.L.: Trajectory planning and control of an underactuated dynamically stable single spherical wheeled mobile robot. In: IEEE International Conference on Robotics and Automation, ICRA 2009. IEEE (2009)

15. Nagarajan, U.: Dynamic constraint-based optimal shape trajectory planner for shape-accelerated underactuated balancing systems, pp. 27–31 (2010)
16. Zafar, M., Hutchinson, S., Theodorou, E.A.: Hierarchical Optimization for Whole-Body Control of Wheeled Inverted Pendulum Humanoids. arXiv preprint arXiv:1810.03074 (2018)
17. Zafar, M., Patel, A., Bogdan, V., Glaser, N., Aguillera, S., Hutchisnon, S.: Online center of mass estimation for a humanoid wheeled inverted pendulum robot. arXiv preprint arXiv:1810.03076 (2018)
18. Dantam, N., Stilman, M.: Robust and efficient communication for real-time multi-process robot software. In: 2012 12th IEEE-RAS International Conference on Humanoid Robots (Humanoids). IEEE (2012)

Learning to Use a Ratchet by Modeling Spatial Relations in Demonstrations

Li Yang Ku[1(✉)], Scott Jordan[1], Julia Badger[2], Erik Learned-Miller[1], and Rod Grupen[1]

[1] University of Massachusetts Amherst, Amherst, MA, USA
{lku,sjordan,elm,grupen}@cs.umass.edu
[2] National Aeronautics and Space Administration, Houston, TX, USA
julia.m.badger@nasa.gov

Abstract. We introduce a framework where visual features, describing the interaction among a robot hand, a tool, and an assembly fixture, can be learned efficiently using a small number of demonstrations. We illustrate the approach by torquing a bolt with the Robonaut-2 humanoid robot using a handheld ratchet. The difficulties include the uncertainty of the ratchet pose after grasping and the high precision required for mating the socket to the bolt and replacing the tool in the tool holder. Our approach learns the desired relative position between visual features on the ratchet and the bolt. It does this by identifying goal offsets from visual features that are consistently observable over a set of demonstrations. With this approach we show that Robonaut-2 is capable of grasping the ratchet, tightening a bolt, and putting the ratchet back into a tool holder. We measure the accuracy of the socket-bolt mating subtask over multiple demonstrations and show that a small set of demonstrations can decrease the error significantly.

1 Introduction

Learning from demonstration (LfD) is an appealing approach to programming robots due to its similarity to how humans teach each other. However, most work on LfD has focused on learning the demonstrated motion [14], action constraints [15], and/or trajectory segments [4,5] and has assumed that object labels and poses can be identified correctly. This assumption may be warranted in well-structured industrial settings, but does not hold, in general, for the kinds of uncertainty and variability common in everyday human environments.

We present an integrated approach that treats identifying informative features as part of the learning process. This gives robots the capacity to manipulate objects without fiducial markers and to learn actions focused on salient parts of the object. Actions are traditionally defined as movements relative to the pose of

Electronic supplementary material The online version of this chapter (https://doi.org/10.1007/978-3-030-33950-0_35) contains supplementary material, which is available to authorized users.

J. Xiao et al. (Eds.): ISER 2018, SPAR 11, pp. 398–410, 2020.
https://doi.org/10.1007/978-3-030-33950-0_35

a landmark; we deviate from this standard and define actions based on informative features. With additional guidance provided by the operator, the features that support actions can be identified automatically. With this approach, the robot can still interact with an object even if (1) the object does not have a global notion of pose, as in the case of an articulated object, or (2) when the object's pose is ambiguous but an affordance of that object can be identified. Two major contributions in this work are as follows.

1. Action demonstrations are classified into three different types based on the interaction between visual features and robot end effectors. This allows robots to repeat tool usage demonstrations by modeling the spatial relations between visual features from the tool and the workpiece.
2. An approach that distills multiple demonstrations of the same action to produce more accurate actions by identifying spatial relations that are consistent across demonstrations.

We show that a challenging tool use task—tightening a bolt using a ratchet—can be learned from a small set of demonstrations using our framework. A different in-hand ratchet pose may result in failure of mating the socket to the bolt if the robot only considers the pose of the hand and the bolt. The proposed approach learns what part of the ratchet should be aligned with the bolt by recognizing consistent spatial relations between features among a set of demonstrations.

2 Related Work

Much research has focused on methods for "learning from demonstration (LfD)," in which robots acquire approximate programs for replicating solutions to sensory and motor tasks from a set of human demonstrations. In work by Calinon et al. [4,5], Gaussian mixture models are used to model multiple demonstrated trajectories by clustering segments based on means and variances. In work by Pastor et al. [14], dynamic movement primitives are used to generalize trajectories with different start and end point. Instead of modeling trajectories in terms of motion invariants, our work focuses on learning consistent perceptual feedback that provides informative guidance for situated actions.

Approaches that learn from multiple demonstrations often require an experienced user to show a variety of trajectories in order to estimate task information. In work by Alexandrova et al. [2], instead of generalizing from multiple examples, the user demonstrates once and provides additional task and feature information via a user interface. The approach in this paper is similar–the user specifies demonstration types and the informative features are identified automatically.

In the work by Phillips et al. [16], experience graphs are built from demonstration to speed up motion planning. A manipulation task such as approaching a door and opening it can be planned in a single stage by adding an additional dimension that represents the degree of opening. However, the demonstrated tasks are restricted to cases where the object can be manipulated in a one dimensional manifold that is detectable. In this work, demonstrations are stored as aspect transition graphs (ATGs). ATGs are directed multi-graphs composed of aspect nodes that represent observations and edges that represent action transitions. Aspect nodes represent observations directly and can, therefore, be used to model a higher dimensional space.

ATGs were first introduced in Sen's work [17] as an efficient way of storing knowledge of objects hierarchically and are redefined as a directed multigraph to capture the probabilistic transition between observations in [12]. In previous work [10], we further demonstrated that ATG models learned from demonstrations can be used to plan sequences of actions to compensate for the robot's reachability constraint. In this work, we extend the ATG representation to model interaction between objects and show that by distilling multiple ATGs learned from demonstrations the accuracy of actions can increase significantly.

In the work by Akgun et al. [1], a demonstrator provides a sparse set of consecutive keyframes that summarizes trajectory demonstrations. Pérez-D'Arpino and Shah [15] also introduced C-Learn, a method that learns multi-step manipulation tasks from demonstrations as a sequence of keyframes and a set of geometric constraints. In our work, aspect nodes that contain informative perceptual feedback play a similar role as keyframes that guide the multi-step manipulation. Instead of considering geometric constraints between an object frame and the end effector frame, relations between visual features and multiple robot frames are modeled.

In the work by Finn et al. [6], states that are represented by visual feature locations on the image plane are learned through deep spatial autoencoders. These features are then filtered and pruned based on feature presence for manipulation tasks learned through reinforcement learning. In our work, features are generated from a network trained on image classification and are selected based on consistency in response and spatial variance across demonstrations.

There has been a lot of work on developing visual descriptors that are robust to viewpoint variations [3,13,19]. Recently, several papers have investigated learning image descriptions using Convolutional Neural Networks (CNNs) [7,18]. In this work, we use the hierarchical CNN features [11] that can represent parts of an object that are informative for manipulation. In the work done by Huang and Cakmak [8], a tool for setting custom landmarks that can represent object parts is introduced. In our work, object parts are represented by a set of features identified based on the demonstration type.

3 Approach

In this section, we describe our approach by teaching the robot to use a ratchet from demonstrations. First, we provide background on the aspect transition graph (ATG) that is used to model demonstrations. Second, we describe how we classify demonstrations into three types. Third, we explain how user demonstrations are used to build ATG models and how these models can be used for planning. Last, we illustrate how multiple ATGs created from demonstrations are merged to create more robust models.

3.1 Aspect Transition Graph Model

In this work, aspect transition graphs (ATG) created from demonstrations are used to represent how actions lead from one observation to another. An aspect transition graph (ATG) is a directed multigraph $G = (\mathcal{X}, \mathcal{U})$, composed of a set of aspect nodes $\mathcal{X}$ connected by a set of action edges $\mathcal{U}$ that capture the probabilistic transition between aspect nodes.

We define an aspect as a multi-feature observation that is stored in the model. In this work, an aspect node stores an aspect representation composed of visual, force, and proprioceptive feedbacks. The visual feedback is represented by hierarchical CNN features introduced in [11]. Instead of representing a feature with a single filter in a certain CNN layer, hierarchical CNN features use a tuple of filter indices to represent a feature such as (f_i^5, f_j^4, f_k^3), where f_i^n represents the ith filter in the nth convolutional layer. These features can represent hierarchcial local structures of an object and be localized in 3D to support actions. The force feedback is based on load cells in Robonaut-2's forearms. Force values are projected to the body frame at 10 Hz and averaged over the preceding one-second interval. The force information is used to distinguish between aspect nodes. The proprioceptive feedback is composed of robot frames calculated based on motor encoders and is used to support actions.

In previous work, a single object was considered per ATG model [10]. In this work, we consider multiple objects and their interactions. An aspect node can be used to represent a particular "view" of an object or a distinctive interaction between objects. For example, two disjoint feature clusters generated by two objects are modeled by two aspect nodes, each representing how the robot perceives them. In contrast, a single feature cluster can span two (partially) assembled objects to focus on object-object interactions. The ATG representation can therefore model object interactions that result in transitions between these two types of aspect nodes.

3.2 Demonstration Types

In our learning from demonstration framework, tasks are demonstrated as a sequence of actions. An action is represented using a controller in the control basis framework [9] and is written in the form $\phi|_\tau^\sigma$, where ϕ is a potential function that describes the error between the current and target robot configuration, σ represents sensory resources allocated, and τ represents the motor resources allocated. The potential functions are formulated as $\phi_V = \sum_{v \in V}(v - g_v)^2$, where v and g_v are visual features and goal locations for these features ($v, g_v \in \mathbb{R}^3$) and $\phi_R = \sum_{r \in R}(r - g_r)^2$, where r and g_r are robot frames and goals for these frames ($r, g_r \in SE(3)$). Each demonstrated action is classified into one of the following three demonstration types based on the interacting features:

(a) The *robot-visual action* ($a_{RV} = \phi_\mathrm{R}|_\tau^{\sigma_\mathrm{V}}$) specifies the target pose of a set of robot frames with respect to a set of 3-D visual feature locations. The left column of Fig. 1 shows an example of executing an a_{RV} action where the goal is to reach the ratchet pre-grasp pose. The yellow and cyan dots are visual feature

locations in 3-D based on hierarchical CNN features [11] and the red and green circles represent the goals for the hand and fingers. The arrows represent the learned offset from features to goal locations.

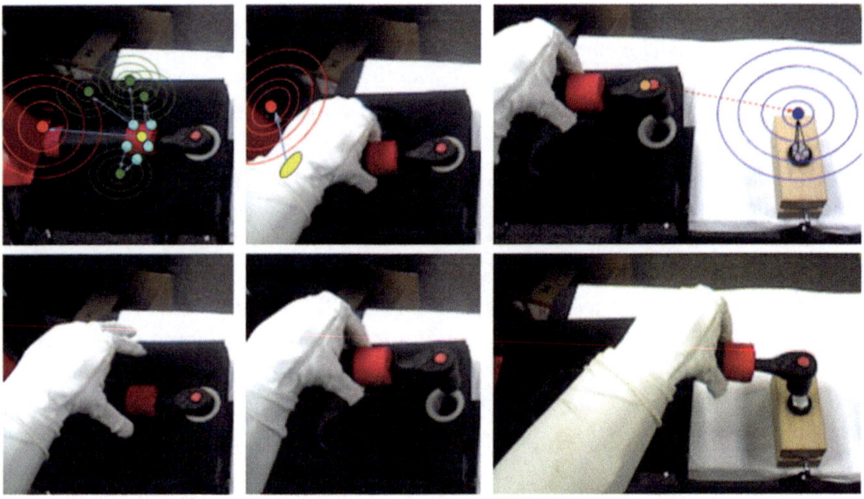

Fig. 1. Examples of the three demonstration types: robot-visual actions a_{RV}, robot-proprioceptive actions a_{RP}, and visual-visual actions a_{VV}.

(b) The *robot-proprioceptive action* ($a_{RP} = \phi_{\text{R}}|_{\tau}^{\sigma_{\text{P}}}$) specifies the target pose of a set of robot frames with respect to a set of current robot frames based on proprioceptive feedback. The middle column of Fig. 1 shows an example of executing an a_{RP} action where the goal is to move the hand relative to the current hand frame so that the grasped ratchet is extracted from the tool holder. The yellow ellipse is the current hand pose and the arrow indicates the reference offset derived from demonstration.

(c) The *visual-visual action* ($a_{VV} = \phi_{\text{V}}|_{\tau}^{\sigma_{\text{V}'}}$) specifies the goal position of a set of controllable visual features relative to another set of visual features on a different object in 3-D. The right column in Fig. 1 shows an example of executing an a_{VV} action where the goal is to place the socket on top of the bolt. The purple dots are features on the bolt used as references for the goal and the orange dot is the feature on the socket. The blue dots are goal positions generated based on relative positions to features indicated by the black arrows. Modeling spatial relations between visual features achieves the same intended outcome even when the in-hand ratchet poses are different. This visuo-servoing approach re-observes after movements until convergence and is therefore more robust to kinematic noise and warped point clouds.

The detected locations of visual features and robot frames are inevitably influenced by noise in the system that may be caused by imperfect sensors or changes in the environment. This makes tasks that require high precision challenging. To accommodate this problem we assume that the references for motor resources τ is generated by adding zero mean noise $N(0, \Sigma)$ to the original reference. By sampling from this distribution during execution, the controller superimposes an additive zero mean search to the motion. Such movement increases the tolerance of the insertion task to uncertainty.

Figure 2 shows the sensorimotor architecture that drives transitions in the ATG model. The perceptual feedback is used to represent aspect nodes and actions are executed based on these sensory resources defined in action edges.

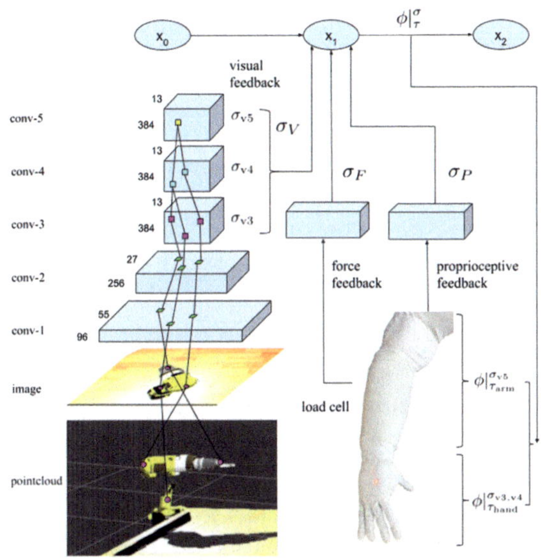

Fig. 2. The sensorimotor architecture driving transitions in the ATG framework. The sensory resources σ_{F} that represent a set of features based on visual and force feedback and σ_{P} that represents a set of robot frames based on proprioceptive feedback are used to parameterize actions $\phi|_{\tau}^{\sigma}$. In this example, the 5th layer hierarchical CNN features σ_{v5} are used to control the arm motors τ_{arm} and the 3rd and 4th layer hierarchical CNN features $\sigma_{\mathrm{v3,v4}}$ are used to control the hand motors τ_{hand}.

3.3 Building and Planning with ATG Models

Each demonstration coupled with information provided by the operator is used to create an ATG model. Demonstrations are performed through teleoperation, in which the user drags interactive markers in a graphical interface to move the robot end effector or change the robot hand configuration. Users indicate intermediate steps for each demonstration and provide its demonstration type described in Sect. 3.2. The user also has the option to add a search movement to the demonstrated action. During the demonstration, an aspect node is created for each observed feature cluster at each intermediate step. A feature cluster can be a single object or multiple objects in contact. Based on the demonstration type selected by the user, the system connects new aspect nodes x_t to aspect nodes x_{t-1} created at the previous time step with action edges that store the demonstrated action a_{t-1}.

During execution, the user selects a goal aspect. Based on the maximum a posteriori (MAP) aspect node, the next action is chosen based on the first action edge on the shortest path from the MAP aspect node to the goal aspect node. The posterior probability is modeled by generalized Gaussian distributions as in [10]. If there is no valid path, the planner guesses possible paths by merging similar aspect nodes from the current ATG to other ATGs until a path exists.

3.4 Learning from Multiple Demonstrations

With a single demonstration, there remain ambiguities regarding the goal. For example, in the action that puts the socket on top of the bolt, it is ambiguous whether the demonstration intends to convey a spatial relationship between the socket and the bolt or some other part of the ratchet and the bolt. With multiple demonstrations, this ambiguity may be resolved by observing consistent relations between features. In this section, we describe how to take multiple demonstrations of the same task to create more robust ATG models. We call these ATGs created from multiple demonstrations *distilled* ATGs.

3.4.1 Identifying Common Features

A set of features are stored in the aspect node to represent the observation of an aspect. Correctly associating the current observations with a memorized aspect node is crucial for implementing transitions to goal status. However, not all features provide the same amount of information. Moreover, some features are more sensitive to lighting changes and some may belong to parts of the visual cluster that may change appearance across examples. With a single demonstration, these kinds of features may be indistinguishable. With multiple demonstrations, common features can be identified by estimating the feature variance across demonstrations.

Given demonstrations of the same task with the same sequence of intermediate steps, our approach looks for features that are consistent across multiple demonstrations. For the observations at each intermediate step, the N most consistent features are chosen. The consistency score is defined as $S_c = n_f/std(f)$, where n_f is the number of times feature f appears among the matched intermediate steps and $std(f)$ is the standard deviation of the value of feature f. We score visual features, proprioceptive features, and force features together with weights of $1, 1, 0.001$, respectively.

3.4.2 Recognizing Consistent Actions

For action edges that represent a *robot-visual action* a_{RV} or a *visual-visual action* a_{VV} in an ATG model, the action reference is specified in terms of a subset of features stored in the aspect node. As result of a single demonstration, features are chosen based on their proximity to robot frames or features controllable by the robot. With multiple demonstrations, a more robust set of features can be identified and used to define the aspect.

For the *robot-visual action* $a_{RV} = \phi_{\mathrm{R}}|_{\tau}^{\sigma_{\mathrm{V}}}$, the top N pairs of robot frames $r \in R$ and visual features $v \in V$ that have the lowest variances in XYZ position offsets are chosen to represent the action. For example, when learning from multiple demonstrations of the action that grasps the ratchet, this approach concludes that features on the ratchet are more reliable than features on the tool holder since the ratchet may be placed at different positions in the tool holder across demonstrations.

For the *visual-visual action* $a_{VV} = \phi_{\mathrm{V}}|_{\tau}^{\sigma_{\mathrm{V}'}}$, the top N pairs of visual features in the tool aspect node $v \in V$ and the target object aspect node $v' \in V'$ that have the lowest variance $var(v, v')$ is selected. $var(v, v')$ is the variance of the XYZ position offsets between feature v and feature v' after the action across demonstrations. For example, the action that places the socket of the ratchet on top of the bolt determines that a consistent spatial relation exists between the features on the socket and those on the bolt after executing the action. Figure 3 shows the top feature pairs identified for constructing a *visual-visual action* from demonstrations. The robot is able to comprehend that the head of the ratchet should be aligned with the bolt autonomously.

Fig. 3. Identifying informative features from multiple demonstrations. The two rows represent two demonstrations that place the socket of the ratchet on top of the bolt. The columns from left to right show the aspect nodes representing the tool, the target object, and the interaction. The green and red circles represent the most informative features identified for this *visual-visual action.*

4 Experiments

In this work, we show that with a small set of demonstrations, Robonaut-2 is capable of performing a ratchet task that involves grasping the ratchet, tightening a bolt, and putting the ratchet back into a tool holder. The complete task sequence is shown in Fig. 4 and can be seen in the supplementary video. We further compare the success rate of mating the socket to the bolt as a function of the number of demonstrations and the size of the feature space. The experimental settings and results are described in the following.

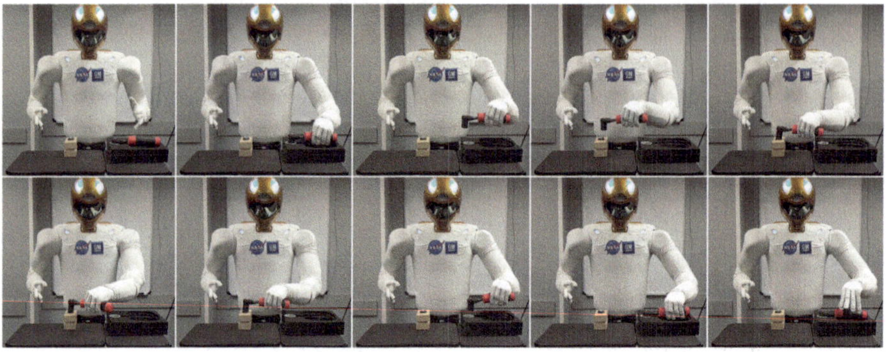

Fig. 4. The ratchet task sequence performed by Robonaut-2. The images from left to right, then top to bottom, show a sequence of actions where Robonaut-2 grasps the ratchet, tightens a bolt on a platform, and puts the ratchet back into a tool holder.

4.1 Demonstrations

Instead of demonstrating the entire ratchet task in one session, we segment the task into shorter sequences of sub-tasks that are easier to demonstrate. The ratchet task is segmented into five different subtasks, *(a)* grasping the ratchet, *(b)* mating socket to the bolt, *(c)* tightening the bolt, *(d)* removing the socket from the bolt, and *(e)* putting the ratchet back into the tool holder. For subtasks *(a)*, two demonstrations are provided. For subtask *(b)* and *(e)* four demonstrations are combined to create the distilled ATG model as described in Sect. 3.4. For subtasks *(c)* and *(d)*, only one demonstration is performed since the features that support these actions are unambiguous.

4.2 Evaluation

The robustness of the framework is tested on the ratchet task based on the ATGs created from demonstrations. During execution, the aspect where the bolt is tightened is first submitted as a goal aspect to the robot. The planner identifies the current aspect node and finds a path to reach the goal aspect. Once the robot finishes tightening the bolt, the aspect where the ratchet is put back to the tool holder is set as the goal aspect. A total of 22 settings are tested. For each setting, the initial location of the tool holder or bolt platform is altered. These initial poses are shown in Fig. 5. The number of successes for each subtask are shown in Table 1. Mating socket with the bolt and placing the ratchet back have 86.3% and 81.8% success rate. 14 out of 24 trials succeeded the complete task.

Table 1. Number of successful trials on subtasks.

Subtask on ratchet	Grasp	Mate	Tighten	Lift	Place	Complete task
Successful trials/total trials	22/24	19/22	18/19	22/22	18/22	14/24

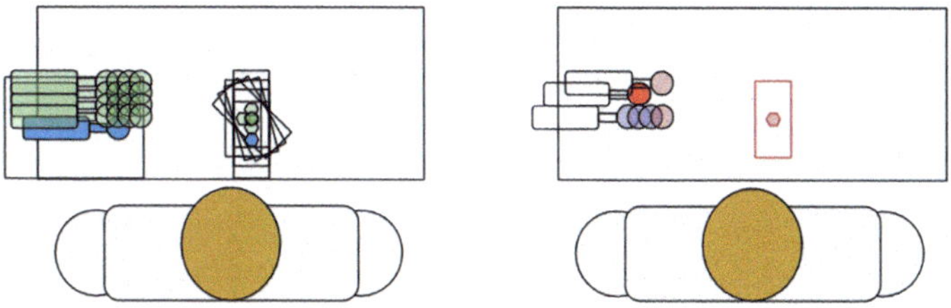

Fig. 5. Top down views of the ratchet task setting. The green objects in the left image are tested poses and the blue objects are the demonstrated pose. The pink objects in the right image shows poses that failed to mate the socket with the bolt and the purple objects show poses that failed to place the ratchet back.

13 corner case settings are further tested on mating the socket with the bolt. This set contains test cases with initial ratchet positions that are close to the sensor and joint limit, in hand ratchet positions that are at opposite ends, and cluttered scenarios. Our approach achieved a similar success rate of 84.6%. Figure 6 shows some of the initial settings.

Fig. 6. Corner case initial settings for mating the socket with the bolt. Note that in the 3rd and 4th image the in-hand ratchet positions are different.

4.3 Comparison

To understand how the number of demonstrations and the size of the visual feature space affect the learned action, we compare the success rates of mating the socket to the bolt under different configurations. In this experiment, we compare the robustness of ATGs created from one to four demonstrations and with hierarchical CNN features in the 3rd and 4th layer. Hierarchical CNN features in the 3rd layer $H^3 = (f_i^5, f_j^4, f_k^3)$, represents a feature with an additional filter f_k^3 and have a feature space $|f^3| = 384$ times larger compared to features in the 4th layer $H^4 = (f_i^5, f_j^4)$, where $|f^3|$ is the number of filters in the conv-3 layer.

Our assumption is that more complex features will require more demonstrations to learn, but may result in more robust actions. For each trial, the robot starts with the grasped ratchet and the bolt placed on the right side of the robot. The trial succeeds if the robot mates the socket to the bolt. We performed 22 trials for each ATG. The results are shown in Fig. 7.

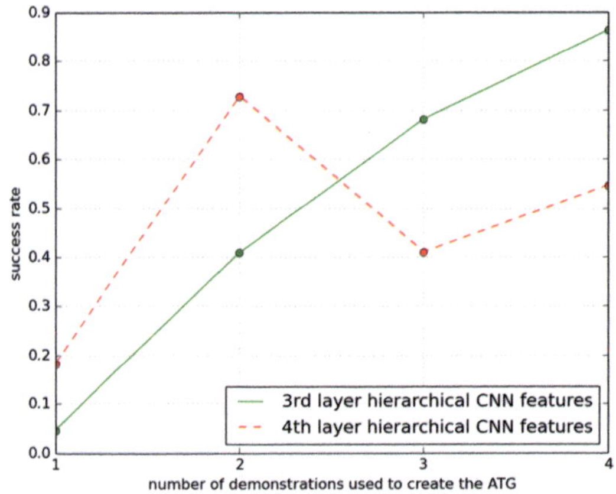

Fig. 7. Success rate versus number of demonstrations and size of feature space

Consistent with our expectations, the success rate of using H^3 features increases with more demonstrations and performs better than H^4 features when more demonstrations are used. The results for using H^4 features however fluctuates with more than two demonstrations. We suspect that this is because H^4 features have a smaller feature space and good features can be found with fewer demonstrations. The up and down in success rate with more demonstrations may be due to imperfect demonstrations and H^4 features that are less precise in location.

5 Conclusion

In this work, we introduced a learning from demonstration approach that learns both actions and features. Categorizing demonstrations into three different types allows the system to define the goal of the task by modeling the spatial relations between features automatically. We show that through multiple demonstrations, informative visual features and relative poses can be identified and used to model actions that are more accurate than models of single demonstrations. This effect is clearly observed in the improvement in success rate over single demonstration models when mating the socket to the bolt. Our experiments also indicate that the larger the feature space is the more demonstrations are needed to achieve robust actions. With this proposed approach, Robonaut-2 is capable of grasping the ratchet, tightening a bolt, and putting the ratchet back into a tool holder with a small set of demonstrations.

Acknowledgements. We are thankful to our colleagues Dirk Ruiken, Mitchell Hebert, Jay Wong, Michael Lanighan, Tiffany Liu, Takeshi Takahashi, and former members of the Laboratory for Perceptual Robotics for their contribution on the control basis framework code base. We are also grateful to Philip Strawser, Jonathan Rogers, Logan Farrell, Kimberly Vana, Evan Laske, and the Robonaut Team for their support on Robonaut-2. This material is based upon work supported under Grant NASA-GCT-NNX12AR16A, the Air Force Research Laboratory and DARPA under agreement number FA8750-18-2-0126, and a NASA Space Technology Research Fellowship. Any opinions, findings, and conclusions or recommendations expressed in this material are those of the authors and do not necessarily reflect the views of the National Aeronautics and Space Administration, the Air Force Research Laboratory, DARPA, or the U.S. Government.

References

1. Akgun, B., Cakmak, M., Yoo, J.W., Thomaz, A.L.: Trajectories and keyframes for kinesthetic teaching: a human-robot interaction perspective. In: Proceedings of the Seventh Annual ACM/IEEE International Conference on Human-Robot Interaction, pp. 391–398. ACM (2012)
2. Alexandrova, S., Cakmak, M., Hsiao, K., Takayama, L.: Robot programming by demonstration with interactive action visualizations. In: Robotics: Science and Systems (2014)
3. Bay, H., Tuytelaars, T., Van Gool, L.: SURF: speeded up robust features. In: Computer Vision–ECCV 2006, pp. 404–417. Springer, Heidelberg (2006)
4. Calinon, S., Billard, A.: A probabilistic programming by demonstration framework handling constraints in joint space and task space. In: IEEE/RSJ International Conference on Intelligent Robots and Systems. IROS 2008, pp. 367–372. IEEE (2008)
5. Calinon, S., Guenter, F., Billard, A.: On learning, representing, and generalizing a task in a humanoid robot. IEEE Trans. Syst. Man Cybern. Part B (Cybern.) **37**(2), 286–298 (2007)
6. Finn, C., Tan, X.Y., Duan, Y., Darrell, T., Levine, S., Abbeel, P.: Deep spatial autoencoders for visuomotor learning. Reconstruction **117**(117), 240 (2015)
7. Fischer, P., Dosovitskiy, A., Brox, T.: Descriptor matching with convolutional neural networks: a comparison to sift. arXiv preprint arXiv:1405.5769 (2014)
8. Huang, J., Cakmak, M.: Flexible user specification of perceptual landmarks for robot manipulation. In: 2017 IEEE/RSJ International Conference on Intelligent Robots and Systems (IROS), pp. 3296–3303. IEEE (2017)
9. Huber, M.: A hybrid architecture for adaptive robot control. Ph.D. thesis, University of Massachusetts Amherst (2000)
10. Ku, L.Y., Learned-Miller, E., Grupen, R.: An aspect representation for object manipulation based on convolutional neural networks. In: 2017 IEEE International Conference on Robotics and Automation (ICRA), pp. 794–800. IEEE (2017)
11. Ku, L.Y., Learned-Miller, E.G., Grupen, R.A.: Associating grasp configurations with hierarchical features in convolutional neural networks. In: 2017 IEEE International Conference on Intelligent Robots and Systems (IROS). IEEE (2017)
12. Ku, L.Y., Sen, S., Learned-Miller, E.G., Grupen, R.A.: Action-based models for belief-space planning. In: Workshop on Information-Based Grasp and Manipulation Planning, at Robotics: Science and Systems (2014)

13. Lowe, D.G.: Distinctive image features from scale-invariant keypoints. Int. J. Comput. Vis. **60**(2), 91–110 (2004)
14. Pastor, P., Hoffmann, H., Asfour, T., Schaal, S.: Learning and generalization of motor skills by learning from demonstration. In: IEEE International Conference on Robotics and Automation. ICRA 2009, pp. 763–768. IEEE (2009)
15. Pérez-D'Arpino, C., Shah, J.A.: C-learn: learning geometric constraints from demonstrations for multi-step manipulation in shared autonomy. In: IEEE International Conference on Robotics and Automation (2017)
16. Phillips, M., Hwang, V., Chitta, S., Likhachev, M.: Learning to plan for constrained manipulation from demonstrations. In: Robotics: Science and Systems, vol. 5 (2013)
17. Sen, S.: Bridging the gap between autonomous skill learning and task-specific planning. Ph.D. thesis, University of Massachusetts Amherst (2013)
18. Simo-Serra, E., Trulls, E., Ferraz, L., Kokkinos, I., Fua, P., Moreno-Noguer, F.: Discriminative learning of deep convolutional feature point descriptors. In: Proceedings of the IEEE International Conference on Computer Vision, pp. 118–126 (2015)
19. Tuytelaars, T., Mikolajczyk, K.: Local invariant feature detectors: a survey. Found. Trends® Comput. Graph. Vis. **3**(3), 177–280 (2007)

Any-Axis Tensegrity Rolling via Symmetry-Reduced Reinforcement Learning

David Surovik[1(✉)], Jonathan Bruce[2], Kun Wang[1], Massimo Vespignani[2], and Kostas Bekris[1]

[1] Computer Science Department of Rutgers University, Piscataway, NJ, USA
{ds1417,kw423,kostas.bekris}@cs.rutgers.edu
[2] NASA Ames Intelligent Robotics Group, Mountain View, CA, USA
{jonathan.bruce,massimo.vespignani}@nasa.gov

Abstract. Tensegrity rovers incorporate design principles that give rise to many desirable properties, such as adaptability and robustness, while also creating challenges in terms of locomotion control. A recent milestone in this area combined reinforcement learning and optimal control to effect fixed-axis rolling of NASA's 6-bar spherical tensegrity rover prototype, SUPERball, with use of 12 actuators. The new 24-actuator version of SUPERball presents the potential for greatly increased locomotive abilities, but at a drastic nominal increase in the size of the data-driven control problem. This paper is focused upon unlocking those abilities while crucially moderating data requirements by incorporating symmetry reduction into the controller design pipeline, along with other new considerations. Experiments in simulation and on the hardware prototype demonstrate the resulting capability for any-axis rolling on the 24-actuator version of SUPERball, such that it may utilize diverse ground-contact patterns to smoothly locomote in arbitrary directions.

1 Introduction

Tensegrity rovers are dynamic truss structures that can deform passively, so as to avoid local stress accumulation, and actively, to locomote [6,13]. Their passive properties aid active control through morphological computation [4], highlighting a close kinship with soft robots. Among other favorable traits, the adaptability and resilience of tensegrity structures makes them appealing for robotic exploration of extreme environments. This has recently led to the second hardware iteration of NASA's self-landing rover prototype, SUPERball (SBv2) [17], which comprises 6 rigid bars and 24 length-actuated elastic cables, shown in Fig. 1.

The same structural properties that motivate use of mobile tensegrities also make them challenging to control due to high system dimensionality, nonlinearity and coupling. Model-based approaches have been applied effectively for structural shape control [14] and can be adapted for locomotion [8]. Nonetheless, an impractical degree of

---

Electronic supplementary material The online version of this chapter (https://doi.org/10.1007/978-3-030-33950-0_36) contains supplementary material, which is available to authorized users.

J. Xiao et al. (Eds.): ISER 2018, SPAR 11, pp. 411–421, 2020.
https://doi.org/10.1007/978-3-030-33950-0_36

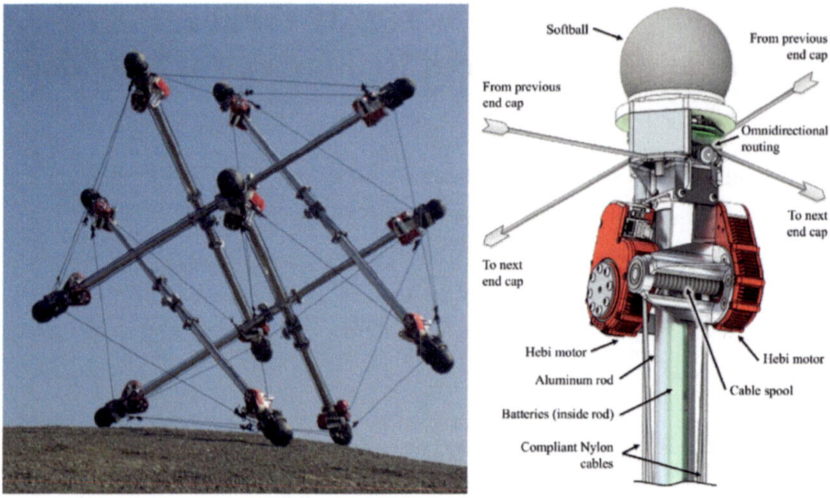

Fig. 1. Left: SUPERball version 2 at NASA Ames Research Center. Right: at either end of a rigid bar (in compression), motorized spools actuate the rest-lengths of elastic cables (in tension) [17].

state knowledge is typically required, and the neglect of the dynamical subtleties of a frequently changing contact state may hinder performance. This is increasingly leading to the application of data-driven approaches, such as evolutionary algorithms and machine learning, for controlling mobile tensegrities [2,3,5,9,12].

1.1 Background

The context and contributions of this paper are better understood by first considering the topology of SUPERball, shown in Fig. 2. The edges of its convex hull consist of the 24 cables (solid black lines) and 6 "virtual" edges (dotted lines). Triangular faces of the hull are bordered by either three cables (Δ faces, light blue) or two (Λ faces, not shaded). The dotted gray box in 3D defines a cut that corresponds to the gray borders in 2D. Thick beige lines are bars, interior to the convex hull.

The previous version of SUPERball (`SBv1`) featured an identical structural topology to `SBv2` but could only actuate half of its cables, shown by the thicker black lines in Fig. 2. This arrangement constrained locomotion to follow one repeating pattern of contact states progressing horizontally on the diagram, i.e., periodic fixed-axis rolling illustrated by the red trajectory. A control policy for generating this motion was obtained for `SBv1` using Guided Policy Search (`GPS`) [18], a reinforcement learning method that will be described in Sect. 2.1. Ultimately, computing this policy required use of `GPS` individually for each of six sub-sequences of the contact pattern before finally merging the corresponding neural networks.

The fully-actuated nature of `SBv2`, along with an increased range of motion of each cable, permits a much broader degree of shape control [17]. This gives rise to the new possibility of any-axis locomotion, i.e., the ability move in arbitrary directions by using

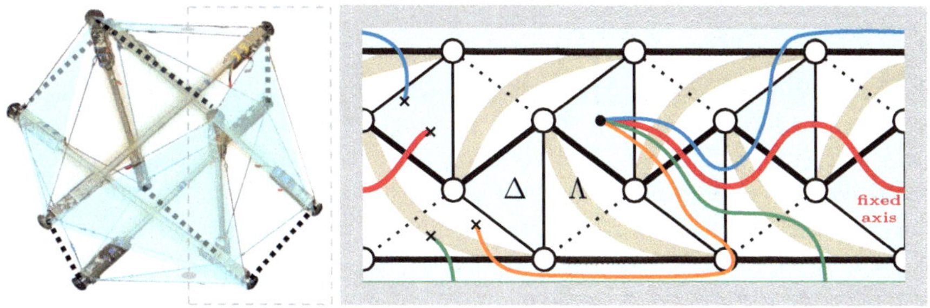

Fig. 2. Topology of SUPERball in 3D and 2D. Colored lines represent paths of the center-of-mass projected downward onto the convex hull of the vehicle during locomotion.

other contact patterns and transition geometries as shown by the green, blue, and orange paths in Fig. 2. Achieving such behaviors under the approach of [18] would require a drastic increase in the number of policy search instances and thus in experimentation costs, among other potential issues. This work incorporates additional features into the policy search pipeline to produce a deployable control policy for any-axis locomotion of SBv2 while avoiding any increase in sample data requirements.

2 Approach

2.1 Prior Tools

Guided Policy Search (GPS) is a reinforcement learning method that leverages optimal control principles to reduce the amount of sampled experimental data required to compute a control policy [7]. This technique fits linear time-varying dynamics $\mathbf{x}_{t+1} = f_i(t, \mathbf{x}_t, \mathbf{u}_t)$ in numerous local regions i and computes optimal linear feedback policies $\mathbf{u}_t = p_i(t, \mathbf{x}_t)$. Then, a single neural network policy $\mathbf{u}_t = \pi(\mathbf{y}_t(\mathbf{x}_t))$ is trained to match the set of local policies given sensor data $\mathbf{y}_t$. Subject to considerations of observability, this allows the comprehensive state knowledge available within a simulator to be leveraged when optimizing a policy that can be deployed under limited sensing. The process, sketched in Fig. 3, is iterated using new samples generated by the neural network in order to converge upon a high-performing global policy.

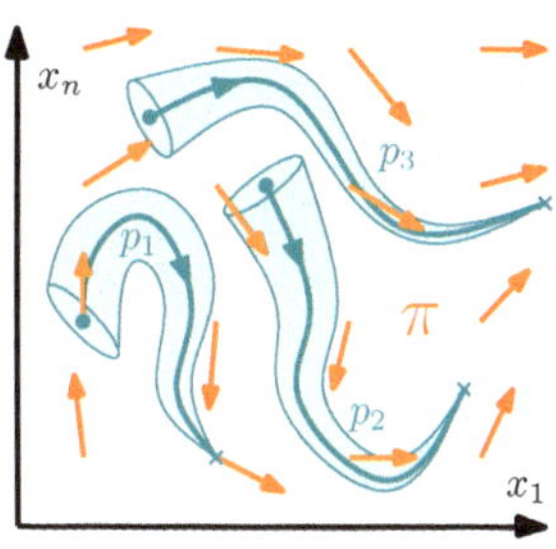

Fig. 3. Closed-loop dynamics; GPS trains a neural network π to match local optimal control laws p_i.

Model-Based Control faces some limitations when applied to tensegrity motion, due to the influence of evolving contact states and the complications of gradient-based operations on a high-dimensional state. Nonetheless, certain simplifications can be used to generate useful controls with reasonable compute effort. Under full knowledge of the vehicle's current shape and assumptions about its contact state, it is possible to solve for cable length changes that displace the center-of-mass enough to cause locomotion via instability [8]. An apriori policy such as this can be used to reduce the required number of GPS iterations relative to an initial null policy.

Exploitation of Vehicle Symmetry furthermore reduces the amount of data necessary per iteration of GPS. Due to the highly regular topology of SBv2, there exist 24 operations H_j that permute the labels of individual structural elements and transform the spatial reference frame without altering the intrinsic dynamics of the system. The policy may then take the form $\mathbf{u} = H_j^{-1}\pi(H_j\mathbf{y})$. With proper selection of a ruleset $j = j(\mathbf{y})$, the required coverage of the observation space $\mathscr{Y}$ is condensed by a factor of 24 to the subset $\mathscr{Y}_R$, as shown in Fig. 4 [15].

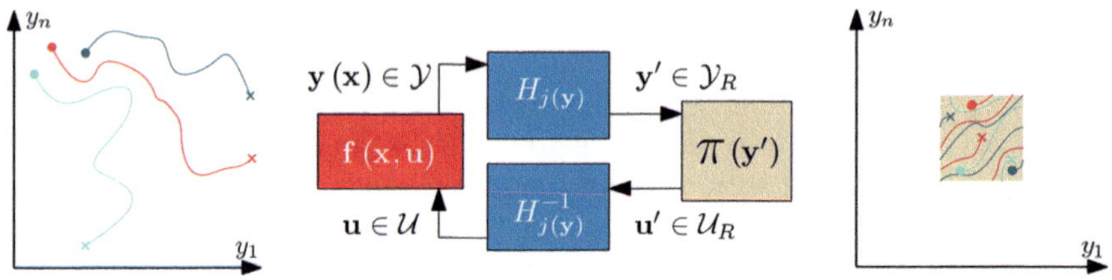

Fig. 4. Trajectories in raw (left) and symmetry-reduced (right) observation spaces. Center: the data-driven feedback policy controls the true state using only symmetry-reduced knowledge.

2.2 Policy Search Pipeline

Overview: The basic procedure for reinforcement learning is to obtain sample data by executing a control policy under various conditions, update the control policy based on how well each sample performed, and repeat. As in prior work [18], NASA's Tensegrity Robotics Toolkit [1] is used as a simulation testbed for generating samples. To bootstrap the learning process, the simple model-based controller [8] is used as the initial policy π_0. Figure 5 gives the pipeline for improving the policy over one iteration of GPS. Use of sample segmentation and of dynamics fitting within a feature space, rather than the full state space, are primary new aspects.

Segmentation and Classification of sample data is necessary for coping with the broad variety of motion under any-axis rolling, where nearby trajectories can rapidly diverge. Trajectories are thus segmented whenever the ground-projection of the center-of-mass crosses an edge of the convex hull (i.e., when a black line is crossed in Fig. 2). Segment

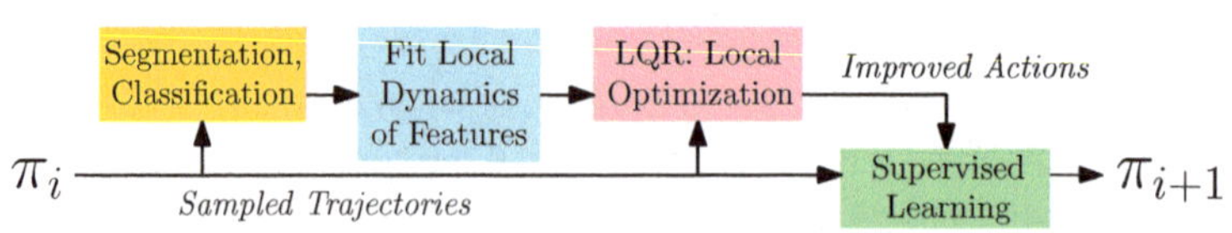

Fig. 5. The flow from sample data to policy improvement over a single iteration of GPS.

categories are then assigned by the type of bottom triangle, Δ or Λ, as well as the relationship between the edge crossed *onto* it and the edge crossed *off of* it.

This process interplays with symmetry reduction in order to enable gradient-based optimization despite the use of a non-differentiable black-box dynamics testbed. For example, left-turning segments on Δ faces exhibit relatively similar dynamics and therefore can contribute to a common local model to be used for computing improvements to the corresponding actions. The prior fixed-axis rolling controller had relied upon short, similar motion samples of a single category to ensure this apriori [11].

Dynamical Feature Space: Even when sample data are well-organized, high dimensionality impedes the fitting of local dynamics models. Lower-dimensional features are instead selected based upon intuition about the system dynamics: the primary external influences are normal and friction forces on contact nodes and gravity acting upon the center-of-mass. The center-of-mass-relative positions of the bottom triangle nodes, (ρ_A, ρ_B, ρ_C) are thus chosen as elements of the feature state χ.

Cost Function: The objective is simply to match the center-of-mass velocity $\bar{\mathbf{v}}$ to a reference $\bar{\mathbf{v}}^*$, giving linear quadratic cost $J = (\bar{\mathbf{v}} - \bar{\mathbf{v}}^*)^T W (\bar{\mathbf{v}} - \bar{\mathbf{v}}^*)$. The weight matrix W penalizes ground-plane components equally and de-weights the vertical. Thus, fitting dynamics of $\chi = (\rho_A, \rho_B, \rho_C, \bar{\mathbf{v}})$ provides enough information to improve the control time series while ignoring less well-behaved state variables.

Observation Space: The supervised learning phase trains on tuples $(\mathbf{y}', \mathbf{u}'')$, where $\mathbf{y}'$ are sensor-derived inputs and $\mathbf{u}''$ are improved associated actions. The observation vector contains the 24 cable rest lengths, the 6 bars' angular velocity vectors, the ID of the bottom triangle, and the commanded direction of motion.

3 Simulated Results

3.1 Single-Shape Controller

Preliminary efforts to train a neural network policy with a model-based supervisor yielded a noteworthy result: an undertrained network produced a constant output, i.e., the policy $\pi_C(H_{j(\mathbf{y})}\mathbf{y}) = \mathbf{u}^*$. This represents single shape that is a broad average of all shapes experienced by the locomoting model-based controller. Embodying this shape causes a transition between support faces, triggering a change in the raw control $\mathbf{u} = H_j^{-1}\mathbf{u}^*$ via the symmetry-based relabeling alone. The resulting sustained motion, seen in Fig. 6, resembles a simple "step-wise" paradigm [16] directable along

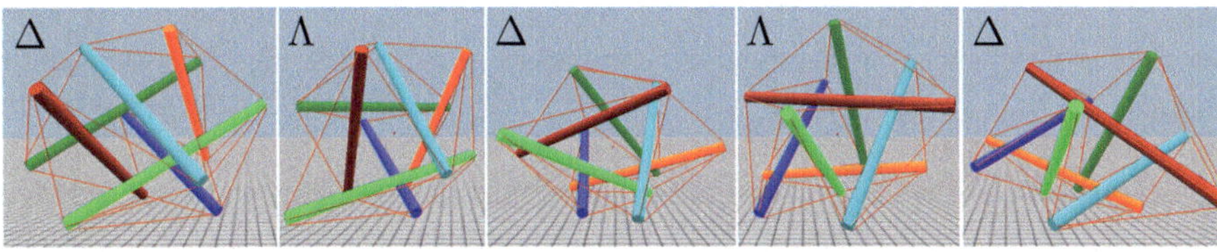

Fig. 6. A sequence of states as the Single-Shape control policy locomotes SBv2 to the right.

any ground contact edge-normal, or the periodic fixed-axis motion of [18] and the red trajectory of Fig. 2. This *Single-Shape* controller thus serves as a useful point of comparison for the desired any-axis control.

3.2 Any-Axis Performance

After basic tuning of the neural network training phase, the application of the policy search update step described by Fig. 5 was repeatedly applied. This iteratively reduces the average sample cost J, corresponding to steady and effective directed locomotion with minimal side-to-side motion. Figure 7 illustrates this improvement with top-down views of typical center-of-mass trajectories for three controllers attempting to visit a sequence of waypoints. A diagram of the robot marks its starting position and size; its center-of-mass must reach the interior of each gray waypoint before proceeding to the next.

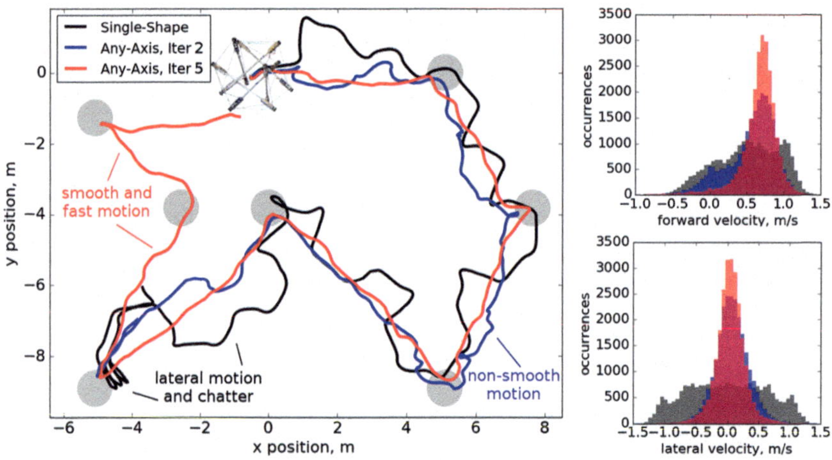

Fig. 7. Iterative performance improvement under policy search. Left: top-down view of center-of-mass trajectories visiting a waypoint sequence. Right: histograms of forward and lateral velocities.

The Single-Shape controller shows zig-zag type motion characteristic of step-wise controllers that move the center-of-mass perpendicular to the forward edge. In some circumstances, such as the bottom-left of the figure, chatter results from its inability to transition over virtual edges. The Any-Axis controller generated by round 2 of policy search locomotes successfully, though it exhibits some irregularity that indicates halting progress. By the round 5 of policy search the controller exhibits relatively smooth paths with few irregularities.

Also shown are velocity distributions aggregated across many trials of each controller in the waypoint scenario. The velocity component toward the next waypoint should ideally be a narrow peak at the target velocity of 0.8 m/s. The Any-Axis-5 controller most closely resembles this distribution, while the Single-Shape controller has a much broader distribution and Any-Axis-2 has a secondary peak at zero caused by stuck states. Lateral velocity shows similar findings, with the exception that zero is the desired value and so target and stuck states are coincident.

3.3 Any-Axis Characteristics

Further illustration of the nature of the Any-Axis controller can help to reveal the capabilities of this policy search pipeline and the possibilities for future results. Figure 8 shows a ground-track path of the Any-Axis controller with locations of the three nodes of the bottom face. Modest movement of these nodes is visible as they either slide along the ground or drift in the air during a ground-edge pivot. Some crossings of edge centers resemble prior $\Delta \to \Lambda \to \Delta$ motions; other crossings occur very close to nodes, while a crossing of a dotted edge indicates a $\Lambda \to \Lambda$ transition.

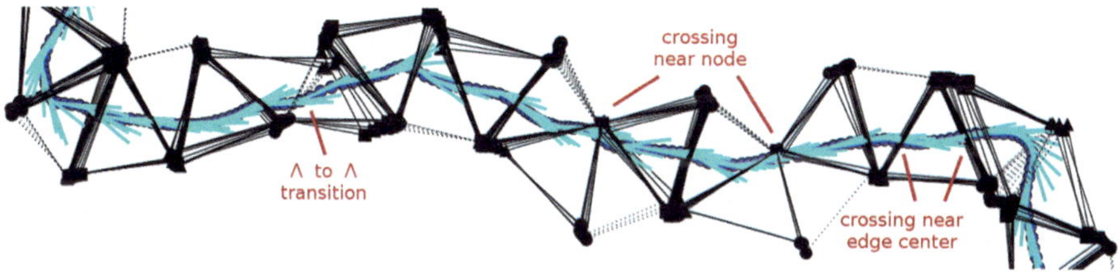

Fig. 8. Center-of-mass and bottom face geometries as the controller locomotes to the right. Light blue lines stem from the center-of-mass location and point in the velocity direction.

Figure 9 more broadly demonstrates this point using two simple geometric measures of the center-of-mass crossing an edge. The crossing *position* describes the center-of-mass location along the edge as a percentage of its total length, while the crossing *angle* relates the center-of-mass velocity direction to the edge direction. Scatter plots of these two values reveal a far wider distribution of behavior for the Any-Axis controller than for the step-wise Single-Shape controller. It is noted, however, that the high-frequency execution of the Single-Shape feedback law in scenarios involving turns does produce some notable variation due to dynamics.

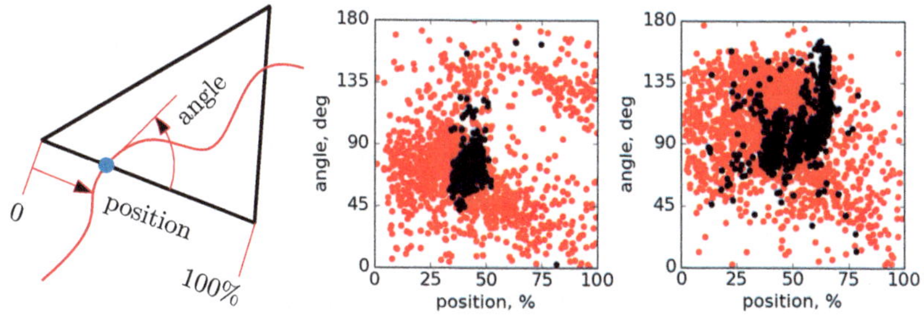

Fig. 9. Left: definition of two quantities describing the geometry of the center-of-mass path over an edge. Center and right: distributions for crossings onto Δ faces and onto Λ faces. Black points: Single-Shape controller; red points: Any-Axis controller.

4 Hardware Experiments

Due to the complex interplay of friction, elasticity, and high dimensionality, the "reality gap" between simulated and hardware behavior of mobile tensegrities is often substantial. Resolving this gap is a deep and profound challenge [10], with limited first steps lying within the scope of this investigation.

Calibration of control inputs (in the form of cable rest lengths) was found to be the most direct route to correcting errors, as the vehicle's shape has a sensitive "tipping-point" relationship with the resulting motion. This was implemented with a choice of two parameters relating hardware and software lengths: the reference length $\hat{u}$, and a scale factor c, such that $\mathbf{u}^*_{HW} = c\left(\mathbf{u}^*_{SW} - \hat{u}\right) + \hat{u}$. Small-scale testing with the Single-Shape controller determined values of $\hat{u} = 0.97m$ and $c = 1.2$ to best approximate intended behavior on hardware (see Fig. 10).

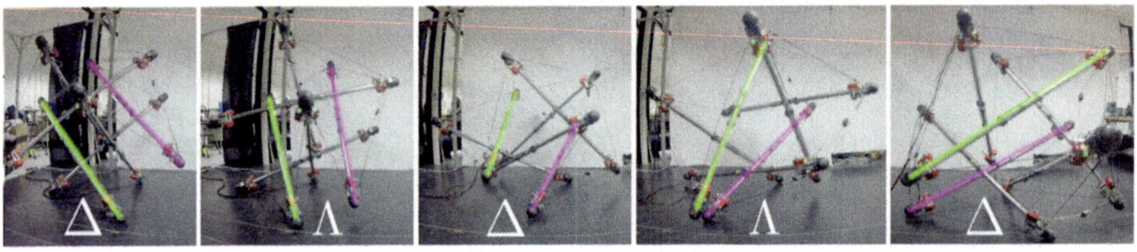

Fig. 10. Execution of the Single-Shape policy on the hardware prototype `SBv2`.

Hardware execution of the Any-Axis neural network policy introduced sensor-related considerations. Locomotion in an arbitrary direction requires knowledge of the vehicle's world frame orientation, or heading, defined via the bottom face. As localization data is not presently available, a simple scheme was introduced to approximate changes in heading based on the sequence of bottom face transitions.

In simulation, this bottom face is computed from node positions, while on hardware it is determined by a classifier that was trained on raw sensor data. As a result, the moment of transition can differ between these two testbeds. Combined with the discontinuous nature of the symmetry-reduced observations, this frequently caused forward/backward chatter of the hardware platform at transition points. Adding a time-delay on the transition detection minimized the occurrence of this issue.

Figure 11 provides the ground contact pattern of a successful hardware trial. Although center-of-mass data could not be obtained, the contact sequence nonetheless reveals directed any-axis locomotion similar to that of the simulated result of Fig. 8. While numerous hardware trials were similarly successful, occasional issues remained. Some transitions resulted in a significantly different heading change than was approximated, causing unintended turns in the trajectory. Some configurations also resulted in "stuck" states on the verge of an intended forward transition.

Footage of the hardware trial is available in the accompanying video. As it reveals, the speed of motion was also found to be significantly less than in simulation. The bar angular rates, important for moderating speed in simulation, could thus be omitted on hardware since the vehicle was not capable of exceeding the desired speed. Friction behavior also differed, with less smooth sliding of bottom nodes.

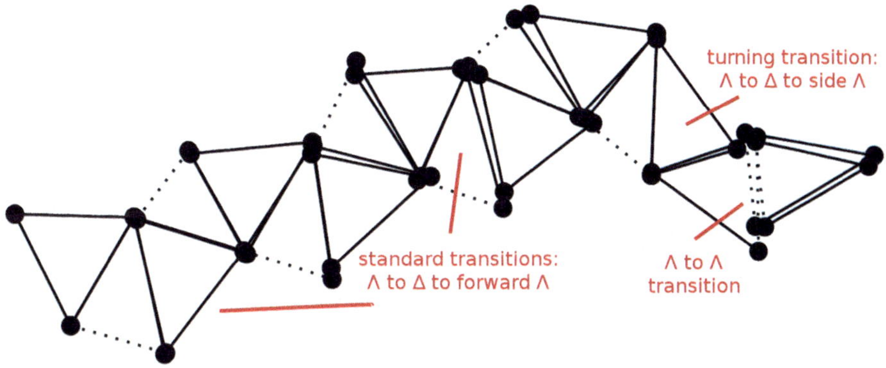

Fig. 11. Bottom face path of Any-Axis locomotion by the hardware prototype SBv2.

5 Discussion

Main Progress: The synthesis of technniques within this paper has produced the first feedback controller to smoothly locomote a 6-bar tensegrity rover in arbitrary directions, independent of orientation. This Any-Axis controller essentially guides state evolution in a variable direction along a 2D+ manifold of the state space, rather than tracking a periodic trajectory or discrete points.

Symmetry-reduced optimization and training required a total of 125,000 time steps sampled at 10 Hz. Previously, without symmetry reduction, 120,000 were needed for training fixed-axis motion with half as many actuators [18]. The broader control manifold is thus learned without significant increase in sample complexity.

Future Directions: Beyond potentially aiding obstacle avoidance, the modified policy search approach may have profound implications for locomotion on unstructured landscapes, a primary motivation for tensegrity use. Traversing rugged features could require both nuanced direction control as well as adaptiveness to non-flat contact geometry, which implies further broadening of the state-space volume navigated by the controller. This would require a corresponding increase in the sample complexity of dynamics fitting and supervised learning, making the applied symmetry reduction and segment classification all the more essential.

Reality Gap: As was seen, many significant differences between the simulation and hardware testbeds resulted in an imperfect transfer of control policies. Control calibrations proved sufficient to enable an initial demonstration of any-axis motion on hardware; however, some occurrences of stuck states and less smooth motion indicate room for more in-depth model identification. Faster and smoother motion, if feasible on hardware actuators, would require more accurate parameter values such as cable elasticity and coefficients of friction. These values might ideally be determined within the simulator testbed via Bayesian optimization, using a loss function of trajectory error relative to physical experiments under identical commands [19]. Finally, additional robustness in the bottom face detection might be obtainable by careful introduction of simulated noise into the symmetry-reduced training pipeline.

Acknowledgment. Supported by NASA ECF grant NNX15AU47G.

References

1. NASA Tensegrity Robotics Toolkit. https://ti.arc.nasa.gov/tech/asr/groups/intelligent-robotics/tensegrity/NTRT/
2. Bliss, T., Iwasaki, T., Bart-Smith, H.: Central pattern generator control of a tensegrity swimmer. Trans. Mech. **18**(2), 586–597 (2013)
3. Iscen, A., Caluwaerts, K., Bruce, J., Agogino, A., SunSpiral, V., Tumer, K.: Learning tensegrity locomotion using open-loop control signals and coevolutionary algorithms. Artif. Life **21**(2), 119–140 (2015)
4. Khazanov, M., Jocque, J., Rieffel, J.: Developing morphological computation in tensegrity robots for controllable actuation. In: 2014 Annual Conference on Genetic and Evolutionary Computation, pp. 1049–1052. ACM, New York (2014). https://doi.org/10.1145/2598394.2605680
5. Kim, K., Agogino, A.K., Toghyan, A., Moon, D., Taneja, L., Agogino, A.M.: Robust learning of tensegrity robot control for locomotion through form-finding. In: IROS (2015)
6. Koizumi, Y., Shibata, M., Hirai, S.: Rolling tensegrity driven by pneumatic soft actuators. In: ICRA, pp. 1988–1993 (2012). https://doi.org/10.1109/ICRA.2012.6224834
7. Levine, S., Finn, C., Darrell, T., Abbeel, P.: End-to-end training of deep visuomotor policies. J. Mach. Learn. Res. **17**(1), 1334–1373 (2016)
8. Littlefield, Z., Surovik, D., Wang, W., Bekris, K.E.: From quasi-static to kinodynamic planning for spherical tensegrity locomotion. In: International Symposium on Robotics Research (ISRR), Puerto Varas, Chile (2017)
9. Mirletz, B., Bhandal, P., Adams, R.D., Agogino, A.K., Quinn, R.D., SunSpiral, V.: Goal directed CPG based control for high DOF tensegrity spines traversing irregular terrain. Soft Robot. **2**, 165–176 (2015)
10. Mirletz, B.T., Park, I.W., Quinn, R.D., SunSpiral, V.: Towards bridging the reality gap between tensegrity simulation and robotic hardware. In: 2015 IEEE/RSJ International Conference on Intelligent Robots and Systems (IROS), pp. 5357–5363. IEEE (2015)
11. Montgomery, W., Ajay, A., Finn, C., Abbeel, P., Levine, S.: Reset-free guided policy search: efficient deep reinforcement learning with stochastic initial states. In: 2017 IEEE International Conference on Robotics and Automation (ICRA), pp. 3373–3380 (2017). https://doi.org/10.1109/ICRA.2017.7989383
12. Paul, C., Valero-Cuevas, F.J., Lipson, H.: Design and control of tensegrity robots for locomotion. TRO **22**(5) (2006). https://doi.org/10.1109/TRO.2006.878980
13. Rovira, A.G., Mirats Tur, J.M.: Control and simulation of a tensegrity-based mobile robot. RAS **57**(5), 526–535 (2009)
14. Sultan, C., Skelton, R.: Deployment of tensegrity structures. Int. J. Solids Struct. **40**(18), 4637–4657 (2003). https://doi.org/10.1016/S0020-7683(03)00267-1
15. Surovik, D.A., Bekris, K.E.: Symmetric reduction of tensegrity rover dynamics for efficient data-driven control. In: ASCE International Conference on Engineering, Science, Construction and Operations in Challenging Environments (2018)

16. Vespignani, M., Ercolani, C., Friesen, J.M., Bruce, J.: Steerable locomotion controller for six-strut icosahedral tensegrity robots. In: IROS (2018)
17. Vespignani, M., Friesen, J.M., SunSpiral, V., Bruce, J.: Design of superball v2, a compliant tensegrity robot for absorbing large impacts. In: IROS (2018)
18. Zhang, M., Geng, X., Bruce, J., Caluwaerts, K., Vespignani, M., SunSpiral, V., Abbeel, P., Levine, S.: Deep reinforcement learning for tensegrity robot locomotion. In: 2017 IEEE International Conference on Robotics and Automation (ICRA), pp. 634–641 (2017). https://doi.org/10.1109/ICRA.2017.7989079
19. Zhu, S., Kimmel, A., Bekris, K.E., Boularias, A.: Fast model identification via physics engines for data-efficient policy search. In: International Joint Conference on Artificial Intelligence (IJCAI), Stockholm, Sweden (2018)

Contacts and Manipulation

Contacts and Manipulation

Jaeheung Park

In this session, a variety of topics were presented, including actuator transparency, contact model and estimation, manipulation planning, and the human-robot interaction for manipulation tasks. The first paper presented the experimental results of the contact space model and the dynamically consistent null space using passive jointed pendulums. The second paper proposed a framework for tracking a desired path of an object held by an adaptive hand. The third paper presented a probabilistic model to enable a robot to obtain knowledge about the workspace from human language and physical interaction. The fourth paper presented a Collision Hypothesis Sets representation for computing a belief of occupancy and proposed a planning and control architecture using this representation to navigate through unknown environments. The fifth paper presented a disturbance observer for floating-base robots using contact information. Finally, the last paper compared the measures of actuators to explore the transparency of the actuators. The session brought interesting and fruitful discussion on the issues on contact modelling, estimation, and manipulation planning, especially in the context of control in contact state dealing with uncertainties.

Experimental Studies of Contact Space Model for Multi-surface Collisions in Articulated Rigid-Body Systems

Shameek Ganguly(✉) and Oussama Khatib

Robotics Laboratory, Stanford University, Stanford, CA 94305, USA
shameekg@stanford.edu, ok@cs.stanford.edu

Abstract. Modeling simultaneous contact and collisions at multiple surfaces in articulated rigid-body systems is computationally challenging because joint and contact constraints must be enforced together. Previously, we have proposed an efficient method to treat multi-surface collisions in the contact space coordinates which represent the relative motion at the contact surfaces [12]. The inertial dynamics of the multi-surface collision is then equivalent to a simultaneous collision between two rigid bodies whose inertia equals the effective inertia of the respective articulated bodies in the contact space. In this paper, we present an experimental study of this model on planar and spatial systems of passive jointed pendulums. Across several frictionless/frictional, single/multi-surface collision events, the model accurately predicted post-collision joint positions and velocities. Furthermore, joint velocity predictions in the dynamically consistent null space were significantly more accurate than those in the contact space. This indicates that the contact space model accurately describes inertial dynamics of multi-surface collisions in articulated rigid-body systems, but a better empirical restitution model can further improve prediction accuracy.

1 Introduction

As a field, robotics is seeing an expansion today from applications in structured manufacturing environments to applications in unstructured dynamic environments such as warehouses, outdoors, and our homes. The ability to reason autonomously about physical interactions with objects is a key requirement for robust manipulation in these environments. A robot working in a kitchen for instance will open several cabinet doors and drawers, grasp various containers, open and close lids on jars, use several tools and utensils to cook, place different objects on counters, shelves, and so on. Each of these interactions involves making and breaking contact several times between the robot or an object the robot is holding, and the environment. Therefore, reasoning about physical interaction critically assumes an understanding of the physics of contacts and collisions in such dynamic scenes.

Robots with rigid links and joints belong to the family of articulated rigid-body mechanisms, whose free space dynamics is well understood. Many simulation systems exist to design and visualize such mechanisms [3,4]. In contrast, the contact dynamics

J. Xiao et al. (Eds.): ISER 2018, SPAR 11, pp. 425–436, 2020.
https://doi.org/10.1007/978-3-030-33950-0_37

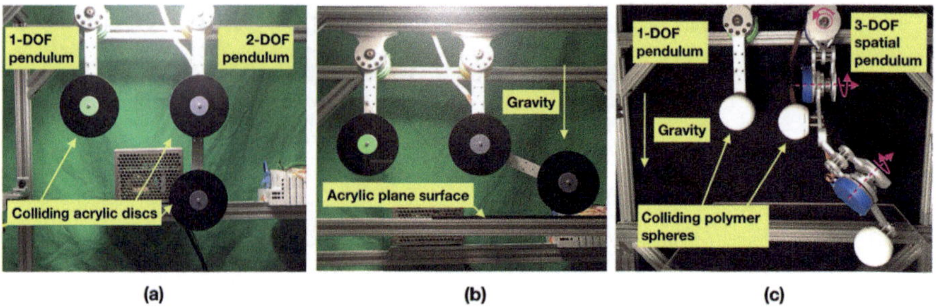

Fig. 1. Designed setups for studying (a) single-surface planar (TP1), (b) two-surface planar (TP2), and (c) single-surface spatial collisions (TS1). Each setup consists of passive pendulum hanging from a fixed structure. The pendulums are allowed to collide by dropping them from different initial configurations. Joint positions, velocities and accelerations are measured at a rate of 4000 Hz using high precision optical joint encoders.

of such systems is more commonly modeled by treating each rigid link as an independent body, and imposing joint constraints through constraint forces or impulses [1,7,14,17]. This approach introduces drifting at the joints, and requires non-physical corrections to produce visually correct results in simulations. The operational space theory was developed to model and control robotic contact manipulation, and has been successfully applied to complex multi-contact tasks [9,11]. Ruspini and Khatib [12] presented a treatment of multi-surface collisions in articulated body systems based on this theory, where the inertial dynamics of the system is represented through a set of contact space coordinates. This removes the need for constraint stabilization, and can be used to interactively simulate large articulated-body systems in real-time [8].

Models used to simulate robotic systems are not only required to be graphically correct but also need to predict outcomes of physical interactions correctly. This is especially true for applications such as transfer reinforcement learning where policies trained on massive datasets generated in simulation are expected to be used to control physical robots [13]. However, a quick review of literature in modeling collisions articulated rigid-body systems reveals a lack of experimental studies in this area. Recent work in studying collisions between a single rigid body and a rigid plane [5,16] suggest several deviations in the behavior of these physical systems from simulation models. Empirical models for restitution have poor physical interpretation, and often fail to capture observed phenomena such as a change in the slip direction at the colliding surface. Conti and Khatib [2] presented an initial investigation into collisions of articulated mechanisms freely floating on an air-bearing surface. Our work extends their work by systematically testing the contact space model on single-surface and multi-surface collisions in articulated rigid-body systems.

2 Contact Space Model

Consider two articulated bodies in contact at time t, touching at two surfaces simultaneously. We assume that the contact patch at each touching surface is modeled as a collection of coincident points. The collection of isolated forces acting at the modeled points on a particular contact patch approximates the true distribution of stresses in the patch. Let ${}^{ij}x$ be the position of contact point c_i located on body j measured with respect to a fixed frame. ${}^{ij}\dot{x}$ denotes the linear velocity of the body at this point. Articulated bodies B_1 and B_2 are defined to be colliding at time t if

$$\begin{aligned} {}^{i1}x(t) = {}^{i2}x(t), \ \forall \ i = 1,2,\ldots,D_C \\ \exists \ k \ s.t. \ {}^{k}n_{12}^{T} \left({}^{i2}\dot{x}(t) - {}^{i1}\dot{x}(t)\right) < 0 \end{aligned} \tag{1}$$

where D_C is the number of contact points and ${}^{k}n_{AB}$ denotes the common surface normal at point c_k directed from body B_1 to body B_2. If the surface normal is not well defined, the collision resolution problem is ill-posed. We exclude this possibility from the Contact Space model. The collision resolution problem is defined as determining a state of the system such that bodies B_1 and B_2 are not penetrating for any time $t+\delta t$ where δt represents a tiny increment in time.

In the Contact State model, the system state is represented by a set of minimal generalized coordinates q and velocities $\dot{q} \in \mathrm{R}^{D_Q}$, where D_Q is the number of degrees of freedom of the system. Using Lagrangian formulation, the equations of motion of the system is given by

$$A(q)\ddot{q} + b(q,\dot{q}) + g(q) = \Gamma \tag{2}$$

where $A(q)$, $b(q,\dot{q})$ and $g(q)$ are the generalized inertia matrix, generalized Coriolis/centrifugal forces and generalized gravity forces respectively for the system. $\ddot{q}$ is the time-derivative of the generalized velocity, that is, the generalized acceleration. Γ is the generalized force acting on the system which can be further differentiated into non-contact generalized forces such as actuator forces/torques, and contact generalized forces. Let Γ_{NC} denote the sum of non-contact generalized forces, and Γ_C the sum of contact generalized forces respectively.

At a point of contact c_i, body B_1 applies a force ${}^{i}F$ on body B_2, and body B_2 applies a force $(-{}^{i}F)$ on body B_1. Then, the relationship between the contact generalized forces and contact forces is given by [6]

$$\Gamma_C = \Sigma_{i=1}^{D_C} \left({}^{i2}J^{T}\,{}^{i}F - {}^{i1}J^{T}\,{}^{i}F\right) = \Sigma_{i=1}^{D_C} \left({}^{i2}J - {}^{i1}J\right)^{T}\,{}^{i}F \tag{3}$$

where ${}^{ij}J$ denotes the linear velocity Jacobian at contact point c_i on body B_j. Since point contact is assumed, we ignore moment components of F. By stacking the forces ${}^{i}F$ into a single vector F_C, and the Jacobian differences into a matrix J_C, we get the relation

$$\Gamma_C = J_C^T F_C \tag{4}$$

Note that the matrix J_C is related to the relative velocities at the contact points as follows.

$$\left({}^{i2}J - {}^{i1}J\right)\dot{q} = {}^{i2}\dot{x} - {}^{i1}\dot{x} \ \ \forall\, i = 1,2,\ldots,D_C \tag{5}$$

Let $\dot{x}_C$ denote the stacked vector of relative velocities at each contact point. Then we have the relationship:

$$\dot{x}_C = J_C\,\dot{q} \tag{6}$$

We define the Contact Space coordinates to be the stacked vector of relative positions ${}^{i2}x - {}^{i1}x$ at each contact point. $\dot{x}_C$ denotes the Contact Space velocity, and J_C denotes the Contact Space Jacobian. The Contact Space is composed of $3D_C coordinates$. Note that the Contact Space coordinates change as a function of time since the points at which the bodies are in contact will change over time.

The concept of Contact Space is inspired by the operational space formulation of robot dynamics used in whole-body control [10]. Similar to the operational space dynamics, the Contact Space dynamics can be deduced to be

$$\ddot{x}_C - \dot{J}_C\dot{q} + J_C A^{-1}(b+g) = J_C A^{-1}\Gamma_{NC} + \Lambda_C^{-1} F_C \tag{7}$$

where $\ddot{x}_C = J_C\ddot{q} + \dot{J}_C\dot{q}$ denotes the Contact Space acceleration and Λ_C denotes the Contact Space inertia matrix.

Rigid body collisions are characterized by a very short duration of collision events. In the Contact Space model, we model collision events to be instantaneous. The system configuration is assumed to remain unchanged through the collision, but the system velocity is assumed to change instantaneously to prevent inter-penetration of the rigid bodies. Since this requires an infinite acceleration and contact force at the instant of collision, the effect of finite centrifugal, Coriolis, gravitational and non-contact forces is ignored. The generalized coordinates equation modeling the system during collision is therefore given by

$$A\left(\dot{q}^+ - \dot{q}^-\right) = J_C^T P_C \tag{8}$$

where P_C is the Contact Space impulse applied during the collision. The superscripts $-$ and $+$ denote quantities before and after the collision event respectively. Similarly, the Contact Space collision model is given by

$$\dot{x}_C^+ - \dot{x}_C^- = \Lambda_C^{-1} P_C \tag{9}$$

Λ_C exists only when the contact forces are non-redundant. The contact forces are said to be redundant when there exist combinations of contact forces that do not produce any change in the motion of the system. An example is a square table standing on four legs of equal height. A contact force of $1\,N$ applied to any two diagonally opposite legs, and $-1\,N$ applied to the remaining two legs produces no net force or moment on the table. It is common for contact forces to be redundant in real scenarios, and the Contact Space method can be generalized to such systems through the LCP complementarity formulation. For the purpose of this paper, we restrict ourselves to experiments where the contact forces are non-redundant, and Λ_C is well-defined.

Combining Eqs. 8 and 9 we get that the post-collision generalized velocities predicted by the model are

$$\begin{aligned} c\dot{q}^+ &= \dot{q}^- + A^{-1}J_C^T\Lambda_C\left(\dot{x}_C^+ - \dot{x}_C^-\right) \\ &= A^{-1}J_C^T\Lambda_C\dot{x}_C^+ + \left(I - A^{-1}J_C^T\Lambda_C J_C\right)\dot{q}^- \\ &= \bar{J}_C\dot{x}_C^+ + N_d\dot{q}^- \end{aligned} \tag{10}$$

where $\bar{J}_C = A^{-1}J_C^T\Lambda_C$ and $N_d = I - A^{-1}J_C^T\Lambda_C J_C$ are the *dynamically consistent right inverse* of J_C, and the corresponding null space projection matrix respectively. I is the $D_Q \times D_Q$ identity matrix. From these definitions, the following conservation law follows:

$$N_d \dot{q}^+ = N_d \dot{q}^- \tag{11}$$

The rank of N_d is equal to the number of degrees of freedom of the system in the null space of the contact constraints, that is, $D_N = rank(N_d) = \max(0, D_Q - 3D_C)$. Therefore, the above equation predicts D_N conserved quantities corresponding to the linearly independent rows of N_d. Intuitively, this is a restatement of the conservation of momentum in the system, in the subspace orthogonal to both the joint and contact constraints.

Equation 10 is insufficient to determine the full post-collision state of the system. Predicting $\dot{x_C}^+$ requires further modeling of the local contact dynamics at each contact point between bodies B_1 and B_2. Consider the component of the relative velocity normal to the surfaces at contact point c_i given by ${}^iv_n = {}^in_{12}\left({}^{i2}\dot{x} - {}^{i1}\dot{x}\right)$. This is the *separation speed* of the bodies at c_i. Let ${}^iv_t = {}^ix_C - {}^iv_n^in$ be the component of the relative velocity tangential to the surfaces at the same point. There are many available models of the rigid multi-body collision resolution problem [7, 15, 17]. All of these models make the following assumptions:

- The normal force at each contact point is non-negative, that is, there is no attraction between the surfaces
- The tangential forces at each contact point are governed by Coulomb's static frictional model

Unfortunately, these laws still do not yield a unique post-collision state of the system, but a set of admissible solution states. Empirical restitution parameters, path restrictions in the impulse space and assumptions such as complementarity are commonly used to find a unique solution state of the system. Note that *any* of these rigid multi-body models that yield physically admissible solutions can be employed in conjunction with the Contact Space model. In the SAI simulation system [8] developed at Stanford Robotics Lab, we employ the following model:

$$\begin{array}{c} find\ \dot{x}_C^+\ \ s.t. \\ \dot{x}_C^+ = \dot{x}_C^- + \Lambda_C^{-1}\, P_C \\ {}^iP_n \geq 0,\ {}^iv_n^+ = -\varepsilon\, {}^iv_n^- \quad OR \quad {}^iP_n = 0,\ {}^iv_n^+ \geq -\varepsilon\, {}^iv_n^- \quad \forall\ i = 1, 2, \ldots, D_C \\ {}^iv_t^+ = 0,\ \mu\, {}^iP_n \geq \|{}^iP_t\| \quad OR \quad \|{}^iv_t^+\| > 0,\ -\mu\, {}^iP_n\, {}^i\hat{s}^+ = {}^iP_t \quad \forall\ i = 1, 2, \ldots, D_C \end{array} \tag{12}$$

where ε is the coefficient of Newtonian restitution, and μ is the coefficient of static friction. ${}^iP_n$ and ${}^iP_t$ represent the (scalar) normal and (vector) tangential components of the impulse respectively at contact point c_i. ${}^i\hat{s} = {}^iv_t^+ / \|{}^iv_t^+\|$ denotes the direction of the tangential velocity vector. Since the purpose of this paper is to experimentally study the Contact Space model, we will not discus the specifics of the rigid multi-body collision solver used here.

Note that although the Contact Space model was derived in this section for a 2-body system, it is easily generalized to a system of more bodies by constructing the contact space coordinates from the relative motion at all points of contacts between pairs of bodies simultaneously in Eq. 5.

2.1 Frictionless Contacts

The Contact Space coordinates are composed of relative motion between bodies at the contact points. In the special case that a contact c_k is frictionless, ${}^kP_t = 0$. As a result, we can eliminate the columns in J_C^T that represent the generalized forces produced by a unit force in the tangent directions at contact point c_k. Therefore, if a collision occurs in the system with $D_{C,f}$ frictional contacts and $D_{C,nf}$ frictionless contacts, the size of J_C can be reduced to $D_X = 3D_{C,f} + D_{C,nf}$ without changing the resolved collision outcome. The overall contact space dimension is given by D_X, and the null space dimension increases to $D_N = \max(0, D_Q - D_X)$.

3 Experiments

To test the accuracy of the Contact Space model, we devised several setups consisting of pendulums with serial kinematic chain structures as shown in Fig. 1. In each experiment, a pair of pendulums hangs from a rigid grounded frame close enough such that they collide often when swinging. To ensure that the pendulums are rigid and joint friction is negligible, the links are constructed out of aluminum and steel precision ball bearings are used at each joint. Unlike real robots, the joints are not actuated. We do so in order to eliminate the effects of drive friction and inertia on the collision outcome. This can be an area of future investigation.

3.1 Experimental Collision Events

We observed collisions in our experimental setup by releasing both pendulums from different heights. The motion of the pendulums is measured by optical encoders at each joint at a resolution of 0.009° and a rate of 4 kHz. Smoothed joint position, velocity and acceleration are estimated offline using a Savitzky-Golay filter. The kinematic and inertial models of the pendulums are obtained from their CAD models.

While the Contact Space models collisions as instantaneous events, we observe on our setup that the separation speed does not change instantaneously but instead over a small interval of time during which the bodies remain in contact. We define an observed collision event as follows:

Observed Collision Event: An interval of time $[t, t + \Delta t_I]$ is said to encompass an observed collision event if

- There are at least two bodies in the system whose surface distance is greater than ρ_I units at time t, but less than ρ_I at time $t + \Delta t_S$
- For any pair of bodies that are within a distance ρ_I of each other at time $t + \Delta t_S$, either
  - the distance between the bodies is larger than ρ_I at time $t + \Delta t_I$, or
  - the separation speed between the bodies is smaller than a predetermined threshold v_{min} at all times in the time interval $[t + \Delta t_I, t + \Delta t_I + \Delta t_{obs}]$.

where Δt_S is the time interval between consecutive encoder samples. The last condition is to quantify the end of inelastic collision events. ρ_I, v_{min} and Δt_{obs} are selected empirically based on measurement noise in the system. The smallest possible time interval that meets the above criteria is considered for each observed collision event. We set $\rho_I = 0.002$ m, $v_{min} = 0.01$ m/s and $\Delta t_{obs} = 0.01$ s for our experiments.

3.2 Test Setups

We performed experiments on three different test setups of colliding pendulums:

TP1: Shown in Fig. 1a, this setup consists of a 1-DOF and a 2-DOF pendulum freely hanging from the grounded support. The pendulums are planar and collisions occur at a single point between the discs mounted on the pendulums. The stiff discs are laser cut out of acrylic, and mounted on bushings to allow them to rotate freely. This is done so as to allow the discs to roll against each other upon contact and minimize the frictional impulse. The 1-DOF pendulum weighs 329 gm whereas the 2-DOF pendulum weighs 655 gm. A total of $S = 78$ collision events were observed, across various drop heights and mounting distances of the pendulums. The observed collision events lasted for a median of 0.023 s.

TP2: This setup is similar to TP1 except that the 2-DOF pendulum is initially resting on an acrylic plane surface (Fig. 1b). This ensures that at the time of collision with the 1-DOF pendulum, there are two points of simultaneous contact. The contact between the 2-DOF pendulum and the plane surface is assumed to be frictionless as well, since the disc rolls freely on it. $S = 66$ collision events were observed on this setup in our experiments, across various drop heights and mounting distances of the pendulums, and across various position and slope angles of the plane surface. Observed collision events lasted for a median of 0.089 s.

TS1: A spatial setup was instrumented as well where a 1-DOF pendulum collides against a freely hanging 3-DOF spatial pendulum (Fig. 1c). The colliding surfaces consist of 3D printed polymer spheres. Friction is non-negligible at the contact points. To our knowledge, this is the first experimental setup to study spatial rigid-body collisions. The 3-DOF pendulum weighs 911 gm. $S = 117$ collision events were observed, across various drop heights, mounting distances and orientations of the pendulums. Observed collision events lasted for a median of 44 ms.

3.3 Evaluation Metrics

Most existing models of simultaneous collision outcome in multibody systems have not been tested experimentally. Common metrics used to compare models are computational time, long term stability and macroscopic qualitative behavior of the system [1, 17]. Past experimental research in rigid-body collisions have focused either on the velocity at the point of contact or the predicted trajectory of the colliding object. In practice, predicting the long term quantitative behavior of robots is rarely of interest.

We use the coefficient of determination, R^2 between quantities predicted by the Contact Space model, and their measured value at the end of the observed collision

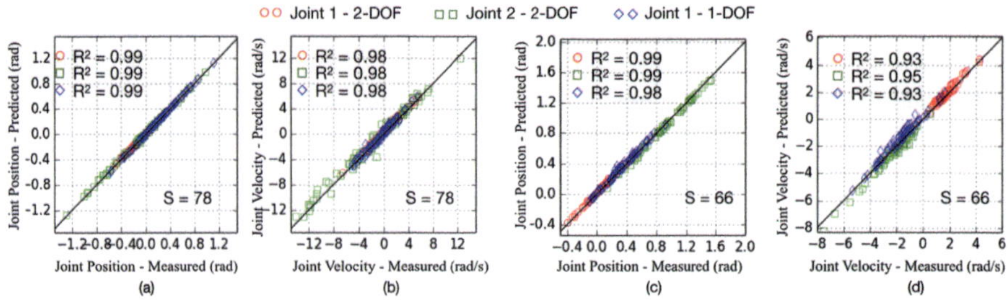

Fig. 2. Figure 4: Prediction of post-collision state of the system in planar experiments. (a) Joint positions for setup TP1, (b) joint velocities for setup TP1, (c) joint positions for setup TP2, and (d) joint velocities for setup TP2.

event as a metric of model accuracy. The R^2 value is computed across several observed collision events performed on a specific setup of pendulums. We investigate the R^2 of the measured vs predicted values at the end of the observed collision event for the following quantities:

1. joint angle (q),
2. joint velocity ($\dot{q}$),
3. separation speed (v_n), and
4. dynamically consistent null space velocity ($N_d\dot{q}$).

Metrics 1 and 2 quantify the accuracy with which the post-collision state of the system is predicted by the Contact Space model whereas metrics 3 and 4 specifically test the accuracy of the model in predicting the contact space and null space velocities after the collision. In computing the predicted state of the system at the end of the observed collision event, we follow a three step process:

1. Starting from the measured state of the system at the beginning of the observed collision event, the equations of continuous motion of the system are integrated till a collision is detected (Eq. 1).
2. The Contact Space model is applied to resolve the collision. This yields a non-colliding state of the system.
3. Starting from the resulting state of the system, the equations of continuous motion of the system are integrated again till the end of the observed collision event.

A forward Euler integrator with time step of 0.00025 s is used in steps 1 and 2 to match the sampling frequency of the system. Decreasing the time step of integration does not have a significant effect on the accuracy.

The restitution parameter ε and friction coefficient μ used in the Contact State Model are estimated for each test setup separately through empirical fitting over the measured post-collision states. $\varepsilon = 0.21$ is used for setup $TP1$ and $\varepsilon = 0.15$ for setup $TP2$. $\varepsilon = 0.28$ and $\mu = 0.14$ are used for setup $TS1$.

4 Results

Figure 2a and b show the predicted vs measured state of the system for the experiments performed on test setup TP1. A high $R^2 = 0.99$ indicates that joint positions are accurately predicted on this setup by the Contact Space Model. Joint velocities are predicted with a high accuracy of $R^2 = 0.98$. Figure 2c and d shows the predicted vs measured state of the system for test setup TP2. While joint positions are predicted with high accuracy ($R^2 > 0.98$), there is an apparent drop of prediction accuracy joint velocities ($R^2 > 0.93$). This is likely due to the fact that the observed collision events occur over a small time interval (median of < 0.090 s). As the joint velocities are limited (< 8 rad/s), the expected change in joint position is small. But joint velocities changes significantly across observed collision events.

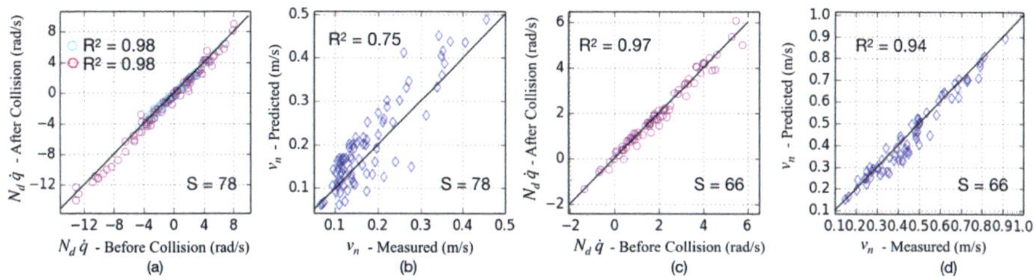

Fig. 3. Figure 4: Prediction of factored post-collision velocity of the system in planar experiments. (a) Dynamically consistent null space velocity components for setup TP1, (b) separation speed for setup TP1, (c) Dynamically consistent null space velocity components for setup TP2, and (d) separation speed for setup TP2.

Performance in metrics 3 and 4 sheds further light on the source of model errors. As the collisions are assumed to be frictionless and to be occurring at a single point for setup TP1, the dimension of the contact space $D_X = 1$, and the dimension of the null space $D_N = 2$. For setup TP2, the dimension of the contact space $D_X = 2$, and the dimension of the null space $D_N = 1$. As shown in Fig. 3, the joint velocity components in the dynamically consistent null space are conserved remarkably well across the observed collision events with $R^2 = 0.98$ for setup TP1 and $R^2 = 0.97$ for setup TP2. On the other hand, the separation speed shows poorer prediction accuracy with $R^2 = 0.75$ for setup TP1 and $R^2 = 0.94$ for setup TP2. This result indicates that while the assumption that the collisions occur at a single point with negligible friction and negligible contact moments is reasonable, the empirical Newtonian restitution model leads to poor predictions of the separation speed both setup TP1 and setup TP2. We also note that the joint velocity prediction is better for single-point collisions as compared to the two-point collisions. This is likely because the joint velocities in the former are largely governed by the null space conservation whereas those in the latter are primarily governed by the restitution law.

To further examine the observation that $N_d\dot{q}$ is conserved across the observed collision interval as predicted by the Contact Space model, we also added a 500 gm mass to the 1-DOF manipulator in test TP2. With the additional mass and fasteners, the mass of the 1-DOF manipulator increases to 845 gm. As shown in Fig. 4a and b, the model remains accurate even with this large change in the mass of the system ($R^2 > 0.97$ with and without the added mass).

Experiments with the spatial test setup show a similar outcome (Fig. 5). $R^2 > 0.99$ for joint position prediction, and $R^2 > 0.88$ for joint velocities. Dynamically consistent null space velocity is conserved to $R^2 = 0.99$ whereas the separation velocity is predicted with an accuracy of $R^2 = 0.88$.

The dynamically consistent null space projection matrix N_d is unique. In robot control literature, the orthogonal null space projection matrix $N_p = \left(I - J_C^+ J_C\right)$ is often used instead to get the smallest possible 2-norm change in $\dot{q}$ to produce the change in contact velocity stipulated by the restitution model. J_C^+ is the Moore-Penrose pseudo-inverse of J_C. Figure 4c shows the value of $N_p\dot{q}$ before and after observed collision events for the TP1 planar single-point experiments. With $R^2 = 0.36$, this hypothesis is unsupported by the experimental data.

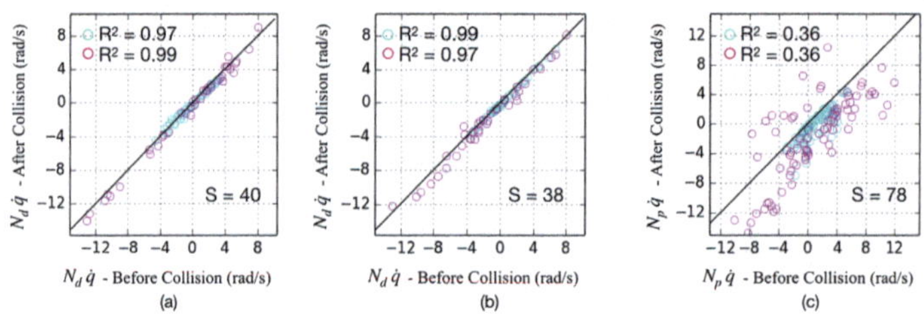

Fig. 4. (a) Dynamically consistent null space velocity components without added mass for setup TP1, (b) dynamically consistent null space velocity components with added mass for setup TP1, (c) orthogonal null space velocity components over all experiment for setup TP1

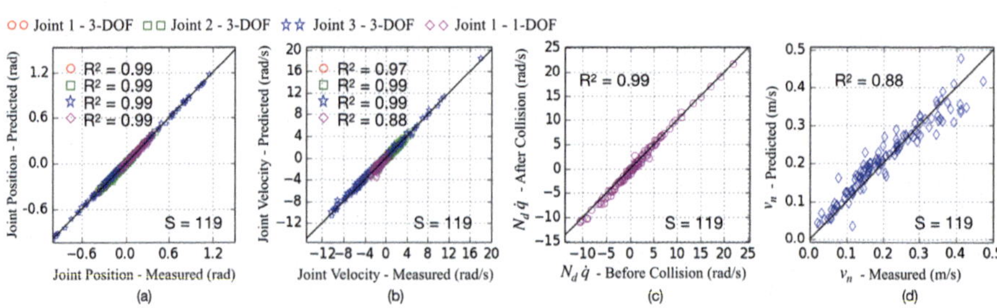

Fig. 5. Prediction of (a) joint positions, (b) joint velocities, (c) dynamically consistent null space velocity, and (d) separation speed for spatial single-surface experiments on setup TS1.

5 Conclusion

Articulated body collisions are complex phenomena with multi-scale dynamics and several constraints. However, the theory of rigid articulated body dynamics along with the contact space model factorizes the resolution problem into the reduced dimension contact space, and the dynamically consistent null space. Conservation of momentum is folded into this formulation and does not need to be handled explicitly, reducing the algorithmic time complexity from the order of number of joints to the order of number of contact surfaces. The experimental results presented above provide strong evidence of the accuracy of the contact space model in predicting physically correct results for real collisions. However, as with state-of-the-art rigid body contact models, inaccuracies are introduced through empirical restitution and friction models that approximate the local interaction at the contact surfaces [5, 16]. Better models, such as data-driven ones, promise to address this issue, and will be explored in future effort.

Acknowledgements. The authors would like to acknowledge the help of Anushka Nair and Chinmay Tyagi in performing the planar collision experiments. They also acknowledge Samir Menon for his insights. Toyota Research Institute (TRI) provided funds to assist the authors with their research but this article solely reflects the opinions and conclusions of its authors and not TRI or any other Toyota entity.

References

1. Baraff, D.: Fast contact force computation for nonpenetrating rigid bodies. In: Proceedings of the 21st Annual Conference on Computer Graphics and Interactive Techniques, pp. 23–34. ACM (1994)
2. Conti, F., Khatib, O.: A framework for real-time multi-contact multi-body dynamic simulation. In: Robotics Research, pp. 271–287. Springer (2016)
3. Delp, S.L., Anderson, F.C., Arnold, A.S., Loan, P., Habib, A., John, C.T., Guendelman, E., Thelen, D.G.: Opensim: open-source software to create and analyze dynamic simulations of movement. IEEE Trans. Biomed. Eng. **54**(11), 1940–1950 (2007)
4. Erez, T., Tassa, Y., Todorov, E.: Simulation tools for model-based robotics: comparison of bullet, havok, mujoco, ode and physx. In: 2015 IEEE International Conference on Robotics and Automation (ICRA), pp. 4397–4404. IEEE (2015)
5. Fazeli, N., Zapolsky, S., Drumwright, E., Rodriguez, A.: Fundamental limitations in performance and interpretability of common planar rigid-body contact models. arXiv preprint arXiv:1710.04979 (2017)
6. Featherstone, R., Orin, D.E.: Springer Handbook of Robotics - Dynamics
7. Flores, P., Ambrósio, J., Claro, J.P., Lankarani, H.: Influence of the contact–impact force model on the dynamic response of multi-body systems. Proc. Inst. Mech. Eng. Part K: J. Multi-body Dyn. **220**(1), 21–34 (2006)
8. Khatib, O., Brock, O., Chang, K.S., Conti, F., Ruspini, D., Sentis, L.: Robotics and interactive simulation. Commun. ACM **45**(3), 46–51 (2002)
9. Khatib, O., Chung, S.Y.: Suprapeds: humanoid contact-supported locomotion for 3D unstructured environments. In: 2014 IEEE International Conference on Robotics and Automation (ICRA), pp. 2319–2325. IEEE (2014)
10. Khatib, O., Sentis, L., Park, J., Warren, J.: Whole-body dynamic behavior and control of human-like robots. Int. J. Humanoid Rob. **1**(01), 29–43 (2004)

11. Khatib, O., Yeh, X., Brantner, G., Soe, B., Kim, B., Ganguly, S., Stuart, H., Wang, S., Cutkosky, M., Edsinger, A., et al.: Ocean one: a robotic avatar for oceanic discovery. IEEE Robot. Autom. Mag. **23**(4), 20–29 (2016)
12. Ruspini, D.C., Khatib, O.: Collision/contact models for dynamic simulation and haptic interaction. In: Robotics Research, pp. 185–194. Springer (2000)
13. Rusu, A.A., Vecerik, M., Rothörl, T., Heess, N., Pascanu, R., Hadsell, R.: Sim-to-real robot learning from pixels with progressive nets. arXiv preprint arXiv:1610.04286 (2016)
14. Stewart, D., Trinkle, J.C.: An implicit time-stepping scheme for rigid body dynamics with coulomb friction. In: IEEE International Conference on Robotics and Automation, 2000, Proceedings, ICRA 20000, vol. 1, pp. 162–169. IEEE (2000)
15. Stewart, D.E.: Rigid-body dynamics with friction and impact. SIAM Rev. **42**(1), 3–39 (2000)
16. Stoianovici, D., Hurmuzlu, Y.: A critical study of the applicability of rigid-body collision theory. Trans.-Am. Soc. Mech. Eng. J. Appl. Mech. **63**, 307–316 (1996)
17. Uchida, T.K., Sherman, M.A., Delp, S.L.: Making a meaningful impact: modelling simultaneous frictional collisions in spatial multibody systems. Proc. R. Soc. A **471**, 20140859 (2015). The Royal Society

Path Planning for Within-Hand Manipulation over Learned Representations of Safe States

Berk Calli[1(✉)], Andrew Kimmel[2], Kaiyu Hang[3], Kostas Bekris[2], and Aaron Dollar[3]

[1] Computer Science Department and Robotic Engineering Program, Worcester Polytechnic Institute, Worcester, MA, USA
bcalli@wpi.edu

[2] Computer Science Department, Rutgers University, New Brunswick, NJ, USA
{andrew.kimmel,kostas.bekris}@cs.rutgers.edu

[3] Mechanical Engineering and Materials Science Department, Yale University, New Haven, CT, USA
{kaiyu.hang,aaron.dollar}@yale.edu

Abstract. This work proposes a framework for tracking a desired path of an object held by an adaptive hand via within-hand manipulation. Such underactuated hands are able to passively achieve stable contacts with objects. Combined with vision-based control and data-driven state estimation process, they can solve tasks without accurate hand-object models or multi-modal sensory feedback. In particular, a data-driven regression process is used here to estimate the probability of dropping the object for given manipulation states. Then, an optimization-based planner aims to track the desired path while avoiding states that are above a threshold probability of dropping the object. The optimized cost function, based on the principle of Dynamic-Time Warping (DTW), seeks to minimize the area between the desired and the followed path. By adapting the threshold for the probability of dropping the object, the framework can handle objects of different weights without retraining. Experiments involving writing letters with a marker, as well as tracing randomized paths, were conducted on the Yale Model T-42 hand. Results indicate that the framework successfully avoids undesirable states, while minimizing the proposed cost function, thereby producing object paths for within-hand manipulation that closely match the target ones.

1 Introduction

Within-hand dexterity enables robots to achieve efficient, human-like manipulation skills by reducing unnecessary large whole-arm motions and other compensatory actions, such as re-grasping. This often requires, however, accurate

---

Electronic supplementary material The online version of this chapter (https://doi.org/10.1007/978-3-030-33950-0_38) contains supplementary material, which is available to authorized users.

J. Xiao et al. (Eds.): ISER 2018, SPAR 11, pp. 437–447, 2020.
https://doi.org/10.1007/978-3-030-33950-0_38

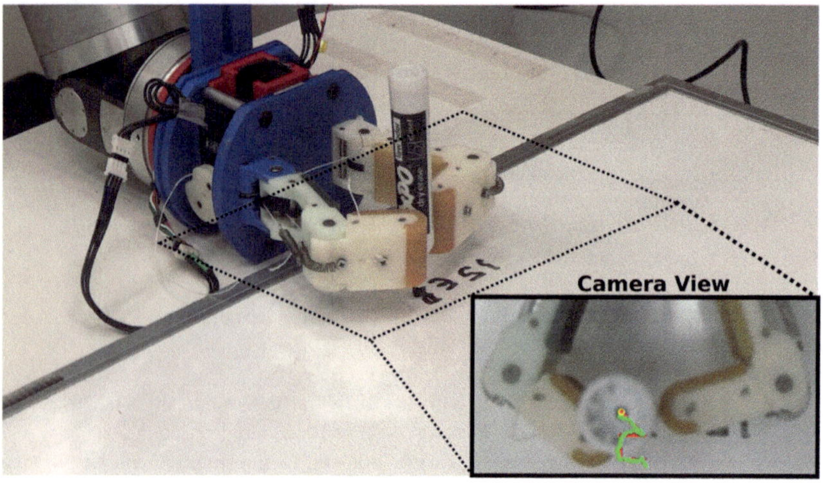

Fig. 1. The Yale T-42 hand writing "ISER".

models of the hand-object system and multi-modal sensory feedback for planning and control [3,8,15]. Underactuated hands, such as the one shown in Fig. 1, passively achieve stable contact thanks to their adaptability. They allow pure vision feedback-based manipulation even without accurate hand-object models [4]. Using a data-driven approach, vision feedback can further enable the prediction of manipulation states, such as rolling, sliding, singularities and object dropping [6].

In tasks, such as handwriting or painting, the object must be moved along a desired path. Directly tracing such a path without considering the undesired states exposes the execution to risks, such as dropping the object. This work focuses on path planning for within-hand manipulation to trace a reference path, while avoiding to drop the object. Accordingly, given a reference path π^+, the objective is to plan a path so as to find path π^*:

$$\pi^* = \arg\min_{\pi} \mathrm{diff}(\pi, \pi^+), \text{ subject to:} \omega(q, \dot{q}) \geq K, \ \forall\ q \in \pi \text{ and } \dot{q} \in \dot{\pi}, \quad (1)$$

where $\mathrm{diff}(\pi, \pi^+)$ is the path difference, $\omega \in [0, 1]$ is the risk estimate given a configuration along the path and its derivative, and K is a risk threshold.

For measuring path difference, this work uses the Dynamic Time Warping (DTW) distance [10,14,16], which minimizes the area between two curves given their time parameterization. DTW is preferred over *Hausdorff* distance [2], which does not reason about the order of states along a path. It is also preferred over the Fréchet distance [1,9,17,18], which can lead to non-robust behavior, where small input variations can significantly distort the output [7].

2 Technical Approach

Planning the joint motions of a robotic hand so as to move a grasped object to a target pose often requires accurate hand-object models, which are difficult to obtain. Instead, the proposed method: (a) utilizes vision feedback for generating velocity references in the task space, (b) projects the velocities to the joint space via rough approximations, and (c) relies upon passive adaptability of the underactuated hands for maintaining stable object contact [4].

Accordingly, the planner reasons only about the configuration of the held object. A regression and classification process is used which evaluates the validity of moving the held object in a particular direction from a given configuration. The objective of the planner is to find a path that minimizes the distance to the reference path, while also ensuring that the paths remains valid (safe) as shown in Fig. 2.

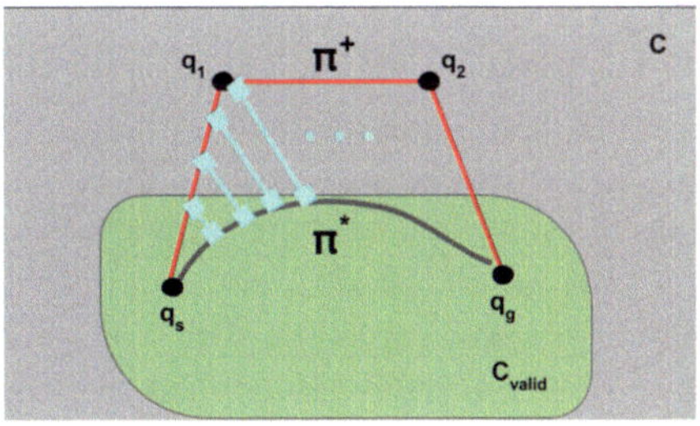

Fig. 2. A reference path π^+ and a path π^* tracking it, where π^+ consists of piecewise straight line segments connecting set points. The valid region $\mathbb{C}_{valid}$ is defined by the classifier. Path π^* must remain within $\mathbb{C}_{valid}$ and minimize the area between it and π^+. The initial and final points q_s, q_g are within $\mathbb{C}_{valid}$.

Classifier: The classifier is based on previous work [6], which uses visual and actuator measurements to detect four manipulation states, i.e. object rolling (normal operation), sliding, near drop, and stuck in a singularity. This work adopts the same framework, and trains a classifier for detecting states that would cause dropping the object. Nevertheless, the approach does not use actuator measurements as features, since these data are not available for offline planning. Instead, the classifier is trained with the object position and its velocity along the trajectory. Therefore, the classifier output ω in Eq. (1) corresponds to the probability of the state being safe given the object's position and reference velocity. The training used state vector machines (SVM) with a radial basis function.

An object configuration $q \in \mathbb{C}$ is valid under the following constraint:

$$\texttt{valid}(q) = \begin{cases} 1, \omega(q, \dot{q}) \geq K \\ 0, \textit{otherwise} \end{cases} \tag{2}$$

where ω is the classifier output, and K is the risk-parameter. The valid configuration space is therefore based on the classifier, $\mathbb{C}_{valid} = \{\forall q \in \mathbb{C} : \texttt{valid}(q) = 1\}$. The risk-parameter K allows for the same classifier to be used in challenging scenarios without requiring it to be retrained. For example, heavier objects can increase the risk that the object gets dropped during within-hand manipulation. In such cases, K should be increased, making the system more conservative and discouraging the object to get close to risky states. Conversely, lowering K would allow exploration of object states closer to the boundaries of $\mathbb{C}_{valid}$, which might be desirable for higher quality paths for lighter objects.

Curve Similarity Through Dynamic Time Warping: Let $q \in \mathbb{C} \subset \mathbb{R}^2$ represent the 2-dimensional configuration of the held object. Given two paths $\pi_1 = \{q_1(t), t \in \{1, T_1\}\}$ and $\pi_2 = \{q_2(t), t \in \{1, T_2\}]\}$, where $q_i(t)$ maps to configurations along the corresponding path, and t represents a discrete time parameterization of the path. Assume that a correspondence map exists $\phi = [\phi_1(t), \phi_2(t)]$ between π_1 and π_2, such that a configuration $q_1(\phi_1(t)) \in \pi_1$ maps to a configuration $q_2(\phi_2(t)) \in \pi_2$ for $t \in \{1, ..., T\}$, where $T = max(T_1, T_2)$. With a slight abuse of notation, let $\mathbf{v}(\phi_t)$ be a vector between corresponding mapped points at time t along π_1, π_2. The similarity measure [14] between these paths is the following:

$$\text{diff}(\pi_1, \pi_2) = \sum_{i=2}^{T} ||\mathbf{v}(\phi_t) - \mathbf{v}(\phi_{t-1})||_2 \tag{3}$$

Traditionally, DTW involves computing the correspondence map ϕ such that Eq. 3 is minimized. Because dynamic programming principles hold for DTW, this lends itself nicely to an incremental search. This also assumes, however, that there are two complete paths available for comparison. In this application, one of these paths is discovered on the fly. To accommodate this, the proposed search process imposes a series of constraints while solving the dual of DTW so as to obtain a path that optimizes Eq. 3.

Optimizing DTW Through Informed Search: Similar to the above discussion, $\mathbb{C}_{valid}$ denotes the valid configuration space of the object, i.e., the manipulator is not in an undesirable state (singularity, dropping or sliding object). This subspace is defined through the classifier. Given a sequence of target object configurations $Q^+ = \{q_1^+, q_2^+, ..., q_T^+\} \in \mathbb{C}$, the *reference path* π^+ is composed of piecewise linear segments connecting subsequent pairs of points in Q^+. The objective then is to compute a continuous curve $\pi : [0, 1] \rightarrow \mathbb{C}_{valid}$, which minimizes Eq. 3 relative to the reference path π^+, i.e. $\text{diff}(\pi, \pi^+)$.

In order to use Eq. 3, a time parameterization over π, π^+ must be employed. This is defined by discretizing the reference path using a fixed-length δ, essentially enforcing a constant velocity over the path. The correspondence map can then be defined as $\phi = [\phi(t), \phi_+(t)]$ for N timesteps, where N stands for the number of δ segments in π^+. Furthermore, the optimization is constrained, such that $\pi(0) = q_1^+, \pi(1) = q_T^+ \in \mathbb{C}_{valid}$, and only *monotonic* paths are explored. The *monotonicity* requirement enforces that corresponding configurations can only be reached by prior time steps, which prevents sudden "jumps" along the reference path.

Computing π can then be accomplished by utilizing an A* search rooted at q_1^+ and terminating at q_T^+. Each search node keeps track of its correspondence map index t, which determines which point on π^+ the node is mapped to. The search proceeds by expanding valid configurations by a δ amount at each iteration, constrained in the 8 cardinal directions for computational efficiency. The search nodes are visited in order according to the evaluation function

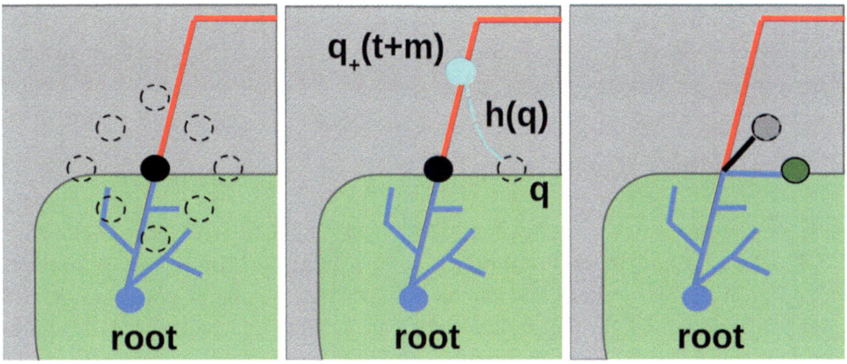

Fig. 3. (Left) iSST selects the black vertex to expand from; 8 possible directions are considered; (Middle) a heuristic is computed for each possible direction using Eq. 4, (Right) directions with the lowest heuristic whose edges lie wholly in the valid configuration space are added to the tree.

$f(q(t)) = g(q(t), q^+(t)) + h(q(t))$, where $q(t) \in \pi$ and $q^+(t) \in \pi^+$. The cost of a node g is then computed from Eq. 3 given the duration of the path that corresponds to the search node. Let $m = \frac{||q(t)-q^+(t)||_2}{\delta}$, then:

$$h(q(t)) = \frac{1}{2}\sum_{k=0}^{m}(||q(t) - q^+(t)||_2 - \delta * k). \tag{4}$$

The heuristic represents the minimum required area to the reference path for a path starting from $q(t)$ to latch on to the reference path $\pi^+(t)$. The best case involves the paths π^+ and π approaching one another at a rate of $2 * \delta$ per time step. This is therefore an admissible heuristic for the search process.

Efficient DTW Optimization: Although the A* process described above manages to produce appropriate solutions, as shown in Sect. 3, there are some drawbacks which warrant the use of an alternative process. Primarily, as the resolution of the search increases (i.e. by decreasing the value of δ), the A* process increasingly depends on the validation of subsequent search nodes to be fast. Despite the classifier calls being relatively quick, they incur a measurable cost, and accordingly the number of calls to the classifier could be viewed as an objective to minimize. Furthermore, A* can only provide solutions up to the selected resolution. To alleviate these issues, an alternative search process is proposed, based on an informed asymptotically optimal sampling-based planner (iSST [12,13]) that does not require access to a steering function [11].

The heuristic described in Eq. 4 can be directly applied in this method, allowing it to reduce the number of validity checks, as well as guiding the search process. Along with randomly sampling in the configuration space, the method also makes use of maneuvers (i.e. guided samples), which in this case will be similar to the A* (8 cardinal directions). An illustration of this is shown in Fig. 3. The initial solution returned by the method, while not guaranteed to be optimal as in

A*, can be improved upon, due to the asymptotically-optimal property of iSST. This allows an anytime performance, which is a desirable property for execution of trajectories on real robots.

3 Simulation Results

The search framework was implemented using both an A* approach, as well as iSST [12]), a heuristically-guided, asymptotically-optimal tree planner that propagates a single control during each iteration. A visual comparison of the expansion process for both is shown in Fig. 4. For each simulated evaluation, a reference path is generated by uniformly sampling the set points with the following constraints: (1) the start and goal are valid, (2) there are at least 4 total set points, and (3) at least 1 of the set points is invalid. These constraints ensure that directly tracing the reference path results in a failure state.

The two alternatives are compared in terms of average success rate, computation time, and solution cost. Both methods are evaluated with two classifiers: (i) *simulated*, which partitions $\mathbb{C}_{valid}$ using a random distribution of obstacles and uses standard collision-checking to determine state validity; and (ii) *trained*, which invokes the trained classifier to determine state validity. Fifty trials with different randomized reference paths were tested. A trial was counted as successful if it found a solution within 60 and 300 s for *simulated and trained* classifiers respectively. The solution cost was computed using Eq. 3.

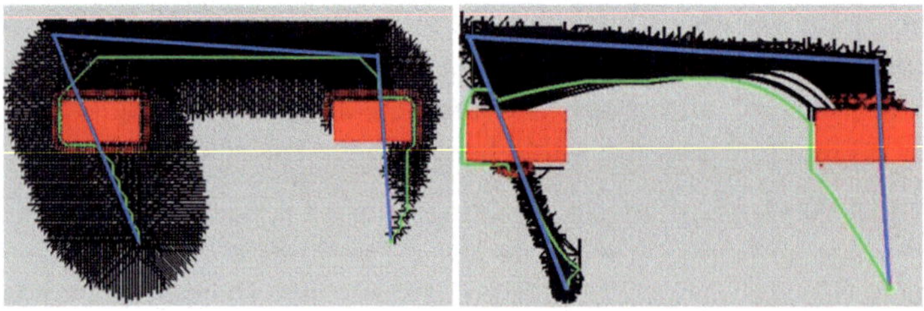

Fig. 4. A visual comparison of A* (left) and iSST (right) search trees with simulated obstacles. The blue line represents π^+, the green line is the solution π, the black lines are the search tree, and the red blocks are invalid regions.

All experiments were run on a single computer with an Intel Xeon E5-1660 CPU and 32 GB RAM, with results shown in Fig. 5. In the *simulated* classifier case, the A* performed the best across all metrics, although iSST remained competitive. In the *trained* classifier case, A* only produced solutions 18% of the time (within the 300 s time limit), compared to iSST's 78%, which requires fewer propagations that are computationally expensive as they require multiple calls to the trained classifier.

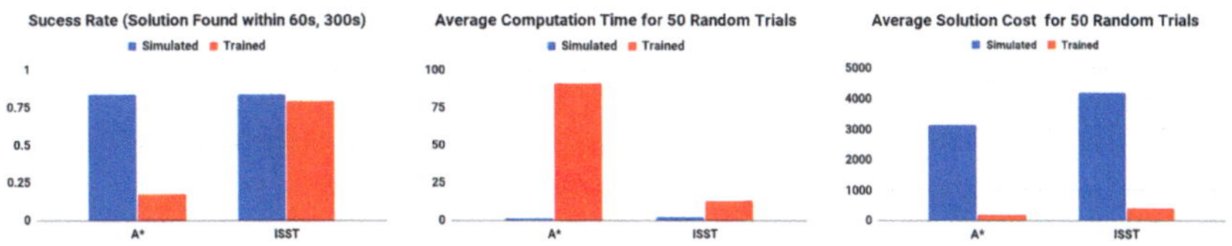

Fig. 5. Results of A* vs. iSST with *simulated and trained* classifiers over 50 randomized reference trajectories. (left) success rate, (middle) average computation time, (right) average cost.

4 Experiments

Given the randomized set points described in Sect. 3, several random reference paths were generated and tested given (a) a naive planner that follows the reference path with a straight line, and (b) the iSST planner that considers the classifier's output for avoiding risky regions. The Model T-42 hand (Fig. 6) is utilized together with a webcam observing it from the top.

Results with Randomized Trajectories: For testing the ability of the algorithm to avoid risky regions, we generated 10 random reference paths, and conducted experiments with the naive planner and the safe iSST planner using a cylindrical object with 2 cm diameter. For all 10 paths, the naive trajectory resulted in dropping the object, whereas the iSST planner successfully reaches the goal by avoiding the risky regions. Three samples from our runs are presented in Fig. 6. It can be seen that the planner diverts from the naive trajectory to stay in the safe regions according to the classifier output, but the distance to

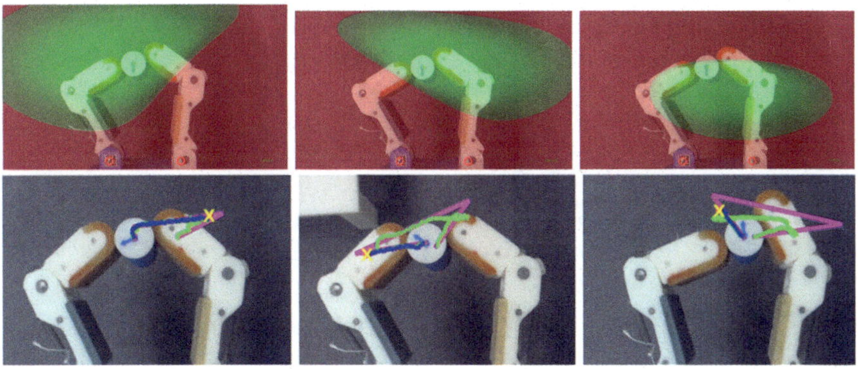

Fig. 6. (Top) The classifier outputs for the exploration direction that caused dropping the object (*green*: normal, *red*: drop), and (Bottom) the generated random paths (pink lines), executed trajectory with the naive planning (blue line), and executed trajectory with I-SST (green line). The object drops for the naive execution (the drop locations are marked with yellow crosses), while iSST completes the trajectory successfully.

the straight path is minimized as much as possible. Despite these deviations, the executed path preserves a notion of similarity to the original reference trajectory.

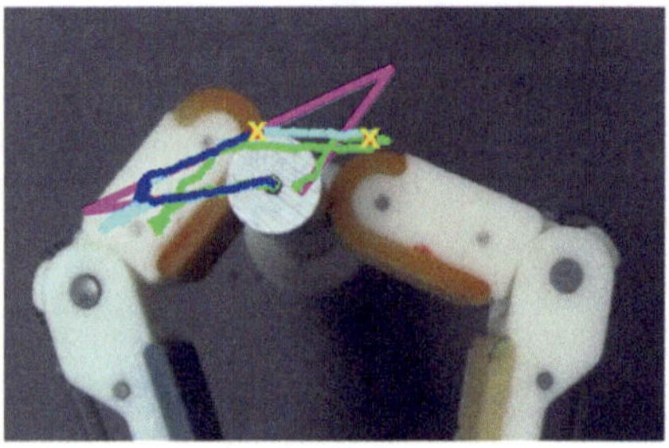

Fig. 7. Manipulation of a heavier object with varying risk parameter (K). Reference trajectory (pink), executed trajectory with $K = 0$ (blue), executed trajectory with $K = 0.6$, and executed trajectory with $K = 0.8$. The object drops for the $K = 0$ and $K = 0.6$ at the locations indicated by yellow crosses.

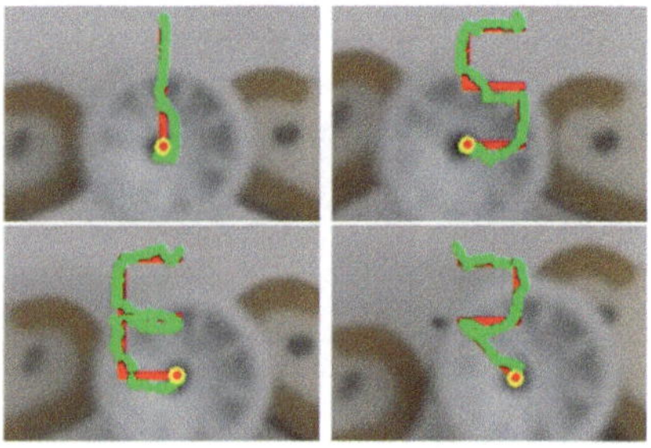

Fig. 8. Camera view while writing the letters "I", "S", "E", "R". Red lines are the generated reference trajectories, green lines are the execution of the trajectories.

Results with Different Weights: In this set of experiments, the same cylindrical object is used, but additional weight is attached to make it six times heavier. Here, we analyze the effectiveness of the risk factor parameter K for handling this scenario, which is significantly different than the data used for training the classifier. We run the experiments with risk factors varying from 0 to 1 with 0.1 increments. The trajectories with $K = 0$, $K = 0.6$ and $K = 0.8$ are presented in Fig. 7. It is seen that the system is able to execute longer segments of the trajectory without failure as the K value is increased. The object is dropped for risk factor values between $K = 0$ and $K = 0.7$, and the whole trajectory is successfully executed for values between 0.8 and 1.0. This indicates that the risk parameter is instrumental to cope with challenging scenarios by generating more conservative, cautious trajectories.

Results with Different Object Geometries: In order to further analyze the robustness of the algorithm, we conducted experiments with two cylinders and two rectangles with 2 cm and 4 cm radius and widths respectively. Four random paths are generated as explained in Sect. 3, and each path is tested with all the objects using the naive approach and the safe iSST planner (16 runs each). The risk parameter K was set to zero. The naive approach was successful only for 1 out of 16 cases, whereas the planner was successful for 15/16. This indicates the effectiveness of the safe planning framework. Nevertheless, for three of the successful iSST executions, even though the object is not dropped, significant out-of-plane rotations were observed for the rectangular objects. This was largely caused by within-hand sliding. In these cases, since $K = 0$, the system operates very close to the vicinity of the risky region, and cannot compensate for the unplanned sliding motions. Since the classifier utilized in this paper does not take sliding into account, the system gets too close to the risky region border. The failed iSST case was with the large rectangle, and was also caused by such sliding motions. This can potentially be fixed by increasing the K value, and obtaining more conservative trajectories. Utilizing a classifier that can detect sliding is another option, which is discussed in Sect. 5. In the successful execution of the naive planner, the sliding motion caused a positive effect by keeping the object close to the palm, resulting in path completion.

Handwriting Experiments: The safe planning framework was used to generate trajectories of the "I", "S", "E" and "R" letters: the feasibility of each letter was checked with our classifier, and the letters are shrunk until safe trajectories were possible. We used a white board and a board marker grasped by the Model T42 hand. A snapshot of the experiment is given in Fig. 1, and the letter trajectories followed by the marker is presented in Fig. 8. Even though the writing operation was successful, we observed that the letters written by the system deviated from the ideal trajectories. This was partially caused by the contact between the pen and the white board, which can be addressed with a more robust image-based controller, i.e. [5]. Nevertheless, the shapes were still well-preserved, and clearly recognizable (Fig. 9). To the best of our knowledge, this is the first execution of handwriting purely implemented with within-hand motions (except for moving between letters, which was done with the robot arm motion).

Fig. 9. "ISER" written by the T42 hand.

5 Conclusions

The proposed method is a crucial part of a vision-based underactuated within-hand manipulation framework. The system does not rely on accurate hand-object models, or force sensors, but it is still able to generate safe and reliable trajectories. The risk factor parameter is an effective tool to make the system cope with challenging cases by generating more conservative trajectories. Nevertheless, we have also seen that unplanned sliding motions may occur during manipulation, which effects system robustness negatively. One solution would be to utilize a classifier that can recognize gripper singularities and sliding along with drop modes as in other related work [6]. Not being able to access actuator loads, velocities and finger positions, however, during offline planning, makes it much more challenging to recognize all the target modes and have a good partitioning of the workspace as in Fig. 6. This issue can be addressed by employing different regression methods and data evaluation.

Another improvement can be achieved by using a probabilistic classifier in a planning under uncertainty framework, which reasons over distributions of safe states. This can potentially provide improved tracking results by allowing the robot to momentarily enter "dangerous" regions for short periods of time. This would entail changing the objective function to operate over expected costs.

Acknowledgements. This work was partially sponsored by NSF IIS-1734492, IIS-1723869, and IIS-1317976.

References

1. Alt, H., Godau, M.: Computing the fréchet distance between two polygonal curves. Int. J. Comput. Geom. Appl. **5**(01n02), 75–91 (1995)
2. Belogay, E., Cabrelli, C., Molter, U., Shonkwiler, R.: Calculating the hausdorff distance between curves. Inf. Process. Lett. **64**, 17–22 (1997)
3. Bicchi, A., Melchiorri, C., Balluchi, D.: On the mobility and manipulability of general multiple limb robots. IEEE Trans. Robot. Autom. **11**(2), 215–228 (1995). https://doi.org/10.1109/70.370503
4. Calli, B., Dollar, A.M.: Vision-based precision manipulation with underactuated hands: simple and effective solutions for dexterity. In: 2016 IEEE/RSJ International Conference on Intelligent Robots and Systems (IROS), pp. 1012–1018 (2016). https://doi.org/10.1109/IROS.2016.7759173
5. Calli, B., Dollar, A.M.: Robust precision manipulation with simple process models using visual servoing techniques with disturbance rejection. IEEE Trans. Autom. Sci. Eng. 1–14 (2018). https://doi.org/10.1109/TASE.2018.2819661
6. Calli, B., Srinivasan, K., Morgan, A., Dollar, A.M.: Learning modes of within-hand manipulation. In: IEEE International Conference of Robotics and Automation (2018, accepted)
7. Efrat, A., Fan, Q., Venkatasubramanian, J.: Curve matching, time warping, and light fields, new algorithms for computing similarity between curves. Math. Imaging Vis. **27**, 203–216 (2007)
8. Hang, K., Li, M., Stork, J.A., Bekiroglu, Y., Pokorny, F.T., Billard, A., Kragic, D.: Hierarchical fingertip space: a unified framework for grasp planning and in-hand grasp adaptation. IEEE Trans. Robot. **32**(4), 960–972 (2016). https://doi.org/10.1109/TRO.2016.2588879
9. Har-Peled, S., Raichel, B.: The fréchet distance revisited and extended. ACM Trans. Algorithms (TALG) **10**(1), 3 (2014)
10. Keogh, E., Ratanamahatana, C.A.: Exact indexing of dynamic time warping. Knowl. Inf. Syst. **7**(3), 358–386 (2005)
11. Li, Y., Littlefield, Z., Bekris, K.E.: Asymptotically optimal sampling-based kinodynamic planning. Int. J. Robot. Res. (IJRR) **35**(5), 528–564 (2016)
12. Littlefield, Z., Bekris, K.E.: Informed asymptotically near-optimal planning for field robots with dynamics. In: Field and Service Robotics, pp. 449–463. Springer (2017)
13. Littlefield, Z., Bekris, K.E.: Efficient and asymptotically optimal kinodynamic motion planning via dominance-informed regions. In: IEEE/RSJ International Conference on Intelligent Robots and Systems (IROS) (2018)
14. Munich, M.E., Perona, P.: Continuous dynamic time warping for translation-invariant curve alignment with applications to signature verification. IEEE Int. Conf. Comput. Vis. **1**, 108–115 (1999)

15. Okamura, A.M., Smaby, N., Cutkosky, M.R.: An overview of dexterous manipulation. In: IEEE International Conference on Robotics and Automation (ICRA), vol. 1, pp. 255–262 (2000). https://doi.org/10.1109/ROBOT.2000.844067
16. Salvador, S., Chan, P.: Toward accurate dynamic time warping in linear time and space. Intell. Data Anal. **11**(5), 561–580 (2007)
17. Solovey, K., Halperin, D.: Efficient sampling-based bottleneck pathfinding over cost maps. In: 2017 IEEE/RSJ International Conference on Intelligent Robots and Systems (IROS), pp. 2003–2009. IEEE (2017)
18. Wenk, C., et al.: Geodesic fréchet distance inside a simple polygon. ACM Trans. Algorithms (TALG) **7**(1), 9 (2010)

Real-Time Human-Robot Communication for Manipulation Tasks in Partially Observed Environments

Jacob Arkin[1(✉)], Rohan Paul[2], Daehyung Park[2], Subhro Roy[2], Nicholas Roy[2], and Thomas M. Howard[1]

[1] University of Rochester, Rochester, USA
jarkin@ur.rochester.edu
[2] MIT, Cambridge, USA

Abstract. In human teams, visual and auditory cues are often used to communicate information about the task and/or environment that may not otherwise be directly observable. Analogously, robots that primarily rely on visual sensors cannot directly observe some attributes of objects that may be necessary for reference resolution or task execution. The experiments in this paper address natural language interaction in human-robot teams for tasks where multi-modal (e.g. visual, auditory, haptic, etc) observations are necessary for robust execution. We present a probabilistic model, verified through physical experiments, that allows robots to acquire knowledge about the latent aspects of the workspace through language and physical interaction in an efficient manner. The model's effectiveness is demonstrated on a mobile and a stationary manipulator in real-world scenarios by following instructions under partial knowledge of object states in the environment.

1 Introduction

Effective human-robot teams coordinate activities via communication to optimally perform tasks. In human teams, visual and auditory cues are often used to communicate information about the task and/or environment that may not otherwise be directly observable. Analogously, robots that primarily rely on visual sensors cannot directly observe some attributes of objects that may be necessary for reference resolution or task execution. Algorithms that fuse information communicated by humans about hidden states of objects with information obtained from physical interactions with those objects, such as force measurements and/or joint torques, enable robots to more robustly perform tasks when posed with inaccurate or insufficient information. Also, minimizing latency in understanding language is crucial for human-robot teams because latency in communication reduces mission tempo and limits the effectiveness of collaboration. A robot with

Electronic supplementary material The online version of this chapter (https://doi.org/10.1007/978-3-030-33950-0_39) contains supplementary material, which is available to authorized users.

J. Xiao et al. (Eds.): ISER 2018, SPAR 11, pp. 448–460, 2020.
https://doi.org/10.1007/978-3-030-33950-0_39

a language model that can anticipate a subset of utterances that a human collaborator will likely communicate can leverage idle system time to proactively generate and ground those expressions in the context of its current world model, thus minimizing the latency between receiving an utterance and understanding its meaning.

The experiments in this paper address natural language interaction in human-robot teams for tasks where multi-modal (e.g. visual, auditory, haptic, etc) observations are necessary for robust execution. Contemporary approaches to probabilistic language understanding in partially known environments deal with uncertainty in both physical locations and semantic labels that is resolved upon observation with visual sensors [4,5,19]. These works target the navigation domain and focus on the alignment of metric and semantic maps wherein the unobserved aspect of the environment is the locations of hidden objects or locations of landmarks that extend beyond the current map. This is in contrast to the work presented here that targets hidden semantic attributes of objects necessary for successful task execution. Approaches that incorporate declarative knowledge [9,11,14] assume that such information is correct and sufficient for task execution, and thus are not robust to situations in which the declared knowledge is incorrectly understood by the robot or factually inaccurate.

We address the problem of realtime interpretation of instructions in scenarios that require knowledge of objects' states involved in task execution that may not be fully observed from the visual modality alone. We incorporate two information sources for estimating the latent aspects of the workspace. First, we use declarative knowledge from the human embedded in descriptions such as, "the gas can on the left is empty" that provide information about one or more of an object's hidden semantic states. Second, while manipulating the object, the robot's force/torque measurements can also inform the belief over the hidden state of the object. We estimate and propagate a belief in the presence of erroneous observations where language and interaction measurements may indicate different states. Further, reasoning with temporally increasing context adds a computational burden, hence mechanisms are needed for online language grounding.

This paper presents a probabilistic model, verified through physical experiments, that allows robots to acquire knowledge about the latent aspects of the workspace through language and physical interaction in an efficient manner. The proposed model builds on three lines of work: (i) efficient language grounding in large symbol spaces [13], where approximation of the complete model is fundamental to efficient inference, (ii) acquiring factual knowledge [14] over a temporally extended visual and linguistic interaction, and (iii) proactively searching for and inferring the meaning of likely phrases given the interaction history and current state of the world [1]. Our approach is robust to noisy or incorrectly asserted semantic information via the ability to incorporate the physical interaction measurements into the internal belief state. We demonstrate the model's effectiveness on a mobile and a stationary manipulator in real-world scenarios following instructions under partial knowledge of object states in the environment.

2 Technical Approach

We introduce a probabilistic graphical model, illustrated in Fig. 1, that allows a robot to follow instructions while remaining resilient to incomplete or inaccurate workspace knowledge. Consider a robot manipulator at a time step t has received sensory observations such as declarative language utterances $\lambda_{0:t}$ communicated by a human that express facts about the world (e.g. "the cup on the left is empty"), observations from a visual sensor $x_{0:t}$, and force/torque measurements $z_{0:t}$ during a physical interaction with an object. The robot's world representation consists of two components: (i) a metric model Υ_t derived from visual observations resulting in object locations and (ii) a symbolic model Γ_t comprised of logical predicates that convey semantic attributes/relationships that may be true for a possible world. A visual sensor can inform the symbolic states by providing observations of entities (blocks, cups, etc.) and spatial relationships (on, inside, front, etc.). Task execution in real-world domains often requires knowledge of the intrinsic object attributes that cannot be determined from visual observations alone, however. For example, executing the instruction "clear away the cups on the table" requires knowledge of the internal states of the cups (full or empty) to decide whether the cup should be trashed or put away in a store.

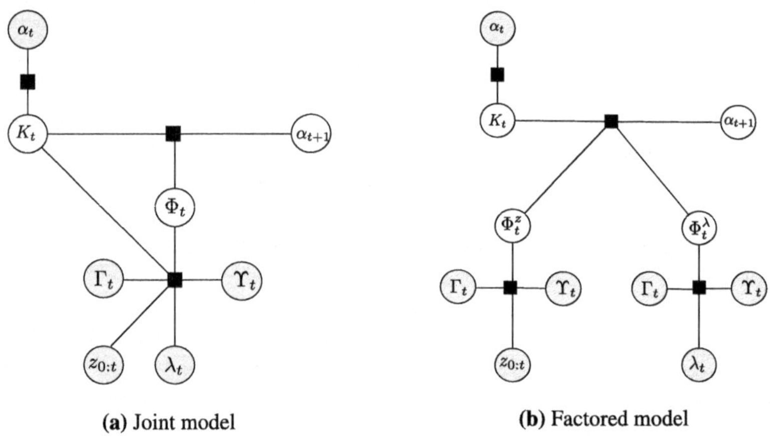

Fig. 1. Probabilistic models for knowledge acquisition over latent object attributes from descriptive language utterances and force measurements during physical interaction. (a) The joint model postulates a single grounding factor for both percepts $z_{0:t}$ and language λ_t resulting in likely correspondences Φ_t propagated to the next time step α_{t+1}. Learning in this joint model is challenging due to the large combined space of possible language utterances and interaction observations. (b) The factored model postulates separate grounding factors that relate percepts $z_{0:t}$ and language λ_t with likely correspondences Φ_t^z and Φ_t^λ respectively. The estimated groundings from both visual and linguistic modalities are fused to inform a posterior belief over the latent semantic state α_t propagated to the next time step.

In order to estimate a belief over the unobserved object states, we incorporate language descriptions from a human (e.g. "the cup on the table is empty") as well as force measurements during the robot's physical interaction (e.g. end-effector force profiles during manipulation). This belief over unobserved object states is included in the probabilistic graphical model as a random variable K_t that represents the state of knowledge about semantic object attributes in the world. For example, if the workspace contains a cup and a box, the knowledge state K_t would include a set of binary variables such as $IsFull(cup)$, $IsMovable(box)$ etc. Due to the diverse and large space of both language and percepts, joint inference about the knowledge state is challenging. Below, we describe our approach for making the learning problem computationally tractable.

2.1 Inferring Latent Semantic States

Our goal is to infer the latent semantic states of the objects (e.g., empty/full etc.) from language utterances and force measurements during physical interaction. Each semantic object state is symbolic in nature and represented as a discrete binary random variable; we assume that object properties are independent. We assume a Bernoulli distribution over the latent semantic states of objects and impose a conjugate Beta distribution prior with hyper-parameter α_t [2]. The distribution over the state K_t is parameterized by α_t and models the current belief over the true likelihood of a symbolic attribute (e.g., $p(IsFull(cup) = 1)$). Formally, the true likelihood over the predicate is $p(IsFull(cup) = 1) \sim Ber(\theta)$ with $\theta \sim Beta(a_t, b_t)$ where $\alpha_t = \{a_t, b_t\}$. The model fuses the factual knowledge provided via language from the human and force/torque measurements acquired from direct physical interaction to estimate a belief over the latent symbolic state of the object. Formally, we pose the problem as inference on a dynamic Bayesian network, incrementally fusing observations to yield a posterior estimate over the latent knowledge state as:

$$p(\alpha_{t+1} | \lambda_{0:t}, z_{0:t}, \Upsilon_t, \Gamma_t, \alpha_t) \tag{1}$$

where α_{t+1} is the hyper-prior over the knowledge predicate at time $t+1$.

The task of inferring symbolic knowledge from language and force/torque measurements can be formulated as estimating correspondences between measurements and probable semantic concepts. Following prior work on language grounding [7,13,14], we introduce a correspondence variable Φ_t to indicate the association between language or percepts and hidden object states. Additionally, the diverse nature of the observations makes direct estimation of Eq. 1 computationally challenging; so, we introduce a factorization wherein the estimation of the knowledge update is conditioned on the current hyper-prior and the state of the observation correspondence variable. As illustrated in Fig. 1, we can update the formulation of the desired distribution accordingly:

$$p(\alpha_{t+1} | \lambda_{0:t}, z_{0:t}, \Upsilon_t, \Gamma_t, \alpha_t) = \sum_{\Phi_t} \overbrace{p(\alpha_{t+1} | \Phi_t, \alpha_t)}^{\text{Knowledge Update}} \; \overbrace{p(\Phi_t | \lambda_{0:t}, z_{0:t}, \Upsilon_t, \Gamma_t)}^{\text{Observation Correspondences}} . \tag{2}$$

However, this formulation suffers from the challenge of jointly inferring a correspondence for the language and force/torque observations. We assume conditional independence between visual and linguistic modalities and factorize the conditional probability of the observation correspondence variable Φ_t; we represent the correspondence variables for language and force/torque observations as Φ_t^λ and Φ_t^z respectively. This factorized model, illustrated in Fig. 1b, is equivalent to Eq. 1 under the stated assumptions and can be expressed as:

$$\sum_{\Phi_t^\lambda}\sum_{\Phi_t^z} \overbrace{p(\alpha_{t+1}|\Phi_t^\lambda, \Phi_t^z, \alpha_t)}^{\text{Knowledge Update}} \overbrace{p(\Phi_t^\lambda|\lambda_{0:t}, \Upsilon_t, \Gamma_t)}^{\text{Lang. Corresp.}} \overbrace{p(\Phi_t^z|z_{0:t}, \Upsilon_t, \Gamma_t)}^{\text{Percept Corresp.}}. \tag{3}$$

The language grounding factor involves estimating probable correspondences Φ_t^λ between the language utterance and semantic concepts that the robot can interpret (e.g., a grounding $IsEmpty(cup)$ inferred from "the cup on the table is empty"). This factor is trained using a structured log-linear model with features extracting linguistic cues and spatial attributes of the current world state [7]. The estimation of semantic attributes from the human's utterance can be viewed as a declarative top-down inference over symbolic knowledge. A second source of symbolic knowledge comes from signatures derived from the robot's physical interaction with the object. The model infers latent object states ($IsFull$, $IsMovable$ etc.) using a hidden Markov model (HMM) with time series observations of the 3-axis end-effector force and arm pose measurements recorded during the physical interaction with the object. The HMM is trained using 30 interaction traces with varied object states and configurations. The inference results in likely correspondences Φ_t^z, a process that can be viewed as a bottom-up source of symbolic knowledge derived from grounding raw force measurements from physical object manipulation.

As shown above in Eq. 3, the two sources of symbolic knowledge, (i) the top-down groundings from descriptive language and (ii) the bottom-up groundings from interaction measurements, can then be fused to inform a posterior belief over the latent knowledge state. The conjugacy property of the Beta-Bernoulli distributions allows closed-form posterior updates marginalizing out the likelihood over the object states. The updated Beta distribution parameters are estimated as:

$$\alpha_{t+1} = \{a_t + n_\lambda^1 + n_z^1, b_t + n_\lambda^0 + n_z^0\}, \tag{4}$$

where $\{n_\lambda^1, n_\lambda^0\}$ and $\{n_z^1, n_z^0\}$ indicate the number of true and false observations derived from language $\Phi_{0:t}^\lambda$ and haptic groundings $\Phi_{0:t}^z$ respectively. The Bayesian treatment can be viewed as state estimation over logical symbolic variables and offers resilience to noisy or contradictory observations from language and physical interactions.

2.2 Proactive Instruction Following and Deliberative Interaction

Given an input language instruction, the robot must select actions μ_{t+1} as per the human's instructions and the robot's current belief over the world state as maintained as per the model described in Sect. 2.1. We extend the Temporal Grounding Graphs [14] for estimating intended manipulation goals from an input instruction. As discussed, the task of interpreting an instruction requires knowledge of hidden attributes for objects involved in the intended plan execution. For example, the interpretation of the instruction "clear the cups on the table" requires knowledge of the latent full/empty binary states of the cups to decide their appropriate destinations in the clearing task. We determine the expected state likelihoods by sampling the Bernoulli distribution from the current Beta prior. The model selects actions based on a confidence measure, computed as normalized entropy of the distribution, over the states of task-relevant objects. An uninformed belief necessitates information gathering actions such as repeatedly manipulating the cup until the latent belief is sufficiently informed to execute the intended action. The model formulation for inferring μ_{t+1} can be seen below:

$$p(\mu_{t+1}|\lambda_t, \Upsilon_t, \Gamma_t, \alpha_t) = \int_{K_t} \sum_{\lambda_t} \overbrace{p(\mu_{t+1}|K_t, \Phi_t^{\lambda})}^{\text{Estimating Actions}} \overbrace{p(K_t|\alpha_t)}^{\text{Belief}} \overbrace{p(\Phi_t^{\lambda}|\lambda_t, \Upsilon_t, \Gamma_t)}^{\text{Language Grounding}} . \quad (5)$$

The language grounding factor acts as a computational bottleneck as it involves a search over a large space of interpretations for an input instruction. Rather than reactively interpreting a full instruction, we proactively compute groundings that are likely to be relevant for future instructions. This results in real-time inference by boot-strapping a novel utterance with estimated groundings (true correspondences) from the set of proactively grounded phrases possessing similar parse structure. Formally, the set of proactive groundings Φ_t^{psg} is determined as a function of the current environment state Υ_t and the space of instructions determined by a grammar G. Since conditional independence is assumed across both individual phrases within the parse tree and individual groundings within the full space of semantic concepts, any given phrase with the same Υ_t will always ground to the same set of symbols regardless of parent phrases in the parse tree. Once the symbols that correspond to a simple phrase have been found, they can be reused within more complex phrases for as long as changes in the environment do not alter their meaning. Further, the set of candidate groundings are estimated by constraining the space of possible sentences from a vocabulary by imposing linguistic production rules from a phrase grammar. We leverage the hierarchical and compositional structure of language to construct proactive grounding sets in a bottom-up manner. The pre-computation of likely groundings reduces the runtime for interpreting a novel instruction by restricting the online computation to salient phrases λ_t^s which are typically smaller than the full set of phrases in the instruction ($|\lambda_t^s| \leq |\lambda_t|$). The resulting model formulation is equivalent to Eq. 5 and can be mathematically described as:

$$\int_{K_t} \sum_{\lambda_t^s} \overbrace{p(\mu_{t+1}|K_t, \Phi_t^{\lambda})}^{\text{Estimating Actions}} \overbrace{p(K_t|\alpha_t)}^{\text{Belief}} \overbrace{p(\Phi_t^{\lambda}|\lambda_t^s, \Upsilon_t, \Gamma_t) p(\lambda_t^s|\lambda_t, \Phi_t^{psg})}^{\text{Proactive Language Gnd.}} . \quad (6)$$

3 Experiments and Results

In order to validate the performance of the proposed system and its components, we designed three independent experiments. The first experimental evaluation targets the proactive symbol grounding component in simulation and quantitatively compares the inference runtime to a reactive baseline. For this experiment, we assumed a sufficiently expressive symbolic representation (see [13]), a grammar, and a corpus of annotated examples used for training. To quantify performance, we trialed different durations of proactive grounding time, increasing from 0 s to 8 s in 2 s intervals, during which the process grounded candidate phases, illustrated in Table 1 as "PSG Duration" (Proactive Symbol Grounding Duration) and "Number of Grounded Phrases" respectively. The row "NLSG Inference Time" (Natural Language Symbol Grounding Time) reports the runtime for a novel utterance; as expected, the runtime decreases as a function of the PSG Duration due to the process generating more matches to phrases in the novel utterance's parse tree and thus reducing the number of phrases to be computed at inference time. We include a trial with 0 s of proactive grounding time to establish a baseline of performance for the natural language symbol grounding process without any boot-strapping by the proactive grounding module.

Table 1. The proactive symbol grounding runtime (PSG) versus natural language symbol grounding (NLSG) runtime for a single instruction.

PSG duration (sec)	0.0	2.0	4.0	6.0	8.0
Number of grounded phrases	0	31	62	102	146
NLSG inference time (sec)	0.21	0.18	0.14	0.13	0.09

The second experimental evaluation used a Baxter Research Robot in a table top setup populated with household objects as shown in Fig. 3. In the first scenario, the robot's workspace contained two coffee cups (with closed lids), a tray and a trash can; the internal state of the cups was hidden with one cup being empty and the other full. We assume that the robot possesses learned background knowledge that empty cups are to be discarded in the trash and full cups are to be placed on the tray. As discussed in Sect. 2.1, the robot also possesses a trained hidden Markov model for classifying signatures from physical interaction with the cups. A plot of the different z-axis force measurements for a full and an empty cup can be seen in Fig. 2. The robot did not have access to the internal state of the cups. The robot was instructed to, "clear away the cups on the table," resulting in a grounded solution referencing the two coffee cups. The grounding model estimated the probable grounding of the sentence as the two cups on the table. The robot picked up each, updating the belief over the internal states according to force/torque sensing. This knowledge allows the robot to estimate the correct location to discard the empty cup in the trash or to place the filled cup on the tray.

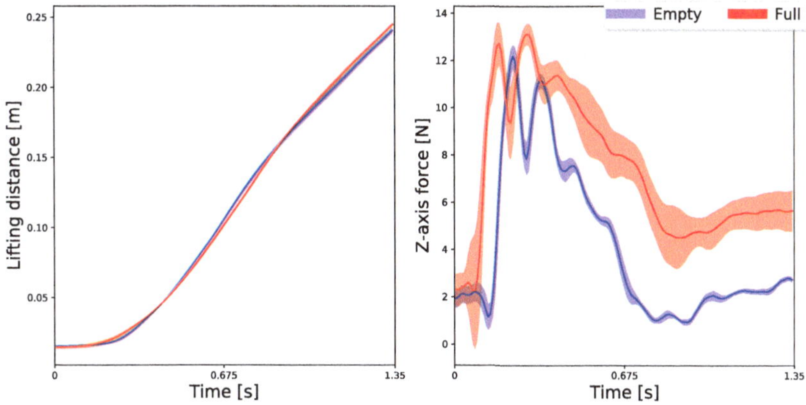

Fig. 2. Lifting distance and z-axis force measurements over time for both full (red) and empty (blue) cups. The patterns of force measurements over lifting distances are modeled by two HMMs which are then leveraged during log-likelihood-based binary classification to infer an object's state.

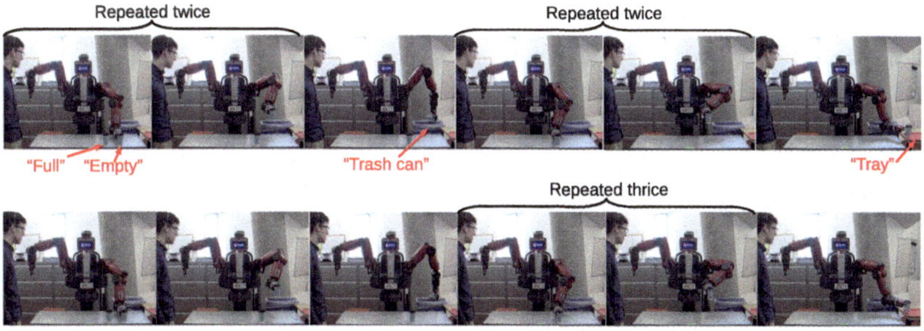

Fig. 3. The second experiment evaluating knowledge acquisition over latent object states from declarative knowledge and physical interaction. The Baxter robot was instructed to "clear away the cups on the table". **Top:** the robot attempts to pick up each cup in turn and infers the latent state of the cups from the time series of interactions. Once the belief is sufficiently informed the robot discards the empty cup in the trash bin and the filled cup on the tray. **Bottom:** the human informs the robot that "the cups on the table are empty" a fact that is true only for only one of the cups. The robot's physical interaction results in a posterior belief correcting the prior resulting from the incorrectly stated fact and correctly accomplished the task of clearing in correct locations.

In a subsequent scenario, the human declared, "the cups on the table are empty," before instructing the robot to "clear away the cups." Contradictory to the initial statement, the actual state of one of the cups is filled and should not be discarded. The robot determined the true state of the cups during interaction, correctly updating its prior belief from force/torque sensing and choosing the correct actions.

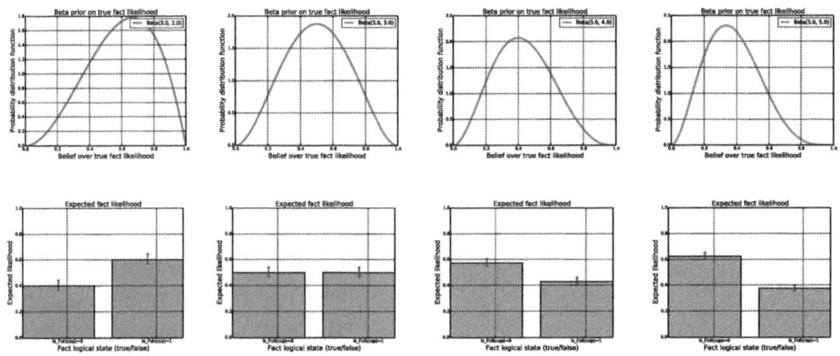

(a) Lang. $IsFull(cup) = 1$ **(b)** Obs. $IsFull(cup) = 0$ **(c)** Obs. $IsFull(cup) = 0$ **(d)** Obs. $IsFull(cup) = 0$

Fig. 4. Temporal evolution of belief over factual knowledge informed by language and interaction. The Beta distribution at time t for the Bernoulli likelihood over factual groundings is plotted in top row. The maximum likelihood true likelihoods for a predicate state appears below. Temporal evaluation from left to right. The initials "Lang." and "Obs." denote estimated groundings obtained from language and time series interaction data respectively. The estimation of the correct belief allows the robot to correctly follow the instruction of clearing the empty cups to the trash and placing the fill cup on the tray.

Figure 4 shows the resulting changes to both the Beta distribution and the expected likelihood of the expressed fact as the robot interacts with one of the cups in the second scenario. The robot first receives a declarative fact from language expressed as "the cups on the table are empty", leading to a posterior update to the Beta hyper-prior for the likelihood using the estimated grounding ($IsFull(cup) =$ True). Upon engaging in a time-series of physically interactions with the cup whose hidden state is actually $IsFull(cup) =$ False, the robot successively updates the latent belief over the symbolic state. The robot interacts with the object until the normalized entropy of the latent distribution is sufficiently informative (set via a threshold). The estimation of the correct belief allows the robot to correctly follow the instruction of clearing the empty cups despite initially receiving an incorrect fact from the human.

In the third experimental evaluation, we tested an integrated system that incorporates both the proactive symbol grounding process for fast inference and the joint use of declarative knowledge and force sensing for updating beliefs about objects' states. We used a Clearpath Husky A200 mounted with a Universal Robots UR5 manipulator in a mobile manipulation setting composed of two cases, as shown in Fig. 5; the internal state of the cases was hidden. The case on the robot's left was full and heavy, and the case on the right was empty and light. We executed three different types of scenarios in this experiment: (i) no declarative knowledge, (ii) accurate declarative knowledge, and (iii) inaccurate declarative knowledge. In one case of (i), the husky was instructed to, "pick up the heavy case" resulting in an ambiguous grounded reference solution. The robot picked up the left case, updating the belief that it was heavy; a second interaction made the robot confident enough to complete the action. In one case of (ii), the human accurately declared, "the case on the left is heavy" followed by, "pick up the heavy case." The robot picked up the left case, updating its belief which reinforced the human's provided fact. A single force/torque interaction and the accurate declared fact made the robot sufficiently confident to complete the action; the fact reduced the number of required interactions. In one case of (iii), the human declared, "the case on the right is heavy" followed by, "pick up the heavy case." The robot picked up the case on the right, updating its belief in contradiction to the human's provided fact. The robot then lifted the left case twice to be sufficiently confident and complete the action. A video demonstrating the experiments with both the Baxter and Husky robots can be found here: https://www.youtube.com/watch?v=JTVJkJavU6g.

4 Related Work

The approach described in this paper and the associated experiments address human-robot collaborative tasks in which multi-modal observations (language, vision, and haptic) are required to sufficiently inform the robot about latent object attributes necessary for task completion. Some contemporary approaches to probabilistic language understanding in partially known environments dealt with uncertainty in the metric location of objects or landmarks that was resolved upon observation with the visual sensors [4,5,12,19]. We address a different element of "partial observability" by inferring the state of object attributes as opposed to hypothesized locations of objects or landmarks that exist beyond the robot's current visual field or internal map of the explored world. We also incorporate a novel knowledge state variable in our graphical model and incrementally update a distribution over that knowledge state rather than reason over a distribution of maps.

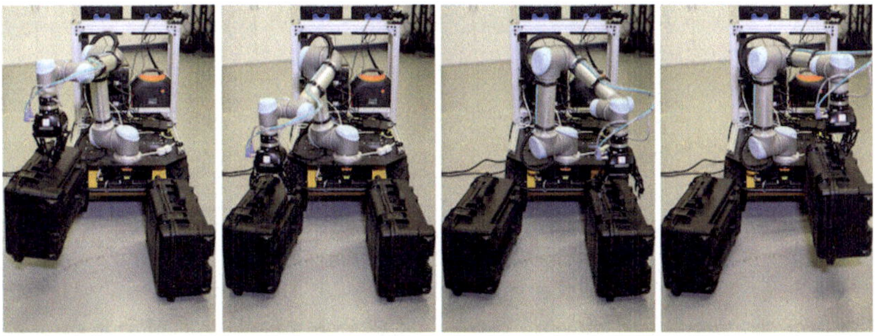

Initial state of the right case is heavy. | Updated belief is uncertain about heavy case. | Interaction with the other case. | Updated belief that the left case is heavy.

Fig. 5. A third experiment incorporating both proactive symbol grounding and updates to beliefs about objects' states via declarative knowledge and force/torque sensing. The Husky robot with a mounted robot arm was inaccurately told, "the case on the right is heavy" before receiving the instruction "pick up the heavy case."

There are other recent approaches for leveraging multi-modal observations to learn object attributes. In some work, human gesture was incorporated as a modality to learn object and relation classifiers, as well as attributes such as color [8,10,20]. Others have incorporated audio and haptic as modalities to learn visually-hidden attributes [3], such as whether a container is full or not based on the sounds produced while picking up and shaking [17]. There is also work to learn behaviorally-grounded or sensorimotor-grounded classifications [6], such as the work by Sinopav et al that uses vision, proprioception, and audio to learn semantic labels for objects via the robot's interaction with them [16].

Our contribution leverages language as a source of information about latent object states by grounding declarative statements from user utterances. Other natural language symbol grounding approaches that incorporate declarative knowledge [9,11,14,15,18] assumed that such information is correct and sufficient for task execution. In contrast, our model incrementally fuses information from language and force/torque interactions making task execution more robust to inaccurate or incorrectly understood declarations.

5 Discussion and Conclusion

This work demonstrated the ability to proactively infer a context for a future instruction from past interactions and world state. The ability to proactively predict the set of likely groundings has a significant improvement in the runtime of grounding a command, contributing towards real-time online symbol grounding in complex workspaces. Further, we observe that the task of interpreting an instruction is crucially tied to the semantic knowledge the robot has acquired about its workspace. In real-world scenarios, the robot only has partial knowledge about its workspace. We demonstrated how both linguistic descriptions from a human and signatures derived from the robot's physical interaction can be used to infer the latent semantic properties of the environment. We showed how a probabilistic model can update and maintain a belief over the latent knowledge state of the world from noisy linguistic and interaction observations. The experiments in this work contribute towards bridging the gap between higher-order inputs from the human such as language and the low-level representation of the robot such as cost functions, interaction forces etc. via grounded learning of semantic concepts.

As part of future work, we intend to demonstrate the effectiveness of the system on additional robot platforms in field experiment settings as part of an effort to both further test the reproducibility of our results and explore how our approach can be integrated into a larger system architecture for complex human-robot teaming scenarios. In addition, we seek to expand this framework to acquiring world knowledge in the context of multi-step plans common in complex missions rather than functioning as a multi-modal strategy for resolving reference ambiguity. To support this goal, we intend to increase the number of concepts expressed in training and consider more complex correlations and background knowledge in state estimation.

Acknowledgements. Authors gratefully acknowledge funding support in part by the Robotics Consortium of the U.S. Army Research Laboratory under the Collaborative Technology Alliance Program (RCTA) and the Toyota Research Institute (TRI) Award Number LP-C000765-SR. We thank our colleagues in the lab for helpful feedback on this paper.

References

1. Arkin, J., Howard, T.M.: Experiments in proactive symbol grounding for efficient physically situated human-robot dialogue. In: Late-breaking Track at the SIGDIAL Special Session on Physically Situated Dialogue (RoboDIAL-18), July 2018
2. Bishop, C.M.: Probability distributions. In: Jordan, M., Kleinberg, J., Schölkopf, B. (eds.) Pattern Recognition and Machine Learning, Chapter 2. Springer-Verlag, New York (2006)
3. Chu, V., McMahon, I., Riano, L., McDonald, C.G., He, Q., Perez-Tejada, J.M., Arrigo, M., Darrell, T., Kuchenbecker, K.J.: Robotic learning of haptic adjectives through physical interaction. Robot. Auton. Syst. **63**, 279–292 (2015)
4. Duvallet, F., Walter, M.R., Howard, T.M., Hemachandra, S., Oh, J., Teller, S., Roy, N., Stentz, A.: Inferring maps and behaviors from natural language instructions. In: Experimental Robotics, pp. 373–388. Springer (2016)
5. Hemachandra, S., Duvallet, F., Howard, T.M., Roy, N., Stentz, A., Walter, M.R.: Learning models for following natural language directions in unknown environments. In: Proceedings of IEEE International Conference on Robotics and Automation (ICRA), Seattle, WA, May 2015
6. Hogman, V., Bjorkman, M., Kragic, D.: Interactive object classification using sensorimotor contingencies. In: 2013 IEEE/RSJ International Conference on Intelligent Robots and Systems (IROS), pp. 2799–2805. IEEE (2013)
7. Howard, T.M., Tellex, S., Roy, N.: A natural language planner interface for mobile manipulators. In: Proceedings of IEEE International Conference on Robotics and Automation (ICRA), pp. 6652–6659. IEEE (2014)
8. Kollar, T., Krishnamurthy, J., Strimel, G.P.: Toward interactive grounded language acqusition. Robot. Sci. Syst. **1**, 721–732 (2013)
9. Kollar, T., Perera, V., Nardi, D., Veloso, M.: Learning environmental knowledge from task-based human-robot dialog. In: Proceedings of IEEE International Conference on Robotics and Automation (ICRA), pp. 4304–4309. IEEE (2013)
10. Matuszek, C., Bo, L., Zettlemoyer, L., Fox, D.: Learning from unscripted deictic gesture and language for human-robot interactions. In: Proceedings of National Conference on Artificial Intelligence (AAAI) (2014)
11. Matuszek, C., FitzGerald, N., Zettlemoyer, L., Bo, L., Fox, D.: A joint model of language and perception for grounded attribute learning. In: Proceedings of the: International Conference on Machine Learning, p. 2012. Edinburgh, Scotland (2012)
12. Oh, J., Howard, T.M., Walter, M.R., Barber, D., Zhu, M., Park, S., Suppe, A., Navarro-Serment, L., Duvallet, F., Boularias, A., Romero, O., Vinkkrov, J., Keegan, T., Dean, R., Lennon, C., Bodt, B., Childers, M., Shi, J., Daniilidis, K., Roy, N., Lebiere, C., Hebert, M., Stentz, A.: Integrated intelligence for human-robot teams. In: Proceedings of the 2016 International Symposium on Experimental Robotics, October 2016

13. Paul, R., Arkin, J., Aksaray, D., Roy, N., Howard, T.M.: Efficient grounding of abstract spatial concepts for natural language interaction with robot platforms. Int. J. Robot. Res. **37**, 1269–1299 (2018)
14. Paul, R., Barbu, A., Felshin, S., Katz, B., Roy, N.: Temporal grounding graphs for language understanding with accrued visual-linguistic context. In: Proceedings of the Twenty-Sixth International Joint Conference on Artificial Intelligence, pp. 4506–4514 (2017)
15. Perera, I.E., Allen, J.F.: Sall-e: Situated agent for language learning. In: Proceedings of the 27th AAAI Conference on Artificial Intelligence, pp. 1241–1247 (2013)
16. Sinapov, J., Schenck, C., Staley, K., Sukhoy, V., Stoytchev, A.: Grounding semantic categories in behavioral interactions: Experiments with 100 objects. Robot. Auton. Syst. **62**(5), 632–645 (2014)
17. Sinapov, J., Stoytchev, A.: From acoustic object recognition to object categorization by a humanoid robot. In: Proceedings of the RSS: Workshop on Mobile Manipulation, p. 2009. Seattle, WA (2009)
18. Thomason, J., Sinapov, J., Svetlik, M., Stone, P., Mooney, R.J.: Learning multimodal grounded linguistic semantics by playing "I Spy". In: IJCAI, pp. 3477–3483 (2016)
19. Walter, M.R., Hemachandra, S., Homberg, B., Tellex, S., Teller, S.: A framework for learning semantic maps from grounded natural language descriptions. Int. J. Robot. Res. **33**(9), 1167–1190 (2014)
20. Whitney, D., Eldon, M., Oberlin, J., Tellex, S.: Interpreting multimodal referring expressions in real time. In: Proceedings IEEE International Conference on Robotics and Automation (ICRA), pp. 3331–3338. IEEE (2016)

Motion Planning for Manipulators in Unknown Environments with Contact Sensing Uncertainty

Brad Saund(✉) and Dmitry Berenson

University of Michigan, Ann Arbor, USA
brad.saund@gmail.com

Abstract. Localization error, sensor noise, and occlusions can lead to an imperfect model of the environment, which can result in collisions between a robot arm and unobserved obstacles when manipulating. The robot must navigate around these obstructions despite not knowing their shape or location. Without tactile sensors, the robot only observes that a contact occurred somewhere on its surface, a measurement containing very little information. We present the Collision Hypothesis Sets representation for computing a belief of occupancy from these observations, and we introduce a planning and control architecture that uses this representation to navigate through unknown environments. Despite the dearth of information, we demonstrate through experiments that our algorithms can navigate around unseen obstacles and into narrow passages. We test in multiple environments in simulation and on a physical robot arm both with and without the aid of a 2.5D depth sensor. Compared to a baseline representation Collision Hypothesis Sets produce an approximately 1.5-3x speed-up and improve the success rate from 40–60% to 100% in tested scenarios with narrow passages.

1 Introduction

Robots rely on sensors to construct models of the world for use in motion planning, but in many practical scenarios sensing limitations result in an incomplete or inaccurate model, resulting in plans that can collide with unobserved obstacles. Sensing limitations occur in manipulation tasks due to limited range and field of view, invalid measurements caused by glare, and insufficient accuracy for motion in tight areas. Furthermore, robots may reach into occluded areas during maintenance and assembly tasks (e.g. reaching into a car engine) and

This research was funded in part by Toyota Research Institute (TRI). This article solely reflects the opinions of its authors and not TRI or any other Toyota entity.

Electronic supplementary material The online version of this chapter (https://doi.org/10.1007/978-3-030-33950-0_40) contains supplementary material, which is available to authorized users.

J. Xiao et al. (Eds.): ISER 2018, SPAR 11, pp. 461–474, 2020.
https://doi.org/10.1007/978-3-030-33950-0_40

household tasks (e.g. reaching deep into a cabinet or behind a box). In the scenarios examined in this paper collisions between the arm and environment can be sensed without damage to the robot and used to inform future plans, but a lack of tactile sensing generates large non-Gaussian uncertainty over the belief of the occupancy of the environment.

The task we consider is to move from a given start configuration through free space to a goal configuration as quickly as possible, however the free space is not known *a priori* and the occupancy must be sensed through contact. We do not require tactile sensing to detect contact since most robots do not possess touch-sensitive skin. Even robots with such skin may pick up objects, effectively extending their kinematic chain with unsensorized geometry. We assume a robot is able to detect contact using joint torque feedback as is available on many robot arms. We further use that joint torque to determine which links may be in contact.

The task of moving to a goal in an unknown environment can be framed as a Partially Observable Markov Decision Process (POMDP), where the belief over occupancy is obtained through noisy collision measurements. In our implementation the workspace, observed collisions, and known obstacles are all stored in voxel grids of size N, thus the size of a belief state is 2^N. A measurement either indicates a configuration along a path did not collide and thus the voxels occupied by the robot do not contain obstacles, or that a collision occurred and thus at least one voxel blocked movement to the new configuration. Measurement uncertainty does not primarily come from sensor noise, but because a measurement only provides a set of voxels where at least one is occupied. The size of the belief space and the measurement uncertainty make this problem intractable for a standard POMDP solver, thus we propose an approach which is specialized to our domain.

Our key contribution is a representation for contact uncertainty we call *Collision Hypothesis Sets (CHS)*, which enables planners to more accurately reason about potential collisions. This paper first reviews prior work on planning in uncertain environments (Sect. 2). The task is then defined (Sect. 3), the CHS representation is motivated and presented (Sect. 4), and the planning and control architecture is then described (Sect. 5). Results are reported for experiments performed in simulation and on a physical robot, comparing the CHS representation to a baseline unified cost grid (Sect. 6). With enough open free space the two methods perform similarly. However, when the robot must enter narrow passages to reach a goal the CHS representation reduced the total execution time by approximately a factor of 1.5 to 3. Furthermore, in narrow passage scenarios where the baseline method only reached the goal in approximately 40–60% of trials, using CHSs produced a success rate of 100%.

2 Related Work

A core component of motion planning is the ability to check the validity of a path. Many motion planning algorithms assume deterministic collision checking but with uncertainty in the robot or environment it is only possible to calculate the probability of collision. With an arbitrary belief Monte Carlo simulations (MCS) can be used to estimate collision probabilities [14], although this approach is computationally expensive and generally not practical for large state spaces, so many approaches apply only to specific belief distributions.

Assuming Gaussian uncertainty over the robot and specific objects enables computation of collision probability for a single configuration [4,20,21]. To estimate the collision probability along a full path, rather than at an individual configuration, [17] propagates only the portion of a Gaussian robot belief that are not in collision. Other forms of uncertainty, such as point cloud measurements from range sensors, may be better modeled with specifically-tailored distributions dependent on distance and incidence angle [1]. However, none of these distributions accurately model the information obtained when a robot contacts the environment.

Localization methods exist that are specifically tailored to model the unusual contact sensing uncertainty. Using a particle filter, a belief consistent with a contact measurement can be updated using rejection sampling [18], or sampling from the contact manifold [11]. Both of these methods require models of objects in the world, which we do not assume are known. Other methods use joint position and torques to estimate the contact location on the robot surface [3, 12], but require accurate torque measurements to produce accurate estimates. Since we wish to sense contact just above the the noise threshold the torque measurements will contain significant noise, thus we use a comparatively simple method that more reliably estimates which links may be in contact.

By treating the probability of collision as a cost, the problem of planning under uncertainty can be framed as an optimal path planning problem. While there exist numerous optimal planners through continuous space, such as RRT* [9] and its many variants, these methods rely on a quick computation of path cost to run efficiently and since computing collision probability is far more expensive than a typical path length cost these methods ultimately explore too few nodes for our problem in a reasonable time.

Rather than attempting to minimize the probability of collision, many algorithms search for a path with some acceptably-low probability of collision. This can be done conservatively by inflating the robot [15], iteratively considering simplified dynamics [5], or in a Probabilistic Roadmap where each edge collision cost is bounded [6]. These methods approximate the full path probability of collision from the probability of collision of the individual states, thereby assuming that probabilities of collision are independent. In our work the occupancy uncertainty is heavily coupled across space, thus this independence assumption does not hold (see Sect. 5 for further discussion).

While executing a plan a robot may learn new information that causes the plan to become invalid or suboptimal, and the robot can replan a new path given this new information. Most prior work on replanning assumes collisions can be sensed at a distance and a primary goal is to avoid collisions [8]. Some methods do allow for collisions and store contact locations sensed using tactile skin on a robotic arm [2,16]. Using a simulated skin a fast planning architecture was demonstrated with a fast local controller that falls back to a slower sampling-based planner when stuck in a local minimum [10]. We use a similar approach for mixing planning and control, but our method does not require tactile skin to sense precise contact points.

3 Problem Statement

Given a robot with configuration space $\mathcal{C}$, define a workspace voxel grid of size $N \times N \times N$ as $\mathcal{W}$ with static workspace occupied voxels $\mathcal{W}_\mathcal{O} \subseteq \mathcal{W}$ and free space $\mathcal{W}_\mathcal{F} = \mathcal{W} \backslash \mathcal{W}_\mathcal{O}$. For any configuration $q \in \mathcal{C}$ the robot occupies a subspace of workspace, defined by the mapping $\mathcal{R}(q) : \mathcal{C} \rightarrow \mathbb{P}(\mathcal{W})$ where $\mathbb{P}$ denotes the powerset. A collision occurs when $\mathcal{R}(q) \cap \mathcal{W}_\mathcal{O} \neq \emptyset$, and is detected by joint torque feedback (discussed in Sect. 4). While the planner does not have access to $\mathcal{W}_\mathcal{O}$ directly, the swept volume of the previously visited configurations $q^i_{visited}$ is known to be free and is stored in $\mathcal{W}_{SV} = \cup_i \mathcal{R}(q^i_{visited}) \subseteq \mathcal{W}_\mathcal{F}$.

A discretized path ξ consists of configurations q_i such that $||q_{i+1} - q_i|| < \delta$ where δ is set based on the desired resolution. A robot at q_i attempting to follow the ith path ξ^{plan}_i takes execution time $T_{exec}(\xi^{plan}_i) > 0$ to arrive at

$$M(q_i, \xi^{plan}_i) = q_{reached} \tag{1}$$

If ξ^{plan}_i collides then $q_{reached} \in \xi^{plan}_i$ is the configuration directly before the first $q \in \xi^{plan}_i$ in collision. After a collision a new ξ^{plan}_{i+1} may be executed.

Define an algorithm $\mathcal{P}$ that takes time t^{plan}_i to produce ξ^{plan}_i, a path that must begin at q_i. $\mathcal{P}$ is aware of past planned paths and visited configurations $Q^{visited}_i = \{q_0, q_1, ..., q_i\}$. We refer to $\mathcal{P}$ as a "planner" if it plans ξ^{plan} reaching the goal, and a "controller" if it computes a short ξ^{plan} towards the goal. A single $\mathcal{P}$ may choose to act as either a planner or controller depending on context.

$$\mathcal{P}(Q^{visited}_i, \xi^{plan}_1, \ldots, \xi^{plan}_{i-1}) = (\xi^{plan}_i, t^{plan}_i) \tag{2}$$

Given a robot in configuration q_{init} and a set of goal configurations Q_{goal} our objective is to choose a $\mathcal{P}$ to reach a goal configuration in the least time.

$$\underset{\mathcal{P},n}{\text{minimize}} \sum_{i=0}^{n-1} t^{plan}_i + T_{exec}(\xi^{plan}_i) \tag{3}$$

$$\text{subject to: } q_0 = q_{init}, q_n \in Q_{goal} \tag{4}$$

$$(\xi^{plan}_i, t^{plan}_i) = \mathcal{P}(Q^{visited}_i, \xi^{plan}_1, \ldots, \xi^{plan}_{i-1}) \tag{5}$$

$$q_{i+1} = M(q_i, \xi^{plan}_i) \tag{6}$$

Since planning time is part of the objective, $\mathcal{P}$ must choose a tradeoff between analyzing all information to choose the best ξ^{plan}_i, and minimizing planning time t_i^{plan}. A controller typically achieves a small t_i^{plan}, while a planner spends more time to produce better ξ_i^{plan}. As ξ^{plan} are executed, $\mathcal{P}$ has access to more knowledge about $\mathcal{W}$, and effectively using this information is key to minimizing total time. We will not solve Eq. 3 computationally, but instead design a $\mathcal{P}$ with desirable qualities and justify our choices with experimental trials.

4 Representing Uncertain Contact Information

When a collision is detected during execution the swept volume of a set of configurations $q_{collision}^i$ along the robot path within some small distance $d_{\mathcal{K}}$ after collision is assumed to contain the point of contact. While in theory the set of points on the robot surface contains the contact point, in practice joint measurement uncertainties, robot geometry uncertainties and approximations, and material compliance require a larger set to guarantee encapsulation of the contact point. A set containing the contact point is constructed $\mathcal{K}_i = \cup_i \mathcal{R}(q_{collision}^i) \setminus \mathcal{W}_{SV}$, the total volume of the robot in the possible collision configurations with the known free space removed (Fig. 1).

In our problem collisions are detected using measured joint torque $\tau_{meas} \in \mathbb{R}^J$, where J is the number of robot joints. Using a mass model of the robot the expected joint torque due to gravity and dynamics τ^{exp} is calculated and used to estimate the external joint torque $\tau^{ext} = \tau^{meas} - \tau^{exp}$. A noise threshold τ^{th} is set for each joint and τ^{ext} triggers a collision detection whenever any joint exceeds its threshold. Joint i exceeding τ_i^{th} implies an external (contact) force on a link after joint i on the kinematic chain. A set of links that must contain a contact $\mathcal{L}_{contact}$ is constructed by first finding the highest i where $\tau_i^{ext} > \tau_i^{th}$, then adding all links downstream from joint i to $\mathcal{L}_{contact}$. Only the links in $\mathcal{L}_{contact}$ are used to create $\mathcal{K}_i$.

4.1 Baseline: Unified Cost Grid

When provided with noisy measurements a common approach to modelling uncertainty extends a binary occupancy map to a Unified Cost Grid (UCG), a voxel grid where each voxel stores its likelihood of occupancy. We compute this likelihood as the count of how many observed collisions could be explained if that voxel were occupied. The UCG representation computes a path cost by summing the likelihood of the voxels in the swept volume of the path.

Unfortunately, by combining all measurements into a single grid the UCG representation loses the information that each collision was caused by at least one occupied voxel. As we show in Sect. 6, this approach fails when entering narrow passages as collisions near the entrance create a high cost for feasible paths. Therefore, we construct a representation that maintains this information.

4.2 Collision Hypothesis Sets

Define a *Collision Hypothesis Set* (CHS) as a set of points in the robot workspace containing at least one point in collision. The $\mathcal{K}_i$ constructed after a collision

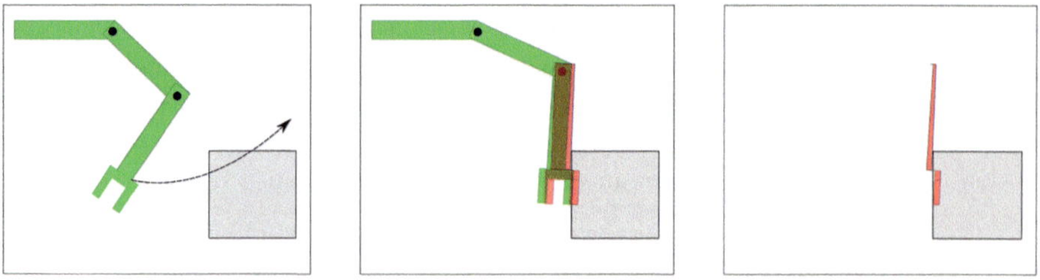

Fig. 1. A plan for the green robot (left) results in a collision with the grey unknown obstacle. A red collision hypothesis set is added using links possibly in collision based on measured joint torques (center). The known free space is removed (right).

are CHSs, since $\mathcal{K}_i \cap \mathcal{W}_{\mathcal{O}} \neq \emptyset$. However, unlike in UCG, the CHS representation never combines $\mathcal{K}_i$s into a unified grid, and instead maintains a set $\mathcal{K}$ of all generated $\mathcal{K}_i$.

Planning will require evaluating the probability of collision for a path using $\mathcal{K}$. Let the swept volume of a path ξ on the robot be $\mathcal{W}_\xi \subseteq \mathcal{W}$. We define the probability of collision of ξ with a single $\mathcal{K}_i$ as

$$p_{collision}(\mathcal{W}_\xi, \mathcal{K}_i) = \frac{|\mathcal{W}_\xi \cap \mathcal{K}_i|}{|\mathcal{K}_i|} \tag{7}$$

Assuming there exists a single occupied voxel uniform randomly selected from $\mathcal{K}_i$, this is precisely the probability the path collides. While this will likely be an underestimate of the true probability, it encourages exploration and further collision measurements will help localize the contact. The probability of collision for a full path is computed assuming independence between $\mathcal{K}_i$s, thus

$$p_{collision}(\mathcal{W}_\xi, \mathcal{K}) = 1 - \prod_i (1 - p_{collision}(\mathcal{W}_\xi, \mathcal{K}_i)) \tag{8}$$

This definition for collision probability captures several key features that the UCG representation lacks. A $\mathcal{K}_i$ with fewer voxels represents a more precise knowledge of where the contact occurred and thus a more precise estimate of workspace occupancy. In addition, a path that moves the robot through an entire CHS is guaranteed to collide, since $\frac{|\mathcal{W}_\xi \cap \mathcal{K}_i|}{|\mathcal{K}_i|} = 1$, representing the information that a CHS contains at least one point in collision. A path ξ that is attempted but blocked due to a detected collision creates a $\mathcal{K}_i$ that lies entirely within $\mathcal{W}_\xi$, so the updated collision probability for ξ will now be 1.

5 Interleaving Planning and Control

To achieve our objective of reaching a goal configuration in minimal time (Eq. 3), we must choose an algorithm $\mathcal{P}$ that compute good motions ξ_i^{plan} in low planning times t_i^{plan}. Local controllers quickly compute locally good ξ_i^{plan} but may get stuck in local minima. Global planning can escape these minima by planning a full path, but requires significant computation time. Our environments include many undetected obstacles and a local controller may produce many collisions before getting stuck, thereby providing more information to the global planner without adding much total time. Thus we use a planner initially and when stuck in cul-de-sacs, and local control otherwise. Our full architecture is presented in Algorithm 1.

Planning: The objective of the planner is to find a path to the goal with the minimal probability of collision. Unfortunately, as discussed in Sect. 2, previous methods that plan over obstacle uncertainty are not applicable when using CHSs, as these planners typically rely on a path cost definition that is purely the sum [9] or maximum [20] of costs of states/edges. Figure 2 illustrates two problems with inferring path collision probability using the costs of only individual states along the path. In Fig. 2a the swept volume of adjacent states overlap significantly, thus summing costs of states could significantly overestimate the probability of collision. In Fig. 2b multiple states collectively intersect the entirety of a CHS thus guaranteeing a collision, but each state individually only intersects a fraction of that CHS, thus taking the maximum over all state costs would significantly underestimate the probability of collision.

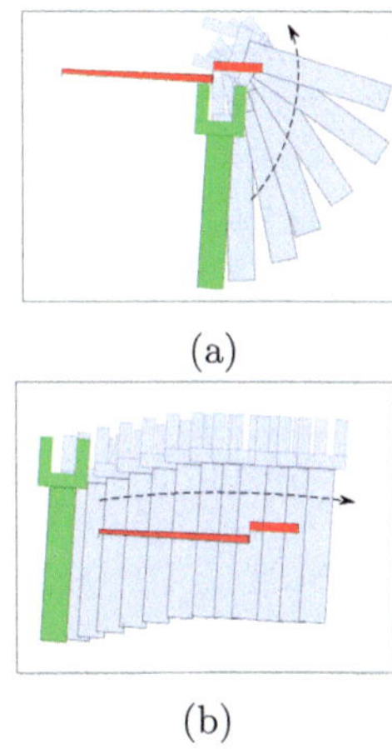

Fig. 2. Plans for the green arm sweep through the blue region. Red: a CHS

Algorithm 1
MainLoop(q_{cur}, q_{goal})

$\mathcal{K} \leftarrow \emptyset$; $\mathcal{W}_{SV} \leftarrow \emptyset$
while $q_{cur} \neq q_{goal}$ **do**
    $\xi \leftarrow$ Planner($q_{cur}, q_{goal}, \mathcal{K}$)
    $q_{cur} \leftarrow$AttemptPath($\xi, \mathcal{K}, \mathcal{W}_{SV}$)
    while $q_{cur} \neq q_{goal}$ **and** ($\xi \leftarrow$ Controller($q_{cur}, q_{goal}, \mathcal{K}$)) $\neq \emptyset$ **do**
        $q_{cur} \leftarrow$ AttemptPath($\xi, \mathcal{K}, \mathcal{W}_{SV}$)

Algorithm 2
AttemptPath($\xi, \mathcal{K}, \mathcal{W}_{SV}$)

for q_i in ξ **do**
    if q_i causes collision **then**
        $\mathcal{K}$.addNew($q_i, \xi, d_{\mathcal{K}}$)
        break
    $q_{cur} \leftarrow q_i$
    $\mathcal{W}_{SV} \leftarrow \mathcal{W}_{SV} \cup \mathcal{R}(q_{cur})$
$\mathcal{K}$.subtract($\mathcal{W}_{SV}$)
return q_{cur}

To plan a path we create PathBiRRT (Algorithm 3), a planner based on bi-directional RRT [13] that ensures the cost of the path generated is below a specified threshold p_{thr}. For the CHS representation the `Cost` function is given by Eq. 8. For the baseline UCG representation, which we compare to in the results, the `Cost` sums the cost of all voxels in $\mathcal{W}_\xi$, i.e. $\sum_i |\mathcal{W}_\xi \cap \mathcal{K}_i|$.

Ideally we desire a `Connect` function that ensures the cost of the entire path to any new node is below p_{thr}. In practice, computing this cost for every explored node is prohibitively expensive, thus we approximate this cost as an accumulation of branch costs computed in the `Connect` function (Algorithm 4). When extending towards q_{target} the full path cost of the branch from q_{near} is calculated (Line 6). The cost from the root to the q_{new} is approximated by accumulating the approximate cost to q_{near} (computed previously) and the cost of the new branch (Line 8). When using CHS the `Accumulate` function is the combination of independent probabilities: $1 - (1 - p_1)(1 - p_2)$. UCG accumulates by adding the costs: $c_1 + c_2$. By computing the cost over full branches this approximation is significantly better than accumulating cost purely based on states, however, this approximation may still over or underestimate the true cost, as separate branches along a path may overlap in $\mathcal{W}$.

PathBiRRT (Algorithm 3) repeatedly calls `Connect` to build a tree from the start and a tree from the goal, generating a potential path ξ when the two trees meet. ξ may exceed p_{thr} due to the approximation error within `Connect` and because ξ is the combination of paths from two trees. The cost of ξ is checked (Line 9) and if it exceeds p_{thr} then the highest cost edge from ξ is pruned along with all child edges, and planning continues.

For planning within fixed time t_{plan}, we provide two methods for setting p_{thr}. The anytime APathBiRRT (Algorithm 5) begins with $p_{thr} = \infty$ and continues searching for lower-cost paths until time runs out. In contrast, IPathBiRRT (Algorithm 6) begins with an optimistic $p_{thr} = c_{init}$. In each iteration IPathBiRRT allocates a fraction ψ_f of the remaining time t_ψ for planning with the current p_{thr}. If a plan is not found within t_ψ, p_{thr} is increased for the next iteration, approaching p_{max} for CHS, or v_{max} for UCG. Once a path is found, IPathBiRRT then invokes APathBiRRT for the remainder of the planning time.

As a benchmark, we also implemented ABiRRT, which iteratively decreases a cost threshold (as in Algorithm 5), but when checking to add a new node (as in Algorithm 4, Line 5–9) only the configuration cost is considered ($\texttt{Cost}(\mathcal{R}(q_{new}), \mathcal{K}) < p_{thr}$) and there is no full path check (Algorithm 3, Line 9). To support the claim that asymptotically optimal planners are not practical for this problem we also compare against RRT* [9], which always computes the full path cost when considering new connections.

Local Control: The local controller samples a specified number n_c of straight-line motions of length d_c uniformly from the half-sphere in $\mathcal{C}$-space that reduce the robot's distance to the goal. From these samples, the controller greedily selects the motion with lowest probability of collision, using Eq. 8. If no motion is found with probability of collision $< p_c$ for CHS, or cost $< \text{nvox}_c$ for UCG, the controller assumes it is stuck (Algorithm 1 Line. 5) in a cul-de-sac and invokes the planner.

Algorithm 3
PathBiRRT($q_{init}, q_{goal}, \mathcal{K}, p_{thr}, t$)

$\mathcal{T}_A$.init(q_{init}); $\mathcal{T}_A[q_{init}]$.approxCost $\leftarrow 0$
$\mathcal{T}_B$.init(q_{goal}); $\mathcal{T}_B[q_{goal}]$.approxCost $\leftarrow 0$
while timeElapsed() $< t$ **do**
    $q_r \leftarrow$ sampleConfig()
    status, q_{new} = Connect($\mathcal{T}_A, q_r, p_{thr}$)
    if status $\neq$ *Trapped* **then**
        if Connect($\mathcal{T}_B, q_{new}, p_{thr}$) = *Reached* **then**
            $\xi \leftarrow$ path($\mathcal{T}_A, \mathcal{T}_B$)
            if Cost($\mathcal{W}_\xi, \mathcal{K}$) $< p_{thr}$ **then**
                return ξ
            else
                e $\leftarrow$ highestCostEdge(ξ)
                if e in $\mathcal{T}_A$ **then**
                    $\mathcal{T}_A$.prune(e)
                else
                    $\mathcal{T}_B$.prune(e)
    swap($\mathcal{T}_A, \mathcal{T}_B$)
return $\emptyset$

Algorithm 4
Connect($\mathcal{T}, q_{target}, \mathcal{K}, p_{thr}$)

$q_{near} \leftarrow$ nearest($\mathcal{T}, q_{target}$)
$\mathcal{W}_{seg} \leftarrow \{\}$
$\xi \leftarrow$interpolate($q_{near}, q_{target}, \delta$)
for q_{new} in ξ **do**
    $\mathcal{W}_{seg} \leftarrow \mathcal{W}_{seg} \cup \mathcal{R}(q_{new})$
    $c_{seg} \leftarrow$ Cost($\mathcal{W}_{seg}, \mathcal{K}$)
    $c_{near} \leftarrow \mathcal{T}[q_{near}]$.approxCost
    $c_{approx} \leftarrow$ Accum(c_{seg}, c_{near})
    if collides(q_{new}) **or** $c_{approx} \geq p_{thr}$ **then**
        if $q_{new} = q_{near}$ **then**
            return {Trapped, q_{new}}
        return {Advanced, q_{new}}
    $\mathcal{T}$.add(q_{new})
    $\mathcal{T}[q_{new}]$.approxCost $\leftarrow c_{approx}$
return {Reached, q_{new}}

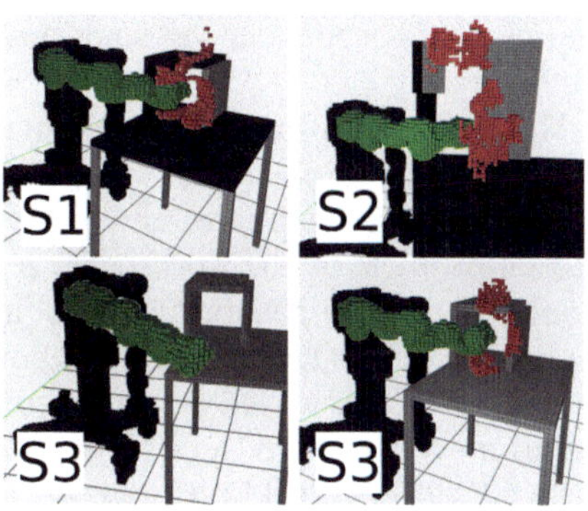

Fig. 3. Simulated scenarios

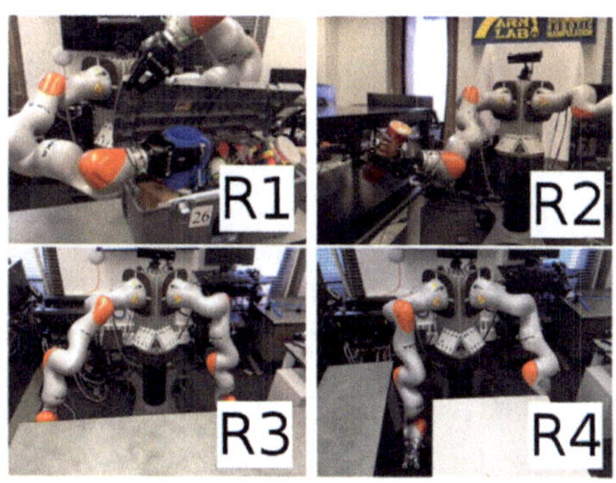

Fig. 4. Physical robot environments

Algorithm 5 APathBiRRT ($q_{init}, q_{goal}, \mathcal{K}, p_{thr} = \infty$)

$\xi_{best} \leftarrow \emptyset$
while timeRemaining()>0 **do**
  $\xi \leftarrow$ PathBiRRT($q_{init}, q_{goal}, p_{thr}$, timeRemaining())
  if $\xi \neq \emptyset$ **then**
    if Cost($\mathcal{W}_\xi, \mathcal{K}$) = 0 **then**
      return ξ
    $p_{thr} \leftarrow$ Cost($\mathcal{W}_\xi, \mathcal{K}$) - ϵ
    $\xi_{best} \leftarrow \xi$
return ξ_{best}

Algorithm 6 IPathBiRRT ($q_{init}, q_{goal}, \mathcal{K}, c_{init}, c_{max}, \psi_f$)

$p_{thr} \leftarrow c_{init}$
while timeRemaining()>0 **do**
  $t_\psi \leftarrow$ timeRemaining() $\cdot \psi_f$
  $\xi \leftarrow$ PathBiRRT($q_s, q_{goal}, p_{thr}, t_\psi$)
  if $\xi \neq \emptyset$ **then**
    return APathBiRRT(q_s, q_{goal}, p_{thr} =Cost($\mathcal{W}_\xi, \mathcal{K}$))
  $\alpha \leftarrow$ timeElapsed()/totalTime()
  $p_{thr} \leftarrow \alpha \cdot c_{max} + c_{init}$
return $\emptyset$

6 Experiments and Results

To demonstrate the advantages of our representation we compared Collision Hypothesis Sets (CHS) to the baseline Unified Cost Grid (UCG) in multiple environments in simulation and on a physical robot using multiple planning approaches[1]. Parameters used in experiments are given in Fig. 5. Voxel grids were implemented on the GPU using GpuVoxels [7]. Planners were implemented using OMPL [19] with modification. Code was run on a computer with i7-7700 processor and a NVidia 1080 Ti GPU. For all planners each path was smoothed using 100 iterations of shortcut smoothing.

Simulation Experiments simulated a 7DOF Kuka iiwa arm in the environments in Fig. 3, shown with red $\mathcal{K}_i$s, grey unobserved obstacles, and black observed obstacles. Rather than simulating joint torque, collisions were determined when the simulated robot moved into an obstacle in the workspace. All links downstream of the true link in collision were used to generate a CHS. In scenario S1, the simplest environment, the robot's goal was to reach inside a box located on a table with some occupancy known from a simulated depth sensor. Scenario S2 was harder as the robot needed to move the entire arm through a narrow slot occluded from the sensor. Scenario S3 was identical to S1 except the robot used no depth sensor information. Each simulation trial allowed 15 min for the robot to reach the goal.

Physical Robot Experiments were conducted on a physical 7DOF Kuka iiwa arm capable of sensing joint torque. A Kinect depth sensor created known obstacle occupancy. Figure 4 shows the physical robot experimental setups. Each physical trial allowed 5 min for the robot to reach the goal. In physical robot scenario R1 the robot placed a pitcher inside a box with the lid occluding the side and back walls from the Kinect. R2 involved placing a cylindrical can on a short shelf. Glare and occlusions resulted in a sparse and noisy occupancy map from the Kinect, both blocking a feasible path to the goal and missing portions

[1] Videos of experiments available at https://www.youtube.com/watch?v=EjCq1Q4nNUc.

	Description	Value
p_c	controller max collision probability	0.9
nvox_c	controller max collision voxels	50
d_c	controller motion length	0.3 radians
n_c	controller number of samples	20
δ	path discretization size	0.14 radians
$d_{\mathcal{K}}$	dist. for $\mathcal{K}_i$ creation	0.05 radians
τ^{th}	torque threshold	[20, 20, 15, 5, 4, 3, 1] Nm
t_{plan}	allowed planning time per iteration	30s
N	voxel grid size	200x200x200
ψ_f	IPathBiRRT phase fraction	1/4
c_{init}	IPathBiRRT initial cost	0.3
p_{max}	IPathBiRRT CHS c_{max}	1
v_{max}	IPathBiRRT UCG c_{max}	400
ϵ	APathBiRRT improvement factor	0.0001

Fig. 5. Experimental parameters

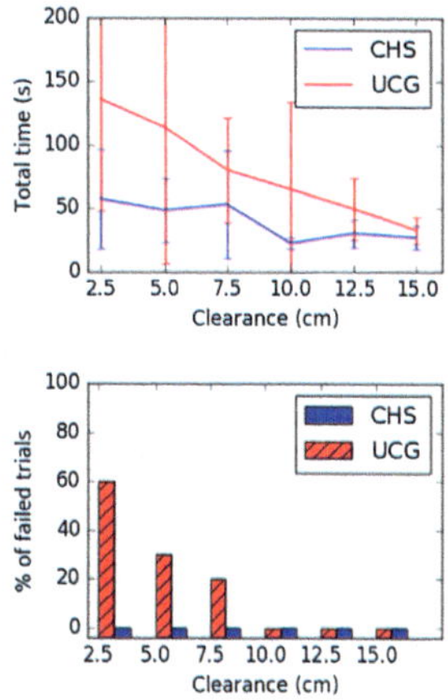

Fig. 6. Total time and failure % for R4, averaged over 10 trials for each clearance with a 5 min. timeout

of the top and bottom of the shelf. To accommodate noise up 30 intersections were allowed between the robot and the Kinect occupancy map.

In scenario R3 the robot arm moved from below to above a table. The sensed table was artificially shifted 5 cm away from the robot, simulating localization or sensor error. R4 tested the behavior moving through a narrow passage between two tables. This gap was adjusted between 15.5 cm to 28 cm, corresponding to a clearance of 2.5 cm to 15 cm for the 13 cm wide robot hand. In R4 Kinect data was not used, thus the robot only sensed obstacles through contact.

Results: Table 1 reports the results for each simulated scenario, comparing CHS and UCG using the proposed APathBiRRT and IPathBiRRT planners as well as RRT* and ABiRRT. RRT* did not reach the goal within the time limit in any scenario. ABiRRT performed far worse than our proposed methods, indicating that considering full path collision probability is superior to only per-configuration collision probability. Table 2 reports the results for each physical scenario, comparing CHS and UCG using IPathBiRRT only, as this planner performed the best in simulation trials. In both simulation and physical trials we observed our CHS formulation outperforms the UCG approach in terms of computation time.

In simpler scenarios (S1, S2, R1) early collisions were navigated better by the local controller, causing fewer lengthy planning iterations. For example in R1 our method required invoking the global planner only once in all ten trials, leading to an average success time of 18.6 s compared to 76.6 s when using a unified

Table 1. Simulated scenarios: successes within 15 min, total time (s), and number of planner calls averaged over 20 trials for each entry. Blue: Proposed methods.

	Planner	S1			S2			S3		
		Succ.	Time	P.Calls	Succ.	Time	P.Calls	Succ.	Time	P.Calls
CHS	APathBiRRT	100%	110	4.1	100%	100	3.4	100%	220	6.5
	IPathBiRRT	85%	69	3.0	100%	64	2.6	100%	180	5.3
	RRT*	0%	-	-	0%	-	-	0%	-	-
	ABiRRT	100%	78	3.5	30%	230	20	60%	200	15
UCG	APathBiRRT	100%	110	4.2	95%	120	3.9	40%	440	18
	IPathBiRRT	55%	71	2.9	90%	130	3.9	60%	280	15
	RRT*	0%	-	-	0%	-	-	0%	-	-
	ABiRRT	100%	113	4.4	80%	250	11	65%	500	22

Table 2. Physical robot experiments: successes within 5 min., total time (s), number of planner calls, planning time (s), and controller time (s) averaged over 10 trials.

	Planner	R1					R2				
		Succ.	Time	P.Calls	Plan	Ctrl	Succ.	Time	P.Calls	Plan	Ctrl
CHS	IPathBiRRT	100%	19	1.1	3.4	15	100%	62	2.1	38	24
UCG	IPathBiRRT	100%	77	2.3	57	19	40%	180	4.8	130	43
	Planner	R3					R4: clearance=2.5cm				
		Succ.	Time	P.Calls	Plan	Ctrl	Succ.	Time	P.Calls	Plan	Ctrl
CHS	IPathBiRRT	100%	100	1.0	32	65	100%	57	1.2	1.5	50
UCG	IPathBiRRT	100%	100	1.0	34	65	40%	140	2.5	50	82

cost grid. In harder scenarios (S3, R2) collisions occurred on multiple sides of narrow passages. Using a unified cost grid, the planner and controller avoided the center of the passage as this area has cost accumulated from multiple collisions, thus only 40% of the trials successfully reached the goal for R2. Paths through the center do not intersect with all voxels from any CHS, thus our approach correctly identified possible paths and succeeded in 100% of trials in R2. Even considering only trials where UCG succeeded our approach still reached the goal in approximately one third of the time on average.

In R3 the table obstacle created a cul-de-sac for the local controller, though there was significant free space for the planner to conservatively avoid the table. Since plans were able to avoid much of the CHSs, accurately modeling collision probability was less important, thus the baseline method and proposed method performed similarly.

In R4 with a large gap between tables both methods found a path quickly. For both methods, successful trials primarily invoked the local controller and rarely needed the global planner. As the gap narrowed both methods required more time to find the opening, however UCG took longer on average and in 60% of trials did not find the opening within the allowed time of 5 min (Fig. 6).

7 Conclusions and Future Work

Most robots do not have touch-sensitive skin, and those that do may manipulate unsensorized objects. The proposed Collision Hypothesis Sets allow reasoning about the knowledge gained when these robots and objects come into contact with the environment. We showed how collision hypothesis sets can be used in controllers and planners to search for paths with minimum probability of collision. We performed simulated and physical robot experiments and found for simpler environments our approach takes less time to reach the goal while for more complex environments our methods succeeds where other approaches fail.

We explained why many existing methods for planning under uncertainty cannot be applied to contact observations and presented two planners that first approximate, then compute the full path cost. However, the planners implemented could be made more efficient and the resulting paths are often significantly suboptimal. Our future work seeks to improve these results.

References

1. Bae, K., Belton, D., Lichti, D.D.: A closed-form expression of the positional uncertainty for 3D point clouds. TPAMI **31**, 577–590 (2009)
2. Bhattacharjee, T., Grice, P., Kapusta, A., Killpack, M., Park, D., Kemp, C.: A robotic system for reaching in dense clutter that integrates model predictive control, learning, haptic mapping, and planning. In: IROS (2014)
3. Bicchi, A., Salisbury, K., Brock, D.: Contact sensing from force measurements. IJRR **12**, 249–262 (1993)
4. Blackmore, L.: A probabilistic particle control approach to optimal, robust predictive control. In: AIAA (2006)
5. Bry, A., Roy, N.: Rapidly-exploring random belief trees for motion planning under uncertainty. In: ICRA (2011)
6. Guibas, L., Hsu, D., Kurniawati, H., Rehman, E.: Bounded Uncertainty Roadmaps for Path Planning. Springer, Heidelberg (2010)
7. Hermann, A., Drews, F., Bauer, J., Klemm, S., Roennau, A., Dillmann, R.: Unified GPU voxel collision detection for mobile manipulation planning. In: IROS (2014)
8. Janson, L., Hu, T., Pavone, M.: Safe motion planning in unknown environments: Optimality benchmarks and tractable policies. CoRR (2018)
9. Karaman, S., Frazzoli, E.: Sampling-based algorithms for optimal motion planning. IJRR **30**, 846–894 (2011)
10. Killpack, M., Kapusta, A., Kemp, C.: Model predictive control for fast reaching in clutter. Auton. Robots **40**, 537–560 (2016)
11. Klingensmith, M., Koval, M., Srinivasa, S., Pollard, N., Kaess, M.: The manifold particle filter for state estimation on high-dimensional manifolds. CoRR (2016)
12. Koonjul, G., Zeglin, G., Pollard, N.: Measuring contact points from displacements with a compliant, articulated robot hand. In: ICRA (2011)
13. Kuffner, J.J., LaValle, S.M.: RRT-connect: an efficient approach to single-query path planning. In: ICRA (2000)
14. Lambert, A., Gruyer, D., Saint Pierre, G.: A fast monte carlo algorithm for collision probability estimation. In: ICARCV, December 2008

15. Lee, A., Duan, Y., Patil, S., Schulman, J., McCarthy, Z., van den Berg, J., Goldberg, K., Abbeel, P.: Sigma hulls for gaussian belief space planning for imprecise articulated robots amid obstacles. In: IROS (2013)
16. Park, D., Kapusta, A., Hawke, J., Kemp, C.: Interleaving planning and control for efficient haptically-guided reaching in unknown environments. IEEE-RAS Humanoids (2014)
17. Patil, S., van den Berg, J., Alterovitz, R.: Estimating probability of collision for safe motion planning under gaussian motion and sensing uncertainty (2012)
18. Saund, B., Chen, S., Simmons, R.: Touch based localization of parts for high precision manufacturing. In: ICRA (2017)
19. Şucan, I., Moll, M., Kavraki, L.: The open motion planning library. IEEE Robot. Autom. Mag. **19**, 72–82 (2012). http://ompl.kavrakilab.org
20. Du Toit, N.E., Burdick, J.W.: Probabilistic collision checking with chance constraints. IEEE Trans. Robot. **27**, 809–815 (2011)
21. van den Berg, J., Wilkie, D., Guy, S.J., Niethammer, M., Manocha, D.: LQG-Obstacles: Feedback control with collision avoidance for mobile robots with motion and sensing uncertainty. In: ICRA (2012)

Contact-Consistent Disturbance Observer for Floating-Base Robots

Hosang Lee[1], Yisoo Lee[1], and Jaeheung Park[1,2(✉)]

[1] Graduate School of Convergence Science and Technology, Seoul National University, Seoul, Republic of Korea
{hslee1989,howcan11,park73}@snu.ac.kr

[2] Advanced Institute of Convergence Technology, Suwon, Republic of Korea

Abstract. In this study, a novel disturbance observer for floating-base robots has been developed using contact information to reduce the estimation errors of joint accelerations. Typically, joint angles are measured using motor encoders and joint accelerations are computed by differentiating them twice in discrete time domain. Therefore, the estimated value inherently has a large numerical error, which is an obstacle for precise on-line estimation of the disturbance. The proposed method computes the joint acceleration error, which does not satisfy the kinematic constraints of the contact position such as a supporting foot of the robot, and incorporates it into the dynamics model to remove its effect on the disturbance torque solution. It is shown through experiments using a legged robot that the proposed contact-consistent observer can estimate the disturbance more accurately than an observer that does not consider contact information.

Keywords: Contact-consistent · Disturbance observer · Floating-base robot · Acceleration estimate · Kinematic constraint

1 Motivation, Problem Statement, Related Work

Safe physical human–robot interaction is essential for the practical use of floating-base robots. They can injure people or lose balance when controlled to strictly follow the given trajectory without considering external force. Therefore, contact detection methods have been developed by using exteroceptive sensors such as tactile [1,2] or force/torque (F/T) sensors [3]. Series elastic actuators have also been utilized to estimate the external force [4]. These methods are intuitive, but require additional costs and increase the complexity of the system.

Strategies for contact detection, isolation, and identification using only proprioceptive sensors have been described in [5]. One way of estimating the external force is to use a dynamics-based disturbance observer model, which was originally developed for the robust control of a fixed-base manipulator [6]. It has been applied to floating-base humanoid robots to estimate the external force [7] or zero-moment-point variation [8,9] using F/T measurements at both feet.

J. Xiao et al. (Eds.): ISER 2018, SPAR 11, pp. 475–484, 2020.
https://doi.org/10.1007/978-3-030-33950-0_41

However, the estimated acceleration required for inverse dynamics inherently has a large numerical error owing to the double differentiation of position in discrete time domain, which is an obstacle for the practical use of the disturbance observer. To reduce this effect, a momentum-based disturbance observer has been developed for manipulators [10,11] and applied to humanoid robots [12–14] by using a first-order low-pass filtered estimate of the acceleration. Still, it is difficult to estimate or reject high-frequency disturbances as the cut-off frequency is typically set low owing to the noise of the acceleration.

The effect of the joint acceleration error can be observed in the contact space. Floating-base robots such as humanoids are usually controlled to keep their supporting foot fixed on the ground, which indicates that their acceleration is almost zero. However, the computed value of acceleration without contact constraints was larger than expected in the experiment, leading to the assumption that most of the estimation error originates from the numerical errors in calculation if the error in the kinematic model is negligible. In this study, a novel disturbance observer model for floating-base robots has been designed by using kinematic constraints on fixed contact positions, such as the supporting foot. The joint acceleration error, which does not satisfy the constraints, is removed from the disturbance torque solution. The proposed method was verified through experiments using a torque-controlled biped robot DYROS RED [15].

2 Technical Approach

2.1 Previous Approaches for Disturbance Observer

The whole-body dynamics of a floating-base robot with k actuated joints considering the disturbance torque can be formulated as follows:

$$A(q)\ddot{q} + C(q,\dot{q})\dot{q} + g(q) + J_c^T f_c = (S^k)^T \Gamma^k + \Gamma_{dis}, \tag{1}$$

where $n = k + 6$ is the total degrees-of-freedom (DOFs) of the system including six virtual joints attached to the base frame. The term q is the $n \times 1$ vector of the joint angle, and f_c is the $c \times 1$ vector of the contact force. The term Γ^k is the $k \times 1$ vector of the commanded torque, and Γ_{dis} is the $n \times 1$ vector of the disturbance torque. The term $A(q)$ is the $n \times n$ matrix of the joint space inertia, $C(q,\dot{q})$ is the $n \times n$ matrix of the Coriolis and centrifugal force, and $g(q)$ is the $n \times 1$ vector of the gravitational torque. The term J_c is the $c \times n$ Jacobian matrix of the contact position and $S^k = [0_{k\times 6}\ I_{k\times k}]$ is the $k \times n$ selection matrix, where I is an identity matrix.

The disturbance torque Γ_{dis} can be obtained by substituting the joint angle q, joint velocity $\dot{q}$, joint acceleration $\ddot{q}$, and F/T sensor measurements f_c in Eq. (1). Typically, q is measured using joint encoders, and $\dot{q}$ and $\ddot{q}$ are computed by differentiating q once and twice, respectively. Therefore, the estimated value of $\ddot{q}$, or $\ddot{q}_{est}$, has a large numerical error, resulting in the noise and inaccuracy of the torque solution Γ_{dis}. To rectify this problem, a momentum-based observer model has been widely used in order not to use $\ddot{q}$ as an input variable. It defines a first-order low-pass filtered estimate r of the disturbance torque Γ_{dis} as follows:

$$r(t) = K[p(t) - \int_0^t \{(S^k)^T \Gamma^k + C^T(q, \dot{q})\dot{q} - g(q) - J_c^T f_c + r(t)\}dt], \quad (2)$$

where $r(t)$ is a residual vector, K is a diagonal positive gain matrix, and $p(t) = A(q)\dot{q}$ is the generalized momentum with $p(0) = 0$. Thus, the residual dynamics is given by

$$\dot{r} = -Kr + K\Gamma_{dis}. \quad (3)$$

The applied low-pass filter can reduce the numerical noise in Γ_{dis}, but it can also cause a large delay and data loss in actual implementation. For example, the cut-off frequency was set to 1.0–2.0 Hz in the previous experiment [11], which is not suitable for high-frequency disturbance estimation or rejection.

2.2 Contact-Consistent Observer Model

The key idea is to estimate the joint acceleration $\ddot{q}_{err}$ using the kinematic constraints on the contact points and remove it explicitly from Eq. (1). Some of the errors in $\ddot{q}_{err}$ can be caused by joint elasticity of the robot, but we assumed that the joint elasticity is large and most of the errors in $\ddot{q}_{err}$ originate from the numerical errors in calculation. By replacing $\ddot{q}$ with $(\ddot{q}_{est} + \ddot{q}_{err})$, Eq. (1) can be rewritten as follows:

$$A(q)(\ddot{q}_{est} + \ddot{q}_{err}) + C(q, \dot{q})\dot{q} + g(q) + J_c^T f_c = (S^k)^T \Gamma^k + \Gamma_{dis}. \quad (4)$$

If the end effector of the robot, such as the supporting foot or hand, is fixed on the ground, the acceleration of the contact point $\ddot{x}_c$ is ideally zero. Thus, the kinematic constraint on the contact point is given by

$$\ddot{x}_c = \dot{J}_c\dot{q} + J_c(\ddot{q}_{est} + \ddot{q}_{err}) = 0, \quad (5)$$

where x_c is the $c \times 1$ vector of the contact point. From Eq. (5), the estimated acceleration of the contact point is defined as $\ddot{x}_{c,est} = \dot{J}_c\dot{q} + J_c\ddot{q}_{est}$.

There are many solutions of $\ddot{q}_{err}$ satisfying Eq. (5) when the number of joints n is greater than that of the contacts c, which is a typical case. Among the feasible solutions, we chose the one using the dynamically-consistent inverse of J_c [16] as follows.

$$\ddot{q}_{err} = -\bar{J}_c\ddot{x}_{c,est}, \quad (6)$$

where $\bar{J}_c = A^{-1}J_c^T(J_cA^{-1}J_c^T)^{-1}$ is the dynamically-consistent inverse of J_c.

Using Eq. (6), the torque owing to the joint acceleration error $A(q)\ddot{q}_{err}$ in Eq. (4) can be rewritten as

$$A(q)\ddot{q}_{err} = -J_c^T \Lambda_c \ddot{x}_{c,est}, \quad (7)$$

where $\Lambda_c = \{J_cA(q)^{-1}J_c^T\}^{-1}$ is the $c \times c$ mass matrix in the contact space. By using $\bar{J}_c$, the torque $A(q)\ddot{q}_{err}$ in Eq. (7) does not affect the null space of the contact position whereas it compensates the force $\Lambda_c \ddot{x}_{c,est}$ corresponding to the estimation error in the contact space.

By substituting $A(q)\ddot{q}_{err}$ in Eq. (7) into Eq. (4), the whole-body dynamics model for the contact-consistent disturbance observer is expressed as follows:

$$A(q)\ddot{q}_{est} + C(q,\dot{q})\dot{q} + g(q) + J_c^T(f_c - \Lambda_c \ddot{x}_{c,est}) = (S^k)^T \Gamma^k + \Gamma_{dis}. \quad (8)$$

The disturbance torque Γ_{dis} obtained from Eq. (8) was passed through the first-order low-pass filter used in Eq. (3) as in the momentum-based disturbance observer.

3 Results

3.1 Experimental Setup

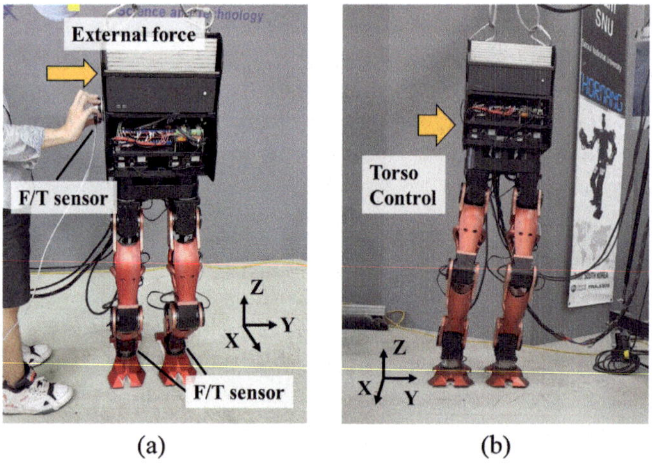

Fig. 1. Experiments using a torque-controlled biped robot DYROS RED. (a) An external force was applied on the torso in the y-direction. The measurements obtained using the F/T sensor at the human hand were compared with the estimation results of the disturbance torque obtained using the contact-consistent observer and the momentum-based observer. (b) The torso was controlled to move along the y-direction without the application of an external force. The estimated disturbance torque was compensated at the joint level using the two observer models.

Figure 1 illustrates two kinds of experiments conducted using a torque-controlled biped robot DYROS RED at a double-support phase. In Experiment 1 shown in Fig. 1(a), the torso was controlled to maintain its initial position and orientation using the proportional–derivative (PD) controller in the task space [17] while an external force was applied on the torso in the y-direction. The external force was measured using the F/T sensor at the human hand, and compared with the estimated value of Γ_{dis} on the virtual joint attached to the torso. The comparison was performed using two different disturbance observer models—the proposed model and the momentum-based model. Furthermore, the estimated value of Γ_{dis} on the right ankle roll (RAR) joint was compared with the measurements obtained using the F/T sensor at the right foot. The control gains for the given controller k_p and k_v were set to 30.0 and 7.5, respectively, and the cut-off frequency of the first-order low-pass filter for both observer models was set to 50.0 rad/s.

In Experiment 2 shown in Fig. 1(b), the torso position was controlled to move in the y-direction by the task PD controller without the application of an external force. Subsequently, the disturbance torque Γ_{dis} on the k-actuated joints consisting of only internal friction and model errors was compensated at the joint level to improve the control performance of the robot. The values of Γ_{dis} on the virtual joints attached to the base were mapped onto the k-actuated joints, and compensated as well. The y-direction trajectories of the torso obtained using the two observer models are shown in the results. The robot control gains k_p and k_v were set to 100.0 and 7.5, respectively, and the cut-off frequency of the first-order low-pass filter for both observer models was set to 5.0 rad/s.

In both cases, the joint angular velocity $\dot{q}$ and acceleration $\ddot{q}$ of the k-actuated joints were computed through the first- and second-order differentiation of the encoder values q without being filtered. The linear and angular velocities of the base were obtained through fusion of IMU measurements attached to the torso and the relative velocity of the torso with respect to the supporting foot. Therefore, the proposed method is more effective when the control frequency was 2.0 kHz. To prevent the supporting foot from being lifted, a balancing controller [17] was applied to the robot.

3.2 Experiment 1: External Force Estimation

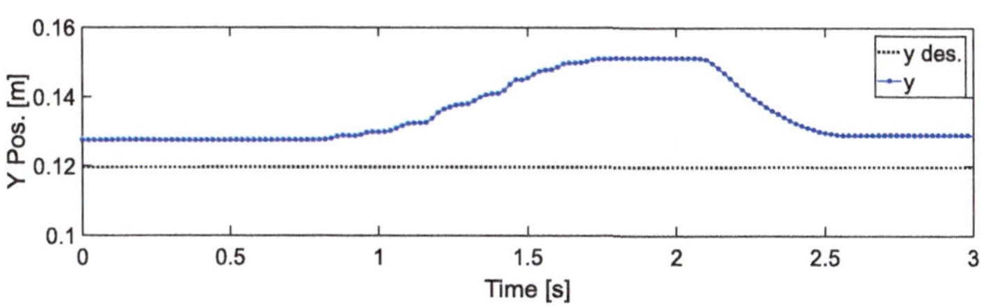

Fig. 2. Torso trajectory in the y-direction when an external force was applied

Figure 2 shows the torso trajectory in the y-direction when an external force was applied. The steady error exists before and after the external force was applied as the control gains were set low. The corresponding acceleration of the contact point on the right supporting foot is shown in Fig. 3, which was computed as

$\ddot{x}_{c,est} = \dot{J}_c\dot{q} + J\ddot{q}_{est}$. The maximum magnitude of $\ddot{x}_{c,est}$ is above 10.0 m/s^2 in the x- and z-directions and above 50.0 m/s^2 in the y-direction, which is too large for slight slippage or chattering of the supporting foot. Thus, these values are considered to originate mostly from the numerical errors in calculation.

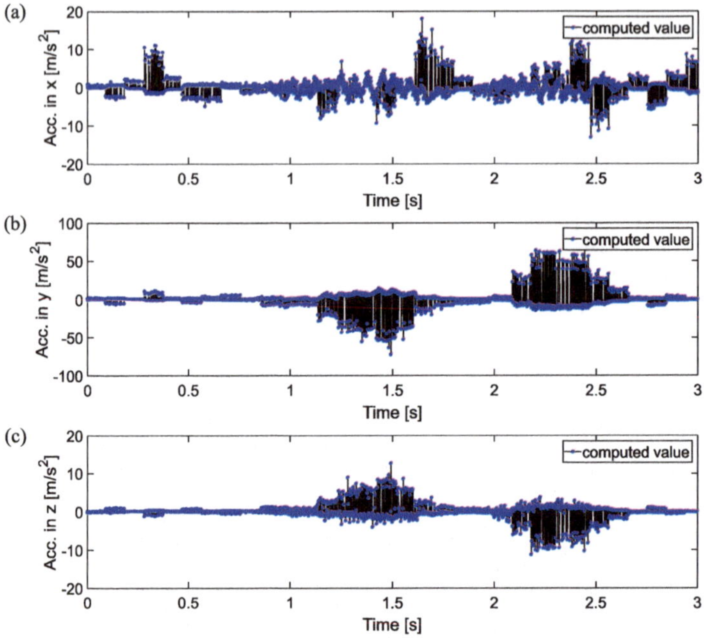

Fig. 3. Computed acceleration of the contact point on the right supporting foot in (a) the x-direction, (b) the y-direction, and (c) the z-direction

Figure 4 shows the estimation results of the disturbance torque Γ_{dis} for the two observer models. Overall, the noise in Γ_{dis} was reduced with the application of the contact-consistent observer model compared with the application of the typical momentum-based observer. To confirm that the values estimated with contact constraints are more accurate than those estimated without constraints, Γ_{dis} of the virtual prismatic joint in the y-direction, F_y, and Γ_{dis} of the virtual rotational joint around the x-axis, M_x, are compared with the measurements obtained using the F/T sensor at the human hand in Fig. 5(a) and (b). For convenience of comparison, Γ_{dis} in Fig. 4 was shifted to the initial value of the F/T measurements. Noise was further reduced in M_x than in F_y because the rotational joints of the base have a greater effect on the acceleration of the contact point than the prismatic joints. In Fig. 5(c), Γ_{dis} on the right ankle roll joint was compared with the measurements obtained using the F/T sensor at the right foot, showing the same result that Γ_{dis} with contact constraints was more accurate than that without constraints.

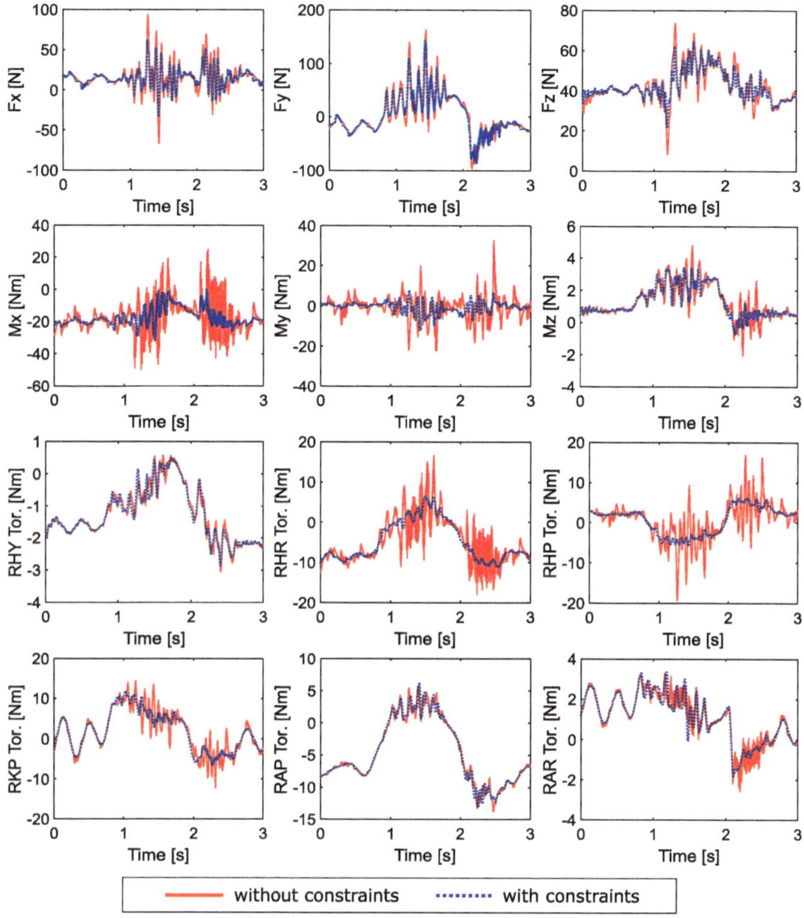

Fig. 4. Estimation results of the disturbance torque on the virtual joints (F_x, F_y, F_z, M_x, M_y, M_z) and the right leg joints depending on the observer model. RHY, RHR, RHP, RKP, RAP, and RAR indicate the right hip yaw/roll/pitch, knee pitch, and ankle pitch/roll joint torque, respectively.

3.3 Experiment 2: Internal Disturbance Rejection

Figure 6 shows the torso trajectories in the y-direction obtained using the two compensation methods of the disturbance torque. The target path was generated to move along the y-direction for 1.5 s. The steady-state error y_{err} without compensation of Γ_{dis} was 5.4 mm, as shown in Fig. 6(a). The error was reduced to 0.2 mm when Γ_{dis} obtained from the contact-consistent disturbance observer was compensated at the joint level, as shown in Fig. 6(b). However, the robot became unstable and the control was deactivated before the target position was reached when Γ_{dis} obtained from the momentum-based observer was compensated with the same cut-off frequency, as shown in Fig. 6(c).

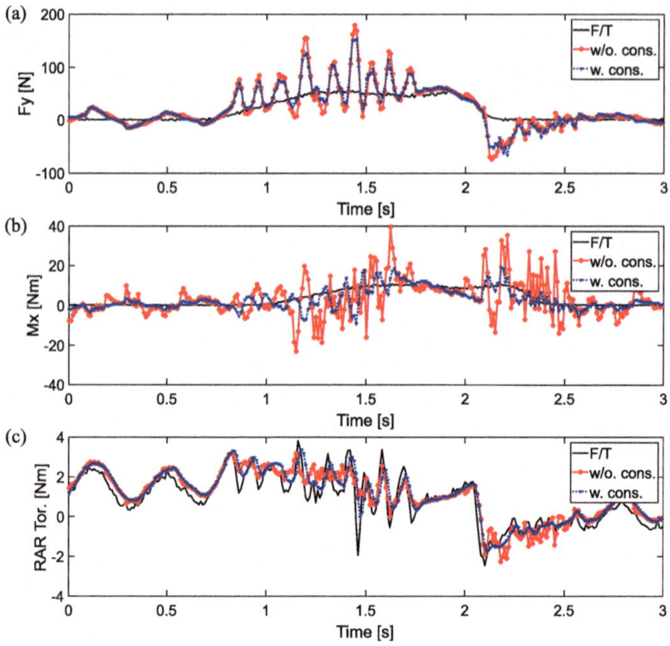

Fig. 5. Comparison between the disturbance torque obtained using the observer (a) F_y, (b) M_x, and (c) RAR and that obtained from the F/T measurements

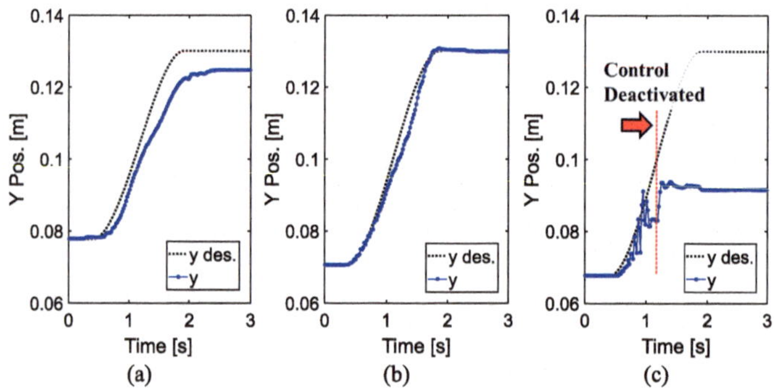

Fig. 6. Torso trajectory in the y-direction (a) without disturbance compensation, (b) with disturbance compensation using the proposed observer model, and (c) with disturbance compensation using the previous observer model

4 Discussion and Conclusion

The proposed method computes the joint acceleration error $\ddot{q}_{err}$ using the kinematic constraints on the contact position and removes it explicitly from the dynamics model to improve the accuracy of the estimated disturbance torque Γ_{dis}. Experimental results have shown that the contact-consistent observer can reduce the noise in Γ_{dis} and increase the cut-off frequency of the low-pass filter for its compensation at the joint level, in contrast to the momentum-based observer which does not consider the contact information.

In this work, the solution of the torque error $A(q)\ddot{q}_{err} = -J_c^T \Lambda_c \ddot{x}_{c,est}$ does not affect the null space of the contact position. In other words, $A(q)\ddot{q}_{err}$ is computed as zero for the joints not involved in the movement of the contact point, and thus, the noise in Γ_{dis} for them is not reduced. For example, $A(q)\ddot{q}_{err}$ on the left leg joints is zero during the right single-support phase. This is due to the lack of the number of contacts c in comparison with that of the total joints n. Thus, the contact-consistent observer is more effective in cases where the number of contacts is large, such as in multi-legged robots or humanoids having multiple contacts with the environment.

In future studies, the velocity and acceleration of the contact position will be measured using external sensors such as IMU and GYRO to verify if the real value of $\ddot{x}_c$ is negligible compared with the computed value. Moreover, the performance of the proposed observer model will be verified when the contact state keeps changing such as in the case of walking.

Acknowledgements. This work was supported by the National Research Foundation of Korea (NRF) grant funded by the Korea government (MSIP) (No. NRF-2015R1A2A1A10055798) and the Technology Innovation Program (10060081) funded by the Ministry of Trade, industry, and Energy (MI, Korea).

References

1. De Maria, G., Natale, C., Pirozzi, S.: Force/tactile sensor for robotic applications. Sens Actuators A Phys. **175**, 60–72 (2012)
2. Lumelsky, V.J., Cheung, E.: Real-time collision avoidance in teleoperated whole-sensitive robot arm manipulators. IEEE Trans. Syst. Man Cybern. **23**(1), 194–203 (1993)
3. Lu, S., Chung, J.H., Velinsky, S.A.: Human-robot collision detection and identification based on wrist and base force/torque sensors. In: Proceedings of the 2005 IEEE International Conference on Robotics and Automation, ICRA 2005, pp. 3796–3801. IEEE (2005)
4. Oh, S., Lee, C., Kong, K.: Force control and force observer design of series elastic actuator based on its dynamic characteristics. In: IECON 2015-41st Annual Conference of the IEEE Industrial Electronics Society, pp. 004639–004644. IEEE (2015)
5. Haddadin, S., De Luca, A., Albu-Schäffer, A.: Robot collisions: a survey on detection, isolation, and identification. IEEE Trans. Rob. **33**(6), 1292–1312 (2017)
6. Morinaga, S., Kosuge, K.: Collision detection system for manipulator based on adaptive impedance control law. In: 2003 Proceedings of the IEEE International Conference on Robotics and Automation, ICRA 2003, vol. 1, pp. 1080–1085. IEEE (2003)
7. Kaneko, K., Kanehiro, F., Morisawa, M., Yoshida, E., Laumond, J.P.: Disturbance observer that estimates external force acting on humanoid robots. In: 2012 12th IEEE International Workshop on Advanced Motion Control (AMC), pp. 1–6. IEEE (2012)

8. Czarnetzki, S., Kerner, S., Urbann, O.: Observer-based dynamic walking control for biped robots. Robot. Auton. Syst. **57**(8), 839–845 (2009)
9. Hawley, L., Suleiman, W.: External force observer for medium-sized humanoid robots. In: 2016 IEEE-RAS 16th International Conference on Humanoid Robots (Humanoids), pp. 366–371. IEEE (2016)
10. De Luca, A., Mattone, R.: Sensorless robot collision detection and hybrid force/motion control. In: Proceedings of the 2005 IEEE International Conference on Robotics and Automation, ICRA 2005, pp. 999–1004. IEEE (2005)
11. Sadeghian, H., Villani, L., Keshmiri, M., Siciliano, B.: Task-space control of robot manipulators with null-space compliance. IEEE Trans. Rob. **30**(2), 493–506 (2014)
12. Ott, C., Henze, B., Lee, D.: Kinesthetic teaching of humanoid motion based on whole-body compliance control with interaction-aware balancing. In: 2013 IEEE/RSJ International Conference on Intelligent Robots and Systems (IROS), pp. 4615–4621. IEEE (2013)
13. Vorndamme, J., Schappler, M., Haddadin, S.: Collision detection, isolation and identification for humanoids. In: 2017 IEEE International Conference on Robotics and Automation (ICRA), pp. 4754–4761. IEEE (2017)
14. Flacco, F., Paolillo, A., Kheddar, A.: Residual-based contacts estimation for humanoid robots. In: 2016 IEEE-RAS 16th International Conference on Humanoid Robots (Humanoids), pp. 409–415. IEEE (2016)
15. Schwartz, M., Hwang, S., Lee, Y., Won, J., Kim, S., Park, J.: Aesthetic design and development of humanoid legged robot. In: 2014 14th IEEE-RAS International Conference on Humanoid Robots (Humanoids), pp. 13–19. IEEE (2014)
16. Khatib, O.: A unified approach for motion and force control of robot manipulators: the operational space formulation. IEEE J. Robot. Autom. **3**(1), 43–53 (1987)
17. Lee, Y., Hwang, S., Park, J.: Balancing of humanoid robot using contact force/moment control by task-oriented whole body control framework. Auton. Robots **40**(3), 457–472 (2016)

Actuator Transparency and the Energetic Cost of Proprioception

Gavin Kenneally[1(✉)], Wei-Hsi Chen[2], and Daniel E. Koditschek[2]

[1] Department of Mechanical Engineering and Applied Mechanics, University of Pennsylvania, Philadelphia, PA, USA
gake@seas.upenn.edu

[2] Department of Electrical and Systems Engineering, University of Pennsylvania, Philadelphia, PA, USA
{weicc,kod}@seas.upenn.edu

1 Introduction

In the field of haptics, conditions for mechanical "transparency" [1] entail such qualities as "solid virtual objects must feel stiff" and "free space must feel free" [2], suggesting that a suitable actuator is able both to do work and readily have work done on it. In this context, seeking actuator transparency has come to mean a preference for minimal dynamics [3] or no impedance [4]. While such general notions seem satisfactory for a haptic interface, actuators with good mechanical transparency are now being used in high-performance robots [5,6] where once again they must be able to do work, but are now also expected to perceive their environment by processing signals related to contact forces in the leg or manipulator when an explicit force sensor is not present. As robotics researchers develop models [7] suitable for programming behaviors that require systematic making and breaking of contact within the environments on which they perform work, actuators must be capable of: (a) generating the high forces at speed needed to accelerate the body during locomotion [5]; (b) robustness to high forces and impacts during locomotion [8]; (c) perceiving high force events quickly, such as touchdown in stance [9]; (d) perceiving contact quickly without exerting significant force on the object, such as in gentle manipulation [10]; and (e) reacting quickly during time-sensitive behaviors [11].

This work aims to describe a quantitative assay of transparency that might, for example, predict the advantage in proprioceptive tasks of an electromagnetic direct-drive (DD) motor (i.e., one without gearbox), relative to actuation schemes consisting of both a motor and a geared reduction. Specifically, we

This work was supported by ONR grant #N00014-16-1-2817, a Vannevar Bush Fellowship held by the last author, sponsored by the Basic Research Office of the Assistant Secretary of Defense for Research and Engineering.

Electronic supplementary material The online version of this chapter (https://doi.org/10.1007/978-3-030-33950-0_42) contains supplementary material, which is available to authorized users.

J. Xiao et al. (Eds.): ISER 2018, SPAR 11, pp. 485–495, 2020.
https://doi.org/10.1007/978-3-030-33950-0_42

explore the prospects for characterizing transparency as revealed by comparing the energetic cost of "feeling" the environment. Our sample proprioceptive task is instantiated by a simple torque estimator in Sect. 2. This scheme is then instrumented in simple contact detection experiments paired with a model to empirically explore the relationships between collision energy and detection time delay in Sect. 3. The actuators are then tested with a feel-cage task to illustrate the advantage of good transparency in Sect. 4.

2 Actuator and Estimator Models

We quantify four sensorimotor subsystems consisting of an actuator and drive electronics with position sensor [12] built using four actuators denoted U10DD, EC60, T2822, and U10GB, as shown in Table 1. These actuators, offering roughly the same thermally allowable continuous torque, are chosen as representatives of distinctly different robot motor sizing strategies. The T2822 has significant gear reduction (51.24:1) similar to [13,14] which provides the benefit of increased mass-specific torque and reduced Joule heating at the cost of decreased transparency. The EC60 has a very modest (4.33:1) planetary gearbox, a strategy also employed by the MIT Cheetah [5] and Hermes [15] machines. The U10DD represents the DD strategy employed in manipulators [16] and legged machines, including Minitaur [6] and the Penn Jerboa [12], which benefit from excellent bandwidth and transparency at the cost of increased Joule heating. Our final comparison abandons the uniform continuous torque constraint to explore directly the decreased transparency caused by fitting a small gear reduction (4.33:1) gearbox to a U10.

2.1 Properties Related to Performing Work

Table 1 shows some useful criteria when choosing which motor to perform work. The motor constant, k_m [16] is found by the equation $k_m = (k_v\sqrt{R_m})^{-1}$, where k_v ($\frac{\text{rad/s}}{\text{V}}$) is the motor velocity constant and R_m is the motor phase to phase resistance. k_m is winding invariant and captures the key trade-off between torque required to work and wasteful Joule heating; for example note the substantial increase in k_m relative to U10DD manifesting the greater thermal efficiency resulting from the addition of the gearbox in U10GB. k_m/m is the mass-normalized motor constant, reflecting the importance of mass budgeting for dynamic machines. Nominal speed ν_n is calculated at an input voltage of $V_{in} = 15\,\text{V}$ but is not winding invariant, thus this is of limited utility when comparing actuators. The maximum continuous torque is experimentally found by $\tau_c = k_\tau i_c$, where k_τ ($\frac{\text{N·m}}{\text{A}}$) is the motor torque constant and i_c is the maximum continuous current, at which the maximum winding temperature reaches 125 °C. As shown in Table 1, the maximum continuous torque value of the first three actuators are similar enough for our purpose of comparison. Peak torque is also measured experimentally, where stall torque and corresponding motor

Table 1. Motor properties

Properties	U10DD[a]	EC60[b]	T2822[c]	U10GB[d]
Physical properties				
Absolute Reduction	N/A	13/3	17576/343	13/3
Backlash (deg)	0.0	0.6	1.0	0.6
Mass m (kg)	0.405	0.730	0.251	0.865
Performing work				
Max Continuous Torque τ_c (N · m)	**2.36**	**1.21**	**2.12**	**10.22**
Motor Constant k_m $(\mathrm{N \cdot m}/\sqrt{\mathrm{W}})$	0.285	0.370	0.617	1.236
k_m/m $(\mathrm{N \cdot m}/\sqrt{\mathrm{W}} \cdot \mathrm{kg})$	0.705	0.506	2.56	1.430
Nominal Speed ν_n(rev/s)	25.00	18.06	8.99	5.77
Peak Torque τ_p (N · m)	8.05	12.35*	11.44	35.3*
Max Continuous Power P_c (W)	92.60	34.29	29.96	92.60
Information acquisition				
Reflected Inertia J_r $(\mathrm{kg \cdot m^2})$	4.47×10^{-4}	2.27×10^{-3}	3.02×10^{-3}	8.39×10^{-3}
Static Friction τ_s (N · m)	0.075	0.061	0.222	0.347
Coulomb Friction τ_{coul} (N · m)	0.032	0.026	0.114	0.253
Viscous Coefficient b_v (N · m/(rad/s))	0.099	0.153	0.835	1.890

[a]T-Motor U10 (direct-drive) [6]
[b]Maxon EC60 Flat with Maxon GP42C (1-stage gearbox, 4.33:1)
[c]Turnigy SK3 2822 with Maxon GP32HP (3-stage gearbox, 51.24:1)
[d]T-Motor U10 with GP52C (1-stage gearbox, 4.33:1)
*Torque exceeds gearbox rating

current i under different PWM duty cycle are measured and fit to the nonlinear torque-current model, $\tau(i) = \tau_p - c_1 e^{-c_2 i}$, to get the peak torque τ_p. It is important to note that the peak torque of EC60 and U10GB exceeds the peak torque rating of their gearboxes, an important consideration for both dynamic maneuvers and long-term wear. Maximum continuous power is calculated at the half point of the motor speed-torque curve, that is $P_c = \frac{1}{4}\tau_c \nu_n$.

2.2 Properties Related to Acquiring Information

While backlash and various frictions all impede backdrivability, reflected inertia is the most critical property to control when designing transparent actuators. With a higher reflected inertia, the actuator is more slowly accelerated by reaction torque from the environment, incurring a longer response time. The inertia of the motors J_m and gearbox J_g are found in the Maxon catalog [17] or by creating detailed CAD of the rotor. The reflected inertia of a system with a gearbox is calculated as $J_r = J_m Gr^2 + J_g$, where Gr is the absolute reduction. Both static friction and backlash act as thresholds for information acquisition since the actuator senses nothing until the torque or deflection (respectively) exceeds a threshold. For purposes of this paper we measure static friction τ_s by hanging incremental weights off a 0.05 m pulley until the motor starts to turn. By collecting terminal velocities from this mass-hanging task with different hanging masses, the remaining first order (constant and linear) terms characterizing general motor drag τ_{drag} can be fitted as $\tau_{drag} = \beta_\nu \dot{\theta} + \tau_{coul}$, as shown in Table 1. For each actuator, there is a very strong linear fit ($r^2 \geq 0.999$). The viscous damping coefficient b_v is a collective effect from mechanical Rayleigh damping and the drag-like effect generated from the back emf of the electrical-magnetic motor (the motor controllers are performing sinusoidal voltage control, this term does not exist with ideal current control.) It is clear that b_v would contribute more to the energy dissipation as $\tau_{drag} \approx b_v \dot{\theta}$, and it would require more energy for a system with larger b_v to acquire information. Table 1 shows that with the addition of gearboxes, b_v increases, thus it would require more energy input to the actuator when acquiring information with a gearbox.

2.3 Torque Estimation

We seek to characterize an actuator's ability to estimate torque using a very simple model that depends only on position and ignores all the terms associated with velocity and acceleration, which would require additional sensing or estimation. Assuming a simple DC motor model [17] with negligible inductance, $V = i_m R_m + G k_e \dot{\theta}$, where V is the motor voltage command, i_m the current in the motor windings, $k_e(\frac{\mathrm{V}}{\mathrm{rad/s}})$ the motor back-emf constant and $G\dot{\theta}$ the motor shaft velocity. Consider a general impedance controller with the form $V = k_s^*(\theta_d - \theta_a) + k_d^*(\dot{\theta}_d - \dot{\theta}_a)$, where $k_s^*(\frac{\mathrm{V}}{\mathrm{rad}})$ and $k_d^*(\frac{\mathrm{V}}{\mathrm{rad/s}})$ are the proportional and derivative gains respectively. The lowercase "d" and "a" of the motor shaft angle θ and its velocity $\dot{\theta}$ correspond to the desired and actual readings. The output torque of the actuator is estimated as being proportional to the winding current and scaled by the gear ratio G, $\hat{\tau} = G k_\tau i_m$, where $k_\tau(\frac{\mathrm{N{\cdot}m}}{\mathrm{A}})$ is the motor torque constant. The estimated torque is

$$\begin{aligned}\hat{\tau} = G k_\tau i_m &= G k_\tau \tfrac{V - G k_e \dot{\theta}_a}{R_m} = \tfrac{G k_\tau}{R_m} V - \tfrac{G^2 k_\tau k_e}{R_m} \dot{\theta}_a \\ &= k_s(\theta_d - \theta_a) + k_d(\dot{\theta}_d - \dot{\theta}_a) - \beta \dot{\theta}_a,\end{aligned}$$

where $k_s = \frac{k_s^* k_\tau G}{R_m}(\frac{\mathrm{N{\cdot}m}}{\mathrm{rad}})$, $k_d = \frac{k_d^* k_\tau G}{R_m}(\frac{\mathrm{N{\cdot}m}}{\mathrm{rad/s}})$ can are the virtual spring constant and the virtual generalized damper, and $\beta = \frac{G^2 k_\tau k_e}{R_m}(\frac{\mathrm{N{\cdot}m}}{\mathrm{rad/s}})$ can be viewed as the generalized electromechanical drag. Since deriving velocity from position measurements requires some filtering, torque estimation schemes using velocity will inevitably introduce signal delay, further slowing response time. By setting $k_d = 0$ and assuming $\beta \dot{\theta}$ to be small, the estimate is now much simpler,

$$\hat{\tau} = k_s(\theta_d - \theta_a), \tag{1}$$

yet still accurate enough to be useful in demanding applications of the sort we target such as [9]. Indeed, this is the same approach as used with series-elastic actuators, except the spring is virtual instead of physical.

3 Contact Detection Experiment, Model, and Data

To quantify the ability of the motor to acquire information, a contact detection experiment is introduced, where the true contact force measured by a load cell and the estimated contact force acquired from the motor's reading are recorded as well as the detection delay t_r and the energy loss ΔE.

3.1 Contact Detection Experiment

We now introduce our proposed transparency measure by relating detection delay, t_r, to the collision energy required for contact detection. A contact detection experiment was performed for all the actuators, wherein a shuttle of known mass m_s installed on an instrumented linear rail is dropped from various heights onto the end of an aluminum link fixed to the actuator output shaft (with overall inertia J_m, arm length r, and actuator angle θ), as shown in Fig. 1a. The actuator was commanded with impedance ($k_s = 2.0\ \frac{\text{N}\cdot\text{m}}{\text{rad}}$, $k_d = 0\ \frac{\text{N}\cdot\text{m}}{\text{rad/s}}$). A load cell, sampled at 10 kHz with a DAQ (NI USB-6210), was mounted on a shuttle of known mass m_s (0.54 kg for U10DD and EC60, 0.71 kg for T2822 and U10GB). The shuttle was dropped from various heights within the range of 0.03 m to 1.1 m onto a 0.1 m actuator arm. The true contact time t_t was recorded when the shuttle hit the actuator arm and the measured force F_μ from the load cell exceeded a threshold of $1\,\text{N}\cdot\text{m}$. A linear potentiometer, sampled at 10 kHz, was attached to the shuttle to record the drop distance, which was then used to specify its velocity, $v_{s,t}$, and, in turn, its kinetic energy, $T = \frac{1}{2}m_s v_{s,t}^2$, at the time of collision, t_t. When the encoder position error reached a threshold of 0.1 rad, chosen by slightly exceeding the deflection of the virtual spring necessary to overcome the actuator with the largest static friction, the time was recorded as the estimated contact t_e, the velocity of the shuttle mass was marked as $\nu_{s,e}$, and the estimated contact force F_{est} was found through Eq. (1) and the linkage length, r. The motor controller outputs the motor's position at 1 kHz, which is synchronized with the DAQ using a digital signal. The detection delay time of the actuator t_r is computed as $t_r = t_e - t_t$, and is recorded in each trial in addition to T, F_μ, the load cell reading, and F_{est} as shown in Fig. 1b.

3.2 Contact Detection Model

Assuming an instantaneous event, we model impact as an inelastic collision between the mass and the actuator arm[1]. The velocity of the shuttle before the contact is $\nu_{s,t}$ and the angular velocity of the mass and the actuator after the collision is $\omega_c = \frac{m_s r \nu_{s,t}}{J_m + m_s r^2}$ [18]. The dynamics after contact can be modeled using the

[1] This was observed literally in experiments with the more backdrivable actuators; for the highly geared cases (where some repeated bouncing was observer), if contact is not detected on the first collision, it will not be detected in subsequent collision as the first collision necessarily dissipates some energy.

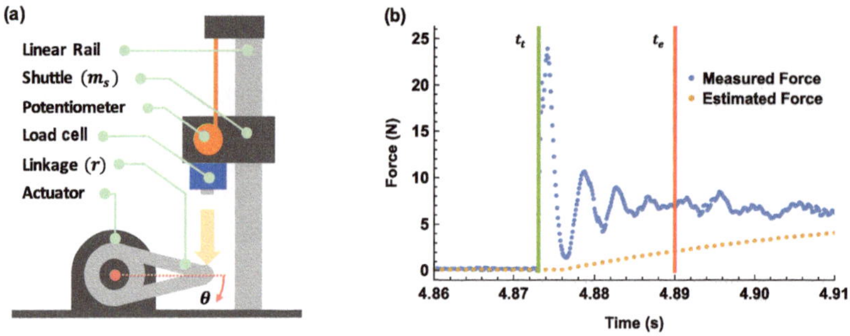

Fig. 1. (a) Setup for the contact detection experiment as described in Sect. 3.1. (b) Single trial of the contact detection experiments in Sect. 3. The green and red vertical lines mark, respectively t_t, and t_e, while the blue and yellow dots plot, respectively, F_μ and F_{est}. The linkage length, r, is 0.1 m.

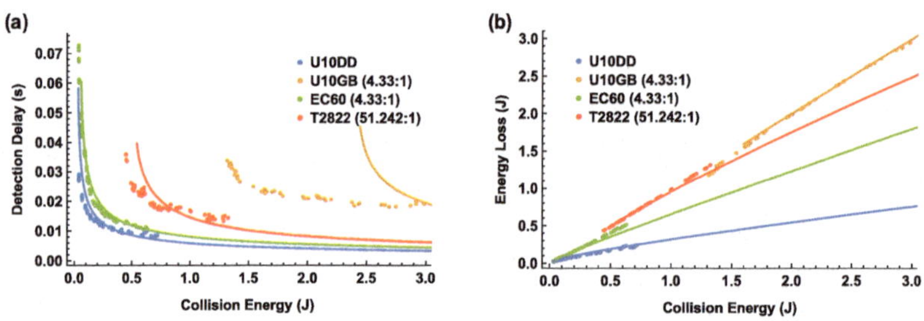

Fig. 2. (a) Detection delay, t_r, as a function of collision energy, T. (b) Energy Loss, ΔE, as a function of collision energy, T. The dotted and solid traces plot, respectively, the experimental data and the modeled predictions.

estimated torque from Eq. (1) (with the desired position $\theta_d = 0$ and the actual position $\theta_a = \theta$), $\hat{\tau} = -k_s\theta$, and the motor general drag $\tau_{drag} = \beta_\nu\dot{\theta} + \tau_{coul}$ as

$$(J_m + m_s r^2)\ddot{\theta} = -k_s\theta - \beta_\nu\dot{\theta} - \tau_{coul}, \tag{2}$$

with the initial condition $\theta(0) = 0$, $\dot{\theta}(0) = \omega_c$ and has the solution

$$\theta(t) = e^{\frac{-t(\beta_\nu+\Delta)}{2J}}\left(\frac{-\tau(\beta_\nu-\Delta)-2Jk_s\omega_c}{2k_s\Delta}\right) + e^{\frac{-t(\beta_\nu-\Delta)}{2J}}\left(\frac{-\tau(\beta_\nu+\Delta)+2Jk_s\omega_c}{2k_s\Delta}\right) - \frac{\tau_{coul}}{k_s}, \tag{3}$$

where $\Delta = \sqrt{\beta_\nu^2 - 4k_s(J_m + m_s r^2)}$. As shown in Fig. 1 all these cases represent the condition in which the system is underdamped, so Δ is a complex number. The trajectories resulting from varied impact velocities, $\nu_{s,t}$ are solved for until $|\theta(t)| \geq 0.1$rad at time $t := t_e = t_t + t_r$. With the assumption of inelastic collision, the angular velocity of the shuttle mass and the actuator at the detected time is

$\dot{\theta}(t_e)$, and the velocity of the shuttle mass is $\nu_{s,e} = \dot{\theta}(t_e)r$. Results of the contact detection experiment and the simulation model are shown in Fig. 2.

Figure 2a shows detection delay t_r as a function of collision energy. The experimental results (dots) match well with the numerical simulation from the model (solid curves). From the energy-detection delay curve for each actuator, we can see that it takes more time for the actuator to detect contact when the collision energy is smaller. Because some of the energy transferred to the actuator at the time of collision is converted into motor shaft kinetic energy, it takes more time to rotate the actuator with less energy. However, if the collision results in a force insufficient to overcome the actuator friction then the shaft will not move, precluding any possibility of contact detection, and this accounts for the vertical asymptote of these curves at very low energies. Figure 2a also shows that when a motor is equipped with a gearbox, more energy is required to trigger contact detection.

For the actuator to detect contact, there must be enough energy to back-drive the actuator, overcome the frictions, in addition to any Joule heating. This energy can be measured as the energy loss ΔE in kinetic energy of the shuttle during the detection delay, or

$$\Delta E = \frac{1}{2}m_s(\nu_{s,t}^2 - \nu_{s,e}^2). \tag{4}$$

Fig. 2b shows the energy loss as a function of its initial collision energy. ΔE is the energy transferred from the shuttle into and that back-drives the actuator, and can be thought as the energetic cost of proprioception. The figure shows that as collision energy increases, the energy loss also increases. This is due to the fact that the viscous friction and damping term of the actuator are both proportional to its velocity, thus the actuator takes more energy away when the shuttle has higher relative velocity with the stationary actuator at the beginning of the contact. As predicted by the model, the experiments validate that the larger viscous friction and inertial terms result in a larger slope for the more geared actuators.

4 Behaviors Using Proprioception

Sensing contact and then re-positioning the actuator is a primitive task embedded in a wide variety of robotics behaviors. We first characterize the performance of these actuators with respect to these tasks, then explore the broader behavioral implications.

4.1 Feel-Cage Task

We illustrate the advantage of good actuator transparency and bandwidth by considering a contact detection and caging task involving balls of varied mass confined to a horizontal rail. Each actuator rotates continuously at 1 rev/s CCW and its maximum error accrued during steady operation due to torque ripple [19] and imperfect commutation is recorded as e_s. The actuator continues to rotate until contact is detected with a position error threshold of $(\epsilon + e_s)$, where the additional error ϵ satisfies the estimated torque error $\tau_{err} = k_s\epsilon = 0.02\,\mathrm{N}\cdot\mathrm{m}$. If contact is detected, the motor rotates CW at 10.5 V open-loop until a position setpoint to kinematically cage the ball. Figure 3 shows the still image of the feel-cage task performed on the U10DD actuator, and the resulting performance of these actuators with respect to this task with different balls is shown in Table 2. As expected, lighter balls are harder for the actuators to feel then cage since a given collision energy results in faster post-collision ball velocity. The U10DD performs best by successfully detecting contact with and then recirculating in time to feel and cage all of the test balls. As predicted by Fig. 2b, the successively more highly geared actuators must impart successively more energy to the balls during collision, propelling them out of the workspace before caging or even detection, in the order predicted by the successively greater slopes of the energy-delay curves.

Table 2. Feel-cage task results

Balls (mass, diameters)\actuators	U10DD	EC60	T2822	U10GB
Rubber ball (0.05 kg, 0.063 m)	feel + cage	feel	no detection	no detection
Lacrosse ball (0.145 kg, 0.064 m)	feel + cage	feel	no detection	no detection
Bocce ball (0.716 kg, 0.073 m)	feel + cage	feel + cage	feel	no detection

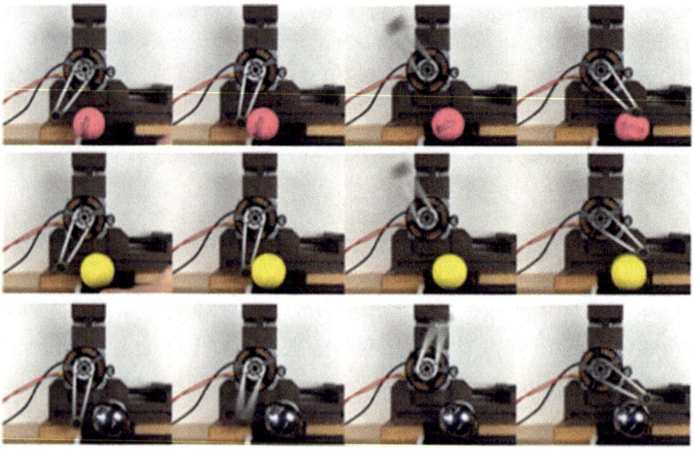

Fig. 3. U10DD performing a feel-cage task for several different balls on a rail. The photo is taken at an interval of 0.05 s

4.2 Broader Behaviors

The feel-cage task is an example of an emerging class of dynamic primitives that rely on proprioception in which the sensorimotor subsystem must:

1. **feel** the environment or object of interest using proprioception (causing a delay t_{feel}, or t_r in our study, dictated by the transparency)
2. **process** these signals, and generate a command (causing a delay, t_{process}, dictated by the digital signal processing speed)
3. **act** as a reaction to the command (causing a delay, t_{act}, dictated by the actuation bandwidth).

The task is successfully accomplished if the total time:

$$t_{\text{total}} = t_{\text{feel}} + t_{\text{process}} + t_{\text{act}}$$

is shorter than the time budget, t_{budget}, dictated by the system's dynamics. We now neglect the second step as t_{process} is typically orders of magnitude faster than the other two (only 1 ms in our case), and independent of the sensorimotor subsystem.

Dynamic door opening [11] fits this class of behaviors, since the robot jumps up and towards the door with a given horizontal component of velocity (imposing an energy budget). The machine must then catch itself by exerting a normal force on the wall (it is too high off the ground to be supported) and perform the remaining elements of the task before falling back to the ground. The robot must **feel** the door handle with a given limb, then use the same limb to **act**, first retracting then engaging with the door handle, all before the robot falls back to the ground. If the transparency is too low, the robot will not perceive the door at all with the given energy budget and just fall back to the ground ($t_{\text{feel}} \geq t_{\text{budget}}$). If the transparency is perfect, but the actuation bandwidth is too low, the robot will feel the door, but will not be able to react fast enough to engage the handle before falling to the ground ($t_{\text{move}} \geq t_{\text{budget}}$). Finally, any combination of transparency and bandwidth that cause $t_{\text{feel}} + t_{\text{act}}$ to take more time than t_{budget} (dictated by the dynamics of the robot and the door handle) will also fail.

High speed locomotion [5,6,20] also has similar constraints, though the energy budget for the task (in this case related to the vertical component of the velocity) is much higher. The actuator in the leg must **feel** touchdown without wasting too much energy (plastic collisions and damping in the leg), or consuming too much time. The leg must then **act** in the remaining stance time to carry out the desired active control (the efficacy of which is dictated by the actuator's bandwidth).

The feel-cage manipulation task can be generalized to consider different kinds of dynamic [21] proprioceptive manipulation in which the manipulator must **feel** the object before it must then **act** on the object to do work on in. In this case, the "energy budget," i.e. the threshold of imparted energy beyond which the task will fail, might arise from any combination of physical events, such as:

- pushing the object out of the workspace (or close enough and with the right velocity that the bandwidth is not sufficient to do the **move** step)
- reaching the manipulator's terminal velocity while contacting the object before it has been sensed, so no further acceleration is possible
- not damaging the object if it is fragile (which can also impose additional constraints on maximum forces).

5 Conclusion

We explore transparency in the context of actuation and compare the measure of three different actuators, each broadly representative of a design strategy. Specifically, we explore the use of the energy-delay curves in Fig. 2b as a quantitative assay for the energetic cost of contact detection. Further, we propose a very simple model which broadly captures this trend of transparency by roughly predicting the experimentally revealed energy-delay curves. In turn, this model suggests the possibility of developing a quantitative metric of "transparency" that would predict these models' energy-delay curves from the key parameters presented in Table 1, most notably the reflected inertia, J_r, and the viscous coefficient b_ν. Such a metric could offer substantial insight when choosing an actuator that will be used to acquire information from the environment. The feel-cage task is an example of an emerging class of dynamic primitives that rely on proprioception in which the actuator must feel the environment, process the information and take action. This simple 1 DoF task inspires various useful higher DoF and highly dynamic tasks such as agile door opening, high speed locomotion, and dexterous manipulation which have similar dynamic constraints dictating their success. We are continuing to explore the extent to which similar dynamically derived metrics of transparency may reveal fundamental robotic design limitations.

References

1. Lawrence, D., Pao, L.Y., Salada, M.A., Dougherty, A.M.: Quantitative experimental analysis of transparency and stability in haptic interfaces, pp. 441–449 (1996)
2. Massie, T.H., Salisbury, J.K.: The phantom haptic interface: a device for probing virtual objects. In: Proceedings of the ASME Winter Annual Meeting, Symposium on Haptic Interfaces for Virtual Environment and Teleoperator Systems, Chicago, IL, vol. 55, pp. 295–300 (1994)
3. Carignan, C.R., Cleary, K.R.: Closed-loop force control for haptic simulation of virtual environments (2000)
4. Lawrence, D.A.: Stability and transparency in bilateral teleoperation. IEEE Trans. Robot. Autom. **9**(5), 624–637 (1993)
5. Seok, S., Wang, A., Chuah, M.Y., Otten, D., Lang, J., Kim, S.: Design principles for highly efficient quadrupeds and implementation on the MIT cheetah robot. In: 2013 IEEE International Conference on Robotics and Automation (ICRA), pp. 3307–3312. IEEE (2013)

6. Kenneally, G., De, A., Koditschek, D.: Design principles for a family of direct-drive legged robots. IEEE Robot. Autom. Lett. **1**(2), 900–907 (2016)
7. Johnson, A.M., Burden, S.A., Koditschek, D.E.: A hybrid systems model for simple manipulation and self-manipulation systems. Int. J. Robot. Res. **35**(11), 1354–1392 (2016)
8. Kim, S., Wensing, P.: Proprioceptive actuation design for dynamic legged locomotion, March 2016. http://meetings.aps.org/Meeting/MAR16/Event/270932
9. De, A., Koditschek, D.E.: Vertical hopper compositions for preflexive and feedback-stabilized quadrupedal bounding, pacing, pronking, and trotting. Int. J. Robot. Res. **37**(7), 743–778 (2018)
10. Wolf, S., Hirzinger, G.: A new variable stiffness design: matching requirements of the next robot generation. In: 2008 IEEE International Conference on Robotics and Automation (ICRA), pp. 1741–1746. IEEE (2008)
11. Topping, T.T., Kenneally, G., Koditschek, D.: Quasi-static and dynamic mismatch for door opening and stair climbing with a legged robot. In: 2017 IEEE International Conference on Robotics and Automation (ICRA), pp. 1080–1087. IEEE (2017)
12. De, A., Koditschek, D.E.: The Penn Jerboa: a platform for exploring parallel composition of templates. arXiv preprint arXiv:1502.05347 (2015)
13. Grimes, J.A., Hurst, J.W.: The design of ATRIAS 1.0 a unique monopod, hopping robot. In: Proceedings of the 2012 International Conference on Climbing and Walking Robots and the Support Technologies for Mobile Machines, pp. 548–554 (2012)
14. Spröwitz, A., Tuleu, A., Vespignani, M., Ajallooeian, M., Badri, E., Ijspeert, A.J.: Towards dynamic trot gait locomotion: design, control, and experiments with Cheetah-cub, a compliant quadruped robot. Int. J. Robot. Res. **32**(8), 932–950 (2013)
15. Wang, A., Ramos, J., Mayo, J., Ubellacker, W., Cheung, J., Kim, S.: The HERMES humanoid system: a platform for full-body teleoperation with balance feedback. In: 2015 IEEE-RAS 15th International Conference on Humanoid Robots (Humanoids), pp. 730–737. IEEE (2015)
16. Asada, H., Youcef-Toumi, K.: Direct-drive Robots: Theory and Practice. MIT Press, Cambridge (1987)
17. Maxon Motors: Maxon motors program 2017/18, e-catalog. http://maxon.blaetterkatalog.ch/b9990/catalog/index.html?data=b9990/b999045&lang=e
18. Landau, L.D., Lifshitz, E.M.: Mechanics, 3rd edn. Butterworth-Heinemann (1982). Chap. 4
19. Piccoli, M., Yim, M.: Anticogging: torque ripple suppression, modeling, and parameter selection. Int. J. Robot. Res. **35**, 148–160 (2016)
20. Duperret, J., Kramer, B., Koditschek, D.E.: Core actuation promotes self-manipulability on a direct-drive quadrupedal robot. In: International Symposium on Experimental Robotics, pp. 147–159. Springer (2016)
21. Dafle, N.C., Rodriguez, A., Paolini, R., Tang, B., Srinivasa, S.S., Erdmann, M., Mason, M.T., Lundberg, I., Staab, H., Fuhlbrigge, T.: Extrinsic dexterity: in-hand manipulation with external forces. In: 2014 IEEE International Conference on Robotics and Automation (ICRA), pp. 1578–1585. IEEE (2014)

Unmanned Aerial Vehicles III

Unmanned Aerial Vehicles III

Jonathan Kelly

This session featured an exciting array of presentations describing both large-scale and small-scale UAV research. Förster et al. discussed a novel approach for cross-wind stabilization of large hybrid airships that incorporate lift cells. Shankar, Elbaum, and Detweiler examined methods for multirotor vehicles to transfer small payloads in mid-air using tethers, while Spitzer et al. reported on the development of an augmented teleoperation system for fast, agile, and stable vision-based flight of multirotor platforms. Zhao et al. explored the complexities of path planning for transformable, multilink aerial robots maneuvering through small openings. Stager and Tanner closed the session by characterizing new models for vehicle-environment interactions that could enable robots to leverage physical contact during dynamic motion. The session highlighted the breadth and depth of ongoing UAV research work, and drew attention to some of the many interesting problems that remain to be explored within the domain.

Hybrid Model Predictive Control for Crosswind Stabilization of Hybrid Airships

Julian F. M. Foerster[1(✉)], Mohamed K. Helwa[1,2], Xintong Du[1], and Angela P. Schoellig[1]

[1] Dynamic Systems Lab, University of Toronto Institute for Aerospace Studies, Toronto, Canada
{julian.forster,mohamed.helwa,xintong.du}@robotics.utias.utoronto.ca, schoellig@utias.utoronto.ca
http://www.dynsyslab.org

[2] Electrical Power and Machines Department, Cairo University, Giza, Egypt

Abstract. A hybrid airship is an aerial vehicle that generates lift by leveraging both buoyancy and aerodynamic principles. The operation of such a vehicle can be limited by its high susceptibility to crosswinds during taxiing, take-off and landing. With the goal to mitigate this issue, this paper proposes a novel controller design for a stabilization system consisting of wing tip thrusters. Due to the response of the vehicle to wind disturbances (e.g. lifting off a wheel during taxiing), modeling it as a hybrid dynamical system is appropriate. A novel, customized hybrid model predictive control (MPC) scheme is proposed for crosswind stabilization. As shown in simulation as well as in experimental results in controlled and realistic environments, the proposed control scheme succeeds in stabilizing the vehicle despite artificial or actual wind disturbances, even in scenarios where simple linear MPC fails. Simultaneously, our approach is computationally efficient enough to run on an onboard computer.

1 Introduction

Hybrid airships are heavier-than-air vehicles that generate a majority of their lift using buoyancy which is caused by a helium inflated wing-shaped envelope [1]. Resulting properties are a high energy efficiency during operation and short take-off and landing distances. Therefore, hybrid airships are very suitable for a number of logistics applications with remote origins and/or destinations, including supplying remote mines, disaster relief missions or, in general, supplying remote but inhabited locations with various goods [2]. However, the range of the vehicles' safe operating conditions can be limited due to a high susceptibility to crosswinds during taxiing, take-off and landing, which occurs due to the large

This work was partially supported by the Zeno Karl Schindler Foundation and the Natural Sciences and Engineering Research Council of Canada (NSERC).

J. Xiao et al. (Eds.): ISER 2018, SPAR 11, pp. 499–510, 2020.
https://doi.org/10.1007/978-3-030-33950-0_43

surface area of the wing envelope. In the worst case, crosswind can lead to the vehicle tipping over.

The goal of this work is to design and implement an automatic counter-gust system (CGS) that stabilizes a hybrid airship despite wind disturbances during ground operations, to extend the set of safe operating conditions. This is achieved by estimating the vehicle attitude and current wind disturbance based on inertial measurement unit (IMU) data and then controlling thrusters that are mounted to the vehicle's wing tips to counteract the disturbance effect. The CGS controller aims to compute optimal control inputs, run autonomously without pilot intervention, be computationally efficient to run on onboard hardware, and be flexible regarding adaption to future hybrid airship designs.

All aircraft need to take precautions when taxiing under wind disturbances. For fixed-wing airplanes, this is done manually by the pilot using the control surfaces [3]. A wind stabilization system for hybrid airships is, for example, Lockheed Martin's air cushion landing system [4], but it is based on suction and thus only applicable when the vehicle is stationary. To the best of the authors' knowledge, there is no automatic wind stabilization system that operates during taxiing for any type of airship or fixed-wing plane.

In contrast, in-flight wind stabilization is quite common for small unmanned flying vehicles [5]. Often, wind disturbance estimates based on sensor data are computed and used to compensate for disturbances [5,6]. In this paper, a similar approach is pursued, although vehicle size and CGS setup are unique in our case. More importantly, the approach in this paper is focussed on wind stabilization during ground operations as opposed to stabilization in flight.

Studying the hybrid aerial vehicle used for experiments in this paper led to modeling it as a hybrid dynamical system. Furthermore, we use predictive control for its ability to optimize the behavior over a prediction horizon. However, most common approaches to *hybrid model predictive control* (HMPC) are not suitable for our application due to their high computational complexity [7] or the lack of flexibility of explicit HMPC methods [8,9]. A very recent approach allows to run nonlinear MPC faster than realtime, using quasi-translations [10]. However, it is based on the assumption that future predictions are almost perfect, which is violated in our case, given that wind disturbances are hard to predict. Therefore, a novel, customized HMPC scheme is proposed in Sect. 2.4, which is the main contribution of this paper.

2 Technical Approach

In this section, the methodology of the CGS design is detailed. Due to space constraints, emphasis is placed on the required estimation and control algorithms.

2.1 Physical Setup

In this paper, a CGS is designed, implemented and evaluated on a 4-m wingspan unmanned hybrid airship. Figure 1 visualizes the physical setup of the vehicle and the CGS. The vehicle body mainly consists of an inflated wing envelope and two inflated pontoons. The tricycle undercarriage landing gear is composed of three wheels, two of which are mounted to the pontoons and one to the fuselage which is attached to the wing envelope. The CGS consists of sensors, a processor and thrusters. To maximize the control authority and thus the lever of the actuators, the thrusters are placed at the wing tips.

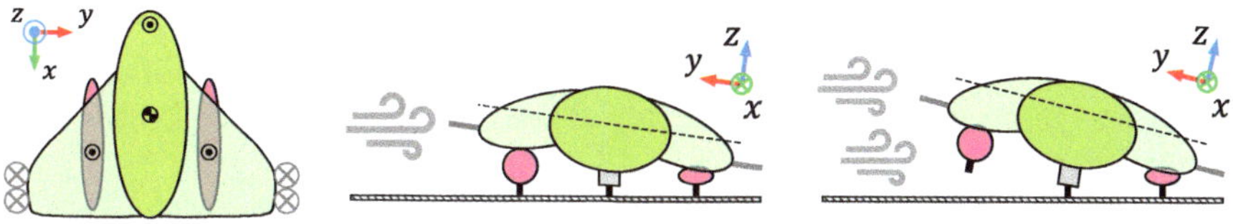

(a) Top-down view. The center of gravity is marked by ⊕.

(b) Front view, vehicle in **mode 1** (all wheels have ground contact).

(c) Front view, vehicle in **mode 2** (back right wheel lifted).

Fig. 1. Schematic views of the hybrid airship studied in this paper, showing wing envelope (green), pontoons (magenta), landing gear positions (black circles), and wing tip thrusters (gray). The orientation of the body frame is shown for each view.

2.2 Hybrid System Model

Observations of the vehicle in operation led to differentiating three different vehicle dynamics, depending on the vehicle state. Therefore, the system is modeled as a hybrid dynamical system, consisting of a set of modes with continuous dynamics describing the system in each mode [11] (see Fig. 2). As the yaw motion during ground operations is controlled manually by the vehicle's pilot, it is assumed to be constant at zero and thus not included in the vehicle state. The state consists of the roll and pitch angles (${}_1\alpha_x$ and ${}_1\alpha_y$) and their angular velocities (${}_1\omega_x$ and ${}_1\omega_y$), both in a body-fixed frame denoted by the leading subscript 1 and visualized in Fig. 1. Frame 1's origin is located on the vehicle's roll axis. The state $\mathbf{x}$ is written as $\mathbf{x} = \left[{}_1\alpha_x \; {}_1\omega_x \; {}_1\alpha_y \; {}_1\omega_y\right]^T$. To obtain the dynamics $\dot{\mathbf{x}} = \left[{}_1\omega_x \; {}_1\dot{\omega}_x \; {}_1\omega_y \; {}_1\dot{\omega}_y\right]^T$, an expression for the angular acceleration in the body frame is needed. In the following, the dynamic modes used in this paper are introduced.

For disturbances below a certain magnitude, the vehicle's three wheels stay on the ground and its wing envelope rotates around its roll axis due to compression of the leeward pontoon. This is visualized in Fig. 1b and referred to as mode 1. The angular acceleration expression for mode 1 is based on Euler's rotation equations for rigid bodies:

$$
{}_1\dot{\boldsymbol{\omega}} = \begin{bmatrix} {}_1\dot{\omega}_x \\ {}_1\dot{\omega}_y \\ * \end{bmatrix} = {}_1\boldsymbol{\theta}_{\text{wing}}^{-1} \left({}_1\boldsymbol{M} - \begin{bmatrix} {}_1\omega_x \\ {}_1\omega_y \\ 0 \end{bmatrix} \times \left({}_1\boldsymbol{\theta}_{\text{wing}} \begin{bmatrix} {}_1\omega_x \\ {}_1\omega_y \\ 0 \end{bmatrix} \right) \right), \tag{1}
$$

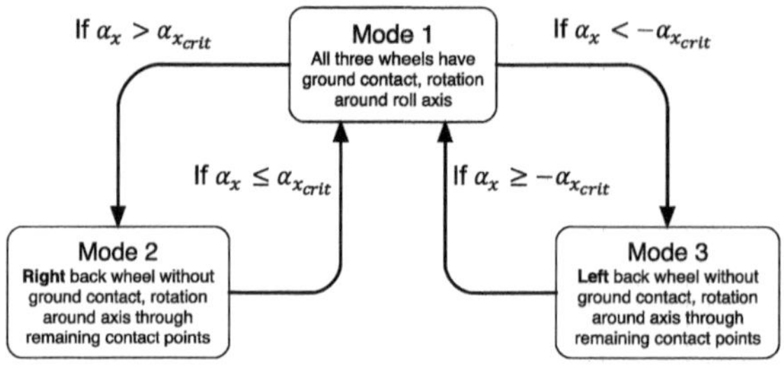

Fig. 2. Overview of the permissible mode switches and the conditions for a mode switch to happen. The occurrence of a switch only depends on the vehicle's roll angle α_x and the constant critical roll angle $\alpha_{x_{\text{crit}}}$.

where ${}_1\dot{\boldsymbol{\omega}}$ is the angular acceleration vector (the last component, marked with $*$, is discarded as justified above), ${}_1\boldsymbol{\Theta}_{\text{wing}}$ is the inertia matrix of the wing only (without pontoons, fuselage and landing gear), ${}_1\boldsymbol{M}$ is the sum of all moments acting on the vehicle and $(.)^{-1}$ and $\times$ denote the matrix inverse and the cross product respectively. The moment ${}_1\boldsymbol{M}$ is composed as follows: ${}_1\boldsymbol{M} = {}_1\boldsymbol{M}_{\text{pontoons}} + {}_1\boldsymbol{M}_{\text{wind}} + {}_1\boldsymbol{M}_{\text{thrust}}$, ${}_1\boldsymbol{M}_{\text{pontoons}} = \left[-k_x \cdot {}_1\alpha_x - d_x \cdot {}_1\omega_x, -k_y \cdot {}_1\alpha_y - d_y \cdot {}_1\omega_y, 0\right]^T$, ${}_1\boldsymbol{M}_{\text{wind}} = {}_1\boldsymbol{b}_g \times (\boldsymbol{R}_{I1}^T {}_I\boldsymbol{F}_{\text{wind}})$ and ${}_1\boldsymbol{M}_{\text{thrust}} = \sum_{i=1}^{4}({}_1\boldsymbol{b}_g + \boldsymbol{R}_{1g}\, {}_g\boldsymbol{r}_{t,i}) \times \boldsymbol{R}_{1g}\, {}_g\boldsymbol{F}_{t,i}$, where ${}_1\boldsymbol{M}_{\text{pontoons}}$, ${}_1\boldsymbol{M}_{\text{wind}}$ and ${}_1\boldsymbol{M}_{\text{thrust}}$ are the moments exerted by pontoons, wind disturbance and wing tip thrusters, respectively, k_i and d_i denote spring and damping coefficients for axis i, ${}_1\boldsymbol{b}_g$ is the vector from the origin of frame 1 to the vehicle's center of mass (CoM), $\boldsymbol{R}_{I1}$ is the rotation matrix between frame 1 and the inertial frame, in which the wind disturbance force ${}_I\boldsymbol{F}_{\text{wind}}$ is expressed, $\boldsymbol{R}_{1g}$ is the rotation matrix between frame 1 and the body frame g with the origin at the CoM ($\boldsymbol{R}_{1g}$ is the identity matrix in this paper), ${}_g\boldsymbol{r}_{t,i}$ is the vector from the CoM to the ith wing tip thruster and ${}_g\boldsymbol{F}_{t,i}$ is the thrust produced by the ith thruster. The produced thrust is obtained via an experimentally determined static thrust map and the input command $\mathbf{u} = \left[u_1, u_2\right]^T$, where $u_1, u_2 \in [0, 1]$ is the starboard/port input command, respectively. A command of zero means idle rotation, and a command of one means full thrust.

For higher disturbances, one of the back wheels can loose ground contact, which occurs at the *critical roll angle*. Consequently, the vehicle rotates around the axis passing through the remaining two contact points, which is visualized in Fig. 1c and referred to as mode 2 or 3, depending on which back wheel is lifted. It is also possible that the front wheel lifts off. However, this behavior is typically related to disturbances other than wind (e.g. potholes) and therefore out of the scope of this paper. To formalize modes 2 and 3, new body fixed reference frames are introduced. Both have their origin in the contact point between front wheel and ground and the x-axis points towards the remaining back wheel contact point (the left one for mode 2 and the right one for mode 3). In the following, equations for mode 2 are given. The ones for mode 3 can be obtained analogously. Similar to mode 1, formalizing the dynamics in mode 2 requires the angular acceleration:

$$ {}_2\dot{\boldsymbol{\omega}} = {}_2\boldsymbol{\theta}^{-1} \left({}_2\boldsymbol{M} - \begin{bmatrix} {}_2\omega_x \\ 0 \\ 0 \end{bmatrix} \times \left({}_2\boldsymbol{\theta} \begin{bmatrix} {}_2\omega_x \\ 0 \\ 0 \end{bmatrix} \right) \right), \tag{2} $$

where the leading subscript denotes the reference frame for mode 2 introduced above and $\boldsymbol{\theta}$ is the inertia matrix of the whole vehicle. The sum of the moments acting in mode 2, denoted ${}_2\boldsymbol{M}$, is composed as follows: ${}_2\boldsymbol{M} = {}_2\boldsymbol{M}_{\text{wind}} + {}_2\boldsymbol{M}_{\text{thrust}} + {}_2\boldsymbol{M}_{\text{gravity}}, {}_2\boldsymbol{M}_{\text{gravity}} = {}_2\boldsymbol{b}_g \times (\boldsymbol{R}_{2I}\ {}_I\boldsymbol{F}_g)$. The moments ${}_2\boldsymbol{M}_{\text{wind}}$ and ${}_2\boldsymbol{M}_{\text{thrust}}$ are defined analogously to the respective moments in mode 1 and ${}_2\boldsymbol{b}_g$ is the vector from the origin of frame 2 to the CoM, $\boldsymbol{R}_{2I}$ is the rotation matrix between frame 2 and the inertial frame, and ${}_I\boldsymbol{F}_g$ is the gravity force, expressed in the inertial frame.

To make the models usable in a discrete time setting, the above continuous time models are linearized and discretized. The operating point for the linearization is the horizontal attitude with zero angular velocities for mode 1 and the critical roll angle with zero angular velocities for modes 2 and 3. Finally, the models for the separate modes are combined into a piecewise affine (PWA) formulation [12], which consists of the three linear discrete time models and for each mode a region of the state and input space in which it is active (see Fig. 2).

2.3 State and Disturbance Estimation

The goal of the CGS controller is to control the wing tip thrusters to compensate for the effect of the disturbance. To achieve this, an estimate of the current disturbance is used by the MPC to account for it over the prediction horizon and to find an optimal control input. Based on our previous work [13], the best way to get a disturbance estimate that offers a reasonable trade-off between accuracy, simplicity and cost is to deduce it from IMU data using a Kalman filter. Other solutions to obtain a disturbance estimate, like adding an anemometer, turned out to be too expensive or technically infeasible. Therefore, apart from estimating the vehicle state $\mathbf{x}[k]$ at a discrete time k, the Kalman filter estimates the current disturbance $d[k]$ by using an augmented state $\tilde{\mathbf{x}}[k] = [\mathbf{x}^T[k], d[k]]^T$. Note that since the main dynamics cause a roll motion, the disturbance ${}_I\boldsymbol{F}_{\text{wind}}$ is reduced to its y component, which is denoted by $d[k]$.

As process model, the system model is used for the state $\mathbf{x}[k]$ (see Sect. 2.2) and the disturbance is modeled as $d[k+1] = d[k] + w[k]$, where $w[k] \sim \mathcal{N}(0, \sigma_d^2)$ is white Gaussian noise with zero mean and standard deviation σ_d. The disturbance model assumes that the disturbance is approximately constant over one sampling interval, which is a common assumption for short term wind prediction [14].

Despite the fact that the vehicle is modeled as a hybrid system (as introduced in Sect. 2.2), a Kalman filter for linear systems is used. In each time step, it uses the model for the mode that the vehicle was in at the last time step according to the previous state estimate. This procedure introduces a small error if a mode switch occured during the current sampling interval. The error is compensated for in subsequent Kalman filter executions.

2.4 Customized Hybrid Model Predictive Control

To control the CGS, MPC is used, mainly due to its prediction capabilities, intuitive tuning, and seamless integration of state and input constraints. Since the system model is hybrid as discussed in Sect. 2.2, HMPC must be used. For HMPC, the optimization problem that must be solved at each time step is a *mixed integer program* (MIP) [15], which is computationally expensive to solve. A preliminary experimental analysis of the system showed that a sampling frequency of at least 10 Hz and a prediction horizon of at least 1 s is needed for adequate performance of the CGS. A benchmark study by Corona and De Schutter [7] shows that solving HMPC for such a long prediction horizon within such a short sampling period is not achievable with today's MIP solvers. Therefore and as the main contribution of this paper, a customized HMPC scheme is proposed that respects the hybrid nature of the vehicle dynamics, and simultaneously can be executed online on a microprocessor.

The proposed, customized HMPC strategy was designed by starting from a brute force approach to HMPC and, subsequently, reducing the necessary computations significantly based on insights into the system behavior. The brute force approach solves at each time step one quadratic program (QP) for each possible switching sequence. A switching sequence is a sequence of modes over the prediction horizon. If a switching sequence is given, the optimization problem to be solved by the controller is a QP. The optimization problem corresponding to a given switching sequence ϑ can be stated as

$$\begin{aligned}
\mathbf{U}^* = \arg\min_{\mathbf{U}} \quad & J(\mathbf{X},\mathbf{U}) = \sum_{i=1}^{i=N_p} \left(\mathbf{x}_{k+i}^T \mathbf{Q} \mathbf{x}_{k+i}\right) \\
& + \sum_{i=0}^{i=N_p-1} \left(\mathbf{u}_{k+i}^T \mathbf{R} \mathbf{u}_{k+i} + (\delta\mathbf{u}_{k+i})^T \mathbf{S} (\delta\mathbf{u}_{k+i})\right) \\
\text{s.t.} \quad & \mathbf{x}_{k+i+1} = \mathbf{L}_{j(i)} + \mathbf{A}_{j(i)} \mathbf{x}_{k+i} + \mathbf{B}_{j(i)} \mathbf{u}_{k+i} \\
& \mathbf{x}_k = \hat{\mathbf{x}}_k \\
& j(i) = \vartheta(i) \\
& \mathbf{X} = (\mathbf{x}_{k+1}^T, \dots, \mathbf{x}_{k+N_p}^T)^T, \mathbf{U} = (\mathbf{u}_k^T, \dots, \mathbf{u}_{k+N_p-1}^T)^T \\
& i \in \{0, \dots, N_p - 1\},
\end{aligned} \tag{3}$$

where J is the cost function, $\delta\mathbf{u}_i = \mathbf{u}_i - \mathbf{u}_{i-1}$, $\mathbf{Q}$, $\mathbf{R}$ and $\mathbf{S}$ are weight matrices of appropriate dimensions for the state, the input and the input change $\delta\mathbf{u}$, respectively, $\{\mathbf{A}_j, \mathbf{B}_j, \mathbf{L}_j\}$ are the matrices of the affine state space model active in mode j, $\mathbf{x}_i$ and $\mathbf{u}_i$ are state and input at time step i and N_p is the prediction horizon length.

With 3 modes, the number of switching sequences for the brute force approach is 3^{N_p}. For $N_p = 15$ as required for our system, the brute force approach would be to solve 14 348 907 QPs per time step, which is computationally infeasible under the constraints of the onboard processor. Instead, only switching sequences with a single mode switch are considered by our proposed controller. Consequently, the number of QPs that need to be solved at each time step is reduced to $N_p + 1$. This simplification is based on experimental observations and the intuition that more than one mode switch during the short prediction horizon (around 1 s) is very rare due to the high inertia of the vehicle.

The proposed algorithm (executed once per sampling period) is shown in Algorithm 1. As inputs, it takes a state estimate $\hat{\mathbf{x}}$ (consisting of the roll and pitch angles and angular velocities) and a scalar disturbance estimate $\hat{d}$. Then, for each possible switching sequence ϑ with at most one mode switch, a QP is formulated and solved. Finally, the control input $\mathbf{u}$ is selected for which the cost c accumulated over the prediction horizon is smallest.

Algorithm 1. Customized HMPC scheme. *Inputs*: state estimate $\hat{\mathbf{x}}$, disturbance estimate $\hat{d}$. *Output*: optimal control input $\mathbf{u}^*$.

```
m ← getCurrentMode(x̂)
if m = 1 then                        ▷ m: current mode
    if x̂^(1) ≥ 0 then                ▷ Roll angle ≥ 0
        n ← 2                        ▷ n: anticipated mode
    else
        n ← 3
else
    n ← 1
i_switch ← N_p                       ▷ i_switch: switching index
c* ← ∞                               ▷ c*: lowest cost so far
while i_switch ≥ 0 do                ▷ ϑ: switching seq.
    ϑ ← (m, ..., m, n, ..., n)       (first i_switch entries m, remaining N_p − i_switch entries n)
    (u, c) ← solveQP(ϑ, x̂, d̂)
    if c < c* then
        c* ← c
        u* ← u
    i_switch ← i_switch − 1
return u*
```

3 Simulation Results

In a first simulation, which is shown in Fig. 3a, the proposed controller is compared to linear MPC (LMPC), which assumes that mode 1 is active throughout the whole state space. The same disturbance is applied for both controllers. While the vehicle tips over when controlled by LMPC, the proposed CHMPC stabilizes the vehicle successfully. The reason is the control input issued by each controller. Because the CHMPC considers the chance of an upcoming mode switch, it anticipates the risk of leaving mode 1. Thus, it increases the thrust earlier than LMPC, which ensures that the vehicle is stabilized successfully.

Furthermore, the performance of CHMPC for the case when more than one mode switch occurs during the prediction horizon is investigated. For this purpose, another controller is introduced: the brute force hybrid MPC (BFHMPC). At each time step, it solves one QP for each possible switching sequence (as defined in Sect. 2.4). As shown in Fig. 3b, the performance of the proposed

CHMPC scheme is very similar to the performance of BFHMPC, despite the fact that CHMPC only anticipates one mode switch.

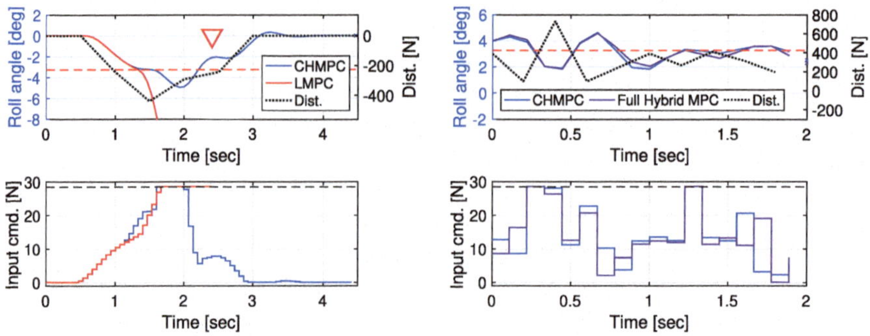

(a) Roll angles during a simulation comparing the proposed customized HMPC (CHMPC) strategy to a standard linear MPC (LMPC). The red dashed line represents the border between mode 1 (above) and mode 3 (below). The LMPC does not manage to stabilize the vehicle against the applied disturbance (red triangle marks tipping over) while the CHMPC does. The reason for this is that the CHMPC anticipates mode 3 and thus increases the thrust earlier to avoid it (see lower plot between 1 and 1.5 sec).

(b) Comparison between the brute force HMPC and the proposed strategy. The disturbance in this example is chosen to force more than one mode switch during the prediction horizon length of one second. Despite the fact that CHMPC is computationally significantly more efficient than brute force HMPC, their performance is comparable.

Fig. 3. Simulation results.

4 Experiments

This section gives details on how the proposed control scheme was tested in experiments and on the results achieved in these experiments.

4.1 Experimental Setup

The main goals of the experiments are to assess the CGS' performance in practice as well as to compare the proposed customized HMPC strategy with the simpler linear MPC approach, which only considers mode 1. As shown in Fig. 3a, simulations predict that the proposed control strategy outperforms linear MPC in certain situations. What remains is to assess if this can also be observed in practice. With these goals in mind, experiments in two different settings were carried out:

First, **indoor experiments** were conducted with the goal to verify estimator accuracy and controller performance in an environment that allows full control over the disturbance acting on the vehicle. This is achieved by attaching known weights to one end of a rope, which is diverted using a pulley system and attached to the vehicle at the other end (see Fig. 4a). The rope is diverted so that it exerts a horizontal force on the vehicle to closely match the effect wind disturbances have.

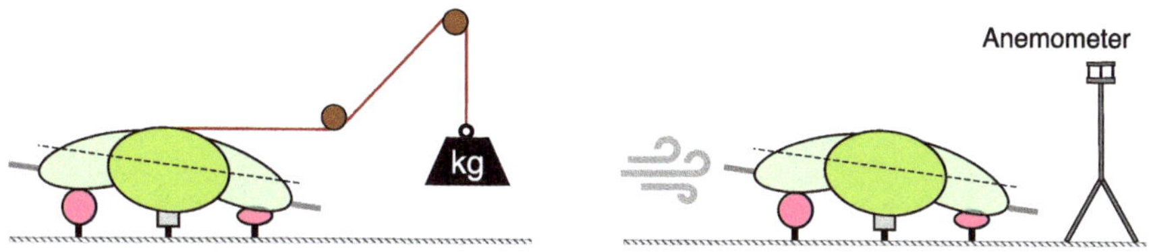

(a) **Indoor experiments**. The vehicle is deflected around its roll axis using a rope-pulley-system and different weights.

(b) **Outdoor experiments**. Realistic wind disturbances act on the vehicle. The current wind speed is measured using an anemometer.

Fig. 4. Schematics showing the indoor and outdoor experimental setups.

Second, **outdoor experiments** were carried out to assess the overall system performance under realistic weather conditions. In order to quantify the disturbance acting on the vehicle at each time step, an ultrasonic anemometer was used during all outdoor experiments (see Fig. 4b). It has a sampling rate of 10 Hz and an accuracy of ±2% below 30 m/s and ±3% between 30 m/s and 70 m/s. Note that the wind speed measurements from the anemometer are solely used as ground-truth for the Kalman filter disturbance estimate and they are not available to the controller.

For these experiments, both the controller and the estimator are implemented in C++, using ROS to manage communication. The algorithms are executed on a *Raspberry Pi 3 Model B* with a *QuadCore* 1.2 GHz CPU and 1 GB of RAM. To seize the full potential of the multi-core hardware, the execution of the control algorithm is parallelized. With this setup, a sampling frequency of 15 Hz and a prediction horizon of 15 time steps is achievable and was selected for all experiments. Data from six gyroscopes and six accelerometers is used as input to the estimator. Two of each are placed on each wing tip and the remaining two accelerometers and gyroscopes are placed on top of the wing envelope above the roll axis in the center of the vehicle.

4.2 Experimental Results and Insights

The estimation accuracy during an indoor experiment is shown in the center plot in Fig. 5a. Both in the case with and without CGS, the applied disturbance is estimated accurately using the proposed Kalman filter.

As illustrated in the top plot in Fig. 5a, the controller reduces the steady-state roll angle by 66% to 68%. The angle is not reduced to zero because the maximum thrust is not large enough to completely counteract the applied disturbance.

In the case without CGS, the vehicle gets very close to a mode switch, which corresponds to one wheel lifting off the ground. A slightly higher disturbance would trigger a mode switch and therefore compromise safe taxiing. In contrast, when the vehicle is controlled by the CGS, the thrusters' capabilities are used optimally to compensate for the disturbance and keep the vehicle in the safe mode 1.

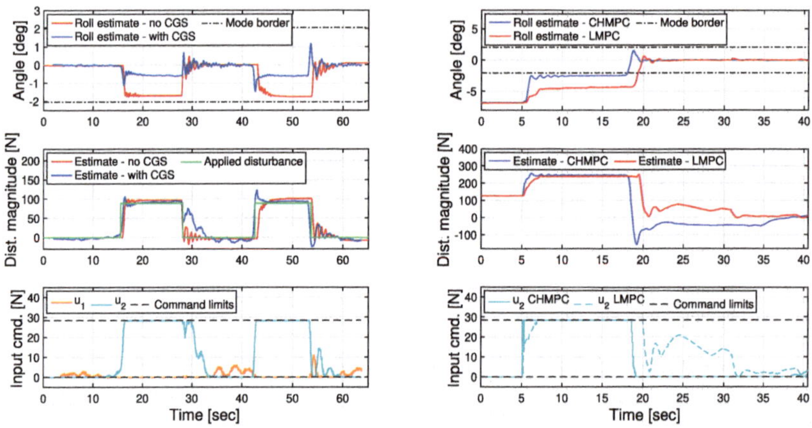

(a) The cases with and without CGS are compared for the case when a square disturbance (20 lb) is applied and released twice. With CGS, the resulting roll angle is reduced by 66% to 68% compared to the case without CGS.

(b) Comparison of proposed controller (CHMPC) to linear MPC (LMPC). The vehicle is manually deflected to a roll angle of 6° (back left wheel raised) before the disturbance is applied and the CGS is activated, both happening simultaneously at 5 sec. The CHMPC anticipates the risk of being in mode 3 and therefore increases the control command faster. Around 18 sec, the disturbance is removed.

Fig. 5. Results from **indoor experiments**.

During further indoor experiments, the proposed HMPC was compared to linear MPC (see Fig. 5b). For this purpose, the vehicle is manually deflected around its roll axis before the disturbance is applied and the CGS is activated. Then, the controller activation and disturbance application happen simultaneously (after 5 s in Fig. 5)b. This procedure allows to observe the controller behavior when the vehicle is not in mode 1, which is where the strength of the proposed controller becomes apparent. The observed performance is similar to what was encountered in the simulation results (Sect. 3): while the LMPC ramps up the counteracting control command gradually, the CHMPC correctly recognizes the risk of being in mode 3 (i.e. having one wheel raised). As a result, the CHMPC stabilizes the vehicle at a roll angle that is 44% lower than the roll angle achieved by the

LMPC. Note that the rope pulley system used in the indoor experiments introduces some friction, which damps any vehicle motion. Without this damping effect, the LMPC is less likely to prevent the vehicle from tipping over.

Finally, during the outdoor experiments, the CGS successfully stabilized the vehicle against wind gusts up to 9 m/s (32.4 km/h), which is the highest wind speed that was observed while the controller was active. The corresponding data is shown in Fig. 6a. Note that the controller manages to keep the roll angle below the critical roll angle (at which a mode switch would occur) at all times. During a run without CGS, a gust with a speed of 10 m/s managed to flip the vehicle over (see Fig. 6b at 68 s). Even before that, at 54 and 59.5 s, and at lower wind speeds the roll angle exceeds the critical roll angle. This illustrates that the CGS is capable of increasing the hybrid airship's stability under realistic crosswind conditions, within the limits of the actuators.

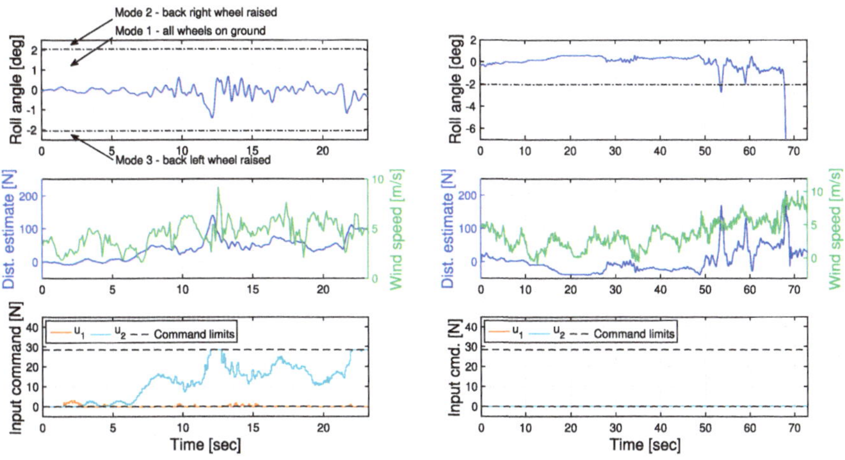

(a) Roll angle estimate, disturbance estimate, crosswind speed measurements and control inputs during an outdoor experiment. There is a correlation between disturbance estimate and wind speed, see for example spike around 13 sec. The CGS manages to stabilize the vehicle against a wind gust of up to 9 m/s.

(b) Data from an experiment run in which the CGS was disabled. At 69 sec, the wind speed (green curve) increases to above 10 m/s. As a result, the vehicle flips over.

Fig. 6. Results from **outdoor experiments**.

5 Conclusions

In this paper, the design and experimental validation of a counter gust system for a hybrid airship were discussed. For this purpose, the vehicle is modeled as a hybrid dynamical system, which challenges the controller design. To address this, the main contribution of this paper is a customized hybrid MPC scheme that, on the one hand, outperforms a linear MPC approach and, on the other hand, was proven to run in real time onboard the vehicle. Future work includes additional in-depth outdoor testing of the system under realistic wind conditions to further prove that the system is reliable in the real-world application it was designed for.

References

1. Khoury, G.A.: Airship Technology, 2nd edn. Cambridge University Press, Cambridge (2012)
2. Jiron, Z.B.: Hybrid airships for lift: a new lift paradigm and a pragmatic assessment of the vehicle's key operational challenges. Technical report, Air University, Maxwell Air Force Base, Alabama, December 2011
3. Airplane Flying Handbook. U.S. Department of Transportation, Federal Aviation Administration, Flight Standards Service (2016)
4. Lockheed Martin Corp.: Air cushion landing system. Accessed 20 Apr 2018. https://www.lockheedmartin.com/en-us/news/features/2016/hovercraft-technology-help-people-remote-parts-of-world.html
5. Waslander, S., Wang, C.: Wind disturbance estimation and rejection for quadrotor position control. In: AIAA Infotech@ Aerospace Conference, p. 1983 (2009)
6. Demitri, Y., Verling, S., Stastny, T., Melzer, A., Siegwart, R.: Model-based wind estimation for a hovering VTOL tailsitter UAV. In: 2017 IEEE International Conference on Robotics and Automation (ICRA), pp. 3945–3952 (2017)
7. Corona, D., De Schutter, B.: Adaptive cruise control for a SMART car: a comparison benchmark for MPC-PWA control methods. IEEE Trans. Control Syst. Technol. **16**(2), 365–372 (2008)
8. Oberdieck, R., Pistikopoulos, E.N.: Explicit hybrid model-predictive control: the exact solution. Automatica **58**, 152–159 (2015)
9. Giorgetti, N., Ripaccioli, G., Bemporad, A., Kolmanovsky, I., Hrovat, D.: Hybrid model predictive control of direct injection stratified charge engines. IEEE/ASME Trans. Mechatron. **11**(5), 499–506 (2006)
10. Maitland, A., McPhee, J.: Fast NMPC with mixed-integer controls using quasi-translations. In: 6th IFAC Conference on Nonlinear Model Predictive Control (2018)
11. Engell, S., Frehse, G., Schnieder, E.: Modelling, Analysis and Design of Hybrid Systems, vol. 279. Springer, Heidelberg (2003)
12. Sontag, E.D.: Nonlinear regulation: the piecewise linear approach. IEEE Trans. Autom. Control **26**(2), 346–358 (1981)
13. Helwa, M.K., Esser, A., Schoellig, A.P.: Estimation-based model predictive control for automatic crosswind stabilization of hybrid aerial vehicles. arXiv preprint arXiv:1810.00046 [cs.SY] (2018)
14. Soman, S.S., Zareipour, H., Malik, O., Mandal, P.: A review of wind power and wind speed forecasting methods with different time horizons. In: North American Power Symposium, pp. 1–8 (2010)
15. Camacho, E.F., Ramirez, D.R., Limon, D., Munoz de la Pena, D., Alamo, T.: Model predictive control techniques for hybrid systems. Ann. Rev. Control **34**, 21–31 (2010)

In-Air Exchange of Small Payloads Between Multirotor Aerial Systems

Ajay Shankar(✉), Sebastian Elbaum, and Carrick Detweiler

Department of Computer Science and Engineering,
University of Nebraska-Lincoln, Lincoln, NE, USA
ashankar@cse.unl.edu

Abstract. Exchanging payloads between multiple vehicles is a challenging problem that finds applications in sustained monitoring and long-range payload transportation. In this paper, we take the problem one step further and explore the design and implementation of a system of aerial vehicles that can exchange a small cargo/payload while air-borne. The proposed approach exploits the hover-capabilities of multirotor vehicles, and works within the constraints and the challenges imposed by such a task (such as proximity and contact forces). It is also portable to a wide range of vehicles, and does not require custom flight maneuvers in order to be viable. A simulation framework is utilized to highlight the key challenges addressed by the system. We present detailed results from experiments conducted outdoors that demonstrate two vehicles exchanging a small 60 g payload unit between them.

1 Introduction

Multirotor unmanned aerial systems (UASs) have the geometry, stability, and maneuverability that makes them attractive platforms for transporting a variety of cargo payloads. They are, however, limited in their endurance. Hence it is natural that a number of applications can benefit by having multiple UASs collaboratively perform tasks. For instance, long-range payload transportation can be achieved by relaying the payload from one UAS to another. Likewise, UASs that collect data from onboard sensors by hovering for prolonged periods of time can benefit by having a second UAS take its place.

Multiple UASs engaging with the same payload has transformative applications, however, it has usually been studied indoors in varied contexts [1–3]. Outdoors, the most straightforward way to transfer a payload carried by one UAS to another is for the former to land and deposit the payload on a reliable terrain, and then for the latter to pick it up. This can be slow, oftentimes infeasible, and does not utilize the full capabilities of a multirotor vehicle/flight.

In this paper, we present the first steps towards transferring payload between two UASs *while* they are both air-borne. A system that allows aerial transfer of payloads can offer potential a speed-up over ground-based systems, while also enabling a transfer over regions where it may be impractical to land. Exchanging

J. Xiao et al. (Eds.): ISER 2018, SPAR 11, pp. 511–523, 2020.
https://doi.org/10.1007/978-3-030-33950-0_44

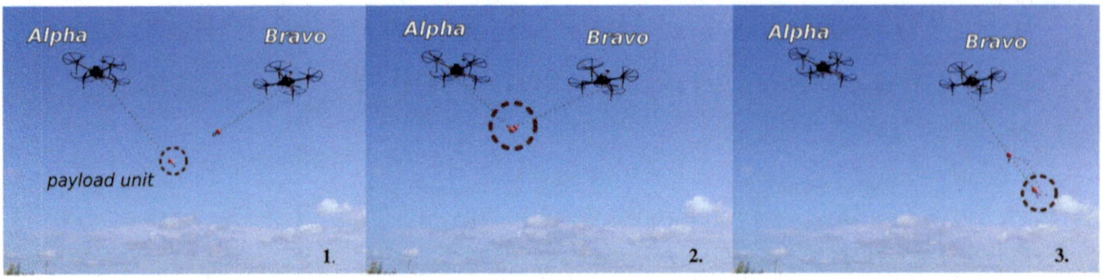

Fig. 1. Snapshots depicting the two vehicles *Alpha* and *Bravo* as they (1) approach each other, (2) intersect the payload strings, and (3) transfer the payload.

payloads mid-flight, however, is challenging due to the rich dynamics and the complexity of the process, as well as the impact of failures which are likely catastrophic for the UASs involved. Our proposed approach involves a controlled swinging of a payload system on both the sender and the receiver vehicles, as shown in Fig. 1. The approach does not require custom flight controllers, which makes it broadly applicable to a wide variety of multirotors.

Numerous examples in the literature discuss multiple UASs collaboratively transporting a payload with a flexible number and geometry of the vehicles [3,4]. Previous work has also demonstrated the potential for two vehicles to exchange payloads by means of a deployed parachute [5]. Adding or removing mass from an aerial vehicle is typically limited to systems that allow swapping batteries [6], and manned aerial refueling [7]. While these involve multiple UASs tethered to the same object (or to each other), they do not specifically address the scenario of one vehicle handing off an object to another vehicle mid-air. Methods for swapping out batteries also typically follow the routine of land–exchange–takeoff. Another class of related work includes robotic arms and grippers that can manipulate objects [8,9]. Accessibility of the payload is one of the key challenges posed by the structural design of the moving parts of a multirotor. While robotic manipulators can work around this by extending outwards, they are usually limited in their reach. Additionally, positioning the payload becomes a greater challenge when it is extended far outwards from the body. The problem is aggravated outdoors as multiple vehicles must maintain precise relative positions. Furthermore, the controller must account for contact forces, which may be harder to estimate in the presence of environmental noise. It is evident that the task of exchanging payloads mid-air is inherently full of challenges, and few have explored this area.

Our proposed approach circumvents several of these concerns and reduces the risk imposed on the UASs. We show that it is possible to exchange payloads in-flight through physical experiments, and present simulations to explore some of the parameter space. We begin by quantifying the primary challenges involved and then defining our problem statement.

1.1 Challenges

Interaction between two simultaneously airborne vehicles outdoors has several challenges which can be grouped into three main classes:

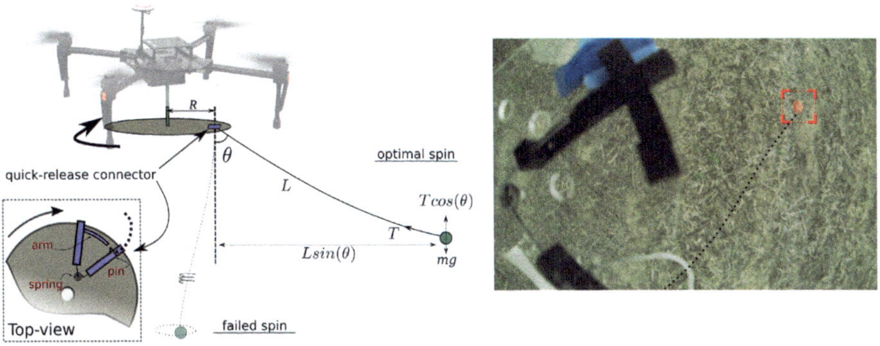

Fig. 2. *Left.* A diagrammatic representation of our approach for extending the payload far out from the UAS. The disc housing is rigidly affixed with a DC motor (not shown) to the body of the vehicle. As the disc spins, the vertical component of the tension in the string causes the payload to move on a conical pendulum with increasing θ. *Right.* The view from the downward facing camera with the payload and the cable highlighted for clarity.

1. *Separation between vehicles:* A mechanism that enables two vehicles to exchange payloads mid-air must maximize the distance of their closest approach. Given environmental uncertainties, hovering in close proximity will increase the risk of collision for the vehicles.
2. *Accuracy in relative positioning:* To perform a payload hand-off, the vehicles will need to maintain extremely precise relative positions. In the absence of extremely accurate state estimates, this may be infeasible to achieve.
3. *Contact forces:* If the payload hand-off is not executed rapidly, the two vehicles in the process will be in contact with each other for a sustained period of time, which makes the joint dynamics very complex. Force-based flight controllers [10,11] may be utilized to achieve some degree of compliance, however, state estimation still remains challenging outdoors.

1.2 Problem Statement

We are interested in developing a system that can enable the transfer of one unit of payload between two UASs, hereafter called *Alpha* and *Bravo*, while in hover. In addition, our goal is to impose minimal constraints on the choice and the design of the vehicles, and be able to function oblivious to the specific characteristics of the flight controller. Particularly, we require a setup that can operate without (and eliminates the need for): (1) having to land, (2) robotic end-effectors or manipulators that require precise positioning, (3) prolonged contact between two end-effectors, or (4) specialized flight controllers for 1, 2 or 3. To be practical, such a mechanism must provide reasonable guarantees of safety, while utilizing the capabilities of a multirotor.

2 Approach

Our approach enables two multirotors to exchange a unit of payload mid-air by extending it outwards from the body of the vehicle through a rotational motion. The extension enables the positioning of the vehicles with a safe separation between them, while the rotational element is designed such that the end-effectors need to be in contact very briefly, and a transfer can happen in spite of a lower precision in positioning.

Figure 2(left) shows a graphical representation of this system mounted underneath a UAS (*Alpha*). The payload is connected to a circular disc by means of a lightweight inelastic string. As a motor spins the disc, the payload is extended outwards as a function of θ, the half-angle of the inscribed conic. In an ideal steady-state, the relationship between the angular velocity of the payload (ω_p) and θ is governed by

$$g \tan(\theta) = (R + L \sin(\theta)) \cdot \omega_p^2,$$

where g denotes the acceleration due to gravity, R is the radius of the disc, and L is the length of the string used.

A critical challenge in practice is the possibility that the disc spins at a high speed while the payload simply dangles almost vertical (shown faded in Fig. 2 left). Due to a flexible string and the two degrees of freedom at the joint between the string and the disc, the payload's rotational velocity (ω_p) will not necessarily equal the disc's rotational velocity (ω_d), and external disturbances (such as wind) can cause the payload to dangle vertical. To recover from such a state, the disc must slow down and ramp up its angular velocity such that the payload spins up "in phase." We reason that feedback is then an essential element in the system design. First, it is critical for a spinning system subject to external disturbances such as the ones introduced by the UAS. Second, feedback helps in designing a system that is more accommodative of variations in R and L. In our implementation we employ visual feedback by means of a machine-vision camera and a compact onboard computer mounted on the UAS, described later.

Quick-Release Connector. One essential component of the design is a reliable mechanism for quickly disconnecting the payload from *Alpha*. There are two conflicting requirements on the design: (1) it must withstand the high tension in the cable during spinning, and (2) it must disengage upon a resistance that is both *external* (not caused by drag or inertia) and acts in the *same* direction as the tension. Our mechanism[1] for addressing these requirements, shown in Fig. 2 (left), consists of a fixed bar and a movable pin that is loaded on a torsion spring. The arm of the pin has an adjustable length so that the sensitivity of the mechanism can be adjusted. During spinning, the tension on the cable compresses the spring to some extent. However, upon external contact, the higher cable tension compresses the spring further in order to dislodge the pin that holds the cable, thereby releasing the payload. The components are made of 3D printed nylon-onyx material, and are produced in our laboratory.

[1] Special thanks to Ashraful Islam for his contributions to the design.

The string is connected to the disc through this quick-release connector which is designed to detach upon contact with any external impact that resists its spin. *Alpha* maintains a hover as the payload spins up, extending outwards at a distance of $R + L\sin(\theta)$ from the center of *Alpha*. A similar mechanism is mounted on the receptor vehicle (*Bravo*). This allows *Bravo* to position itself laterally adjacent to *Alpha*, and "snatch" the payload by having their strings intersect.

Visual Feedback. We have emphasized the importance of feedback regarding the states of the disc and the payload in the reliability and the performance of the system. This can be accomplished by several means; however, in our implementations, we use visual feedback for its flexibility and ease of design. We do so by utilizing a downward facing camera mounted under the body of the vehicle (see Fig. 2, right), and a vision processing algorithm that segments out the payload and the disc using color. The control input to the motor is a commanded rotational rate that increases with time. The algorithm measures the actual rotational rate of the disc (ω_d) and the payload (ω_p), and generates a commanded $\omega_d \pm \Delta\omega_d$ so as to minimize $(\omega_d - \omega_p) \bmod 2\pi$. In essence, the goal of this proportional controller is to ramp up the speeds of the disc and the payload while ensuring that they remain in phase. A pseudo-code for the algorithm is listed in Algorithm 1. The function s (line 3) represents the scale at which the commanded rate grows with time, and k is constant for proportionality.

Algorithm 1: Pseudo-code listing that controls the speed of the disc in order to ensure that the payload can be spun up reliably.

```
1 Procedure CTRL-DISC-SPEED:
      Input  : ω_d, ω_p, t
      Output: Commanded speed (cmd_rpm)
2     while mission_active do
3         ω ← s(t)
4         if ¬ IN-PHASE() then
5             Δω ← (ω_d − ω_p)  mod 2π
6             ω ← ω + k ∗ Δω
7         end
8         cmd_rpm ← ω
9     end

1 Function IN-PHASE():
2     return |t_{d,ref} − t_{p,ref}| ≤ t_{thresh,ref}
```

To assess whether the disc and the payload are spinning in phase, the algorithm also measures $t_{d,ref}$ and $t_{p,ref}$, the times at which the disc and the payload cross an arbitrary reference line (ref) in the image plane. The helper function in Algorithm 1 uses a small constant ($t_{thresh,ref}$) to ensure that $t_{d,ref} \approx t_{p,ref}$. The variable *mission_active* is unset once the payload is disengaged from *Alpha*.

The proposed approach is generic and portable to a wide-range of multirotors. Generic slung-load control has been studied in the literature [12,13], however, in order to conform to the problem statement we outlined, our system design functions completely oblivious to the choice of aircraft and flight controllers. Furthermore, our method also addresses the three primary challenges outlined in Sect. 1.1 – (1) the length of the two cables allows the vehicles to be separated well apart, (2) the vehicles can be in relative motion while transferring the payload, and (3) the quick-release connector minimizes the the interaction time between the two vehicles.

3 Simulation Study

To evaluate certain key behavioral characteristics of the system, we use a physics-based simulation framework designed in MATLAB Simulink®. Particularly, we investigate two matters: (1) the angular acceleration profile that must be commanded to the motor to ensure that the payload can be spun up, and (2) the value of using state feedback in this process. In the our simulations, we model the string as a series of short elements interlinked via elastic ball joints that mimic its experimentally observed physical characteristics. The motor and the disc are likewise modeled as physical objects connected via a revolute joint. We then simulate the rotation of the disc by gradually increasing the torque applied by the motor. Note that we do not attempt to model physical factors such as damping, friction, cable elasticity, etc.

Acceleration Profile. To spin the payload up, we command an angular acceleration, $\dot{\omega}_d(t)$, for the disc in the form of applied torque. Intuitively, $\dot{\omega}_d(t)$ must not be a step function if a sustained "in-phase" spin is desired. The failed spin shown as a faded illustration in Fig. 2 (left) is one possible consequence of modeling $\dot{\omega}_d(t)$ as a step function. We hypothesize that a linear ramp (corresponding to $s(t) = c_1 t + c_2$, for some c_1 and c_2, in Algorithm 1) would yield better results. However, the parameters of the ramp function are not known. Later on, we also consider the possibility that deviations from a steady ramp may be necessary.

In order to empirically find a suitable $\dot{\omega}_d(t)$, we utilize MATLAB's Global Optimization Toolbox combined with our Simulink model to optimize its response. We first constrain the simulation to run only for a finite time period (T). This time period is further divided into n smaller intervals, and an applied torque is computed for each of these intervals. The optimization problem is given an initial guess in the form of a ramp function (c_1, c_2), and an n-element vector for applied torque is sought. The objective function minimizes

$$J(\theta(t), \phi(t)) = \int_0^T \phi(t)dt + \big(\frac{\pi}{2} - \theta(T)\big),$$

constrained to $0 \leq \theta(t) < \frac{\pi}{2}$. The variable $\phi(t)$ in the cost denotes the angular phase difference modulo 2π. Figure 3 shows the results from one sample optimization routine which was seeded with an initial guess of $c_1 = 0.026$ and $c_2 = 0.2$. The snapshots on the left graphically show the response of the system (comprised of a disc and a payload connected by a string) for the initial guess for $\dot{\omega}_d(t)$ and for the optimized $\dot{\omega}_d(t)$. The figures on the right also show the corresponding responses plotted. Note how small modifications to the linear ramp can produce dramatic differences in the response of the system.

In practice, it may be hard to solve for a desired $\dot{\omega}_d(t)$, even if all system parameters were known before hand. As mentioned before, environmental disturbances will further necessitate adjustments to $\dot{\omega}_d(t)$ in real-time, which further justifies the use of a feedback system. We also note that a mathematical model of this system is immensely complex, and therefore a reactive procedure listed in Algorithm 1 is well suited for our application.

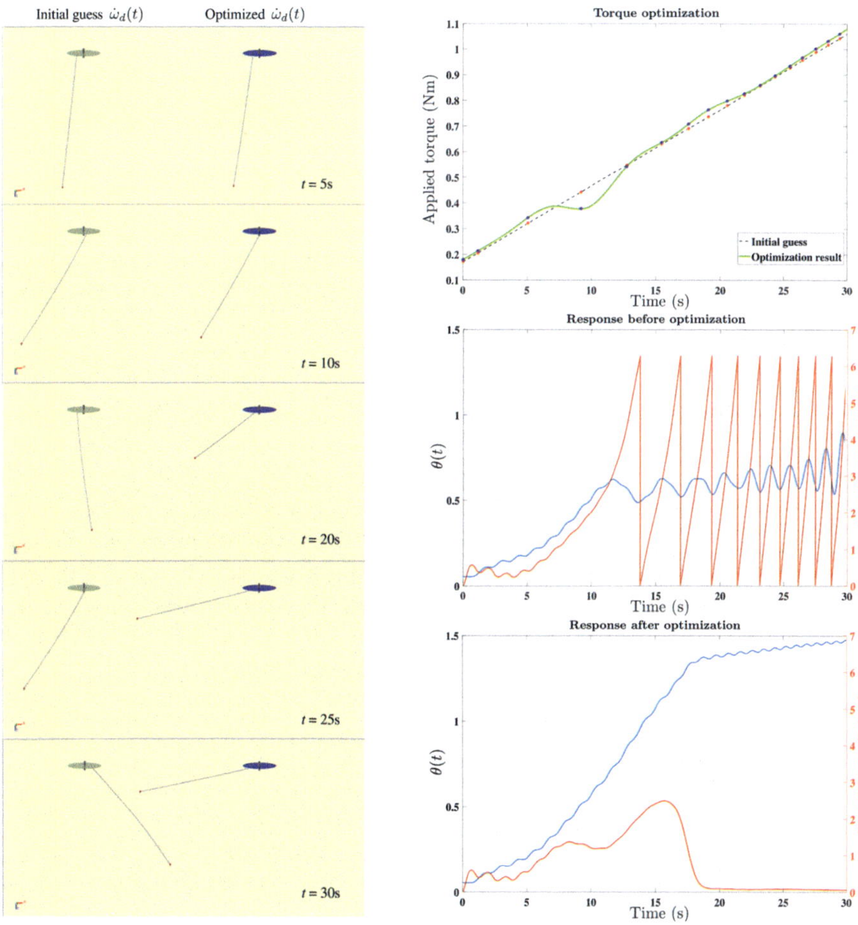

Fig. 3. Snapshots of the results from one optimization routine.

Feedback. In general, a particular acceleration profile, $\dot{\omega}_d(t)$, applied to systems with different combinations of the radius of the disc (R) and the length of the string (L) will not yield good results. In order to demonstrate the usefulness of feedback, we first define the $\dot{\omega}_d(t)$ that works well with our selected values of R and L. Then, the performance of the system is evaluated over a range of values of R and L. Since our intent is to maximize the separation between the vehicle and the payload, we inspect the maximum value of θ attained by the system.

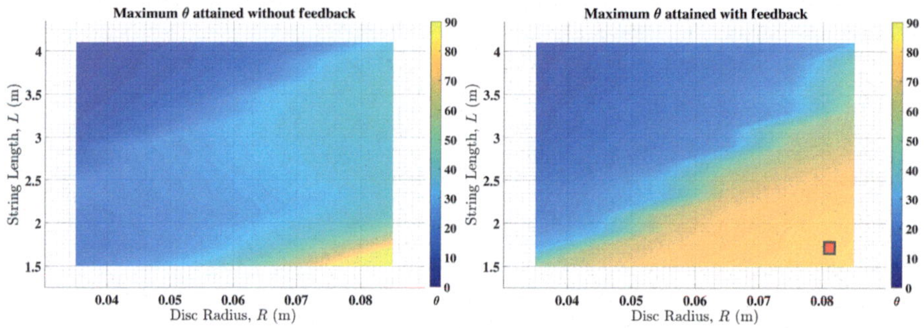

Fig. 4. Maximum θ attained by the string when the system is simulated over a range of values of R and L, (a) without, and (b) with feedback. The addition of feedback improves the tolerance of the system to variations in these parameters. The marked point (red) shows the values we use in our implementation.

The simulation is initialized at $\theta = 0$ (vertical cable) and $\omega_d = 0$. Figure 4 shows the maximum θ attained by the string when, for different values of R and L, (a) the same $\dot{\omega}_d(t)$ is applied to the system (no feedback), and (b) the selected $\dot{\omega}_d(t)$ is adjusted according to a feedback law. Clearly, without feedback, the string attains near-horizontal angles only for a very small range of values. However, the addition of a feedback control increases this range significantly, thereby improving the tolerance to disturbances. We also emphasize here that, in practice, near-horizontal angles are generally unattainable outside of simulations because of physical imperfections of the systems involved.

4 Experiment Design

Through field experiments conducted outdoors, we are interested in assessing (1) how well the system can induce and sustain a spin in the presence of environmental disturbances and autopilot behavior, and (2) the feasibility of the actual transfer mechanism between two vehicles.

4.1 Hardware Implementation

We use a DJI Matrice 100 [14], a medium-scale quadcopter as our aerial platform. For our tests, we employ a solid payload that weighs approximately 60 g, connected to the disc using a lightweight nylon string of length (L) 1.6 m.

The disc is cut out of acrylic and has a total radius (R) of 92 mm. It is spun by a 6 V brushless DC motor, which is rigidly affixed to the body of the vehicle. A downward-facing machine-vision camera (MatrixVision BlueFox, with a resolution of 752×480 px and approx. 110° horizontal field of view) mounted on the vehicle provides video feed to an Odroid XU4 (also on the vehicle). The Odroid observes the states of the disc and the payload and updates the rotational speeds commanded to the motor. We note here that entire setup is housed completely onboard the vehicle. Our approach is also independent of the choice of the multirotor and the flight controller, and does not require any specialized maneuvers by the vehicle, other than hovering.

Indoor and outdoor experiments are conducted to assess *Alpha*'s ability to induce and sustain the spin on a payload unit, and then transfer it to *Bravo*. Preliminary indoor tests are performed in unstructured environments that have no provisions for the UASs to maintain a position hold. The goal here is to ascertain that (1) the spinning mechanism can withstand disturbances induced by a non-stationary platform, and (2) the proposed system can function despite the vehicles not maintaining strict relative positions.

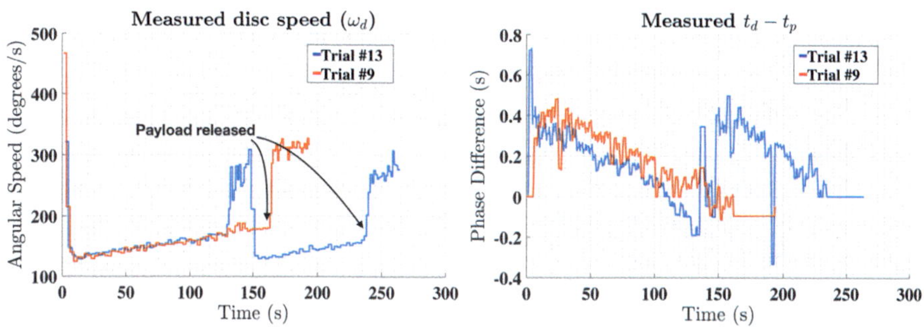

Fig. 5. The measured rotational speed of the disc, and the phase offset between the disc and the payload from two sample experiments (recorded on *Alpha*). When the payload is released, the disc has zero resisting torque which causes the spike in its speed. The phase difference starts at a positive value (the disc is "dragging" the payload) and then gradually approaches zero.

Both vehicles have identical builds, except for the quick-release connector (which is mounted only on *Alpha*), and *Bravo*'s payload (a dummy object which is similar to *Alpha*'s actual payload). In our experiments, both vehicles are piloted manually to approximately the same altitude above ground, while they begin spinning their respective attachments. As the payload approaches a steady spin, *Bravo* proceeds to approach *Alpha* horizontally in order to let their strings intersect. As the payload disengages from the quick-release connector on *Alpha*, it is trapped in the string affixed to *Bravo*. Figure 1 shows frames from an actual experiment in which *Bravo* (right) (a) approaches *Alpha* that is already spinning the payload, (b) interrupts the spin, thereby intercepting the payload and finally (c) moves away with the payload.

5 Results

We begin by investigating how well the systems can spin up the payload outdoors. As listed in Table 1, the time taken to spin up the payload is fairly consistent for most of our trials. Figure 5 (left) shows the gradually increasing measured rotation speeds of the disc from two sample tests. The figure on the right shows the phase offset between the disc and the payload (measured in seconds). In the beginning, this offset is positive, and as the payload gets in phase with the disc, this offset approaches zero. Starting from completely vertical, the time taken to get the payload up to its maximal extent is similar for the tests (unless influenced by strong external factors). However, as seen in Fig. 5, the time that *Alpha* continues to spin the payload can be different for the tests. This can be either because *Bravo* does not intercept the payload right away, or because the contact force does not disengage the payload in the first attempt. In such a scenario (trial #13), *Alpha* simply restarts its spin. The sharp increase in the rotational speed of the disc is caused by the lack of resisting torque (load) on the disc after the payload is released (or restarts from vertical).

In a complete autonomous mission, the vehicles will need to hover adjacent to each other for some finite amount of time before the transfer is successful.

Table 1. The total mission time, split into spin time and the waiting time (for *Alpha*), along with the approximate payload separation and number of transfer attempts made in 16 flights.

Trial	Spin time (s)	Wait time (s)	Total (s)	Dist (m)	# attempts	Notes
1	117	20	137	1.3	1	–
2	131	118	249	1.2	2	Drop
3	117	119	236	1.0	1	
4	120	44	164	1.3	1	Drop
5	150	120	270	1.3	2	
6	143	137	279	1.2	2	No release
7	120	34	154	1.2	1	
8	117	21	138	1.2	1	Drop
9	120	44	164	1.4	1	Drop
10	115	26	141	1.2	1	
11	120	34	154	1.4	1	Drop
12	118	67	186	1.4	1	
13	146	96	241	1.0	2	
14	117	36	153	1.3	1	
15	117	18	135	1.2	1	
16	485	110	594	1.0	4	
Mean	147.0	65.29	212.29	1.24	–	
Median	120.0	44.15	164.15	1.23	–	

Two properties of the system that are of interest are, then, the time it takes for *Alpha* to spin the payload up to its maximum extent (starting from completely vertical), and the approximate distance between the payload and *Alpha*'s center. Table 1 lists these two variables, along with *Alpha*'s waiting time for 16 sample tests we performed outdoors. All but one of the tests resulted in a successful disengagement of the payload from *Alpha*; and out of these, *Bravo* was able to successfully retain the payload in ten missions. The table also lists the number of attempts the system needed to make before the payload was disengaged. Majority of the missions took one attempt; however, trial #16, which suffered strong gusts of wind required 4 attempts before both vehicles reached a steady state of spin. For the rest of the missions that took more than one attempt, both *Alpha* and *Bravo* simply reset their spinning mechanisms. Note that in manual missions, the waiting time is dependent on the readiness of *Bravo*. Also, as we note in trial #6, there's a possibility that the payload does not release at all if it collides with *Bravo*'s dummy object (instead of the strings intersecting).

6 Discussion

One of the key insights from our simulations is the relevance of feedback in the mechanism that spins the payload. For a system that must work with a range of parameters (R, L, and the weight of the payload), and in different environmental conditions, it is necessary to have a controller that can adapt the rate of spin even if many of those parameters are known beforehand.

Even though our method only requires the vehicles to hover, it is imperative that the torques exerted due to the spinning motion remain within critical limits. While spinning, the payload's momentum may induce rotations on the body of the vehicle. Figure 6 shows the 3-axis gyroscope readings from the vehicle's onboard IMU during one representative test mission. The spinning mechanism is initiated at approximately 160 s and is active throughout the highlighted region. We notice that as time progresses (i.e., the rotation speed increases), the effect becomes increasingly pronounced. At the peak (approx. 430 s), the measured rotation rate is close to 8°/s. Even though it is noticeable, the effect is small (approx. 2.5%) compared to the maximum angular velocity that our particular vehicle and flight controller can attain. We also note that since this is, in general, a function of the mass and the rotational speed of the

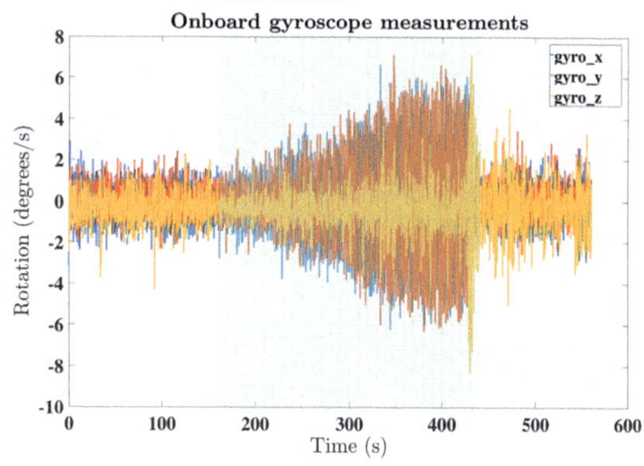

Fig. 6. The progression of 3-axis gyroscope measurements as the payload is spun up (shaded area between 160 s to 430 s).

payload, an appropriate choice of vehicle can be made by considering its hover performance.

Surprisingly, the motion of *Alpha* did not have a significant impact on the way the payload is spun up. We expected that deviations from a perfect hover would deteriorate the spin; however, from our preliminary tests, which were performed indoors (GPS-denied, no position hold) we found that the performance of the system does not change significantly. It was empirically determined that due to environmental disturbances such as wind, only the rate at which the rotational speed of the disc is increased needed to be altered (the factor s in Algorithm 1). This value was decreased, so as to allow the system to spin up less aggressively. Our experiments were conducted in winds of up to 3 m/s (the range within which the vehicles do not exhibit significant drifts in altitude). Since the string will not be perfectly horizontal, the shape of the conic inscribed by it allows a difference in altitude of approximately 1 m. Nevertheless, we acknowledge that one limitation of our method is that a significant drift in altitude by one vehicle may pose a threat to the other.

7 Conclusion

This paper presents an approach for two multirotor UASs to exchange their cargo payloads in air. We demonstrate the system's capabilities through outdoor field experiments, and present a detailed analysis of the results. We noted earlier that our proposed approach is contained entirely onboard the UAS, and has no dependencies on external resources such as a motion capture system or custom flight controllers. In the future, we wish to make the process entirely autonomous, and towards that end, we have already conducted some field trials with missions that have shared autonomy.

References

1. Su, K., Shen, S.: Catching a flying ball with a vision-based quadrotor. In: International Symposium on Experimental Robotics, pp. 550–562. Springer (2016)
2. Ritz, R., Müller, M.W., Hehn, M., D'Andrea, R.: Cooperative quadrocopter ball throwing and catching. In: 2012 IEEE/RSJ International Conference on Intelligent Robots and Systems (IROS), pp. 4972–4978. IEEE (2012)
3. Sreenath, K., Kumar, V.: Dynamics, control and planning for cooperative manipulation of payloads suspended by cables from multiple quadrotor robots. In: Robotics: Science and Systems (2013)
4. Jiang, Q., Kumar, V.: The inverse kinematics of cooperative transport with multiple aerial robots. IEEE Trans. Robot. **29**(1), 136–145 (2013)
5. Shankar, A., Elbaum, S., Detweiler, C.: Towards aerial recovery of parachute-deployed payloads. In: 2018 IEEE/RSJ International Conference on Intelligent Robots and Systems (IROS). IEEE (2018)
6. Wang, M.: Systems and methods for UAV battery exchange, US Patent 9,346,560, 24 May 2016

7. Smith, R.K.: Seventy-Five Years of Inflight Refueling. Ohne Verlagsangabe, Washington (1998)
8. Kim, S., Choi, S., Kim, H.J.: Aerial manipulation using a quadrotor with a two DOF robotic arm. In: 2013 IEEE/RSJ International Conference on Intelligent Robots and Systems (IROS), pp. 4990–4995. IEEE (2013)
9. Jimenez-Cano, A., Martin, J., Heredia, G., Ollero, A., Cano, R.: Control of an aerial robot with multi-link arm for assembly tasks. In: 2013 IEEE International Conference on Robotics and Automation (ICRA), pp. 4916–4921. IEEE (2013)
10. Parra-Vega, V., Sanchez, A., Izaguirre, C., Garcia, O., Ruiz-Sanchez, F.: Toward aerial grasping and manipulation with multiple UAVs. J. Intell. Robot. Syst. **70**(1–4), 575–593 (2013)
11. Augugliaro, F., D'Andrea, R.: Admittance control for physical human-quadrocopter interaction. In: 2013 European Control Conference (ECC), pp. 1805–1810. IEEE (2013)
12. Palunko, I., Fierro, R., Cruz, P.: Trajectory generation for swing-free maneuvers of a quadrotor with suspended payload: a dynamic programming approach. In: 2012 IEEE International Conference on Robotics and Automation (ICRA), pp. 2691–2697. IEEE (2012)
13. Guerrero, M., Mercado, D., Lozano, R., García, C.: Passivity based control for a quadrotor UAV transporting a cable-suspended payload with minimum swing. In: 2015 IEEE 54th Annual Conference on Decision and Control (CDC), pp. 6718–6723. IEEE (2015)
14. DJI Matrice 100. http://www.dji.com/matrice100. Accessed Aug 2018

Fast and Agile Vision-Based Flight with Teleoperation and Collision Avoidance on a Multirotor

Alex Spitzer(✉), Xuning Yang, John Yao, Aditya Dhawale, Kshitij Goel, Mosam Dabhi, Matt Collins, Curtis Boirum, and Nathan Michael

Robotics Institute at Carnegie Mellon University, 5000 Forbes Ave, Pittsburgh, PA 15213, USA
{aspitzer,xuningy,johnyao,adityand, kgoel1,mdabhi,mcollin1,cboirum,nmichael}@andrew.cmu.edu

Abstract. We present a multirotor architecture capable of aggressive autonomous flight and collision-free teleoperation in unstructured, GPS-denied environments. The proposed system enables aggressive and safe autonomous flight around clutter by integrating recent advancements in visual-inertial state estimation and teleoperation. Our teleoperation framework maps user inputs onto smooth and dynamically feasible motion primitives. Collision-free trajectories are ensured by querying a locally consistent map that is incrementally constructed from forward-facing depth observations. Our system enables a non-expert operator to safely navigate a multirotor around obstacles at speeds of 10 m/s. We achieve autonomous flights at speeds exceeding 12 m/s and accelerations exceeding 12 m/s^2 in a series of outdoor field experiments that validate our approach.

1 Introduction

Autonomous aerial vehicles with onboard vision-based sensing and planning have been shown to be capable of performing fast and agile maneuvers. However, long-horizon planning required to achieve a task has proven to be a challenge, particularly with limited onboard compute. We propose a fully integrated vision-based multirotor platform that allows minimally trained operators to teleoperate the vehicle at high speeds with agility, while remaining safe around obstacles in unstructured outdoor environments. At high speeds, the environment around the vehicle changes quickly, and is subject to dynamic obstacles and lighting conditions. Our multirotor architecture integrates the following to achieve agile

A. Spitzer and X. Yang—authors contributed equally.

Electronic supplementary material The online version of this chapter (https://doi.org/10.1007/978-3-030-33950-0_45) contains supplementary material, which is available to authorized users.

J. Xiao et al. (Eds.): ISER 2018, SPAR 11, pp. 524–535, 2020.
https://doi.org/10.1007/978-3-030-33950-0_45

and collision-free flight in these scenarios: (1) an extension of motion primitive-based teleoperation [12] to generate jerk-continuous local trajectories, a crucial component to prevent instability in agile flights, and (2) efficient local mapping for collision avoidance purposes using a KD-tree.

Many prior works in high speed flight exploit the field-of-view (FOV) of stereo cameras for fast collision checking for autonomous flights, including [4,5], where trajectories are constrained to be inside the FOV of the depth sensor with a max range of 10 m. In [8], trajectories generated by RRT* are checked for collisions directly in the disparity space. Lopez and How [7] presents aggressive flight on a 1.2 kg MAV, achieving a velocity of 3 m/s. These methods achieve fast collision checking by circumventing the need to construct a local map and checking for collisions in the sensor's FOV. This however, limits the range of motions the vehicle can perform. Approaches with local map generation using a laser range finder [1] and a monocular RGB camera [3] have been shown to achieve maximum velocities of 1.8 m/s and 1.5 m/s respectively, but data processing limits the update rate of the local maps. In our approach, we give a minimally trained operator full control of the vehicle and show that fast and agile flights can be achieved with a human-in-the-loop while maintaining safety.

We perform a series of high speed collision avoidance trials in both indoor and outdoor environments with untrained operators. In our experiments, our hexarotor attains speeds exceeding 12 m/s and accelerations exceeding 12 m/s^2. We are able to safely avoid obstacles at speeds up to 10 m/s and accelerations of 8 m/s^2, while retaining a local map.

2 Technical Approach

2.1 Smooth Motion Primitive-Based Teleoperation

Aggressive multirotor flights demand large angular velocities and large angular accelerations, which are directly related to the jerk and snap of the reference position [9]. Thus to avoid incurring large tracking error due to discontinuous trajectories, we extend *forward-arc motion primitives* [12] to generate trajectories that retain differentiability up to jerk and continuity up to snap. From the resulting trajectories, we can calculate desired vehicle attitude, angular velocity, and angular acceleration for use as feedforward terms in the controller.

The motion primitives are parameterized as follows. We define a *local frame* $\mathcal{L}$ to be a fixed z-axis aligned frame, taken at a snapshot in time. The motion primitive definition will be provided in the local frame at the time at which an input is issued, and can be freely transformed into a fixed global frame or body frame for control purposes. An action $\mathbf{a}$ specified by a continuous external input, such as a joystick, generates a single motion primitive. For a multirotor, we define an action as $\mathbf{a} = \{v_x, v_z, \omega\}$ in the local frame at which the input is issued, where v_x is the x-velocity (i.e., the forward velocity of the vehicle), v_z is the z-velocity, and ω is the yaw rate. Let $\mathbf{x}$ denote a vector containing the position and yaw of the vehicle, i.e. $\mathbf{x} = [x,\ y,\ z,\ \theta]$. Then, the endpoint velocities are defined by the unicycle model [10]. The unicycle model dynamics are given by $\dot{\mathbf{x}}(\mathbf{a},\tau) = [v_x\cos(\omega\tau) \quad v_x\sin(\omega\tau) \quad v_z \quad \omega]^\top$, where $\tau \in [0,T]$. The position endpoints are unconstrained and we enforce all higher order derivatives above velocity to be zero.

A regeneration step k occurs when a new input is received from the joystick, or the previous trajectory $\boldsymbol{\gamma}(\mathbf{a}_{k-1},T)$ finishes executing. Alternatively, a fixed regeneration rate can be chosen in order to accommodate changes in the environment for collision avoidance. Suppose regeneration step k occurs at time t_k. Then, a library of dynamically feasible motion primitives is generated in the local frame $\mathcal{L}_{t_k}$ specified by the reference state at time t_k, i.e. $\mathbf{x}_{\text{ref}}(t_k) = \boldsymbol{\gamma}(\mathbf{a}_{k-1}, t_k - t_{k-1})$, given a set of discretized actions $\mathbf{a}_k$. Each motion primitive $\boldsymbol{\gamma}$ is a vector of four 8th order polynomials that specify the trajectory along the position coordinates x, y, z and yaw coordinate θ. Given an action $\mathbf{a}_k$ at regeneration step k, each motion in the motion primitive library is generated in frame $\mathcal{L}_{t_k}$ according to

$$\begin{aligned} \boldsymbol{\gamma}(\mathbf{a}_k,\tau) &= \sum_{i=0}^{8} \mathbf{c}_i \tau^i \\ \text{s.t. } \boldsymbol{\gamma}^{(j)}(\mathbf{a}_k,0) &= \mathbf{x}_{\text{ref}}^{(j)}(t_k) \text{ for } j = 0,1,2,3,4 \\ \dot{\boldsymbol{\gamma}}(\mathbf{a}_k,T) &= \dot{\mathbf{x}}(\mathbf{a}_k,T) \\ \boldsymbol{\gamma}^{(j)}(\mathbf{a}_k,T) &= 0 \text{ for } j = 2,3,4 \end{aligned} \tag{1}$$

where $\{\cdot\}^{(j)}$ specifies the jth time derivative. Note that all constraints are appropriately transformed into $\mathcal{L}_{t_k}$. The duration of the trajectory T and the maximum x, z, and yaw velocities are specified according to the desired operational range. We further allow the operator to freely rotate the motion primitive library with respect to $\mathcal{L}_{t_k}$'s z-axis to allow for sideway slalom motions.

The result of having snap-continuous trajectories (see Fig. 1) ensures that we have smoothness in error dynamics, thus minimizing instabilities and tracking error.

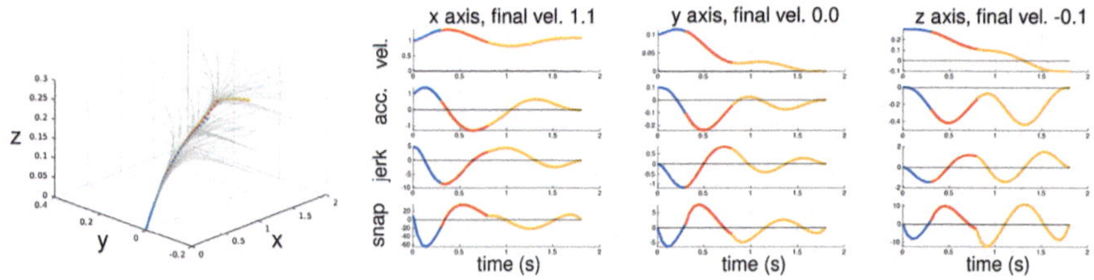

(a) (b)

Fig. 1. (a) A trajectory composed of 3-segments of motion primitives that switches to a new motion primitive at arbitrary points along the trajectory that have non-zero higher-order-derivative terms. The discarded trajectory is shown in dotted lines. (b) Higher-order time derivatives (velocity, acceleration, jerk, and snap) of the three segment trajectory, showing that the trajectories are differentiable up to jerk and continuous in snap at the switching points. At the end of the trajectory, all higher order derivatives are zero except for the user specified velocity.

2.2 Pruning and Trajectory Selection

At every time step, a family of motion primitives, called the motion primitive library, is created. The motion primitive library is constructed by discretizing the continuous input along each action dimension, such that each action $\mathbf{a}_i \in A$ is selected from a convex set $A := \{\mathbf{a} \in [\mathbf{a}_{\text{min}}, \mathbf{a}_{\text{max}}]\}$ with size $N_1 \times N_2 \times \cdots \times N_n$ where N_i is the dimension of the space of each input. An example of the discretization is shown in Fig. 2.

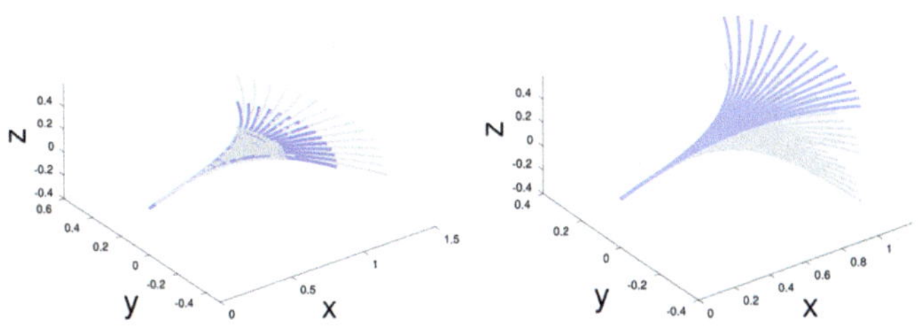

(a) (b)

Fig. 2. Motion primitive library with (a) variation in linear velocity and with (b) variation in z velocity

At every time step, the operator input is mapped to the closest input in the action space, as defined by the Euclidean norm. A priority queue that minimizes input distance from the selected input $\mathbf{a}_{\text{joystick}}$ to each input in the action space $\mathbf{a}_i \in A$ is used to iterate through the action space until a feasible, collision-free trajectory is found. This results in having the operator input mapped to the feasible motion primitive in the library that is parameterized by the closest discretized action, i.e. $\boldsymbol{\gamma}(\mathbf{a}_i) = \text{argmin} \|\mathbf{a}_{\text{joystick}} - \mathbf{a}_i\|$. A trajectory is deemed collision-free if the minimum distance between any point along the trajectory and the surrounding environment is above the sum of the vehicle size and an operator specified collision radius. Algorithm 1 describes teleoperation with reactive collision avoidance in detail.

The effect of this pruning algorithm is that the vehicle exhibits natural behavior in the presence of obstacles. If a pillar is in front of the vehicle, then the vehicle chooses a motion primitive some angle away and avoids the obstacle. If a wall is present, then the vehicle will choose linear velocities that gradually decrease until the vehicle is stopped.

2.3 Local Map Representation Using KD-Trees

We present a local mapping framework that generates a spatially consistent local map of the robot surroundings represented as voxel grids. The local map is generated by retaining only the depth sensor measurements obtained at poses that lie

Algorithm 1. Snap-continuous Motion Primitives based Teleoperation and Collision Checking with KD-Tree Local Map

Given KD-Tree local map L, collision radius r, vehicle radius r_v
Receive input $\mathbf{a}_{\text{joystick}}$
Generate the minimum input distance queue according to $d_i = \|\mathbf{a}_{\text{joystick}} - \mathbf{a}_i\|$
while $\mathbf{a}_i$ is infeasible **do**
  Pop the top action element off of the minimum input distance queue $\mathbf{a}_i$
  Generate $\gamma_i = \gamma(\mathbf{a}_i)$ according to (1)
  for $\tau = 0 : T$ discretized at some $\triangle t$ **do**
    Query L for the closest point p to $\gamma_i(\mathbf{a}, \tau)$
    if $\|p - \gamma_i(\mathbf{a}, \tau)\| \geq r + r_v$ **then**
      Set $\mathbf{a}_i$ to feasible
      Set $\gamma = \gamma_i$
    end if
  end for
end while

in the vicinity of the vehicle's current pose. This enables trajectories to span in the space observed by all of the retained sensor measurements. Sequential sensor measurements obtained using a stereo imaging sensor often contain redundant information about the surroundings of the robot. The novelty of information in an incoming sensor measurement at the resolution of the voxel grids is dictated by the rotation and translation displacement of the robot between the current frame and the previous frames. To enforce spatial locality, we dynamically select anchor frames and transform subsequent sensor measurements that contain novel information about the surroundings in to the anchor frames. In order to efficiently incorporate only novel information in the local map, we classify each incoming sensor measurement into one of the following categories: a KeyFrame (*KF*), a SubFrame (*SF*), or a BufferFrame (*BF*) (see Table 1). The local map (L) is updated in a locally consistent coordinate frame according to the type of sensor frame ($F \in \{KF, SF, BF\}$).

A sensor measurement that is more than α_k meters away from the current *KF* is classified as a *KF*. Sensor measurements that are not new *KF*s, but are either more than α_s meters in position or β_s degrees in heading away from the previous *SF*, are classified as *SF*s. Sensor measurements that are neither *KF*s nor *SF*s are classified as *BF*s, which do not contain sufficient novel information about the surroundings, but are registered to L to account for dynamic changes in the scene. *BF*s are in L only for the iterations in which they are observed.

A new local map L is spawned every time a new *KF* is detected. L consists of all the *SF*s registered to the current *KF*, the current *BF* and all the *SF*s registered to the previous *KF*, resulting in a voxel grid map representing the occupied space centered around the vehicle. This allows us to only update the KD-Tree that contains the voxel centers of the *KF* to which the incoming sensor scan is registered. Because a high fidelity map is not essential for collision avoidance, each sensor measurement is downsampled into an occupancy grid with a fixed

Table 1. Frame Classification

Frame type	Definition	Initialization condition for frame F
KF (anchor)	Sensor frames to which subsequent SFs and BFs are registered	$T(KF_{i-1}, F) > \alpha_k$
SF	Sensor frames that provide novel information about the vehicle surroundings	$T(SF_{i-1}, F) > \alpha_s \Vert H(SF_{i-1}, F) > \beta_s$
BF	Sensor frames that are registered for one time step to accommodate the dynamic changes in the surroundings when the vehicle has not been displaced by a sufficient distance	$F \neq (KF, SF)$

$T(a, b)$ is the translational distance between a and b; $H(a, b)$ is the heading rotational distance between a and b.

voxel size before registering it to L. Voxel centers in the map are then arranged in a KD-Tree to minimize the time complexity of nearest neighbor queries.

The algorithmic efficiency and independence of the definition of the local map from the sensor model enables our approach to incorporate measurements from multiple depth sensors with widely separate FOVs.

2.4 State Estimation

We use VINS-Mono [11], a tightly-coupled visual-inertial odometry framework that has been shown to perform favorably when compared to other state of the art open source state estimation algorithms [2]. VINS-Mono jointly optimizes vehicle motion, feature locations, camera-IMU extrinsics, and IMU biases over a sliding window of monocular images and preintegrated IMU measurements. Because our local planning strategy does not require a globally consistent map, we disable the loop closure functionality of VINS-Mono to reduce its computational footprint. We run an auxiliary state estimator during takeoff in order to provide smooth state estimates when the vehicle has not yet experienced sufficient motion excitation for VINS-Mono to initialize. The auxiliary state estimator is an unscented Kalman filter that fuses downward rangefinder altitude observations, downward optical flow, and IMU measurements to estimate vehicle velocity, position, and attitude.

3 Implementation

Hardware. We experimentally evaluate our proposed approach on a 3.8 kg hexarotor that fits within a 20 cm × 60 cm × 80 cm volume (Fig. 3a). The hexarotor has an average flight time of 7 min and a power to weight ratio of 3. Two Intel RealSense D435 depth cameras are used for mapping: one facing forwards and one facing upwards at a 45° angle, which aids obstacle avoidance while accelerating forwards. A downward facing Matrix Vision mvBlueFOX-MLC200w is used as the RGB camera input for VINS-Mono and the NVIDIA Tegra TX2 is used for computation. The hexarotor uses a cascaded PD control architecture as in [9] with jerk and snap references used to compute feedforward angular velocities and accelerations.

Teleoperation. Motion primitives are generated with an angular velocity bound of 2 rad/s. There are 25 discretizations for the v_x action, 11 for ω, and 5 for v_z for a total of 1375 motion primitives generated per trajectory iteration. Trajectories are generated at 25 Hz, and the local map is updated at 30 Hz.

We follow [6] and choose the maximum velocity of the motion primitives to be such that the vehicle can always safely stop given a known constant maximum acceleration, known sensor range and known sensor and mapping rates. Since our vehicle has a power to weight ratio of 3, we assume a conservative maximum acceleration at 6 m/s^2, and assume a worst case sensor and mapping rate of 10 Hz. With a sensor range of 10 m, this allows for a maximum motion primitive velocity of 10.37 m/s.

(a) (b) (c)

Fig. 3. (a) The hexarotor vehicle used for aggressive flights in (b) outdoor environments and (c) a dimly lit garage. The overlays of the vehicle positions over time depict the trajectories the vehicle took to avoid pillars in its way.

4 Experiments and Results

4.1 Experiments in Aggressive High Speed Flights

We conduct seven aggressive flight experiments in order to test our framework, including four outdoor experiments over an area of 40 m × 20 m with obstacles over grass (Fig. 3b), and two indoor experiments in a dimly lit garage (Fig. 3c). The experiments are described in Table 2. The local map used for all experiments uses $\alpha_k = 2$ m, $\alpha_s = 0.2$ m, $\beta_s = 0.1$ rad and a voxel size of 0.2 m.

Table 2. Experiment descriptions and parameters.

Experiment	Description	**T** (s)	$\mathbf{v}_x^{\max}$ (m/s)	$\omega^{\max}$ (rad/s)	r (m)
Outdoor-1	High speed teleoperation outside	2.2	12.0	2.0	N/A
Outdoor-2, 3	Aggressive teleoperation with collision avoidance outside	2.0	5.0	1.0	0.8
Indoor-1, 2	Aggressive teleoperation with collision avoidance in a dimly lit garage	1.5	3.0	1.5	0.4
Outdoor-4	High speed, aggressive teleoperation with collision avoidance outside	1.3[a]	7.0	2.0	0.5
Outdoor-5	High speed, aggressive teleoperation with collision avoidance outside	1.5[a]	10.0	2.0	0.2

T denotes the duration of the motion primitive, $\mathbf{v}_x^{\max}$ is the maximum desired speed, $\omega^{\max}$ is the maximum yaw rate, and r is the collision radius.
[a]Motion primitive duration increased adaptively as a linear function of the desired velocity change

High Speed Flight. In **Outdoor-1**, we execute straight line trajectories that hit a maximum desired speed of 12 m/s using the teleoperation system. Figure 4 shows the speed and acceleration attained during three runs. The vehicle achieves a maximum acceleration of 13.5 m/s^2.

High Speed and Aggressive Flight with Collision Avoidance. Experiments **Outdoor-2-5**, and **Indoor-1,2** stress test our collision avoidance algorithm at high speeds while maintaining aggressiveness in the commanded trajectories. In both environments, the operator repeatedly tries to fly the vehicle at the maximum speed, as indicated in Table 2, into an obstacle. Figure 6 shows the speeds and accelerations attained during the six experiments. The vehicle successfully reaches speeds of 10 m/s and accelerations of 8 m/s^2 in the outdoor environment and speeds of 3 m/s and accelerations of 5 m/s^2 in the indoor environment. Figure 6 also shows regions where the operator's selected trajectory would bring the vehicle closer than r meters to an obstacle. In all such cases, the operator's

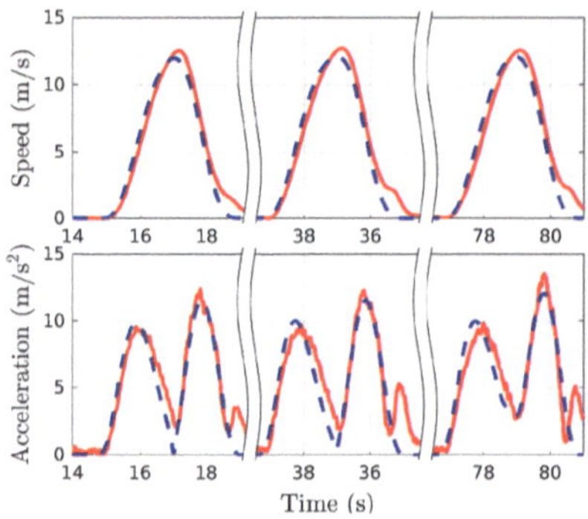

Fig. 4. Desired (dashed blue) vs. estimated (solid red) speed and acceleration achieved during high speed flight in experiment **Outdoor-1**. A maximum acceleration of $13.5\,\mathrm{m/s^2}$ was attained.

trajectory is pruned and a collision-free trajectory is selected. Figure 5 shows an example instance of motion primitive pruning to avoid a collision, along with the local map generated by the vehicle.

Computational Performance Analysis. Table 3 shows that our trajectory generation and pruning time is faster than the user input rate (10 Hz) and that our map generation time is faster than the sensor input rate (30 Hz), enabling real time operation. Furthermore, Table 4 indicates that the in-flight computational footprint of the proposed system is less than 75% of the available onboard capacity.

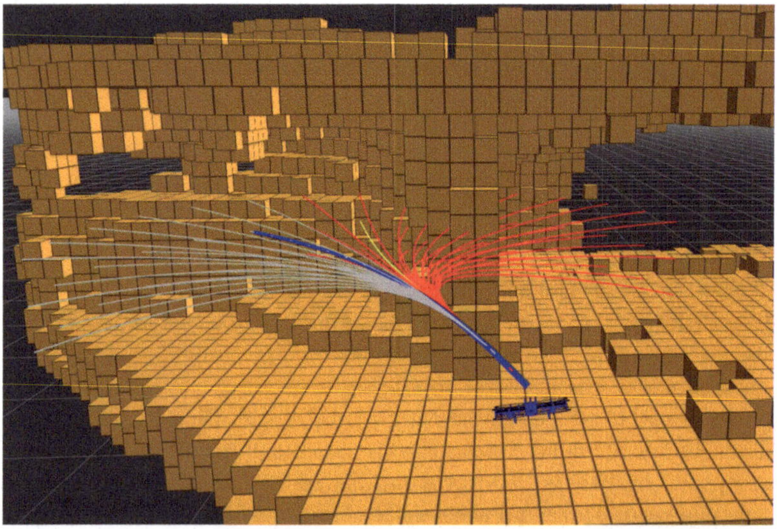

Fig. 5. A snapshot of the map and motion primitive library during an indoor experiment when the user selected trajectory (yellow) is not chosen to avoid a collision.

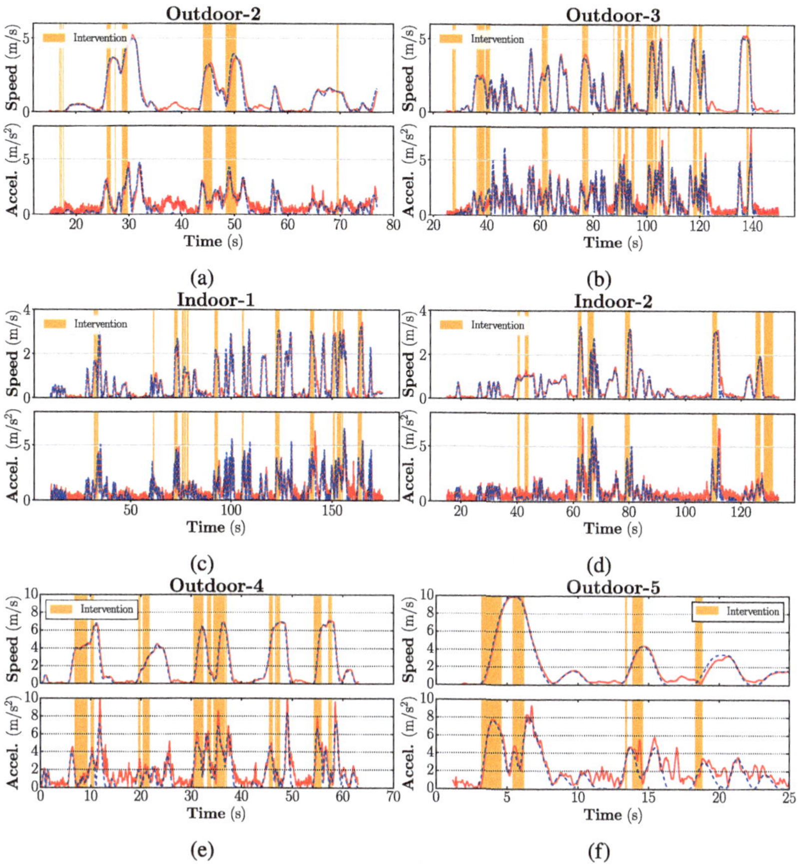

Fig. 6. Desired (dashed blue) vs. estimated (solid red) speed and acceleration achieved during our six collision avoidance experiments. During orange regions, the operator's trajectory was not selected in order to keep a minimum distance to surrounding obstacles. The system accurately tracks trajectory references and avoids obstacles while reaching speeds of up to 10 m/s, and accelerations up to 8 m/s^2.

Table 3. Execution time (ms) and std. dev. per iteration for safe teleoperation during some of our experiments

Experiment	Trajectory generation	Trajectory pruning	Local map generation
Outdoor-2	0.47 ± 0.072	29.37 ± 24.03	5.72 ± 5.54
Outdoor-3	0.48 ± 0.071	23.65 ± 15.42	5.84 ± 7.95
Indoor-1	0.50 ± 0.125	8.34 ± 5.19	11.67 ± 14.83
Indoor-2	0.51 ± 0.219	15.73 ± 7.23	12.71 ± 16.78

Table 4. CPU usage (%, out of a total available 600%) and std. dev. on a 6 core NVIDIA TX2

Experiment	Safe teleoperation	VINS Mono	Total
Outdoor-2	32.92 ± 17.59	67.45 ± 10.87	395.72 ± 43.53
Outdoor-3	26.70 ± 19.98	56.19 ± 19.47	359.89 ± 60.87
Indoor-1	57.38 ± 21.25	64.21 ± 3.42	450.40 ± 43.14
Indoor-2	58.76 ± 24.09	63.82 ± 6.41	456.47 ± 36.83

5 Conclusion and Future Work

In this work, we present a system architecture designed for collision-free, agile autonomous flight through cluttered environments. Our hexarotor vehicle is capable of flying at speeds exceeding 12 m/s and with accelerations exceeding $12\,\mathrm{m/s}^2$, as shown in **Outdoor-1**. The teleoperation and collision avoidance system presented has been shown to consistently avoid obstacles while traveling at speeds up to 10 m/s in an outdoor environment (**Outdoor-5**) and in a dimly lit garage (**Indoor-1**, **Indoor-2**).

Although we've attained relatively high levels of performance, there are limitations to the architecture. Currently, motion primitive durations are chosen by the operator. If the motion primitive duration is set too low, the vehicle will not be able to meet the desired acceleration and performance will suffer. Consequently, it is difficult to attain high speeds while remaining responsive at low speeds. A natural extension of the approach is to choose motion primitive durations in a principled manner to ensure that generated references are dynamically feasible. We also believe that chaining together motion primitives to generate more complex reference paths is an exciting avenue for future work. This would enable the vehicle to execute more complicated maneuvers, particularly in cluttered environments.

References

1. Chen, J., Liu, T., Shen, S.: Online generation of collision-free trajectories for quadrotor flight in unknown cluttered environments. In: Proceedings of the IEEE International Conference on Robotics and Automation, Stockholm, Sweden (2016)
2. Delmerico, J., Scaramuzza, D.: A benchmark comparison of monocular visual-inertial odometry algorithms for flying robots. In: Proceedings of the IEEE International Conference on Robotics and Automation, Brisbane, Australia (2018)
3. Dey, D., Shankar, K.S., Zeng, S., Mehta, R., Agcayazi, M.T., Eriksen, C., Daftry, S., Hebert, M., Bagnell, J.A.: Vision and learning for deliberative monocular cluttered flight. In: Field and Service Robotics, pp. 391–409. Springer (2016)
4. Florence, P., Carter, J., Tedrake, R.: Integrated perception and control at high speed: evaluating collision avoidance maneuvers without maps. In: Workshop on the Algorithmic Foundations of Robotics, San Francisco, USA (2016)

5. Florence, P.R., Carter, J., Ware, J., Tedrake, R.: NanoMap: fast, uncertainty-aware proximity queries with lazy search over local 3D data. In: Proceedings of the IEEE International Conference on Robotics and Automation, Brisbane, Australia (2018)
6. Liu, S., Watterson, M., Tang, S., Kumar, V.: High speed navigation for quadrotors with limited onboard sensing. In: 2016 IEEE International Conference on Robotics and Automation (ICRA), pp. 1484–1491 (2016). https://doi.org/10.1109/ICRA.2016.7487284
7. Lopez, B.T., How, J.P.: Aggressive 3-D collision avoidance for high-speed navigation. In: Proceedings of the IEEE International Conference on Robotics and Automation, Singapore, pp. 5759–5765 (2017)
8. Matthies, L., Brockers, R., Kuwata, Y., Weiss, S.: Stereo vision-based obstacle avoidance for micro air vehicles using disparity space. In: Proceedings of the IEEE International Conference on Robotics and Automation, pp. 3242–3249, Hong Kong, China (2014)
9. Mellinger, D., Kumar, V.: Minimum snap trajectory generation and control for quadrotors. In: 2011 IEEE International Conference on Robotics and Automation (ICRA), pp. 2520–2525. IEEE (2011)
10. Pivtoraiko, M., Nesnas, I.A., Kelly, A.: Autonomous robot navigation using advanced motion primitives. In: Proceedings of the IEEE Aerospace Conference, pp. 1–7. Big Sky (2009)
11. Qin, T., Li, P., Shen, S.: Vins-Mono: a robust and versatile monocular visual-inertial state estimator. IEEE Trans. Robot. **34**(4), 1004–1020 (2018). https://doi.org/10.1109/TRO.2018.2853729
12. Yang, X., Sreenath, K., Michael, N.: Online adaptive teleoperation via motion primitives for mobile robots. In: Special Issue on Learning for Human-Robot Collaboration, Autonomous Robots 2018. Springer (2018)

Path Planning Based on Differential Kinematics for Passing Through Small Opening by Transformable Multilinked Aerial Robot

Moju Zhao(✉), Tomoki Anzai, Fan Shi, Kei Okada, and Masayuki Inaba

JSK Lab Engineering Building 2, The University of Tokyo,
7-3-1 Hongo, Bunkyo-ku, Tokyo, Japan
chou@jsk.imi.i.u-tokyo.ac.jp
http://www.jsk.t.u-tokyo.ac.jp/index.html

Abstract. In this paper, we propose a path planning method allowing a transformable multilinked aerial robot to pass through small openings. The majority of the related studies have used under-actuated and unfoldable multirotors to enable passing through narrow gaps by performing an aggressive motion to compensate for disadvantage of underactuation. However, this approach is challenging in an unknown environment since an aerial robot needs to run-up and decelerate to perform agile maneuvers. Furthermore, the unfoldable frame cannot allow the robot to squeeze through small openings. To overcome these limitations, we have developed a transformable aerial robot with a multilinked structure in our latest work, along with a motion planning algorithm using a sampling-based method. In spite of the probabilistic completeness, the sampling-based method cannot guarantee the real-time property since the dimension of the configuration space related to the multilinked model is high. Thus, in this work, we apply differential kinematics to achieve instantaneous generation of the valid path by finite numerical integration. To validate our approach, we present an experiment employing a multilinked aerial robot composed from four links to perform online motion planning for a tight horizontal opening and track the desired squeezing trajectory instantaneously.

Keywords: Transformable multirotor · Path planning · Differential kinematics

---

Electronic supplementary material The online version of this chapter (https://doi.org/10.1007/978-3-030-33950-0_46) contains supplementary material, which is available to authorized users.

J. Xiao et al. (Eds.): ISER 2018, SPAR 11, pp. 536–548, 2020.
https://doi.org/10.1007/978-3-030-33950-0_46

1 Introduction

Recent studies on multirotor aerial robots have demonstrated the ability of the robots to perform autonomous exploration in unknown environments [1]. The collision avoidance in these studies is considered as one of the important issues in path planning. In the case of traversing through small openings in a complex environment, the collision avoidance problem becomes more challenging since the free space between the area of an opening and the robot's body is small. Previous studies on maneuvering in a complex environment have used under-actuated multirotors to enable dynamic passing through narrow openings by rotating the robot's body [2,3]. However, the under-actuated property cannot allow the aerial robot to move slowly while the state is not in the near-hover condition. This is the reason why these previous studies have focused only on the aggressive maneuvering. Furthermore, the following two crucial preconditions are required. First, a sufficient space is required for running-up before passing and decelerating after passing; otherwise, the robot might collide with the unknown objects after exiting the opening. In this regard, the majority of the previous studies have assumed that no obstacles lie behind the opening [4], which is not common in real-life scenarios. Second, the majority of small openings are assumed to be narrow vertical gaps, and thus, the aerial robot can fly through a gap by rotating its body. However, if the opening is completely closed and the area is smaller than the arbitrary cross section of the multirotor model, then it is entirely impossible for the robot to pass through.

To become free from these two preconditions, the current studies focus on the following two features.

Fully/Over-Actuation: If more than six actuators are employed on a rigid frame, it is possible to achieve control of the full pose in $\mathsf{SE}(3)$ and hover at arbitrary attitude. For example, an omni-directional aerial robot composed from a long rod and six asymmetrically aligned rotors is proposed by Park [5], whereas in [6], one DoF force vectoring mechanism is applied for each rotor and a quadrotor model is constructed. These special robots can slowly traverse through narrow vertical gaps because of the fully/over-actuated platform, reducing the risk to collide with unknown obstacles in a cluttered environment. However, these robots are still required to meet the second precondition since they cannot traverse through closed openings smaller than the cross section of the robot's model.

Folding Mechanism: Special folding mechanisms are proposed to negotiate with narrow gaps or small openings. Riviere [7] has developed a quad-rotor to enable folding in a narrow and elongated configuration, which allows flying through narrow vertical gaps. However, it is associated with aggressive motion, and the tight horizontal openings are not possible to negotiate. As a different mechanical design drawing inspiration from the ability of a snake to slowly pass through a small hole by changing its shape, a two-dimensional multilinked structure was applied for a multirotor in our previous work [8] allowing passing through narrow vertical gaps. However, this multilinked robot is an under-actuated model, indicating the difficulty to negotiate with completely vertical openings.

Fig. 1. (a) A quad type transformable aerial robot **DRAGON: D**ual-rotor embedded multilink **R**obot with the **A**bility of multi-de**G**ree-of-freedom aerial transformati**ON**; (b) the aerial transformation.

To improve versatility of the multilinked robot, we developed an advanced transformable aerial robot called **DRAGON** (**D**ual-rotor embedded multilink **R**obot with the **A**bility of multi-de**G**ree-of-freedom aerial transformati**ON**) in our latest work [9] as shown in Fig. 1; it integrates the advantages of over-actuation and the folding mechanism.The subsequent issue is how to find a valid path for this multilinked aerial robot to traverse arbitrary openings. In addition to collision avoidance, other constraints such as joint angle limitations and the flight stability should be taken into consideration. Among various searching method, the sampling-based method [10] is one of the most common algorithms because of the probabilistic completeness. In our previous work [11], the sampling-based method with optimization property called RRT* [12] was applied to search the transformation path However, the crucial problem with sampling-based methods is a long time required to find a solution. In [11], a tight vertical opening in the middle between a ground floor and a closed ceiling was designed, and the time to search path using a high-performance computer was around 10[h] for the four-links model. This is a huge handicap to perform the online processing inside the robot while flying.

Thus, in this paper, we present a different path searching method based on differential kinematics [13], which aims to ensure immediacy of online path planning before performing the desired motion. The main feature of the proposed method is the numerical integration of infinitesimal motion. Each infinitesimal motion is calculated on the basis of differential kinematics related to the multilinked structure. Although optimization for each segment (i.e., infinitesimal motion) is performed, such numerical integration cannot ensure global optimization and the solution is not complete. This means, the path planning might be stuck in the local minimum point. Hence, in the case of squeezing through a small opening, we focus on the fact that all points of the multilinked model must pass the opening (i.e., the cross section of the opening) once, and an appropriate Cartesian error (i.e., target infinitesimal motion) can be further added at the point next to the point just on the cross section. This is similar to the motion of gradually passing a string through a tiny hole by pinching the part that is close to the hole. Nevertheless, this planning method is limited to the environment with single-layer opening or gap.

2 Technical Approach

As proposed in our previous studies [9,11], aerial transformation is assumed to be slow enough, so that at each time point, the multilinks can be regarded as a special rigid body, the rotor configuration and inertial parameters of which vary. According to this assumption, multilinked dynamics and the proposed motion planning algorithm are derived subsequently.

2.1 Kinematics and Dynamics

The counter rotating dual-rotor unit is employed as depicted in Fig. 2(a) to counteract the drag moment and gyroscopic moment by applying the same speed to each rotor. The dual-rotor unit can be further tilted by the 2DoF vectoring mechanism (i.e., the first rotating axis ϕ around the main rod and the second axis θ around the dual-rotor connection rod). The integrated thrust force f_{T_i} generated by the dual-rotor unit is then used instead of the individual rotor thrust force. As shown in Fig. 2(b), the vectored thrust force of the i-th rotor module ${}^{\{CoG\}}\boldsymbol{f}_i$ w.r.t. the frame $\{CoG\}$ can be defined as:

$$ {}^{\{CoG\}}\boldsymbol{f}_i = f_{T_i}\,{}^{\{CoG\}}\boldsymbol{b}_i = f_{T_i}\,{}^{\{CoG\}}R_{\{L_i\}}(\boldsymbol{q})\,{}^{\{L_i\}}R_{\{F_i\}}(\theta_i, \phi_i)\boldsymbol{b} \tag{1} $$

where $\boldsymbol{q} \in \mathcal{R}^{2(N-1)}$ denotes the joint angles, and ${}^{\{CoG\}}\boldsymbol{f}_i$ can be divided into the norm f_{T_i} and the unit vector ${}^{\{CoG\}}\boldsymbol{b}_i$. The total wrench can be further summarized as follows:

$$ \begin{pmatrix} {}^{\{CoG\}}\boldsymbol{f} \\ {}^{\{CoG\}}\boldsymbol{\tau} \end{pmatrix} = \begin{pmatrix} \displaystyle\sum_{i=1}^{N} {}^{\{CoG\}}\boldsymbol{f}_i \\ \displaystyle\sum_{i=1}^{N} {}^{\{CoG\}}\boldsymbol{p}_i \times {}^{\{CoG\}}\boldsymbol{f}_i \end{pmatrix} = \begin{pmatrix} B \\ P \times B \end{pmatrix} \boldsymbol{f}_T = Q\boldsymbol{f}_T \tag{2} $$

$$ B = \left({}^{\{CoG\}}\boldsymbol{b}_1\ {}^{\{CoG\}}\boldsymbol{b}_2 \cdots {}^{\{CoG\}}\boldsymbol{b}_N \right); P = \left({}^{\{CoG\}}\boldsymbol{p}_1\ {}^{\{CoG\}}\boldsymbol{p}_2 \cdots {}^{\{CoG\}}\boldsymbol{p}_N \right); $$

$$ \boldsymbol{f}_T = \left(f_{T_1}\ f_{T_2} \cdots f_{T_N} \right)^{\mathrm{T}} $$

where the i-th column of $P \times B$ is defined as ${}^{\{CoG\}}\boldsymbol{p}_i \times {}^{\{CoG\}}\boldsymbol{b}_i$. The configuration matrix Q varies because of the joint angles $\boldsymbol{q}$; it is also influenced by the control input of the rotor tilt angles $\boldsymbol{\varphi} \in \mathcal{R}^N$ and $\boldsymbol{\theta} \in \mathcal{R}^N$.

Given that the flight motion to pass through an opening is considered to be slow, the multilinked model can be approximated as a single rigid body with variable inertial parameters under the assumption that all joints rotate slowly. Thus, the dynamics can be simplified as follows:

$$m_{\sum}{}^{\{W\}}\ddot{\boldsymbol{r}}_{\{CoG\}} = {}^{\{W\}}R_{\{CoG\}}{}^{\{CoG\}}\boldsymbol{f} - m_{\sum}\boldsymbol{g} \quad (3)$$
$${}^{\{CoG\}}I_{\sum}(\boldsymbol{q})^{\{CoG\}}\dot{\boldsymbol{\omega}}_{\{CoG\}} + {}^{\{CoG\}}\boldsymbol{\omega}_{\{CoG\}} \times {}^{\{CoG\}}I_{\sum}(\boldsymbol{q})^{\{CoG\}}\boldsymbol{\omega}_{\{CoG\}} = {}^{\{CoG\}}\boldsymbol{\tau} \quad (4)$$

where the inertia tensor ${}^{\{CoG\}}I_{\sum}(\boldsymbol{q})$ is variable. Regarding this approximated dynamics model, the corresponding control method is derived in the near-hover condition proposed in our previous works [9,11].

2.2 Path Planning Based on Differential Kinematics

Numerical Iteration of Optimal Infinitesimal Motion. The path to traverse through a gap can be regarded as a sequence composed of discrete states. Each state can be defined as $\boldsymbol{\xi}_k = ({}^{\{W\}}R_{\{L_1\}}(k), {}^{\{W\}}\boldsymbol{r}_{\{L_1\}}(k), \boldsymbol{q}(k))$. The first two components denote to the orientation and the position of the first (i.e., root) link $\{L_1\}$ w.r.t the world frame $\{W\}$, and the last one is the vector of the joint angles. The main planning procedure based on differential kinematics can be summarized as the repetition of the following two steps:

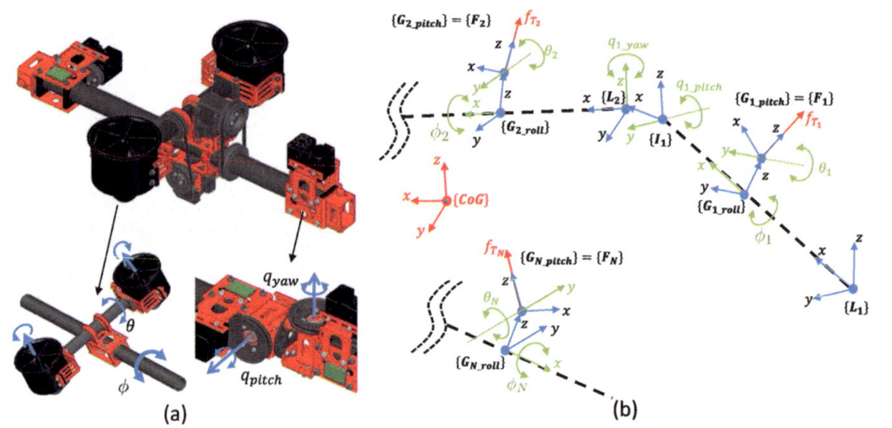

Fig. 2. (a) The link rotors is validated as follows-rotor unit and the joint module; (b) the kinematics model of the multilinked structure. The green arrows in (b) indicate the rotation axes associated with both the joint angles (q_{i_yaw}, q_{i_pitch}) and tilt angles (θ_i, ϕ_i). The origin of the frame $\{CoG\}$ is the center of gravity of the multilinked structure.

1. Find optimal infinitesimal motion $\delta\boldsymbol{\xi}$ from following optimization problem:

$$\min_{\delta\boldsymbol{\xi}} \quad f(\delta\boldsymbol{\xi}, \delta\boldsymbol{x}_k, \boldsymbol{\xi}_k) \quad (5)$$
$$s.t. \quad \underline{\boldsymbol{b}}(\boldsymbol{\xi}_k) \leq g(\delta\boldsymbol{\xi}, \boldsymbol{\xi}_k) \leq \overline{\boldsymbol{b}}(\boldsymbol{\xi}_k) \quad (6)$$

where $f()$ and $g()$ are the generalized cost and constraint functions, respectively, and $\boldsymbol{\xi}_k$ and $\delta \boldsymbol{x}_k$ are the current state and target infinitesimal value (e.g., Cartesian motion of a target point), respectively.

2. Update the state by adding the infinitesimal motion to the current motion:

$$\boldsymbol{\xi}_{k+1} = \boldsymbol{\xi}_k + \delta \boldsymbol{\xi} \tag{7}$$

Generally, the optimization problem (Eqs. 5 and 6) is highly nonlinear. However, the majority of the nonlinear functions can be approximated by quadratic forms around the infinitesimal motion. Hence, we derive the cost and constraint terms to quadratic form, allowing to obtain a quadratic programming (QP) problem.

Infinitesimal Cartesian Motion for the Link Across the Opening. The solution of the optimization problem (Eqs. 5 and 6) can only provide the local minimum of each infinitesimal motion, implying that the completeness of the global solution is not guaranteed. For example, if we consider the end of last link as the reference point $\boldsymbol{P}_e$ and add the infinitesimal Cartesian motion ($\delta \boldsymbol{x}_{\boldsymbol{P}_e}$) at this reference point, the general framework of inverse kinematics [13] can be applied to the traversing problem in this work. However, such Cartesian motion at the $\boldsymbol{P}_e$ might make the iteration process stuck when encountering an obstacle. Thus, an appropriate guidance is necessary to enable passing through a small opening in this work, and we draw the inspiration from the motion of gradual passing of a string through a tiny hole by pinching the string's part that is close to the hole. This is formulated by updating the reference point $\boldsymbol{P}_k$ on every iteration based on the rule as shown in Fig. 3. The frame $\{O\}$ is aligned at the center of the cross section of the opening, whereas the reference point $\boldsymbol{P}_k$ is next to the point $\boldsymbol{P}_{cross}$ that is just on the cross section. The target infinitesimal Cartesian motion $\delta \boldsymbol{x}_k$ can be further obtained, as a vector from the reference point $\boldsymbol{P}_k$ to the origin of the frame $\{O\}$.

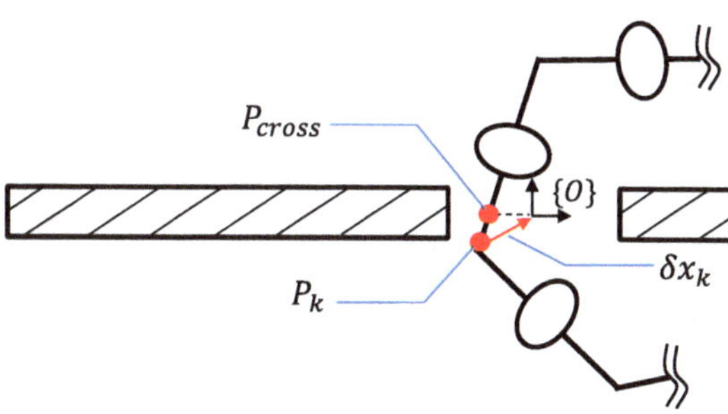

Fig. 3. The desired Cartesian motion $\delta \boldsymbol{x}_k$ at the point $\boldsymbol{P}_k$ for incremental passing through an opening.

Quadratic Form Costs and Constraints. The linear relationship between the infinitesimal change of the state $\delta\boldsymbol{\xi}$ and the Cartesian motion $\delta\boldsymbol{x}_k$ as illustrated in Fig. 3 can be written as $\delta\boldsymbol{x}_k = J(\boldsymbol{\xi}_k)\delta\boldsymbol{\xi}_k$, where $J(\boldsymbol{\xi}_k)$ is the Jacobian obtained from the multilinked kinematics. Although this can be considered as an equality constraint, the following translational cost is introduced aiming to relax the influence of this Cartesian error and balance with other constraints:

$$\min_{\delta\boldsymbol{\xi}} \left[(J(\boldsymbol{\xi}_k)\delta\boldsymbol{\xi} - \delta\boldsymbol{x}_k)^{\mathrm{T}} W_1 (J(\boldsymbol{\xi}_k)\delta\boldsymbol{\xi} - \delta\boldsymbol{x}_k) + \delta\boldsymbol{\xi}^{\mathrm{T}} W_2 \delta\boldsymbol{\xi}\right] \tag{8}$$

where the positive definite diagonal weight matrices W_1 and W_2 balance the cost of the Cartesian error and the robot's differential motion (i.e., the twist of root link and the joint velocity).

On the other hand, the constraints for this optimization problem can be classified into following six types:

1. **State bounds.** The state bounds can be described as $\boldsymbol{\xi} \in [\boldsymbol{\xi}_{min}, \boldsymbol{\xi}_{max}]$. We apply the velocity damper [14] to set the following inequality constraint:

$$\delta\xi_j^- \leq \delta\xi_j \leq \delta\xi_j^+, \quad j \in \{1, \cdots, n\} \tag{9}$$

$$\delta\xi_j^- = \begin{cases} \delta\xi_j^{min} \frac{(\xi_j - \xi_j^{min}) - \xi_F}{\xi_R - \xi_F} & (\text{if} \quad \xi_j - \xi_j^{min} < \xi_R) \\ \delta\xi_j^{min} & (\text{otherwise}) \end{cases} \tag{10}$$

$$\delta\xi_j^+ = \begin{cases} \delta\xi_j^{max} \frac{(\xi_j^{max} - \xi_j) - \xi_F}{\xi_R - \xi_F} & (\text{if} \quad \xi_j^{max} - \xi_j < \xi_R) \\ \delta\xi_j^{max} & (\text{otherwise}) \end{cases} \tag{11}$$

where ξ_j^{min} and ξ_j^{max} are the lower and upper limits of the j-th state, respectively, and $\delta\xi_j^{min}$ and $\delta\xi_j^{max}$ are the lower and upper limits of the infinitesimal change of the corresponding state, respectively. ξ_R and ξ_F denote the ranges where the state is restricted and forbidden, respectively.

2. **Collision avoidance.** The Velocity damper is applied again to set the collision avoidance constraint. The constraint can be written as:

$$\delta d \geq -\epsilon \frac{d_k - d_F}{d_R - d_F} \qquad \text{if} \quad d_k < d_F \tag{12}$$

where d_k is the shortest distance between two objects, ϵ is the damping gain parameter, d_R and d_F play the same role as described in Eqs. 10 and 11. Considering that $\boldsymbol{P}_1$ and $\boldsymbol{P}_2$ are the nearest points on each object and $\boldsymbol{n}$ is the unit vector pointing from $\boldsymbol{P}_1$ to $\boldsymbol{P}_2$, δd can be further written as:

$$\delta d = \boldsymbol{n}(J_{\boldsymbol{P}_1} - J_{\boldsymbol{P}_2})\delta\boldsymbol{\xi} \tag{13}$$

where $J_{\boldsymbol{P}_1}$ and $J_{\boldsymbol{P}_2}$ are the Jacobian of $\boldsymbol{P}_1$ and $\boldsymbol{P}_2$, respectively. The object referred here could be a robot link or an obstacle in the environment.

3. **Hovering thrust force.** Besides the avoidance of singular forms, it is also necessary to ensure that thrust force is always within the valid range generated by the thrust rotor. As the desired motion is considered slow, the hovering thrust can be employed. As proposed in our previous work [11], the hovering thrust can be calculated as:

$$\boldsymbol{f}_{T_s}(\boldsymbol{\xi}) = Q(\boldsymbol{\xi})^{\mathrm{T}}(Q(\boldsymbol{\xi})Q(\boldsymbol{\xi})^{\mathrm{T}})^{-1}\boldsymbol{g} = Q(\boldsymbol{\xi})^{\#}\boldsymbol{g} \tag{14}$$
$$\boldsymbol{g} = \begin{bmatrix} 0 \; 0 \; mg \; 0 \; 0 \; 0 \end{bmatrix}^{\mathrm{T}}$$

where the matrix Q is corresponding to Eq. 2 and mg denotes the gravity force from the robot. The bounds for the hovering thrust are introduced by $\boldsymbol{f}_{T_s} \in [\boldsymbol{f}_{min}, \boldsymbol{f}_{max}]$, and the same velocity damper method as presented in Eq. 10 is applied to calculate the bounds for the infinitesimal motion ($\delta \boldsymbol{f}_{T_s}^-$, $\delta \boldsymbol{f}_{T_s}^+$). Given the infinitesimal motion, we extract the first-order term from the Taylor expansion of Eq. 14 to achieve linearization. Thus, the related constraint can be described as:

$$\delta \boldsymbol{f}_{T_s}^- \le \frac{\partial Q_3^{\#}(\boldsymbol{\xi}_k)}{\partial \boldsymbol{\xi}} mg\delta\boldsymbol{\xi} \le \delta \boldsymbol{f}_{T_s}^+ \tag{15}$$

where $Q_3^{\#}(\boldsymbol{\xi}_k)$ denotes the third row vector of this matrix.

4. **Stability margin.** If all links are aligned in a straight line, i.e., all joints are 0 [rad], the configuration becomes singular because the motion around this line is uncontrollable. However, the singular form cannot be detected by the calculation of the hovering thrust state from Eq. 14, because the matrix Q is always of the full rank. On the basis of the decoupled control methods proposed by [9], we simplify this problem by focusing only on the controllability of the roll and pitch motion regarding the CoG frame, and the major components of the moment are generated from $\sum_{i=1}^{N} {}^{\{CoG\}}p_{i_y} f_{T_i}$ and $\sum_{i=1}^{N} {}^{\{CoG\}}p_{i_x} f_{T_i}$ respectively. Thus, the stability margin for the roll and pitch motion is associated with the planar distribution of 2D points (${}^{\{CoG\}}p_{i_x}$, ${}^{\{CoG\}}p_{i_y}$). In terms to statistics, the stability margin corresponds to the correlation of these points, and the lower the correlation is, the higher the stability margin will be. The following variance-covariance matrix is further introduced:

$$\begin{pmatrix} \frac{1}{N}\sum_{i=1}^{N}(p_{i,x}-\bar{p}_x)(p_{i,x}-\bar{p}_x) & \frac{1}{N}\sum_{i=1}^{N}(p_{i,x}-\bar{p}_x)(p_{i,y}-\bar{p}_y) \\ \frac{1}{N}\sum_{i=1}^{N}(p_{i,x}-\bar{p}_x)(p_{i,y}-\bar{p}_y) & \frac{1}{N}\sum_{i=1}^{N}(p_{i,y}-\bar{p}_y)(p_{i,y}-\bar{p}_y) \end{pmatrix} = T^{\mathrm{T}} \begin{pmatrix} S_1 & 0 \\ 0 & S_2 \end{pmatrix} T \tag{16}$$

where $\bar{p}_x$ and $\bar{p}_y$ are the averages of $p_{i,x}$ and $p_{i,y}$ respectively, S_1 and S_2 ($S_1 < S_2$) are the eigenvalues. We focus on the smaller eigenvalue S_1, since it reflects the correlation of all points. Then, the following dimensionless value is introduced to present the stability margin:

$$\gamma_{-} < \frac{\sqrt{S_1}}{l_{link}} \tag{17}$$

where l_{link} is the length of the single link, and γ_{-} is the dimensionless threshold. The same linearization presented in Eq. 15 can be applied.

5. **Overlapping of rotors in different links.** If there are differences in the height among the links, the rotors in different links may overlap with each other. In this case, air flow from the upper rotor would significantly degrade the performance of the lower one, indicating the necessity to avoid such form. Moreover, since the rotor vectoring control actively changes the tilt angle of rotors (e.g. <0.1$[rad]$), air flow does not always blow vertically. Therefore, a circular truncated cone is introduced to consider the diffusion of air flow and the overlapping of the i-th rotor and j-th rotors is validated as follows:

$$\sqrt{(p_{r_i,x} - p_{r_j,x})^2 + (p_{r_i,y} - p_{r_j,y})^2} > \|p_{r_i,z} - p_{r_j,z}\| tan\beta + 2r_{rotor} \tag{18}$$

Note that, ${}^{\{CoG\}}\boldsymbol{p}_{r_i}$ is the true position of rotor, which is different from the thrust generation point ${}^{\{CoG\}}\boldsymbol{p}_i$. The first term on the right-hand side represents the diffusion influence of the rotor max tilt angle β, whereas r_{rotor} in the second term denotes the radius of the rotor. If we consider all combination of rotors (${}_{2N}\mathrm{C}_2 - N$), the corresponding constraint matrix would become very large as the link number increases. Hence, only the closest pair is selected to derive the corresponding linearized inequality presented in Eq. 15.

6. **Inclination of links.** The calculation of the desired tilt angles ϕ_i and θ_i for the i-th dual-rotor unit involves the inverse trigonometric function tan^{-1}. This indicates the unsolvable case for the i-th dual-rotor unit due to $tan^{-1}(0,0)$ as the i-th link is vertical towards the world frame $\{W\}$. Such region can be avoided by calculating the pitch angle ${}^{\{W\}}\boldsymbol{\alpha}_{\{L_1\},y}$ of the i-th link according to forward kinematics by the given parameters $({}^{\{W\}}\boldsymbol{\alpha}_{\{L_1\}}, \boldsymbol{q})$. Then, the following inequality is applied and the same linearization around the infinitesimal motion $\delta\boldsymbol{\xi}$ is performed:

$${}^{\{W\}}\boldsymbol{\alpha}_{\{L_1\},y}(\boldsymbol{\xi}) < \alpha_{+} \tag{19}$$

To summarize, the proposed path planning method can be represented as:

$$\begin{gathered} \min_{\delta\boldsymbol{\xi}} \left[(J(\boldsymbol{\xi}_k)\delta\boldsymbol{\xi} - \delta\boldsymbol{x}_k)^{\mathrm{T}} W_1 (J(\boldsymbol{\xi}_k)\delta\boldsymbol{\xi} - \delta\boldsymbol{x}_k) + \delta\boldsymbol{\xi}^{\mathrm{T}} W_2 \delta\boldsymbol{\xi} \right] \\ s.t. \quad Eq.9, Eq.12, Eq.15, Eq.17, Eq.18, Eq.19 \\ \boldsymbol{\xi}_{k+1} = \boldsymbol{\xi}_k + \delta\boldsymbol{\xi} \end{gathered} \tag{20}$$

Besides the iterative update of the state $\boldsymbol{\xi}$, the reference point $\boldsymbol{P}_k$ is also changed after each iteration as depicted in Fig. 3. If the valid point $\boldsymbol{P}_k$ can no longer be assigned, i.e., the last reference point $\boldsymbol{P}_k$ is the end of the last link, it can be considered that the motion of squeezing through the opening has been completed.

3 Experimental Results

The performance of the proposed planning method is validated using the four-link aerial robot shown in Fig. 1. The main component of the experimental environment is a small horizontal opening, the width of which is 0.6[m], as depicted in Fig. 4. Although the size of this opening is smaller than the size of the transformable robot under the normal form (0.8[m]), some small under-actuated multirotors can still pass through this opening. However, the essential point is the relative difference between the opening and the robot. Besides, we assume the existence of the floor and ceiling, leading to a more complex environment compared with the related studies. The experimental procedure involves online planning by the on-board processor inside the robot. The configuration of the target opening is considered well known; thus, object recognition is omitted. The opening width in the planning problem is set as 0.5[m] to provide enough margin (i.e., 0.05[m]) for collision avoidance. The CPU of the on-board processor is Intel Atom x7-8700 Quad-Core. Besides path planning, flight control is also performed by the same on-board processor, which partially depends on the external motion capture system to localize the position and the yaw angle of the baselink (i.e., link2).

The snapshots in Fig. 5 demonstrate the successful result of passing through the opening. As the proposed planning method can only generate a discrete path composed from key poses (i.e., $\boldsymbol{\xi}(k), (k = 1, \cdots, n)$), the subsequent process to obtain a continuous path (i.e., $\boldsymbol{\xi}(t)$) is necessary. The uniform B-spline curve [15] is applied in this work, and the implementation detail can be found in our previous work [11]. Online planning is performed after the takeoff; the total computation time is about 5.8[s]. The green path in Fig. 5 is the desired trajectory of the $\{CoG\}$ frame, which is a part of the result of $\boldsymbol{\xi}(t)$.

To guarantee the assumption that the flight motion is slow, the duration of the whole motion sequence is 80[s], and the details of the path tracking control method can be also found in our previous work [11]. To validate the trackability towards the desired path, we consider the CoG motion because the robot can be regarded as a rigid body at each time point. Figure 6(a)–(b) and (c)–(d)

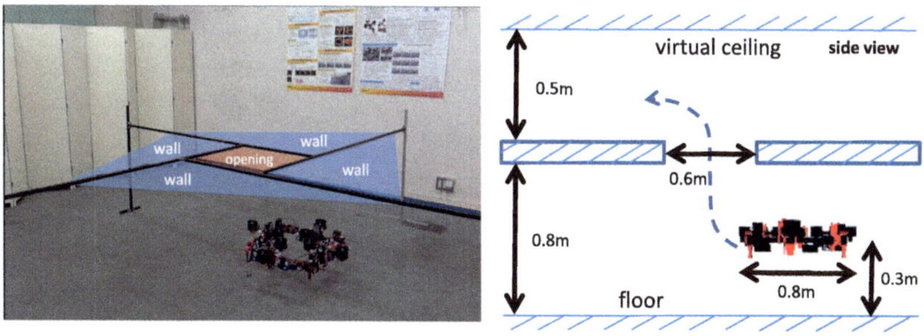

Fig. 4. The configuration of the tight horizontal opening compared with the robot. The ceiling and the floor are also taken into consideration during planning.

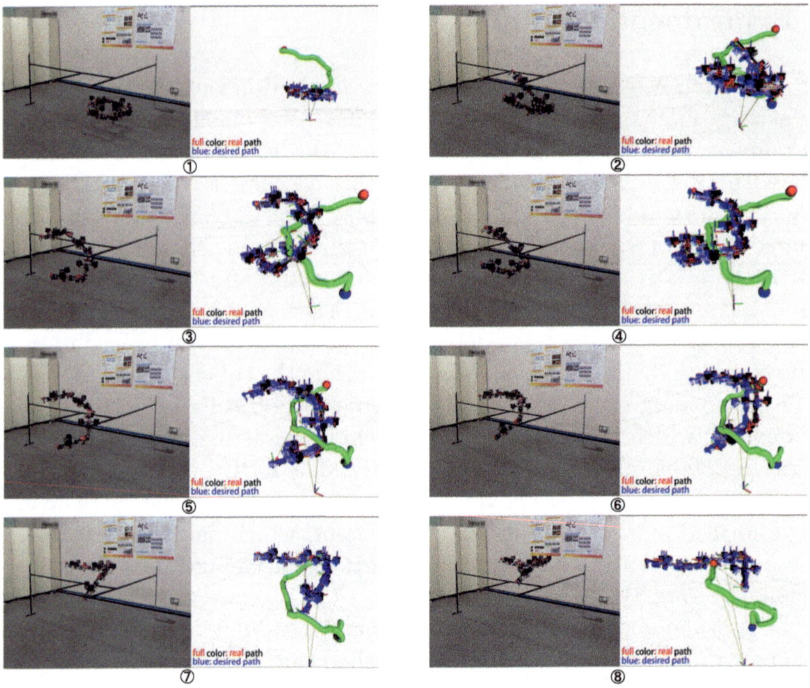

Fig. 5. Snapshots of the motion sequence of passing through the horizontal opening with aerial transformation. The right images in each figure unit show the difference between the desired path and the real path.

show the rotational and translational trackability regarding the $\{CoG\}$ frame, respectively, whereas Fig. 6(e) plots the joint angles. The RMS tracking errors with regard to the CoG motion are r_x: 0.042[m], r_y: 0.063[m], r_z: 0.015[m], α_x: 0.023[rad] α_y: 0.015[rad], α_z: 0.030[rad]. Hence, the difference between the true and desired trajectories is relatively small. As shown in Fig. 6(b) and (d), the tracking errors of the horizontal and yaw motions are relatively large, i.e., the maximum errors are 0.085[m], 0.11[m] and 0.12[rad], respectively. We consider the main reason for this is that these states are controlled by the rotor tilting control method as proposed in [9], and this controller cannot guarantee enough trackability for the desired motion speed requested by the planned path. One of the straightforward solutions is to enlarge the motion duration (e.g., 160[s]). However, the flight time for this robot is limited to 150[s], thus, improvement of the mechanical design to extend the flight time is required. Nevertheless, feasibility of the proposed planning method and the corresponding flight motion required to pass through a complex opening has been confirmed.

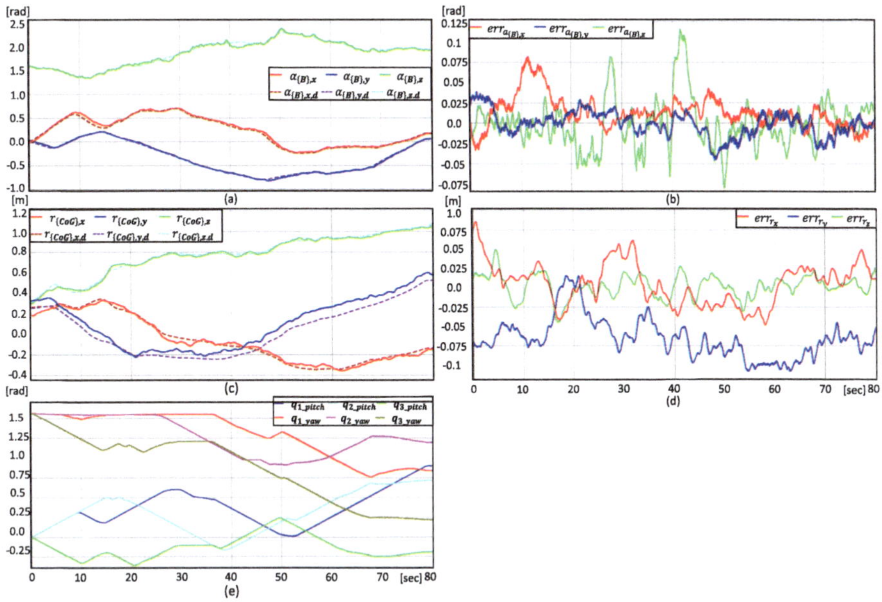

Fig. 6. The trajectories of the CoG motion of the multilinked model ($\boldsymbol{\alpha}_{\{B\}}(t)$, $\boldsymbol{r}_{\{CoG\}}(t)$), along with the joint motion $\boldsymbol{q}(t)$. The dashed lines show the desired path for the corresponding states.

4 Conclusions and Future Work

In this paper, we presented a motion planning method based on differential kinematics, allowing a multilinked aerial robot to navigate in a complex environment, e.g., to pass through a small horizontal opening between a ground floor and a closed ceiling. Online planning executed on the onboard processor during flying demonstrated the immediacy of our proposed method. A stable motion to squeeze through a tight horizontal opening by the transformable aerial robot composed of four links also confirmed the validity of our proposed method.

Our future work involves improvement of the dynamic path trackability without the assumption of the rigid body. We will also focus on completeness while searching a valid solution for more complicated environments (e.g., multi-layer openings), and a possible combination of our method with the sampling based methods will be taken into consideration.

References

1. Papachristos, C., Khattak, S., Alexis, K.: Uncertainty-aware receding horizon exploration and mapping using aerial robots. In: 2017 IEEE International Conference on Robotics and Automation (ICRA), pp. 4568–4575, May 2017
2. Mellinger, D., Michael, N., Kumar, V.: Trajectory generation and control for precise aggressive maneuvers with quadrotors. Int. J. Robot. Res. **31**(5), 664–674 (2012)

3. Loianno, G., Brunner, C., McGrath, G., Kumar, V.: Estimation, control, and planning for aggressive flight with a small quadrotor with a single camera and IMU. IEEE Robot. Autom. Lett. **2**(2), 404–411 (2017)
4. Falanga, D., Mueggler, E., Faessler, M., Scaramuzza, D.: Aggressive quadrotor flight through narrow gaps with onboard sensing and computing using active vision. In: 2017 IEEE International Conference on Robotics and Automation (ICRA), pp. 5774–5781, May 2017
5. Park, S., Her, J., Kim, J., Lee, D.: Design, modeling and control of omni-directional aerial robot. In: 2016 IEEE/RSJ International Conference on Intelligent Robots and Systems (IROS), pp. 1570–1575, October 2016
6. Oosedo, A., et al.: Flight control systems of a quad tilt rotor unmanned aerial vehicle for a large attitude change. In: 2015 IEEE International Conference on Robotics and Automation (ICRA), pp. 2326–2331 (2015)
7. Riviere, V., Manecy, A., Viollet, S.: Agile robotic fliers: a morphing-based approach. Soft Robot. **5**(5), 541–553 (2018)
8. Zhao, M., Kawasaki, K., Okada, K., Inaba, M.: Transformable multirotor with two-dimensional multilinks: modeling, control, and motion planning for aerial transformation. Adv. Robot. **30**(13), 825–845 (2016)
9. Zhao, M., Anzai, T., Shi, F., Chen, X., Okada, K., Inaba, M.: Design, modeling, and control of an aerial robot dragon: a dual-rotor-embedded multilink robot with the ability of multi-degree-of-freedom aerial transformation. IEEE Robot. Autom. Lett. **3**(2), 1176–1183 (2018)
10. LaValle, S.M., Kuffner, J.J.: Randomized kinodynamic planning. In: Proceedings of the 1999 IEEE International Conference on Robotics and Automation, vol. 1, pp. 473–479 (1999)
11. Zhao, M., Shi, F., Anzai, T., Chaudhary, K., Chen, X., Okada, K., Inaba, M.: Flight motion of passing through small opening by dragon: transformable multilinked aerial robot. In: 2018 IEEE/RSJ International Conference on Intelligent Robots and Systems (IROS), October 2018
12. Karaman, S., Frazzoli, E.: Incremental sampling-based algorithms for optimal motion planning. In: Proceedings of Robotics: Science and Systems, Zaragoza, Spain, June 2010
13. Whitney, D.E.: The mathematics of coordinated control of prosthetic arms and manipulators. ASME J. Dyn. Syst. Meas. Control **94**(4), 303–309 (1972)
14. Kanehiro, F., Morisawa, M., Suleiman, W., Kaneko, K., Yoshida, E.: Reactive leg motion generation method under consideration of physical constraints. J. Robot. Soc. Jpn. **28**(10), 1251–1261 (2010)
15. de Boor, C.: On calculating with B-splines. J. Approx. Theory **6**(1), 50–62 (1972)

Mathematical Models for Physical Interactions of Robots in Planar Environments

Adam Stager(✉) and Herbert G. Tanner

Department of Mechanical Engineering, University of Delaware, Newark, DE 19711, USA
{astager,btanner}@udel.edu

Abstract. This paper reports on the development of a collision capable omni-driven platform called Omnipuck with an accompanying reflection model capturing uncertainty present after collision events. Path planning and navigation strategies focus almost entirely on avoiding collisions, but this platform was designed to specifically benefit by colliding with obstacles in the environment. Collisions are modeled as partially reflected diffusions characterizing boundary interaction behaviors including uncertainty with experimentally validated parameters. Experiments highlighting convergence time demonstrate the potential benefits of using boundaries to partially recover impact energy in planning. The results herein lay the foundation for path planning and navigation strategies that exploit collisions.

Keywords: Reflection · Diffusion · Modeling · Robot design

1 Introduction

Planning algorithms and navigation strategies have aimed almost entirely at avoiding collisions, with the obstacles and workspace boundaries serving to impose constraints on robot paths [1]. In the past, a robots structural and internal components were expensive, fragile, and in general not considered dispensable, which clearly motivated the urge for collision avoidance. Today, however, new methods for manufacturing and the emergence of affordable, smaller scale electronics have enabled the development of a new class of robots that can either withstand collisions, or be considered dispensable when they do not always survive them. There has even been evidence that robotic missions can benefit from contact with the environment [2,3]. Yet efforts to understand, implement, and exploit such behaviors are still at their infancy.

Boundary following is one exception, having its roots in early bug algorithms where the presence of obstacles actually allows provable convergence [4]. In fact, insects commonly collide with their boundaries continuing with no ill-effect [5]. Several aerial robots have been developed demonstrating similar behavior, with intentional mechanisms for adding robustness to collisions [2,6]. Early

J. Xiao et al. (Eds.): ISER 2018, SPAR 11, pp. 549–558, 2020.
https://doi.org/10.1007/978-3-030-33950-0_47

work demonstrates the potential for physical robot-environment interactions as a basis for new strategies in motion planning [7].

Compared to aerial robots, ground mobile robots offer more flexibility for designers to make them robust to collisions, add shielding, or even for employing mechanisms to exploit boundary interactions. Small-scale platforms in particular, have been shown to benefit from colliding with their boundary; traveling on new, sometimes improved paths after an impact [3], or offering a foothold to improve post-collision mobility [8]. These platforms, however are plagued with uncertainty and are difficult to control reliably [3].

Despite progress toward developing robust, collision resistant platforms, mathematical models that capture impact behavior, and planners that intensionally incorporate these behaviors are currently under-developed. One model that has been used to capture reflections in terms of a coefficient of restitution [9] provides a simplified proof of concept, but does not go as far as capturing the uncertainty that is inherent in the robot-boundary interaction. A *stochastic* model capturing the mechanics of collision, recovery, as well as failure, could be particularly useful in exploiting such phenomena in low-friction or high speed, time critical scenarios [10].

This work identifies such a model, based on an existing mathematical framework for partially reflected diffusions [11]. We construct a ground mobile robot called Omnipuck as an experimental testbed for validating the hypothesis that the mathematics of partially reflected diffusions fits well in the context of robot-environment collisions. Omnipuck is designed to benefit from impact events not only by being collision resistant, but also by storing/releasing potential energy during collisions. The results reported herein have utilized the Omnipuck to match probability distribution parameters and thus compare analytical results to experimental reflection data. The results suggest that the partially reflected diffusion model shows promise in describing complex physical robot-environment interaction phenomena, as well as predicting not only the aftermath of collisions (excluding the possibility of failure, at this moment) but also the evolution of the system during the process.

2 Technical Approach

Post collision behavior is studied as a partially reflected diffusion [11], capturing the position of the robot after surviving a collision in the form of a distribution. An absorption parameter controls the rate at which sample paths reaching a boundary are not reflected back—this percentage models the rate at which collisions lead to failure. The stochastic model seems appropriate for capturing not only the possible variability in the motion of the robot during the collision phase, but also the uncertainty regarding both the contact point, as well as the location of the robot after reflection.

Partially reflected diffusions [11] are characterized by a drift term $a(x,t)$ capturing the expectation of the motion, and a diffusion term $\sigma(x,t) > 0$ describing the amount of underlying uncertainty. In practice, these terms —especially the latter— are determined experimentally. In this work, we assume decoupled motion along each direction, an assumption allowing the analysis to proceed as a single-dimensional model for the motion along each direction. This single dimensional analysis allows for a direct comparison between the formalism of partially reflected diffusions and the robots position as it evolves over time away from the collision surface.

If x expresses the state (read: position) of a particle along a positive semi-axis (with zero marking the workspace boundary), then the stochastic motion of the particle may be modeled as,

$$\mathrm{d}x = a'(x,t)\,\mathrm{d}t + \sqrt{2\sigma'(x,t)}\,\mathrm{d}w. \tag{1}$$

The solution of (1) with partially absorbing boundaries has been defined [11] as the limit of the solutions of a Markov jump process $x_{\Delta t}$, which for $s \leq t \leq s+T < \infty$ and $\Delta t = \frac{T}{N}$ for $N \in \mathbb{N}$, is generated by the Euler scheme,

$$\begin{aligned} x_{\Delta t}(s) &= x \\ x_{\Delta t}(t+\Delta t) &= x_{\Delta t}(t) + a'\big(x_{\Delta t}(t),t\big)\,\Delta t + \sqrt{2\sigma'\big(x_{\Delta t}(t),t\big)}\,\Delta w(t,\Delta t) \end{aligned} \tag{2}$$

where the random variables Δw are normally distributed with zero mean and variance σ'. In the Euler scheme of (2), the absorption at the boundary is modeled through the boundary conditions. Let R denote the probability of reflection, i.e., the event that the robot bounces off undamaged, and define [11]

$$P \triangleq \lim_{\Delta t \to 0} \frac{1-R}{\sqrt{\Delta t}}.$$

Then the boundary conditions for (2) are activated whenever it is detected that the next iterate of (2) has crossed the (zero) boundary, namely when

$$x_{\Delta t(t)} + a'\big(x_{\Delta t}(t),t\big)\,\Delta t + \sqrt{2\sigma'\big(x_{\Delta t}(t),t\big)}\,\Delta w(t,\Delta t) < 0$$

in which case it is reset to some other positive value

$$x_{\Delta t}(t+\Delta \mathrm{t}) := \tag{3}$$
$$\begin{cases} -(x_{\Delta t}(t) + a\big(x_{\Delta t}(t),t\big)\,\Delta t + \sqrt{2\sigma\big(x_{\Delta t}(t),t\big)}\,\Delta w(t,\Delta t)) & \text{w.p } 1 - P\sqrt{\Delta t} \\ \text{process terminates} & \text{w.p. } P\sqrt{\Delta t} \end{cases}$$

When the process terminates in (3) it is understood as a catastrophic failure of the robot resulting from the collision.

The Fokker-Planck equation (FPE) associated with (3) is [11]

$$\frac{\partial p(y,t|x,s)}{\partial t} = -\frac{\partial\, a(y,t)\,p(y,t|x,s)}{\partial y} + \frac{\partial^2\, \sigma(y,t)\,p(y,t|x,s)}{\partial y^2} \tag{4}$$

with initial condition $p(y,t|x,s) \to \delta(y-x)$ as $t \downarrow s$, and radiation boundary condition

$$a(0,t)\,p(0,t|x,s) - \frac{\partial\, \sigma(0,t)\, p(0,t|x,s)}{\partial y} = -\kappa\, p(0,t|x,s)$$

with $\kappa = \frac{P\sqrt{\sigma(0,t)}}{\sqrt{\pi}}$ denoting the reactive coefficient related to the probability of absorption at the boundary $y = 0$ [11].

In one dimension, and for constant a and σ, the FPE (4) has an explicit solution with condition $x(0) = x_0$ initialized at the moment of impact [11],

$$p(x,t|x_0) = \frac{1}{\sqrt{4\pi\sigma t}}\left[\exp\left\{-\frac{(x-x_0-at)^2}{4\sigma t}\right\} + \exp\left\{-\frac{ax_0}{\sigma} - \frac{(x+x_0-at)^2}{4\sigma t}\right\}\right] - \frac{2\kappa + a}{2\sigma}\exp\left\{\frac{ax + \kappa\left[x + x_0 + (\kappa + a)t\right]}{\sigma}\right\}\operatorname{erfc}\left[\frac{x + x_0 + (2\kappa + a)t}{\sqrt{4\sigma t}}\right]. \quad (5)$$

The goal now is to express the distribution describing the location of the robot *after* impact, and as soon as it is capable of reassuming control of its motion, by a probability distribution function of the form (5), with time t counting from the time of impact (Fig. 1).

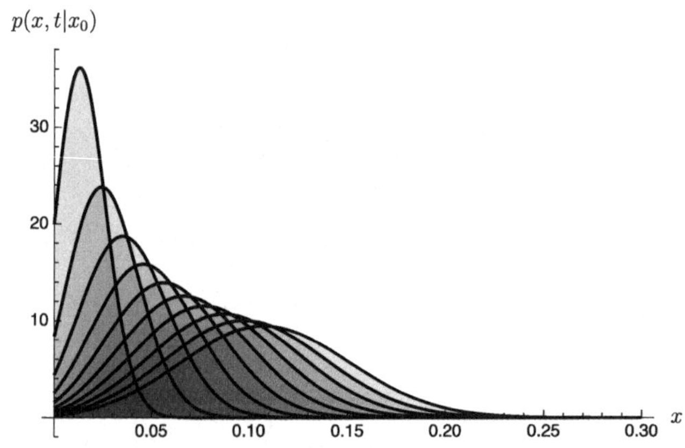

Fig. 1. How the density given in (5) changes over time. As time t increases, the superimposed peak of $p(x,t|x_0)$ shifts to the right of the boundary $x_0 = 0$. At any time instant the experimentally validated $p(x,t|x_0)$ describes the position of the robot away from the boundary.

In order to validate the hypothesis that (5) can capture the outcome of workspace boundary collisions, at least in a single-dimension, we constructed a collision-tolerant, omni-directional ground mobile robot (Fig. 2).

Omni-directional wheels where chosen to maximize collision effects by allowing the robot to resume post-collision motion in any direction on the Cartesian plane. A single stage 10:1 gearbox fit to each of the robots three motors increases torque while providing minimal friction and rolling resistance. A specially designed ring surrounding the robot incorporates counteracting elastic bands, hovering in equilibrium around the robot until a collision event occurs. Acting as a protective shield the reflection ring also allows Omnipuck to store and redirect impact energy. The Omnipuck was designed by a combination of 3D printed and off-the-shelf components affording flexibility for iteration of the design and reducing the time and cost of assembling the platform. The diameter of the platform is 0.18 m and with a mass of 350 g it can collide with stationary obstacles in its environment non-destructively. The omni-wheels can be used in multiple configurations; in this particular case, three wheels were chosen because they maintain consistent ground contact and require only three motors controlling the origin of a body centered robot frame in $\mathbb{R}^2$.

(a) (b) (c)

Fig. 2. The Omnipuck robot constructed to study workspace boundary reflections. (a) an Atmega 328 micro-controller sits at the core of the Omnipuck, controls motors through PWM, and takes accelerometer input for collision detection; (b) a reflection ring is held in tension between three elastic bands providing robustness to collisions while storing/redirecting impact energy; (c) experiments were conducted on a smooth flat $4' \times 8'$ table made of medium-density fiberboard. A 3D printed jig helps automatically align the robot before manually verifying its initial state with Vicon.

3 Results

To experimentally validate the potential for partially reflected diffusions to capture reflection behaviors a set of head-on collisions were captured using a Vicon Motion Capture system. A "there-and-back" experiment exemplifies the effect of leveraging reflection surfaces to improve convergence times in planning and provides insight to the practical benefits of capturing uncertainty in the reflection model.

3.1 Experimental Setup

The parameters of the reflection model (5) are estimated experimentally, based on a body of trajectory data collected from 500 head-on collisions between the Omnipuck and a stationary boundary as pictured in Fig. 2(c). The robot is positioned using a 3D printed jig to an initial distance of $x = 2.3\,\text{m}$ relative to the reflection boundary where $x = 0$, and then commanded to travel in a straight line perpendicular to a synthetic wall. Specifically the pre-collision control law was constructed as a -wall directed- vector field,

$$\dot{x} = dX + p(x_0 - x - ((x_0 - x)dX + (y_0 - y)dY)dX)$$
$$\dot{y} = dY + p(y_0 - y - ((x_0 - x)dX + (y_0 - y)dY)dY),$$

where (x_0, y_0) are the initial point, $[dX, dY]$ is a unit vector in the direction of the reflection surface and $p = 1.5$ is a tuning parameter. Field lines drive the robot into the boundary and enforce the desired collision angle simultaneously. Commands are updated at 20 Hz and result in an average impact velocity of 1.3 m/s.

An on-board accelerometer is used to detect collision and immediately switch motors into a free-spinning mode allowing the robot to roll freely with minimal resistance. The moment a collision is detected dictates the initial time $t = 0$ and sets the collision models initial state as $x(0) = 0$ on the boundary. The motion of the robot off of the boundary proceeds with x increasing positively along the positive collision axis. Collision data is collected until the Omnipuck comes to a complete rest for a total of 500 sample paths so that multiple post-collision time instances can be compared.

3.2 Model Fitting

The proposed collision model is meant to capture the distribution of post-collision positions as a function of time. Zero terminal collisions were reported so that $\kappa = 0$ (pure reflection), leaving a and σ to be determined as parameters in the underlying distribution. Although experimental data determines pure reflection, it is possible during any one trial to experience an absorbing situation which could be factored into the reflection model as $\kappa > 0$.

Figure 3 shows three snapshots from the complete post-collision time sequence. Reflection distance is captured on the x-axis at each of the three time instances, representing the number of trials at each position. Data is shifted from the frequency domain where empirical probability distribution functions are overlaid. Originally the aim was to capture the entire post-collision behavior with a single analytically represented distribution. Analytical solutions arising from constant coefficients can capture short intervals of the post-collision event, however, solving for drift and diffusion parameters for each time instant results in theoretical distributions that are well matched with the corresponding experimental data. In Fig. 3 one can see how this empirical density function compares to the theoretical model (5) when parameterized with the estimated values of a and σ.

In comparison, reflection models generally consider only the average reflection behavior and assume exact collision velocity for determining post-collision behavior. Models such as the partially reflected diffusion (4), however are able to capture uncertainty not only due to surface interaction and frictional ground effects, but also the effect of variations in impact velocity. This is particularly useful in considering the motion of small-mobile robots - especially ground mobile robots - and is an important next step as robots perform tasks in real-world environments where such deviations undoubtedly play a role.

3.3 Experimental Results

In order the emphasize the proposed benefits of including collisions in planning the time to complete a basic "there-and-back" maneuver is compared for free-space and reflected scenarios. Although this experiment is a simplistic case it is not unmotivated. Swimmers, parcor runners, and many competitive activities have naturally adopted similar simple, but effective behaviors to improve times and generate desirable maneuvers.

Similarly to the setup described in Sect. 3.1 the Omnipuck is positioned in a vector field originally directed in the negative x direction. In the free-space situation a half-plane perpendicular to the direction of the vector field, 1.0 m from the robot's initial state, acts as an imaginary boundary inverting the vector field once the robots position crosses. In comparison, for reflecting experiments, a stationary boundary acts in place of the half-plane with the collision event acting as an indicator function to switch the directionality of the driving vector field. This effectively acts as a bang-bang controller with each experiment concluding when the robot crosses another half-plane coincident with the starting position.

The free-space situation results in a significant increase in convergence time, with an average time to return nearly 25% greater than in the reflected scenario. Also, visible in Fig. 4 are the high accelerations in the reflected experiments, represented by a nearly vertical slope at the point of impact compared to a more gradual transition from negative to positive velocity in the free-space case. Free-space planning results in motor input saturation and insufficient friction leading to overshoot and poor tracking. Because the effects of motor saturation and friction can be challenging to model reliably, planners often enforce constraints reducing the obtainable velocities and further increasing the convergence time. These experiments demonstrate an example of how boundaries can be used to reduce convergence time, allowing for high speeds even in situations with dramatic direction changes.

4 Main Experimental Insights

The preliminary analysis conducted suggests that the effect of physical robot-environment interactions on a robots position can be captured in the form of probability distributions with experimentally validated parameters capable of

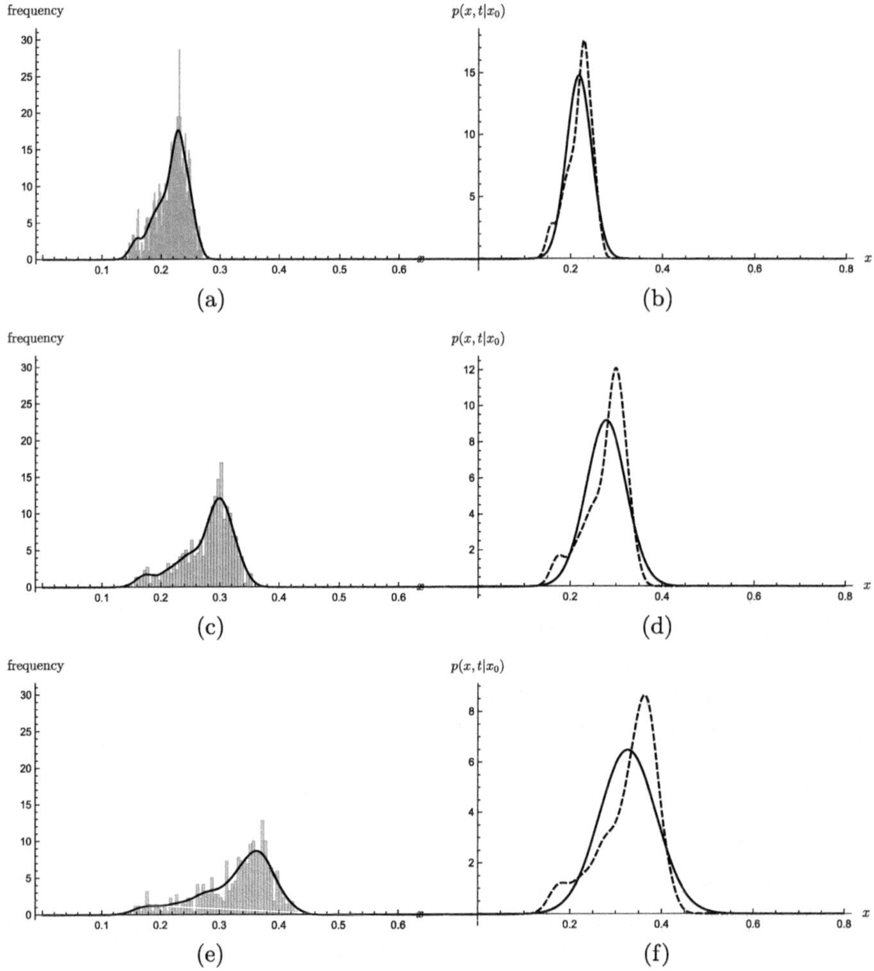

Fig. 3. Three post collision snapshots at times $t = 0.10, 0.20, 0.30$ show the progression of the post impact position and how it can be captured with an accompanying distribution. (a) a histogram of the positions recorded at time $t = 0.10$, and an empirical density function that is constructed based on the histogram; (b) comparison of the empirical density (dashed) with the parameterized model (5) with $a = 2.1633$ and $\sigma = 0.00369405$. (c) a histogram of the positions recorded at time $t = 0.20$, and an empirical density function that is constructed based on the histogram; (d) comparison of the empirical density (dashed) with the parameterized model (5) with $a = 1.37348$ and $\sigma = 0.00481413$. (e) a histogram of the positions recorded at time $t = 0.30$, and an empirical density function that is constructed based on the histogram; (f) comparison of the empirical density (dashed) with the parameterized model (5) with $a = 1.06867$ and $\sigma = 0.0065008$.

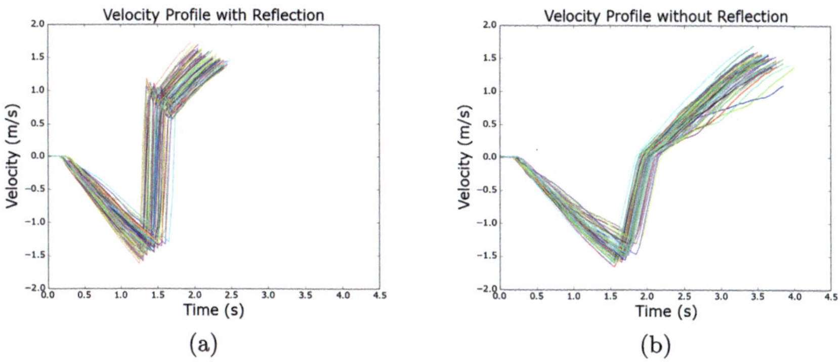

Fig. 4. Plots of the velocity profile from the reflected and free-space scenarios show faster return times and higher average speed for 65 experimental trials each. (a) the reflecting boundary results in a return time of 2.814 s; (b) the free-space change in direction results in a return time of 3.665 s.

describing not only the location of the robot after impact, but also the possibility for terminal, catastrophic collisions. The theoretical model suggested shows promise in its ability to describe this phenomenon probabilistically, making it a crucial component for practically including these behaviors in motion planners that consider uncertainty. When incorporated into a planner, this model can help rule out prohibitive reflections. For example reflections with high probability of carrying the robot into a second undesirable collision can be avoided.

An interesting next step is understanding how these positional distributions relate to coefficients of restitution and how they can be used to model post-collision velocity distributions. Also from a planning perspective it would be useful to develop a set of rules that can help in determining which boundaries are good candidates for intensional reflection. These developments can unearth a new paradigm; planning strategies that seek to exploit collisions between robust (or dispensable) robot designs and their surrounding environment, when the trade-offs between the risk of failure and the probability of improving mission objectives through collisions appear favorable.

References

1. LaValle, S.M.: Planning Algorithms. Cambridge University Press, Cambridge (2006)
2. Briod, A., Kornatowski, P., Klaptocz, A., Garnier, A., Pagnamenta, M., Zufferey, J.-C., Floreano, D.: Contact-based navigation for an autonomous flying robot. In: Proceedings of the IEEE/RSJ International Conference on Intelligent Robots and Systems, pp. 3987–3992 (2013)
3. Karydis, K., Zarrouk, D., Poulakakis, I., Fearing, R.S., Tanner, H.G.: Planning with the STAR(s). In: Proceedings of the IEEE/RSJ International Conference on Intelligent Robots and Systems, pp. 3033–3038 (2014)

4. Lumelski, V.J., Stepanov, A.A.: Dynamic path planning for a mobile automaton with limited information on the environment. IEEE Trans. Autom. Control **31**(11), 1058–1063 (1986)
5. Briod, A.: Observation of insect collisions. Ph.D. dissertation, Harvard University (2008)
6. Mulgaonkar, Y., Cross, G., Kumar, V.: Design of small, safe and robust quadrotor swarms. In: Proceedings of the IEEE International Conference on Robotics and Automation, pp. 2208–2215 (2015)
7. Mulgaonkar, Y., Makineni, A., Guerrero-Bonilla, L., Kumar, V.: Robust aerial robot swarms without collision avoidance. IEEE Robot. Autom. Lett. **3**(1), 596–603 (2018)
8. Haldane, D.W., Plecnik, M.M., Yim, J.K., Fearing, R.S.: Robotic vertical jumping agility via series-elastic power modulation. Sci. Robot. **1**(1), eaag2048 (2016)
9. Garwin, R.L.: Kinematics of an ultraelastic rough ball. Am. J. Phys. **37**(1), 88 (1969)
10. Karaman, S., Frazzoli, E.: High-speed motion with limited sensing range in a poisson forest. In: Proceedings of the IEEE Conference on Decision and Control, pp. 3735–3740 (2012)
11. Singer, A., Schuss, Z., Osipov, A., Holcman, D.: Partially reflected diffusion. SIAM J. Appl. Math. **68**(3), 844–868 (2008)

Human-Robot Interaction

Human-Robot Interaction

Berk Calli

Human robot interaction received a great deal of interest from the audience. First, Leung and Schmerling gave an inspirational presentation of their paper on an application of HRI for autonomous vehicles to ensure safety with minimal intervention. Safety was again the focus of the second paper, presented by Joseph, which proposed a model-based dissipated energy calculation to be used to constrain a robotic system. The third paper was about human-based study on how to represent human intent using elliptical models and discussed obtaining the model parameters. Next, the fourth paper presented by Fitter addressed the opposite problem: how to express robot's intent to humans. She presented us her extensive study using mechanical turk. Penultimately, the fifth paper presented by Clabaugh addressed how to effectively use socially assistive robots to aid children with autism spectrum disorders. She proposed techniques for measuring children's attention and stressed the challenges of generating such models due to limited amount of available data. Nakamura presented the last paper of the session on a backdrivable and force controllable humanoid robot design. He discussed how compliant behavior can be achieved using hydrostatic mechanisms and viscoelasticity controllers. This session was full of questions and interesting discussions throughout.

On Infusing Reachability-Based Safety Assurance Within Probabilistic Planning Frameworks for Human-Robot Vehicle Interactions

Karen Leung[1(✉)], Edward Schmerling[2], Mo Chen[4], John Talbot[3], J. Christian Gerdes[3], and Marco Pavone[1]

[1] Department of Aeronautics and Astronautics, Stanford University, Stanford, CA 94305, USA
{karen17,pavone}@stanford.edu
[2] Institute for Computational and Mathematical Engineering, Stanford University, Stanford, CA 94305, USA
schmrlng@stanford.edu
[3] Department of Mechanical Engineering, Stanford University, Stanford, CA 94305, USA
{john.talbot,cgerdes}@stanford.edu
[4] School of Computing Science, Simon Fraser University, Burnaby, BC V5A 1S6, Canada
mochen@cs.sfu.ca

Abstract. Action anticipation, intent prediction, and proactive behavior are all desirable characteristics for autonomous driving policies in interactive scenarios. Paramount, however, is ensuring safety on the road—a key challenge in doing so is accounting for uncertainty in human driver actions without unduly impacting planner performance. This paper introduces a minimally-interventional safety controller operating within an autonomous vehicle control stack with the role of ensuring collision-free interaction with an externally controlled (e.g., human-driven) counterpart. We leverage reachability analysis to construct a real-time (100 Hz) controller that serves the dual role of (1) tracking an input trajectory from a higher-level planning algorithm using model predictive control, and (2) assuring safety through maintaining the availability of a collision-free escape maneuver as a persistent constraint regardless of whatever future actions the other car takes. A full-scale steer-by-wire platform is used to conduct traffic weaving experiments wherein the two cars, initially side-by-side, must swap lanes in a limited amount of time and distance, emulating cars merging onto/off of a highway. We demonstrate that, with our control stack, the autonomous vehicle is able to

M. Chen—Work performed as a postdoctoral scholar at Stanford University.
K. Leung and E. Schmerling—Contributed equally to this work. This work was supported by the Office of Naval Research (Grant N00014-17-1-2433), by Qualcomm, and by the Toyota Research Institute ("TRI"). This article solely reflects the opinions and conclusions of its authors and not ONR, Qualcomm, TRI, or any other Toyota entity. The authors would like to thank Thunderhill Raceway Park for accommodating testing.

J. Xiao et al. (Eds.): ISER 2018, SPAR 11, pp. 561–574, 2020.
https://doi.org/10.1007/978-3-030-33950-0_48

avoid collision even when the other car defies the planner's expectations and takes dangerous actions, either carelessly or with the intent to collide, and otherwise deviates minimally from the planned trajectory to the extent required to maintain safety.

1 Introduction

Decision-making and control for mobile robots is typically stratified into levels. A high-level planner, informed by representative dynamics of a robot and its environment, might be responsible for selecting a coarse trajectory plan, which is implemented through a low-level controller that respects more accurate models of the robot's dynamics and control constraints. While additional components may be required to flesh out a robot's full control stack from model to motor commands, selecting the right "division of responsibilities" is fundamental to system design.

One consideration that defies clear classification, however, is how to ensure a mobile robot's safety when operating in close proximity with a rapidly evolving and stochastic environment. Safety is a function of uncertainty in both the robot's dynamics and those of its surroundings; high-level planners typically do not replan sufficiently rapidly to ensure split-second reactivity to threats, yet low-level controllers are typically too short-sighted to ensure safety beyond their local horizon. Hand-designed artificial potential fields have been employed at the planning level [1] to encourage the selection of safer nominal trajectories, while reactive collision avoidance techniques, e.g., based on online optimal control [2] or precomputed emergency maneuver libraries [3], have been applied at the controller level to avoid previously unseen static obstacles, but these approaches do not account for interactive scenarios in which another sentient agent is a key environmental consideration.

In this work we implement a control stack for a full-scale autonomous car (the "robot") engaging in close proximity interactions with a human-controlled vehicle (the "human"). Freedom of motion is essential for the planner to carry out the driving task while conveying future intent to the other vehicle. Our primary tool for designing a controller that does not needlessly impinge upon the planner's choices is Hamilton-Jacobi (HJ) reachability. Backward HJ reachability, in particular, has been studied extensively and applied successfully in a variety of safety-critical interactive settings [4–9] due to its flexibility with respect to system dynamics, and its optimal (i.e., non-overly-conservative) avoidance maneuvers stemming from its equivalence to an exhaustive search over joint system dynamics. Previously reachability-based controllers have largely been used in a switched fashion: the robot is normally allowed to apply any control, but switches to the optimal avoidance controller when near safety violation. While this is sometimes effective [10], to the best of our knowledge there has not been any work that explicitly addresses the integration of reachability-based safety controllers as a component within a robot's control stack, i.e., with safety as a constraint upon a primary planning objective.

Statement of Contributions: The contributions of this paper are twofold. First, we propose a method for formally incorporating reachability-based safety within an existing optimization-based control framework. The main insight that enables our approach is the recognition that, near safety violation, the set of safety-preserving controls often contains more than just the optimal avoidance control. Instead of directly applying this optimal avoidance control when prompted by reachability considerations, as in a switching control approach, we quantify the set of safety-preserving controls and pass it to the broader control framework as a constraint. Our intent is to enable truly minimal intervention against the direction of a higher-level planner when evasive action is required. Second, we evaluate the benefits and performance of this safe control methodology in the context of a probabilistic planning framework for the traffic weaving scenario studied at a high level in [11], wherein two cars, initially side-by-side, must swap lanes in a limited amount of time and distance. Experiments with a full-scale steer-by-wire vehicle reveal that our combined control stack achieves better safety than applying a tracking controller alone to the planner output, and smoother operation (with similar safety) compared with a switching control scheme; in our discussion we provide a roadmap towards improving the level of safety assurance in the face of practical considerations, i.e., unmodeled dynamics, as well as towards generalizations of the basic traffic weaving scenario.

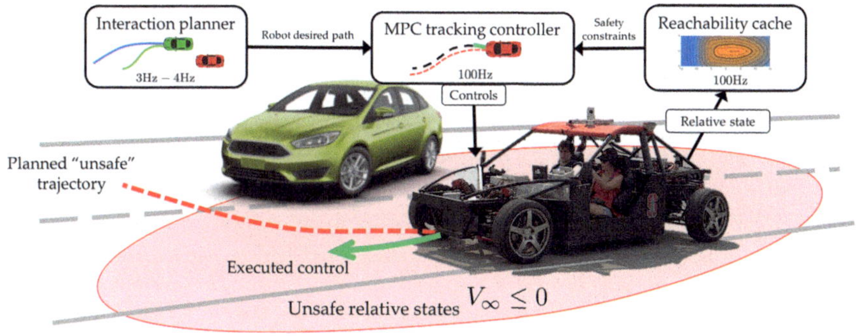

Fig. 1. Decision-making and control stack for human-robot pairwise vehicle interactions. Our contribution in this work is the integration of safety-ensuring control constraints, derived from a HJ backward reachable set computed and cached offline, into a model predictive controller's tracking optimization problem. A high-level interaction planner produces nominal trajectories for the robot car (foreground); the low-level safe tracking controller executes controls that minimally deviate from the planner's choice if the vehicles approach the set of unsafe relative states.

2 Technical Approach

To bring safe traffic weaving to the road, we combine a model-based interaction planner [11] with a real-time trajectory tracking model predictive control (MPC) controller inspired by [12], modified to include an additional invariant set constraint detailed in this section. An outline of our control framework is shown in Fig. 1.

Interaction Planner: The traffic-weaving interaction planner [11] uses a predictive model of future human behavior to select a desired trajectory for the robot car to follow, updated at ~3 Hz. We extend this work by using a hindsight optimization policy [13] instead of the limited-lookahead action policy used in [11] in order to encourage more information-seeking actions from the robot. Though this interaction planner optimizes an objective that weighs safety considerations (e.g., distance between cars) relative to other concerns (e.g., control effort), it reasons anticipatively with respect to a *probabilistic* interaction dynamics model. That is, safety is not enforced as a deterministic constraint at the planning level.

MPC Tracking Controller: The MPC tracking controller, operating at 100 Hz, computes optimal controls to track a desired trajectory by solving a quadratic program (QP) at each iteration. This optimization problem is based on a single track vehicle model (also known as the bicycle model) and incorporates friction and stability control constraints (in addition to state and control constraints), while minimizing a combination of tracking error and steering effort. We incorporate additional constraints computed from HJ reachability analysis into this QP to ensure that at each control step the robot car does not enter a state of possible inevitable collision.

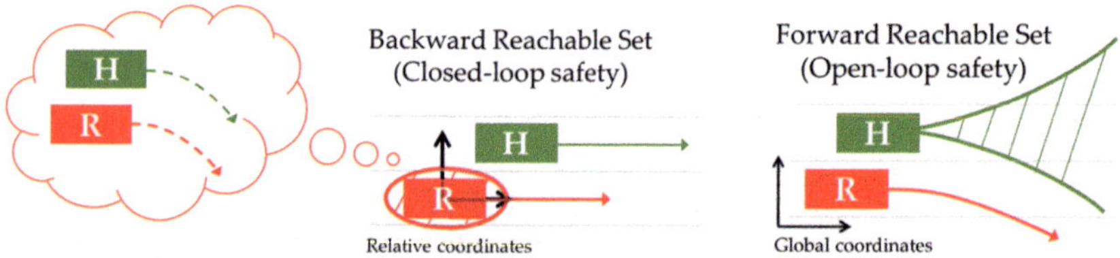

(a) Joint dynamics are propagated *backwards* in time. The sets are not overly-conservative even for long time horizons. However, they are computationally expensive to compute.

(b) Human dynamics are propagated *forward* in time (overly-conservative in general).

Fig. 2. Illustration of backward and forward reachable sets (shaded) to be avoided.

2.1 Backward Reachability Analysis

We use infinite time horizon backward reachability analysis to compute additional safety constraints for the MPC tracking controller. Critically, *backward reachability* is computed backwards in time allowing for closed-loop robot reactions to avoid future collisions (see Fig. 2a), while *forward reachability* is computed forwards in time assuming open-loop control for the robot (see Fig. 2b) which leads to an overly-conservative robot outlook. Though more expensive to compute, backward reachability solutions may be precomputed offline and cached for online use. We briefly review relevant HJ reachability definitions in the remainder of this section; see [14] for a more in-depth treatment.

Let the relative dynamics of the robot and human be given by $\dot{x}_{\mathscr{R}} = f(x_{\mathscr{R}}, u_{\mathrm{R}}, u_{\mathrm{H}})$, and the collision set be $\mathscr{T}$. The BRS $\mathscr{A}(t)$ represents the set of "avoid states" from which there exists an adversarial policy $u_{\mathrm{H}}(\cdot)$ for the human such that any robot policy $u_{\mathrm{R}}(\cdot)$ would lead to the relative state trajectory $x_{\mathscr{R}}(\cdot)$ being inside $\mathscr{T}$ within a time horizon $|t|$ (note that $t < 0$ when propagating backwards in time):

$$\mathscr{A}(t) := \{\bar{x}_{\mathscr{R}} \in \mathbb{R}^n : \exists u_{\mathrm{H}}(\cdot), \forall u_{\mathrm{R}}(\cdot), \exists s \in [t, 0], x_{\mathscr{R}}(t) = \bar{x}_{\mathscr{R}} \wedge \dot{x}_{\mathscr{R}} = f(x_{\mathscr{R}}, u_{\mathrm{R}}, u_{\mathrm{H}}) \wedge x_{\mathscr{R}}(s) \in \mathscr{T}\} \tag{1}$$

Assuming optimal human actions, $\mathscr{A}(t)$ can be computed by defining a value function $V(t, x_{\mathscr{R}})$ which obeys the Hamilton-Jacobi-Isaacs (HJI) partial differential equation [9,15]; the solution $V(t, x_{\mathscr{R}})$ gives the BRS as its zero sublevel set: $\mathscr{A}(t) = \{x_{\mathscr{R}} : V(t, x_{\mathscr{R}}) \leq 0\}$.[1] When the control capabilities of the human are no greater than those of the robot, one can take the limit $t \to -\infty$ and obtain the infinite time horizon BRS $\mathscr{A}_\infty$ with corresponding value function $V_\infty(x_{\mathscr{R}})$.[2] Illustrative slices of the value function and the BRS for the vehicle-vehicle relative dynamics considered in this work are shown in Fig. 3.

The optimal robot avoidance control offers the greatest increase in V assuming optimal (worst-case) actions by the human:

$$u_R^* = \arg\max_{u_{\mathrm{R}}} \min_{u_{\mathrm{H}}} \nabla V(x_{\mathscr{R}}) \cdot f(x_{\mathscr{R}}, u_{\mathrm{R}}, u_{\mathrm{H}}) \tag{2}$$

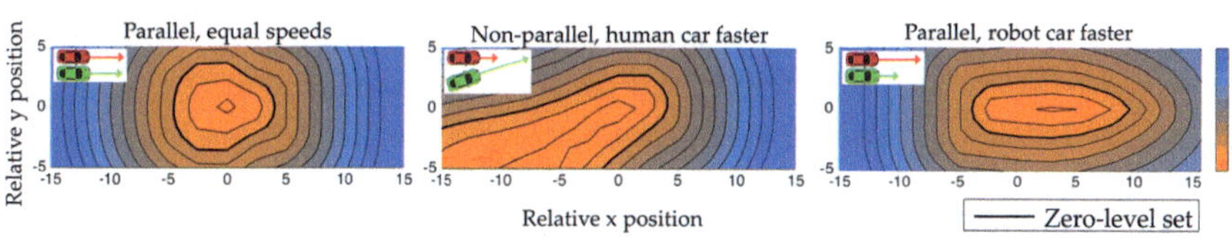

Fig. 3. Contour plots of slices of the HJI value function for the relative dynamics (4); slices show V as a function of x_{rel} and y_{rel} with all other states held fixed. Left: The pear-shaped BRS stems from the fact the robot car can swerve its front more rapidly than its rear, e.g., to avoid collision. Middle: If the human car is traveling faster (in this case at an angle), it is unsafe for the robot to be in front of the human. Right. If the robot car is traveling faster, it should not be directly behind the human as collision may be unavoidable if the human brakes abruptly.

[1] The HJI PDE is solved starting from the boundary condition $V(0, x_{\mathscr{R}})$, the sign of which reflects set membership of $x_{\mathscr{R}}$ in $\mathscr{T}$; see Sect. 5 for discussion of specific choices of $V(0, x_{\mathscr{R}})$.

[2] For ease of notation going forward we will often write $V := V_\infty$.

Previous applications of HJI solutions have switched to this control when near the boundary of the BRS [10], i.e., when safety is nearly violated. In an interactive scenario where, for example, we may want to let a robot planner convey intent by nudging towards the human car to the extent that is safe, we prefer a less extreme control strategy. Rather than switching to the optimal avoidance controller (2), we add containment in the set of safety-preserving controls defined below as an additional constraint:

$$\mathscr{U}_R(x_{\mathscr{R}}) = \{u_{\rm R} : \min_{u_{\rm H}} \nabla V(x_{\mathscr{R}}) \cdot f(x_{\mathscr{R}}, u_{\rm R}, u_{\rm H}) \geq 0\}. \tag{3}$$

$\mathscr{U}_R(x_{\mathscr{R}})$ represents the set of robot controls that ensure the value function is nondecreasing. We compute the BRS offline using the BEACLS toolkit [16], and cache the value function V and its gradient ∇V. Online we employ a safety buffer $\epsilon > 0$ so that when the condition $V(x_{\mathscr{R}}) \leq \epsilon$ holds, indicating that the robot is nearing safety violation, we add a half-space constraint $\mathscr{U}_R(x_{\mathscr{R}}) = \{u_{\rm R} : M_{\rm HJI} \cdot u_{\rm R} + b_{\rm HJI} \geq 0\}$ to the list of MPC constraints in the tracking control problem. This constraint is a linearized approximation of (3); specifically, for the current relative state $\tilde{x}_{\mathscr{R}}$, current robot control $\tilde{u}_R$, and optimal, i.e., worst-case, human action defined analogously to (2) $u^*_{\rm H}$, $M_{\rm HJI} = \nabla V \frac{\partial f}{\partial u_{\rm R}}(\tilde{x}_{\mathscr{R}}, u^*_{\rm H}, \tilde{u}_R)$ and $b_{\rm HJI} = \nabla V \cdot f(\tilde{x}_{\mathscr{R}}, u^*_{\rm H}, \tilde{u}_R) - M_{\rm HJI} \cdot \tilde{u}_R$.

3 Methodology

In this section we describe how we tailor the technical approach described above to the particular application of ensuring safety for human-robot pairwise vehicle interactions. We detail the relative dynamics model for HJI computation and outline the MPC tracking controller; all the details of the traffic weaving interaction planner can be found in [11]. All code for the planner, MPC tracking controller, and BRS computation is available at https://github.com/StanfordASL/safe_traffic_weaving.

3.1 Relative Dynamics

We use a six-state single track model to describe the robot car's dynamics (see [12] for details) and assume the human car obeys a dynamically extended unicycle model. This represents a compromise between model fidelity and the number of state dimensions in the relative dynamics; reducing the latter is essential since solving the HJI PDE suffers greatly from the curse of dimensionality.[3]

[3] The computation becomes notoriously expensive past five or more states without compromising on grid discretization or employing some decoupling strategy. We use a grid size of $13 \times 13 \times 9 \times 9 \times 9 \times 9 \times 9$ for our 7D system uniformly spaced over $(x_{\rm rel}, y_{\rm rel}, \psi_{\rm rel}, U_{x{\rm R}}, U_{y{\rm R}}, v_{\rm H}, r_{\rm R}) \in [-15, 15] \times [-5, 5] \times [-\pi/2, \pi/2] \times [1, 12] \times [-2, 2] \times [1, 12] \times [-1, 1]$; computing the BRS with this discretization takes approximately 16 hours on a 3.6 GHz octocore AMD Ryzen 1800X CPU.

We define the relative dynamics (4) in a coordinate system centered on and aligned with the robot, where R_θ is the counterclockwise rotation matrix over θ. x_{rel}, y_{rel} and ψ_{rel} are the relative position and heading angle between the human and robot, while the other states (longitudinal, lateral and angular velocities) are needed to completely describe the human and robot car dynamics. The robot control inputs $u_{\text{R}} = [\delta,\ F_x]$ are steering angle and longitudinal tire force. We can distribute the tire force to the front and rear tires $F_x = F_{x_{\text{f}}} + F_{x_{\text{r}}}$ via a fixed mapping. The human control inputs $u_{\text{H}} = [\omega,\ a]$ are yaw rate and longitudinal acceleration. The robot car experiences a longitudinal drag force $F_{x_{\text{drag}}}$, and d_{f} and d_{r} are the distances from the center of mass to the front and rear axles respectively.

$$\begin{bmatrix} x_{\text{rel}} \\ y_{\text{rel}} \end{bmatrix} = R_{-\psi_{\text{R}}} \begin{bmatrix} x_{\text{H}} - x_{\text{R}} \\ y_{\text{H}} - y_{\text{R}} \end{bmatrix} \qquad \psi_{\text{rel}} = \psi_{\text{H}} - \psi_{\text{R}} \qquad \begin{bmatrix} \dot{x}_{\text{rel}} \\ \dot{y}_{\text{rel}} \\ \dot{\psi}_{\text{rel}} \\ \dot{U}_{x\text{R}} \\ \dot{U}_{y\text{R}} \\ \dot{v}_{\text{H}} \\ \dot{r}_{\text{R}} \end{bmatrix} = \begin{bmatrix} v_{\text{H}} \cos\psi_{\text{rel}} - U_{x\text{R}} + y_{\text{rel}} r_{\text{R}} \\ v_{\text{H}} \sin\psi_{\text{rel}} - U_{y\text{R}} - x_{\text{rel}} r_{\text{R}} \\ \omega - r_{\text{R}} \\ \frac{1}{m}(F_{x_{\text{f}}} \cos\delta - F_{y_{\text{f}}} \sin\delta + F_{x_{\text{r}}} + F_{x_{\text{drag}}}) + r_{\text{R}} U_{y\text{R}} \\ \frac{1}{m}(F_{y_{\text{f}}} \cos\delta + F_{y_{\text{r}}} + F_{x_{\text{f}}} \sin\delta) - r_{\text{R}} U_{x\text{R}} \\ a \\ \frac{1}{I_{zz}}(d_{\text{f}} F_{y_{\text{f}}} \cos\delta + d_{\text{f}} F_{x_{\text{f}}} \sin\delta - d_{\text{r}} F_{y_{\text{r}}}) \end{bmatrix} \tag{4}$$

We assume that the robot and human car share the same power, steering, and friction limits. These limits are chosen to reflect the same physical constraints assumed for the MPC tracking controller's dynamic bicycle model [12]. Due to its simpler dynamics representation, the human car has a transient advantage in control authority over the robot car (it may change its path curvature discontinuously, while the robot may not), but by equating the steady state control limits we ensure that the infinite time horizon BRS computation converges.

3.2 The MPC+HJI Tracking Controller

Both the trajectory tracking objective and safety-preserving control constraint rely on optimizing over the robot steering and longitudinal force inputs simultaneously. The associated tracking optimization problem is non-convex; we apply an approximate variant of sequential quadratic programming (SQP) in which we time-discretize and linearize the dynamics over a lookahead horizon and solve one corresponding QP at each MPC step. We interpolate along each solution trajectory to compute the linearization nodes for the QP at the next MPC step.[4]

We use the `ForwardDiff.jl` automatic differentiation (AD) package implemented in the Julia programming language [17] to linearize the trajectory tracking dynamics as well as the HJI relative dynamics to facilitate the computation of M_{HJI} and b_{HJI}. We call the Operator Splitting Quadratic Program (OSQP) solver [18] through the `Parametron.jl` modeling framework [19]; this combination of software enables us to solve the following MPC optimization problem at 100 Hz.

[4] Only a single iteration of SQP is solved for the tracking problem at each MPC step, rather than the usual iteration until convergence. Since the tracking problems are so similar from one MPC step to the next, we find that this approach yields sufficient performance for our purposes.

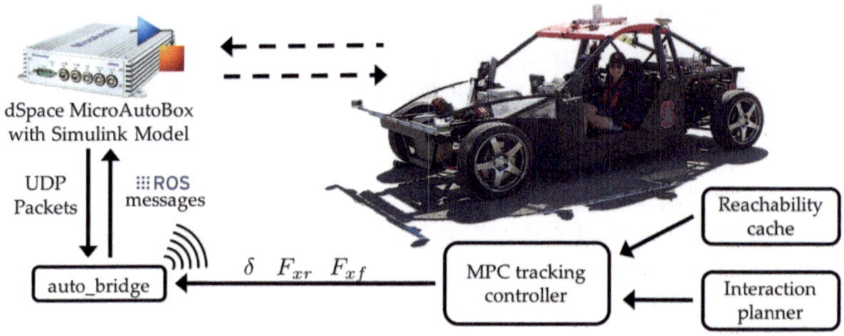

Fig. 4. X1: a steer-by-wire experimental vehicle platform. It is equipped with three LiDARs (one 32-beam and two 16-beam), a differential GPS/INS which provides pose estimates accurate to within a few centimeters as well as high fidelity velocity, acceleration, and yaw rate estimates. ROS is used to interface the planning/control stack described in this work with a dSpace MicroAutoBox onboard X1 which handles sensing and control at the hardware level.

The QP is given by (5). Let $q_k = [\Delta s_k, U_{x,k}, U_{y,k}, r_k, \Delta\psi_k, e_k]^T$ be the state of the robot car with respect to a nominal trajectory at discrete time step k. Δs_k, e_k and $\Delta\psi_k$ denote longitudinal, lateral, and heading angle error; $U_{x,k}$, $U_{y,k}$, and r_k are body-frame longitudinal and lateral velocity, and yaw rate respectively. Let $u_k = [\delta_k, F_{x,k}]$ be the controls at step k and let $A_k q_k + B_k^- u_k + B_k^+ u_{k+1} + c_k = q_{k+1}$ denote linearized first-order-hold dynamics. We adopt the varying time steps method and stable handling envelope constraint (H_k, G_k) from [12]; $\sigma_{\beta,k}$, $\sigma_{r,k}$, and σ_{HJI} are slack variables to ensure the existence of a feasible solution. The constraint $M_{\mathrm{HJI}} u_j + b_{\mathrm{HJI}} \geq -\sigma_{\mathrm{HJI}}$ is added only when $V(x_{\mathcal{R}}) \leq \epsilon$. Although HJI theory suggests that applying this constraint on the next action alone is sufficient, we apply it over the next 3 timesteps (30 ms lookahead) to account for the approximations inherent in our QP formulation.

$$
\begin{aligned}
\underset{q,u,\sigma,\sigma_{\mathrm{HJI}},\Delta\delta,\Delta F_x}{\text{minimize}} \quad & \sum_{k=1}^{T} \Delta s_k^T Q_{\Delta s} \Delta s_k + \Delta\psi_k^T Q_{\Delta\psi} \Delta\psi_k + e_k^T Q_e e_k + \Delta\delta_k^T R_{\Delta\delta} \Delta\delta_k + \\
& \Delta F_{x,k}^T R_{\Delta F_x} \Delta F_{x,k} + W_\beta \sigma_{\beta,k} + W_r \sigma_{r,k} + W_{\mathrm{HJI}} \sigma_{\mathrm{HJI},k} \\
\text{subject to} \quad & \delta_{k+1} - \delta_k = \Delta\delta_k, \quad \Delta\delta_{\min} \leq \Delta\delta_k \leq \Delta\delta_{\max}, \quad \delta_{\min} \leq \delta_k \leq \delta_{\max} \\
& F_{x,k+1} - F_{x,k} = \Delta F_{x,k}, \quad V_{\min} \leq U_{x,k} \leq V_{\max}, \quad F_{x,\min} \leq F_{x,k} \leq F_{x,\max} \\
& \sigma_{\beta,k} \geq 0, \quad \sigma_{r,k} \geq 0 \\
& H_k \begin{bmatrix} U_{y,k} \\ r_k \end{bmatrix} - G_k \leq \begin{bmatrix} \sigma_{\beta,k} \\ \sigma_{r,k} \end{bmatrix}, \quad A_k q_k + B_k^- u_k + B_k^+ u_{k+1} + c_k = q_{k+1} \\
& q_1 = q_{\mathrm{curr}}, \quad u_1 = u_{\mathrm{curr}}, \quad M_{\mathrm{HJI}} u_j + b_{\mathrm{HJI}} \geq -\sigma_{\mathrm{HJI}} \\
& \text{for } j = 1, ..., T_{\mathrm{HJI}}, \quad k = 1, ..., T
\end{aligned}
\tag{5}
$$

4 Experiments

4.1 Experimental Vehicle Platform

X1 is a flexible steer-by-wire, drive-by-wire, and brake-by-wire experimental vehicle developed by the Stanford Dynamic Design Lab (see Fig. 4). To control X1, desired steering (δ) and longitudinal tire force (F_{x_f}, F_{x_r}) commands are sent to the dSpace MicroAutoBox (MAB) which handles all sensor inputs except LiDAR (handled by the onboard PC) and implements all low level actuator controllers. We use the Robot Operating System (ROS) to communicate with the MAB; the planning/control stack described in this work is running onboard X1 on a consumer desktop PC running Ubuntu 16.04 equipped with a quadcore Intel Core i7-6700K CPU and an NVIDIA GeForce GTX 1080 GPU.

4.2 Results

To evaluate our proposed control stack—a synthesis of a high-level probabilistic interaction planner with the MPC+HJI tracking controller—we perform full-scale human-in-the-loop traffic-weaving trials with X1 taking on the role of the robot car.[5] We investigate and evaluate the effectiveness of our control stack by allowing the human car to act carelessly (i.e., swerving blindly towards the robot car) during the experiments. We compare our proposed controller (MPC+HJI) against a tracking-only MPC controller (MPC) and a controller that switches to the HJI optimal avoidance controller when near safety violation (switching). To ramp up towards testing with two full-scale vehicles in the near future, we investigate two types of human car: (1) a virtual human-driven car and (2) a 1/10-scale LiDAR-visible human-driven RC car (see Fig. 5).

Fig. 5. 1/10-scale RC car with LiDAR- visible mast.

4.3 Virtual Human-Driven Vehicle

To ensure a completely safe experimental environment, our first tier of experimentation uses a joystick-controlled virtual vehicle for the human car and allows the robot control stack to have perfect observation of the human car state. Experimental trials of the probabilistic planning framework for the MPC+HJI and switching controllers are shown in Fig. 6a and b respectively, along with a simulated comparison between the two safety controllers and the tracking-only

[5] We scaled the highway traffic-weaving scenario (mean speed ~28 m/s) in [11] down to a mean speed of ~8 m/s by shortening the track (reducing longitudinal velocity by a constant) and scaling time by a factor of 4/3 (with the effect of scaling speeds by 3/4 and accelerations by 9/16).

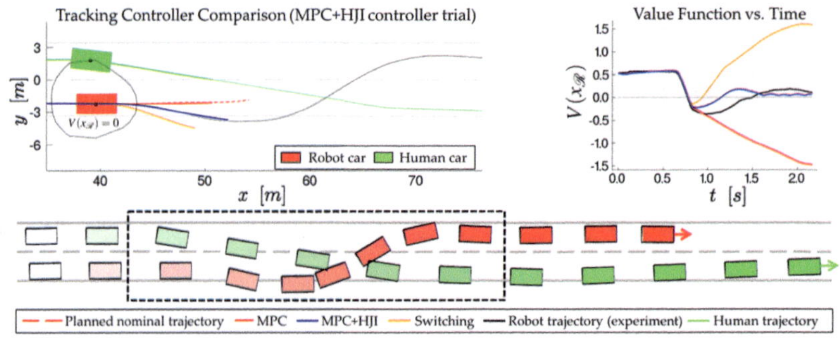

(a) Controller comparison on experimental data where the robot car uses a MPC+HJI controller.

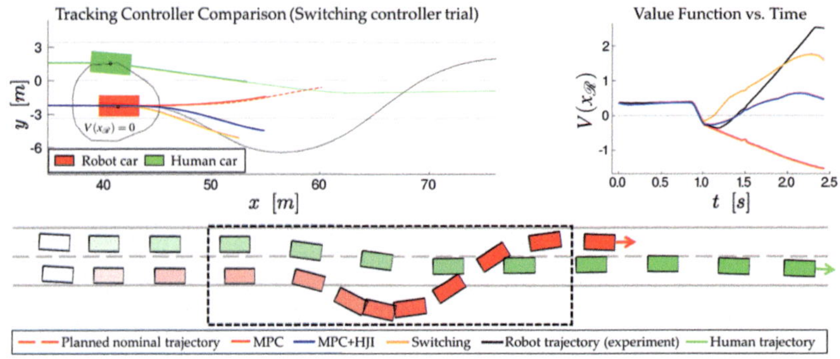

(b) Controller comparison on experimental data where the robot car uses a switching controller.

Fig. 6. Controller comparison on planner trajectories from X1/virtual human-driven vehicle experiments. Top left: A close up view of the different simulated controller trajectories when $V(x_{\mathscr{R}})$ first drops below $\epsilon = 0.05$. Top right: The corresponding evolutions of $V(x_{\mathscr{R}})$. Bottom: A time-lapse view of the traffic-weaving experiment, with the dashed box corresponding to top left figure.

controller.[6] As expected, we see that the MPC+HJI controller represents a middle ground between the tracking-only MPC which does not react to the human car's intrusion, and the switching controller which arguably overreacts with a large excursion outside the lane boundaries. Evident from Fig. 6a, our proposed controller tries to be minimally interventional—the robot car swerves/brakes but only to an extent that is necessary.

4.4 1/10-Scale Human-Driven Vehicle

To begin to investigate the effects of perception uncertainty on our safety assurance framework we use three LiDARs onboard X1 to track a human-driven RC

[6] For comparative purposes, the controllers were simulated with the displayed nominal trajectory held fixed, but in reality, the nominal trajectory in these experiments was updated at ~4 Hz.

car, and implement a Kalman filter for human car state estimation (position, velocity, and acceleration). Even with imperfect observations, we show some successful preliminary results (an example is shown in Fig. 7) at mean speeds of 4 m/s, close to the limits of the RC car + LiDAR-visible mast in crosswinds at the test track. We observe similar behavior as in the virtual human car experiments, including the fact that the value function dips briefly below zero before the MPC+HJI controller is able to arrest its fall; we discuss this behavior in the next section.

5 Discussion

Beyond the qualitative confirmation of our design goals, our experimental results reveal two main insights.

Takeaway 1: The reachability cache is underly-conservative with respect to robot car dynamics and overly-conservative with respect to human car dynamics.

In all cases—hardware experiments as well as simulation results—the HJI value function V dips below zero, indicating that neither the HJI+MPC nor even the optimal avoidance switching controller are capable of guaranteeing safety in the strictest sense. The root of this apparent paradox is in the computation of the reachability cache used by both controllers as the basis of their safety assurance. Though the 7-state relative dynamics model (4) subsumes a single-track vehicle model that has proven successful in predicting the evolution of highly dynamic vehicle maneuvers [2,12], the way it is employed in computing the value function V omits relevant components of the dynamics. In particular, when computing the optimal avoidance control (2) as part of solving the HJI PDE, we assume total freedom over the choice of robot steering angle δ and longitudinal force command F_x, up to maximum control limits. This does not account for, e.g., limits on the steering slew rate (traversing $[-\delta_{\max}, \delta_{\max}]$ takes approximately 1 s), and thus the value function is computed under the assumption that the robot can brake/swerve far faster than it actually can.

Fig. 7. Time-lapse of pairwise vehicle interaction: X1 with 1/10-scale human-driven RC car. The RC car (green trajectory) nudges into X1 (red trajectory), which swerves gently to avoid.

We note that simply tuning the safety buffer ϵ is insufficient to account for these unmodeled dynamics. In Fig. 6 we see that V may drop from approximately 0.5 (the value when the two cars traveling at 8 m/s start side-by-side in lanes) to -0.3 in the span of a few tenths of a second. Selecting $\epsilon > 0.5$ might give enough time for the steering to catch up, but such a selection would prevent the robot car from accomplishing the traffic weaving task even under nominal conditions, i.e., when the human car is equally concerned about collision avoidance. This is because the safety controller would push the robot car outside of its lane from the outset to maintain the buffer. This behavior follows from wide level sets associated with the transient control authority asymmetry (recall that in the HJI relative dynamics the human car may adjust its trajectory curvature discontinuously), assumed as a conservative safety measure as well as a way to keep the relative state dimension manageable.

The simplest remedy for both of these issues is to increase the fidelity of the relative dynamics model by incorporating additional integrator states $\dot{\delta}$ for the robot and $\dot{\omega}$ for the human. Naively increasing the state dimension to 8 or 9, however, might not be computationally feasible (even offline) without devising more efficient HJI solution techniques or choosing an extremely coarse discretization over the additional states.[7] We believe that simulation, accounting for slew rates, could be a good tool to prototype such efforts, noting that as it stands we have relatively good agreement between simulation and experimental platform in our testing.

Takeaway 2: Interpretability of the value function V should be a key consideration in future work.

In this work the terminal value function $V(0, x_{\mathscr{R}})$ is specified as the separation/penetration distance between the bounding boxes of the two vehicles, a purely geometric quantity dependent only on x_{rel}, y_{rel}, and ψ_{rel}. Recalling that V $(:= V_\infty)$ represents the worst-case eventual outcome of a differential game assuming optimal actions from both robot and human, we may interpret the above results through the lens of worst-case outcomes, i.e., a value of -0.3 may be thought of as 30 cm of collision penetration assuming optimal collision seeking/avoidance from human/robot. When extending this work to cases with environmental obstacles (e.g., concrete highway boundaries that preclude large deviations from lane), or multi-agent settings where the robot must account for the uncertainty in multiple other parties' actions, for many common scenarios it may be the case that guaranteeing absolute safety is impossible. Instead of avoiding a BRS of states that might lead to collision, we should instead treat the value function inside the BRS as a cost. In particular we should specify more contextually relevant values of $V(0, x_{\mathscr{R}})$ for states in collision, e.g., negative kinetic energy or another notion of collision severity as a function of the velocity states $U_{x_{\text{R}}}$, $U_{y_{\text{R}}}$, v_{H}, and r_{R} in addition to the relative pose. This would lead to a controller that prefers, in the worst case, collisions at lower speed, or perhaps "glancing blows" where the velocities of the two cars are similar in magnitude

[7] By literature standards we already use a relatively coarse discretization grid for solving the HJI PDE; associated numerical inaccuracies may be another source of the observed safety mismatch.

and direction. Computationally we could use such a value function in an MPC formulation similar to that presented in this work, with static obstacles either incorporated in $V(0, x_{\mathcal{R}})$ or represented as additional MPC constraints as done in [12], and accommodating multiple other agents by considering them pairwise with the robot and taking the minimum over the corresponding value functions.

6 Conclusions

We have investigated a control scheme for providing real-time safety assurance to underpin the guidance of a probabilistic planner for human-robot vehicle-vehicle interactions. By essentially projecting the planner's desired trajectory into the set of safety-preserving controls whenever safety is threatened, we preserve more of the planner's intent than would be achieved by adopting the optimal control with respect to separation distance. Our experiments show that with our proposed minimally interventional safety controller, we accomplish the high level objective (traffic weaving) despite the human car swerving directly onto the path of the robot car, and accomplish this relatively smoothly compared to using a switching controller that results in the robot car swerving more violently off the road. We note that this work represents only a promising first step towards the integration of reachability-based safety guarantees into a probabilistic planning framework. We have already discussed the concrete modifications to this controller we believe are necessary to improve the impact of these guarantees; further study should also consider better fitting of the planning objective at the controller level. That is, instead of performing a naive projection, i.e., the one that minimizes trajectory tracking error, it is likely that a more nuanced selection informed by the planner's prediction model would represent a better "backup choice" in the case that safety is threatened. We recognize that ultimately, guaranteeing absolute safety on a crowded roadway may not be realistic, but we believe that in such situations value functions derived from reachability may provide a useful metric for near-instantly evaluating the future implications of a present action choice.

References

1. Wolf, M.T., Burdick, J.W.: Artificial potential functions for highway driving with collision avoidance. In: Proceedings of IEEE Conference on Robotics and Automation (2008)
2. Funke, J., Brown, M., Erlien, S.M., Gerdes, J.C.: Collision avoidance and stabilization for autonomous vehicles in emergency scenarios. IEEE Trans. Control Syst. Technol. **25**, 1204–1216 (2017)
3. Arora, S., Choudhury, S., Althoff, D., Scherer, S.: Emergency maneuver library – ensuring safe navigation in partially known environments. In: Proceedings of IEEE Conference on Robotics and Automation (2015)

4. Bokanowski, O., Forcadel, N., Zidani, H.: Reachability and minimal times for state constrained nonlinear problems without any controllability assumption. SIAM J. Control Optim. **48**, 4292–4316 (2010)
5. Chen, M., Hu, Q., Fisac, J., Akametalu, K., Mackin, C., Tomlin, C.: Reachability-based safety and goal satisfaction of unmanned aerial platoons on air highways. AIAA J. Guid. Control Dyn. **40**, 1360–1373 (2017)
6. Dhinakaran, A., Chen, M., Chou, G., Shih, J.C., Tomlin, C.J.: A hybrid framework for multi-vehicle collision avoidance. In: Proceedings of IEEE Conference on Decision and Control (2017)
7. Gattami, A., Al Alam, A., Johansson, K.H., Tomlin, C.J.: Establishing safety for heavy duty vehicle platooning: a game theoretical approach. IFAC World Congr. **44**, 3818–3823 (2011)
8. Margellos, K., Lygeros, J.: Hamilton-Jacobi formulation for reach-avoid differential games. IEEE Trans. Autom. Control **56**, 1849–1861 (2011)
9. Mitchell, I.M., Bayen, A.M., Tomlin, C.J.: A time-dependent Hamilton-Jacobi formulation of reachable sets for continuous dynamic games. IEEE Trans. Autom. Control **50**, 947–957 (2005)
10. Fisac, J.F., Akametalu, A.K., Zeilinger, M.N., Kaynama, S., Gillula, J., Tomlin, C.J.: A general safety framework for learning-based control in uncertain robotic systems. J. IEEE TAC **64**(7), 2737–2752 (2018)
11. Schmerling, E., Leung, K., Vollprecht, W., Pavone, M.: Multimodal probabilistic model-based planning for human-robot interaction. In: Proceedings of IEEE Conference on Robotics and Automation (2018)
12. Brown, M., Funke, J., Erlien, S., Gerdes, J.C.: Safe driving envelopes for path tracking in autonomous vehicles. Control Eng. Pract. **61**, 307–316 (2017)
13. Yoon, S., Fern, A., Givan, R., Kambhampati, S.: Probabilistic planning via determinization in hindsight. In: Proceedings of AAAI Conferrence on Artificial Intelligence (2008)
14. Chen, M., Tomlin, C.J.: Hamilton-Jacobi reachability: some recent theoretical advances and applications in unmanned airspace management. Robot. Auton. Syst. Annu. Rev. Control **1**, 333–358 (2018)
15. Fisac, J.F., Chen, M., Tomlin, C.J., Sastry, S.S.: Reach-avoid problems with time-varying dynamics, targets and constraints. In: Hybrid Systems: Computation and Control (2015)
16. Tanabe, K., Chen, M.: BEACLS: Berkeley Efficient API in C++ for Level Set methods. https://github.com/HJReachability/beacls
17. Revels, J., Lubin, M., Papamarkou, T.: Forward-mode automatic differentiation in Julia (2016). https://arxiv.org/abs/1607.07892
18. Stellato, B., Banjac, G., Goulart, P., Bemporad, A., Boyd, S.: OSQP: An operator splitting solver for quadratic programs (2017). https://arxiv.org/abs/1711.08013
19. Koolen, T., Contributors: Parametron.jl: Efficiently solving parameterized families of optimization problems in Julia. https://github.com/tkoolen/Parametron.jl

Experimental Validation of an Energy Constraint for a Safer Collaboration with Robots

Lucas Joseph[1,2,3(✉)], Vincent Padois[1], and Guillaume Morel[1,3]

[1] Sorbonne Université, CNRS UMR 7222, Institut des Systèmes Intelligents et de Robotique, ISIR, 75005 Paris, France
{joseph,padois,morel}@isir.upmc.fr
[2] General Electric Healthcare, 78530 Buc, France
[3] INSERM, U1150, Agathe-ISIR, 75005 Paris, France

Abstract. In this paper a platform dedicated to the measure of the energy dissipated by a robot during an impact is presented. This platform is used to validate a controller that explicitly constrains the robot kinetic energy thanks to a linear quadratic problem formulation. This constraint is used to ensure safety at the control level by computing intrinsically safe commands that comply with ISO norm 15066 on collaborative robot safety. This platform has been designed to validate this approach in case of a collision with an obstacle and is experimented on a KUKA LWR4+ robot.

1 Introduction

The further adoption of robots for collaborative tasks with an operator must go through an enhancement of the user safety. To that extent several research projects have been conducted. They led to the introduction of lightweight robots with a reduced mass and torque sensing capabilities [1], to new variable impedance actuators [2] and solutions to detect and avoid obstacles with exteroceptive sensors [3].

Collaborative robots are designed to work closely with humans. To perform their tasks, these robots have torque/force capacities that make them potentially dangerous for human-robot collaboration. A first measure to improve safety in the vicinity of these robots consists in avoiding unwanted contact with humans. Different obstacle avoidance algorithms have been proposed in the literature. [4] expose a technique called elastic bands that updates the robot trajectory on-line to define a collission-free path. Potential fields methods, first introduced by [5], create virtual forces around obstacles. The magnitude of these forces is linked to the distance between the robot and an obstacle. Whether obstacle algorithm are implemented by mean of control or through path planning, in cluttered environments with unpredictable human motions, they cannot guaranty that no contact can happened between a fixed base robot and a human.

J. Xiao et al. (Eds.): ISER 2018, SPAR 11, pp. 575–583, 2020.
https://doi.org/10.1007/978-3-030-33950-0_49

If contact cannot be avoided, it should at least be safe for the operator. A common approach is to implement algorithms to detect and react to a collision. Many detection methods exist. They range from comparing the robot actual torques to the theoretical ones necessary to perform its task to observing the change of energy in the system [6]. For a complete review on collision detection algorithms, one may refer to [7]. Once a collision is detected, the work of [8] proposes to stop the robot or to switch its control mode to a zero-gravity torque control. Alternatively, [9] develops a control solution to realise motion and/or force control at any point on the robot and propose to control the force exerted by a robot at the point where a collision occurred.

As effective as these techniques can be, detecting and reacting to a collision is not sufficient to ensure safety. If at the moment of impact the robot is moving too fast or with an important load, the collision by itself might be fatal. Before reacting to a collision, the robot should be actuated in a way that is intrinsically safe. ISO Norm 15066 [10] proposes a new definition of a robot dangerousness during transient contact, *i.e.* an impact. The norm proposes to quantify a robot kinetic energy and provides energy limits to respect in order to avoid severe human injuries.

In a previous work [11], a robot kinetic energy is expressed as a constraint inside a quadratic programming (QP) problem. This constraint ensures that the robot does not exceed a defined limit considered as dangerous. It shows that by correctly expressing the constraint, it is possible to ensure safety in different cases of collision. However, the kinetic energy computed through this method is based on a theoretical dynamic model of the robot and does not guaranty to be representative of the true robot kinetic energy.

To that extent, a platform has been developed to measure the energy transferred by a robot during contact. This platform is used in this work to validate the approach proposed in [11] by comparing the energy measured by the platform to the one computed with the robot model.

In a first section, the QP formulation is explained and the kinetic energy constraint is detailed. In a second section the experimental set-up for the experiments is detailed. In a third section several experiments are realised and experimental results are presented to validate the correct enforcement of the proposed energy constraint. The last section concludes on this work and proposes further improvements that are being investigated.

2 Constrained Convex Problem Formulation

The control of a robot can be expressed as an optimization problem where cost functions define the robot tasks and constraints modelize the robot actuation limits [12]. The optimization problem requires expressing tasks and constraints as functions of the control variable. In this work, the control variable is the joint torque, $\boldsymbol{\tau} \in \mathbb{R}^n$, where n is the robot number of degree of freedom (dof). This section focuses on the expression of a quadratic programming problem to ensure safety by defining a kinetic energy constraint.

2.1 Tasks

The robot tasks are formulated in terms of commanded task space acceleration $\dot{\boldsymbol{v}}^* \in \mathbb{R}^6$, where $\boldsymbol{v} \in \mathbb{R}^6$ is the end-effector twist expressed at some point of interest. The tracking of this task is ensured by a PID controller. A robot task can thus be expressed as:

$$\boldsymbol{T}(\boldsymbol{\tau}) = \dot{\boldsymbol{v}}^* - \dot{\boldsymbol{v}}(\boldsymbol{\tau}) \tag{1}$$

Using the robot equation of motion, it is possible to express tasks as a function of the robot torque such that

$$\boldsymbol{T}(\boldsymbol{\tau}) = \dot{\boldsymbol{v}}^* - (\dot{J}\dot{\boldsymbol{q}} + JM^{-1}(\boldsymbol{\tau} - \boldsymbol{b}(\boldsymbol{q}, \dot{\boldsymbol{q}}) - \boldsymbol{g}(\boldsymbol{q}))) \tag{2}$$

where $\boldsymbol{q}$ and $\dot{\boldsymbol{q}} \in \mathbb{R}^n$ are respectively the joint configuration and its derivative, J and $\dot{J} \in \mathbb{R}^{6\times n}$ represent the robot Jacobian and its derivative expressed at the end-effector point. $\boldsymbol{b}(\boldsymbol{q}, \dot{\boldsymbol{q}}) \in \mathbb{R}^n$ is the centrifugal and Coriolis induced joint torque, $\boldsymbol{g}(\boldsymbol{q}) \in \mathbb{R}^n$ the gravity induced joint torque and $M(\boldsymbol{q})$ is the $n \times n$ joint-space inertia matrix.

2.2 Constraints

Quadratic programming allows expressing constraints as inequalities. These constraints prevent the robot from reaching forbidden states. The robot intrinsic constraints, *i.e.* its bounds on joint positions, velocities and torques are one of the constraints that should be considered when controlling a robot. These constraints can all be expressed as inequalities and using a proper discretization scheme such as in [12] they can overall be written under the general form

$$C(\boldsymbol{q}, \dot{\boldsymbol{q}})\boldsymbol{\tau} \leq \boldsymbol{h} \tag{3}$$

with $C(\boldsymbol{q}, \dot{\boldsymbol{q}})$ the constraint matrix and $\boldsymbol{h}$ its associated limit vector. The specific case of the energetic constraint is dealt with in Sect. 2.4.

2.3 Quadratic Programming Formulation

The control of the robot requires finding the optimal command input, $\boldsymbol{\tau}^{opt}$, satisfying the constraints and minimizing the tasks. This problem can be expressed as a quadratic programming optimization which is defined as

$$\begin{aligned} \boldsymbol{\tau}^{opt} = &\underset{\boldsymbol{\tau}}{\text{argmin}} \frac{1}{2}\left(||\boldsymbol{T}(\boldsymbol{\tau})||^2 + \varepsilon||\boldsymbol{r}||^2\right) \\ &s.t. \quad C(\boldsymbol{q}, \dot{\boldsymbol{q}})\boldsymbol{\tau} < \boldsymbol{h} \end{aligned} \tag{4}$$

where $||.||$ is the euclidean norm, $\boldsymbol{r}$ is a regularization term ensuring the uniqueness of the optimized solution and $\varepsilon \ll 1$ is a weighting term allowing monitoring the impact of the regularization on the main tasks performances.

For a kinematically redundant robot the choice of $\boldsymbol{r}$ is important as it affects the robot behaviour when the defined tasks do not fully constrain the robot. $\boldsymbol{r}$ is a task using all degrees of freedom. A common regularization task consists in minimizing the control variable, so here $\boldsymbol{r} = \|\boldsymbol{\tau}\|$. In [11], the controller is implemented on a redundant robot and the regularization task is set to minimize $||\boldsymbol{\tau} - \boldsymbol{g}(\boldsymbol{q})||^2$. This regularization task minimizes the difference between the computed torque and the gravity induced external torque. It allows performing free motions in the null-space of the robot main tasks by physically interacting with it.

2.4 The Kinetic Energy Constraint

In its discrete form, the kinetic energy, $E_{c,k}$, of the robot end-effector can be written:

$$E_{c,k} = \frac{1}{2}\boldsymbol{v}_k^T \Lambda_k\left(\boldsymbol{q_k}\right)\boldsymbol{v}_k \tag{5}$$

where, $\Lambda_k\left(\boldsymbol{q_k}\right) = \left(J_k(\boldsymbol{q_k})M_k^{-1}(\boldsymbol{q_k})J_k^T(\boldsymbol{q_k})\right)^{-1} \in \mathbb{R}^{6\times 6}$ is the operational-space inertia matrix and $\boldsymbol{v}_k \in \mathbb{R}^6$ its current velocity.

Considering a time period, Δt, between two discrete time k and $k+1$, the robot velocity at the next time step can be approximated to

$$\boldsymbol{v}_{k+1} = \boldsymbol{v}_k + \dot{\boldsymbol{v}}_k^c \Delta t, \tag{6}$$

where $\dot{\boldsymbol{v}}_k^c$ is the current control acceleration. Considering that the modification of the configuration of the robot during the time interval Δt is small enough, it can be shown that the difference of kinetic energy between two successive discrete times can be written:

$$E_{c,k+1} = E_{c,k} + \underbrace{\left(\boldsymbol{v}_k \Delta t + \frac{1}{2}\dot{\boldsymbol{v}}_k^c \Delta t^2\right)^T}_{\Delta \boldsymbol{x}} \underbrace{\Lambda_k \dot{\boldsymbol{v}}_k^c}_{\boldsymbol{F}_k^c}. \tag{7}$$

On the one hand, $\Delta\boldsymbol{x}$ represents the expected variation of pose given an initial operational state $\{\boldsymbol{x}_k; \boldsymbol{v}_k\}$ and the commanded acceleration trajectory $\dot{\boldsymbol{v}}_k^c$. On the other hand, $\boldsymbol{F}_k^c = \Lambda_k \dot{\boldsymbol{v}}_k^c$, is the equivalent control wrench.

This formulation can be interpreted as the provisional kinetic energy in Δt seconds. This energy is expressed as a function of the robot torque using the equation of motion. It is then inserted as a constraint in the optimization problem so that $E_{c,k+1}(\boldsymbol{\tau}) \leq E_c^{lim}$. In [11], it is shown that this constraint can be used to limit the energy that can be transferred from the robot to an obstacle during transient contact and also to limit wrenches applied by the robot when it entraps an obstacle against a fixed surface. However, further validations are required to asses the correct respect of this energy limit.

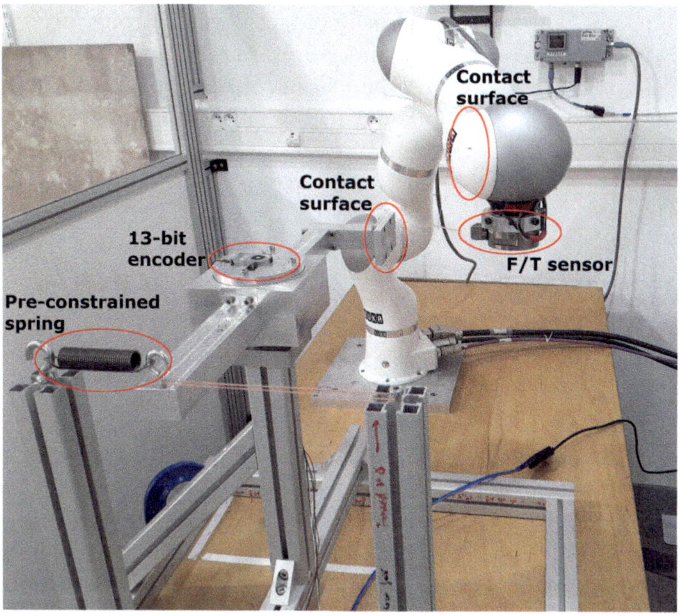

Fig. 1. Platform to measure the energy transferred by a robot during an impact. The elongation of the spring is measured by an encoder to determine the potential energy accumulated when an impact occurs.

3 Experimental Setup

This work proposes to use an external device to quantify the energy transferred from the robot to an obstacle during an impact. This platform, depicted in Fig. 1, measures the potential energy stored in a spring when a contact occurs. If the friction effects are negligible, this potential energy is equal the energy transferred by the robot at the impact.

This platform consists in a rigid structure on which is mounted a rotating beam. A spring is attached on one side of the beam and is pre-constrained. The robot enters in collision with the other side. Ball bearings are used to reduce friction at best. A 13-bit encoder is placed at the structure centre of rotation and a micro-controller queries its position at 500 Hz. The resulting elongation, Δx, of the spring is obtained using geometric relations. The spring stiffness, k, is measured beforehand using a micrometric table and a dynamometer with a 0.05 N precision. The potential energy, E_p loaded in the spring is equal to $\frac{1}{2}k\Delta x^2$. The robot and the platform are mounted on a fix and rigid table. A bracing is mounted on the beam carrying the encoder to reduce any deformation of the structure during the collision.

The control framework proposed in Sect. 2 is tested on a KUKA LWR4+. Torque commands are sent to the robot at 1 kHz using the KUKA Fast Research Interface through an OROCOS component [13]. The QP problem is implemented using the qpOASES library [14]. When contact occurs, strong inertial efforts are applied on the robot. An ATI Gamma sensor is used to record these forces. A threshold is applied to determine the instant of collision. The kinetic energy of the robot at that instant is compared to the maximum potential energy stored in the platform spring.

4 Experimental Results

This section provides experimental results obtained with the platform described previously. First the measurements of the platform are compared with the model based kinetic energy of the robot. In a second part, an experiment is realised to verify the respect of the kinetic energy constraint.

In the two experiments, the robot moves along a straight line with a target position located behind the platform. The robot trajectory tracking task is performed using a PID controller. When a contact is detected, the task is cancelled so that it only remains the regularization task, leaving the robot in gravity compensation mode. This is to ensure that the PID controller does not integrate the error resulting from the collision and thus inject more energy to the platform.

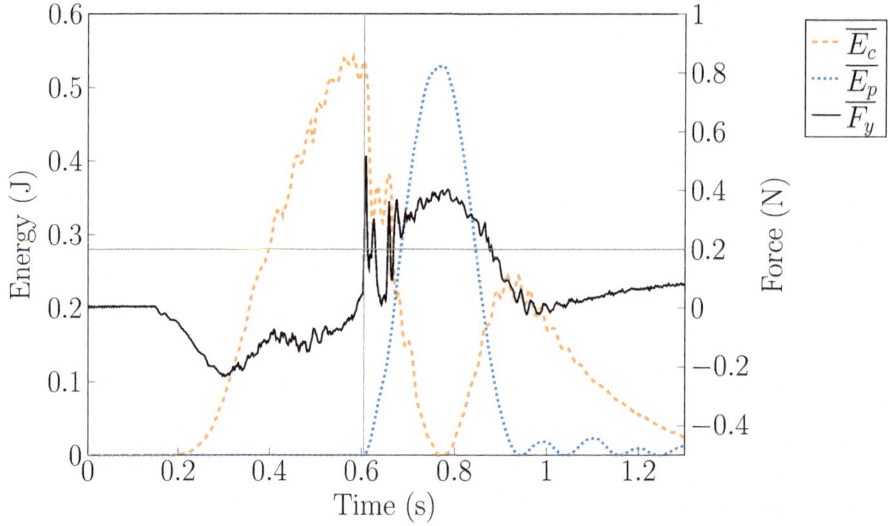

Fig. 2. Energy dissipation during an impact without a constraint on the kinetic energy.

Figure 2 compares the energy of the robot and the platform when a collision occurs. The orange curve represents $\overline{E_c}$, the average kinetic energy of the robot during the experiment over 10 trials. The blue curve represents the average potential energy stored in the platform spring, $\overline{E_p}$. In dark is the average force along the y-axis, $\overline{F_y}$, recorded by the ATI sensor. It can be seen that prior to contact, the forces recorded by the ATI are close to zero. When the collision occurs, a spike in the force is recorded. At the same instant a sudden drop of the robot kinetic energy can be observed as well as an accumulation of potential energy stored in the spring. Overall, the robot average kinetic energy at the instant of contact is 0.516 J. The average potential energy stored in the spring rises up to 0.528 J. It results a difference of 3.17% and a standard deviation over 10 trials of 0.012 J between the model based kinetic energy and the measurement of the platform. This experiment thus validates the model-based computation of the kinetic energy.

In a second experiment, depicted on Fig. 3., the kinetic energy limit, E_c^{lim}, is set at 0.2 J, bellow the 0.5 J required to correctly execute the task. The horizon time step Δt is set at 15 ms. Once again, the experiments are realised 10 times. The red curve represents the kinetic energy limit. The green curve represents the average predicted kinetic energy, $\overline{E_{c,k+1}}$, computed with the robot dynamic model. It is the energy, expressed in (7), that is constrained inside the QP solver. The predicted energy being expressed relatively to a horizon of time Δt, it represents the energy that the robot would have in that horizon of time. The QP solver finds torque commands to ensure that it never exceeds the limit E_c^{lim} (in red). Consequently, the resulting real kinetic energy is inferior to the predicted one. When the collision occurs, this current kinetic energy is on average equal to 0.168 J. The potential energy stored in the spring after impact rises up to 0.170 J. Over 10 trials, the standard deviation of the error is 0.002 J.

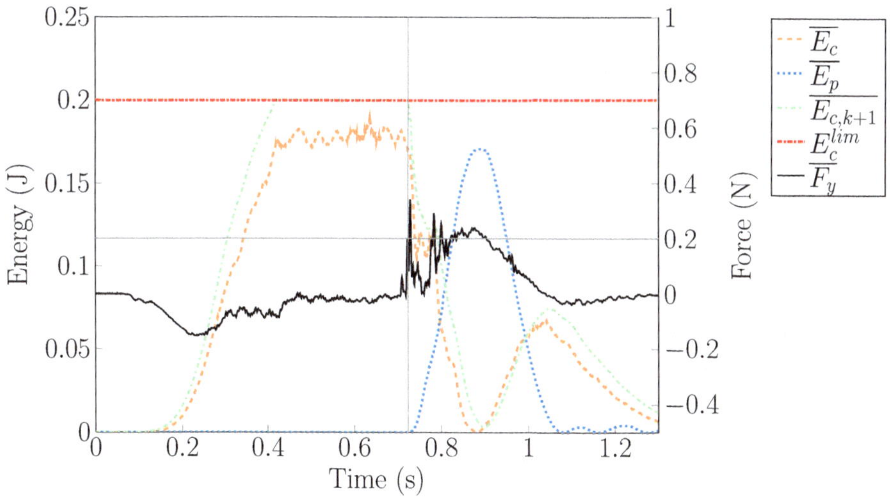

Fig. 3. Energy dissipation during an impact with a limited kinetic energy of 0.2 J.

The resulting energy that is constrained through this method is always inferior to the limit and validates the proposed solution. To be closer to the limit, it would require choosing a horizon of time Δt as close to the robot control period as possible. However, a too small time step could induce important variation of the robot commanded torques at the next time step which may lead to oscillation and commands that are not feasible by the actuators. Constraint compatibility should be taken into account as thoroughly detailed in [15].

5 Conclusion

Torque sensing robots offer new ways to interact with a robot and to detect collisions. However, depending on the robot current velocity and the tools it carries, the detection of a collision may not be sufficient to ensure safety. Ensuring safety thus also requires considering the case of unwanted contact. To alleviate this problem, ISO Norm 15066 proposes to limit the energy that can be transferred from the robot to a human body.

This work proposes a control solution that directly constrains the robot kinetic energy through a constrained convex optimization formulation. This kinetic energy constraint is expressed according to the theoretical model of the robot and does not mandatorily represents its real kinetic energy. To validate this model, a platform has been designed to record the energy transferred from the robot during an impact. The results show that the model based kinetic energy is similar to the real one. This platform is then used to show the correct enforcement of the kinetic energy constraint. It experimentally proves that the formulation of a kinetic energy constraint can efficiently prevent the robot from transferring a dangerous amount of kinetic energy during an impact, thus improving the safety of a robotic solution.

When the robot kinetic energy is not sufficient to realise the tasks, the QP solver finds torque solutions reducing the trajectory tracking performances. This induces a decrease of the robot velocity. The only way to maximise the robot velocity while satisfying this constraint is to minimize the robot operational inertia matrix in the direction of motion. In the work of Mansfeld *et.al.* [16], a gradient descent method is used to reactively compute, at each time step, a joint configuration minimizing the robot perceived mass in the direction of motion. Such algorithms could be used in the control framework presented in this paper and enforced using the regularisation task. It is possible to define a regularisation task that pushes the robot toward a desired position minimizing its perceived mass towards an obstacle. Ongoing work is being performed to test this implementation and yields promising results.

Acknowledgement. This work has been partially funded by ANRT under CIFRE grant n° 2015/0288.

References

1. Albu-Schäffer, A., Haddadin, S., Ott, C., Stemmer, A., Wimböck, T., Hirzinger, G.: The DLR lightweight robot: design and control concepts for robots in human environments. Ind. Robot Int. J. **34**, 376–385 (2007)

2. Bicchi, A., Tonietti, G.: Fast and "soft-arm" tactics. IEEE Robot. Autom. Mag. **11**(2), 22–33 (2004)
3. Flacco, F., Kroeger, T., De Luca, A., Khatib, O.: A depth space approach for evaluating distance to objects. J. Intell. Robot. Syst. **80**(1), 1–21 (2014)
4. Quinlan, S., Khatib, O.: Elastic bands: connecting path planning and control. In: Proceedings IEEE International Conference on Robotics and Automation, pp. 802–807 (1993)
5. Khatib, O.: Inertial properties in robotic manipulation: an object-level framework. Int. J. Robot. Res. **14**(1), 19–36 (1995)
6. De Luca, A., Albu-Schäffer, A., Haddadin, S., Hirzinger, G.: Collision detection and safe reaction with the DLR-III lightweight manipulator arm. In: IEEE International Conference on Intelligent Robots and Systems, pp. 1623–1630 (2006)
7. Haddadin, S., De Luca, A., Senior Member: Robot collisions: detection, isolation, and identification. IEEE Trans. Robot. **33**(6), 1292–1312 (2017)
8. Haddadin, S., Albu-Schäffer, A., De Luca, A., Hirzinger, G.: Evaluation of collision detection and reaction for a human-friendly robot on biological tissues. In: 6th IARP/IEEE-RAS/EURON Workshop on Technical Challenges for Dependable Robots in Human Environments (2008)
9. Magrini, E., Flacco, F., De Luca, A.: Control of generalized contact motion and force in physical human-robot interaction. In: ICRA, pp. 2298–2304 (2015)
10. ISO/TS 15066. Robots and robotic devices – collaborative robots (2016)
11. Joseph, L., Padois, V., Morel, G.: Towards X-ray medical imaging with robots in the open: safety without compromising performances. In: IEEE International Conference on Robotics and Automation (ICRA) (2018)
12. Rubrecht, S., Padois, V., Bidaud, P., De Broissia, M.: Motion safety and constraints compatibility for multibody robots. Auton. Robots **32**(3), 333–349 (2012)
13. Bruyninckx, H.: OROCOS: design and implementation of a robot control software framework. In: ICRA (2002)
14. Ferreau, H.J., Kirches, C., Potschka, A., Bock, H.G., Diehl, M.: qpOASES: a parametric active-set algorithm for quadratic programming. Math. Program. Comput. **6**(4), 327–363 (2014)
15. Meguenani, A.: Safe control of robotic manipulators in dynamic contexts. Université Pierre et Marie Curie - Paris VI, Theses (2017)
16. Mansfeld, N., Djellab, B., Rald, J., Beck, F., Ott, C., Haddadin, S.: Improving the performance of biomechanically safe velocity control for redundant robots through reflected mass minimization. In: IROS, pp. 5390–5397 (2017)

A Study of Reaching Motions for Collaborative Human-Robot Interaction

Sara Sheikholeslami[1], Gilwoo Lee[2], Justin W. Hart[3], Siddhartha Srinivasa[2], and Elizabeth A. Croft[4(✉)]

[1] Department of Mechanical Engineering, University of British Columbia, Vancouver, BC, Canada
sara.sheikholeslami@ubc.ca
[2] Paul G. Allen School of Computer Science & Engineering, University of Washington, Seattle, USA
[3] Faculty of Computer Science, University of Texas at Austin, Austin, TX, USA
[4] Faculty of Engineering, Monash University, Clayton, Australia
elizabeth.croft@monash.edu

Abstract. In human-human interactions, individuals naturally achieve fluency by anticipating the partner's actions. This predictive ability is largely lacking in collaborative robots, leading to inefficient human-robot interactions. Fluent meshing in human-robot collaboration requires the robot to make its intentions clear to its human collaborator. We propose a unified generative model of human reaching motions that allows the robot to **(a)** infer human intent, and then **(b)** plan its motion to be legible, or intent-expressive. We conducted a study on human reaching motion and constructed an elliptical motion model that is shown to yield a good fit to empirical data. In future studies, we plan to confirm the effectiveness of this model in predicting human intent and conveying robot intent for achieving fluency in human-robot handovers.

1 Introduction

We focus on single-arm reaching motions as a natural communication channel in tightly coupled physical human-robot collaborative tasks, like a human-robot handover scenario shown in Fig. 1. Such collaboration requires legible coordination of the specific behavior of human-robot handovers: as the human reaches out to handover the water-bottle, the robot should be able to tell early on and reach out to receive the bottle (Fig. 1).

Similar to human-human interactions, attaining fluency in human-robot interactions requires the collaborative robot to be able to infer the intentions of its human counterparts in order to determine the next appropriate action [1–3]. Likewise, in taking actions, the robot must balance two, often juxtaposed, objectives: **(1)** moving in a *predictable* (expected) way that the human trusts and understands [4], and **(2)** moving in a *legible* (intent-expressive) way that conveys its intent to the human - it enables the inference of intentions [5].

J. Xiao et al. (Eds.): ISER 2018, SPAR 11, pp. 584–594, 2020.
https://doi.org/10.1007/978-3-030-33950-0_50

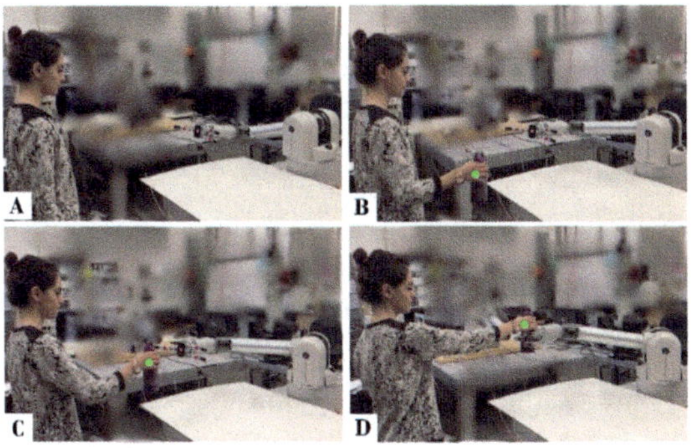

Fig. 1. An example of a human-to-robot handover in a shared-workspace collaboration: Human initial hand motion (**A, B**) is used to make predictions about the remainder of the path, generating a suitable trajectory for the robot's motion to grasp the bottle (**C, D**).

Few studies have sought to develop algorithms for the prediction of goal-directed human reaching motions [6], and generating legible robot motions [5]. These studies often model human-like motion as complex and time-expensive cost functions capturing different aspects of how the human observer(s) expects the robot to move [7,8]. Such trajectory optimization techniques present two major challenges in high dimensional space: **(1)** learning the cost function based on context as the efficiency of robot motion has different interpretations for different observers [9,10]; and **(2)** finding an efficient robot trajectory in real time that minimizes such non-convex cost functions, often subject to local minima [11]. In more ambiguous situations with many possible reach targets, such as a collaborative assembly tasks, the robot is faced with a much more challenging burden of conveying its intent. Modeling fast predictions of how the human observer expects the robot to move to be intent-expressive is an active research challenge that benefits from investigating some regularities of biological motion control [12].

Work in neuroscience [13,14] and biomechanics [15,16] suggests that the central nervous system (CNS) formulates motor commands for human arm reaching movements in terms of spatial trajectories of the hand rather than joint angles [14,17]. Findings indicate that the hand follows a predictable path with smooth and symmetric bell-shaped velocity profiles [18], motivating our investigation of reaching motion geometry to model human reaching motions.

2 Research Goals and Hypotheses

Our overarching goal is to create a single, unified model for describing reaching motions that can be solved in real-time for both motion prediction and motion production of collaborative robots in various shared human-robot collaborative work-space paradigms. This requires **(a)** understanding and predicting how people coordinate their reaching motions, **(b)** modeling these mechanisms, and **(c)** incorporating the models into robots to produce trajectories that synchronize naturally with human motion. This work primarily addresses **(a)** and **(b)**, setting out to better understand and construct models of human reaching motions. To this end, we proposed the following Goal (**G1**) and Hypothesis (**H1**):

- **G1:** Understanding and modeling unconstrained goal-directed human reaching movements.
- **H1:** For human reaching movements, the hand trajectory follows a predictable ellipse-shaped path with smooth velocity profiles.

If **H1** holds true, an elliptical fit to the motion early in the path can be predictive of the remainder of the path, and, subsequently, models of velocity and acceleration can be used to determine the final timing and location of the reach trajectory early in the process. This predictive ability is necessary to attaining fluent meshing in human-robot team activities.

3 A Study of Legible Motion

To better understand and construct models of human reaching motions, a high-fidelity Vicon motion capture system was used to track human participants reaching movements in a simulated box packing task. The task was designed to represent a possible warehouse scenario with the goal of generating a sample of reaching movements that would be naturally occurring in the application domain.

Central Insight: *Human unconstrained reaching motion tends to follow an arc shaped trajectory that can be modeled by an ellipse.*

3.1 Task

Participants were asked to fulfill the role of a warehouse packer in a shipping operation while we recorded their movements. Each participant sat at a desk. Surrounding them were a ramped shelf with five columns of colored ball supplies, a ball holder with a single space for intermediate ball placement, and a rack consisting of a $3x3$ square grid with a total of 9 possible spaces for final packing, Fig. 2. The experimental task proceeded in the following phases:

P1. *Picking*, in which products are staged and scanned to be packed. This phase involves unshelving and staging a ball by removing a ball from the ramp and placing it on the ball-holder; followed by

P2. *Packing*, in which products are packed for shipment. This phase involves unracking the current ball from the ball-holder and arranging it on an available spot on the rack.

The *Picking* and *Packing* tasks were repeated 9 times, such that the rack was full. Participants were instructed to pick and pack balls in any order they desired as the final arrangement of the balls on the rack was not important—The

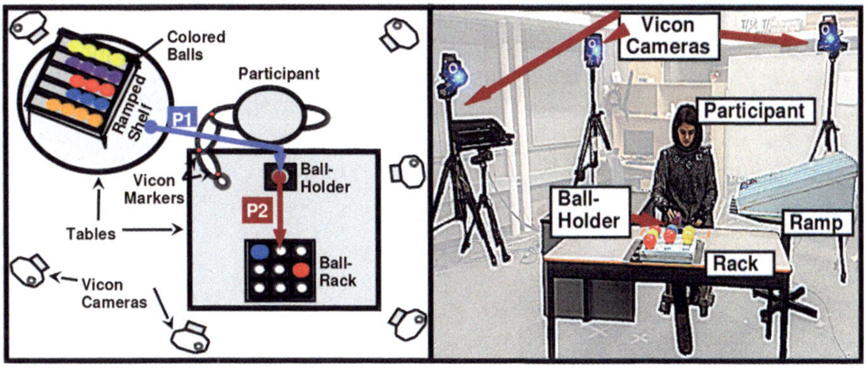

Fig. 2. Simulated Box Packing task designed with the goal of understanding and modeling point-to-point human reaching movements. **P1** corresponds to trajectories collected from the *Picking* phase, i.e. from ramped shelf to ball-holder, and **P2** corresponds to trajectories collected from the *packing* phase, i.e. from ball-holder to ball-rack.

goal was to allow participants to focus their attention on completing the task at hand (see discussion of limitations in Sect. 6). After, in order to prepare for another round of testing, participants reset the setup by removing balls from the rack and placing them back onto the shelf.

Each trial was repeated 15 times, yielding a total of 135 unshelving and racking motion samples (**P1**), and 135 unracking and packing motion samples (**P2**) per participant. For the study, 10 participants were recruited.

3.2 Data Annotation

The Vicon motion capture system used in this study included six cameras that performed data processing to create real-time raw $2D$ marker data. To minimize distortion of the movement and improve tracking accuracy (i.e., fill in any gaps in the trial data), a total of 19 markers were attached to each participant's upper body to measure the kinematics of their motion, and construct a calibrated $3D$ model of their upper-body skeletal structure[1].

The results herein are based on the reconstructed motion of the wrist marker, placed above the head of the ulnar bone, which we take as representing the movement of the arm in space. Two different categories of reach trajectories were collected from each trial: (**P1**) corresponding to motion samples from the ramp to the ball-holder during *Picking*, and (**P2**) corresponding to motion samples from the ball-holder to the rack during *Packing*. The start and end of each reach sample was determined through simple heuristics such as the wrist position, velocity, and distance from the target.

[1] See Vicon Nexus Documentation at https://docs.vicon.com.

4 Building Elliptical Models of Human Reaching Motion

Ellipses arise as second degree curves generated by the intersection of a plane and a cone. Therefore, to justify an elliptical fit to the reach data, the reach trajectory must be planar (Sect. 4.1). After verifying planarity, an ellipse is fit to the motion points projected onto the best fit plane; the fit error is measured as the sum of the squared orthogonal distances between the found ellipse and the motion points (Sect. 4.2).

4.1 Modeling Planarity of Reaching Motion

To compute the best fit plane to a motion trajectory, principal component analysis (PCA) can be applied to each motion trajectory, providing an intuitive visualization and approximation of a new orthogonal basis along which the motion points are mainly distributed (cf. Fig. 3). In the context of this study, the covariance matrix Σ is a 3 by 3 positive-definite or positive-semidefinite matrix constructed from the $3D$ coordinates of a single reach sample as [19]:

$$\Sigma = \frac{(\mathbf{X} - \mathbf{X}_{centroid})(\mathbf{X} - \mathbf{X}_{centroid})^T}{n} \tag{1}$$

where n is the number of the points collected in one reaching motion, and $\mathbf{X}$ is a 3 by n matrix with all the 3D coordinates of the points row-stacked. $\mathbf{X}_{centroid}$ is the mean coordinate, i.e., the centroid of the reach points.

Eigenvectors of Σ (i.e. the principal components (PCs)) point to 3 mutually perpendicular directions along which the reach points, n, are mainly distributed, and subsequently, construct the basis of the best fitting plane. The corresponding Eigenvalues of Σ indicate the mean spread of the points along each PC. Therefore, in practice, due to the presence of noise in laser scanning data, a reach trajectory can be assumed planar only if the smallest PC, approximating the mean deviation of the points normal to the best fit plane surface, is found statistically insignificant compared to the other two non-trivial PCs constructing the plane of the motion.

4.2 Modeling Reaching Motion as an Ellipse

The equation of a an ellipse in cartesian coordinates is a polynomial of degree two with five conic coefficients, defined as $Ax^2 + Bxy + Cy^2 + Dx + Ey + F = 0$. For a sample trajectory projected onto the best fit plane as described in Sect. 4.1, each transformed motion position $(x; y)$ places one constraint on these conic coefficients. Therefore, the best fit conic to the motion points can be modeled as the null vector of the matrix obtained from stacking the constraints from n wrist positions as it moves through space [20].

The type of the conic section, whether elliptical or not, can be determined from the sign of the invariant discriminant of the conic section. For a conic to be elliptical, 2 conditions must satisfy [21]:

$$B^2 - 4AC < 0 \tag{2a}$$

$$A \neq C \tag{2b}$$

For elliptical conics, determined by the above conditions, The fit error is measured as the mean of the squared orthogonal distances between the found ellipse and the motion points (Sect. 5.2).

5 Results

Results presented in this section are based on **P1** and **P2** movements collected from 9 participants. For each participant, the number of sampled trajectories is indicated as **N**, and correspond to the sample of trajectories that we were able to reconstruct from the scanned data due to the noise inherent in the Vcon system (e.g., Fig. 4).

5.1 Analyzing Planarity of Human Reaching Motions

Principal component analysis (PCA) was performed to each of the collected reaching trajectories from **P1** and **P2** for all participants to model the orthogonal bases defining the plane of the motion (Fig. 4). Motion planarity is approximated by the smallest principal component which corresponds to the mean offset of the fit motion plane with respect to the horizontal X-Y plane of the camera.

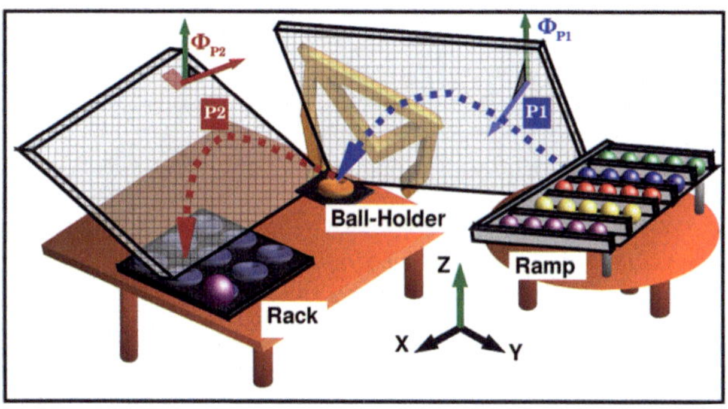

Fig. 3. The 3D view of the box packing experiment. For each phase, (**P1**, and **P2**) , ϕ represents the angle between the normal vector of the best fit plane to each reach sample, and the Z-axis of the global coordinate system.

A. *The PCA Solution:* The covariance matrix Σ, defined in Sect. 4.1, has two degrees of freedom comparing the spread of points along the transformed basis of the fit plane: the ratios $\{eigenvalue3 : eigenvalue2 : eigenvalue1)\}$, or equivalently the 3 principal components (PCs) less one for scale. Figure 4 provides a visualization of the mean of the smaller of the two ratios, comparing the minor planar distribution of motion (i.e. along the vertical axis of the fit plane, PC_2) to its out of plane spread (PC_3).

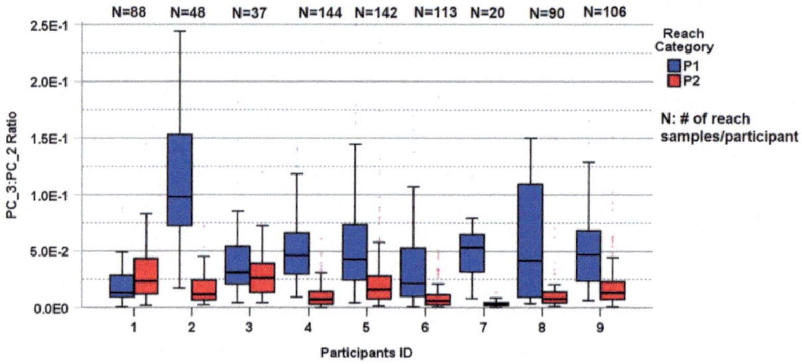

Fig. 4. Motion planarity is evaluated from the PCA solution, providing a comparison of the motion distribution along the minor axis of the modeled motion plan, approximated by PC_2, to the deviation of the motion perpendicular to the plane (PC_3).

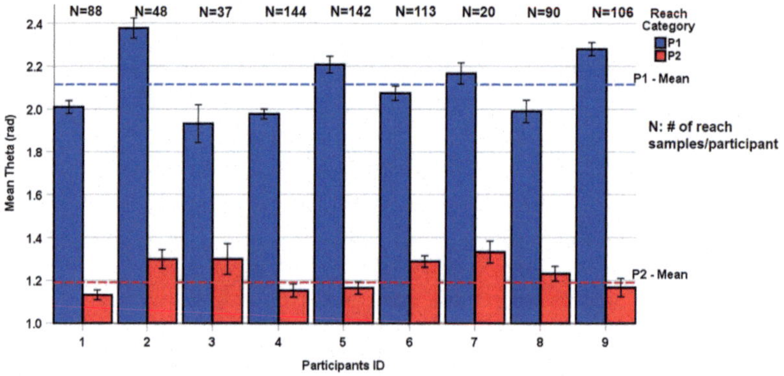

Fig. 5. ***Theta*** represents the angle between the normal vector of the best fit plane to each reach sample, and the Z-axis of the global coordinate system.

A *repeated measures ANOVA*, with a *Greenhouse-Geisser* correction, to account for the violation of sphericity assumption, was applied to the PCA solution of 1574 reconstructed trajectories from 9 participants (collected from Sect. 3) to evaluate the planarity of the sampled motions. Results of this analysis determined that the mean eigenvalue ratios were statistically significantly different ($F(1.368, 3233.32) = 12425.89$, $p < 0.0001$). Further, for each motion category, **P1** and **P2**, a *Bonferroni* analysis was performed to provide pairwise comparisons of the motion distribution along each principal direction. This post hoc test revealed that both motion categories, the distribution of the points perpendicular to the surface of motion, approximated as smallest eigenvalue of Σ (PC_3), is significantly negligible compared to the distribution of the points along PC_1 ($p < 0.0001$), or PC_2 ($p < 0.0001$), corresponding to the major and minor axis of motion plane, respectively. These experimental results indicate that human reaching motions can be accurately fitted to a plane, regardless of the constraints imposed during the experimental task (Sect. 3).

B. *Modeling the Motion Plane:* The smallest PC represents the vector normal to the plane of the motion, forming an inclination angle ϕ with the vertical Z-axis of the camera frame (Fig. 4). Our experimental results indicate that the fit planes to both **P1** and **P2** reaching movements are at an oblique angle to the vertical Z-axis of the camera frame, represented as inclination angles ϕ_1 and ϕ_2, respectively (Fig. 4). Analysis of *Independent-samples t-tests* was performed to the measures of ϕ from 1574 trajectories across **P1** ($N = 788$) and **P2** ($N = 788$) motion categories; ϕ_1 was found to be significantly smaller than ϕ_2 for all participants ($t(1492.61) = 83.69$, $p < 0.0001$). This analysis indicates that the participants consistently moved more horizontally to the camera X-Y plane to perform P2 reaching movements compared to P1 reaching movements (Fig. 5).

5.2 Evaluating the Fit of Elliptical Models to Human Reaching Motion

Analysis of the planar reach trajectories, transformed using PCA (Sect. 4.1), indicates that human reaching movements can be modeled as an ellipse 89% of the time, for both **P1** and **P2** motion categories ($N : 788, M : 88.35\%, SD : 11.50\%$, and $N = 788, M : 89.64\%.SD : 6.76\%$, respectively), see Fig. 6. Supporting **H1**, evaluation of the performance of an ellipse on approximating the motion path determined that the model can fit the sampled data within a mean accuracy of ± 0.31 mm for **P1**, and ± 0.23 mm for **P2** motion trajectories, see Fig. 7.

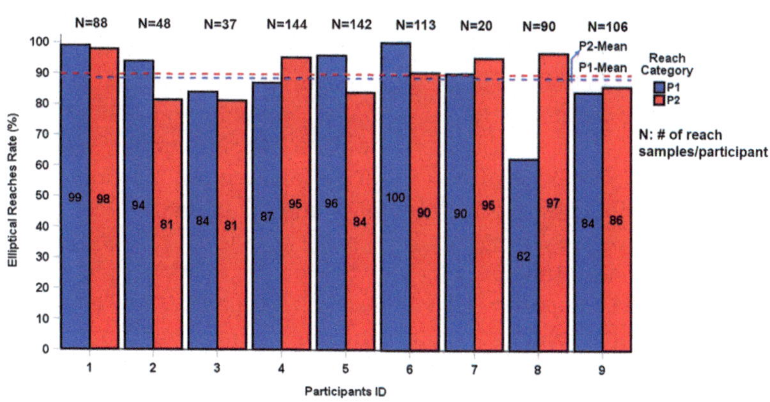

Fig. 6. Rate of reach trajectories for both (**P1**) and (**P2**) data (Sect. 3.1) that fit to an Elliptical Conic collected for the box-packing task across the participants.

Figure 8 illustrates two different views of the 3D coordinates of a sample trajectory collected from a random participant overlaid by the corresponding predictions of the elliptical model. The predicted ellipse closely fits the reach.

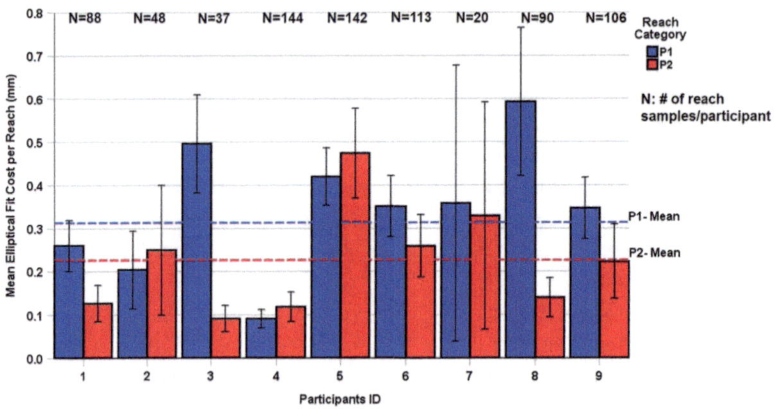

Fig. 7. The fit error measured as the sum of the squared orthogonal distances between the found ellipse and the motion points. The bars indicate the margin of error for a 95% confidence interval.

6 Discussion

This work proposes a mathematically simple and very efficient elliptical motion model of human movements that involves geometric fitting to the motion points, given the entire sampled trajectory. The analysis of our experimental data indicates that such model exists, and is able fit the motion within a mean accuracy few millimeters (Sect. 5.2).

A. Implications: Collaborating agents often share spaces, parts, tools and equipment and, in the normal course of their work, encounter conflicts when accessing such shared resources. To resolve such conflicts, each agent's selection of anticipatory actions must have a temporal dependency on the actions of their collaborating counterpart, and be based on **early** and **fast** predictions of the future state of their collaborator's motion. Existing motion planning techniques for these contexts primarily focus on learning and optimizing complex cost functions of biological motion control, rendering online and fast predictions of human-motion very challenging. The proposed elliptical motion model (**H1**) has potential to alleviate some of these issues due to its simple formulation. Future studies will evaluate the effectiveness of the elliptical model in predicting human intent to robot, and generating legible motion conveying robot intent for achieving fluency in human-robot handovers.

B. Limitations and Future Work: The presented model is preliminary, and developed based on the limited data sampled from 9 individuals. To test the validity of the elliptical model (**H1**), the experiment was designed to minimize the influence of any external constraints, such as visual information or proprioceptive disturbances, on determining the model of an ongoing movement. Our analysis only focused on the hand positional data, and did not include other types of motion, nor other important aspects, such as orientation, grasp condition, or the relative coordination of the gaze, head and hand.

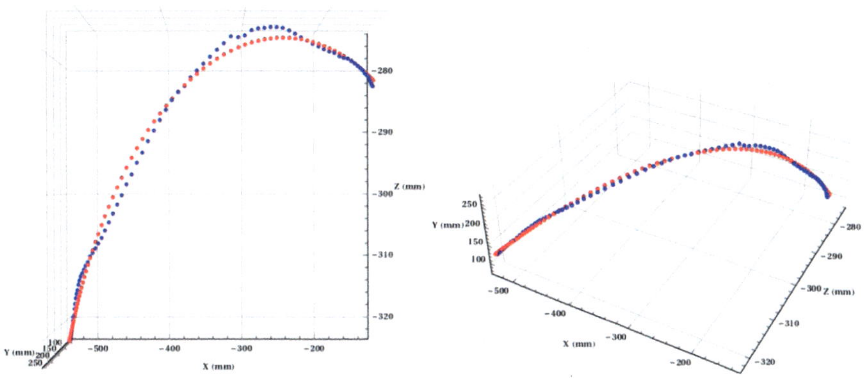

Fig. 8. A demonstration of the current state of the developed model applied to a sample reaching trajectory collected from the Vicon cameras: Top (*Left*) and Isometric (*Right*) views of the cartesian position of the wrist marker, in mm, as it moves through space (blue), and the corresponding elliptical model approximated to the wrist trajectory (red).

Future investigations will aim to validate and refine the current model by generating a more exhaustive exploration of the application domain, incorporating other constraints and variations. Our current analysis is investigating the effect of introducing visual cues on the onset of the movement by incorporating data collected from an analogous box-packing task (Sect. 3.1); however, instead of providing no guidelines to participants, the experiment was modified such that, during each phase, two computer monitors provided pictographic instructions on the completion of the task. Another limitation and exciting area of future work that our current analysis 5 lacks is to evaluate and compare motion consistency among individuals based on the measures of the inclination angle of the computed motion plane, and/or the fit ellipse parameters, such as the axis ratio.

6.1 Summary

This work has aimed to understand and construct models of human reaching motions to facilitate prediction of the timing and location of the end of human arm reaching motions in tasks such as grasping or object handover. Incorporating the elliptical model (**H1**) into robotic devices can improve human-robot interaction fluency by (1) providing faster predictions of human motion, enabling the robot to make quick inferences, and (2) generating time-efficient and human-like robot trajectories that conveys intent, enabling the human collaborator to focus their attention on completing the task at hand, rather than controlling, or understanding the robot.

References

1. Tomasello, M., Carpenter, M., Call, J., Behne, T., Moll, H.: Behav. Brain Sci. **28**(5), 675–691 (2005)
2. Dragan, A.D., Bauman, S., Forlizzi, J., Srinivasa, S.S.: In: Proceedings of the Tenth Annual ACM/IEEE International Conference on Human-Robot Interaction (ACM), pp. 51–58 (2015)
3. Cakmak, M., Srinivasa, S.S., Lee, M.K., Kiesler, S., Forlizzi, J.: In: Proceedings of the 6th International Conference on Human-robot Interaction. ACM, New York, NY, USA, , HRI 2011, pp. 489–496 (2011)
4. Beetz, M., Stulp, F., Esden-Tempski, P., Fedrizzi, A., Klank, U., Kresse, I., Maldonado, A., Ruiz, F.: Autonom. Robots **28**(1), 21 (2009)
5. Dragan, A.D., Lee, K.C.T., Srinivasa, S.S.: In: 2013 8th ACM/IEEE International Conference on Human-Robot Interaction (HRI), pp. 301–308 (2013)
6. Hoffman, G., Breazeal, C.: IEEE Trans. Robot. **23**(5), 952 (2007)
7. Lasseter, J.: In: ACM Siggraph Computer Graphics, vol. 21, pp. 35–44. ACM (1987)
8. Flash, T., Hogan, N.: J. Neurosci. **5**(7), 1688 (1985)
9. Argall, B.D., Chernova, S., Veloso, M., Browning, B.: Robot. Auton. Syst. **57**(5), 469 (2009)
10. Abbeel, P., Ng, A.Y.: In: Proceedings of the Twenty-First International Conference ON Machine Learning, p. 1. ACM (2004)
11. Dragan, A.D., Gordon, G.J., Srinivasa, S.S.: In: Robotics Research, pp. 309–326. Springer (2017)
12. Sciutti, A., Sandini, G.: IEEE Trans. Neural Syst. Rehabil. Eng. **25**(12), 2295 (2017)
13. Flash, T., Hogans, N.: J. Neurosci. **5**, 1688 (1985)
14. Harris, C., Wolpert, D. M.: Signal-dependent noise determines motor planning. Nature **394**(6695), 780 (1998)
15. Wu, G., van der Helm, F.C., et al.: J. Biomech. **38**(5), 981 (2005)
16. Todorov, E., Jordan, M.: Nat. Neurosci. **5**, 1226 (2002)
17. Morasso, P.: Exp. Brain Res. **42**(2), 223 (1981)
18. Collewijn, H., Erkelens, C.J., Steinman, R.M.: J. Physiol. **404**(1), 157 (1988)
19. Nurunnabi, A., Belton, D., West, G.: ISPRS Annals of Photogrammetry, Remote Sensing and Spatial Information Science, vol. I-3, pp. 269–274 (2012)
20. Hartley, R.I., Zisserman, A.: Multiple View Geometry in Computer Vision, 2nd edn. Cambridge University Press, Cambridge (2004). ISBN: 0521540518
21. Ayoub, A.B.: Math. Mag. **66**(5), 322 (1993)

Generating Expressive Light Signals for Appearance-Constrained Robots

Elizabeth Cha[1], Naomi T. Fitter[1,3(✉)], Yunkyung Kim[2], Terrence Fong[2], and Maja Matarić[1]

[1] University of Southern California, Los Angeles, CA 90089, USA
echa@usc.edu
[2] NASA Ames Research Center, Mountain View, CA 94035, USA
[3] Oregon State University, Corvallis, OR 97331, USA
naomi.fitter@oregonstate.edu

Abstract. Our goal is to enable robots to communicate information about their internal state using expressive light signals. Since nonverbal cues are typically given context by a robot's other actions, we combined light signals, varying their color, pattern, and frequency, with robot base motion and investigated the effects of these nonverbal signals on human observers' attribution of certain task-related properties (e.g., urgency and safety) to the robot. The results show that variations in light signal parameters affect human attribution of common notification properties. Using these findings, we present a first step towards systematically generating new light signals that can convey different state-related properties relevant for human-robot interaction, validated in a video-based study.

1 Introduction and Background

Robots are complex machines that require a high degree of expressivity to convey the many facets of their state [3,4]. Recent work has suggested the use of light signals as a robot communication modality because LEDs are affordable, reliable, salient, and versatile [3,4,14,16]. Modulating light signal parameters, such as color, position, duration, and frequency, can be used to communicate a wide range of state information for appearance-constrained robots [10,14].

Although light signals have been employed by many devices, there are few universal guidelines or principles to drive their design [10,14]. As a result, many signals are simplistic and not particularly iconic, requiring humans to learn them by rote [10]. As robots take on different forms and purposes, their signaling needs will continue to grow [4]. Learning new signals will become increasingly difficult, especially as humans reach the limits of their perceptual capability and memory.

The current method of creating specialized signals for each robot requires extensive time and effort from human designers. In contrast, this work aims to enable more automatic generation of light signals by leveraging existing human perceptions that are formed from prior experiences. A greater understanding of these perceptions not only facilitates intuitive signal design but also prevents

J. Xiao et al. (Eds.): ISER 2018, SPAR 11, pp. 595–607, 2020.
https://doi.org/10.1007/978-3-030-33950-0_51

design inconsistencies in which the robot inadvertently expresses information that conflicts with its communication objectives.

The primary contribution of this work is a exploratory user study investigating the effects of manipulating light signal parameters on a robot's expressiveness. Since robots typically have primary goals involving motion, we coupled the robot's light signals with base motion to mimic the realistic task-oriented settings that robots will operate in. The goal of this study is to support intuitive signal design and modulation of signal parameters to create a rich set of robot communication behaviors. The gained insights also facilitate the creation of common light signaling conventions across different robot platforms.

We focus primarily on perceptions of functional properties, rather than social or affective ones, as many current robots are used for task-oriented applications. These properties are general communication attributes (e.g., urgency), rather than concrete state information (e.g., low power). As prior work has shown that light signals may be too abstract to iconically represent precise information without additional training or context [10], the scope of this study was to determine the effects of light signal design characteristics on these attributes.

A secondary contribution of this work is the application of the initial study findings to several types of appearance-constrained robots and applications. We propose a model for generating robot light signals for human-robot interaction (HRI) and validated it in a video-based study with four different appearance-constrained robot platforms. As much of current HRI research still takes place in laboratory settings, it is important to understand how robot communication behaviors are affected by the context of these scenarios. This knowledge also facilitates the creation of standard nonverbal signals.

2 Technical Approach

The key insight of this work is that consideration of human users' existing perceptions or biases can be leveraged to design more intuitive light signals for robots. To better understand how the design of these signals affects humans' attribution of desired properties to a robot, we conducted an exploratory video-based user study that varied parameters of a robot's light signal and motion. The signal parameters and measured properties were derived from past works [13,15].

Since light signaling for robot communication is a nascent area of research, formulating well-informed hypotheses is still challenging. Instead, we explored several signal parameters and measured properties with the goal of building a foundation that will allow for generating light signals more systematically and for hypothesis testing in future research.

2.1 Robot

This study employed a Parrot AR drone, equipped with an array of 80 individually programmable Adafruit Neopixel RGBW LEDs embedded in the robot's outer foam hull and surrounded by white filter sheets for light diffusion (Fig. 1).

Fig. 1. The light signal patterns used in the exploratory video-based user study.

2.2 Manipulated Variables

We manipulated the robot's *light signal parameters* (color, pattern, and pattern frequency) [2,10] and its *base motion parameters* (path and speed) [8,15,18]. Figure 2 shows a table of the resulting conditions.

Light Signal Parameters: To better understand how existing signals affect human perceptions, we chose to study three common and fundamental light signal parameters: *color*, *pattern*, and *pattern frequency* [2,10,15].

1. Color- Color is a common differentiator for light signals on individual devices. Consequently, standards for its usage have emerged in certain fields, especially transportation (e.g., traffic signals). We included five common colors in this study: *green*, *red*, *orange*, *blue*, and *white* [10].
2. Pattern- Past work has investigated light signal patterns on both point light sources and light arrays [2,10]. We investigated four basic and frequently utilized patterns, as shown in Fig. 1. *Steady* is a static light signal in which all lights in the array are on. *Blink* consists of the entire array turning on and off at a constant frequency. *Beacon* consists of a set of three lights moving across the array [10,15]. *Wipe* is visually similar to a progress bar in which subsequent lights are turned on to fill the array.
3. Pattern Frequency- Since the steady pattern is static, it does not have a pattern frequency. For the other three patterns above, we varied the frequency or the speed of the pattern repeating itself to be either *slow* or *fast* with the fast condition about four times faster than the slow condition.

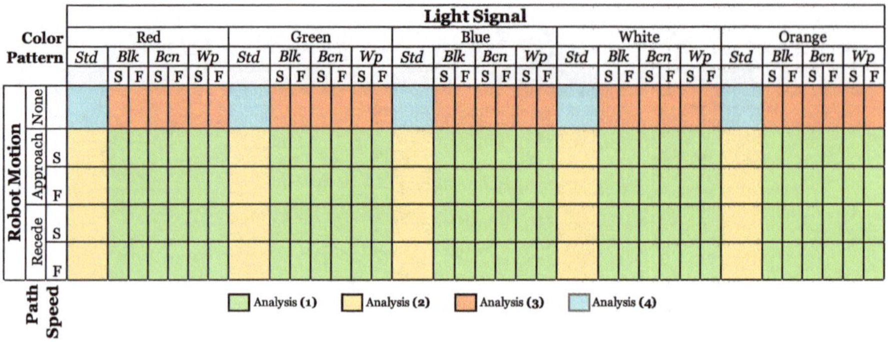

Fig. 2. The variables manipulated in the study: base motion parameters (path, speed) on the left and light signal parameters (color, pattern, pattern frequency) on the top. Cell coloring indicates separate analyses.

Motion Parameters: We manipulated two motion parameters: *path* and *speed*. Although these parameters were inspired by work in HRI and animation, few researchers to date have explored their effects on light signal perception.

1. Motion Path- Motion path is defined as the direction of the robot's base motion and has three levels, *no path*, *approaching*, and *receding*. In the no path condition, the robot hovers at a fixed altitude.
2. Motion Speed- We varied the robot's motion speed to be either *slow* or *fast*. The robot reaches a maximum velocity of $0.5\frac{m}{sec}$ in the fast condition and a maximum velocity of $0.15\frac{m}{sec}$ in the slow condition.

2.3 Participant Allocation

We generated a total of 175 video conditions, shown in Fig. 2. Since a fully within-subjects design was infeasible with the high number of conditions, participants were randomly assigned 7 of the video conditions (partially within-subjects), with a total of 18 participants per video condition [18].

We recruited participants using the Amazon Mechanical Turk platform. Participants were located in the United States, were at least 18 years of age, and had an approval rating of at least 95%. A total of 450 participants (48% male and 52% female; $M = 39.4$ years, $SD = 12.5$ years) participated in the study.

2.4 Dependent Measures

To assess how the light and motion parameters affect human attribution of desired task-related properties, we identified 3 dependent measures: *urgency*, *error*, and *difficulty to ignore*. These attributes and their measurement scales are derived from notification systems literature [13]. We used a unipolar, 5-point Likert scale ranging from "Not at all" to "Extremely".

		Light Signal			Robot Motion		Interaction Effect		
		Color	Pattern	Frequency	Path	Speed	Color, Pattern	Pattern, Frequency	Color, Pattern, Frequency
Notification Attributes	Urgency	✓	✓	✓	✓	✓	✓		
	Error	✓				✓			
	Difficulty to Ignore	✓	✓	✓	✓	✓		✓	

Fig. 3. Analysis **(1)** results: Check marks indicate a significant mean effect for the manipulated variable (column) and dependent measure (row).

1. Urgency- Urgency is defined as the need for immediate attention. Notification systems and alarm literature often explore how to convey varying levels of urgency in order to elicit appropriate reactions from the user.
2. Error- Since robots are embodied agents that can physically interact with the world, it is important for human safety and task success to consider the degree at which the robot is operating or functioning correctly.
3. Difficulty to Ignore- This property measures the salience of a signal is and is shown to affect the success and speed of eliciting a human response [5].

3 Results

In this section, we present an overview of the statistical analyses performed on the data gathered in the initial user study. From this data and the results of our analyses, we generated a decision tree for generating robot light signals.

3.1 Analyses

First, we performed a multivariate analysis to confirm that the items used as dependent measures were not highly correlated. Due to the unbalanced nature of study, we performed multiple analyses that examined subsets of our nested data. For each subset of data, we employed a factorial repeated measures analysis of variance (ANOVA). We present highlights of the analyses, primarily analysis **(1)** (Fig. 3) and include results that showed statistical significance, were surprising, or provided other interesting insights. We will refer to each of these analyses shown by the colored boxes in Fig. 2 using the numbers in the figure key.

Urgency: In analysis **(1)**, we found a significant main effect for all of the manipulated variables: light signal color ($F(3, 2159) = 5.24, p < 0.001$), pattern ($F(2, 2159) = 18.1, p < 0.001$), and pattern frequency ($F(1, 2159) = 71.87, p < 0.001$) and motion path ($F(1, 2159) = 9.84, p = 0.002$) and speed ($F(1, 2159) = 68.87, p < 0.001$). We also found an interaction effect between the light signal pattern and its frequency ($F(2, 2159) = 6.64, p = 0.001$).

Color was expected to have a strong effect on urgency, due to its strong usage in transportation. A post hoc analysis using Tukey HSD revealed, however, that only red and green were rated significantly different, with red perceived as the most urgent and green perceived as the least. We found similar results in analyses **(2)** and **(3)**, suggesting that color alone does not convey urgency.

Past research exploring perceptions of urgency in alarm and alert signal design have primarily looked at acoustic properties [1,9]. We found that varying similar signal characteristics (e.g., blinking vs. beeping cycle frequency) across modalities had the same effects on urgency, as shown in the post hoc analysis for **(1)**; all dynamic patterns were rated as significantly different with blink the most urgent and wipe the least.

As the robot's overall goal is ambiguous in the videos used in the study, we did not expect that its motion path would affect perceptions of urgency. However, approaching motions towards the camera were rated as significantly more urgent than receding motions. This could be due to participants framing the term urgency relative to themselves. The effect size of the motion path was the smallest ($\eta_p^2 = 0.005$) of our manipulated variables.

For the interaction effect found between the light signal pattern and frequency in analysis **(1)**, a post hoc analysis showed that conditions with blink and beacon patterns at the faster frequency had significantly higher ratings than the other conditions. Conditions with the fast pattern frequency were rated on average higher than those with the slow frequency; within these groups, the blink pattern was rated the most urgent and wipe the least. This suggests that the pattern frequency ($\eta_p^2 = 0.018$) has a stronger effect than the pattern ($\eta_p^2 = 0.035$).

Error: As high urgency scenarios are often associated with off-nominal conditions, such as errors, we expected similar findings for error and urgency. However, we found significant effects for only a subset of the dependent measures. In analysis **(1)**, a significant main effect was found for light signal color ($F(3, 2159) = 5.85, p < 0.001$) and motion speed ($F(1, 2159) = 29.85, p < 0.001$).

We expected that participants would draw on their experiences with colors used to signify error (e.g., red stop signs, emergency stop buttons, alarm lights). A post hoc analysis showed significant differences in the conditions with red and orange lights compared to conditions with green and blue lights. Surprisingly, white light was not found to be significantly different than any other color. As white is often a default light color, these results suggest that participants may have a more neutral or diverse perception towards its usage.

We also found that in the absence of motion (analyses **(3)** and **(4)**) color has no significant effect. This was surprising as static red or orange lights, symbols, and signs are often found in everyday life. As many of these instances combine color with a symbol, sound, or text, people may expect the signal to be less ambiguous when conveying important information such as errors (e.g., stop sign).

Surprisingly, we did not find any effect for signal pattern or frequency. We expected that light patterns with some variation, such as those resembling a police beacon or blinking fire alarm, would have a significant effect. The mean results showed that beacon was the highest-rated pattern for error and wipe was the lowest. The fast pattern frequency was also rated higher than the slow.

Difficulty to Ignore: We expected that signals with larger variations in their parameters over time will be more salient. This was confirmed in analysis **(1)**, as we found a significant main effect for light signal pattern ($F(2, 2159) = 9.8, p < 0.001$) and pattern frequency ($F(1, 2159) = 11.26, p = 0.001$). Color was shown

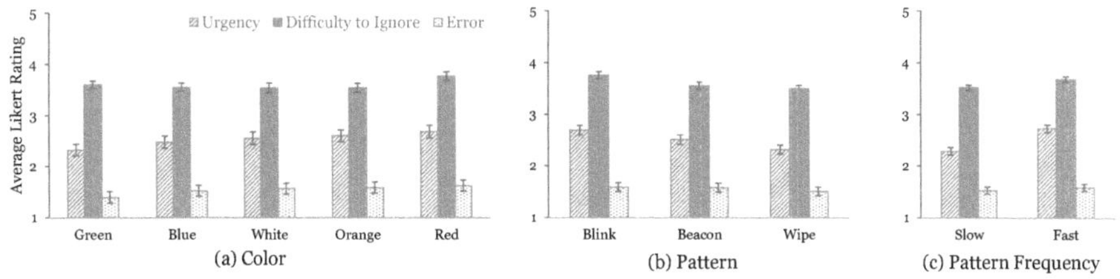

Fig. 4. The average Likert ratings (including 95% confidence intervals) for the urgency, difficult to ignore, and error measures in analysis **(1)**.

only to be significant ($F(3, 2159) = 3.11, p = 0.015$) in analysis **(1)**, indicating the weakness of color without motion or a light signal pattern for salience.

The post hoc analysis revealed that red and green were rated significantly differently than the other colors. Since past work has not shown color alone to have a significant effect on salience [17], this could suggest that participants rated light colors commonly utilized by other signals (e.g., traffic lights) higher. This could be because they are better trained to notice these signals or it could be an effect of utilizing videos rather than in-person interactions.

We expected that a higher number of LEDs changing over a set period of time would cause more apparent motion and increase saliency [6]. This was confirmed by the post hoc analysis for **(1)** which showed that the blink pattern was rated significantly higher than the beacon and wipe patterns. A significant mean effect was also found for pattern frequency, with the faster frequency rated more difficult to ignore.

Both motion parameters, path and speed, were found to have significant effects. We expected that participants would have more difficulty ignoring the approaching motion as the robot moves in the direction of the camera and ends in close proximity. This finding was confirmed, suggesting that motion biases found in prior work [7,11] also apply to robot motion when a dynamic light signal is present, as shown in analysis **(2)**. We also found that the faster motion was rated as significantly more difficult to ignore.

3.2 Light Signal Design Decision Tree

As a first step towards a more systematic approach for signal design, we propose a decision tree for light signal generation based on common parameters found in notification systems literature [12,13] (Fig. 5). Since the focus of this work is light signal design and motion acts as an implicit cue, only light signal parameters are included in the decision tree.

We employed the ID3 decision tree algorithm for each of the three notification attributes: urgency, error, and difficulty to ignore. The robot attributes were not included in the decision tree as their results were mixed and not as readily related with the signal design parameters used to split the tree.

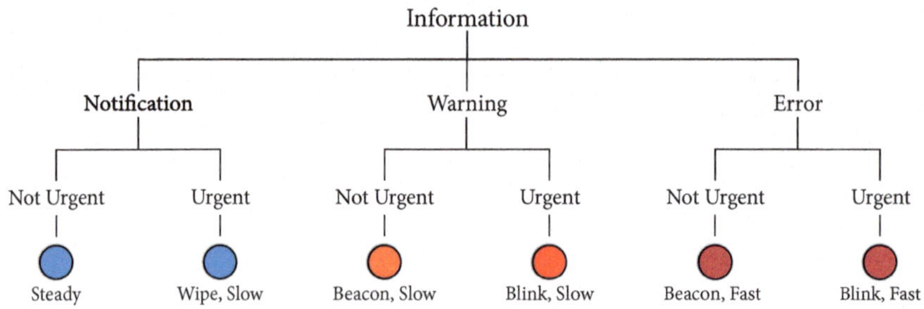

Fig. 5. A decision tree for light signal generation using the signal parameters investigated in the exploratory user study.

Signal Design Parameters. The decision tree includes two signal design parameters: *information type* and *urgency*. Our study findings suggested that light signals are better suited for conveying broader categories of information. Therefore, the goal of this decision tree is to generate a set of generalizable signals well suited for many HRI scenarios.

The *signal type* is defined as a notification, warning, or error. Notification is defined as information from the robot that may be beneficial for the human. Warnings and errors indicate off-nominal states with errors preventing successful operation of the robot. *Urgency* only had two levels, with the assumption that signals that are not urgent requires less attention from a human observer.

Notification Signal Design: Our study results showed two distinct color groupings: (1) red, orange and (2) green, blue, white. For notifications, we employed blue signals as green was rated more salient than blue or white, and comments from participants indicated that green might inadvertently express a positive state (e.g., operating correctly).

Since non-urgent notification signals are used for information that is not critical to the human or constrained by time, they require low levels of awareness or attention from the human interactor [12,13]. The steady light pattern was rated the lowest in salience, urgency, and error, making it well suited for non-urgent signals.

Since the wipe pattern was rated the next lowest for salience and error, it was chosen to represent urgent notifications. Although pattern frequency was shown to provide the highest information gain for urgency, the faster frequency increased salience significantly, leading us to choose the low frequency pattern.

Warning Signal Design: Color provided the highest information gain for the error measure with red and orange indicating the highest levels of error. Since orange (or yellow) is often employed for warning signals in other applications, we employed orange signals for the warning signal type.

We also assumed that warnings require a larger level of human attention, suggesting the need for higher saliency signals. We chose the beacon and blink signal

Fig. 6. The four appearance-constrained robots—the Astrobee, Fetch mobile base, Ohmni telepresence robot, and autonomous car—used in the validation study.

patterns, which were rated highest in saliency and error. We chose to employ only the slower pattern frequency for warning signals which are by definition less critical than error signals.

Error Signal Design: As red signals were rated significantly higher for urgency, error, and salience, red was employed for both types of error signals. Similar to the warning signal design, beacon and blink were chosen for their higher ratings of error and saliency. We assumed that all errors require higher levels of human awareness, and consequently, employed only the faster pattern frequency for error signals.

4 Validation Study

We conducted a video-based study validating the decision tree across four HRI contexts. These contexts involve different appearance-constrained, non-humanoid robots (Fig. 6) performing their respective tasks (e.g., autonomous car driving) while interacting with humans. For each robot, we generated videos of three different scenarios in which the robot employs light signals for communication (Fig. 7) and had participants rate the robot and its signals using several measures.

4.1 HRI Contexts

We chose four types of appearance-constrained robots, focusing on platforms and applications already found in the real world (Fig. 6). We added LED arrays to the models of the robots that do not already possess communicative lights.

Free-Flying Robot: We used the NASA Astrobee, a free-flying robot designed to perform tasks on the International Space Station (ISS) and operate in close proximity with human astronauts at all times.

Industrial Mobile Base: We used the Fetch mobile base, a robot that is significantly shorter than adult humans (16in. tall) and typically employed to transport loads around industrial environments, such as factories and warehouses.

Telepresence Robot: We used an OhmniLabs telepresence robot that is 48in. tall and designed to provide human operators with face-to-face communication capabilities and mobility in a remote environment.

Type		Scenario	Signal Type	Urgency
Free-Flying Robot	1	Waiting for new task	Notification (85%)	Not Urgent (90%)
	2	Issue while docking	Warning (55%)	Not Urgent (90%)
	3	Out of control motion	Error (60%)	Urgent (80%)
Industrial Mobile Base	1	Waiting for a person to pass	Notification (80%)	Urgent (10%)*
	2	Robot is physically stuck	Warning (45%)	Urgent (70%)
	3	Overheating	Error (45%)	Urgent (80%)
Telepresence Robot	1	Remote operator logging on	Notification (100%)	Urgent (0%)*
	2	Low battery	Warning (45%)	Urgent (65%)
	3	Loss of connection	Error (45%)	Not Urgent (65%)
Autonomous Car	1	Operating autonomously	Notification (65%)	Not Urgent (80%)
	2	Slowing down for pedestrian	Warning (70%)	Not Urgent (66%)
	3	Sensor damage	Error (65%)	Not Urgent (55%)

Fig. 7. Percentages of correct responses for the forced-choice measures: the asterisk indicates a majority of participants chose the incorrect response.

Autonomous Car: We used a Prius model taken from open source ROS repositories, modified with simulated sensors to mimic an autonomous vehicle capable of sensing the environment and navigating without human input.

4.2 Participant Allocation

We recruited 80 participants (67.5% male and 32.5% female; $M = 34.5$ years, $SD = 11.5$ years) using the Amazon Mechanical Turk platform. We used a partially within-subjects design; each participant saw all three videos (randomized and counterbalanced) for one robot platform.

4.3 Dependent Measures

1. Description- Participants were asked to describe the light signal and what they believe it indicates. This item was used as a manipulation check for the videos and to check whether participants could infer the precise meanings behind the signals.
2. Signal Type- Participants chose the signal type—"notification", "warning", or "error"—they believed best represented the robot's light signal.
3. Urgency- Participants chose whether they believed the robot's light signal to be urgent or not.
4. Signal Usage- Participants rated on a 5-point Likert scale how helpful they found the robot's light signal in understanding the scenario they observed.

To avoid biasing participants responses, we provided only minimal information about the robot platforms, the simulated environments, and the tasks the robots were performing.

4.4 Analysis

We present the results of the validation experiment, primarily focusing on counts of the forced-choice measures describing the light signals. We refer to the scenarios representing the notification, warning, and error signals for each of the robot platforms by numbers (1–3), as shown in Fig. 7.

Signal Type- The signal type fixed-choice measure was chosen correctly for all scenarios. We found that despite participants choosing the notification signal type correctly most often, their comments showed they were uncertain about what precise information the robot was trying to notify them of. Although the warning and error scenarios had more mixed results, participants were able to infer the precise cause of the warning and error signals more often in their descriptions of the videos.

Urgency- The urgency fixed-choice measure was chosen correctly for all scenarios, except for those with urgent, notification signals. 90% of participants chose non-urgent for the notification scenarios (1). Comments suggest that participants may never classify notifications as urgent due to their lower criticality.

Signal Usage- For all four robot types, participants rated the error signal most helpful and the notification the least helpful. Since notifications encompass a wider range of information, participants had more trouble inferring what the robot was notifying them of. We also found that participants had more difficulty understanding the scenarios with the Astrobee free-flying robot and the telepresence robot due to their lack of familiarity with these types of robot.

5 Conclusion

Our goal in this work is to enable appearance-constrained robots to communicate information about their state using expressive light signals. First, we conducted an exploratory video-based user study on Amazon Mechanical Turk to identify how variations in certain light signal parameters affect human perception of notification attributes.

The rich set of overall light parameters and the human ratings from the initial exploratory user study gave us the information needed to generate distinct notification, warning, and error signals using a decision tree. The results of this study are a promising first step for transforming the generation of light signals from a specialized task for individual human designers to an automated and standard process across a multitude of robot platforms.

We conducted a second video-based study validating the light signal decision tree. The results demonstrated that our signal designs were reasonable and illuminated certain distinctions that are difficult for human onlookers. In the future, this work could be further validated by studies in the real world, but the presented studies and findings supply the initial step needed to automatically generate light signals that can convey state-related properties of non-humanoid robots.

This work may guide roboticists who are interested in providing appearance-constrained robots the expressive abilities needed to make the world a easier and safer place for humans and robots to coexist.

References

1. Arrabito, G.R., Mondor, T.A., Kent, K.J.: Judging the urgency of non-verbal auditory alarms: a case study. Ergonomics **47**(8), 821–840 (2004)
2. Baraka, K., Paiva, A., Veloso, M.: Expressive lights for revealing mobile service robot state. In: Robot 2015: Second Iberian Robotics Conference, pp. 107–119. Springer, Cham (2015)
3. Baraka, K., Rosenthal, S., Veloso, M.: Enhancing human understanding of a mobile robot's state and actions using expressive lights. In: IEEE International Symposium on Robot and Human Interactive Communication, pp. 652–657. IEEE (2016)
4. Cha, E., Kim, Y., Fong, T., Matarić, M.J.: A survey of nonverbal signaling methods for non-humanoid robots. Found. Trends Robot. **6**(4), 211–323 (2018)
5. Cha, E., Matarić, M.: Using nonverbal signals to request help during human-robot collaboration. In: IEEE/RSJ International Conference on Intelligent Robots and Systems, pp. 5070–5076. IEEE (2016)
6. Cohn, T.E.: Method and apparatus for enhancing visual perception of display lights, warning lights and the like, and of stimuli used in testing for ocular disease, 20 January 1998. US Patent 5,710,560 (1998)
7. Dautenhahn, K., Walters, M., Woods, S., Koay, K.L., Nehaniv, C.L., Sisbot, A., Alami, R., Siméon, T.: How may I serve you? a robot companion approaching a seated person in a helping context. In: ACM/IEEE International Conference on Human-Robot Interaction, pp. 172–179. ACM (2006)
8. Dragan, A.D., Lee, K.C.T., Srinivasa, S.S.: Legibility and predictability of robot motion. In: ACM/IEEE International Conference on Human-Robot Interaction, pp. 301–308. IEEE (2013)
9. Guillaume, A., Drake, C., Rivenez, M., Pellieux, L., Chastres, V.: Perception of urgency and alarm design. In: International Conference on Auditory Display. Elsevier (2002)
10. Harrison, C., Horstman, J., Hsieh, G., Hudson, S.: Unlocking the expressivity of point lights. In: ACM Conference on Human Factors in Computing Systems, pp. 1683–1692. ACM (2012)
11. Lewis, C.F., McBeath, M.K.: Bias to experience approaching motion in a three-dimensional virtual environment. Perception **33**(3), 259–276 (2004)
12. McCrickard, D.S., Chewar, C.M., Somervell, J.P., Ndiwalana, A.: A model for notification systems evaluation-assessing user goals for multitasking activity. ACM Trans. Comput. Hum. Interact. **10**(4), 312–338 (2003)
13. Pousman, Z., Stasko, J.: A taxonomy of ambient information systems: four patterns of design. In: Working Conference on Advanced Visual Interfaces, pp. 67–74. ACM (2006)
14. Song, S., Yamada, S.: Effect of expressive lights on human perception and interpretation of functional robot. In: Extended Abstracts of the 2018 CHI Conference on Human Factors in Computing Systems, p. LBW629. ACM (2018)
15. Szafir, D., Mutlu, B., Fong, T.: Communicating directionality in flying robots. In: ACM/IEEE International Conference on Human-Robot Interaction, pp. 19–26. ACM (2015)

16. Szafir, D.J.: Human interaction with assistive free-flying robots. PhD thesis, The University of Wisconsin-Madison (2015)
17. Wool, L.E., Komban, S.J., Kremkow, J., Jansen, M., Li, X., Alonso, J.M., Zaidi, Q.: Salience of unique hues and implications for color theory. J. Vis. **15**(2), 10 (2015)
18. Zhou, A., Hadfield-Menell, D., Nagabandi, A., Dragan, A.D.: Expressive robot motion timing. In: ACM/IEEE International Conference on Human-Robot Interaction, pp. 22–31. ACM (2017)

Month-Long, In-Home Socially Assistive Robot for Children with Diverse Needs

Caitlyn Clabaugh(✉), Shomik Jain, Balasubramanian Thiagarajan, Zhonghao Shi, Leena Mathur, Kartik Mahajan, Gisele Ragusa, and Maja Matarić

University of Southern California, Los Angeles, CA 90089, USA
{clabaugh,bthiagar,zhonghas,lmathur,kmahajan, ragusa,mataric}@usc.edu, shomikja@gmail.com

Abstract. Socially assistive robotics (SAR) has the potential to close the gap between need and access to individualized, longitudinal support for children with autism spectrum disorders (ASD) that supplements and augments that of therapist, clinicians, and other caregivers. To this end, we developed a fully autonomous SAR system designed for month-long, in-home interventions with children with ASD. In this paper, we investigate a single case study with two siblings; one is typically developing and the other is affected by ASD. To better understand their help-seeking behaviors and engagement patterns, we define and analyze measures of participants' attentiveness. Our analysis shows that there are strong correlations between attentiveness and challenge level for both participants. We discuss how attentiveness can be used to optimize challenge level, detect help-seeking behaviors, and improve robot feedback for individuals with diverse needs.

1 Introduction

The Center for Disease Control estimates that approximately 1 in 59 children in the United States have been diagnosed with autism spectrum disorders (ASD), a constellation of developmental disorders characterized by delays and developmental difficulties in communication and social skills [1]. Regular therapeutic intervention significantly impact children on the autism spectrum [10]. For instance, occupational therapists deliver individualized lessons for developing turn-taking, perspective-taking, joint attention, and communication skills through play, storytelling, and engaging in social groups [15]. Though paramount, these services are not universally accessible nor equitably affordable [8].

Emerging research in socially assistive robotics (SAR) seeks to reduce the disparity between need and access to ASD therapeutic interventions with individualized human-robot interaction (HRI) to supplement and augment the work of clinicians and caregivers [6]. SAR for children with ASD holds great potential as a broadly replicable means for targeting, scaffolding, and tracking the development of critical social skills [12]. Specifically, SAR has been shown to effectively

J. Xiao et al. (Eds.): ISER 2018, SPAR 11, pp. 608–618, 2020.
https://doi.org/10.1007/978-3-030-33950-0_52

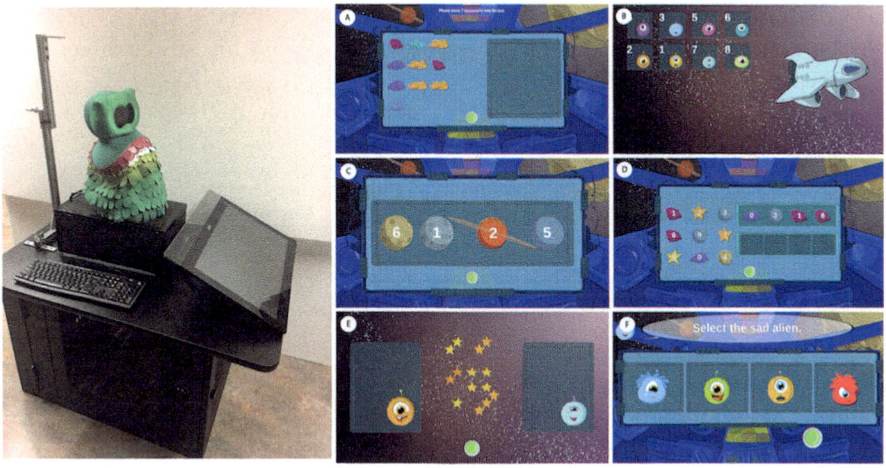

Fig. 1. The in-home SAR system (left) and educational space-themed games (right).

mediate joint attention [11], increase interaction complexity [5], motivate users [7], and elicit social interaction [11].

However, many fundamental research challenges remain in enabling long-term, personalized SAR for children with ASD in real-world environments [12]. With these goals in mind, we developed and deployed a fully autonomous SAR system for month-long, in-home interventions with children with ASD. Building on our work in [4], we consider a single case study of the SAR system in the home of two siblings with diverse needs. In this paper, we define two nonverbal measures of human attentiveness to better understand participants' help-seeking behaviors and engagement patterns. Our analysis shows strong correlations between both participants' attentiveness and level of challenge. This work demonstrates the potential of applying attentiveness to SAR toward optimizing challenge level, detecting help-seeking behaviors, and improving robot feedback for individuals with diverse needs.

2 In-Home Socially Assistive Robot for Children with ASD

For this research, we specifically developed a SAR system to support long-term, in-home interventions with children with ASD. We designed the system shown in Fig. 1 to be simple, accessible, and secure. The child-robot interaction centers around a set of space-themed education games designed for ordering and sequencing skills and numeracy [2]. Throughout the intervention, children play the games on a touchscreen tablet while the Stewart-platform robot [13] named Kiwi provides verbal feedback with a set of affective behaviors.

The system records all touch interactions with the tablet. Additionally, at the top of the monitor, a USB camera records a frontal view of the participant interacting with the system. For the convenience and safety of the participating families, the SAR system boots up and shuts down, halting all data recording, with one simple button. All equipment is provided and assembled by the research team at the start of a study. The only requirement from participants is a power outlet and sufficient physical space in the home.

2.1 Semi-Daily Interaction

We deployed the SAR system in the homes of the participating families for one month each. Child participants were encouraged but not required to interact with the system about five times per week. Since the SAR system is fully autonomous and we are interested in studying it in real-world in-home settings, it was left to the participants to choose when to begin and end sessions with the robot. Child participants were encouraged to play at least one of the games per session; the games are described below and are shown in Fig. 1.

1. **Pack Moon-rocks:** The player is instructed to drag 1–10 moon-rocks into a box.
2. **Select Galaxy:** The player is instructed to select the galaxy with more or fewer stars.
3. **Select Planet:** The player is instructed to select the planet with a particular number.
4. **Feed Space Pets:** The player is instructed to evenly divide a set of stars between two "alien pets".
5. **Pets on Spaceship:** The player is instructed to drag numbered "alien pets" into a spaceship in increasing or decreasing order.
6. **Organize Moon-rocks:** The player is instructed to separate and organize moon-rocks based on sprite and number.
7. **Organize Space Objects:** The player is instructed to separate and organize various space-themed objects based on sprite and number.
8. **Pattern Matching:** The player is instructed to copy a pattern with the space objects provided.
9. **Pattern Completion:** The player is instructed to complete a pattern with the provided space objects.
10. **Identify the Alien Emotion:** The player is instructed to determine the emotion of one or more "alien friends" based on their facial expression.

Each of the above games has five difficulty levels; more challenging levels introduce math, add distracting elements onto the screen, or ask the child to keep track of multiple items (e.g., match the pattern based on the object and the number).

2.2 Single-Case Study of Siblings with Diverse Needs

The SAR system and educational games described above are designed for multiple long-term, in-home deployments. Given that children with ASD vary widely in terms of need, our study follows a single-subject design such that each child participant serves as his/her own baseline [9]. Under the University of Southern California IRB study UP-16-00755, we collected data on child participants' semi-daily interactions, including a short questionnaire completed by their parents and/or guardians. We also conducted pre- and post- assessment using two math subtests from the Wechsler Individual Achievement Test Second Edition (WIAT-II), Numerical Operations and Math Reasoning [14], as well as a longitudinal assessment of each child participants' retention of the intervention context.

Our first case study was conducted in the home of a family with two dissimilarly affected siblings. The family consisted of one primary guardian with full-time employment and two sons (study participants). Child participant 1 (P1) was a 6 year old boy affected with ASD, and child participant 2 (P2) was a 5 year old typically developing (TD) boy. While dissimilarly affected and of slightly different ages, the two participants were in the same year at school. The child with ASD scored in the mild to moderate range of ASD. Developmental assessments for the ASD affected child indicated language skills of approximately a 2.5 to 3-year-old, and typical cognition and social- emotional skills of a 2-year-old. Accordingly, his primary areas of need were in communication and social behaviors. His TD sibling was performing essentially like typically developing 5-year-olds across developmental domains.

Without strict enforcement of the guidelines described in Sect. 2.1, we found that the participating family voluntarily adhered to the guidelines throughout the intervention of one month. Over that period, each child participant interacted with the SAR system a total of 18 times: on average, 8–9 min per day and 4–5 days per week. The siblings remained engaged throughout the intervention despite the fact that game difficulty was randomly selected in this pilot study. By randomizing the game difficulty, we were able to observe how each child reacted and performed at various challenge levels. In particular, we were interested in investigating the children's attentiveness to the SAR system (defined in Sect. 3, below) relative to game difficulty, performance, and help-seeking behavior.

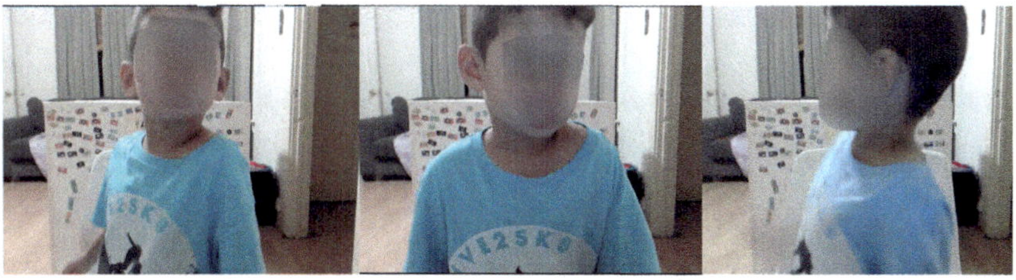

Fig. 2. Participant's VFOA directed at the robot (left), tablet (center), and away (right)

3 Measures of Attentiveness

Visual focus of attention (VFOA), commonly classified using eye gaze and head pose, has been used as an important nonverbal cue related to attention and engagement in human-robot interaction (HRI) and human-computer interaction (HCI) research [3]. In this work, we build upon VFOA as an indicator of (1) attentiveness with respect to the SAR system and (2) potential help-seeking behaviors. Specifically, we define two measures of *attentiveness*: Continuous Focus (CF) and Transitions of Focus (TF).

Definition 1 – ***Continuous Focus (CF)*** *is the length of time an individual's VFOA is continuously directed toward a particular entity.*

Definition 2 – ***Transitions of Focus (TF)*** *are transitions in an individual's VFOA between entities.*

3.1 Annotating Visual Focus of Attention

During interaction, the system recorded video data (without audio) of the participant's face and key game events such as game type and difficulty, robot dialogue and instruction, and the participant's performance. Two human coders were used per video to annotate the child's VFOA as directed at the (1) robot, (2) tablet, or (3) away (as shown in Fig. 2), and presence of other people in the environment (camera frame). Inter-rater reliability among annotators was measured using Fleiss' kappa; a threshold of $k \geq 0.7$ was maintained for all videos.

We define CF as the length of time (in seconds) that a participant's VFOA is continuously directed at the (1) robot, (2) tablet, or (3) away. Given that participants occasionally left the system for extended periods of time, we consider the median as the participant's central tendency of CF. Additionally, we analyze participants' TF, specifically the percentage of transitions that are (1) within-system and (2) outside-system. We define *within-system TF* as transitions

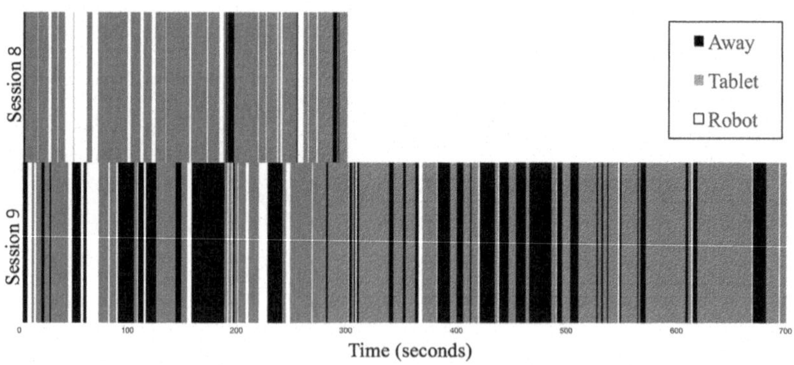

Fig. 3. Visualization of CF of P1 in sessions 8 and 9. CF "away" is greater in session 9 with the highest difficulty level than in session 8 with the lowest difficulty level.

between "robot" and "tablet", and *outside-system TF* as transitions between "away" and the system ("robot" or "tablet"). We use these measures to indicate the participants' attentiveness with respect to the SAR system and relate them to potential help-seeking behaviors, specifically examining longer intervals of CF "away" and increased outside-system TF.

3.2 Visualizing Continuous Focus

To introduce how we analyze these measures of attentiveness, we examine sessions 8 and 9 for P1. These were the only two sessions for P1 that the parent reported the child participant to be "Anxious", "Frustrated", "Tired", and "Distracted" in a post-session questionnaire. We observed longer intervals of CF "away" and increased outside-system TF in session 9 than in session 8. This difference can be attributed to the difficulty level; session 8 was played at the lowest difficulty and session 9 was at the highest. In addition, despite playing twice as long in session 9 than in session 8, P1 completed half the amount of games in session 9 compared to session 8, and the game performance was significantly worse ($p < 0.001$).

When considering CF of P1 during these two sessions, we found the median was 1.64s for "away", 8.45s for "tablet", and 1.73s for "robot" in session 8, as compared to 4.11s "away", 6.29s "tablet", and 1.38s "robot" in session 9. Through a visualization of P1's VFOA in Fig. 3, we observed longer intervals where CF was directed "away" in session 9 than in session 8.

3.3 Visualizing Transitions of Focus

When examining TF of P1 during the same two sessions, 8 and 9, we found that 78.0% were within-system and 22.0% were outside-system in session 8, as compared to 30.8% within-system and 62.2% outside-system in session 9. Through a visualization of P1's TF in Fig. 4, we observed more outside-system TF in session 9 than in session 8.

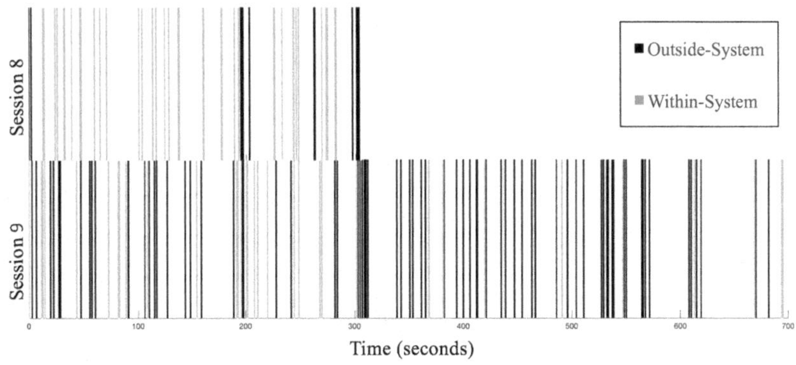

Fig. 4. Visualization of TF of P1 in sessions 8 and 9. Outside-System TF are greater in session 9 with the highest difficulty level than in session 8 with the lowest difficulty level.

3.4 Implications and Limitations

Our visual analysis of CF and TF indicates that the activities were possibly too challenging in session 9, wherein the child's CF "away" and outside-system TF increased as a result of seeking help from others in the environment. In contrast, the activities may have been too easy in session 8, so the child potentially became disengaged and turned off the system. This demonstrates the correlation between the measures of attentiveness and the game difficulty level, discussed further in the following section.

There are limitations to these measures because the video recording was only capturing the participant's face. For example, if the parent was standing behind the SAR system, it was possible for VFOA to be marked as "away" when the participant was receiving help from the parent and was still engaged with the game.

4 Analysis of Attentiveness

In the previous section, we considered the relationship between attentiveness and game difficulty through the specific examples of sessions 8 and 9 for P1. In this section, we our analysis on both participants' attentiveness measures over the multiple, repeated interactions. In particular, we observed (1) longer intervals of CF "away" and (2) more outside system TF in sessions with higher difficulty levels, likely as a result of the participants seeking help from others in the environment. We also found that the performance of both participants decreased as game difficulty increased. Our analysis shows the potential of applying attentiveness measures to the SAR system to inform game difficulty selection, detect help-seeking behaviors, and improve robot feedback.

4.1 Continuous Focus and Level of Difficulty

As shown in Fig. 5, we found a positive correlation between the game difficulty level and CF "away" (P1: $r_s = 0.70$, $p = 0.20$; P2: $r_s = 0.70$, $p = 0.20$). We postulate that difficulty level 3 had the lowest value for CF "away" because difficulty was selected randomly and level 3 was played the least. By excluding difficulty level 3, we found a perfect positive correlation between difficulty and CF "away" ($r_s = 1.00$) for both participants.

4.2 Transitions of Focus and Level of Difficulty

Our analysis also revealed a positive correlation between the game difficulty level and outside-system TF (P1: $r_s = 0.70$, $p = 0.20$; P2: $r_s = 0.70$, $p = 0.20$). As game difficulty increased, a larger percentage of TF were outside-system, compared to within-system, as depicted in Fig. 6. In sessions with higher difficulty, it is likely that the child's outside-system TF increased as a consequence of seeking help from others in the home environment. In fact, as also indicated in Fig. 6,

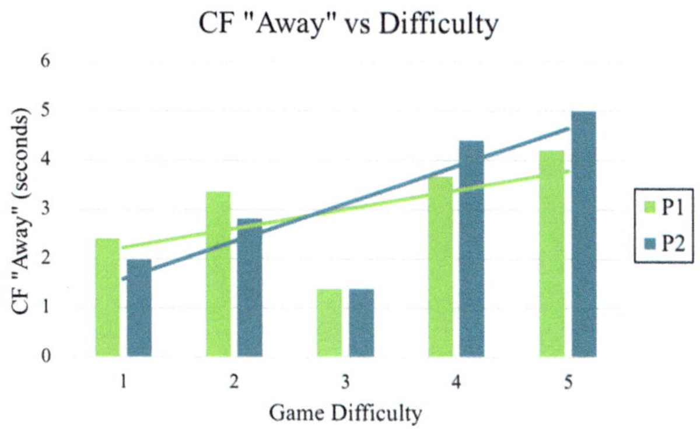

Fig. 5. CF "away" significantly increases with game difficulty for both participants (P1: $r_s = 0.70$, $p = 0.20$; P2: $r_s = 0.70$, $p = 0.20$).

the positive correlation between difficulty and outside-system TF is more pronounced when multiple people were present in the environment (P1: $r_s = 1.00$, $p =$ undefined; P2: $r_s = 0.90$, $p = 0.02$).

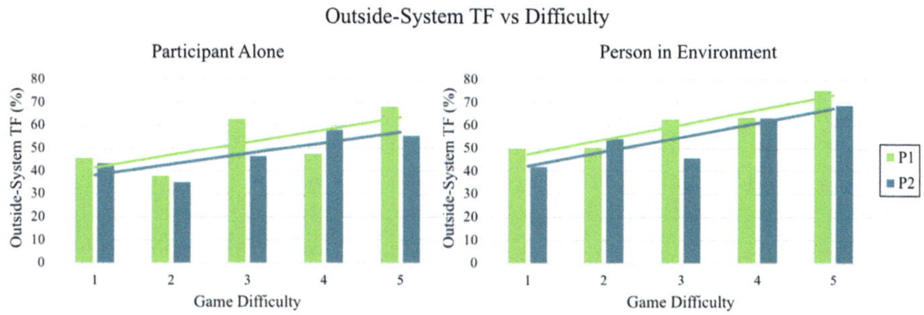

Fig. 6. Outside-system TF significantly increase with game difficulty (P1: $r_s = 0.70$, $p = 0.20$; P2: $r_s = 0.70$, $p = 0.20$) and are even higher when another person is present in the environment (P1: $r_s = 1.00$, $p =$ undefined; P2: $r_s = 0.90$, $p = 0.02$).

4.3 Attentiveness as an Indicator of Need

There were no statistically significant differences between P1 and P2 for either measure of attentiveness ($p > 0.05$). Relative to CF, we found that the two participants sustained attention on each VFOA category for similar lengths of time. Additionally, P1 and P2 had comparable percentages of outside-system versus within-system TF. These findings indicate that both participants, whether affected with ASD or TD, demonstrated consistent nonverbal cues for attention.

Relative to performance, there was a negative correlation between the game difficulty level and number of games successfully completed for both participants (P1: $r_s = -0.90$, $p = 0.02$; P2: $r_s = -0.90$, $p = 0.02$). However, TD P2 answered

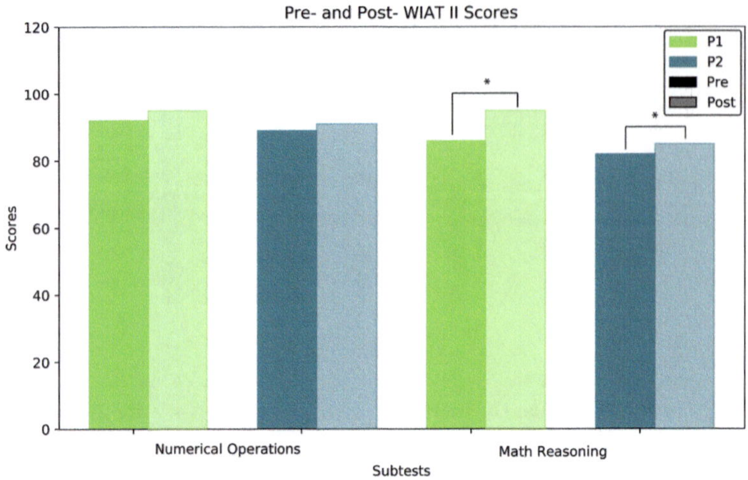

Fig. 7. Both participants showed learning gains according to a pre-/post-intervention assessment that used the WIAT II numerical operations and math reasoning subtests. We found that both participants' math reasoning skills improved significantly ($*p < 0.05$).

correctly on the first attempt significantly more frequently ($p < 0.05$) and made fewer mistakes ($p < 0.05$) than his sibling with ASD, P1. This indicates that P1 may have needed more intermediate feedback.

By conducting a pre- and post- assessment using the WIAT II, we found that both participants' numerical operation skills improved and math reasoning skills significantly improved, as shown in Fig. 7. When both participants were assessed for retention of game knowledge 30 days after the close of the intervention, the child with typical development, P2, retained a greater amount of knowledge from the intervention (87.4%) than his sibling with ASD, P1 (64.9%).

Together, this evidence indicates that P1 and P2 may have differences in need. More specifically, P1 may benefit from more intermediate assistance and feedback. For both participants, we observed longer intervals of CF "away" and more outside-system TF in sessions with higher difficulty. This indicates that both participant exhibited similar help-seeking behaviors; P1 exhibited these with a higher frequency as he was in need of more assistance. The application of measures of attentiveness to the SAR system could inform difficulty selection and intermediate feedback, personalized to each participant's unique needs.

5 Conclusion and Future Work

We developed a fully autonomous SAR system deployed in a real-world home for 30 days with two siblings: one child with ASD and the other typically developing. Using a single case study approach, our research shows the potential and desirability of applying multimodal features to select the appropriate game difficulty level, detect help-seeking behaviors, and improve robot feedback.

We defined two measures of attentiveness and analyzed how they relate to game difficulty and performance per participant. In future work, we plan to utilize multimodal features to detect help-seeking behaviors and incorporate a machine learning model in the SAR system that could offer intermediate support to participants while games are in progress. Additionally, to further enable personalization and long-term adaptation of the SAR system, we are developing reinforcement learning algorithms to adapt the game difficulty level and feedback to each participant's individual needs. Our implementation seeks to maximize game difficulty while minimizing the probability that a user will make mistakes. These computational approaches to multimodal, long-term adaptation could enable SAR to better support and understand the complex challenges facing individuals with special needs.

Another important insight from our work is the multi-party nature of real-world, in-home environments. Of the entire month-long intervention, only 50.8% of sessions for P1 and 61.5% for P2 were purely dyadic. Thus, it is insufficient to restrict research applications of SAR to purely dyadic interaction or isolate them into highly-controlled laboratory settings. The long-term family-situated nature of this research pushes the boundaries of SAR, HRI, and broader robotics research.

Additionally, we found that after the first week of interaction, the participants' parent became comfortable with leaving the child participants to interact more independently with the SAR system. The parent identified this as a potential benefit: freeing time to focus on other tasks while the child remained socially and educationally engaged. Both child participants also responded that they would like the robot to stay longer; one said: "I want Kiwi to stay for three years!".

In summary, this work built upon visual focus of attention research to define two nonverbal measures of attentiveness, and showed how these measures relate to engagement and help-seeking in the context of a month-long, in-home, fully autonomous SAR intervention for a child with ASD and his typically developing brother. As we look forward to conducting many more long-term studies with our SAR system, we believe that, especially in the context of autism spectrum disorders, the single case study approach ensures that each participating family's data receives deep analysis and attention, in order to design SAR systems that can adapt to the unique particularities of each user and family.

Acknowledgements. This research was supported by National Science Foundation grant IIS-1139148.

Disclosure. Co-author Maja Matarić is co-founder and chief science officer of Embodied, Inc.

References

1. Baio, J., Wiggins, L., Christensen, D.L., Maenner, M.J., Daniels, J., Warren, Z., Kurzius-Spencer, M., Zahorodny, W., Rosenberg, C.R., White, T., et al.: Prevalence of autism spectrum disorder among children aged 8 years-autism and developmental disabilities monitoring network, 11 sites, united states, 2014. MMWR Surveill. Summ. **67**(6), 1 (2018)
2. Clabaugh, C., Ragusa, G., Sha, F., Matarić, M.J.: Designing a socially assistive robot for personalized number concepts learning in preschool children. In: Joint IEEE International Conference on Development and Learning and Epigenetic Robotics (ICDL-EpiRob), pp. 314–319. IEEE (2015)
3. Clabaugh, C., Ram, T., Matarić, M.J.: Estimating visual focus of attention in dyadic human-robot interaction for planar tasks. In: International Conference on Social Robotics (2015)
4. Clabaugh, C., Becerra, D., Deng, E., Ragusa, G., Matarić, M.J.: Month-long, in-home case study of a socially assistive robot for children with autism spectrum disorder. In: Companion of the 2018 ACM/IEEE International Conference on Human-Robot Interaction, pp. 87–88. ACM (2018)
5. Dautenhahn, K.: Robots as social actors: Aurora and the case of autism. In: Proceedings CT99, The Third International Cognitive Technology Conference, August, San Francisco, vol. 359, p. 374 (1999)
6. Feil-Seifer, D., Matarić, M.J.: Defining socially assistive robotics. In: 9th International Conference on Rehabilitation Robotics, ICORR 2005, pp. 465–468. IEEE (2005)
7. Kim, E.S., Berkovits, L.D., Bernier, E.P., Leyzberg, D., Shic, F., Paul, R., Scassellati, B.: Social robots as embedded reinforcers of social behavior in children with autism. J. Autism Dev. Disord. **43**(5), 1038–1049 (2013)
8. Lavelle, T.A., Weinstein, M.C., Newhouse, J.P., Munir, K., Kuhlthau, K.A., Prosser, L.A.: Economic burden of childhood autism spectrum disorders. Pediatrics **133**(3), e520–e529 (2014)
9. Odom, S.L., Brown, W.H., Frey, T., Karasu, N., Lee Smith-Canter, L., Strain, P.S.: Evidence-based practices for young children with autism: contributions for single-subject design research. Focus Autism. Dev. Dis. **18**(3), 166–175 (2003)
10. Ospina, M.B., Seida, J.K., Clark, B., Karkhaneh, M., Hartling, L., Tjosvold, L., Vandermeer, B., Smith, V.: Behavioural and developmental interventions for autism spectrum disorder: a clinical systematic review. PloS One **3**(11), e3755 (2008)
11. Robins, B., Dautenhahn, K., Te Boekhorst, R., Billard, A.: Robotic assistants in therapy and education of children with autism: can a small humanoid robot help encourage social interaction skills? Univ. Access Inf. Soc. **4**(2), 105–120 (2005)
12. Scassellati, B., Admoni, H., Matarić, M.: Robots for use in autism research. Annu. Rev. Biomed. Eng. **14**, 275–294 (2012)
13. Short, E., Short, D., Fu, Y., Matarić, M.J.: Sprite: Stewart platform robot for interactive tabletop engagement, Tech. rep., Department of Computer Science, University of Southern California (2017)
14. Wechsler, D.: Wechsler Preschool and Primary Scale of Intelligence, 2nd edn. Psychological Corporation, San Antonio (1989)
15. White, S.W., Keonig, K., Scahill, L.: Social skills development in children with autism spectrum disorders: a review of the intervention research. J. Autism Dev. Disord. **37**(10), 1858–1868 (2007)

Experimental Validation of Resolved Viscoelasticity Control on Hydrostatically Driven Humanoid Hydra

Ko Yamamoto(✉), Tianyi Ko, Kazuya Murotani, and Yoshihiko Nakamura

University of Tokyo, 7-3-1 Hongo, Bunkyo-ku, Tokyo 113-8656, Japan
{yamamoto.ko,kang,murotani,nakamura}@ynl.t.u-tokyo.ac.jp
http://www.ynl.t.u-tokyo.ac.jp

Abstract. A backdrivable or force-controllable actuator is one of the key elements needed to develop a humanoid capable of interacting with humans. Recently, we developed the humanoid Hydra using electro hydrostatic actuators, which have high backdrivability and force sensitivity. Moreover, one of the authors proposed resolved viscoelasticity control (RVC) approach, which achieves robust balance control by transforming task-space viscoelasticity into joint viscoelasticity. However, the RVC was validated only through forward dynamics simulation. Therefore, in this study, experimental validation of the RVC method using Hydra was achieved. Further, robust standing balance control was realized in the experiments.

Keywords: Humanoid robot · Whole-body balance control · Hydraulic actuator

1 Introduction

For a humanoid to move in a real-life environment and interact with humans, it is required to have a softness in its body. A backdrivable or force-controllable actuator is one of the key elements to realize such a humanoid. Recently, humanoid robots with torque-sensing actuators [3] or series elastic actuators

This work was partially supported by New Energy and Industrial Technology Development Organization (NEDO) the International R&D and Demonstration Project on Robotic Field/Research and Development of Disaster-Response Robot Open Platform (FY2014-FY2015), NEDO Core Technology Development of Next Generation Robots, Innovative Robot Element Technology, "Field Actuation Technology using Compact Hydraulic Actuators and Fuel Cell/Rechargeable Battery Hybrid Power Supply", and JSPS KAKENHI Grant Number 18K19802.

Electronic supplementary material The online version of this chapter (https://doi.org/10.1007/978-3-030-33950-0_53) contains supplementary material, which is available to authorized users.

J. Xiao et al. (Eds.): ISER 2018, SPAR 11, pp. 619–628, 2020.
https://doi.org/10.1007/978-3-030-33950-0_53

[12] were developed. Hydraulic actuation has received significant attention ever since the humanoid Atlas, developed by Boston Dynamics, employed a servovalve hydraulic actuator. Although a servovalve hydraulic actuator has high torque density, it lacks backdrivability. Kaminaga et al. [5] developed the humanoid Hydra by using electro hydrostatic actuators (EHAs). An EHA achieves high backdrivability by directly connecting a pump and hydraulic actuator without a servovalve.

Such an actuator has the potential to realize sophisticated whole-body control of a humanoid. Popular approaches include inverse dynamics control [8] and operational space formulation [9]. One of the authors proposed the resolved viscoelasticity control (RVC) approach, which transforms task-space viscoelasticity, including the center of gravity (COG) viscoelasticity, into joint viscoelasticity. It also set a desired value of whole-body viscoelasticity, which was realized in the null-space of the task-space viscoelasticity. A previous study extended the RVC by considering the structure-variability of a humanoid [16] and multiple viscoelasticities [15]; this study was able to realize walking on an uneven terrain and a slope. However, these validations were only made by forward dynamics simulation.

The purpose of this study is to experimentally validate the RVC method by using Hydra. Although the concept of a humanoid with EHAs was also proposed by Alfayad et al. [1], there has not been a significant amount of reported results. In this study, robust standing balance control is realized in the experiments. The rest of this paper is organized as follows. Section 2 reviews the RVC method, and Section 3 describes the specification of Hydra and its architecture of joint force control using EHAs. Then, Sect. 4 reports three experimental results: static response against slow disturbance, dynamic response against impulsive disturbance, and null-space compliance validation. Finally, Sect. 5 summarizes the experimental validations.

2 Resolved Viscoelasticity Control [16]

2.1 Outline of RVC

In this section, the basic formulation of the RVC is explained for the single support phase of a humanoid, which is modeled as an open kinematic chain. The following formulation can be extended to the double support phase, in which the whole-body structure includes a closed kinematic chain [16].

Figure 1 shows the outline of the RVC. A robot with an open kinematic chain is considered with n degrees of freedom (DOFs) of actuated joints. It is assumed that $\boldsymbol{\theta}$ and $\boldsymbol{\tau} \in \mathbb{R}^n$ denote the joint angle and torque vector, respectively. Additionally, it is assumed that joint viscoelasticity is given as follows:

$$\boldsymbol{\tau} = {}^{ref}\boldsymbol{\tau} + \boldsymbol{K}_\theta \Delta\boldsymbol{\theta} + \boldsymbol{D}_\theta \Delta\dot{\boldsymbol{\theta}} \tag{1}$$

where ${}^{ref}*$ denotes a reference value of a quantity $*$, and $\Delta * = {}^{ref}* - *$ denotes the error from the reference. $\boldsymbol{K}_\theta, \boldsymbol{D}_\theta \in \mathbb{R}^{n\times n}$ denote the stiffness and viscosity

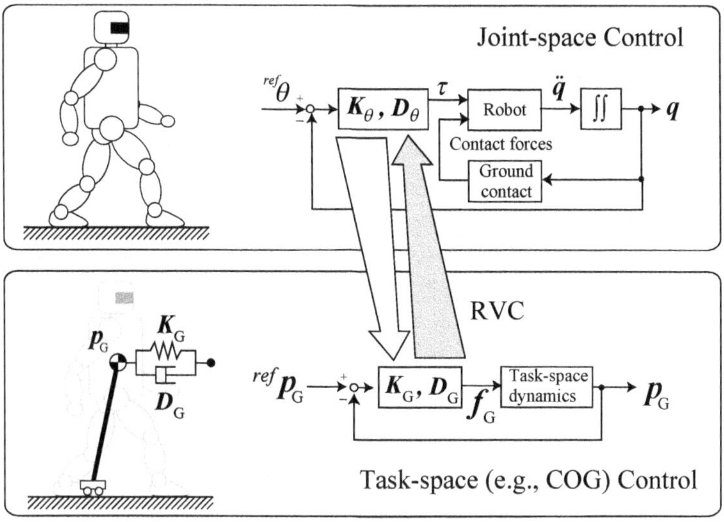

Fig. 1. Outline of the RVC. The mapping relationship from joint-space viscoelasticity to task-space viscoelasticity is given by $\boldsymbol{K}_i^{-1} = \boldsymbol{J}_i \boldsymbol{K}_\theta^{-1} \boldsymbol{J}_i^T$. The RVC treats its inverse problem by considering redundant DOFs.

matrices. Typically, $\boldsymbol{K}_\theta$ and $\boldsymbol{D}_\theta$ are set to be diagonal. In this study, we consider a wider class of $\boldsymbol{K}_\theta$ and $\boldsymbol{D}_\theta$, which are positive-definite and symmetric matrices.

Furthermore, the viscoelasticity of a feature quantity $\boldsymbol{p}_i \in \mathbb{R}^m$ in task space is considered as follows:

$$\boldsymbol{f}_i = {}^{ref}\boldsymbol{f}_i + \boldsymbol{K}_i \Delta \boldsymbol{p}_i + \boldsymbol{D}_i \Delta \dot{\boldsymbol{p}}_i \tag{2}$$

where $\boldsymbol{f}_i \in \mathbb{R}^m$ denotes force acting at $\boldsymbol{p}_i$, and $\boldsymbol{K}_i, \boldsymbol{D}_i \in \mathbb{R}^{m \times m}$ denotes stiffness and viscosity matrices, respectively. With respect to $\boldsymbol{p}_i$, we can select a link position, orientation, the COG, angular momentum around a point, or a combination of them.

Hereafter, the study focuses on stiffness for simplicity of explanation. The virtual work principle gives the following relationship.

$$\boldsymbol{K}_i^{-1} = \boldsymbol{J}_i \boldsymbol{K}_\theta^{-1} \boldsymbol{J}_i^T \tag{3}$$

where $\boldsymbol{J}_i(\boldsymbol{\theta}) \in \mathbb{R}^{m \times n}$ denotes the Jacobian matrix. For example, $\boldsymbol{J}_G (i = G)$ is the COG Jacobian [11] when we consider the COG $\boldsymbol{p}_G$ as a task.

The RVC treats the inverse problem of (3). The solution of (3) is not uniquely determined when there are redundant DOFs, that is $m < n$. A general solution of (3) is given as follows [2]:

$$\boldsymbol{K}_\theta^{-1} = \boldsymbol{J}_i^\sharp \boldsymbol{K}_i^{-1} \boldsymbol{J}_i^{T\sharp} + (\boldsymbol{Y} - \boldsymbol{J}_i^\sharp \boldsymbol{J}_i \boldsymbol{Y} \boldsymbol{J}_i^T \boldsymbol{J}_i^{\sharp T}) \tag{4}$$

where $\boldsymbol{J}_i^\sharp$ represents the pseudo-inverse of $\boldsymbol{J}_i$, and $\boldsymbol{Y}$ denotes an arbitrary matrix. If $\boldsymbol{Y}$ is set as a desired value of the joint compliance, then it implies that the following optimization problem is solved [16].

$$\min \frac{1}{2} \|\boldsymbol{Y} - \boldsymbol{K}_{\theta}^{-1}\|^2 \quad \text{subject to Eq. (3)} \tag{5}$$

Considering the inverse of (4), we obtain joint stiffness $\boldsymbol{K}_{\theta}$. Similarly, joint viscosity $\boldsymbol{D}_{\theta}$ is also obtained. The resultant $\boldsymbol{K}_{\theta}$ and $\boldsymbol{D}_{\theta}$ become functions of the joint angle $\boldsymbol{\theta}$. It can be said that the RVC computes the joint viscoelasticity as the robot posture changes, so that pre-designed task space viscoelasticity is realized in whole-body kinematics.

2.2 COG Viscoelasticity for Balancing [14]

For humanoid balancing, an appropriate setting of COG viscoelasticity is important. In this section, the COG viscoelasticity is derived from the linear inverted pendulum model (LIPM) [4], in which dynamics between the COG $\boldsymbol{p}_G = [x_G \; y_G \; z_G]^T$ and zero moment point (ZMP) [13] $\boldsymbol{p}_Z = [x_Z \; y_Z \; 0]^T$ are considered. The dynamics in the x and y directions of the LIPM are represented as the following state equation.

$$\dot{\boldsymbol{x}} = \begin{bmatrix} \boldsymbol{O} & \boldsymbol{E} \\ \omega^2 \boldsymbol{E} & \boldsymbol{O} \end{bmatrix} \boldsymbol{x} + \begin{bmatrix} \boldsymbol{O} \\ -\omega^2 \boldsymbol{E} \end{bmatrix} \boldsymbol{u}, \quad \omega := \sqrt{\frac{\ddot{z}_G + g}{z_G}} \tag{6}$$

where the state and input vectors are defined as $\boldsymbol{x} = [x_G \; y_G \; \dot{x}_G \; \dot{y}_G]^T$ and $\boldsymbol{u} = [x_Z \; y_Z]^T$, and $\boldsymbol{O}$ and $\boldsymbol{E}$ denote the zero and unit matrices, respectively.

In order to maintain the balance, we consider a feedback controller that causes the COG to converge to its reference ${}^{ref}\boldsymbol{p}_G$. For motion in the x and y directions, the following state feedback is applied.

$$\boldsymbol{u} = {}^{ref}\boldsymbol{u} + \boldsymbol{F}({}^{ref}\boldsymbol{x} - \boldsymbol{x}) \tag{7}$$

where $\boldsymbol{F}$ denotes the state feedback gain. In this study, we employ $\boldsymbol{F}$ as a feedback controller of the capture point [7] because the stabilizable region was maximized with respect to the ZMP constraint [10]. Then, the following feedback controller is employed in the z-axis direction.

$$f_z = mg + k_z \Delta z_G + d_z \Delta \dot{z}_G \tag{8}$$

where f_z denotes the vertical element of the ground reaction force, m denotes total mass, and k_z and d_z denote gain parameters.

The relationship among the COG viscoelasticity, $\boldsymbol{K}_G, \boldsymbol{D}_G$, and the COG feedback gain, $\boldsymbol{F}$, is represented as follows [14]:

$$\boldsymbol{K}_G = \begin{bmatrix} -m\omega^2(\boldsymbol{E} + \boldsymbol{F}_k) & \begin{matrix} 0 \\ 0 \end{matrix} \\ 0 \; 0 & k_z \end{bmatrix}, \; \boldsymbol{D}_G = \begin{bmatrix} -m\omega^2 \boldsymbol{F}_d & \begin{matrix} 0 \\ 0 \end{matrix} \\ 0 \; 0 & d_z \end{bmatrix} \tag{9}$$

where $\boldsymbol{F}_k, \boldsymbol{F}_d \in \mathbb{R}^{2\times 2}$ denote element matrix of $\boldsymbol{F}$, i.e., $\boldsymbol{F} = [\boldsymbol{F}_k \; \boldsymbol{F}_d]$.

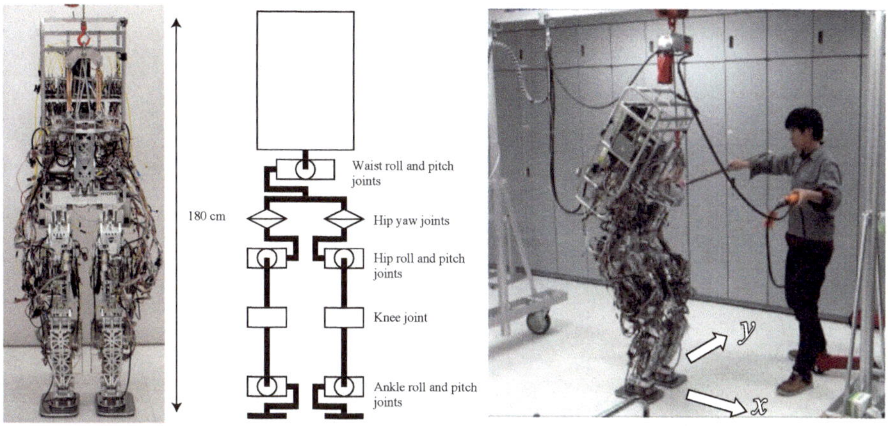

Fig. 2. External and joint configuration of Hydra. The lower-body version (with 14 DOFs, 180 cm, 74 kg) is used in this study.

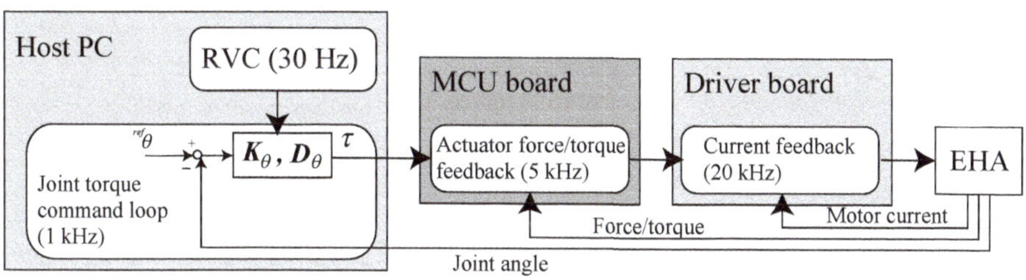

Fig. 3. Implementation of triple-loop feedback for accurate joint torque control.

3 Hydrostatically Driven Humanoid Hydra [5]

The validity of the RVC was verified by experiments with the humanoid Hydra [5]. Figure 2 shows an external view, joint configuration, and experimental setup of Hydra. In this study, the lower-body version (180 cm, 74 kg) was used for validation. The total DOF is 14: 2 DOFs for waist and 6 DOFs for each leg. All joints are actuated by EHAs.

To realize joint torque control with high accuracy, we implemented tripe-loop feedback control: current feedback, actuator force/torque feedback, and joint torque command loop. As shown in Fig. 3, the lowest-level control is the current feedback, which runs at 20 kHz on the driver board. Outside the current feedback control, the force/torque feedback runs at 5 kHz on the microcontroller unit (MCU) board. Finally, the joint torque command loop runs at 1 kHz on the host PC, which commands joint torque to the MCU board through the EtherCAT.

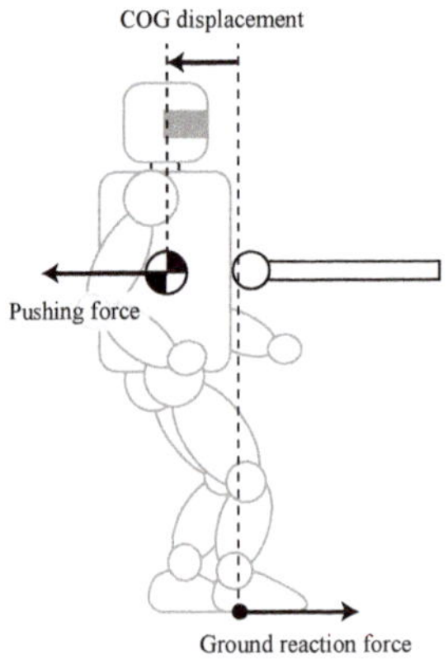

(a) Relationship between pushing force and ground reaction force

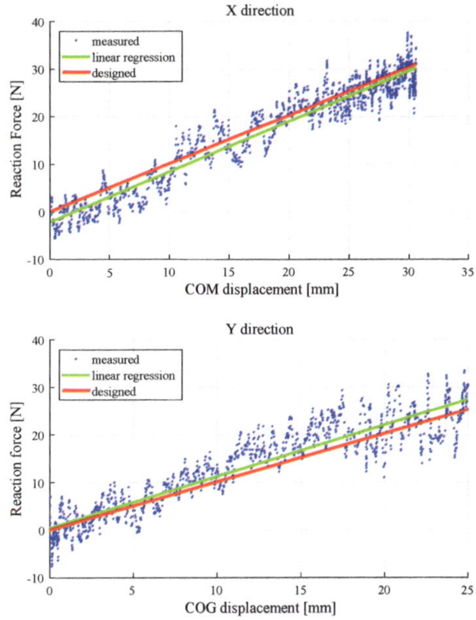

(b) Experimental result

Fig. 4. Results of static response against slow and constant disturbances. (a) It is considered that the pushing force is equivalent to the horizontal element of the ground reaction force. (b) Relationship between the pushing force and the COG displacement in x- and y-directions. The red line implies the pre-designed COG stiffness in the RVC ($k_x = k_y = 1.008$ N/mm). The blue points indicate the measured data, and the green line indicates the linear regression of the measured data. From the linear regression, the estimated stiffnesses in the x- and y-directions are 1.055 and 1.079, respectively. It is observed that the pre-designed COG stiffness is regenerated with a good accuracy.

The RVC is implemented as a ROS node on the host PC, which is updated at 30 Hz. This means that the RVC sometimes calculates the joint viscoelasticity $\boldsymbol{K}_\theta$ and $\boldsymbol{D}_\theta$ using past joint angle data. Nevertheless, the calculated $\boldsymbol{K}_\theta$ and $\boldsymbol{D}_\theta$ can stabilize the current state because they result in positive-definite matrices. In the following experiments, the value of joint viscosity $\boldsymbol{D}_\theta$ was scaled by multiplying with 0.1 because Hydra has potential joint viscosity.

4 Experimental Results

4.1 Static Response Against Slow Disturbances

To validate the RVC, particularly its standing balance performance, we added external forces to Hydra and evaluated its responses.

First, we performed an experiment for evaluating disturbance response against *static* pushing force. In the experiment, Hydra was pushed slowly and constantly, as shown in the video accompanying the manuscript. After we released the pushing force, the COG goes back to its initial position by the RVC. By slowly pushing the robot, we can ignore the effect of the COG viscosity and consider the effect of the stiffness only. In this static condition, it is considered as illustrated in Fig. 4(a) that the pushing force is equivalent to the horizontal element of the ground reaction force, which was measured by foot force sensors. Figure 4(b) shows the relationship between the pushing force and the COG displacement from its initial position. In the figure, the red line implies the pre-designed COG stiffness in the RVC ($k_x = k_y = 1.008$ N/mm). The blue points indicate the measured data, and the green line indicates the linear regression of the measured data. From the linear regression, the estimated stiffnesses in the x- and y-directions are 1.055 and 1.079 N/mm, respectively. It is observed that the pre-designed COG stiffness is regenerated with a high accuracy.

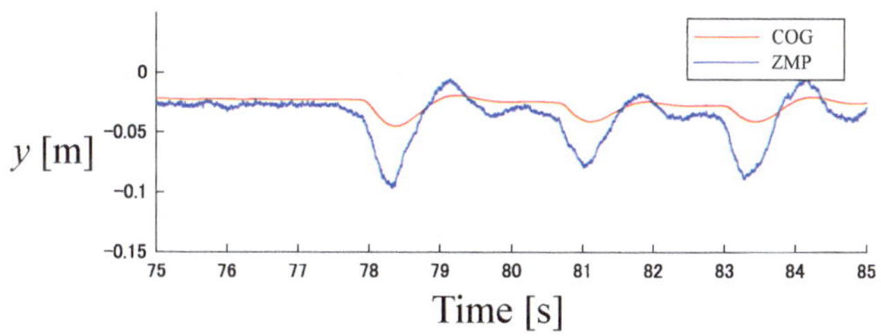

(a) Experimental results

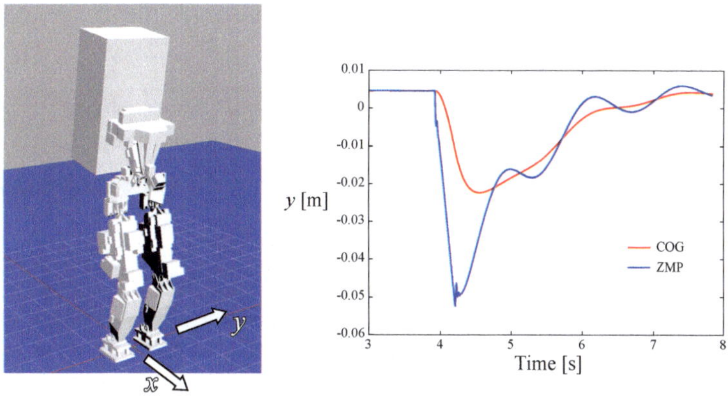

(b) Forward dynamics simulation

Fig. 5. Results of dynamic response against impulsive disturbances: (a) the results of the COG and ZMP in the experiment, and (b) ideal response in forward dynamics simulation. It is observed that the ZMP moves so that the COG is stabilized.

4.2 Dynamic Response Against Impulsive Disturbances

Second, we performed an experiment for evaluating disturbance response against *impulsive* pushing force, as shown in the video accompanying the manuscript.

Figure 5(a) shows the results of the COG and ZMP in the experiment. It is observed that the ZMP (blue line) moves so that the COG (red line) is stabilized. This type of response is typical of the LIPM-based feedback control, which was mentioned in Sect. 2.2. For comparison, we evaluated a similar response by forward dynamics simulation. Figure 5(b) shows the results of the COG and ZMP in the simulation. From $t = 4$ s, an impulsive force was added with 30 N for 0.3 s. We can observe that the response of the COG and ZMP is similar to the experimental results, in particular between 78 s and 80 s of Fig. 5(a), although the time and space scales are different.

In general, it is difficult to perfectly match the experimental and simulation results because of modeling errors. In particular, an ideal impulse force can be added in the simulation, whereas a disturbance force was added manually in the experiment. Moreover, the potential joint viscosity in Hydra is not modeled in the simulation. Nevertheless, these results show a similar response of the COG and ZMP, which reveals that the RVC can realize robust balancing control by regenerating the feedback controller of the COG.

4.3 Validation of Null-Space Compliance

Finally, an experiment validated the null-space compliance. As shown in Fig. 6, we manually moved different parts of Hydra in different directions. In Fig. 6(a),

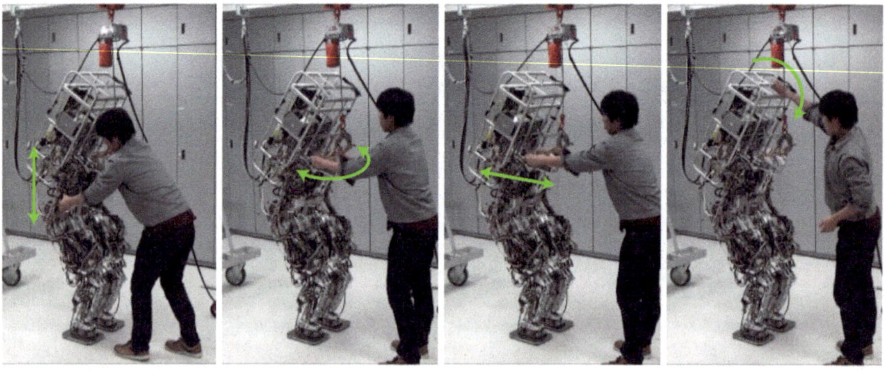

(a) (b) (c) (d)

Fig. 6. Validation of null-space compliance. Different parts of Hydra are manually moved in different directions. In the case of (a), the robot does not largely move in the vertical direction because the COG viscoelasticity is set in this direction. In the case of (b), (c), and (d), the body link is moved around the yaw axis, in the x-direction, and around the pitch axis, respectively. In these cases, the robot moves largely because of the null-space compliance.

the waist link is moved in the vertical direction. The robot does not largely move in this case because the COG viscoelasticity is set in this direction. In Figs. 6 (b), (c), and (d), the body link is moved around the yaw axis, in the x-direction, and around the pitch axis, respectively. In these cases, the robot moves largely because of the null-space compliance. This experiment reveals that the RVC contributes to the realization of the null-space compliance.

5 Conclusion

In this study, we experimentally validated the RVC method using the humanoid Hydra. The following findings were obtained through the experiments.

1. The experiment on the static response to slow and constant disturbances showed that the RVC also realized COG stiffness with high accuracy as assumed in the task-space. When setting the pre-designed COG stiffness to be 1.008 N/mm in both the x- and y-directions, the estimated stiffnesses were 1.055 N/mm in the x-direction and 1.079 N/mm in the y-direction.
2. The dynamic response to impulsive disturbances experiment showed that the RVC realized COG stabilization control as assumed in the forward dynamics simulation.
3. The RVC realized null-space control in the real robot. High compliant motion was realized without losing COG stability.

One of the advantages of the RVC is that it does not require real-time computation. In the experiment, the RVC was updated at 30 Hz. Nevertheless, a robust standing balance was realized by the RVC. The computational burden of the RVC is much smaller in the total system, and we can allocate the remaining resources to other functions.

The RVC also has a practical advantage in performing experiments. Because the output of the RVC corresponds to gain matrices in joint position control, we can perform experiments by comparing them with the gains of a traditional joint PD controller. This is a practical advantage when developing a humanoid because basic control is usually tested with a joint PD controller in an early stage of development.

Recently, we achieved walking motion using the RVC on Hydra, which will be reported in [6]. The experiment demonstrated that the walking motion was stabilized by the RVC even when there was a disturbance. Future works will address modeling errors in the hardware or uncertainties in the environment.

References

1. Alfayad, S., Ouezdou, F.B., Namoun, F., Gheng, G.: High performance integrated electro-hydraulic actuator for robotics - part i: principle, prototype design and first experiments. Sens. Actuators, A **169**(1), 115–123 (2011)
2. Athans, M.: The matrix minimum principle. Inf. Control **11**(5–6), 592–606 (1968)

3. Englsberger, J., et al.: Overview of the torque-controlled humanoid robot TORO. In: IEEE-RAS International Conference on Humanoid Robots, pp. 916–923 (2015)
4. Kajita, S., et al.: Biped walking pattern generation by a simple three-dimensional inverted pendulum model. Adv. Rob. **17**(2), 131–147 (2003)
5. Kaminaga, H., Ko, T., Masumura, R., Komagata, M., Sato, S., Yorita, S., Nakamura, Y.: Mechanism and control of whole-body electro-hydrostatic actuator driven humanoid robot hydra. In: International Symposium on Experimental Robotics (2016)
6. Ko, T., Yamamoto, K., Murotani, K., Nakamura, Y.: Compliant biped locomotion of hydra, an electro-hydrostatically driven humanoid. In: Proceedings of IEEE-RAS International Conference on Humanoid Robots (2018, under review)
7. Pratt, J., Tedrake, R.: Velocity-Based Stability Margins for Fast Bipedal Walking, pp. 299–324. Springer, Heidelberg (2006)
8. Righetti, L., Schaal, S.: Quadratic programming for inverse dynamics with optimal distribution of contact forces. In: Proceedings of IEEE-RAS International Conference on Humanoid Robots, pp. 538–543 (2012)
9. Sentis, L., Park, J., Khatib, O.: Compliant control of multicontact and center-of-mass behaviors in humanoid robots. IEEE Trans. Robot. **26**(3), 483–501 (2010)
10. Sugihara, T.: Standing stabilizability and stepping maneuver in planar bipdedalism based on the best COM-ZMP regulator. In: Proceedings of IEEE International Conference on Robotics and Automationm, pp. 1966–1971 (2009)
11. Sugihara, T., Nakamura, Y.: Whole-body cooperative balancing of humanoid robot using COG Jacobian. In: Proceedings of IEEE/RSJ International Conference on Intelligent Robots and Systems, pp. 2575–2580 (2002)
12. Tsagarakis, N.G., et al.: Compliant humanoid COMAN: optimal joint stiffness tuning for modal frequency control. In: Proceedings of IEEE International Conference on Robotics and Automation, pp. 665–670 (2013)
13. Vukobratovic, M., Stepanenko, J.: On the stability of anthropomorphic systems. Math. Biosci. **15**, 1–37 (1972)
14. Yamamoto, K.: Humanoid motion analysis and control based on COG viscoelasticity. Adv. Robot. **31**(7), 341–354 (2017)
15. Yamamoto, K.: Resolved multiple-viscoelasticity control for a humanoid. IEEE Robot. Autom. Lett. **3**(1), 44–51 (2017)
16. Yamamoto, K.: Robust walking by resolved viscoelasticity control explicitly considering structure-variability of a humanoid. In: Proceedings IEEE International Conference on Robotics and Automation, pp. 3461–3468 (2017)

Robot Design and Prototypes

Robot Design and Prototypes

Torsten Kroeger

The five papers of this session address hardware concepts and sensor prototypes with the goals of improving 3D printing, quadruped robots, low-cost 3D scanning, and perception and control for manipulation. Yan *et al.* present framework for designing 3D printers based on a novel bi/stable mechanical actuator that allows for inexpensive and rapid integrated electromechanical prototyping. Harper *et al.* introduce a set of algorithms to improve models for motion planning and control of quadruped robots, which were validated using high-accuracy motion tracking systems. Borrmann *et al.* presents a wheeled device that deploys a 2D laser scanner used to generate high-precision 3D maps of outdoor environments. Saldarriaga and his team leverage a foundational method to analytically determine optimal damping coefficients of impedance controllers of industrial robot arms that take into account end-effector tools and application-specific parameters. For applications of industrial manipulation, Brucker *et al.* introduce a deep learning approach that trains models with synthetic data only.

During the panel discussion, the audience and the presenters focused on the advantages and disadvantages of traditional model-based approaches for perception and actuation with respect to data-driven model-free approaches. The conclusion was that one will always benefit from both, and the current and future challenges for the robotics research community are in finding the right interfaces and abstractions to get the best out of both approaches.

Towards Autonomous Printable Robotics: Design and Prototyping of the Mechanical Logic

Wenzhong Yan(✉), Angela L. Gao, Yunchen Yu, and Ankur Mehta

The Laboratory for Embedded Machines and Ubiquitous Robots, University of California, Los Angeles, Los Angeles, CA 90095, USA
wzyan24@ucla.edu

Abstract. New strategies for controlling printable robots are needed for inexpensive and rapid integrated electromechanical prototyping. In this work, we propose a printable, easy-to-fabricate controller—what we refer to as the mechanical logic—which contains a bistable mechanism that acts as a mechanical switch which generates an oscillating binary electrical signal. The integration of the mechanical logic with conductive actuators endows it with clocking and current switching functions; thus, powered by a constant electrical energy source, the mechanical logic is able to autonomously control a sequence of actuations. To demonstrate this technique, we designed and fabricated prototypes of the mechanical logic, whose raw materials (not including the power supply) cost approximately 40 cents. Under a constant-current (0.55 A) power source, our mechanical logic was able to induce current oscillation with an average period of around 3.8 s. This property of the mechanical logic could ultimately be incorporated into onboard controllers for printable robots and devices.

Keywords: Printable robotics · Autonomous robots · Printable controllers · Design and prototyping

1 Introduction

The origami-inspired cut-and-fold technique allows for the application of 2D fabrication tools towards building 3D objects. Based on this technique, a new class of printable robots has been created. These printable robots are inexpensive, easy to design and manufacture, and amenable to rapid prototyping [1]. Therefore, they can be utilized for numerous real-life applications [2–5]. Nevertheless, in order to achieve autonomy, each printable robot still requires an on-board electrical controller, e.g. microcontroller, and its accessories, which raise the cost and the complexity of the resulting system. In this paper, therefore, we take a step towards inexpensive and easy

Electronic supplementary material The online version of this chapter (https://doi.org/10.1007/978-3-030-33950-0_54) contains supplementary material, which is available to authorized users.

J. Xiao et al. (Eds.): ISER 2018, SPAR 11, pp. 631–644, 2020.
https://doi.org/10.1007/978-3-030-33950-0_54

electromechanical prototyping of robots: we present an inexpensive control method for fully printable robots, which allows for fast and inexpensive fabrication of a building block for robotics control. The mechanical logic, as shown in Fig. 1, has the potential to build an autonomous robot, once integrated with onboard power sources, printable structures and transducers.

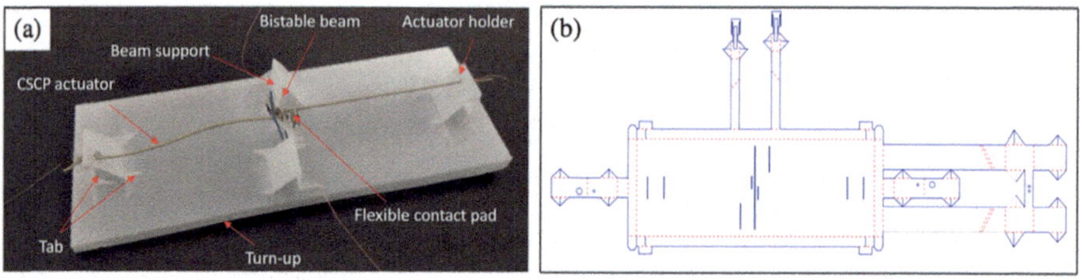

Fig. 1. (a) A prototype of the proposed mechanical logic (110 × 40 × 4 mm, 1.75 g). Power is provided through tethers. (b) The corresponding 2D fold pattern of the origami structure. Blue lines represent cutting while red dash lines represent scoring. Actuators and other accessories are not shown here.

The presented approach allows for "printing" controlling methods for robots. The mechanical logic possesses clocking and current switching functions that are dictated by two factors, a switching mechanism and a time delay. The switching action is accomplished with an origami precompressed bistable beam [6]. The deflection of this bistable beam enables it to store and release energy, giving rise to its two distinct stable positions. This characteristic makes it a great candidate for applications with two distinct working states, such as an on/off switch. Hence, bistable components can transform a continuous actuation into a sequence of discrete, transient motions. The time delay is realized with conductive supercoiled polymer (CSCP) actuators [7] that gradually contract in response to a steady Joule heating resulting from a constant power supply. The CSCP actuator is chosen for its special properties: high conductivity, low cost and easy fabrication.

Towards "printing" complete robotic systems, this paper focuses on a printable method for robots' controlling components. The contributions of this paper include:

- an electromechanical mechanism which generates an oscillating switched current from a constant electrical power supply;
- the low-cost design of a mechanical logic system which combines a bistable beam made with an origami-inspired strategy with CSCP actuators; and
- a printable control logic device featuring the proposed design, modeled, manufactured, and verified by experiments.

2 Related Work

Origami. Origami was introduced to the robotics community in order to achieve easy, rapid and inexpensive robot prototyping. A wide variety of 3D geometries have been created out of origami structures by patterning 2D flat materials and assembling them by folding [8]. This technique has been applied to manufacturing varied printable robots. However, these printable robots still require custom or off-the-shelf, non-printable hardware, such as microcontrollers and electromechanical transducers, to achieve autonomy [9]. There has been an attempt to create printable elements to replace these electrical hardware [10], however, this work primarily focused on printing passive components and sensors, such as resistors and capacitors.

Controller. Recently, Wehner et al. [11] developed an Octobot that incorporates a microfluidic logic [12] as an onboard controller to realize autonomy in fully soft robots. Similarly, Rothemund et al. [13] designed a soft, elastomeric valve that behaves as a mechanical "switch" which controls air flow for soft robots. When integrated into a feedback pneumatic circuit, the valve generates periodic motion using air from a single source of constant pressure, and acts as a pneumatic oscillator. Still, these works are not applicable to electromechanical robots, as they focused on entirely fluidic applications.

Inspired by these works, we came up with a printable controller that incorporates a bistable beam and CSCP actuators to achieve autonomy in origami-inspired, electrically powered printable robots, further reducing cost and increasing accessibility to the public.

3 Design and Modeling

3.1 Overall Design and Mechanism

To implement the current-oscillating function, a bistable mechanism and conductive actuators are incorporated into an integrated design. Bistable mechanisms are ideal for switches because power is only needed for switching the beam between its stable states but not for maintaining it at one state. As shown in Fig. 2, the bistable mechanism functions as a double-pole, single-throw switch for both Loop A and Loop B in opposite phases. The actuators serve two functions: one of which is to complete the circuitry, while the other is to drive the snap-through motion of the bistable mechanism.

If the exerted force generated by the actuator exceeds the activation force required by the bistable mechanism, a snap-through motion of the bistable mechanism can be guaranteed. Under this circumstance, once power is supplied, the mechanical logic starts to oscillate: (1) initially, Loop A is closed. The corresponding actuator begins to heat up and gradually contract along its axial direction; (2) upon reaching its critical displacement, the bistable mechanism snaps to the other stable position; Loop A is now open while Loop B is closed; (3) Actuator A begins to cool down while Actuator B begins to heat up. The previous steps then execute for the opposite loops; after this, the entire actuation process is completed. As the system repeats this entire process periodically, the mechanical logic behaves similarly to an electrical oscillator. The supplied power heats the actuators, and the thermal energy of these actuators subsequently dissipates into the atmosphere. Hence, to achieve continuous periodic oscillation, the time needed for the bistable mechanism to snap through must exceed the heat dissipation time of the actuators.

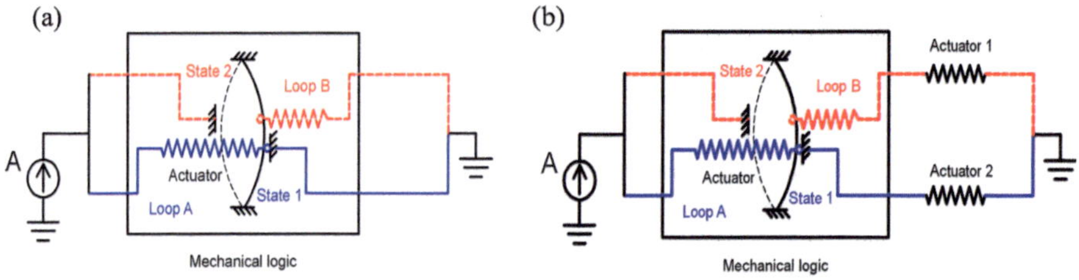

Fig. 2. A schematic of the mechanical logic and one of its possible applications. (a) The mechanism of the mechanical logic. Two loops are connected and disconnected alternatively when the bistable mechanism switches between the two different stable states, inducing current oscillation. (b) A possible configuration of a control-actuation system with the mechanical logic connecting with other actuators.

3.2 Function Implementation

(1) CSCP Actuator. In our system, the actuators are expected to implement the functions of both conductivity and actuation. CSCP actuators are chosen, since they are conductive and able to convert current into power and induce a temperature change, which can finally transform into their axial contraction. They are also low cost and easy to use.

The CSCP actuator is formed by twisting conductive polymer threads [14, 15] until they form coils and then annealing the coils thermally. The output force of the actuator can be described as follows [7]:

$$F = k(x - x_0) + b\dot{x} + c(T - T_0) \tag{1}$$

where x and x_0 are the loaded and unloaded length of the actuator. k and b are the mean stiffness and mean damping coefficient of the actuator, respectively. T is the temperature of the actuator, while T_0 is the room temperature (i.e. 25 °C), and c (a constant) is the mean slope to compensate the temperature rise of the actuator. For static or quasi-static analysis, the second (damping) term of Eq. (1) can be neglected.

The maximum strain of this kind of actuator is about 10% [14]. Under this limitation, the actuator's length (which results in the inverse change of k and b simultaneously) could alter the output force, which leads to the variation of the oscillation period of the mechanical logic. The power supply, which determines the temperature T, could also control the rate of change of the contraction force generated by the actuator, and therefore adjust the oscillation frequency of the mechanical logic. In conclusion, a longer CSCP actuator or higher power supply can increase the oscillation frequency of the mechanical logic.

(2) Bistable Beam Switch. To design a mechanical logic capable of providing control for printable robots, we must come up with an easily fabricated mechanism with a switching function. In this work, a precompressed bistable beam is incorporated into the mechanical logic as a switch, since generating an axial load required for a pre-compressed beam with origami folding is relatively easy.

The beam is supported at its two terminals and subjected to a transverse force applied at its midpoint. Figure 3 is a typical force-displacement diagram of a pre-compressed bistable beam, where state 1 and state 2 are its two distinct equilibrium states. The switching force, F_s, is the minimum force required for the bistable beam to switch from state 1 to state 2; the switch-back force, F_{sb}, is the minimum force required for the beam to switch back to state 1 from state 2. Both F_s and F_{sb} are referred to as critical forces, while the corresponding positions (u_s and u_{sb}) are critical displacements. u_{tr} refers to the distance its midpoint travels when the beam switches from one state to another. The snap-through occurs once the beam reaches its switching point, making it suitable as a switch, especially for our mechanical logic.

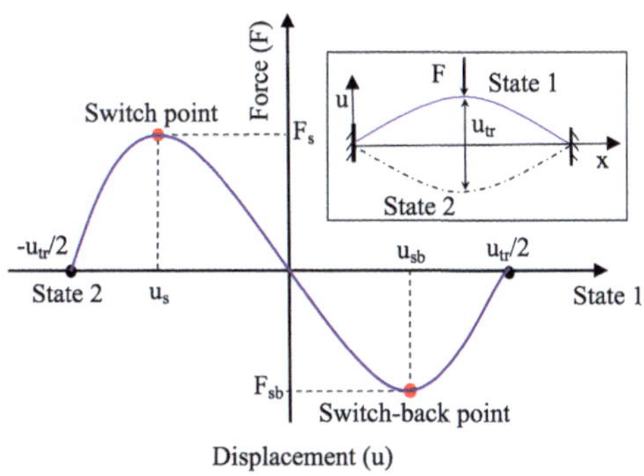

Fig. 3. A typical force-displacement curve of a precompressed bistable beam's middle point, adapted from [16].

The force-displacement curve of the bistable beam, which dictates its switching behavior, can be modeled using linear approximation. This curve is linearized around the beam's three equilibrium positions, and is thus composed of three linear curves [17].

$$F = \left\{ \begin{array}{ll} k_1\left(u + \frac{u_{tr}}{2}\right), & u \leq -u_0 \\ k_2 u, & -u_0 < u < u_0 \\ k_1\left(u - \frac{u_{tr}}{2}\right), & u \geq u_0 \end{array} \right\} \qquad \begin{array}{l} k_1 = \frac{EI}{l^3}\left[\frac{10-\pi^2}{16\pi^4} + \frac{EI}{d_0\pi^2 lEA - 4\pi^4 EI}\right]^{-1} \\ k_2 = -\frac{(2.86\pi)^2 EI}{l^3}\left[\frac{1}{4} - \frac{1}{2.86\pi}\tan\left(\frac{2.86\pi}{4}\right)\right]^{-1} \\ u_0 = \frac{k_1 u_{tr}}{2(k_1 - k_2)} \end{array} \tag{2}$$

where k_1 is the spring constant of the beam at its unstable equilibrium position, and k_2 is the spring constant at its two stable equilibrium positions. *E, A, l,* and *I* refer to the beam's Young's modulus, cross-section area, span, and moment of inertia, respectively. Moreover, d_0 is defined as the difference between the original length l_o and the span of the beam l.

The critical force and corresponding critical displacement are determined by the intersections of these linear curves, which, in turn, depend on the aforementioned parameters of the beam. Therefore, by carefully choosing the material (i.e. E) and the dimensional parameters of the bistable beam (i.e. A, I, l and l_0), we can control its behavior by predetermining its critical points (i.e. the critical force and the travel u_{tr}). Note that the actuation position on the bistable beam in our final design deviates slightly from its middle point; thus, the transverse force needed to facilitate the snap-through motion is expected to be smaller than the modeling result [18].

3.3 Mechanical Characterization

Since the mechanical logic is symmetrical, we only need to consider the actuation process on one side. To ensure a snap-through, the force generated by the actuator must exceed the activation force of the bistable beam. Mechanical testing is done using an Instron machine with an embedded heat chamber, which controls the temperature of actuators. To simulate the static behavior of the CSCP actuator, a simplified quasi-static model derived from Eq. (1) is established. Meanwhile, the force-displacement curve of the bistable beam is obtained from linear approximation [17].

For our characterization, a bistable beam with certain geometric parameters was chosen; we then explored the influence of the length of the CSCP actuator on its exerted force, in order to facilitate the mechanical logic design. The parameters of the beam and actuators are listed as follows: The length, width, thickness, and span of the beam are 14.9 mm, 3.0 mm, 0.1325 mm, and 14.0 mm, respectively. The Young's modulus of the beam is 3.0 GPa; k and c of the CSCP actuator (50 mm) are 186.8 ± 13.0 N · m and 28.6 ± 1.7 mN/°C. k of other actuators could be calculated, since it is inversely proportional to their lengths.

Modeling and experimental results are shown in Fig. 4. Figure 4(a) presents the forces generated by the actuators during contraction when their temperature reaches our predetermined working temperature. By changing the length of the actuator, we can tune the attenuation of the output force against its contraction. As shown in the graph, the contraction force of a longer actuator tends to decrease much more slowly. With this theory, we are able to choose an actuator (of appropriate length) that can generate enough force to activate a bistable beam with certain geometry.

In our current work, we assume that the maximum working temperature of the actuator is 50 °C (lower than its Glass Transition Temperature and melting point of the bistable beam). As shown in Region I of Fig. 4(b), at 50 °C, the force generated by the CSCP actuator is always greater than the activation force of the bistable beam, guaranteeing a snap-through motion. We do not need the force provided by this actuator to exceed the activation force of the beam in Region II of this graph, since in this region, the bistable beam is characterized by negative stiffness behavior. After the beam snaps through and quickly settles in its other equilibrium state, the actuator on the opposite side is expected to function in the same way, resulting in the snap-back motion of the beam.

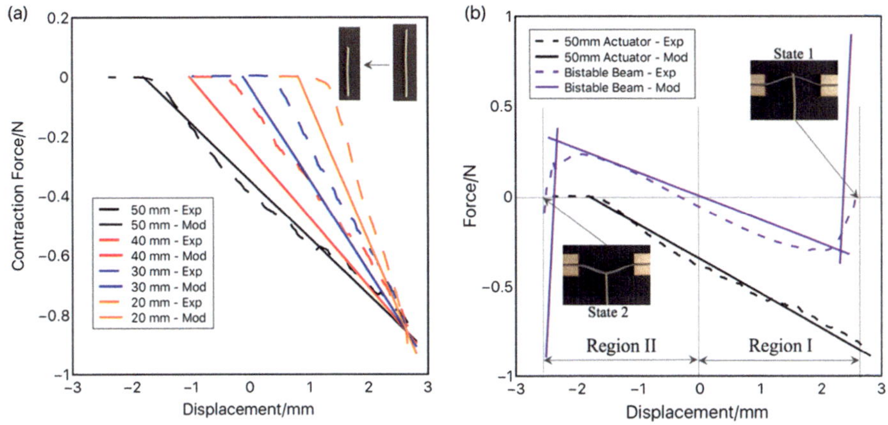

Fig. 4. Mechanical characterization of the mechanical logic, including modeling and experimental results. (a) Parametric study of the influence of the length of the CSCP actuator on its contraction force at 50 °C. The range of the length investigated is between 20 and 50 mm, in 10 mm intervals. (b) The force-displacement curves of the chosen bistable beam and actuator (50 mm). Activation is guaranteed when the contraction force generated by the actuator is greater than the beam's activation force in region I. In region II, the bistable beam can snap back if there is another actuator pulling on it in the opposite direction (not shown in the graph).

Hence, we have demonstrated that the CSCP actuator can be integrated into the mechanical logic by showing that the force it generates exceeds the activation force of the bistable beam. Moreover, by varying the length of the actuator, we can control its output force (see Fig. 4) and thus the activation process of the bistable beam. Actuators with lengths from 40 mm to 50 mm can all trigger the bistable beam used in our experiment to snap through. These actuators of different lengths correspond to different activation times of the bistable beam, resulting in different oscillation periods of corresponding mechanical logics. Eventually, we chose 50 mm-long actuators for building our mechanical logic with the predetermined bistable beam, guaranteeing a highly fault-tolerant snap-through motion of the selected bistable beam. In other words, this actuator will be able to activate the bistable beam even if its working temperature deviates from the predetermined working temperature (50 °C).

4 Fabrication

4.1 Origami-Inspired Structure

The 2D planar design, shown in Fig. 1(b), was accomplished in AutoCAD and cut using a Silhouette Cameo paper cutter to pattern a flat sheet. After cutting and folding, the final 3D model was created. There were several challenges in our origami-inspired structure fabrication, including the structure fastening, bistable beam formation, and flexible contact pads design.

(1) Fastening Method for the Origami Structure. To construct 3D structures out of 2D sheets, a fastening method was needed. To simplify fabrication and reduce cost, a mechanical tab was created. This approach obviates the need for fasteners or adhesives during assembly. Two triangle ribs function together as a mechanical stopper to fix the position of the origami structure, as shown in Fig. 5 (Tab A). In some cases, folding the ribs to 90° against their attachment could force the structure itself to stand vertically against the base; in other words, this way of folding helps to shape the geometry of origami structures. Especially for the bistable beam formation, this approach not only helped to straighten the support of the beam, but also greatly increased the stiffness of the boundary of the beam (e.g. Tab B in Fig. 5). This enhanced stiffness is necessary for the symmetric operation of the bistable beam, as explained in the experiment section below.

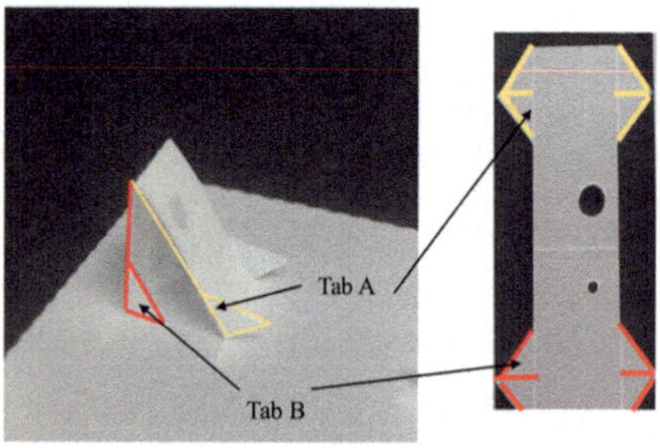

Fig. 5. A mechanical tab. Tab A is a typical mechanical tab, which can be used to fasten the structure. Tab B can help to keep the structure vertical against the base and increase its stiffness.

(2) Bistable Beam Formation. In order to make the desired bistable beam out of a planar sheet, we needed to create a compliant beam with rigid supports which apply axial compression on it and result in buckling. This required that the flat sheet material itself should have flexibility. A low-cost .005" polyester (PET) sheet was selected. We used a z-shape folding pattern and geometry confinement to form the precompressed bistable beam. The z-shape folding pattern (Fig. 6(a)) resulted in a first-step compression on the beam by decreasing its axial span. Then the beam fed through a slot in the support structure. The side walls of this slot could further confine the beam (Fig. 6(b)), giving rise to the final form of the desired bistable beam.

Due to the single-point confinement from each side wall, the bistable beam slightly bends upwards, with its symmetry broken. In order to keep the structure simple enough, we decided not to add accessories to balance the tilt. Instead, we compensated for the bending by differentiating the heights of the fastening tabs, as shown in Fig. 6 (a). The discrepancy in the heights of tabs made the beam bend downwards, counteracting the initial error and therefore improving the symmetry of the bistable beam.

(3) Contact Pad Design. In order to maintain the electrical connection in the loops before the bistable beam snaps through and realize an instant disconnection immediately thereafter, we designed a flexible contact pad, as shown in Fig. 7(a). The pad was made out of a nonconductive PET sheet, and a piece of aluminum foil was attached on

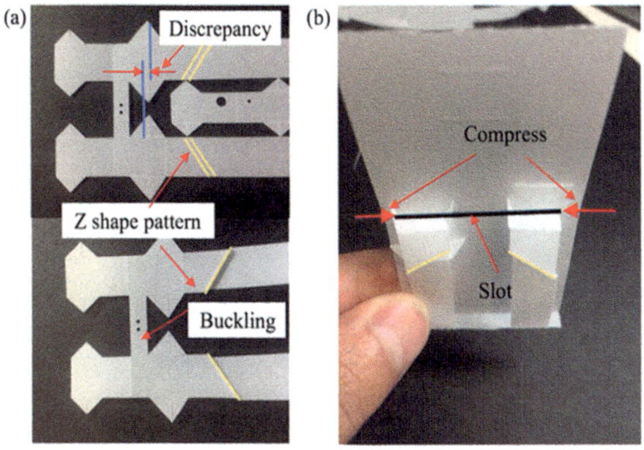

Fig. 6. A precompressed bistable beam formation. (a) z-shape folding is used to preliminarily compress the beam. Note that the discrepancy is exaggerated in the graph. (b) The axial dimension of the bistable beam is further confined by the sidewalls of the slot.

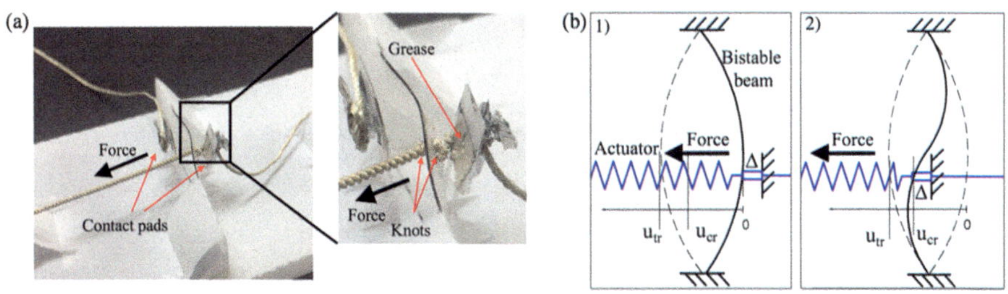

Fig. 7. A flexible contact pad. (a) The side view of the contact pad. (b) The arrangement of the bistable beam and the contact pad. (1) At the beginning, the flexible pad is at $u = -\Delta$. Δ is the length of the overhanging part of the actuator. (2) The actuator gradually pulls the beam leftwards and finally forces it to reach its critical displacement at $u = u_{cr}$. Meanwhile, the contact pad is at $u = u_{cr} - \Delta$.

it to endow it with conductivity. As shown in Fig. 7(b), the critical displacement of the bistable beam is u_{cr}. If we settle the contact pad at the position $u = u_{cr} - \Delta$, its flexibility will ensure electrical connection between the actuator and contact pad before the bistable beam snaps through.

4.2 Actuator Fabrication and Connectivity

Details of the CSCP actuators used in our paper can be found in [7]. Owing to the small contact area, the high contact resistance introduced on the connection between the actuator and the contact pad would lead to overheating. The overheating at the contact point would further increase its resistance or even melt the actuator, eventually destroying the conductivity of the circuit. In our work, we temporarily used conductive grease (Volume Electrical Resistivity 0.01 Ω-cm) to solve this problem. Before the experiment, a tiny volume of the conductive grease was spread on the terminal of actuator, shown in Fig. 7.

5 Experiments and Results

5.1 Testing Apparatus

The total cost of materials to build a mechanical logic with our approach is about 40 cents. After being fixed on the platform, the mechanical logic was powered with a constant current (0.55 A) by a laboratory power supply, as shown in Fig. 2. In addition, a cooling fan was added to control the period of oscillation by balancing the actuation rate and the heat dissipation rate of the actuator, as shown in Fig. 8.

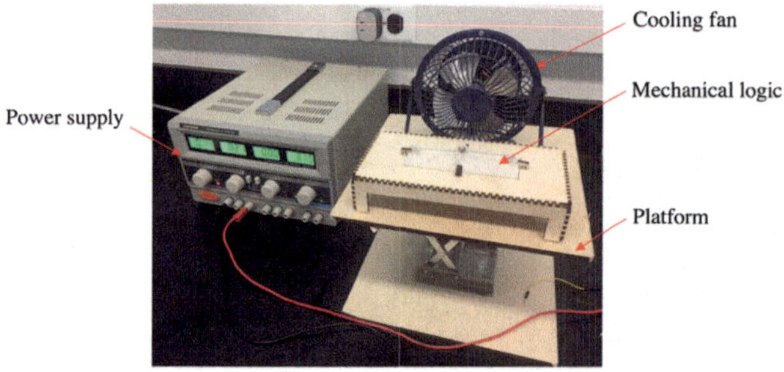

Fig. 8. The test rig for the mechanical logic. The position of the fan is fixed to maintain the same cooling condition of actuators for different tests.

5.2 Main Results

The test was run for around 20 s, during which the mechanical logic went through four complete oscillation cycles, along with one additional snap-through process at the very end. The result of the first cycle is represented in Fig. 9. At the very beginning, shown in Fig. 9(a), Loop A was connected, and then the corresponding Actuator A started to heat up. Once the contraction of Actuator A drove the bistable beam to its critical displacement (this transition mode was noted at around 3.7 s, as indicated in Fig. 9(b)), the bistable beam snapped through and disconnected Loop A at about 3.8 s. At the same time, Loop B was connected. Then Actuator B heated up and contracted while Actuator A cooled down and relaxed. The system was back in a transition mode at around 5.8 s (see Fig. 9(d)), but oriented in the opposite direction. Thereupon, the bistable beam jumped back to its original position, completing the first oscillation at about 5.9 s. Thanks to the symmetrical design, the bistable beam continued this motion periodically, resulting in current oscillation.

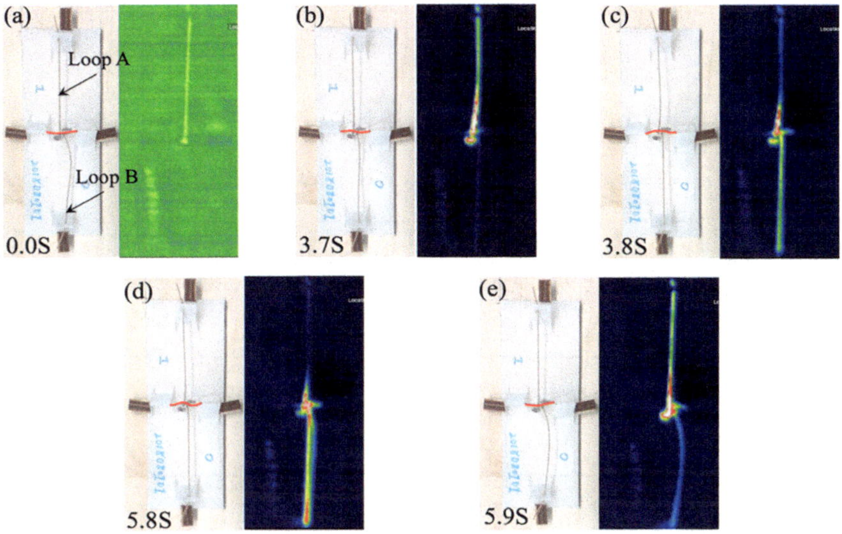

Fig. 9. The first oscillation cycle of the mechanical logic. For each figure from (a) to (e), the left part is the image of the mechanical logic and the right part is the corresponding thermal image from Seek Thermal Compact.

The average oscillation frequency (for four complete cycles) of this device was therefore about 0.27 Hz. This oscillation frequency is determined by the response time of the snap-through motion of the bistable beam. This response time depends on the bistable beam's dimension and the CSCP actuator's dynamic response, which, in turn, is determined by the input power and other parameters, according to the thermo-electro-mechanical model from [7].

The currents running through both actuators, as well as these actuators' temperatures, are plotted in Fig. 10. The current graphs were created by plotting the readings from the multimeters over time, while the temperature changes were recorded with a thermal camera (Seek Thermal Compact). In order to synchronize these two key variables (i.e. current and temperature) of the mechanical logic, we aligned two different curves by manually matching the critical moments when the current was first detected and when the temperature started to change. As shown in Fig. 10, two currents oscillated alternatively with an average period of about 3.75 s (the average of four complete cycles). Theoretically, decreasing the critical displacement of the bistable beam by changing its dimensions can allow it to snap through within a shorter time; thus, the oscillation frequency will increase correspondingly. In addition, the frequency of the current oscillation can be boosted by increasing the input power.

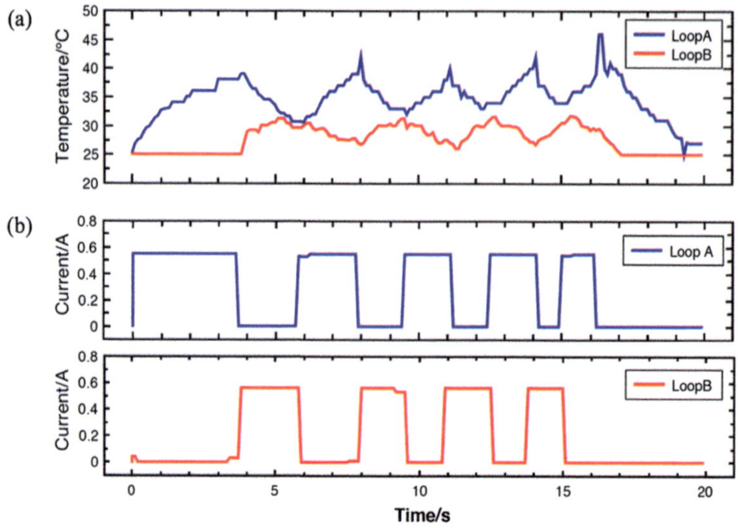

Fig. 10. Curves that show key variables (temperature and current) as a function of time. (a) Temperature curves of two actuators in both Loop A and Loop B. (b) Curves of currents through both Loop A and Loop B.

5.3 Discussion

(1) Period Attenuation. As shown in Fig. 10, the period of current oscillation decreased over time. This phenomenon could have resulted from the temperature rise of the CSCP actuators at the beginning of each cycle, which introduced a small initial contraction and further lessened the time needed to activate the snap-through of the bistable beam. Improving the stability of the CSCP actuators should eliminate this attenuation of the oscillation period. Still, this phenomenon implies a potential approach to tune the oscillation frequency by prestressing the actuators against the bistable beam. However, over-stressing of the actuators would destroy the bistability of the beam. Besides prestressing, changing the operating current, the length of the actuator, or the initial rise of the bistable beam could adjust the final characteristics of the oscillation, as mentioned above.

(2) Thermal Asymmetry. The temperature curves of the two actuators are not symmetric, even though the structure of the system is. The different resistance at the two connections between the contact pads and the terminals of the actuators may account for this lack of symmetry. This hypothesis was verified by the thermal images (see Fig. 9) that each display a hot spot only at the connection between Actuator A and the corresponding contact pad. As the image shows in Fig. 10, Actuator A had higher maximum temperatures (around 41 °C) than Actuator B (about 33 °C) in all four cycles. Fortunately, this thermal asymmetry did not interfere with the functionality of the mechanical logic, though it altered the oscillation frequency. To resolve this issue of asymmetry, we plan to develop a more robust and controllable mechanism that connects the terminals of the actuators and the contact pads.

(3) Oscillation Duration. The oscillation of the mechanical logic terminated after four complete cycles, while a longer oscillation duration is desired. The temperature of both actuators gradually fell back to room temperature; meanwhile, the current across both actuators dropped to zero. These phenomena might be explained by the disrupted conductivity at the connections between the actuators and the contact pads. In fact, a post-experiment examination of these connections verified this explanation: the connection between Actuator A and its corresponding contact pad was broken. Therefore, in order to obtain oscillations with longer duration, a better design of the connecting mechanism is needed.

6 Insights and Future Work

We demonstrated an entirely printable controller that, while using an external constant current power source, was able to generate periodic mechanical and electrical oscillations through the integration of a bistable beam and conductive actuators. The bistable beam acted as a switch, controlling electrical current flow through the actuators, which in turn drove the switching of the bistable beam. This resulting oscillation can be adapted for use in more complicated designs to sequence further switched actuators.

This was the first trial where an origami-inspired printable structure was incorporated into an autonomous printable controller for a robot. Thanks to the fabrication methods we adopted, the manufacturing of this controller was easy and inexpensive, leading to minimal cost of building a completely autonomous printable robot.

Our future work will include combining the mechanical logic with a printable robot body and additional actuators to create an integrated printable robot. The combination of the snap-through instability of a bistable beam with the contraction of CSCP actuators enables the mechanical logic to function on the component level, making it less sensitive to other components of the robots (e.g., the actuators) and more predictable in its behavior. All of these advantages of the mechanical logic imply its easy integration into a printable robot. For example, we can use a worm-like robot [1] as a basic configuration, replace its NiTi coil actuators with conductive actuators (e.g. CSCP actuators), and add our mechanical logic as its controller, as shown in Fig. 1(b). With constant-current (or constant-voltage) power, the worm-like robot can undergo autonomous peristaltic locomotion.

To fabricate autonomous, untethered, and printable robots, the mechanical logic may also be combined with onboard energy sources (e.g. button battery). Furthermore, the printable energy sources (e.g. paper battery [19]) may ultimately be involved in designing a fully printable robot. We believe that our printable controller will be the foundation of manufacturing entirely printable, inexpensive, and autonomous robots, as well as a step towards making robots ubiquitous, accessible for every member of society.

Acknowledgments. This work is partially supported by NSF under grant #1752575.

References

1. Onal, C.D., Wood, R.J., Rus, D.: An origami-inspired approach to worm robots. IEEE/ASME Trans. Mechatron. **18**(2), 430–438 (2013)

2. Mehta, A.M., Rus, D., Mohta, K., Mulgaonkar, Y., Piccoli, M., Kumar, V.: A scripted printable quadrotor: rapid design and fabrication of a folded MAV, pp. 203–219. Springer International Publishing, Cham (2016)
3. Martinez, R.V., Fish, C.R., Chen, X., Whitesides, G.M.: Elastomeric Origami: programmable paper elastomer composites as pneumatic actuators. Adv. Funct. Mater. **22**(7), 1376–1384 (2012)
4. Min, C.C., Suzuki, H.: Geometrical properties of paper spring. In: Mitsuishi, M., Ueda, K., Kimura, F. (eds.) Manufacturing Systems and Technologies for the New Frontier, pp. 159–162. Springer, London (2008)
5. Kim, S.J., Lee, D.Y., Jung, G.P., Cho, K.J.: An origami-inspired, self-locking robotic arm that can be folded flat. Sci. Robot. **3**(16), eaar2915 (2018)
6. Camescasse, B., Fernandes, A., Pouget, J.: Bistable buckled beam and force actuation: experimental validations. Int. J. Solids Struct. **51**(9), 1750–1757 (2014)
7. Yip, M.C., Niemeyer, G.: On the control and properties of supercoiled polymer artificial muscles. IEEE Trans. Rob. **33**(3), 689–699 (2017)
8. Demaine, E.D., Demaine, M.L.: Recent results in computational origami. In: Origami 3: Proceedings 3rd International Meeting of Origami Science, Math, and Education (OSME2001), pp. 3–16. A K Peters, Monterey (2001)
9. Mehta, A.M., DelPreto, J., Shaya, B., Rus, D.: Cogeneration of mechanical, electrical, and software designs for printable robots from structural specifications. In: Intelligent Robots and Systems (IROS), pp. 2892–2897 (2014)
10. Miyashita, S., Meeker, L., Goldi, M., Kawahara, Y., Rus, D.: Self-folding printable elastic electric devices: resistor, capacitor, and inductor. In: Conference on Robotics and Automation (ICRA), pp. 1446–1453 (2014)
11. Wehner, M., Truby, R.L., Fitzgerald, D.J., Mosadegh, B., Whitesides, G.M., Lewis, J.A., Wood, R.J.: An integrated design and fabrication strategy for entirely soft, autonomous robots. Nature **536**, 451 (2016)
12. Mosadegh, B., et al.: Integrated elastomeric components for autonomous regulation of sequential and oscillatory flow switching in microfluidic devices. Nat. Phys. **6**(6), 433–437 (2010)
13. Rothemund, P., Ainla, A., Belding, L., Preston, D.J., Kurihara, S., Suo, Z., Whitesides, G.M.: A soft, bistable valve for autonomous control of soft actuators. Sci. Robot. **3**(16), eaar7986 (2018)
14. Yip, M.C., Niemeyer, G.: High-performance robotic muscles from conductive nylon sewing thread. In: Conference on Robotics and Automation (ICRA), pp. 2313–2318 (2015)
15. Zhang, D., Rahmat-Samii, Y.: Integration of electro-textile RF coil array with magnetic resonance imaging (MRI) system: Design strategies and characterization methods. In: 2018 International Workshop on Antenna Technology (iWAT), pp. 1–3 (2018)
16. Palathingal, S., Ananthasuresh, G.K.: Design of bistable pinned-pinned arches with torsion springs by determining critical points. In: Mechanism and Machine Science, pp. 677–688 (2017)
17. Vangbo, M.: An analytical analysis of a compressed bistable buckled beam. Sens. Actuators, A **69**(3), 212–216 (1998)
18. Cazottes, P., Fernandes, A., Pouget, J., Hafez, M.: Bistable buckled beam: modeling of actuating force and experimental validations. J. Mech. Des. **131**(10), 101001 (2009)
19. Hu, L., Choi, J.W., Yang, Y., Jeong, S., La Mantia, F., Cui, L.F., Cui, Y.: Highly conductive paper for energy-storage devices. Proc. Nat. Acad. Sci. **106**(51), 21490–21494 (2009)

Model Refinement for Terrain Responsive Planning on a Dynamic Running Quadruped

Mario Harper[1(✉)], David Balbuena[2], Justin Larson[2], Camilo Ordonez[2], Gordon Erlebacher[1], Emmanuel Collins[2], and Jonathan E. Clark[2]

[1] Department of Scientific Computing, Florida State University, Tallahassee, FL 32306, USA
myharper@fsu.edu

[2] Department of Mechanical Engineering, Florida A&M University-Florida State University, Tallahassee, FL 32306, USA
jeclark@fsu.edu

Abstract. Running legged robots present several challenges when motion planning, these challenges often stem from the difficulty in predicting motion due to the innate complexity of the mechanical system, persistent effects of terrain and foot-ground interactions. While reasonable approximations of the inherent motion models of the Ghost Robotics Minitaur platform can be learned through data-driven approaches, system mechanical robustness and the requisite experimental time discourages running a full battery of experiments to determine a unique model for each considered terrain. This paper discusses the development of turning maneuvers on the quadruped robot Minitaur and the approach taken to adapt a learned model for new terrains. The prediction model was generated through data collected in indoor experiments using a VICON motion capture system and adapted through use of a correction factor for new terrains. Resulting motion planning differences show that the addition of this factor improves the planning model and reduces the computational burden of replanning.

1 Introduction

Generally, legged animals and robots are able to demonstrate several advantages over traditional wheeled or track-based platforms when navigating through unstructured terrains [18], specifically leveraging discrete foot placement [15] and large ground/obstacle clearance [2]. Planning the motions to navigate using these tools has been achieved quasi-statically [5,20], typically by choosing to restrict behavior to statically stable gaits such as walking. However, to achieve more practically useful velocities, legged robots need gaits which have flight phases

Electronic supplementary material The online version of this chapter (https://doi.org/10.1007/978-3-030-33950-0_55) contains supplementary material, which is available to authorized users.

J. Xiao et al. (Eds.): ISER 2018, SPAR 11, pp. 645–654, 2020.
https://doi.org/10.1007/978-3-030-33950-0_55

rendering them statically unstable [3,8,17], this presents significant difficulty for current motion planning techniques.

Most legged robots currently motion plan based on foothold selection coupled with a high-bandwidth controller [6,7,19]. While this approach gives both flexibility and reliability of locomotion on unstructured terrains [1], it has not been applied to dynamic running systems. Control schemes for complex, deliberative motion robots are fast but often fails to consider the kinodynamic constraints of dynamic running. Although sufficiently complex, high-speed controllers can potentially be used [16], this requires a high level of accurate state information. It is computationally more feasible to generate a motion planning trajectory that can consider the constraints and can accurately predict motions of a fast legged robot.

An important aspect of a motion prediction model is in the terrain effects of a legged robot. Work has been done to model foot-terrain interactions that involves calculation of the normal and tangential forces on the foot [9]. This requires detailed information about the terrain geometry and mechanical properties and thus is not suitable for high speed locomotion over unknown terrain. It is particularly difficult to understand the terrain interactions (as illustrated in Fig. 1) that impact the motion characteristics of a legged robot. This is true even in the case of relatively flat terrains. As an example, soft surfaces absorb more impact and reduce body velocity but tends to increase stability. Conversely, hard terrains with high friction tend to increase the expected body velocity while compounding the effects of body pitch and yaw.

Fig. 1. Initial experiments on different terrains. The robot is commanded to move forwards for 4.7 m (obstacles at 5 m) and turn to the left for 1m. Attempting to follow this trajectory results in collision on some terrains and negligible turns on others.

In this work, we use a motion planning algorithm that is well suited for fast legged robots called Sampling-Based Model Predictive Optimization (SBMPO) [14]. This algorithm provides the benefits of computation speed by using a heuristics-based search method and directly utilizes a learned kinodynamic model [11]. The models can be viewed as the prediction model of how the control space maps to the robot velocities.

This research shows a hardware implementation of the SBMPO planning algorithm on the Minitaur platform. We detail a new model for turning on the Minitaur platform along with introducing a method to map that model to various terrains that accounts for persistent terrain effects.

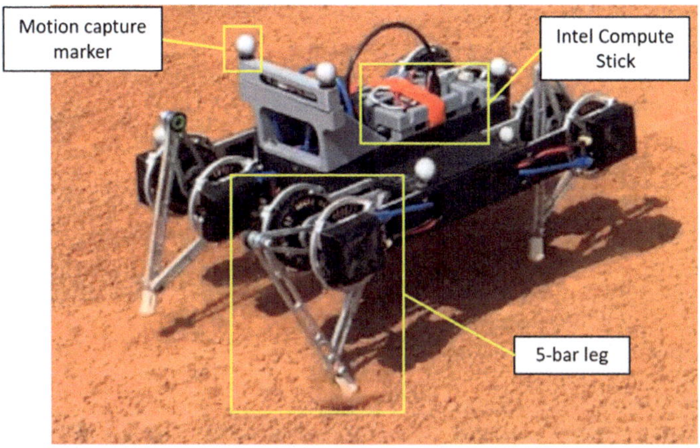

Fig. 2. Modified Minitaur platform. This is the Ghost Robotics Minitaur platform used in this research. It has been modified to add perception and onboard computing.

2 Platform

The experimental platform considered for this work is a modified Minitaur platform. The 8.2 kg Minitaur uses directly driven symmetric 5-bar leg design (Fig. 2) for each of its four legs to produce sagittal plane control of each point foot. Locomotion is achieved by prescribing a clock based trajectory of the end effector (regulated through soft PD gains which produce virtual spring damper systems within each leg), which for this study is operated in a trotting gait at a driving frequency of up to 4.5 Hz.

The Minitaur platform was modified to increase on-board computing power necessary to run the SMBPO algorithm while minimizing payload by adding an m5-series Intel Compute Stick which was chosen for its small form-factor and power requirements.

The trajectory shape for each leg (see Fig. 3) is constrained such that it can be defined by 3 parameters: the horizontal stroke length x_s, vertical stroke length y_s, and offset length x_o. To achieve straight-line running, all four legs are given the same trajectory, with values shown in Fig. 3.

The approach taken in this paper is very similar to approaches taken to learn forward predictive models for skid-steered robots. For example, this approach was used to learn a forward predictive model for the Crusher UGV (Unmanned Ground Vehicle) [4]. The Minitaur may be viewed as a special case of skid-steered vehicle as each leg is locked in a fixed position relative to the sagittal plane. Turning motion is achieved by differential movements of the legs during trotting on opposing sides of the robot. It is hence not surprising that the turning dynamics of this type of robot bear resemblance to those observed for wheeled and tracked robots [13].

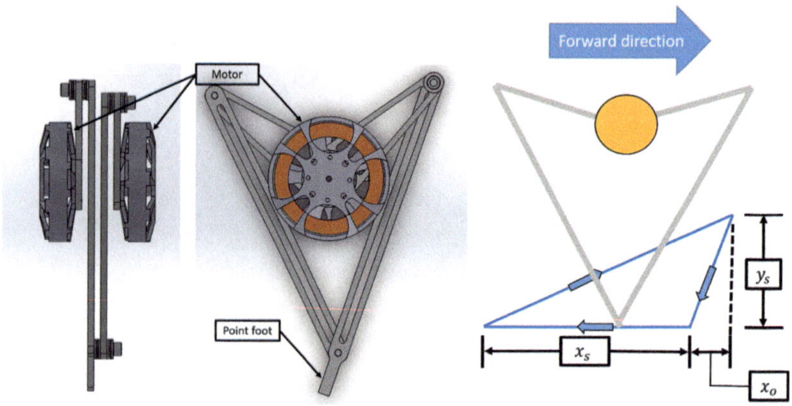

Fig. 3. Leg trajectory parameters. This profile was chosen for simplicity, but can be optimized for speed [12]. The values chosen were $x_s = 12$ cm, $y_s = 7$ cm, and $x_o = 1$ cm.

2.1 Turning

In order to navigate obstacles, we developed a model for turning on the Minitaur platform. This model takes into account the kinematic limitations of the platform, where the 2-DOF legs are constrained to operate in a plane fixed relative to the sagittal plane. To induce a turn, these trajectory parameters are modified for the legs corresponding to the turn side according to the following equations:

$$y_s = y_s \cdot (1 - TG_y)$$

$$x_s = x_s \cdot (1 - TG_x)$$

where $TG_i \in [0, 1)$ is a non-dimensional *turn gain* that reduces the stroke in the i direction.

To understand the behavior of Minitaur when using this model to turn, a sweep of both TG_y and TG_x was done where the robot was instructed to run forward until a steady state velocity was reached and then turn with the specified turn gains. Linear velocity during the forward running and turning motions was collected and then the turn radius of the turning portion of the trajectory was determined using a parametric surface curve fitting technique. It was found that when varying only TG_y, a turn radius as low as 4.3 m could be achieved, while varying only TG_x would give a turn radius as low as 5.3 m. However, varying both TG_y and TG_x together, we were able to achieve a turn radius as small as

1.8 m. Varying both turn gains together made it possible to achieve a smaller turn radius than by varying a single turn gain alone. It was also found that an increase in turn gain comes at a cost. This is in the form of a reduction in velocity, where it was observed that the forward velocity of the robot could be reduced by up to 50%.

3 Experimental Process

With a stable gait and a model for turning implemented, a series of motion-data collection experiments were conducted to determine the prediction model relating the control parameters (TG_x, TG_y) to angular velocity (ω) and forward velocity ($v_{forward}$). Based on the results obtained when developing the turning model, we chosen to keep TG_x fixed at 0.25 while allowing TG_y to vary. This reduces the complexity of the model and still gives a full range of turning behaviors (turn radius of 2.4 m to 15 m).

The robot was given a straight line running command for 15 steps (distance sufficient to reach a steady state velocity) and then commanded to turn using a prescribed given TG for 27 steps. Data was collected on 9 different TG values, ranging from 0.1 to 0.9 incremented by 0.1. Negative TG which corresponds to right-turns were not collected as numerous preliminary gait-development experiments showed that the robot behaved in a near-symmetrical manner.

The synchronized robot control log and VICON data was fed into a neural network which mapped the input TG into body velocities (ω, $v_{forward}$). Due to the desire to adapt and train the network online, a deep feed-forward network was chosen. This network has two hidden layers with 15 and 8 units respectively, giving the capacity to learn and fit more inputs/outputs in the future without needing significant training time or computational resources.

Some of the motion information captured during these experiments were not able to be reflected in the simplified learned model. In particular, transition velocities which occurred on the step when a new command was executed was significantly different to that of the steady state. Training accuracy decreased when trying to consider these transition points due to a lack of data, however, command transitions remains an interesting consideration when exploring further data-collection experiments in the future.

3.1 Laboratory Validation

Once the Minitaur prediction model was learned, experiments were conducted to verify the model's accuracy. These experiments were run on tile floor in a 6 m square area and VICON motion capture equipment was used to gather motion data at a rate of 300 Hz. For each experiment, the start position, end position, and obstacle locations in the field were given to SBMPO, which returned a feasible trajectory using the learned prediction model. Two trials were run for 14 different scenarios. However, only 8 scenarios were recorded as 6 cases were invalidated due to the limits of the motion capture field. Examples of these planned paths are shown in Fig. 4.

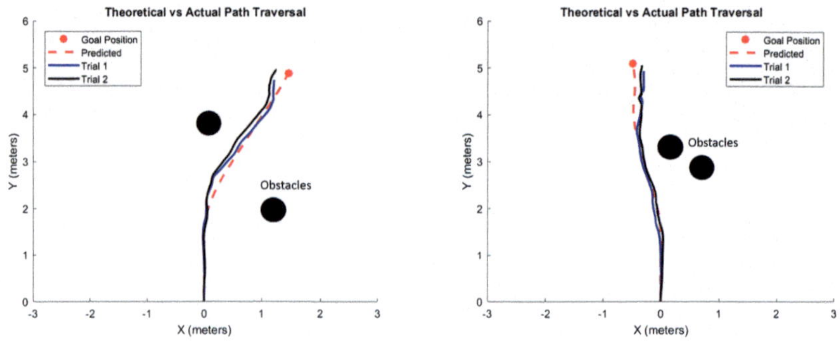

Fig. 4. Example trajectory executed by Minitaur. The average per step error for all 8 trials was 5.6 cm. These trials illustrate both left and right turn behaviors of the Minitaur robot.

3.2 Terrain Correction

While there were many types of terrains available to test, three terrains were of particular interest in this research: grass, dirt, and asphalt. These terrains are not only relatively uniform, but they are readily accessible in almost all parts of the world. Grass and dirt are relatively compliant and soften the foot impact of the stride, absorbing some of the forces slowing down the robot while asphalt tends to have the opposite effect. Despite the interesting characteristics of other terrains (such as tall grass and rocky substrate), these were avoided to reduce the risk of the damaging the exposed motors.

Due to the different friction and compliance of asphalt, dirt, and grass, the prediction model derived from laboratory experiments was no longer sufficiently accurate for motion planning. Therefore, it was hypothesized that a simple scaling factor α can be experimentally determined to give an approximation of the new terrain such that:

$$\omega_i(TG_y) = \alpha_i * \omega(TG_y)$$

where ω is from the original motion model $i \in \{grass,\ dirt,\ asphalt\}$ Rather than running a series of lengthy data collection experiments to develop a prediction model for each terrain, a diagnostic trajectory was instead created to determine this scaling factor. This trajectory is comprised of a prescribed series of maneuvers and was run 3 times on different terrains. The distances covered during the maneuvers were measured by hand. The data generated (Table 1) was used to calibrate the scaling factor α.

Table 1. Distance traveled by Minitaur during diagnostic trajectory. Using the distance traveled on lab floor of 660 cm, the following scaling factors were calculated.

Terrain	Distance (cm)	Standard deviation (cm)	Scaling factor
Asphalt	522	18.06	0.79
Grass	452	13.62	0.68
Dirt	510	4.87	0.77

4 Results

The terrain factor was included in the Minitaur motion model enabling the SBMPO algorithm to generate terrain aware trajectories. As can be seen, trajectories based on grass, asphalt, and dirt (shown in Fig. 5) exhibit different behaviors than the laboratory test case.

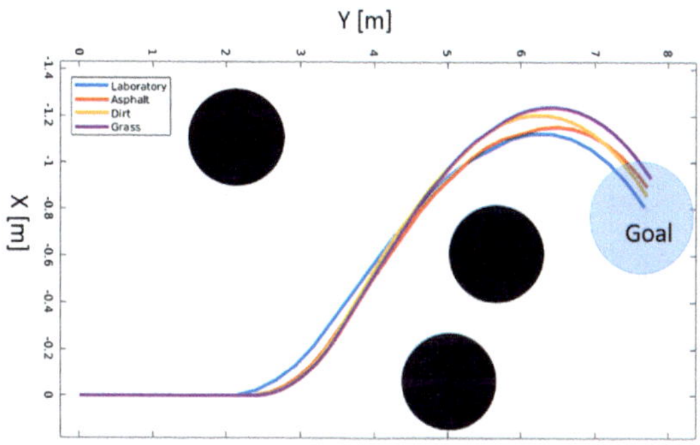

Fig. 5. Differences in terrain factors. Motion planning results compare the nominal model derived experimentally in laboratory conditions with models adjusted for asphalt, dirt and grass. The goal region is defined as being a region around the point $(-0.8, 7.5)$.

In particular, the loss in forward velocity required more robot steps in order to satisfy the goal criterion (to be within 1 m of the goal position) requiring the planner to use slightly more computation time. This increased time is a serious consideration when dealing with limited computational resources onboard the robot, effectively reducing the maximum planning horizon.

Moving from hard surfaces was also found to dampen the effective turn-radius of the robot, causing the planner to become more conservative, often preventing exploration between close obstacles. Particularly on grass where terrain effects inhibited sharp turns, the grass-aware trajectories tended to have the greatest robot to obstacle clearance.

Validation of the motion planner and terrain factor was conducted by running the Minitaur on the terrain of interest with the appropriate terrain factor. These trajectories were executed using no state feedback term, allowing us to directly assess the accuracy of the planning motion model.

It is clear from the final position error (shown in Fig. 6) that simply using the terrain factor does not fully account for how the motion model changes when learned and implemented on new terrain. It can be observed that perturbations to the robot's heading angle caused large errors in the final position. However, the distance covered by the robot was greater with the addition of the terrain factor which indicates that our approach accounts for some of the differences in the model caused by motion differences due to terrain interactions.

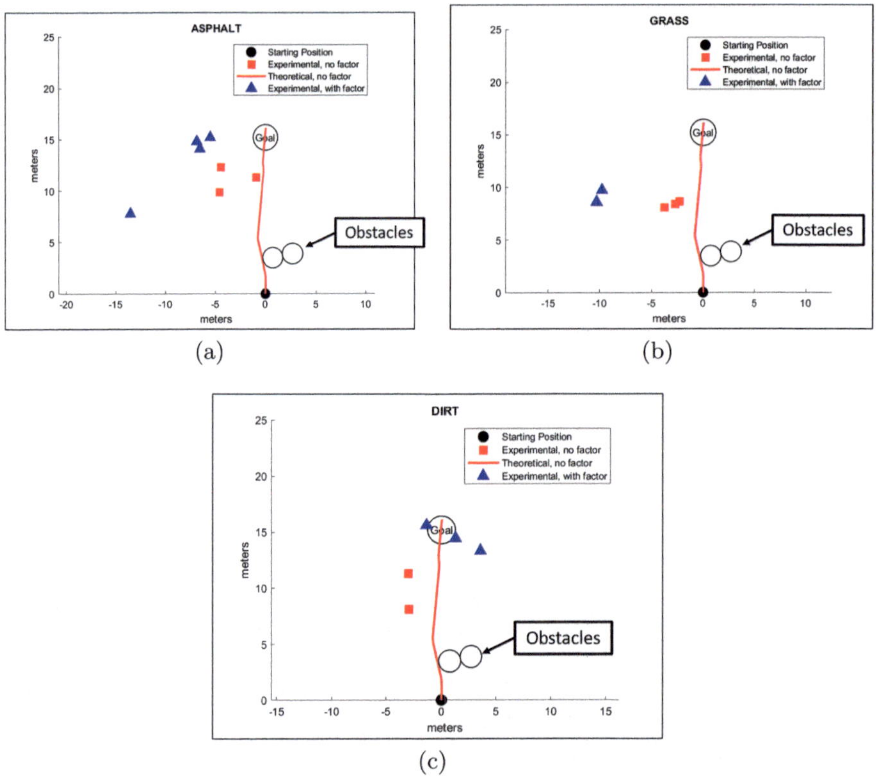

Fig. 6. Terrain factor results. Red squares indicate the final position using the laboratory derived model with no terrain factor. Blue triangles show the new final positions when the terrain factor was used in motion planning.

5 Conclusions

There are several conclusions that can be seen from this research. First, we demonstrate a new model for turning on the Minitaur platform. This method perturbs a trajectory-based gait to tune the turn radius and has been shown to be repeatable and consistent in turning behaviors allowing the motion models to be generated with low error and reduced data.

This research also experimentally validates the use of a simple scaling factor to account for some of the persistent differences in locomotion introduced while running on new terrain. While the initial motion prediction model may be data-intensive, subsequent terrains only requires a simple diagnostic trajectory to tune the terrain factor. While this factor alone was not enough to account for stochastic perturbations that impacted robot heading angle, it was able to account for the additional traversal distance needed on new terrains. This terrain factor is also expected to reduce the computational burden of replanning as we expect fewer pose errors due to its use.

Future work will involve increasing the perception and sensing capabilities to enable robot odometry. The onboard odometry will be used to make trajectory correction and replanning decisions onboard for long-distance planning and heading angle correction. Replanning capabilities have already been developed and tested to extend the SBMPO algorithm to work in real-time through parallel programming [10] to allow for rapid course correction. Further experiments are being conducted to verify and improve terrain factors while SBMPO will be extended to adapt to new terrain surfaces during locomotion.

Acknowledgement. This work was supported by the collaborative participation in the Robotics Consortium sponsored by the U.S. Army Research Laboratory under the Collaborative Technology Alliance Program, Cooperative Agreement DAAD 19-01-2-0012. The U.S. Government is authorized to reproduce and distribute reprints for Government purposes not withstanding any copyright notation thereon.

References

1. Atkeson, C.G., Babu, B.P.W., Banerjee, N., Berenson, D., Bove, C.P., Cui, X., Gennert, M.: No falls, no resets: Reliable humanoid behavior in the DARPA robotics challenge. In: 2015 IEEE-RAS 15th International Conference on Humanoid Robots (Humanoids). IEEE (2015)
2. Bermudez, F.L.G., Julian, R.C., Haldane, D.W., Abbeel, P., Fearing, R.S.: Performance analysis and terrain classification for a legged robot over rough terrain. In: 2012 IEEE/RSJ International Conference on Intelligent Robots and Systems (IROS), October 2012, pp. 513–519. IEEE (2012)
3. Birkmeyer, P., Peterson, K., Fearing, R.S.: DASH: a dynamic 16 g hexapedal robot. In: IEEE/RSJ International Conference on Intelligent Robots and Systems. IROS 2009, October 2009, pp. 2683–2689. IEEE (2009)
4. Bode, M.: Learning the forward predictive model for an off-road skid-steer vehicle, Tech. Rep. Robotics Institute, Pittsburgh. CMU-RITR-07-32 (2007)
5. Clary, P., Morais, P., Fern, A., Hurst, J.: Monte-Carlo Planning for Agile Legged Locomotion. In: ICAPS 2018, pp. 446–450 (2018)
6. Feng, S., Whitman, E., Xinjilefu, X., Atkeson, C.G.: Optimization-based full body control for the DARPA robotics challenge. J. Field Robot. **32**(2), 293–312 (2015)
7. Johnson, M., Shrewsbury, B., Bertrand, S., Wu, T., Duran, D., Floyd, M., Carff, J.: Team IHMC's lessons learned from the DARPA robotics challenge trials. J. Field Robot. **32**(2), 192–208 (2015)
8. Kim, S., Clark, J.E., Cutkosky, M.R.: iSprawl: design and tuning for high-speed autonomous open-loop running. Int. J. Robot. Res. **25**, 903–912 (2006)
9. Ding, L., Gao, H., Deng, Z., Song, J., Liu, Y., Liu, G., Iagnemma, K.: Foot-terrain interaction mechanics for legged robots: modeling and experimental validation. Int. J. Robot. Res. **32**, 1585–1606 (2013). https://doi.org/10.1177/0278364913498122
10. Harper, M., Ordonez, C., Collins, E., Erlebacher, G.: Parallel approach to motion planning in uncertain environments. In: Proceedings of SPIE Unmanned Systems Technology XX. International Society for Optics and Photonics (2018)
11. Harper, M., Pace, J., Gupta, N., Ordonez, C., Collins, E.: Kinematic modeling of a RHex-type robot using a neural network. In: Proceedings of SPIE, Unmanned Systems Technology XIX, International Society for Optics and Photonics (2017). https://doi.org/10.1117/12.2262894

12. Austin, M., Brown, J., Geidel, K., Wang, W., Clark, J.: Gait design and optimization for efficient running of a direct-drive quadrupedal robot. Unmanned Syst. Technol. **XIX**, 10195 (2017)
13. Ordonez, C., Nikhil Gupta, N., Collins, E.G., Clark, J.E.: Power modeling of the XRL hexapedal robot and its application to energy efficient motion planning. Adapt. Mob. Robot., 689–696 (2012)
14. Chuy, O., Collins, E., Dunlap, D., Sharma, A.: Sampling-based direct trajectory generation using the minimum time cost function. In: Jaydev, P., Desai, G.D., Oussama, K., Kumar, V. (eds.) Experimental Robotics, vol. 88, pp. 651–666. Springer Tracts in Advanced Robotics. Springer (2013)
15. Plagemann, C., Mischke, S., Prentice, S., Kersting, K., Roy, N., Burgard, W.: Learning predictive terrain models for legged robot locomotion. In: IEEE/RSJ International Conference on Intelligent Robots and Systems, September 2008. IROS 2008, pp. 3545–3552. IEEE (2008)
16. Siegwart, R., Nourbakhsh, I., Scaramuzza, D.: Introduction to Autonomous Mobile Robots. MIT Press (2011)
17. Spenko, M., Haynes, G.C., Saunders, J., Cutkosky, M.R., Rizzi, A.A., Full, R.J., Koditschek, D.E.: Biologically inspired climbing with a hexapedal robot. J. Field Robot. **25**, 223–242 (2008)
18. Watson, J.T., Ritzmann, R.E.: Leg kinematics and muscle activity during treadmill running in the cockroach, Blaberus discoidalis: I. slow running. J Comp. Physiol. A **182**, 11–22 (1997)
19. Wermelinger, M., Fankhauser, P., Diethelm, R., Krüsi, P., Siegwart, R., Hutter, M.: Navigation planning for legged robots in challenging terrain. In: 2016 IEEE/RSJ International Conference on Intelligent Robots and Systems (IROS), pp. 1184–1189 (2016)
20. Zucker, M., Ratliff, N., Stolle, M., Chestnutt, J., Bagnell, J.A., Atkeson, C.G., Kuffner, J.: Optimization and learning for rough terrain legged locomotion. Int. J. Robot. Res. **30**(2), 175–191 (2011)

RADLER - A RADial LasER Scanning Device

Dorit Borrmann(✉), Sven Jörissen, and Andreas Nüchter

Chair of Robotics and Telematics, University of Würzburg,
Am Hubland, 97074 Würzburg, Germany
{borrmann,joerissen,nuechter}@informatik.uni-wuerzburg.de

Abstract. In recent years a wide range of 3D measurement devices for various applications has been proposed. Each of them has its own benefits and limitations. This paper proposes a modified unicycle with a 2D laser scanner attached to the wheel axle, thus creating a radial 3D scanning pattern. This novel low-cost device combines the advantages of wheeled scanning equipment with those of wearable or hand-held devices. After presenting the hardware setup and the sensor integration, the results are evaluated using three test scenarios and a terrestrial laser scanner for comparison.

1 Introduction

In recent years, optical 3D measurement devices have undergone a rapid development. With increased speed and accuracy on the one hand and decreased weight and hardware costs on the other hand they are on the way of becoming the standard measurement tool in many disciplines. For measuring large areas in a timely manner mobile scanning solutions are the method of choice. Laser scanners as well as camera systems mounted on cars, trolleys, backpacks or aerial vehicles have been developed depending on specific requirements.

Mobile mapping systems consisting of sensors mounted on cars are the state of the art for mapping urban environments. However, they are limited to areas that are accessible by car. Robotic solutions [10] or solutions with scanners mounted on carts, like the viametris iMMS [11,12], the Google Street View Trolley [5], or the NavVis 3D Mapping Trolley [9] are applicable in smaller alleys. At stairs as well as dirt or gravel roads these systems still meet their limits. Airborne laser scanning is not restricted to specific terrain and thus has advantages, but it is not available in roofed environments or tunnels or gives unsatisfying results in areas with a lot of trees or bushes. Backpack mounted systems, also known as personal laser scanning, such as "The Cartographer" by Google [4], the Zebedee 3D sensor system [1], the Leica Pegasus:Backpack [8] or our own backpack mobile mapping system [6], have been presented as ideal solutions to overcome these issues for indoor mapping. While all of them perform great for said requirements and tasks, all come with disadvantages such as high weight of the backpack, high hardware costs or low range (Zebedee 3D, 15–30 m).

Electronic supplementary material The online version of this chapter (https://doi.org/10.1007/978-3-030-33950-0_56) contains supplementary material, which is available to authorized users.

J. Xiao et al. (Eds.): ISER 2018, SPAR 11, pp. 655–664, 2020.
https://doi.org/10.1007/978-3-030-33950-0_56

A different approach was recently presented in form of the "Classical Mechanics Scanner" [7]. Although it is simple and easy to operate, steering the wheel is not possible since the localization is only performed in an intrinsic manner, which requires the path to be straight. Additionally, it needs to be regularly pushed for a continuous movement, making the operation exhausting for long scans. Therefore, the field of application is very limited, especially considering challenging and rough terrain. While camera systems meet their limits in changing lighting conditions and featureless environments, the main drawback of full 3D laser scanners is still the respectable price range and the weight. Inspired by the concept of a surveyor's wheel, this paper presents a low-cost 3D scanning solution, RADLER (a RADial LasER scanning device), that consists of a 2D laser scanner mounted on the axle of a unicycle (cf. Figure 1). The wheel rotation creates a radial 3D scanning pattern. Moving on the ground leads to a smoother trajectory than handheld or backpack systems. Nevertheless, with the large pneumatic wheel, RADLER can be operated on rough terrain and even stairs if the operator stands above the wheel. Additionally, it is lightweight, highly portable and easy to operate.

Fig. 1. First author operating RADLER (https://en.wikipedia.org/wiki/Shandy#Radler) (RADial LasER scanning device).

2 Technical Approach

RADLER is a low-cost scanning device that combines the advantages of wheeled 3D measurement equipment with those of hand-held or wearable devices. It consists of a modified unicycle with an attached handle for easy maneuverability, as seen in Fig. 1. It is light-weight and moves smoothly on the ground due to the large pneumatic wheel, which also allows it to be used in rough terrain and even on stairs, when the operator is rather pulling than pushing.

2.1 Hardware Setup

The sensors, namely a SICK LMS141 LiDAR sensor, a PhidgetSpatial Precision 3/3/3 IMU, a Raspberry Pi 3 single board computer (SBC) with display and a Phidget rotary encoder ISC3004, are mounted on a base plate that is connected to the wheel axle as seen in Fig. 2. Counterweights on the other side help to keep the wheel in balance during operation. For best stability, the weights move the center of gravity onto the wheel (cf. red line in Fig. 2(b)). The rotary encoder is fixed with respect to the unicycle frame

(a) Sensor plate (b) Center of gravity (c) Wheel encoder

Fig. 2. The sensor setup of RADLER. (1) SICK LMS141 LiDAR sensor, (2) PhidgetSpatial Precision 3/3/3 IMU, (3) Raspberry Pi 3 with Display, (4) Cable Box, (5) Phidget optical rotary encoder ISC3004 and encoder high speed.

via a support structure and connected via the PhidgetEncoder High Speed to the SBC through USB. A coupler releases the axial stress from the encoder during rotation. The LMS141 with a maximum operating range of 40 m at a scanning frequency of 25/50 Hz and resolution of 0.25/0.5° is mounted with its scanning plane parallel to the wheel's axle and connected to the SBC via Ethernet. The Raspberry Pi 3 is configured as a WiFi access point, such that easy communication via *ssh* is possible. The wheel encoder is used for simple odometry and later on fused with the IMU to get a precise trajectory. RADLER is powered through a 11.1 V 1000 mAh lithium polymer battery, which is also directly mounted on the base plate, giving 11.1 V for the laser scanner and 5 V via a DC-DC converter for the SBC, that allows an operation time of about 40 to 50 min.

2.2 Sensor Integration

The Raspberry Pi 3 integrates the sensors using ROS (the Robot Operating System). The laser scanner rotates around the wheel axis to create a radial 3D point cloud while moving forward at the same time. This ensures multiple measurements between full 360° rotations to provide enough overlapping data needed by our continuous-time SLAM algorithm [2]. The quality of the resulting 3D point cloud is thus heavily dependent on the quality of the laser scanner pose when unrolling the 2D scan slices.

Let the unicycle coordinate system $\mathscr{U}$ be located in the center of the axis with x pointing forward, z pointing upward and y pointing away from the sensor plate. Let $T_{\mathscr{U} \to \mathscr{G}}$ be the transformation from $\mathscr{U}$ into the global coordinate system $\mathscr{G}$, L the circumference of the tire, and C the counts per revolution, i.e., the encoder count for a full 360° wheel rotation, then the change in pitch angle $\Delta\vartheta$ of the unicycle and the forward movement $\Delta x_{\mathscr{U}}$ of the unicycle are determined from the wheel encoder counts c_t between time steps $t-1$ and t as

$$\Delta\vartheta = \frac{-2\pi}{C}\cdot c_t \quad \text{and} \quad \Delta x_{\mathscr{U}} = \frac{L}{C}\cdot c_t. \tag{1}$$

IMUs are prone to drifting effects. To reduce the impact of the drift on the map quality, only the yaw ψ and roll φ angles are considered. Let $R_{\mathscr{I}\to\mathscr{G}}$ describe the orientation of the IMU in the global coordinate system, then

$$\left(p_x\ p_y\ p_z\right) = R_{\mathscr{I}\to\mathscr{G}}\cdot\left(0\ 0\ 1\right)^T \tag{2}$$

describes the z-axis of $\mathscr{I}$ in the global coordinate system. Thus the orientation of the unicycle is determined as

$$\psi = \operatorname{atan2}(p_y, p_x) \quad \text{and} \quad \varphi = \angle\left(p_x\ p_y\ p_z\right),\left(p_x\ p_y\ 0\right). \tag{3}$$

The pitch angle is determined in a similar matter using the y-axis in the global coordinate system as

$$\vartheta = \operatorname{atan2}(q_y, q_z) \quad \text{with} \quad \left(q_x\ q_y\ q_z\right) = R_{\mathscr{I}\to\mathscr{G}}\cdot\left(0\ 1\ 0\right)^T. \tag{4}$$

ϑ is used to calibrate the wheel encoder as described in the next section.

Let $T_{\mathscr{L}\to\mathscr{U}}$ be the transformation from the scanner coordinate system $\mathscr{L}$ into the unicycle coordinate system, then the initial transformation for any point $p_{\mathscr{L}}$ in scanner coordinates into the global coordinate system is given as

$$p_{\mathscr{G}} = T_{\mathscr{I}\to\mathscr{G}}\cdot T_{\mathscr{L}\to\mathscr{U}}\cdot p_{\mathscr{L}}, \tag{5}$$

where the rotational part of $T_{\mathscr{I}\to\mathscr{G}}$ is given by $(\varphi, \vartheta, \psi)$ and the translational part is given by

$$\Delta x_{\mathscr{G}} = \Delta x_{\mathscr{U}}\cdot\cos\psi, \quad \Delta y_{\mathscr{G}} = \Delta x_{\mathscr{U}}\cdot\sin\psi, \quad \Delta z_{\mathscr{G}} = 0. \tag{6}$$

The initial transformation does not consider any differences in height and is limited by the accuracy of the encoder and the IMU. Algorithmic solutions are necessary to reduce the impact of these limitations. Here the continuous-time SLAM approach from [2] is used. As this approach is based on the ICP (iterative closest point) algorithm, it benefits from the design of RADLER. The radial rotation allows the 2D laser scanner to measure in front and behind simultaneously, increasing the density and the amount of overlap in the scene.

2.3 Sensor Calibration

For accurate odometry, it is important to determine C and L from (1) as precise as possible. While the circumference of the tire L can be easily measured, the counts per revolution C of the encoder need to be calibrated. To this end, Madgwick fused and filtered quaternions from the IMU data are used. Figure 3 shows an exemplary plot of the calculated angular pitch position of the IMU versus the encoder ticks. Detecting peaks in the angular positions allows for calculating the difference in encoder ticks per revolution. These differences are filtered based on the median and then averaged to calculate the number of ticks per revolution. Finally, the offset corresponding to the initial orientation of the encoder is determined. The respective pose of the wheel is then calculated with odometry for movements along the wheel plane and IMU data for the orientation of the wheel as described before. To compensate for systematic errors in the wheel encoder due to environmental influences such as temperature, the automatic calibration procedure is repeated for each experiment.

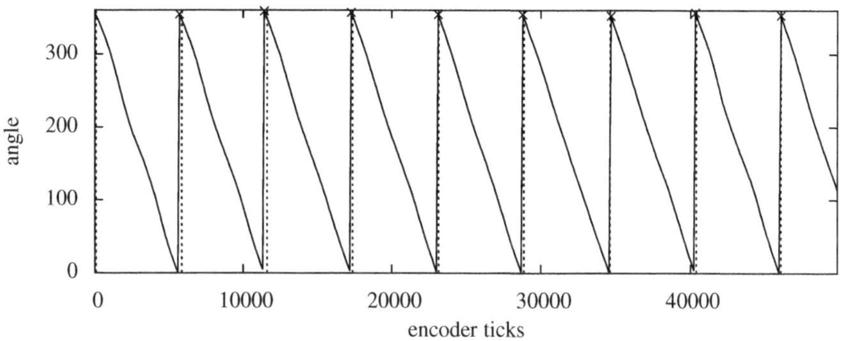

Fig. 3. Determination of encoder RPM with IMU data. The continuous line shows the IMU angles, the crosses mark the detected peaks while the dashed lines represent the zero-crossings based on the calibrated encoder.

3 Experimental Results

To evaluate the performance of RADLER, three test scenarios in different environments were performed. The first and second experiments were conducted in the 1st floor of the University of Würzburg's computer science ("The Hallway") and old mathematics building ("The Circle"), respectively, the third experiment at the Randersacker chapel[1] ("The Chapel"). Scans obtained with a Riegl VZ-400 laser scanner serve as ground truth for "The Hallway" and "The Chapel".

1. "The Hallway" Being the simplest movement consisting of a straight line over approximately 200 m on a smooth surface with no slope. The data set results in around 3.5 million points. Typically for an office building, the hallway consists of multiple doors and glass facades, yielding problems for laser scanners due to reflections and ghost projections (partly visible in Figs. 4(top) and 5(top)).
2. "The Circle" Again, the surface was smooth without slope, a full circle there and back was conducted for this experiment to compensate the limited field of view of the unicycle. The IMU drift is strongly present, but was removed by the continuous-time SLAM. The full data set results in around 6.3 million points.

[1] http://www.wuerzburgwiki.de/wiki/Maria-Schmerz-Kapelle_(Randersacker).

Fig. 4. Ground truth data acquired with a Riegl VZ-400 terrestrial laser scanner. Top view from "The Hallway" and two perspectives from "The Chapel". The grey values represent the reflectance of the materials.

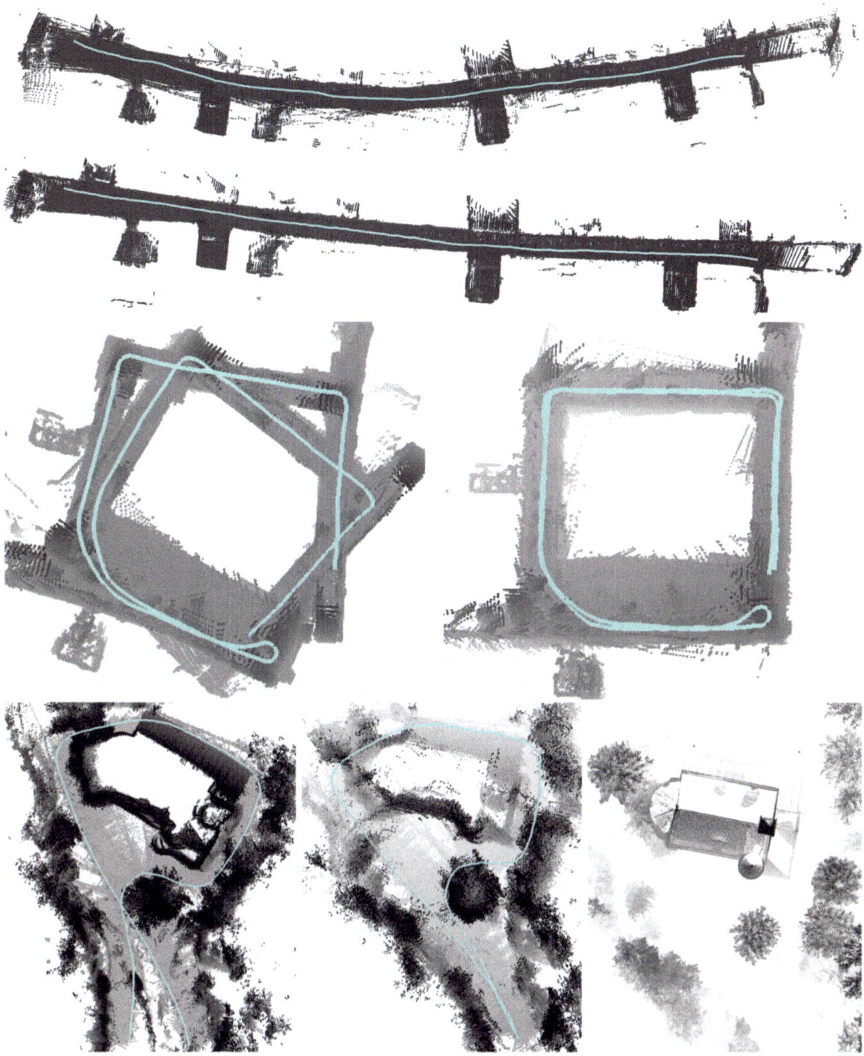

Fig. 5. Topview showing the results from the experiments: first, the initial trajectory, second, the results from the continuous-time SLAM algorithm. From top to bottom: "The Hallway", "The Circle" and "The Chapel". The noisy appearance of the indoor data sets results from reflecting surfaces such as glass. The initial trajectory of "The Chapel" data set is planar. After correction, the true 3D structure of the surface becomes clear when looking at the shading. As the chapel is lower than the starting position the shading by height has less impact on the roof of the chapel which is therefore lighter in color. The third image shows the reference data set collected with a Riegl VZ-400 laser scanner.

Fig. 6. Registered point clouds for "The Hallway" and two perspectives from "The Chapel", respective to Fig. 4. The color represents the shown scale, ranging from 0 m (blue) to 1 m (red) point-to-point distance.

3. "The Chapel" Being the last data set, it is also the most challenging with a gravel-like surface and slopes. Starting at the highest point and descending to the level surroundings of the chapel, a final ascend back to the starting point was done. Since the initial trajectory was only 2D, the differences in elevation were corrected using the SLAM algorithm and the global consistency was improved. The data set consists of around 3.5 million points.

Figure 4 shows the Riegl scans from "The Hallway" as a top view and "The Chapel" from two different sides. The gray values of the points represents the reflectance of the materials. The results shown in Fig. 3 were obtained after about 150 iterations of the continuous time-SLAM algorithm. To calculate the point-to-point distance, the RADLER data was first matched to the ground truth using the ICP algorithm, implemented in 3DTK [3]. The points are colored with the point-to-point distance on the shown color scale, ranging from 0 m to 1 m.

"The Hallway" compares well to the ground truth, especially in the first three quarters of the way from left to right in Fig. 3(top). Black color marks points with a point-to-point distance greater than 1 m. They almost exclusively appear in the last quarter, which is a result of the large glass front at the end of "The Hallway". RADLER scans mostly towards one side causing one wall to be sparsely covered. Without a loop closure the ICP algorithm contracts the data towards one side and thus bending the trajectory.

Figure 5 shows top views of the last two experiments with the originally obtained trajectory as well as the improved trajectory after few iterations of the continuous-time SLAM. The closed loop helps to improve the trajectory significantly, compared to the previously mentioned deviation for "The Hallway". Having the longest trajectory, "The Circle" still converges nicely to the rectangular shape of the building. Blurry edges in the top view show the necessity of further improvements in calibration. For longer trajectories a single offset and static RPM values appear not to be sufficient. A solution is here the integration of the IMU values using a Kalman filter.

"The Chapel" also compares well to the ground truth with less outliers compared to 'The Hallway'. Though the rough terrain does not seem to influence the quality of the output in a negative way, the closed loop within the path made a proper correction of the trajectory with our SLAM algorithm possible. The major drawback in the outdoor scene comes from the limited range of the laser scanner and the lower density at increasing distances. The roof of the chapel is also hardly covered by points. To increase the density, the unicycle could be moved much slower than in the current experiments or the trajectory could be followed twice. It would also be an option to exchange the LMS141 for a model with a greater maximum measuring distance, but this would increase the cost of the hardware drastically.

4 Conclusions

This paper presents a new, low-cost way of generating 3D point clouds without the utilization of a classical 3D laser scanner. A modified unicycle with a 3D laser profiler, an IMU and a wheel encoder was used to generate 3D point clouds and the results were optimized using out continuous-time SLAM algorithm. The experiments and the discussion show, that the results are already promising and meaningful 3D point clouds are generated. But needless to say, a lot of work remains to be done. In the future, we plan on implementing a Kalman filter for sensor fusion of IMU and wheel encoder to improve the accuracy of the odometry. Additionally, a GPS sensor could be integrated to provide a better initial trajectory for our continuous-time SLAM algorithm. Also, the mounting system of the sensors needs to be improved in order to overcome some drawbacks a classical unicycle provides, such as the clearance within the main axis.

Acknowledgements. The authors thank Rahul Bhandari for the mechanical design and construction of RADLER.

References

1. Bosse, M., Zlot, R., Flick, P.: Zebedee: design of a spring-mounted 3-d range sensor with application to mobile mapping. IEEE Trans. Rob. **28**(5), 1104–1119 (2012)
2. Elseberg, J., Borrmann, D., Nüchter, A.: Algorithmic solutions for computing precise maximum likelihood 3D point clouds from mobile laser scanning platforms. Remote Sens. **5**(11), 5871–5906 (2013)
3. Nüchter, A., et al.: 3DTK — The 3D Toolkit, August 2018. http://threedtk.de/
4. Lardinois, F.: TC. Google Unveils The Cartographer, Its Indoor Mapping Backpack, January 2015. http://techcrunch.com/2014/09/04/google-unveils-the-cartographer-its-indoor-mapping-backpack/
5. Google. Street View Trolley (2015). http://maps.google.com/intl/en/maps/about/behind-the-scenes/streetview
6. Lauterbach, H.A., Borrmann, D., Hess, R., Eck, D., Schilling, K., Nüchter, A.: Evaluation of a backpack-mounted 3D mobile scanning system. Remote Sens. **7**(10), 13753–13781 (2015)
7. Lehtola, V.V., Virtanen, J.-P., Kukko, A., Kaartinen, H., Hyyppä, H.: Localization of mobile laser scanner using classical mechanics. ISPRS J. Photogrammetry Remote Sens. **99**, 25–29 (2015)
8. Leica. Leica pegasus:backpack, May 2015. www.leica-geosystems.com/de/Leica-PegasusBackpack_106730.htm
9. NavVis. M3 trolley (2015). https://www.navvis.com/explore/trolley/
10. Nüchter, A., Elseberg, J., Borrmann, D.: Irma3D – an intelligent robot for mapping applications. In: Proceedings of the 3rd IFAC Symposium on Telematics Applications (TA 2013), Seoul, Korea, pp. 119–124 (2013)
11. Thomson, C., Apostolopoulos, G., Backes, D., Boehm, J.: Mobile laser scanning for indoor modelling. ISPRS Annal. Photogrammetry Remote Sens. Spat. Inf. Sci. II-5/W2, 289–293 (2013)
12. VIAmetris. Mobile Mapping Technology, January 2015. http://www.viametris.com/

Design of Damping Matrices for Cartesian Impedance Control of Robotic Manipulators

Carlos Saldarriaga(✉), Nilanjan Chakraborty, and Imin Kao

Stony Brook University, Stony Brook, NY, USA
{carlos.saldarriaga,nilanjan.chakraborty,imin.kao}@stonybrook.edu

Abstract. Impedance control of robotic manipulators has been widely used in manipulation tasks for controlling the interaction behavior between a robotic end-effector and the environment. One of the main challenges in the implementation of impedance control schemes (even in "purely" stiffness control) is the choice of appropriate damping factors of the Cartesian damping matrix for the intended task. Although most tasks are described in the end-effector (or task) space, the required torques for the controller must be applied in the robot joint space. Thus, the attempt to control the end effector dynamics gives rise to dynamic behaviors in the joint space. We show that by analyzing the vibratory dynamics of the robot in joint space and planning out the suppression of the vibration modes of each of the robot joints, we can search for the Cartesian damping matrix parameters in a principled manner. Our method is especially useful for redundant robots and the main contribution of this paper is a novel, general, analytical, and practical method to determine the damping factors. We present experimental results on initial implementation of our algorithm on a 7-DoF robot arm.

1 Introduction

Controlling the dynamics of interaction between a robotic end effector and the environment is useful in a variety of applications including robotic assembly, robotic machining (e.g., deburring, grinding), and service robotics (e.g., wiping). A pre-dominant approach for controlling the dynamics of interaction is impedance control [7,9] and its variations [12], where the dynamics of interaction is controlled by proper choice of Cartesian stiffness and damping matrices. However, there is no principled method to select the Cartesian damping matrices due to the coupling of multi-dimensional mass, damping, and stiffness matrices that determine the overall dynamic response during interaction (although limited work in this context has been done in [1,3,4,6]). Therefore, in this paper,

Electronic supplementary material The online version of this chapter (https://doi.org/10.1007/978-3-030-33950-0_57) contains supplementary material, which is available to authorized users.

J. Xiao et al. (Eds.): ISER 2018, SPAR 11, pp. 665–674, 2020.
https://doi.org/10.1007/978-3-030-33950-0_57

our goal is to present a principled method to compute the Cartesian damping matrix and evaluate it experimentally.

For a given robotic manipulation task, and a specified stiffness matrix in the Cartesian space, it is imperative to compute a damping matrix in the Cartesian space, such that undesirable vibrations at the end-effector are minimized, while ensuring that the task is successfully completed.

We present a novel method to determine the Cartesian damping matrix to achieve the desired damping characteristics in interaction dynamics. The analysis done to select the elements of the Cartesian damping matrix is in the joint space of the robot. Note that any attempt to control the end effector dynamics gives rise to dynamic behaviors in the joint space. Our key contribution is that by analyzing the vibratory dynamics of the robot in joint space and planning out the suppression of the vibration modes of each of the robot joints, we can search for the Cartesian damping matrix parameters in a principled manner, akin to the one-DOF mass-damper-spring system with prescribed damping ratio. Using joint space analysis of the interaction dynamics is challenging, especially for redundant robots, since the dynamics exhibit coupled behavior of the rigid body (RB) mode of motion, which belongs to the null space of the joint stiffness matrix, as well as vibratory dynamics. We use a novel analytical method to separate out the RB mode and form a system of equations that allows us to study only the vibratory joint dynamics. The method presented here does not limit the Cartesian stiffness and damping matrices to be diagonal. To the best of our knowledge, such a method for computing the Cartesian damping matrix based on the analysis of joint space dynamics has not been presented in the past.

2 Related Work

The extant literature on choice of damping matrices either makes simplifying assumptions or are based on trial-and-error. In [3], the authors propose a simultaneous diagonalization of positive definite matrices, which is generally not feasible, all this under the assumption of quasi-static motion, and the method ends up solving decoupled equations. Furthermore, their dual eigenvalue problem in Cartesian space, accounts for displacements and rates of change of displacements in both position and orientation in the same matrices, consequently it should not have a correct physical interpretation. The methods of [4] and [6] make similar assumptions. There is no systematic method to select damping matrices with sound theoretical background.

In [1] the damping matrix is assumed to be a linear combination of the mass and stiffness matrices, constraining it to a proportional damping fitted by a scalar function, where the complex modes are neglected. The method that we propose does not assume quasi-static motion nor the damping to be a linear combination of the mass and stiffness matrices, and since it is entirely developed in the robot joint space, there are no physical unit inconsistencies.

3 Impedance Control: Choosing Elements of the Damping Matrix Based on Dynamic Response

In Cartesian impedance control, an n-DoF robotic manipulator has a certain level of desired dynamic behavior when interacting with the environment [7,9]. The equations of motion of an n-DoF redundant manipulator is given by

$$\mathbf{M}(\mathbf{q})\ddot{\mathbf{q}}(t) + \mathbf{G}(\mathbf{q}, \dot{\mathbf{q}})\dot{\mathbf{q}}(t) + \mathbf{v}(\mathbf{q}) = \boldsymbol{\tau}_{\mathbf{m}} + \boldsymbol{\tau}_{\mathbf{ext}} \tag{1}$$

where $\mathbf{q}$ is the vector of joint angles, $\mathbf{M}$ is the manipulator's mass matrix, $\mathbf{G}$ the matrix containing the centrifugal and Coriolis terms and $\mathbf{v}$ is the gravity vector term. Choosing $\tau_{\mathbf{m}}$ as $-\mathbf{K}(\mathbf{q})\mathbf{q}(t) - \mathbf{C}(\mathbf{q})\dot{\mathbf{q}}(t) + \mathbf{v}(\mathbf{q}) + \mathbf{G}(\mathbf{q}, \dot{\mathbf{q}})\dot{\mathbf{q}}(t)$, the system in Eq. (1) becomes:

$$\mathbf{M}(\mathbf{q})\ddot{\mathbf{q}}(t) + \mathbf{C}\dot{\mathbf{q}}(t) + \mathbf{K}\mathbf{q}(t) = \boldsymbol{\tau}_{\mathbf{ext}} \tag{2}$$

This equation is now comparable to the equation describing a mass, damper and spring system. In order to perform Cartesian impedance control, note that

$$\mathbf{M_C}\ddot{\mathbf{x}}(t) + \mathbf{C_C}\dot{\mathbf{x}}(t) + \mathbf{K_C}\mathbf{x}(t) = \mathbf{f} \tag{3}$$

where, $\mathbf{K_C}$ is the Cartesian stiffness matrix, $\mathbf{C_C}$ is the damping matrix and $\mathbf{M_C}$ is the Cartesian mass matrix. The values of $\mathbf{K_C}$ can be chosen as necessary for the application, and the prospective initial damping matrix $\mathbf{C_C}$ can be approximated or based on relatively low values. For a given configuration, $\mathbf{M}$ in the joint space can be computed [11], and the desired joint stiffness $\mathbf{K}$ and damping $\mathbf{C}$ matrices can be mapped from the Cartesian end-effector space [5,14] using the manipulator Jacobian $\mathbf{J}$:

$$\mathbf{K} = \mathbf{J}^T\mathbf{K_C}\mathbf{J} + \left[\left(\frac{\partial \mathbf{J}^T}{\partial q_1}\mathbf{f}\right)\left(\frac{\partial \mathbf{J}^T}{\partial q_2}\mathbf{f}\right)\cdots\left(\frac{\partial \mathbf{J}^T}{\partial q_n}\mathbf{f}\right)\right] + \mathbf{J^T}\mathbf{C}_C\dot{\mathbf{J}} \tag{4}$$

$$\mathbf{C} = \mathbf{J}^T\mathbf{C_C}\mathbf{J} \tag{5}$$

3.1 Dynamic Response

For sake of finding a solution to Eq. (2), we assume that the external forces $\mathbf{f}$ and the changes in the Jacobian $\mathbf{J}$ of the manipulator are negligible, so only the first term in Eq. (4) is used to map the stiffness matrix from Cartesian end-effector space to robot joint space.

Since we are not limiting to non-redundant robotic manipulators, in general we consider the redundant case, which are unconstrained systems, as described in the theory of analysis of vibration [8,10]. An unconstrained system has freedom to move without being constrained to an inertial frame. Such unconstrained system has RB motion, as if the entire system were a rigid-body moving without relative motions. This RB mode of motion of an n-DOF system is defined as $\mathbf{u}_0$, which is an nx1 vector of constant elements, with the associated frequency of $\omega_0 = 0\, rad/s$. If the system is constrained (non-redundant), the removal of the RB mode(s) would not be necessary and the method presented here becomes mathematically simpler; the robot response would only include the non-rigid-body (NRB) part of the motion. The characteristic of an unconstrained, positive semi-definite system, corresponding to the general case of a redundant manipulator, is a singular stiffness matrix, $\mathbf{K}$. The rigid-body mode $\mathbf{u}_0$ belongs to the null space of $\mathbf{K}$ [13], i.e.,

$$\mathbf{u}_0 \in N(\mathbf{K}) \Rightarrow \mathbf{K}\mathbf{u}_0 = 0 \tag{6}$$

This means that in the rigid body mode there is no net force due to stiffness and consequently, there is no net potential energy change from stiffness, neither stored nor released from the system.

3.2 Methodology to Modulate the Dynamic Response of a General System with Damping

In Eq. (2), because of the damping term $\mathbf{C}$, the system is not a positive-definite conservative system, and therefore the classical modal analysis cannot be employed to obtain a solution for the equation. Thus, the method presented here becomes essential to solve and find a desired dynamic response of a manipulator performing impedance control. The general response (of vibration) of the robot can be expressed as a linear combination of all the bases (modes, or modal vectors), $\mathbf{u}_i$, in the following

$$\mathbf{q}(t) = c_0(t)\mathbf{u}_0 + c_1(t)\mathbf{u}_1 + ... + c_{n-1}(t)\mathbf{u}_{n-1} = \mathbf{q}_{RB}(t) + \mathbf{q}_{NRB}(t) \tag{7}$$

where $\mathbf{q}_{RB}(t) = c_0(t)\mathbf{u}_0$ is the rigid body (RB) mode or modes that belong to the null space of the joint stiffness matrix $\mathbf{K}$, and $\mathbf{q}_{NRB}(t) = \sum_{i=1}^{n-1} c_i(t)\mathbf{u}_i$ is the non-rigid body (NRB) mode, in which all modal vectors, $\mathbf{u}_i$, are normalized.

3.2.1 Rigid-Body Motion in the Null Space of the Stiffness Matrix

Equation (6) renders the RB mode which is normalized, called $\mathbf{u}_0$. Since this motion is unique, we need to determine it using the dynamics of the system. With the normalized RB mode, we can write the dynamic response of the rigid-body as follows

$$\mathbf{q}_{RB}(t) = \mathbf{u}_0\beta(t) \tag{8}$$

where $\beta = \beta(t)$ is a function of time. Substituting Eq. (8) into (2) to obtain

$$\begin{gathered}\mathbf{M}\mathbf{u_0}\ddot{\beta}(t) + \mathbf{C}\mathbf{u_0}\dot{\beta}(t) + 0 = \boldsymbol{\tau}_{\mathbf{ext}}\\ \Rightarrow \mathbf{u_0^T}\mathbf{M}\mathbf{u_0}\ddot{\beta}(t) + \mathbf{u_0^T}\mathbf{C}\mathbf{u_0}\dot{\beta}(t) = \mathbf{u}_0^T\boldsymbol{\tau}_{\mathbf{ext}}\end{gathered} \tag{9}$$

Given that in general both $\mathbf{K}$ and $\mathbf{C}$ matrices go through similar transformations via the Jacobian matrix $\mathbf{J}$, the second term in Eq. (9) also vanishes. When $\mathbf{u}_0$ is normalized, this equation can be simplified to

$$\ddot{\beta}(t) = \mathbf{u}_0^T\boldsymbol{\tau}_{ext} \tag{10}$$

Equation (10) can now be solved for $\beta(t)$, and it is a scalar equation if there is only one RB mode. After $\beta(t)$ has been solved, the RB response $\mathbf{q}_{RB}(t)$, pertaining to the null space of the joint stiffness matrix $\mathbf{K}$ can be obtained by using Eq. (8).

3.2.2 Non-Rigid-Body Motion

To determine the oscillatory motions, we remove the known RB mode q_n by imposing upon the state vector $\mathbf{q}$ a constraint matrix, $\mathbf{S}$ of dimensions $n\times(n-1)$, to ensure that all solutions q(t) are free of the RB mode [10], such that $\mathbf{q}=\mathbf{S}\mathbf{q}'$, where $\mathbf{q}'$ is the reduced state vector of $\mathbf{q}$. Substituting $\mathbf{q}=\mathbf{S}\mathbf{q}'$ in Eq. (2), we obtain

$$\mathbf{S}^T\mathbf{M}(\mathbf{q})\mathbf{S}\ddot{\mathbf{q}}'+\mathbf{S}^T\mathbf{C}\mathbf{S}\dot{\mathbf{q}}'+\mathbf{S}^T\mathbf{K}\mathbf{S}\mathbf{q}'=\mathbf{S}^T\tau_{\mathbf{ext}} \tag{11}$$

$$\Longrightarrow \mathbf{M}'(\mathbf{q})\ddot{\mathbf{q}}'+\mathbf{C}'\dot{\mathbf{q}}'+\mathbf{K}'\mathbf{q}'=\mathbf{S}^T\tau_{\mathbf{ext}} \tag{12}$$

where $(\cdot)'=\mathbf{S}^T(\cdot)\mathbf{S}$. The equation of motion with the $\mathbf{M}'$, $\mathbf{C}'$ and $\mathbf{K}'$ matrices is now a positive definite system, with a non-singular stiffness matrix in a new set of coordinates $\mathbf{q}'$. The system can now be solved and the NRB response can be determined. Defining $\mathbf{x}(\mathrm{t})=[\mathbf{q}'(t)^T\ \dot{\mathbf{q}}'(t)^T]^T$, Eq. (11) can be re-arranged in the form of a linear system equation

$$\dot{\mathbf{x}}(t)=\mathbf{A}\mathbf{x}(t)+\mathbf{B}\mathbf{S}^{\mathbf{T}}\tau_{\mathbf{ext}}(t) \tag{13}$$

where matrices $\mathbf{A}$, $\mathbf{B}$ have size $2(n-1)\times 2(n-1)$, $2(n-1)\times(n-1)$ respectively.

$$\mathbf{A}=\begin{bmatrix}\mathbf{0} & \mathbf{I}\\ -(\mathbf{M}')^{-1}\mathbf{K}' & -(\mathbf{M}')^{-1}\mathbf{C}'\end{bmatrix},\ \mathbf{B}=\begin{bmatrix}\mathbf{0}\\ (\mathbf{M}')^{-1}\end{bmatrix} \tag{14}$$

This linear system has a solution of the form

$$\begin{aligned}\mathbf{x}(t)&=\mathbf{X}e^{\Lambda t}\mathbf{Y}^{\mathbf{T}}\mathbf{x}(0)+\int_0^t\mathbf{X}e^{\Lambda(t-\tau)}\mathbf{Y}^T\mathbf{B}\mathbf{S}^{\mathbf{T}}\mathbf{Q}(\tau)d\tau\\ &=\mathbf{X}e^{\Lambda t}\mathbf{Y}^{\mathbf{T}}\mathbf{x}(0)+\int_0^t\mathbf{X}e^{\Lambda\tau}\mathbf{Y}^T\mathbf{B}\mathbf{S}^{\mathbf{T}}\mathbf{Q}(t-\tau)d\tau\end{aligned} \tag{15}$$

where $\mathbf{X}$ and $\mathbf{Y}$ are the right and left eigenvectors, respectively, and $\mathbf{Y}^{\mathbf{T}}\mathbf{X}=\mathbf{I}$, $\boldsymbol{\Lambda}$ is the matrix that contains all pairs of eigenvalues of the system. The first part of Eq. (15) corresponds to the free vibration response, and the second part to the forced vibration response of the robotic manipulator. The NRB part of the solution, including the generalized coordinates and generalized velocities, can be found in $\mathbf{x}(t)=[\mathbf{q}'(t)\ \dot{\mathbf{q}}'(t)]^T$. Finally, the NRB part of the vibration response $\mathbf{q}_{NRB}(t)$ is obtained after mapping it back using the constraint matrix $\mathbf{S}$.

After solving the system for the oscillatory modes of the reduced-state vector $\mathbf{q}'$ by applying the dual eigenvalue analysis, it is necessary to combine the solutions of both RB and NRB modes as in Eq. (7). From there, the Cartesian damping matrix can be chosen by iteratively analyzing the effect of every element of this (not necessarily diagonal) matrix on the complete response, where every eigenvalue λ_i of the reduced system represents each mode of vibration. Then, the damping ratios ζ_i can be obtained and consistently improved as necessary and physically plausible. Note that,

$$\lambda = a \pm ib; \quad \omega_d = b; \quad \zeta = cos(\gamma) \tag{16}$$

where a and b are the real and imaginary parts of the eigenvalue, respectively, and γ is the angle between them. By observing the complete response (RB+NRB) at each of the joints of the manipulator, it can be established which of the vibration modes ω_d is being excited and which of these damping ratios ζ_i they correspond to. Then, by changing the elements of the Cartesian damping matrix $\mathbf{C_C}$, we aim to improve (increase) the corresponding ζ's.

4 Validation: Choice of Damping Matrix

The variables in Eq. (16) determine the dynamic response of a system. Once the mass, $\mathbf{M}$, and stiffness matrix, $\mathbf{K}$, have been chosen for a system, the natural frequencies ω_n are established. The selection of the damping matrix $\mathbf{C}$ determines the damping ratios ζ's, and in order to modulate them we use our proposed method.

4.1 Use of the Methodology to Choose the Elements of $\mathbf{C_C}$ Matrix Based on the Dynamic Response

In order to validate the method to design damping matrices, experiments on a Baxter 7 DoF robotic manipulator have been performed. By solving and obtaining the analytical equations using the method described in Sect. 3, and taking into consideration the necessary characteristics of the 7 DoF Baxter manipulator, we were able to have a certain idea of how the damping elements affect the response of the joints. We were also able to match the frequencies from a plot of a Fast Fourier Transform with the ones obtained from the complete response of the manipulator using our method. All this was done offline, without running the robot (Fig. 1).

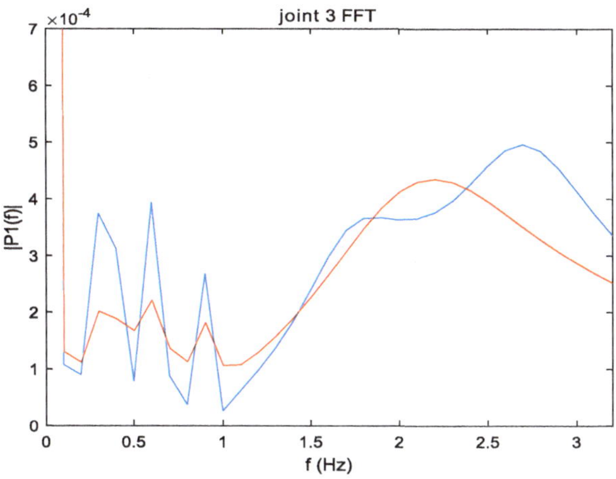

Fig. 1. Frequency plot from the vibration response of one of the robot joints before (in blue) and after (in red) improving the ζ's.

4.2 Experimental Validation

The chosen Baxter 7 DoF robot configuration was: [0; −0.55; 0; 0.75; 0; 1.26; 0] rad. In order to apply the desired torques at every joint of the manipulator, a program using torque control mode of the Baxter robot was implemented on MATLAB using the Robotics System Toolbox interface with ROS.

As shown in [2], a stiffness matrix was chosen such that a convenient stiffness feasibility region was generated for the experiment, taking into consideration the Jacobian at the given configuration, and the maximum and minimum torque limitations at each of the joints of the robot. The chosen Cartesian stiffness matrix for the experiments was $\mathbf{K}_C = diag[300, 300, 300, 10, 10, 10]$. The units are N/m for the first three linear stiffness terms and Nm/rad for the last three angular stiffness terms.

4.2.1 Experiments

The Cartesian stiffness matrix was kept identical for each set of experiments, and the mass matrix for the given robot configuration, needed for the design method described in Sect. 3, was computed as in [11]. The gravity compensation terms $\mathbf{v}(\mathbf{q})$ as in Eq. (1) are being automatically applied by the Baxter robot, so there is no need of computing them nor adding them to the computed torques. The set of experiments consisted in applying an external disturbance at the end-effector of the robot, while performing combined Cartesian stiffness and damping control at the chosen robot configuration using different damping matrices. The applied equation for combined stiffness and damping control of the robotic manipulator was

$$\delta\tau = \mathbf{K} \times \delta\mathbf{q} + \mathbf{C} \times \delta\dot{\mathbf{q}} \tag{17}$$

where **K** and **C** are the joint stiffness and damping matrices of the robot, computed at every iteration as in Eq. (4). The Cartesian stiffness and damping matrices were kept constant by computing (mapping) a set of both matrices to the joint space at each robot configuration, using these matrices in the joint space the torques were computed and applied using Eq. (17). After the experiment was done and the corresponding set of data was collected, the response was compared when using the Cartesian damping parameters from the presented method versus another set of damping parameters randomly chosen, or parameters selected by trial-and-error.

4.2.2 Results

This theoretical model enables one to arbitrarily change any element of the Cartesian damping matrix $\mathbf{C}_C$ individually in order to improve the response based on the analytical output of damping ratios. This innovative method has never been shown nor published before. The plots from the obtained data show how damped the response of each joint of the manipulator was, and how these results convey with the Cartesian displacement.

One of the many sets of experiments performed on the robot was done using the Cartesian stiffness matrix $\mathbf{K}_C = diag[300, 300, 300, 10, 10, 10]$ N/m, N m/rad, and an initial Cartesian damping matrix based on some previous trial-and-error $\mathbf{C}_C = diag[0.8, 0.8, 0.8, 0.5, 0.5, 0.5]$ N s/m, N m s/rad. After solving Eq. (2) and obtaining the free vibration response of the Baxter robotic manipulator analytically as in the method described in Sect. 3, a more suitable set of Cartesian damping parameters (generating better damping ratios) was chosen. From the dual eigenvalue analysis we were able to see which eigenvalues needed to be improved in terms of damping ratio ζ, and that when damping

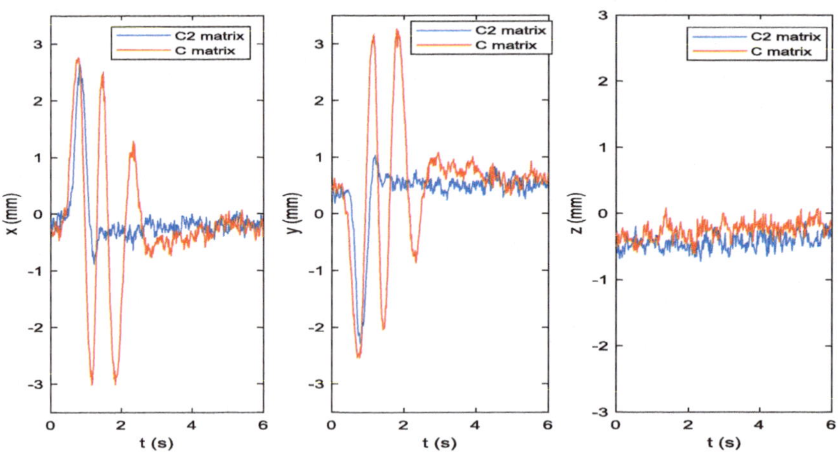

Fig. 2. Displacements in x, y and z directions after applying a disturbance on the end effector for two different damping matrices while performing combined stiffness and damping control

$\mathbf{C}_{C2} = diag[4.5, 5, 5, 1.5, 3.5, 1]$ one of the most critical eigenvalues in the NRB response showed a damping ratio $\zeta = 0.2$ for the corresponding excited mode, showing a much improved dynamic response of the manipulator, and experimentally validating the method.

As shown in Fig. 2, while performing combined Cartesian stiffness and damping control and applying an external disturbance on the end-effector, we can clearly establish the differences between the robot response when using a sound theory-based selected Cartesian damping matrix and an approximated one from trial-and-error.

5 Discussion and Conclusion

The proposed new and general analytical method based on the theory of vibration, akin to the mass-damper-spring system, enables us to choose the elements of the Cartesian damping matrix to directly affect the dynamic behavior of the manipulator. The systematic analysis and independent choice of damping characteristics improve the outcomes of Cartesian impedance control of robotic manipulators with damping.

Before any hardware implementation, just by observing the responses of the robot in simulation, a set of Cartesian damping parameters for the Cartesian impedance control of robotic manipulators can be selected. A sound method based on vibration theory was used as a guide to correctly estimate and damp out the vibration modes of the robot when interacting with the environment.

From the described experimental results on a 7-DoF manipulator, we can see from the actual joint displacement responses, the vibration of the joints can be damped out by using our proposed methodology for selecting the Cartesian damping matrix. In our method, we iteratively update the values of the Cartesian damping matrix based on the joint responses, to arrive at the desired dynamic behavior.

We found out that the expected damped response of the manipulator is more obvious at certain robot configurations than at others. We believe that it is due to the manipulability and the dexterity corresponding to that specific configuration. In addition, the imperfect modeling of the gravity compensation torques of the Baxter robot may affect the dynamic responses.

The method presented here gives a closed-form solution to the vibration of a robotic manipulator for end-effector Cartesian impedance control. It solves and obtains the complete response by first removing the rigid-body modes of motion; these modes belong to the null-space of the joint stiffness matrix, with no change in the potential energy due to stiffness. Once they are removed, only the oscillatory motions are observed, and by solving the dual eigenvalue problem of this system, the actual effect of the damping parameters on the robot can be established and thanks to the data from the corresponding eigenvalues of this reduced-state equations, these damping elements can successively improve the response of the manipulator.

Initially, there were many ROS topics accessed by our code that were not absolutely necessary, this increased the working time per cycle of the system.

After disposing of non-essential access to certain topics, we were able to increase the control frequency of the robot which enhanced the stability and accuracy of the controller.

Another key observation that arose from the experiments was that the damping ratios cannot all be over-damped. This would make the manipulator unnecessarily slow, may make it reach the maximum torques sooner than expected and increase the control effort. Appropriate amount of accommodation can be considered by the presented method that accounts for the eigenvalues of the system and their corresponding damping ratios.

References

1. Adhikari, S.: Damping modelling using generalized proportional damping. J. Sound Vib. **293**(1), 156–170 (2006)
2. Ajoudani, A., Tsagarakis, N.G., Bicchi, A.: On the role of robot configuration in Cartesian stiffness control. In: 2015 IEEE International Conference on Robotics and Automation (ICRA), pp. 1010–1016 (2015)
3. Albu-Schaffer, A., Ott, C., Frese, U., Hirzinger, G.: Cartesian impedance control of redundant robots: recent results with the DLR-light-weight-arms. In: IEEE ICRA (2003)
4. Anderson, R.J.: Dynamic damping control: implementation issues and simulation results. In: Proceedings of IEEE ICRA, vol. 1, pp. 68–77 (1990)
5. Chen, S.F., Kao, I.: Conservative congruence transformation for joint and Cartesian stiffness matrices of robotic hands and fingers. Int. J. Robot. Res. **19**(9), 835–847 (2000)
6. Fiala, J., Wavering, A.J.: Experimental evaluation of Cartesian stiffness control on a seven degree-of-freedom robot arm. J. Intell. Rob. Syst. **5**(1), 5–24 (1992)
7. Hogan, N.: Impedance control: an approach to manipulation: Part I - Theory, Part II - Implementation, Part III - Applications. J. Dyn. Syst. Meas. Contr. **107**(1), 1–24 (1985)
8. Kao, I.: Lecture notes from MEC 532 Vibration and Control (2010). Department of Mechanical Engineering, Stony Brook University
9. Khatib, O.: A unified approach for motion and force control of robot manipulators: the operational space formulation. IEEE J. Robot. Autom. **3**(1), 43–53 (1987)
10. Meirovitch, L.: Fundamentals of Vibrations. McGraw-Hill, New York (2001)
11. Murray, R., Li, Z., Sastry, S.: A Mathematical Introduction to Robotic Manipulation. Taylor & Francis, Boca Raton (1994)
12. Ott, C.: Cartesian Impedance Control of Redundant and Flexible-Joint Robots, 1st edn. Springer Publishing Company, Heidelberg (2008)
13. Strang, G.: Introduction to Linear Algebra, 4th edn. Wellesley-Cambridge Press, Wellesley (2009)
14. Tsagarakis, N.G., Morfey, S., Cerda, G.M., Zhibin, L., Caldwell, D.G.: Compliant humanoid coman: optimal joint stiffness tuning for modal frequency control. In: 2013 IEEE International Conference on Robotics and Automation, pp. 673–678 (2013)

6DoF Pose Estimation for Industrial Manipulation Based on Synthetic Data

Manuel Brucker[1], Maximilian Durner[1], Zoltán-Csaba Márton[1(✉)], Ferenc Bálint-Benczédi[2], Martin Sundermeyer[1], and Rudolph Triebel[1]

[1] German Aerospace Center (DLR), Institute of Robotics and Mechatronics, Münchner Str. 20, Oberpfaffenhofen, 82234 Weßling, Germany
{manuel.brucker,maximilian.durner,zoltan.marton,martin.sundermeyer, rudolph.triebel}@dlr.de

[2] Institute for Artificial Intelligence, University of Bremen, Bremen, Germany
balintbe@cs.uni-bremen.de

Abstract. We present a perception system for mobile manipulation tasks. The primary design goal of the proposed system is to minimize human interaction during system setup which is achieved by several means, such as automatic training data generation, the use of simulated training data, and 3D model based geometric matching. We employ a state-of-the art deep-learning based bounding box detector for rough localization of objects and a Point Pair Feature based matching algorithm for 6DoF pose estimation. The proposed approach shows promising results on our recently published dataset for industrial object detection and pose estimation. Furthermore, the system's performance during four days of live operation at the Automatica 2018 trade fair is analyzed and failure cases are presented and discussed.

1 Introduction and State of the Art

In this study, we evaluate a perception system for manipulation tasks that was designed based on our experiences developing autonomous robots and as part of the RobDREAM project (http://robdream.eu/). The perception system is integrated into our Autonomous Industrial Mobile Manipulator (AIMM) [7,8], which is used to deliver parts to different workcells (see Fig. 1) in a shop floor logistics and kitting scenario.

Recently, deep learning (DL) has dominated vision research, and considerable work has been conducted in object detection, i.e., the simultaneous inference of bounding boxes and semantic labels of objects [6,18,19,23–25]. However, for industrial use cases (and other practical tasks), the lack of labeled training data must be addressed. Here, we present our experiences with robot-aided dataset generation and using simulations of the scenario to obtain synthetic views.

Given the previously reported mixed generalization performance of photo-realistically rendered object views (with background) for object detection and

J. Xiao et al. (Eds.): ISER 2018, SPAR 11, pp. 675–684, 2020.
https://doi.org/10.1007/978-3-030-33950-0_58

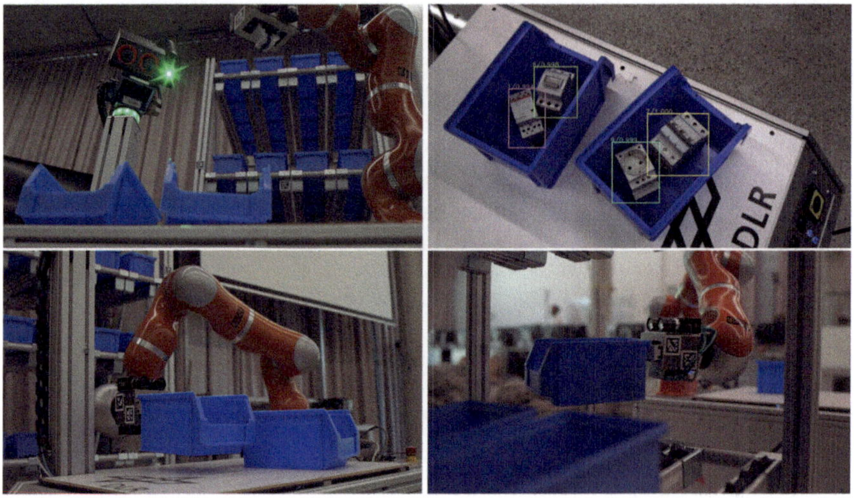

Fig. 1. AIMM platform use case: Small Load Carriers (SLCs) are detected on a workbench and moved to the robot's transport area. The content of the SLCs is detected and placed in the corresponding flow rack.

viewpoint estimation [20,21,27], we investigated the effect of using different combinations of real and UnrealCV [33] data to analyze the benefits of considering synthetic images when training a deep-learned object detector.

Although DL approaches for pose estimation exist [4,16,22,28,29], if reliable depth data is available and sub-framerate speed is sufficient, methods based on Point Pair Features (PPF) [13,31], have demonstrated more robust pose estimation performance on multiple datasets. Note that a PPF-based method won the recent SIXD Challenge (http://cmp.felk.cvut.cz/sixd/challenge_2017/). Typically, runtimes in the range of seconds are required. Here, we use such an approach [17], which is optimized to a runtime of approximately 0.5 s.

Our goal is to minimize human interaction throughout the system setup process or, where human interaction is unavoidable, make the setup process more intuitive and less expert-dependent.

2 Technical Approach

Figure 2 shows an overview of the perception pipeline.

2.1 Dataset Creation

We created a freely available dataset (www.dlr.de/rm/thr-dataset) [10] that will be included in the next iteration of the SIXD Challenge [15].

In addition, we use an annotation tool to annotate ground truth execution data [10]. Assuming the camera position is known relative to the world model,

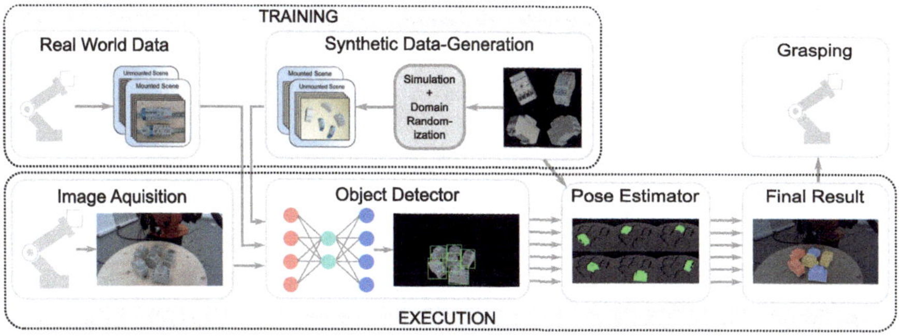

Fig. 2. Perception pipeline flowchart for the training and execution phase.

only a single viewpoint per scene needs to be annotated. With this tool, a non-expert can produce the ground truth for image segmentation, classification, and pose estimation. The ground truth annotation tool was tested in an independent user study conducted by Tecnalia in which expert and non-expert users reported on its usefulness. The results of this user study will be publicly available (http://robdream.eu/). While there remain possibilities for further automation, we found the current data acquisition and ground truth annotation setup was sufficient to label more than 6000 frames with correct segmentation and 6DoF pose information. To obtain more stable prior ground truth estimations, the information of several viewpoints can be fused to diminish drift effects of the camera position [17]. This approach has been used for the Top Hat Rail (THR) dataset and we plan to integrate it on our mobile platform.

2.2 Synthetic Data Generation

We used a previously reported system [1] to generate synthetic training data for object detection using photo-realistic rendering. Here, scene generation is based on variations of the training scenes from the THR dataset. Scenes can be extracted automatically from previously described experience logs [10] and an environment model using the RoboSherlock [2,3] perception framework. We use UnrealCV [33] through the RobCog project (http://www.robcog.org), which offers extensions to communicate through ROS and a wrapper for UnrealCV to stream images over the network. The plugin developed for the RobCoG system offers basic interfaces for spawning and moving objects, as well as the virtual camera. These interfaces are implemented as standard ROS communication schemas. We generated nine variations of the original two test scenes from the THR dataset (five unmounted and four mounted) such that the appearance of each of the nine objects from the database is evenly distributed with a total of 2880 images per variation.

2.3 Object Detection and Pose Estimation

We generate 2D bounding boxes using RetinaNet [18] with a ResNet50 [12] backbone at a resolution of 800 pixels, as well as YOLOv2 [23] for comparison. In addition we use the corresponding depth data as input to the pose estimation method.

For pose estimation, we use a variant of the well-known generic PPF-based pose estimation method from depth data [9], as presented in the *SceneParser* framework [17]. The method is based on extracting parameters from multiple combinations of model surface point pairs and their corresponding normals [32]. The extracted parameters are then used as keys in per object hash tables to quickly find similar point pairs for a given candidate pair. Candidate point pairs are sampled randomly, and similar model point pairs are retrieved from the hash tables. For each combination of candidate and model point pairs, a rigid transformation can be calculated when considering the corresponding normals. Since the hash table keys are not unique, many possibly conflicting hypotheses will be generated. Therefore, another processing step is necessary, in which hypotheses are clustered and only those that are supported by a sufficient amount of candidate/model pairs are processed further. Then, quality values for the remaining hypotheses are calculated by rendering the objects in their corresponding poses, followed by a pixel-wise comparison of the resulting depth buffer to the acquired depth data. If the quality of the highest rated hypothesis exceeds a threshold, it is considered to explain the sensor data sufficiently. Finally, an ICP step is used for local registration.

3 Experiments and Results

The following two subsections focus on evaluating the detection and pose estimation components using our public benchmark dataset and during the Automatica live demo days (Fig. 3). The task of the vision system for the Automatica scenario includes marker less handling of the SLCs, automatic gathering of labeled training data of complex scenes for the detector, detection of the THR components in an SLC, and verification of the SLCs deliveries to different workstations.

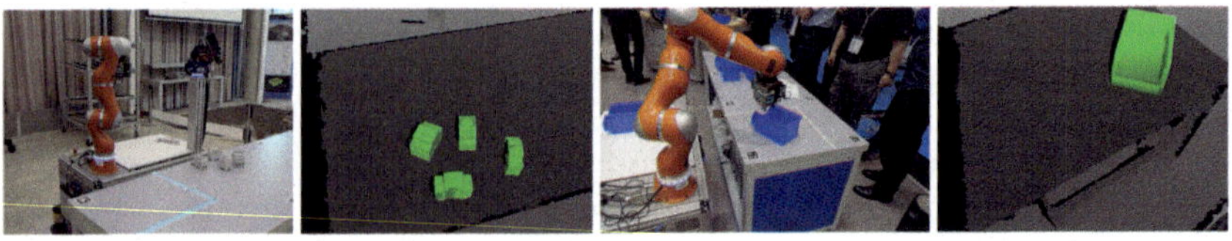

Fig. 3. AIMM detecting THR-elements and SLCs in the lab and at the Automatica fair. Objects are rendered in the estimated poses used for grasping.

Table 1. YOLOv2 and RetinaNet mAP (@0.5IOU) values on the THR dataset test scenes - M and U indicate mounted and unmounted object scenes, respectively.

Training	Real Objects			Real Objects + 2 Real Scenes			Unreal Scenes		
Testing	All	M	U	All	M	U	All	M	U
YOLOv2	0.0576	0.0795	0.0646	0.6765	0.5544	0.7657	0.4828	0.3663	0.6392
RetinaNet	0.1471	0.0699	0.2342	0.7587	0.6563	0.8782	0.6307	0.5349	0.7163
Training	Unreal Scenes + Real Objects			Unreal Scenes + 2 Real Scenes			Unreal Scenes + 2 Real Scenes + Real Objects		
Testing	All	M	U	All	M	U	All	M	U
YOLOv2	0.5614	0.4196	0.7451	0.7602	0.6895	0.8129	0.8118	0.7568	0.8476
RetinaNet	0.6872	0.5939	0.7689	0.8404	0.7367	0.9091	0.8428	0.7368	0.9120

3.1 Benchmarking

The resulting mean average precision (mAP) of the bounding-box detection system [11] are shown in Table 1. Testing always happens on the real test scenes from the THR dataset [10]. Since the training scenes from the dataset only contain four types of objects, we train the network to detect only those objects using all other object types as negative examples.

The pose estimation results for detections with a confidence value greater than 0.5 are shown in Fig. 4. In the reported results, we differentiate between results for mounted and unmounted scenes due to the step change in the difficulty between such scenes.

3.2 Robotic Experiments

For the Automatica shop floor logistics demo, we used our AIMM robot with a stereo pair on its gripper (for in-hand detection as a verification step) and an *rc_visard* by the DLR spinoff RoboCeption GmbH combined with a pattern projector, which was used for the main detection and pose estimation tasks.

During the four days, our system was triggered for 1103 frames, resulting in 1282 SLC detections and corresponding poses. Since these resulted in successful grasps, we assume that most of the estimated poses were sufficiently accurate. Due to the different grasping failure detection steps, we were able to robustly handle the small amount of incorrect percepts, as well as the (intentionally introduced) human disturbance before, during, and after the vision system was triggered. In these cases, the same scenes were captured repeatedly until a grasp was successful, and no operator interaction was required. As we did not have ground truth for the scenes, we evaluated the detection and pose estimation jointly by reporting the average distance between the real edges detected in the RGB image and the rendering of the detected object in the estimated pose [26], as shown in Fig. 5, where errors of up to approximately 6–7 pixels seem visually appealing. Note that errors up to 8–9 pixels would still result in

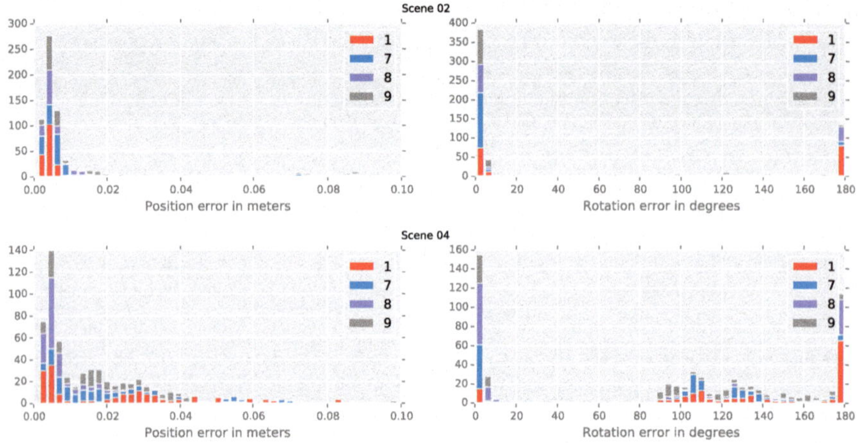

Fig. 4. Translational and rotational errors for objects 1, 7, 8, and 9 in the THR dataset test scenes. Top: Scene 2 with unmounted objects spread on a plane. Bottom: Scene 4 with objects mounted on a rail. High rotational errors of approximately 180° are due to the geometric (near-)symmetries of objects 1 and 8.

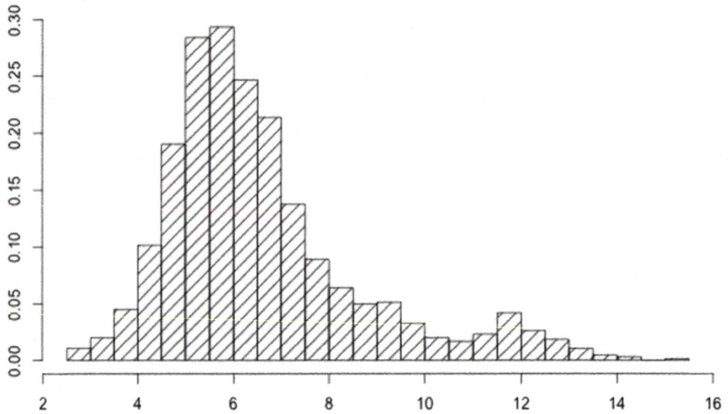

Fig. 5. For each of the 1282 SLC detections at Automatica, we rendered the best 6DoF match of the 3D model, ran Canny edge detection on it, and computed the edge pixels' mean distance to RGB edges. The distances were capped at 25 pixels to avoid false associations. The two peaks correspond to correct and flipped detections.

successful grasps most of the time. The small peak near 12 pixels is due to 90- or 180-degree flipped poses that sometimes occur, or due to occlusions/stacking.

While most detections resulted in pose estimates that were sufficiently accurate for successful grasping, Fig. 6 highlights several interesting cases. For each detection, 3D-based pose estimation was performed inside the bounding box without the use of prior knowledge about the expected poses. Besides misdetections, the most common mistakes were 180° (and rarely 90°) flips. As with

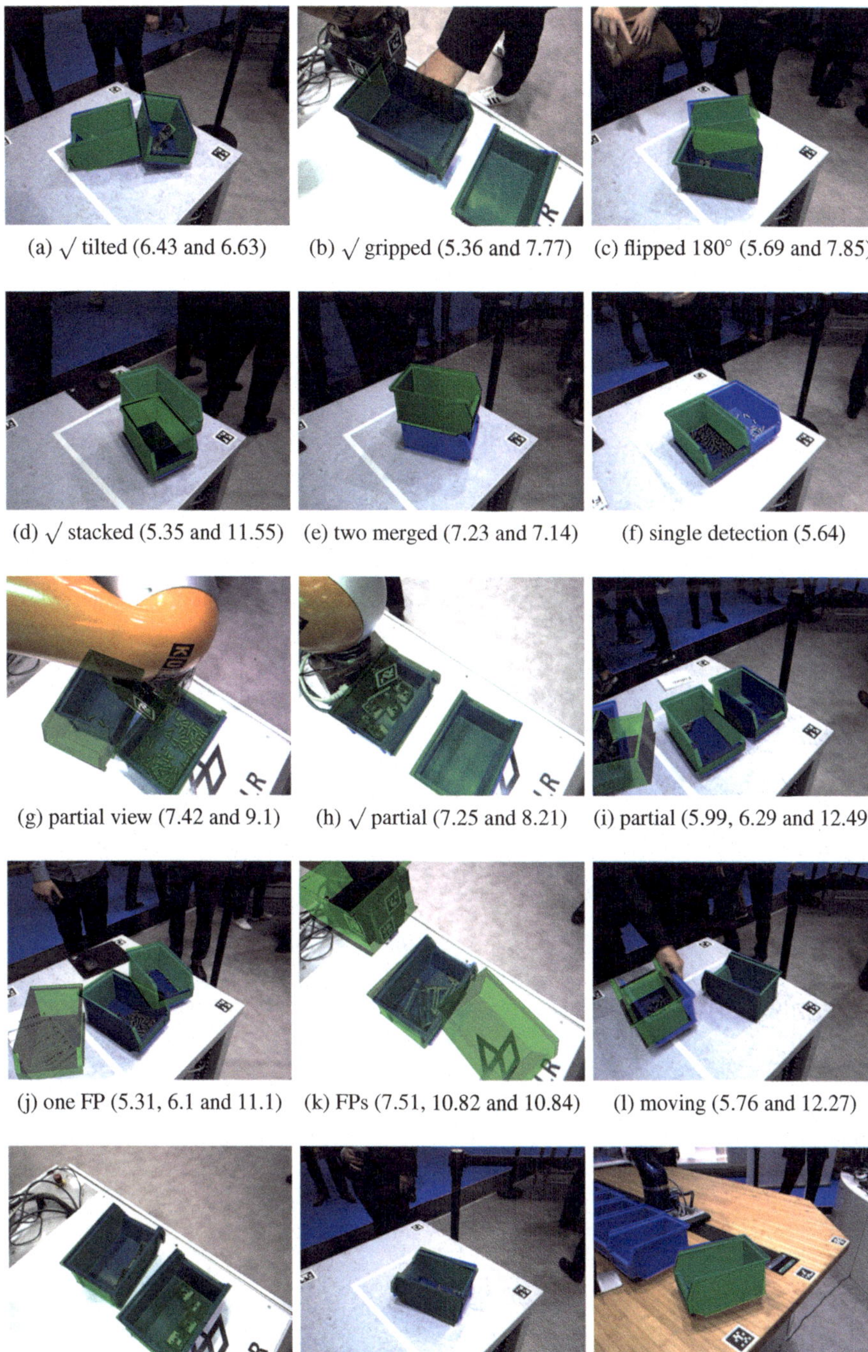

(a) √ tilted (6.43 and 6.63) (b) √ gripped (5.36 and 7.77) (c) flipped 180° (5.69 and 7.85)

(d) √ stacked (5.35 and 11.55) (e) two merged (7.23 and 7.14) (f) single detection (5.64)

(g) partial view (7.42 and 9.1) (h) √ partial (7.25 and 8.21) (i) partial (5.99, 6.29 and 12.49)

(j) one FP (5.31, 6.1 and 11.1) (k) FPs (7.51, 10.82 and 10.84) (l) moving (5.76 and 12.27)

(m) √ with 3.74 and 6.05 (n) × misaligned with 7.31 (o) × misaligned with 7.64

Fig. 6. Example of success (√) and failure (×) cases with mean edge pixel errors.

THR elements, where geometrically similar objects are stacked s.t. their exact boundary and translational errors are difficult to detect while matching, detection and pose estimation errors are common. Perfectly aligned side-by-side SLCs typically result in a single detection and a correct pose. Similarly, in the case of stacked ones, the top (and occasionally the bottom) SLC is matched reliably.

4 Conclusions and Discussion

The experimental results show that for 2D object detection simulated data is, as of now, insufficient and a Domain Adaptation (DA) [5] step is necessary, i.e., using synthetic images combined with real data. By introducing real images during the training phase, the network is less sensitive to color nuance differences between the rendered and the real images. However, by simply generating realistic synthetic data, we achieved a 10% increase in performance on real data.

Improvements of the 2D object detection trained using only synthetic data could be obtained by generating more variations, applying more advanced transfer learning approaches, or by employing additional Domain Randomization (DR) [14,30]. Furthermore, DA can be facilitated using deeper integration of autonomous modeling and ground truth annotation in the mobile robotic system.

For pose estimation, we can relatively often observe confusion with the geometrically symmetric view at 180° rotation because the camera image is not used. Therefore, to improve the results, we plan to add a camera-based classification step [10]. In addition, we plan to fuse our current detection pipeline with our recent DL-based one [28], as well as introduce a plausibility check based on the environment model and non-interpenetration [26].

Note that the THR dataset is freely available online, the UnrealCV extension scenes will be published soon, and the scenes and detections logged during the live demo will be released to the community.

Acknowledgements. We thank the entire team of the mobile manipulation Automatica demo at DLR, especially Andreas Dömel for coordination, Peter Lehner for the robotic manipulation, and Sebastian Riedel for the logging component. We also thank Amrutha Saseendran for her help with the detector, as well as the teams of the other three Factory of the Future demos.

References

1. Balint-Benczedi, F., Beetz, M.: Variations on a theme: 'it is a strange kind of memory that only works backwards'. In: International Conference on Intelligent Robots (IROS). IEEE (2018, under review)
2. Balint-Benczedi, F., Marton, Z.-C., Durner, M., Beetz, M.: Storing and retrieving perceptual episodic memories for long-term manipulation tasks. In: 18th International Conference on Advanced Robotics (ICAR). IEEE (2017). Best Paper Finalist
3. Beetz, M., Balint-Benczedi, F., Blodow, N., Nyga, D., Wiedemeyer, T., Marton, Z.-C.: RoboSherlock: unstructured information processing for robot perception. In: IEEE International Conference on Robotics and Automation (ICRA), Seattle, Washington, USA (2015). Best Service Robotics Paper Award

4. Brachmann, E., Michel, F., Krull, A., Ying Yang, M., Gumhold, S., Rother, C.: Uncertainty-driven 6D pose estimation of objects and scenes from a single RGB image. In: Proceedings of the IEEE Conference on Computer Vision and Pattern Recognition, pp. 3364–3372 (2016)
5. Csurka, G.: Domain adaptation for visual applications: a comprehensive survey. arXiv preprint arXiv:1702.05374 (2017)
6. Dai, J., Li, Y., He, K., Sun, J.: R-FCN: object detection via region-based fully convolutional networks. In: Advances in Neural Information Processing Systems, pp. 379–387 (2016)
7. Dömel, A., Kriegel, S., Brucker, M., Suppa, M.: Autonomous pick and place operations in industrial production. In: 12th International Conference on Ubiquitous Robots and Ambient Intelligence, p. 356. IEEE (2015). Best Video Paper Award
8. Dömel, A., Kriegel, S., Kaßecker, M., Brucker, M., Bodenmüller, T., Suppa, M.: Towards fully autonomous mobile manipulation for industrial environments. Int. J. Adv. Robot. Syst. (2017)
9. Drost, B., Ulrich, M., Navab, N., Ilic, S.: Model globally, match locally: efficient and robust 3D object recognition. In: IEEE Computer Society Conference on Computer Vision and Pattern Recognition, pp. 998–1005, June 2010
10. Durner, M., Kriegel, S., Riedel, S., Brucker, M., Márton, Z.-C., Bálint-Benczédi, F., Triebel, R.: Experience-based optimization of robotic perception. In: 18th International Conference on Advanced Robotics (ICAR), pp. 32–39. IEEE (2017). Best Paper Finalist
11. Everingham, M., Van Gool, L., Williams, C.K.I., Winn, J., Zisserman, A.: The PASCAL visual object classes challenge 2012 (VOC2012) Results (2012). http://www.pascal-network.org/challenges/VOC/voc2012/workshop/index.html
12. He, K., Zhang, X., Ren, S., Sun, J.: Deep residual learning for image recognition. In: Proceedings of the IEEE Conference on Computer Vision and Pattern Recognition, pp. 770–778 (2016)
13. Hinterstoisser, S., Lepetit, V., Rajkumar, N., Konolige, K.: Going further with point pair features. In: European Conference on Computer Vision (2016)
14. Hinterstoisser, S., Lepetit, V., Wohlhart, P., Konolige, K.: On pre-trained image features and synthetic images for deep learning. arXiv:1710.10710 (2017)
15. Hodaň, T., Matas, J., Obdržálek, Š.: On evaluation of 6D object pose estimation. In: Hua, G., Jégou, H. (eds.) European Conference on Computer Vision, vol. 9915, pp. 606–619. Springer, Cham (2016)
16. Kehl, W., Manhardt, F., Tombari, F., Ilic, S., Navab, N.: SSD-6D: making RGB-based 3D detection and 6D pose estimation great again. In: Proceedings of the IEEE Conference on Computer Vision and Pattern Recognition, pp. 1521–1529 (2017)
17. Kriegel, S., Brucker, M., Marton, Z.C., Bodenmüller, T., Suppa, M.: Combining object modeling and recognition for active scene exploration. In: International Conference on Intelligent Robots and Systems, Proceedings of IEEE/RSJ International Conference on Intelligent Robots and Systems, pp. 2384–2391, November 2013
18. Lin, T.-Y., Goyal, P., Girshick, R., He, K., Dollár, P.: Focal loss for dense object detection. arXiv preprint arXiv:1708.02002 (2017)
19. Liu, W., Anguelov, D., Erhan, D., Szegedy, C., Reed, S., Fu, C.-Y., Berg, A.C.: SSD: single shot multibox detector. In: Leibe, B., Matas, J., Sebe, N., Welling, M. (eds.) European Conference on Computer Vision, vol. 9905, pp. 21–37. Springer, Cham (2016)

20. Mitash, C., Bekris, K.E., Boularias, A.: A self-supervised learning system for object detection using physics simulation and multi-view pose estimation. In: IEEE/RSJ International Conference on Intelligent Robots and Systems (IROS), pp. 545–551. IEEE (2017)
21. Movshovitz-Attias, Y., Kanade, T., Sheikh, Y.: How useful is photo-realistic rendering for visual learning? In: Hua, G., Jégou, H. (eds.) European Conference on Computer Vision, pp. 202–217. Springer, Cham (2016)
22. Rad, M., Lepetit, V.: BB8: a scalable, accurate, robust to partial occlusion method for predicting the 3D poses of challenging objects without using depth. arXiv preprint arXiv:1703.10896 (2017)
23. Redmon, J., Farhadi, A.: YOLO9000: better, faster, stronger. arXiv preprint arXiv:1612.08242 (2016)
24. Redmon, J., Farhadi, A.: YOLOv3: An incremental improvement. arXiv (2018)
25. Ren, S., He, K., Girshick, R., Sun, J.: Faster R-CNN: towards real-time object detection with region proposal networks. In: Advances in Neural Information Processing Systems, pp. 91–99 (2015)
26. Schmidt, T., Hertkorn, K., Newcombe, R., Marton, Z., Fox, D.: Depth-based tracking with physical constraints for robot manipulation. In: IEEE International Conference on Robotics and Automation (ICRA), pp. 119–126, May 2015. Best Paper Award finalist
27. Su, H., Qi, C.R., Li, Y., Guibas, L.J.: Render for CNN: viewpoint estimation in images using CNNS trained with rendered 3D model views. In: Proceedings of the IEEE International Conference on Computer Vision, pp. 2686–2694 (2015)
28. Sundermeyer, M., Marton, Z.-C., Durner, M., Brucker, M., Triebel, R.: Implicit 3D orientation learning for 6D object detection from RGB images. In: The European Conference on Computer Vision (ECCV), September 2018
29. Tekin, B., Sinha, S.N., Fua, P.: Real-time seamless single shot 6D object pose prediction. arXiv preprint arXiv:1711.08848 (2017)
30. Tobin, J., Fong, R., Ray, A., Schneider, J., Zaremba, W., Abbeel, P.: Domain randomization for transferring deep neural networks from simulation to the real world. In: IEEE/RSJ International Conference on Intelligent Robots and Systems (IROS) (2017)
31. Vidal, J., Lin, C.-Y., Martí, R.: 6D pose estimation using an improved method based on point pair features. arXiv preprint arXiv:1802.08516 (2018)
32. Wahl, E., Hillenbrand, U., Hirzinger, G.: Surflet-pair-relation histograms: a statistical 3D-shape representation for rapid classification. In: Fourth International Conference on 3-D Digital Imaging and Modeling (3DIM), pp. 474–481. IEEE (2003)
33. Weichao, Q., Fangwei, Z., Yi, Z., Siyuan, Q., Xiao, Z., Kim, T.S., Wang, Y., Yuille, A.: UnrealCV: virtual worlds for computer vision. In: ACM Multimedia Open Source Software Competition (2017)

Mapping and Pose Estimation

Mapping and Pose Estimation

Seth Hutchinson

This session covered a range of topics, from low-level issues related to calibration, to higher level issues related to navigation in the context of pose estimation. The paper by Nobre and Heckman dealt with the problem of sensor calibration and presented a fast and accurate solution that could handle drift n calibration parameters in unobservable in the parameter space. The paper by Zhen and Scherer addressed the problem of generalizing LiDAR-based SLAM algorithms across heterogeneous sensor systems, specifically the case of applying both 2D and 3D LiDAR in the Cartographer SLAM library. The paper by Heiden et al. also considered the problem of heterogeneous sensors, but in a sensor fusion context. They described a method in which cells in a grid-based map were augmented with information that facilitated effective fusion of data from disparate sensors. The paper by Sartipi and Roumeliotis dealt with the problem of concurrently computing the transformation between multiple gravity-aligned maps, given a set of common feature observations, i.e., the alignment problem for visual-inertial maps. They formulated this as an optimization problem that could be expressed in terms of a system of multivariate polynomials and provided an efficient solution for cases when the number of maps was small. Finally, the paper by Manderson et al. addressed the problem of robust, vision-based odometry for underwater robots, presenting a method that could autonomously optimize the robot trajectory in real time based on visual feedback.

Navigation in the Service of Enhanced Pose Estimation

Travis Manderson(✉), Ran Cheng, David Meger, and Gregory Dudek

School of Computer Science, McGill University, Rm 318, 3480 Rue University, Montreal, QC H3A 0E9, Canada
{travism,rancheng,dmeger,dudek}@cim.mcgill.ca

1 Introduction

This paper addresses robust vision-based odometry for underwater robotics by autonomously adjusting the robot trajectory in real-time to optimize the quality of ongoing visual feedback. It is well-known that accurate Visual Odometry (VO) depends on both the presence of sufficient smooth surfaces with manageable reflectance functions and the availability of detectable rigid structures or appearance variations. More generally, some locations in the world are suited to good VO performance while others are not (Fig. 1). Our system finds trajectories that pass over such usable visual content by evaluating a localization quality metric (quality score) on points from a forward-facing camera's image. By planning forward-looking paths that optimize an associated quality score, a downward-facing camera running VO is able to localize our robot more accurately. While our approach is based on a particular underwater vehicle and imaging arrangement, its essential elements can be generalized readily.

Fig. 1. Example coral environment where sandy regions will lead to poor visual odometry estimates.

Fig. 2. Our vision-based robot swimming over a coral reef.

The robotic vehicle used in this work is a fully-autonomous marine system that uses vision for collision avoidance, navigation, and image understanding

Electronic supplementary material The online version of this chapter (https://doi.org/10.1007/978-3-030-33950-0_59) contains supplementary material, which is available to authorized users.

J. Xiao et al. (Eds.): ISER 2018, SPAR 11, pp. 687–701, 2020.
https://doi.org/10.1007/978-3-030-33950-0_59

on a single vehicle (Fig. 2). This vehicle is a variant of the "Aqua" class of six-flipper swimming robots [10,15] configured with two forward-facing cameras and a downward-facing camera (Fig. 6) as well as an on-board GPU module (NVidia Jetson TX2) that is used to run a Deep Convolutional Neural Network (DCNN). The platform is targeted to observational tasks such as coral reef mapping and health assessment.

Underwater robots operating in complex environments such as coral reefs often rely on accurate odometry for tasks like autonomous surveying and collision avoidance. VO is appealing for pose estimation due to its simplicity and energy efficiency, as well as the fact that cameras are exteroceptive sensors (at least in the subset of underwater regions where there is good visibility). We use vision as our primary sensing modality because it is passive, low energy, compatible with our very compact vehicle, and can be used in a shallow-water reef environment, which is acoustically complex. On the other hand, coral reefs are a challenging environment for vision-based algorithms due to: (1) poor visual conditions caused by low light, water turbidity, and floating artifacts, (2) featureless regions that appear plain or uniform (such as sand), (3) caustics caused by refractive effects at the water surface, and (4) non-rigid objects.

One way to evaluate coral reef health is by determining the amount of live coral, which we have previously measured using image data [27]. Aqua is small, lightweight, highly maneuverable, and capable of swimming autonomously in close proximity (tens of centimeters) to coral without collision [25]. Our long-term goal is to deploy Aqua to perform fully-autonomous geometric coverage of underwater environments for surveying tasks such as coral reef health. To this end, we require reliable and accurate vision-based odometry. Although many visual-odometry systems exist, they rely on visual texture to compare photometric differences between frames to estimate change in pose, and when uncoupled from trajectory planning, they are likely to encounter situations where poor texture results in incorrect pose estimates.

2 Related Work

Numerous authors have demonstrated the importance of active motion planning for navigation [2,22,31]. Information-theoretic reasoning is often utilized [6,12,38] to guide the robot along mapping paths that optimize environment coverage and localization accuracy, both with laser-based [5,33] and visual maps [7,36]. Gaze control can improve localization quality by capturing images with visually-salient content [13], with many features visible [20] and with reliable local features [9].

Recently, methods that learn to predict visual reliability from self-supervised robot performance data have been considered. Gurau *et al.* [16] utilize object detector performance along previously visited paths to train an introspective predictor of test-time recognition rates at nearby locations, yielding confidence estimates for self-driving behavior. Our own previous work trained a texture classifier capable of making a greedy decision to orient a single camera towards the most reliable visual content [26,28]. This current paper extends that work significantly, training more flexible deep models through the use of a realistic simulator in conjunction with real open-ocean data, creating an association between two cameras and planning longer navigation paths. This use of simulation data for vision-based algorithm development is a direction we have also explored in the context of marine target tracking [23].

Robot localization in underwater environments is often based upon acoustic or computer vision sensing [19,21,32,35]. Particularly notable is the Direct Sparse Odometry (DSO) [11] algorithm, which we employ in our own work. DSO is one of a family of methods, reaching back to the early work of Horn, that exploit a combination of feature point extraction and tracking to compute 3D structure from motion.

In many domains, the tight coupling of visual cues and trajectories has been considered - for example, to allow a previously-taken path to be replayed [8] or to allow servo-control to be coupled to visual cues [3]. Other authors have considered learning per-viewpoint detection rates as a guide for motion during object search [29]. Active guidance is also a crucial factor for a team of cooperatively exploring robots with communication constraints [4].

3 Approach

Our approach to planning trajectories for accurate navigation is to use the wide field of view of forward-looking cameras to predict eventual VO quality for distant points when observing such points in the environment from the downward-facing camera of the robot. In short, the front facing cameras allow us to see and estimate what will be soon approaching in various parts of the visual field. Since the overall goal is to swim in close proximity over coral, the downward-facing camera provides better VO estimates than running on the forward-facing images.

3.1 Visual Odometry

Visual odometry refers to the processes of computing pose updates for the vehicle from information captured by the camera(s). While several powerful methods exist for this computation, they all depend on a visually suitable environment including good illumination, a suitable number of visual features at the right range of scale, constraints on the rigidity of the material being observed, and other attributes. The suitability of the environment for good VO depends on illumination, vehicle speed, geometry, material properties of the local environment, and other features.

We apply a variant of the DSO algorithm [11] algorithm as our Visual Odometry estimator, with enhancements (outlined below) to permit an estimate of tracking reliability at each frame. More precisely, we track each frame in a sliding window and compute a Hessian for each point being tracked (referred to as

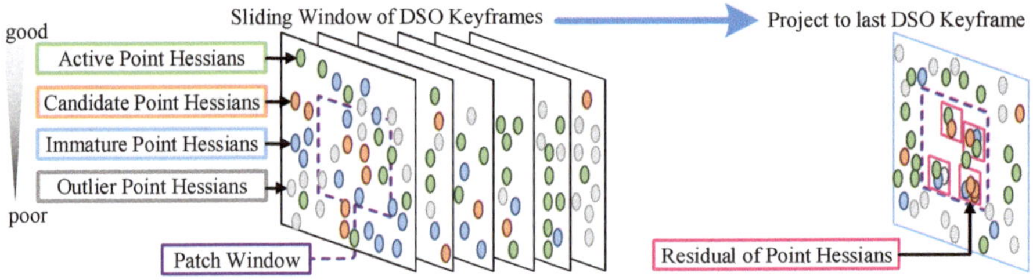

Fig. 3. Pipeline of the DSO quality score prediction.

a Point Hessian (H)) as shown in Fig. 3. The Point Hessian is a data-structure in DSO that contains the Hessian of the temporally tracked inverse depth and intensity values. The residual (or difference between the estimated and measured value) of the Point Hessian dictates to which class it belongs: "Active", "Candidate", "Immature" or "Outlier" point, as described in the DSO literature.

We introduce a measure for tracking quality, b, which is derived from the classification of the Point Hessians in the current frame and the entropy [34] of those in a center region of the current frame. The entropy, E at frame k is then defined as the sum of normalized probabilities that a Point Hessian belongs to a particular class, c:

$$E(k) = \sum_{i=0}^{N^H} \sum_{c \in C} \frac{P_c ln(P_c)}{N^C}, \quad C = \{\text{'Active', 'Candidate', 'Immature', 'Outlier'}\}$$

where P_c is probability of each class: $P(c \mid H(i))$, and $H(i)$ is one of N^H Point Hessians in the current frame and N^C is the total number of classes (5). We define a quality score, g, per frame, k, as a normalized sum of all weighted Point Hessians in a centered patch in the frame:

$$g(k) = \frac{N^H}{\eta} \sum_{c \in C} \frac{1}{w_c n(c, k)}$$

where η is the maximum weight over the previous frames $(max(g(m)),\ m \in (0 : k-1))$, w_c is a weight associated for each class (with poorer classes being penalized more as shown in Fig. 3), and $n(c, k)$ is the number of Point Hessians in class c within a window at the center of the image. Finally, the quality score is:

$$b(t) = \sum_{k=t-N^w}^{t} (g(k) - E_t(k))$$

where N^w is the number of frames in the sliding window (approximately 10 in practice).

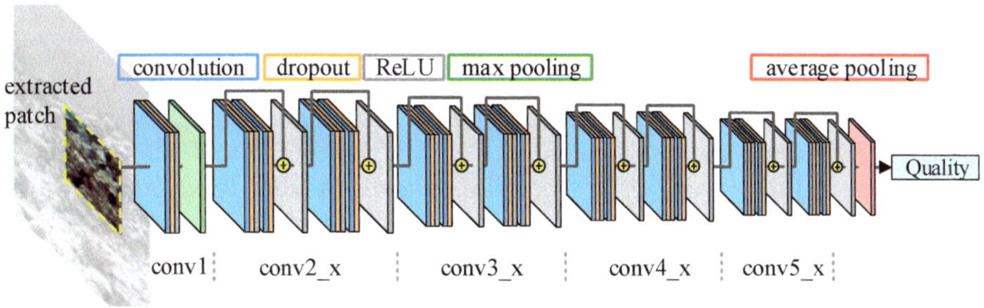

Fig. 4. Deep Convolutional Neural Network architecture, which is based on the Resnet-18 model. The input into the network is an image patch extracted from the whole image and the output is a quality score.

3.2 Predicting Localization Quality

Our quality score has a diagnostic form in relation to DSO. It requires visual odometry to be running real-time and must actually experience potential loss of visual quality for the quality score to drop. We aim to prevent the robot from ever experiencing such circumstances at deployment time, and therefore we utilize a learned classifier trained to predict localization quality in advance by analyzing images from a forward-looking camera. This approach allows our system to be pro-active and select paths such that localization performance is never reduced at deployment time.

Our target trajectory predictor is based on a DCNN that predicts the quality score for a set of points observed from a forward-looking image (using a small patch centered at these points), and chooses a heading based on the highest quality points. The DCNN is based on the Resnet-18 architecture [17,18] as shown in Fig. 4, which performs well on an auxiliary GPU (NVidia Jetson TX2) installed in our vehicle. We introduce dropout layers [14,37] to prevent over-fitting and use the mean squared error cost function for training.

Finally, the robot is controlled using a tuned PID [25] controller that takes temporally averaged image scores and their location to control the yaw and pitch of the robot. A low-level gait controller [30] translates these angular commands to motions of the six flippers.

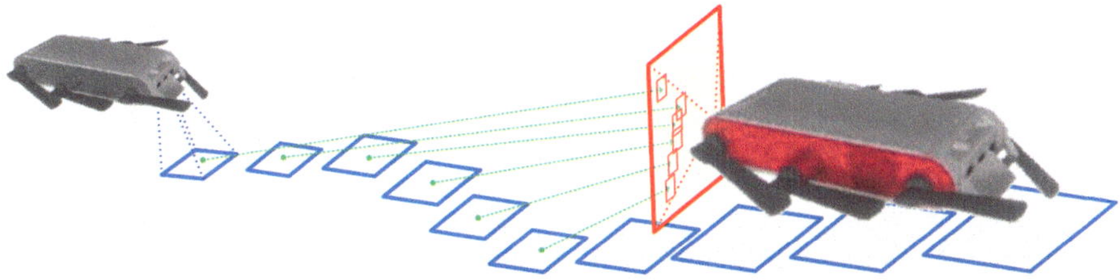

Fig. 5. Image patches are generated by back projecting downward-facing image locations to the front camera image frames.

3.3 Training

Our predictive quality network is trained in a self-supervised fashion by moving the robot over a wide range of terrain and correlating the localization quality with patches in the forward-looking camera via the robot's self-calibration. For each image in the training data set, training labels for the DCNN are generated by back-projecting the relative 3D location of the downward-facing images into the history of front images, as shown in Fig. 5. For each point projected into the front image, a patch is extracted, along with the quality score described above.

The back-projection of a downward facing image at time t into a forward facing image at time t' is given by:

$$u_{t',t}, v_{t',t} = \pi(T_w^c(t')\mathbf{P_b}(\mathbf{t})) \quad (1)$$

where π is a function that projects a 3D point relative to the front camera into the image plane, T_w^c is the transformation from a point in the world to the front camera, P_b is the position of the downward-facing image in world coordinates, and $u_{t',t}$ and $v_{t',t}$ are the pixel locations at time t' of the downward-facing patch at time t.

The robot is controlled using a PID [25] controller that takes temporally averaged image scores and their location to control the yaw and pitch of the robot. A low-level gait controller [30] translates these angular commands to motions of the six flippers.

4 Experiments

Overall, our method processes images in real time to maintain a navigation estimate and close planning and control loops at a number of levels. We perform accurate VO based on the downward looking "Back Camera" images while estimating localization quality in the region ahead of the robot by filtering the "Front Camera" images using a DCNN, whose output feeds longer horizon path planning for the purpose of guiding the robot to regions that support accurate localization.

Our experimental approach is composed of two components: (1) a preliminary proof-of-concept study in the open ocean and (2) a series of rigorous simulation studies allowing repeatable quantitative assessment in a state-of-the-art simulation environment. While our primary focus is achieving reliable performance in the open ocean, the simulation is critical as it allows access to large volumes of reliable ground truth that is challenging to collect and imprecise in field conditions.

Open Ocean Experiment. This experiment served to: (1) generate quality scores in real conditions to validate the methodology and generate initial training data for the DCNN used proceeding experiments, and (2) test the integrated hardware (NVidia Jetson TX2) running DCNN inference alongside the VO algorithm.

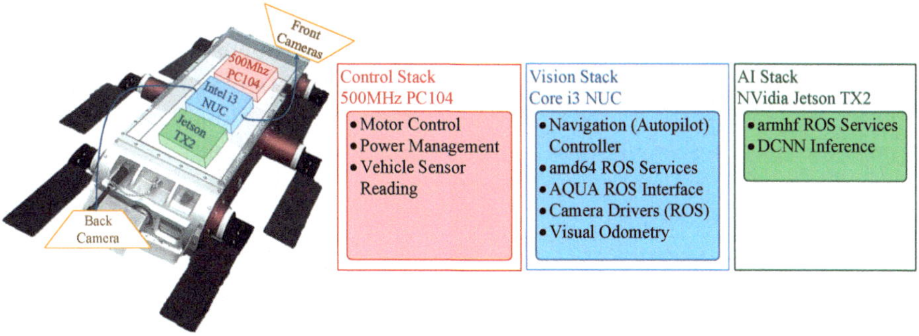

Fig. 6. Block diagram of the Aqua robot and the three computers. The three computers are connected by high-speed Ethernet and the cameras are connected by USB 3.0 to the Intel NUC computer.

Our experimental platform is an Aqua robot (Fig. 6) where the cameras are operated by three efficient, yet powerful computers that are networked using Gigabit Ethernet. The PC104 computer (Control Stack) uses the real-time Xenomai extension to the Linux operating system and performs low-level motor control, power management and vehicle orientation and depth estimation from an internal Inertial Measurement Unit (IMU) and depth sensor. An Intel i3 NUC computer (Vision Stack) running Robot Operating System (ROS) captures the images (at 15 Hz) from the global-shutter and hardware-synchronized cameras, runs the VO algorithm and performs the high-level navigation tasks. An NVidia Jetson TX2 GPU compute module (AI Stack) which is also running ROS, employs TensorFlow [1] for vision-based inference using the DCNN. The inference results are transmitted to the Vision Stack where they are used in the navigation planner.

Fig. 7. An example comparison of the real and simulated underwater ocean environments. The left image is the real underwater environment taken during our open-ocean experiment and the right image is an example taken from our simulator, showing its close resemblance to the real environment.

Simulation Experiments. We developed a simulator with two objectives: (1) perform end-to-end validation of our system beginning with image data characteristic of an ocean environment to control outputs to move the robot, and (2) quantitatively evaluate the effectiveness of our system using the ground truth available. Our simulator is based on Unreal Engine 4 and involves a kinetic model for control of the Aqua robot and a visual-inertial sensor model to obtain vision and inertial data. Ground truth, images, and control are communicated using custom ROS2 and ROS1 messages with the same definition used on the Aqua robot. This approach allows us to use the exact inference and control code in both simulation and real-world. We have also built a customizable ocean environment, which also allows us to vary the environmental conditions such as the water turbidity, water current, and brightness. We have used this simulator to tune the inference model as well as control parameters. In separate work, we have demonstrated that this simulator is faithful enough to train vision-based methods that then perform well in genuine marine imagery [23,24]. As seen in Fig. 7, the simulated environment bears close resemblance to the real underwater, ocean environment.

5 Results

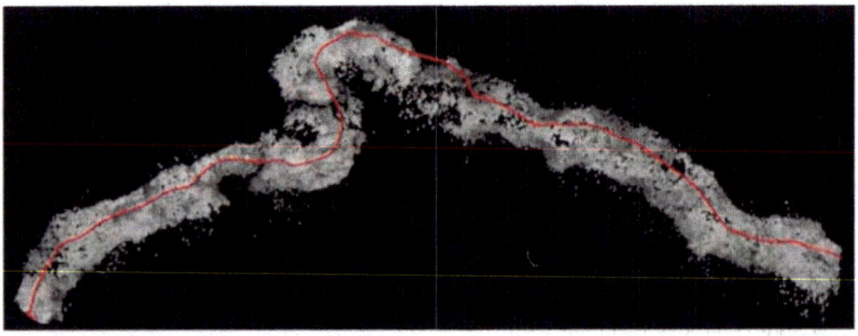

Fig. 8. Illustrative reconstruction from a deployment in the Caribbean.

We ran DSO and our quality score scoring system on our preliminary open-ocean data as well as simulated data as described in Sect. 4. We collected ~74 min (in segments of two to 10 min) of data from autonomous swimming in the open ocean near coral over the course of five days (on the coast of Barbados) at varying locations under varied weather conditions (and visibility) and swimming motions. We successfully ran DSO on this dataset and validated the quality score by examining the distribution of scores, visually inspecting obvious good and poor image patches for VO, and verifying that VO failures were preceded with extremely low quality scores. An illustrative path and reconstruction on our preliminary open-ocean data is shown in Fig. 8.

Table 1. DCNN classification accuracy on simulated validation dataset

Patch Size	40 px	60 px	80 px	100 px	120 px
Accuracy	90.59%	91.46%	93.13%	92.56%	92.37%

Our simulated dataset consisted of 51,892 test images, 21,567 validation images, and 10,104 test images. Our Resnet based DCNN predicted the quality score as one of five classes. The model was initialized with the pre-trained weights after training on ImageNet [17] and the parameters of the model were: batch size: 25, l2 regularization weight: 1e–5, momentum: 1e–4, initial learning rate: 1e–5 and learning rate decay: 1e–3.

Table 1 shows the classification accuracy of our DCNN on a validation dataset collected using our simulator over a number of patch sizes. All patch sizes resulted in accuracy over 90%. The smaller patch sizes allowed our DCNN to perform faster while being more granular in our estimate. A patch size of 80×80 pixels resulted in the maximum accuracy of 93.13%. We suspect that the larger patch sizes performed slightly worse because they were capturing a larger region of the scene and may have contained ambiguous textures such as sand and coral in the same image patch.

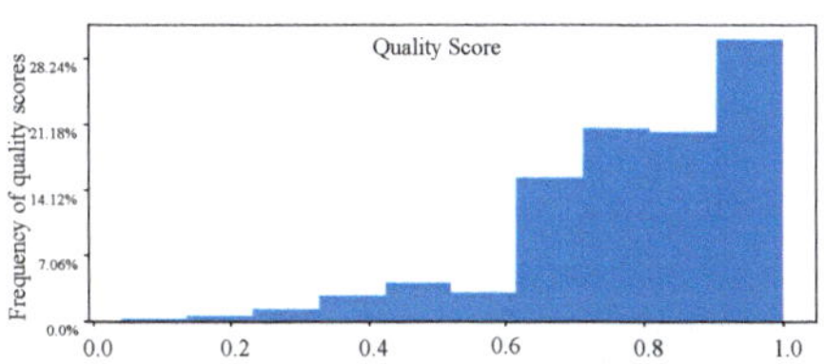

Fig. 9. Sample score histogram swimming with trajectory prediction (this is a good result since mass is concentrated on the right).

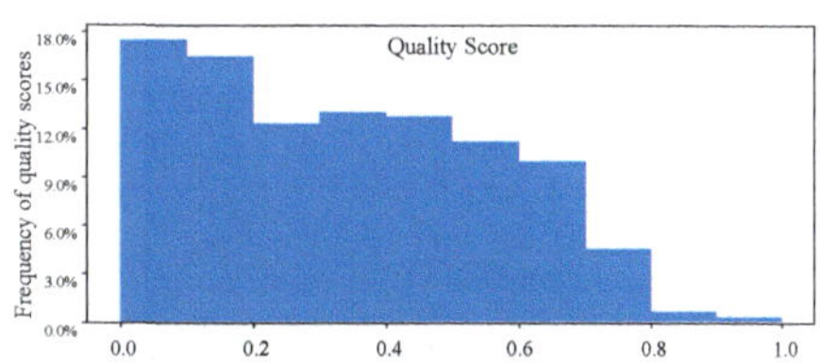

Fig. 10. Sample score histogram swimming without trajectory prediction (this is a less useful result since mass is concentrated on the left).

The effectiveness of our trajectory predictor can be seen in Figs. 9 and 10, which show the distribution of quality scores collected from autonomous swimming in our simulator with and without our trajectory prediction. The concentration of high quality scores sampled when running our trajectory prediction (as compared to running without it) demonstrates the effectiveness of our quality score estimator and trajectory planner. The quality scores on our preliminary open-ocean experiment is similar to Fig. 10.

Fig. 11. Three simplified examples of the image patch classification from the DCNN inference (note that the class overlay visualization removes much of the image texture). The numbers represent the unscaled score prediction, where the highest number (4) represents the highest quality score. In each example, it can be seen that the highest quality prediction is over coral regions and a much lower quality is predicted for seagrass, sand and water.

To quantify the robustness of our approach, we performed several simulated experiments using our DCNN trajectory prediction controller and a naive waypoint controller. Our DCNN trajectory prediction controller navigated the robot towards regions containing the highest quality scores while moving forward at a constant velocity. A simplified example of the DCNN inference is shown in Fig. 11 for three regions of the map (note that the class overlay visualization removes much of the image texture). Each experiment using the predictive controller

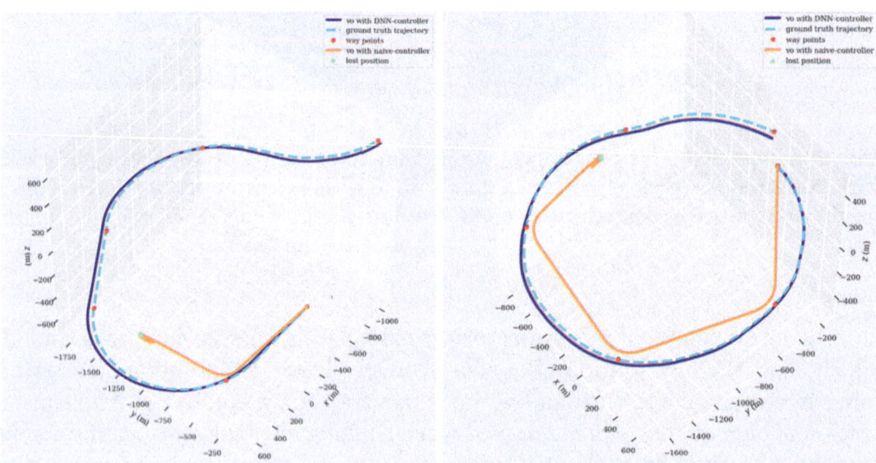

Fig. 12. Two example paths the robot swam with and without trajectory planning. In both cases, it can be seen that the VO path estimate (purple line) closely matches the ground-truth path (blue line). In the left example, the naive waypoint controller, reached the first waypoint, but DSO become lost before reaching the second waypoint (path represented by the orange line). Similarly, in the right example, using the naive waypoint controller, DSO gets lost before reaching the fourth waypoint.

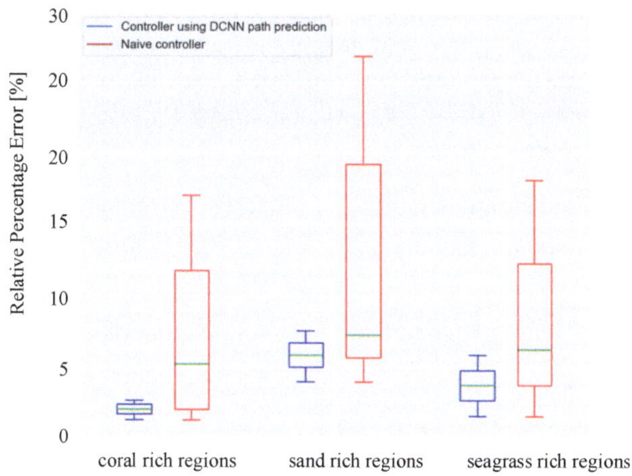

Fig. 13. Comparison of relative percentage error in coral, sand and seagrass environments. Twenty experiments were performed in each of the three simulated environments which resulted in a total distance travelled of 162 km. The mean relative percentage error (RPE) in the coral rich environments was **2.35%** and 8.50% using our trajectory prediction controller and the native waypoint controller respectively. Likewise, **6.75%** and 17.50% for the sand rich regions, and **4.15%** and 11.25% for the seagrass rich regions. In all cases, our trajectory prediction controller outperformed the naive waypoint controller in both the mean error and the total variance.

resulted in a path that we compared against the ground-truth to quantitatively measure the relative error. For each path that the robot took using the predictive controller, we sampled a number of points to create a set of waypoints along the path. We then navigated the set of waypoints using a naive waypoint controller.

The naive waypoint controller moved at a constant forward velocity and used a proportional controller to iteratively point the robot towards the current waypoint until it was reached. To prevent VO error due to rapid rotations, we limited the angular rate of rotation. Once the current waypoint was reached within one meter, the robot navigated toward the next waypoint, repeating until all were reached, or DSO was lost (where the relative pose estimator was unable to converge due to a lack of image texture). An example of two of these experiments is shown in Fig. 12. In each example, the light blue line is the ground-truth path, the purple line is the VO path estimate of the robot while running the trajectory prediction controller, and the orange line is the path of the robot using the naive waypoint controller.

In total, we performed 60 experiments. These were divided evenly between three environments: (1) regions where the majority of the environment contained a mix of **coral**, (2) regions that contained a majority of **seagrass**, and (3) regions that contained a large amount of **sandy** bottom. Figure 13 shows average position error (depicted as the relative percentage error) for the trajectory prediction controller (blue whisker) and the naive waypoint controller

Table 2. Ten example experiments from the set of 60 with the best results for the native waypoint controller showing the total distance travelled, terrain type, and the relative position error for both our trajectory prediction controller and the naive waypoint controller. In all cases, it can be seen that our trajectory prediction controller results in less error over the total path.

Experiment index	Configuration		Relative percentage error	
	Distance	Terrain	DCNN trajectory prediction	Naive waypoint controller
1	714 m	Coral region	3.6%	12.8%
2	2156 m	Coral region	7.6%	18.1%
3	1703 m	Sand region	11.2%	30.4%
4	1442 m	Seagrass region	6.9%	11.7%
5	1756 m	Sand region	7.5%	17.3%
6	815 m	Seagrass region	14.6%	22.1%
7	1677 m	Sand region	9.1%	16.2%
8	561 m	Sand region	5.3%	17.9%
9	2329 m	Seagrass region	6.3%	12.4%
10	1503 m	Coral region	3.7%	6.8%

(orange whisker) for each environment. It can be seen that in all three cases, on average, the trajectory prediction controller outperformed the naive waypoint controller.

Finally, Table 2 tabularizes the 10 experiments with the lowest relative percentage error in the path using the naive waypoint controller. Note that in all cases, our trajectory prediction controller resulted in less position error, demonstrating the robustness of our method.

6 Conclusion

This paper has presented a real-time method for robust underwater navigation that uses medium-to-long term planning to optimize task completion in the context of using visual odometry. This addresses the important but often under-appreciated issue of picking routes in the world where estimation and observability are optimized. We have validated our method on real-world data using an underwater vehicle in the open ocean and have performed extensive quantitative assessments by running the method in a photo-realistic simulator that simulates both the environment of the ocean and the robot physics. In that case, the use of our controller substantially improves localization and waypoint attainment. Our DCNN predicts a quality score for an image region that represents the performance of the VO at the corresponding position in the world with 93.13% accuracy. We simulated 162 km of autonomous swimming in several environments comparing our trajectory predictive controller against a naive waypoint controller. Our method outperformed all of the 10 best results using the naive controller by as much as 3.5 times.

In future work, we plan to conduct additional underwater sea trials in even more visually challenging and complex environments with the long-term goal of robustly building 3D models of underwater environments and performing autonomous surveys of coral reef health in close proximity. One of the primary challenges will be to identify a good source for ground-truth in these real-world conditions on our robot. We suspect that using simple sub-trajectories (such as constant forward translation), we can selectively simplify the correspondence problem between cameras to facilitate the acquisition of reliable training data.

References

1. Abadi, M., Barham, P., Chen, J., Chen, Z., Davis, A., Dean, J., Devin, M., Ghemawat, S., Irving, G., Isard, M., et al.: TensorFlow: a system for large-scale machine learning. In: OSDI, vol. 16, pp. 265–283 (2016)
2. Aloimonos, J.: Purposive and qualitative active vision. In: 1990 Proceedings of 10th International Conference on Pattern Recognition, vol. 1, pp. 346–360. IEEE (1990)
3. Azrad, S., Kendoul, F., Nonami, K.: Visual servoing of quadrotor micro-air vehicle using color-based tracking algorithm. J. Syst. Des. Dyn. **4**(2), 255–268 (2010)
4. Banfi, J., Li, A.Q., Basilico, N., Rekleitis, I., Amigoni, F., et al.: Multirobot online construction of communication maps. In: Proceedings of the International Conference on Robotics and Automation (2017)
5. Bourgault, F., Makarenko, A., Williams, S., Grocholsky, B., Durrant-Whyte, H.: Information based adaptive robotic exploration. In: Proceedings of the IEEE/RSJ International Conference on Intelligent Robots and Systems (IROS) (2002)
6. Carrillo, H., Reid, I., Castellanos, J.A.: On the comparison of uncertainty criteria for active SLAM. In: IEEE International Conference on Robotics and Automation (ICRA), pp. 2080–2087. IEEE (2012)
7. Chaves, S.M., Kim, A., Eustice, R.M.: Opportunistic sampling-based planning for active visual SLAM. In: IEEE/RSJ International Conference on Intelligent Robots and Systems (IROS 2014), pp. 3073–3080. IEEE (2014)
8. Chen, Z., Birchfield, S.T.: Qualitative vision-based path following. IEEE Trans. Robot. **25**(3), 749–754 (2009)
9. Davison, A.J., Murray, D.W.: Simultaneous localization and map-building using active vision. IEEE Trans. Pattern Anal. Mach. Intell. **24**(7), 865–880 (2002)
10. Dudek, G., Jenkin, M., Prahacs, C., Hogue, A., Sattar, J., Giguere, P., German, A., Liu, H., Saunderson, S., Ripsman, A., Simhon, S., Torres-Mendez, L.A., Milios, E., Zhang, P., Rekleitis, I.: A visually guided swimming robot. In: IEEE/RSJ International Conference on Intelligent Robots and Systems (IROS), Edmonton, AB, Canada, pp. 1749–1754 (2005)
11. Engel, J., Koltun, V., Cremers, D.: Direct sparse odometry. arXiv:1607.02565 (2016)
12. Fox, D., Burgard, W., Thrun, S.: Active Markov localization for mobile robots. Robot. Auton. Syst. **25**(3–4), 195–207 (1998)
13. Frintrop, S., Jensfelt, P.: Attentional landmarks and active gaze control for visual SLAM. IEEE Trans. Robot. **24**(5), 1054–1065 (2008)

14. Gal, Y., Hron, J., Kendall, A.: Concrete dropout. In: Advances in Neural Information Processing Systems 30 (NIPS) (2017)
15. Giguere, P., Girdhar, Y., Dudek, G.: Wide-speed autopilot system for a swimming hexapod robot. In: Canadian Conference on Computer and Robot Vision (CRV) (2013)
16. Gurau, C., Rao, D., Tong, C.H., Posner, I.: Learn from experience: probabilistic prediction of perception performance to avoid failure. Int. J. Robot. Res. **37**, 981–995 (2017)
17. He, K., Zhang, X., Ren, S., Sun, J.: Deep residual learning for image recognition. In: Proceedings of the IEEE Conference on Computer Vision and Pattern Recognition, pp. 770–778 (2016)
18. He, K., Zhang, X., Ren, S., Sun, J.: Identity mappings in deep residual networks. In: Leibe, B., Matas, J., Sebe, N., Welling, M. (eds.) European Conference on Computer Vision, pp. 630–645. Springer, Cham (2016)
19. Hidalgo, F., Bräunl, T.: Review of underwater SLAM techniques. In: 6th International Conference on Automation, Robotics and Applications (ICARA), pp. 306–311. IEEE (2015)
20. Jensfelt, P., Kristensen, S.: Active global localization for a mobile robot using multiple hypothesis tracking. IEEE Trans. Robot. Autom. **17**(5), 748–760 (2001)
21. Johnson-Roberson, M., Pizarro, O., Williams, S.B., Mahon, I.: Generation and visualization of large-scale three-dimensional reconstructions from underwater robotic surveys. J. Field Robot. **27**(1), 21–51 (2010)
22. Juliá, M., Gil, A., Reinoso, O.: A comparison of path planning strategies for autonomous exploration and mapping of unknown environments. Auton. Robots **33**(4), 427–444 (2012)
23. Koreitem, K., Li, J., Karp, I., Manderson, T., Shkurti, F., Dudek, G.: Synthetically trained 3D visual tracker of underwater vehicles. In: MTS/IEEE OCEANS, Charleston, SC, USA (2018)
24. Manderson, T., Dudek, G.: GPU-assisted learning on an autonomous marine robot for vision based navigation and image understanding. In: MTS/IEEE OCEANS, Charleston, SC, USA (2018)
25. Manderson, T., Gamboa Higuera, J., Cheng, R., Dudek, G.: Vision-based autonomous underwater swimming in dense coral for combined collision avoidance and target selection. In: Proceedings of the IEEE/RSJ International Conference on Intelligent Robots and Systems (IROS) (2018)
26. Manderson, T., Holliday, A., Dudek, G.: Gaze selection for enhanced visual odometry during navigation. In: Proceedings of the Conference on Computer and Robot Vision (CRV) (2018)
27. Manderson, T., Li, J., Cortés Poza, D., Dudek, N., Meger, D., Dudek, G.: Towards autonomous robotic coral reef health assessment. In: Wettergreen, D., Barfoot, T. (eds.) Field and Service Robotics, pp. 95–108. Springer, Cham (2016)
28. Manderson, T., Shkurti, F., Dudek, G.: Texture-aware SLAM using stereo imagery and inertial information. In: Computer and Robot Vision. IEEE (2016)
29. Meger, D., Gupta, A., Little, J.J.: Viewpoint detection models for sequential embodied object category recognition. In: Proceedings of the International Conference on Robotics and Automation (ICRA) (2010)
30. Meger, D., Shkurti, F., Poza, D.C., Giguère, P., Dudek, G.: 3D trajectory synthesis and control for a legged swimming robot. In: Proceedings of the IEEE/RSJ International Conference on Intelligent Robots and Systems (2014)

31. Mihaylova, L., Lefebvre, T., Bruyninckx, H., Gadeyne, K., De Schutter, J.: A comparison of decision making criteria and optimization methods for active robotic sensing. In: Dimov, I., Lirkov, I., Margenov, S., Zlatev, Z. (eds.) International Conference on Numerical Methods and Applications, pp. 316–324. Springer, Heidelberg (2002)
32. Ribas, D., Ridao, P., Tardós, J.D., Neira, J.: Underwater SLAM in man-made structured environments. J. Field Robot. **25**(11–12), 898–921 (2008)
33. Roy, N., Burgard, W., Fox, D., Thrun, S.: Coastal navigation-mobile robot navigation with uncertainty in dynamic environments. In: 1999 Proceedings of IEEE International Conference on Robotics and Automation, vol. 1, pp. 35–40. IEEE (1999)
34. Roy, N., Gordon, G., Thrun, S.: Finding approximate POMDP solutions through belief compression. J. Artif. Intell. Res. **23**, 1–40 (2005)
35. Sáez, J.M., Hogue, A., Escolano, F., Jenkin, M.: Underwater 3D SLAM through entropy minimization. In: 2006 Proceedings of IEEE International Conference on Robotics and Automation, ICRA 2006, pp. 3562–3567. IEEE (2006)
36. Sim, R., Dudek, G.: Effective exploration strategies for the construction of visual maps. In: Proceedings of the IEEE/RSJ International Conference on Intelligent Robots and Systems (IROS) (2003)
37. Srivastava, N., Hinton, G., Krizhevsky, A., Sutskever, I., Salakhutdinov, R.: Dropout: a simple way to prevent neural networks from overfitting. J. Mach. Learn. Res. **15**(1), 1929–1958 (2014)
38. Stachniss, C., Grisetti, G., Burgard, W.: Information gain-based exploration using Rao-Blackwellized particle filters. In: Robotics: Science and Systems, vol. 2, pp. 65–72 (2005)

A Unified 3D Mapping Framework Using a 3D or 2D LiDAR

Weikun Zhen(✉) and Sebastian Scherer

Carnegie Mellon University, Pittsburgh, USA
{weikunz,basti}@andrew.cmu.edu

Abstract. Simultaneous Localization and Mapping (SLAM) has been considered as a solved problem thanks to the progress made in the past few years. However, the great majority of LiDAR-based SLAM algorithms are designed for a specific type of payload and therefore don't generalize across different platforms. In practice, this drawback causes the development, deployment and maintenance of an algorithm difficult. Consequently, our work focuses on improving the compatibility across different sensing payloads. Specifically, we extend the Cartographer SLAM library to handle different types of LiDAR including fixed or rotating, 2D or 3D LiDARs. By replacing the localization module of Cartographer and maintaining the sparse pose graph (SPG), the proposed framework can create high-quality 3D maps in real-time on different sensing payloads. Additionally, it brings the benefit of simplicity with only a few parameters need to be adjusted for each sensor type.

1 Introduction

As one of the most fundamental problems of autonomous robots, LiDAR-based SLAM has been an active research area for many years. Recent advancements in LiDAR sensing and SLAM techniques have led to the fast growth of robot applications in many industrial fields such as autonomous inspection of civil engineering facilities. Usually, specialized systems are developed according to the particular requirements in different situations. For example, different Unmanned Aerial Vehicles (UAVs) (see Fig. 1) are used to inspect tunnels in our project depending on the scale of the operating environment. The robots are equipped with different LiDARs and typically different SLAM algorithms would be required. However, maintaining multiple algorithms needs significant effort which is especially undesirable in the field. Consequently, in this work we propose a *unified* mapping framework that handles multiple types of LiDARs including (1) a fixed 3D LiDAR, (2) a rotating 3D/2D LiDAR.

Electronic supplementary material The online version of this chapter (https://doi.org/10.1007/978-3-030-33950-0_60) contains supplementary material, which is available to authorized users.

J. Xiao et al. (Eds.): ISER 2018, SPAR 11, pp. 702–711, 2020.
https://doi.org/10.1007/978-3-030-33950-0_60

Fig. 1. *Left*: A customized DJI M100 drone carrying a rotating Hokuyo laser scanner. *Right*: A customized DJI M600 drone equipped with a rotating VLP-16.

The problem can be defined as follows: given a sequence of laser scans collected from a LiDAR sensor of any type, the algorithm will compute the motion of the sensor and build a 3D map in the meantime. As stated before, our work is based on the Cartographer SLAM (Hess et al. 2016) which contains a foreground localization component and a background SPG refinement component. Originally designed to work with stationary 3D LiDARs, it doesn't generalize to rotating 2D LiDARs since directly accumulating 2D scans using the IMU in the localization component will introduce distortion to the map. To accommodate this problem, we apply a different localization method that has two major advantages. First, every single scan is matched to the map to compute a more accurate pose than pure IMU-based methods. Second, the pose is computed regardless of the LiDAR types, allowing the framework to be generalizable across different platforms. With a unified framework, identical parameter-tuning strategies can be shared between systems which significantly simplifies the set-up procedure of multiple platforms during field tests. Additionally, we show that only a few parameters need to be adjusted when switching platforms such as local map resolution, number of accumulated scan and so on. More details will be discussed in the experiments.

The rest of this paper is structured as follows. Section 2 summarizes the related work on LiDAR-based SLAM problem. The proposed method is described in detail in Sect. 3. Experiments and results are presented in Sect. 4. Finally, conclusions and insights are discussed in Sect. 5.

2 Related Work

There has been a vast of research on LiDAR-based SLAM over past decades. Classic probabilistic approaches such as Kalman filters (Castellanos and Tardos 2012; Montemerlo et al. 2002) and particle filters (Dellaert et al. 1999; Doucet et al. 2000; Grisetti et al. 2007) infer the distribution of the robot state and the map based on measurements which are characterized by sensor noise models. Thrun et al. (2002) does a comprehensive review on the techniques. Those work establishes the theoretical fundamentals of the SLAM problem and has achieved great success in robustness, accuracy and efficiency. However, most of these approaches are limited to using fixed 2D LiDARs to solve the planar SLAM problem. Although in principle these algorithms are generalizable to 3D, the computational cost could become intractable as the dimension increases.

In 3D situations, 2D LiDARs may be mounted on a rotating motor (Bosse and Zlot 2009; Zlot and Bosse 2014; Zhang and Singh 2014) or a spring (Bosse et al. 2012) to build 3D maps. The additional degree of freedom significantly enlarges the sensor FOV, which, on the other hand, makes sequential scan matching impossible due to a lack of overlap. To account for this issue, a smooth continuous trajectory (Anderson and Barfoot 2013; Tong and Barfoot, 2013) may be used to represent robot motion instead of a set of pose nodes. However, the smooth motion assumption does not always hold true.

More recently, as 3D ranging technology becomes widely used, methods to achieve real-time, large-scale and low-drift SLAM have emerged using accurate 3D LiDARs. Martín et al. (2014) developed a Differential Evolution-based scan matching algorithm that is shown to be of high accuracy in three dimensional spaces and contains a loop-closure algorithm which relies on surface features and numerical features to encode properties of laser scans. Zhang and Singh (2014) extract edge and planar features from laser scans and then adopt an ICP method (Chen and Medioni 1992) for feature registration. An extension is presented in their later work (Zhang and Singh 2015) where visual data is fused with range data to further reduce drifts. Although they do not compute the loop-closure, the generated map is of high accuracy even after travelling for several kilometers. Hess et al. (2016) introduced the Cartographer SLAM where a local odometry relying on scan matching estimates the poses and meanwhile an SPG is updated and optimized regularly to refine pose estimates and generate consistent maps. Although existing methods vary in specific techniques, most share a similar pipeline, which estimates the pose using ICP or its variants as front-end while solves an SPG or trajectory optimization problem as the back-end.

3 Approach

3.1 Localization

The localization module (shown in Fig. 2) combines an Error State Kalman Filter (ESKF) with a Gaussian Particle Filter (GPF) to estimate robot states inside a prior map. The GPF, originally proposed by Bry et al. (2012), converts raw laser scans to a pose measurement, which frees the ESKF from handling 2D or 3D range data directly. This is a key factor that ensures compatibility. More specifically, the ESKF (illustrated in Fig. 3) numerically integrates IMU measurements to predict robot states and uses a pseudo pose measurement to update the prediction. In the GPF illustrated in Fig. 4, a set of particles (pose hypotheses) are sampled according to the prediction, then weighted, and finally averaged to find the posterior belief. By subtracting the prediction from the posterior belief, the pseudo pose measurement is recovered and used to update the ESKF. Finally, we refer the readers to our previous work (Zhen et al. 2017) for more details.

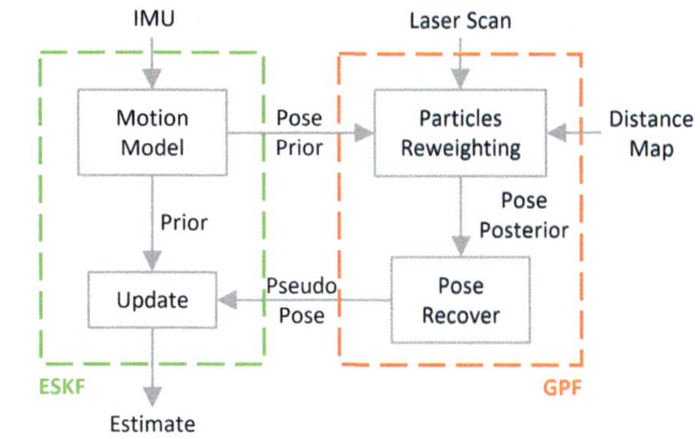

Fig. 2. Pipeline of the localization algorithm that takes the sensor readings and a distance map as input and outputs the pose estimate.

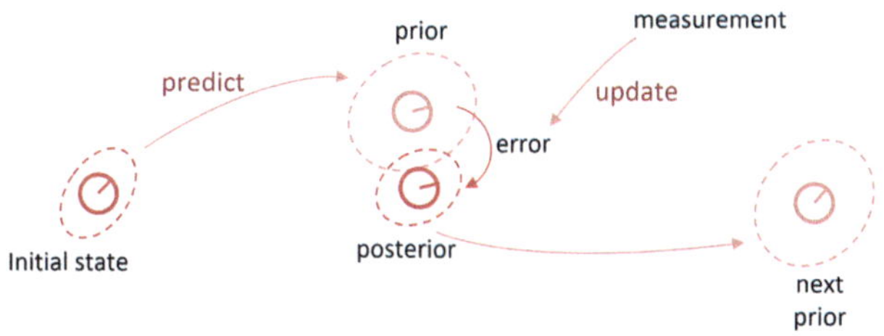

Fig. 3. An illustration of the ESKF in 2D. The circle denotes the robot with a bar indicating its orientation. The dashed ellipse represents the position uncertainty. Here the orientation uncertainty is omitted for simplicity.

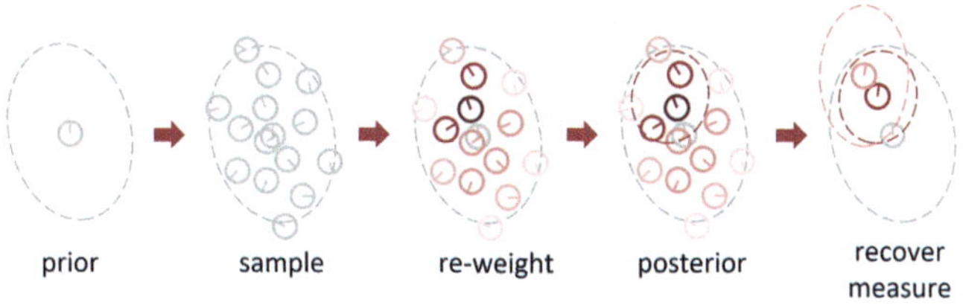

Fig. 4. An illustration of the GPF in 2D. Circles and ellipses share the same meaning as in Fig. 3. Differently, the darker color means a higher weight, namely higher probability of a hypothesis to be true.

3.2 Submaps and Distance Maps

A local occupancy grid map is defined as a submap by Cartographer. Since a different localization method is used, we need to adjust the submap management scheme so that the submaps can be accessed by the localization module. As shown in Fig. 5, there exist two stages of scan accumulation. On the first stage, N scans are accumulated to form a 3D scan, then matched and inserted into the active submaps. The active submaps include a matching submap (blue) and a

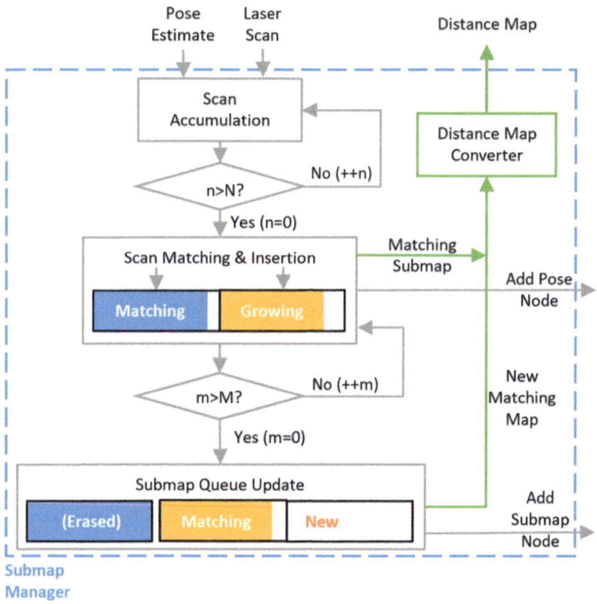

Fig. 5. The pipeline modified based on Cartographer to manage submaps and distance maps. Modifications are highlighted in green.

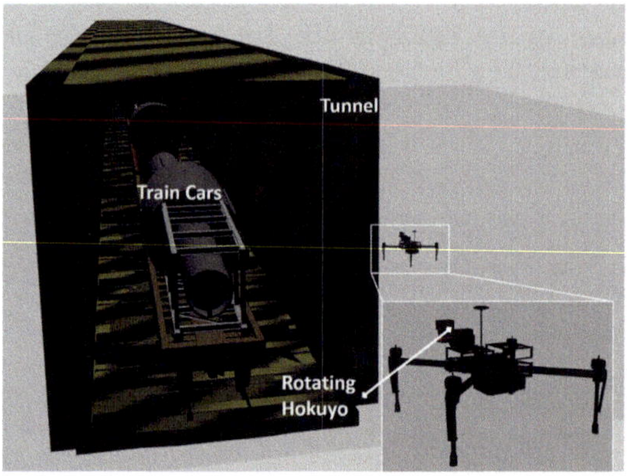

Fig. 6. The tunnel is of size $108\,\text{m} \times 5.8\,\text{m} \times 7\,\text{m}$ ($l \times w \times h$) and is built based on the DOE-PUREX nuclear tunnel wherein 8 train cars are loaded with radioactive waste. The simulated robot shares the identical sensing setup with the DJI M100. The rotating Hokuyo LiDAR is inserted from Gazebo and IMU measurements are generated by adding white noise and biases to the ground truth.

growing submap (yellow). The formed 3D scan is matched and inserted into both submaps. On the second stage, if M 3D scans are inserted into the matching submaps, the growing submap is switched to be the new matching submap and the old matching submap is erased. Meanwhile, a new submap (orange) is created and starts growing. During the two stages, whenever a 3D scan is formed or a new submap is created, new pose nodes are added to the SPG.

The adjustments are done by adding an octomap (Hornung et al. 2013) beside the original grid map. The formed 3D scan is inserted into the octomap and corresponding distance map is updated immediately. The octomap library provides an efficient method to detect changes so that the distance map can be computed efficiently. Additionally, updating octomap and distance map uses multi-threading techniques to avoid time delay caused by the distance map conversion.

4 Experiments

In this section, experiments conducted in simulation and real-world are discussed. For simulation, we focus on the rotating 2D LiDAR payload which is believed to be the most challenging case compared with other types of configurations. For the real-world tests, different configurations including stationary 3D LiDAR, rotating 2D/3D LiDAR are used.

4.1 Simulated Tunnel Test

The main task of the simulated robot is to fly through a train car tunnel (see Fig. 6) and map its interior. The fly-through takes about 15 min since we keep a relatively low velocity (1.13 m/s at maximum) and frequently dwell so that enough laser points are collected to build submaps. Moving too fast will result in unstable localization since the submap is not well observed. This issue can be addressed using a 3D LiDAR which quickly scans many points from the environment.

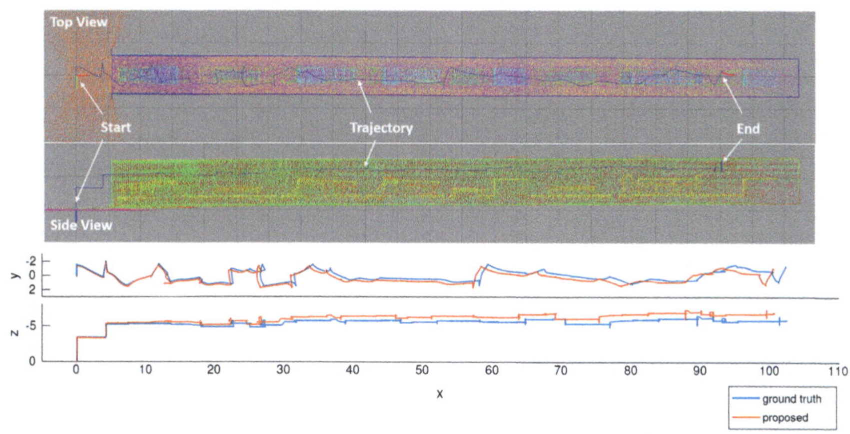

Fig. 7. *Up:* Built tunnel maps along the flight through test. *Down:* Comparison of the ground truth (blue) and the estimated trajectory (red).

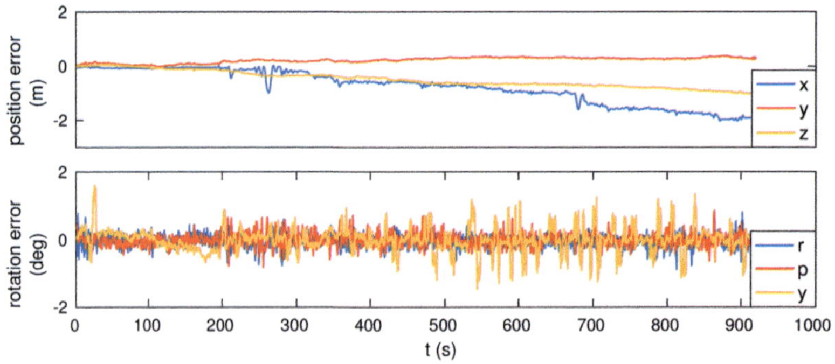

Fig. 8. A plot of pose estimation errors. The rotation error is computed by $e_{\text{rotation}} = \log\left(q_{\text{estimated}}^{-1} \cdot q_{\text{groundtruth}}\right)$, where $q \in S(3)$ is a unit quaternion. The $\log(\cdot)$ function maps a unit quaternion to an angle axis vector in $so(3)$.

The built map is visualized in Fig. 7 (voxel filtered with resolution 0.1 m). In simulation, we are able to compare the estimated poses with the ground truth. From Fig. 8, the maximum position error in three axes is observed to be 2.0 m, 0.37 m and 1.11 m near the end of the flight. Particularly, drift along x-axis is the largest, which is because the number of points on train cars to estimate x is relatively small than that on side walls or ceiling to estimate y and z. In other words, x-axis is under-constrained. The total traversed distance is 165 m and the translational drift rate is 1.34%. The rotation estimation, differently, is more consistent and the averaged error in roll, pitch and yaw are 0.14°, 0.15°, 0.24°. There are error peaks in yaw due to occasional low quality scan matching but can be quickly recovered. The rotational drift rate is 0.003°/m.

4.2 Real World Test

Real-world experiments are carried out on multiple platforms: (1) fixed VLP-16 (10 Hz, range 100 m) with an i7 (2.5 Ghz) computer, (2) rotating VLP-16, (3) rotating Hokuyo (40 Hz, range 30 m) with a 2.32 GHz ARM processor.

The first experiment is conducted inside a corridor loop (see Fig. 9 left). In this test, the VLP-16 is put horizontally and the rotating speed is set to be zero so that the LiDAR is stationary. We found that although the VLP-16 measures 3D structures, its 20° FOV is still not enough to reliably estimate height. The main reason is that inside the narrow corridor, most laser points come from side walls instead of the ground and ceiling. As a result, a larger drift in height is observed when the robot revisits the origin and more time is needed to detect the loop-closure.

The second test is carried out around the patio on the CMU campus. Again the VLP-16 is used and the motor rotating speed is set to be 30 rpm. Since this is a larger area, the distance map used for localization has a coarser resolution of 0.3 m and is constrained within a 40 m × 40 m × 40 m bounding box.

Fig. 9. Three tests are conducted in indoor and outdoor environments. *Left*: Test with a fixed 3D LiDAR in a hallway loop. *Middle*: Test with a rotating 3D LiDAR around the CMU patio. *Right*: Test with a rotating 2D LiDAR in a corridor loop.

Table 1. Parameters that need to be changed when switching platforms.

Localization related Params	Mapping related Params
Accelerator noise σ_a	# of scans per accumulation N
Gyroscope noise σ_g	# of scans per submap M
Distance map resolution	# of scans per optimization I
Max range of distance map	

In the last test, the robot maps a narrow hallway with 1.1 m width at minimum. To ensure enough laser points are collected, the robot is manually carried and moved slowly (≈0.5 m/s). This time only small drifts in height is observed before closing the loop. This is because by rotating the LiDAR, the robot obtains wider FOV, which could significantly improve the mapping performance.

It is important to point out that only a few parameters (listed in Table 1) are changed in the above 3 cases. For localization related parameters, σ_a and σ_g characterize the noise level of IMU. Distance map resolution are chosen according to the scale of environment. The maximum range of distance map sets a limit on the distance the robot will see. For mapping related parameters, M and N are as described in Fig. 5 and I governs how often the background SPG got optimized.

5 Conclusions

In this paper, the proposed algorithm is shown to allow different LiDAR configurations to be handled in a unified framework with only a few parameters need to be tuned, which simplifies the development and application process. Some key insights obtained from the experiments are:

- The FOV of a LiDAR matters. A fixed 3D LiDAR is simple to set up but has quite a limited vertical FOV, which results in unreliable height estimation. In our experiments, the LiDAR has to be pitched down to capture more ground points. A rotating LiDAR has significantly wider FOV and is observed to be more robust to different environments. However, the rotating motor has to be carefully designed to ensure continuous data streaming and accurate synchronization. For example, an expensive slip-ring mechanism is needed to achieve continuous rotation.

- Moving speed is critical in the case of rotating 2D LiDAR. In our tests, a low speed is necessary so that the laser scanner can accumulate enough points to update a submap. Moving too fast may lead to unstable localization. In the case of 3D LiDAR, a low moving speed is not a crucial requirement.
- The choice of submap resolution will affect memory usage, computational complexity and mapping accuracy. From our experience, low resolution submap has low memory usage and is faster to query data from. However, that will sacrifice the map accuracy. On the other hand, higher resolution consumes more memory but doesn't necessarily improve the map accuracy. Therefore, the resolution has to be chosen carefully through trial and error.

Acknowledgement. The authors are grateful to Sam Zeng, Yunfeng Ai and Xiangrui Tian for helping with the UAV development and the mapping experiments, to Matthew Hanczor and Alexander Baikovitz for building the DOE-PUREX tunnel model. This work is supported by the Department of Energy under award number DE-EM0004478.

References

Anderson, S., Barfoot, T.D.: Towards relative continuous-time slam. In: 2013 IEEE International Conference on Robotics and Automation (ICRA), pp. 1033–1040. IEEE (2013)

Bosse, M., Zlot, R.: Continuous 3d scan-matching with a spinning 2d laser. In: IEEE International Conference on Robotics and Automation, ICRA 2009, pp. 4312–4319. IEEE (2009)

Bosse, M., Zlot, R., Flick, P.: Zebedee: design of a spring-mounted 3-d range sensor with application to mobile mapping. IEEE Trans. Robot. **28**(5), 1104–1119 (2012)

Bry, A., Bachrach, A., Roy, N.: State estimation for aggressive flight in GPS-denied environments using onboard sensing. In: 2012 IEEE International Conference on Robotics and Automation (ICRA), pp. 1–8. IEEE (2012)

Castellanos, J.A., Tardos, J.D.: Mobile Robot Localization and Map Building: A Multisensor Fusion Approach. Springer Science & Business Media, New York (2012)

Chen, Y., Medioni, G.: Object modelling by registration of multiple range images. Image Vis. Comput. **10**(3), 145–155 (1992)

Dellaert, F., Fox, D., Burgard, W., Thrun, S.: Monte Carlo localization for mobile robots. In: Proceedings of 1999 IEEE International Conference on Robotics and Automation, vol. 2, pp. 1322–1328. IEEE (1999)

Doucet, A., De Freitas, N., Murphy, K., Russell, S.: Rao-Blackwellised particle filtering for dynamic Bayesian networks. In: Proceedings of the Sixteenth Conference on Uncertainty in Artificial Intelligence, pp. 176–183. Morgan Kaufmann Publishers Inc. (2000)

Grisetti, G., Stachniss, C., Burgard, W.: Improved techniques for grid mapping with Rao-Blackwellized particle filters. IEEE Trans. Robot. **23**(1), 34–46 (2007)

Hess, W., Kohler, D., Rapp, H., Andor, D.: Real-time loop closure in 2D LiDAR SLAM. In: 2016 IEEE International Conference on Robotics and Automation (ICRA), Stockholm, Sweden. IEEE (2016)

Hornung, A., Wurm, K.M., Bennewitz, M., Stachniss, C., Burgard, W.: OctoMap: an efficient probabilistic 3d mapping framework based on octrees. Auton. Robots **34**(3), 189–206 (2013)

Martín, F., Triebel, R., Moreno, L., Siegwart, R.: Two different tools for three-dimensional mapping: DE-based scan matching and feature-based loop detection. Robotica **32**(1), 19–41 (2014)

Montemerlo, M., Thrun, S., Koller, D., Wegbreit, B., et al.: FastSLAM: a factored solution to the simultaneous localization and mapping problem. In: AAAI/IAAI, pp. 593–598 (2002)

Thrun, S., et al.: Robotic mapping: a survey. In: Exploring Artificial Intelligence in the New Millennium, vol. 1, pp. 1–35 (2002)

Tong, C.H., Barfoot, T.D.: Gaussian process Gauss-Newton for 3d laser-based visual odometry. In: 2013 IEEE International Conference on Robotics and Automation (ICRA), pp. 5204–5211. IEEE (2013)

Zhang, J., Singh, S.: LOAM: LiDAR odometry and mapping in real-time. In: Robotics: Science and Systems Conference (RSS), pp. 109–111 (2014)

Zhang, J., Singh, S.: Visual-lidar odometry and mapping: low-drift, robust, and fast. In: 2015 IEEE International Conference on Robotics and Automation (ICRA), IEEE, pp. 2174–2181 (2015)

Zhen, W., Zeng, S., Scherer, S.: Robust localization and localizability estimation with a rotating laser scanner. In: 2017 IEEE International Conference on Robotics and Automation (ICRA), Singapore, Singapore. IEEE (2017)

Zlot, R., Bosse, M.: Efficient large-scale 3d mobile mapping and surface reconstruction of an underground mine. In: Field and Service Robotics, pp. 479–493. Springer (2014)

Efficient Alignment of Visual-Inertial Maps

Kourosh Sartipi(✉) and Stergios I. Roumeliotis(✉)

Google Inc., Mountain View, CA 94043, USA
{sartipi,stergiosr}@google.com

Abstract. In this paper, we address the problem of concurrently computing the transformation between multiple, gravity-aligned maps given common point feature observations. In particular, we formulate the problem as a minimization of the distances between shared features expressed with respect to different maps. We show that by marginalizing the maps' relative positions, the KKT conditions of the resulting minimization problem correspond to a system of multivariate polynomial equations (in the sin and cos of the relative yaw angles) which can be solved analytically for the case of a few maps. Furthermore, and to improve the efficiency when considering numerous maps, we present a fast, iterative process for determining the unknown relative map orientations.

1 Introduction

Creating an accurate 3D map within a GPS-denied area is required in a wide range of applications such as virtual and augmented reality. A camera and an inertial measurement unit (IMU) sensor pair is ideal for such scenarios because of their availability in most mobile devices. Their low computational resources, however, limit the size of the maps that one can create in real time using visual and inertial measurements. For this reason, many recent approaches to simultaneous localization and mapping (SLAM) employ a fast front-end based on a filtering algorithm (*e.g.*, [21]) while mapping is usually performed either offline or at a lower rate on a server.[1]

In order to reduce the dependence on pre-computed maps or the existence of a cloud infrastructure, recent work has focused on performing on-device mapping. One such popular method is ORB-SLAM [13], where the front-end tracks the camera position and orientation (pose) in real-time, while the back-end computes the local map and processes loop closures. To limit the time for optimizing over a large map, [15] divides it into multiple overlapping submaps, where each

[1] For a review of methods employing offline maps or maps from a remote server, please see [3, 7] and references therein.

Electronic supplementary material The online version of this chapter (https://doi.org/10.1007/978-3-030-33950-0_61) contains supplementary material, which is available to authorized users.

J. Xiao et al. (Eds.): ISER 2018, SPAR 11, pp. 712–724, 2020.
https://doi.org/10.1007/978-3-030-33950-0_61

of them is optimized independently, and then aligned by iteratively optimizing only the variables involved in the inter-map measurements. In the aforementioned methods, new loop-closures could cause updates propagating through the entire system and introducing delays to the map optimization as the state size increases. These limitations motivate multi-map approaches which bound the time for creating each map.

Early such approaches (*e.g.*, [20]) employ variants of the extended Kalman filter (EKF) to estimate the device's pose and a local map of the surrounding features, while maintaining a global map containing the history of observed features. These methods, however, have quadratic, in the number of features, memory requirements due to storing the covariance matrix for the global map, rendering them unsuitable for mobile devices. To reduce the memory and processing cost, other approaches, such as [16] only update the local map in the exploration phase and postpone applying the loop-closure corrections between the maps. These updates, however, could potentially propagate through all the maps, delaying the response of the system.

Alternatively, Cunningham *et al.* [6] employ a smoothing approach to take advantage of the sparse structure of the problem in a multi-device setting, where they jointly optimize the common features and the transformations between the maps. As new measurements are processed, however, it is often required to update all the maps, which leads to significant delays. In [7], although each map is optimized independently (and hence in parallel), its last map-merging step has cost cubic in the number of features *common* to the maps, thus limiting the size of the area that can be mapped on a mobile device. A different process was introduced in [10,12] where the concept of "anchor nodes" is employed to represent the transformations between different maps in a pose-graph setting. The system is then optimized incrementally for all the poses and map transforms, which is computationally expensive as it involves all the states in the system.

In [3], two distributed optimization schemes on pose graphs are proposed with different communication bandwidth requirements where each robot updates its states based on information from other robots, as well as its own observations. Employing this method for a centralized, multi-map setup on a mobile device is not desirable however, since all submaps require updates.

In the aforementioned approaches [6,7,10,12], new measurements could potentially cause the estimates of all maps to be updated, thus incurring a high processing cost. As it was shown in [15], however, the local estimates of the maps are typically unaffected by the map-merging and joint optimization process. Hence, given accurate maps and their common feature correspondences, one only needs to compute the *transformations* between the maps, which will allow aligning them.

Estimating the 6 degrees of freedom (d.o.f.) transformation between *only* two maps was first solved in closed-form by Horn [9]. This method was extended to *multiple* maps in [1], where a least squares problem for minimizing the distance between common map features was formulated, then the relative map positions were eliminated and the resulting cost function of only map orientations was minimized in an iterative manner. Other methods such as [11,19] employ a similar minimization approach by employing rotation matrices instead of unit quaternions to represent orientation. Alternatively, Chaudhury *et al.* [2] relax the non-linear, constrained least-squares minimization to a semidefinite program and show improved convergence properties. These algorithms, however, do not take advantage of the problem's structure for cases where the accelerometer measurements render the direction of gravity observable [8]. Specifically, the roll and pitch angles of all map frames are known, reducing the unknown degrees of freedom between relative maps to four (three for position and one for rotation about gravity). Furthermore, they do not incorporate the uncertainty of common feature constraints, which as we will show, improves accuracy. In summary, the main contributions of this paper are:

- We introduce a computationally efficient method for adjusting multiple gravity-aligned maps given common point feature constraints. Moreover, we incorporate the uncertainty in the feature estimates so as to improve accuracy, and introduce a relaxation to our approach to further reduce the processing cost.
- We validate the efficiency and accuracy of our algorithms quantitatively using room and building-scale datasets comprising multiple maps.

2 Technical Approach

Consider N gravity-aligned 3D feature maps with known feature correspondences and unknown transformations between the maps created through a process similar to the algorithm of [7]. Let $\mathcal{F}_{ij}$ be the set of common features between maps i and j, ${}^1\mathbf{p}_i$ and ${}^1_i\mathbf{C}$ the position and orientation of map i with respect to map 1, respectively, where ${}^1_i\mathbf{C} = \mathbf{C}(\mathbf{e}_3, {}^1\theta_i)$ with $\begin{bmatrix}\mathbf{e}_1 & \mathbf{e}_2 & \mathbf{e}_3\end{bmatrix} = \mathbf{I}_3$ in which $\mathbf{I}_3$ is the 3×3 identity matrix, and ${}^i\mathbf{f}_m$ the position of feature m in map i. We define the cost function to be minimized as

$$\mathcal{C} = \sum_{i,j=1}^{N} \sum_{\mathbf{f}_m \in \mathcal{F}_{ij}} ||{}^1\mathbf{p}_j - {}^1\mathbf{p}_i + {}^1_j\mathbf{C}{}^j\mathbf{f}_m - {}^1_i\mathbf{C}{}^i\mathbf{f}_m||^2_{\boldsymbol{\Omega}_{ijm}} \tag{1}$$

where $\mathcal{F}_{ii}$ is an empty set and $\boldsymbol{\Omega}_{ijm}$ is the covariance associated with each cost term.

We start by stating that at its minimum, the cost function $\mathcal{C}$ in (1) will satisfy $\frac{\partial \mathcal{C}}{\partial {}^1\mathbf{p}_k} = 0$, for $k = 2, \ldots, N$, which results in

$$\sum_{i=2}^{N} \mathbf{A}_{ki}{}^1\mathbf{p}_i - \sum_{i=2}^{N} \mathbf{B}_{ki}\mathbf{K}{}^1\mathbf{u}_i - {}^1\mathbf{a}_k = \mathbf{0} \tag{2}$$

where

$$\mathbf{A}_{ki} \triangleq \begin{cases} -\sum_{\mathbf{f}_m \in \mathcal{F}_{ik}} \boldsymbol{\Omega}^{-1}_{ikm} & k \neq i \\ \sum_{i=1}^{N} \sum_{\mathbf{f}_m \in \mathcal{F}_{ik}} \boldsymbol{\Omega}^{-1}_{ikm} & k = i \end{cases} \tag{3}$$

$$\mathbf{B}_{ki} \triangleq \begin{cases} \sum_{\mathbf{f}_m \in \mathcal{F}_{ik}} \boldsymbol{\Omega}_{ikm}^{-1} \mathbf{L}({}^i\mathbf{f}_m) & k \neq i \\ -\sum_{l=1}^{N} \sum_{\mathbf{f}_m \in \mathcal{F}_{lk}} \boldsymbol{\Omega}_{lkm}^{-1} \mathbf{L}({}^k\mathbf{f}_m) & k = i \end{cases} \tag{4}$$

$${}^1\mathbf{u}_i \triangleq \begin{bmatrix} \cos({}^1\theta_i) \\ \sin({}^1\theta_i) \end{bmatrix}, \quad \mathbf{K} \triangleq \begin{bmatrix} \mathbf{e}_1 \; \mathbf{e}_2 \end{bmatrix}, \quad \mathbf{L}(\begin{bmatrix} x \\ y \\ z \end{bmatrix}) \triangleq \begin{bmatrix} x & -y & 0 \\ y & x & 0 \\ 0 & 0 & z \end{bmatrix} \tag{5}$$

$${}^1\mathbf{a}_k \triangleq \sum_{\mathbf{f}_m \in \mathcal{F}_{1k}} \boldsymbol{\Omega}_{1km}^{-1} \, {}^1\mathbf{f}_m + \sum_{i=2}^{N} \mathbf{B}_{ki} \mathbf{e}_3. \tag{6}$$

Defining $\mathbf{A}$, $\mathbf{B}$ and $\mathbf{a}$ as the matrices and vector whose block elements are $\mathbf{A}_{ki}$, $\mathbf{B}_{ki}, \mathbf{K}$ and ${}^1\mathbf{a}_k$, respectively, it is easy to show that

$${}^1\mathbf{p}_k = {}^1\mathbf{t}_k + \sum_{i=2}^{N} \mathbf{T}_{ki} \, {}^1\mathbf{u}_i \tag{7}$$

where $\mathbf{T}_{ki}$ and ${}^1\mathbf{t}_k$ are 3×2 matrices and 3×1 vectors, respectively, and are defined as

$$\mathbf{A}^{-1}\mathbf{a} = \begin{bmatrix} {}^1\mathbf{t}_2 \\ \vdots \\ {}^1\mathbf{t}_N \end{bmatrix}, \quad \mathbf{A}^{-1}\mathbf{B} = \begin{bmatrix} \mathbf{T}_{22} & \dots & \mathbf{T}_{2N} \\ \vdots & \ddots & \vdots \\ \mathbf{T}_{N2} & \dots & \mathbf{T}_{NN} \end{bmatrix}. \tag{8}$$

Substituting (7) into (1) yields a cost function in the orientations *only*:

$$\mathcal{C} = \sum_{i,j=1}^{N} \sum_{\mathbf{f}_m \in \mathcal{F}_{ij}} ||\mathbf{r}_{ijm} + \sum_{k=2}^{N} \mathbf{Q}_{ijmk} \, {}^1\mathbf{u}_k||^2_{\boldsymbol{\Omega}_{ijm}} \tag{9}$$

with

$$\mathbf{r}_{ijm} \triangleq {}^1\mathbf{t}_j - {}^1\mathbf{t}_i + \mathbf{L}({}^j\mathbf{f}_m)(\mathbf{e}_3 + \delta_{1j}\mathbf{e}_1) - \mathbf{L}({}^i\mathbf{f}_m)(\mathbf{e}_3 + \delta_{1i}\mathbf{e}_1) \tag{10}$$

$$\mathbf{Q}_{ijmk} \triangleq \mathbf{T}_{jk} - \mathbf{T}_{ik} + \delta_{jk}\mathbf{L}({}^j\mathbf{f}_m)\mathbf{K} - \delta_{ik}\mathbf{L}({}^i\mathbf{f}_m)\mathbf{K} \tag{11}$$

where δ_{ij} is the Kronecker delta and we defined ${}^1\mathbf{t}_1 \triangleq \mathbf{0}_{3\times 1}$ and $\mathbf{T}_{1k} \triangleq \mathbf{0}_{3\times 2}$. Furthermore, expanding (9) will result in an expression of the form:

$$\mathcal{C} = \mathcal{C}_0 + \bar{\mathbf{v}}^T \bar{\mathbf{u}} + \bar{\mathbf{u}}^T \bar{\mathbf{W}} \bar{\mathbf{u}} \tag{12}$$

where $\bar{\mathbf{u}} \triangleq \begin{bmatrix} {}^1\mathbf{u}_2^T \dots {}^1\mathbf{u}_N^T \end{bmatrix}^T$, $\mathcal{C}_0 \triangleq \sum_{i,j=1}^{N} \sum_{\mathbf{f}_m \in \mathcal{F}_{ij}} \|\mathbf{r}_{ijm}\|^2_{\boldsymbol{\Omega}_{ijm}}$ and

$$\bar{\mathbf{v}} \triangleq \begin{bmatrix} \mathbf{v}_2 \\ \vdots \\ \mathbf{v}_N \end{bmatrix}, \quad \bar{\mathbf{W}} \triangleq \begin{bmatrix} \mathbf{W}_{22} & \dots & \mathbf{W}_{2N} \\ \vdots & \ddots & \vdots \\ \mathbf{W}_{N2} & \dots & \mathbf{W}_{NN} \end{bmatrix} \tag{13}$$

$$\mathbf{v}_k \triangleq 2 \sum_{i,j=1}^{N} \sum_{\mathbf{f}_m \in \mathcal{F}_{ij}} \mathbf{Q}_{ijmk}^T \boldsymbol{\Omega}_{ijm}^{-1} \mathbf{r}_{ijm} \tag{14}$$

$$\mathbf{W}_{kl} \triangleq \sum_{i,j=1}^{N} \sum_{\mathbf{f}_m \in \mathcal{F}_{ij}} \mathbf{Q}_{ijmk}^T \boldsymbol{\Omega}_{ijm}^{-1} \mathbf{Q}_{ijml}. \tag{15}$$

Note that (12) can be minimized *analytically* subject to the constraints ${}^1\mathbf{u}_k^T {}^1\mathbf{u}_k = 1, k = 2, \ldots, N$, at a cost exponential in the number of maps. Specifically, by defining $\nu_{2i-3} = \cos({}^1\theta_i)$ and $\nu_{2i-2} = \sin({}^1\theta_i)$, for $i = 2, \ldots, N$, it is straightforward to show that (12) becomes a quadratic with quadratic constraints of the form $\nu_{2i-3}^2 + \nu_{2i-2}^2 = 1, i = 2, \ldots, N$. Formulating the Lagrangian $\mathcal{L} = \mathcal{C} + \sum_{i=2}^{N} \omega_i(\nu_{2i-3}^2 + \nu_{2i-2}^2 - 1)$ of (12) and computing the KKT conditions will result into a set of quadratic equations whose solution has complexity exponential in the number of maps N [5].

Thus, following such an approach is of practical use for only a small number of maps. Alternatively, we employ the first-order Taylor series expansion, which results in $\bar{\mathbf{u}} \simeq \hat{\bar{\mathbf{u}}} + \mathbf{G}\tilde{\boldsymbol{\psi}}$ in which $\hat{\bar{\mathbf{u}}}$ is the evaluation of vector $\bar{\mathbf{u}}$ around the current estimate obtained analytically from pairwise map matches, and

$$\tilde{\boldsymbol{\psi}} \triangleq \begin{bmatrix} {}^1\tilde{\theta}_2 \\ \vdots \\ {}^1\tilde{\theta}_N \end{bmatrix}, \quad \mathbf{G} \triangleq \begin{bmatrix} \mathbf{J}\,{}^1\hat{\mathbf{u}}_2 & & & \\ & \mathbf{J}\,{}^1\hat{\mathbf{u}}_3 & & \\ & & \ddots & \\ & & & \mathbf{J}\,{}^1\hat{\mathbf{u}}_N \end{bmatrix}, \quad \mathbf{J} \triangleq \begin{bmatrix} 0 & -1 \\ 1 & 0 \end{bmatrix} \tag{16}$$

where ${}^1\tilde{\theta}_k = {}^1\theta_k - {}^1\hat{\theta}_k, k = 2, \ldots, N$, denotes the error between the true map's, ${}^1\theta_k$, and estimated, ${}^1\hat{\theta}_k$, orientations. Substituting (16) in (12) results in the error cost function:

$$\mathcal{C}' = \mathcal{C}_0' + (\bar{\mathbf{v}} + 2\bar{\mathbf{W}}\hat{\bar{\mathbf{u}}})^T \mathbf{G}\tilde{\boldsymbol{\psi}} + \tilde{\boldsymbol{\psi}}^T \mathbf{G}^T \bar{\mathbf{W}} \mathbf{G} \tilde{\boldsymbol{\psi}} \tag{17}$$

in which $\mathcal{C}_0' = \mathcal{C}_0 + \mathbf{v}^T \hat{\bar{\mathbf{u}}} + \hat{\bar{\mathbf{u}}}^T \bar{\mathbf{W}} \hat{\bar{\mathbf{u}}}$ is constant. To minimize (17), we set $\frac{\partial \mathcal{C}'}{\partial \tilde{\boldsymbol{\psi}}} = \mathbf{0}$, then

$$\mathbf{G}^T \bar{\mathbf{W}} \mathbf{G} \tilde{\boldsymbol{\psi}} = -\mathbf{G}^T \left(\bar{\mathbf{W}} \hat{\bar{\mathbf{u}}} + \frac{1}{2} \bar{\mathbf{v}} \right). \tag{18}$$

After finding $\tilde{\boldsymbol{\psi}}$ from (18) and updating the current estimates of the map orientations ${}^1\theta_i^{\oplus} = {}^1\theta_i^{\ominus} + {}^1\tilde{\theta}_i, i = 2, \ldots, N$, we compute the map positions from (7) and initiate a new iteration until convergence ($||\tilde{\boldsymbol{\psi}}||/(N-1) \leq 10^{-5}$). Note that solving (18) requires $\mathcal{O}(N^3)$ operations, hence the overall algorithm's complexity is set by computing the $\bar{\mathbf{W}}$ matrix in (12), which is $\mathcal{O}(N^4 M)$ with $M \triangleq \max |\mathcal{F}_{ij}|$. The matrix $\bar{\mathbf{W}}$ and vector $\bar{\mathbf{v}}$, however, depend only on feature positions and term covariances, and hence need to be computed only *once* in the iterative minimization process.

Note that in the special case where all the covariances are equal to the same scalar times identity (*i.e.*, $\boldsymbol{\Omega}_{ijm} = \sigma^2 \mathbf{I}_3$), we can simplify the algorithm further, achieving lower processing at the expense of accuracy. Specifically, the cost function to minimize becomes

$$\mathcal{C}' = \sum_{i,j=1}^{N} \sum_{\mathbf{f}_m \in \mathcal{F}_{ij}} \left|\left| {}^1\mathbf{p}_j - {}^1\mathbf{p}_i + {}^1_j\mathbf{C}^j\mathbf{f}_m - {}^1_i\mathbf{C}^i\mathbf{f}_m \right|\right|^2. \tag{19}$$

Following the same process as for minimizing (1), it is straightforward to show that

$${}^1\mathbf{p}_k = \sum_{l=2}^{N} \gamma_{kl}{}^1\mathbf{a}'_l - \sum_{i,l=2}^{N} \gamma_{kl}\, {}^1_i\mathbf{C}^i\mathbf{b}'_l \tag{20}$$

where

$${}^i\mathbf{a}'_k \triangleq \sum_{f_m \in \mathcal{F}_{ik}} {}^i\mathbf{f}_m, \quad {}^i\mathbf{b}'_k \triangleq \begin{cases} -{}^i\mathbf{a}_k & i \neq k \\ \sum_{j=1}^{N} {}^i\mathbf{a}_j & i = k \end{cases}, \quad \lambda_{ki} \triangleq \begin{cases} -|\mathcal{F}_{ki}| & k \neq i \\ \sum_{j=1}^{N} |\mathcal{F}_{jk}| & i = k \end{cases} \tag{21}$$

$$\boldsymbol{\Lambda} = \begin{bmatrix} \lambda_{22} & \dots & \lambda_{2N} \\ \vdots & \ddots & \vdots \\ \lambda_{N2} & \dots & \lambda_{NN} \end{bmatrix}, \quad \boldsymbol{\Lambda}^{-1} \triangleq \begin{bmatrix} \gamma_{22} & \cdots & \gamma_{2N} \\ \vdots & \ddots & \vdots \\ \gamma_{N2} & \cdots & \gamma_{NN} \end{bmatrix}. \tag{22}$$

Equation (20) can be further simplified by defining ${}^1\mathbf{t}'_k \triangleq \sum_{l=2}^{N} \gamma_{kl}{}^1\mathbf{a}'_l$ and ${}^i\mathbf{t}'_k \triangleq -\sum_{l=2}^{N} \gamma_{kl}{}^i\mathbf{b}'_l$:

$${}^1\mathbf{p}_k = {}^1\mathbf{t}'_k + \sum_{i=2}^{N} {}^1_i\mathbf{C}^i\mathbf{t}'_k. \tag{23}$$

Computing ${}^i\mathbf{a}'_k$ and ${}^i\mathbf{b}'_k$ requires $\mathcal{O}(N^2M)$ operations, while finding $\boldsymbol{\Lambda}^{-1}$ has $\mathcal{O}(N^3)$ cost. Note, however, that as compared to computing $\mathbf{A}_{ki}$ and $\mathbf{B}_{ki}$ in (2), which involve 3×3 matrix multiplications and additions, determining ${}^i\mathbf{a}'_k$ and ${}^i\mathbf{b}'_k$ involves only vector additions. Furthermore, calculating $\boldsymbol{\Lambda}^{-1}$ requires inverting a $(N-1)\times(N-1)$ matrix, compared to $\mathbf{A}^{-1}$ in (3) which is a $3(N-1)\times3(N-1)$ matrix.

Substituting (23) into (19) and rearranging terms results in the following cost function involving only orientations:

$$\mathcal{C}' = \sum_{i,j=1}^{N} \sum_{f_m \in \mathcal{F}_{ij}} ||\mathbf{r}'_{ijm} + \sum_{l=2}^{N} {}^1_l\mathbf{C}^l\mathbf{q}_{ijm}||^2 \tag{24}$$

where

$$\mathbf{r}'_{ijm} \triangleq {}^1\mathbf{t}'_j - {}^1\mathbf{t}'_i + \delta_{j1}{}^j\mathbf{f}_m - \delta_{i1}{}^i\mathbf{f}_m \tag{25}$$

$$ {}^{l}\mathbf{q}_{ijm} \triangleq {}^{l}\mathbf{t}'_{j} - {}^{l}\mathbf{t}'_{i} + \delta_{jl}{}^{j}\mathbf{f}_{m} - \delta_{il}{}^{i}\mathbf{f}_{m} \tag{26} $$

and we have defined ${}^{1}\mathbf{t}'_{1} \triangleq \mathbf{0}_{3\times 1}$. Note that the third element of the 3×1 vectors in (24) will not be affected by the rotation matrices since all rotations are around the z-axis. Hence, we can disregard the last element of the vectors $\mathbf{r}'_{ijm} + \sum_{l=2}^{N} {}^{1}_{l}\mathbf{C}^{l}\mathbf{q}_{ijm}$ in (24) to reduce computations. This is equivalent to defining $\bar{\mathbf{r}}_{ijm} \triangleq \mathbf{K}^{T}\mathbf{r}'_{ijm}$, ${}^{l}\bar{\mathbf{q}}_{ijm} \triangleq \mathbf{K}^{T l}\mathbf{q}_{ijm}$ and ${}^{1}_{l}\mathbf{R} \triangleq \mathbf{K}^{T}{}^{1}_{l}\mathbf{C}\mathbf{K}$ and minimizing the following modified cost function instead:

$$ \mathcal{C}'' = \sum_{i,j=1}^{N} \sum_{f_m \in \mathcal{F}_{ij}} ||\bar{\mathbf{r}}_{ijm} + \sum_{l=2}^{N} {}^{1}_{l}\mathbf{R}^{l}\bar{\mathbf{q}}_{ijm}||^2 \tag{27} $$

which can be written as

$$ \mathcal{C}'' = \mathcal{C}''_0 + \mathbf{g}^T\bar{\mathbf{u}} + \bar{\mathbf{u}}^T\bar{\mathbf{H}}\bar{\mathbf{u}} \tag{28} $$

where we have employed the identity ${}^{1}_{k}\mathbf{R}^{T}{}^{1}_{k}\mathbf{R} = \mathbf{I}_2$ and defined

$$ \mathcal{C}''_0 \triangleq \sum_{i,j} \sum_{f_m \in \mathcal{F}_{ij}} \left\{ ||\bar{\mathbf{r}}_{ijm}||^2 + \sum_{l=2}^{N} ||{}^{l}\bar{\mathbf{q}}_{ijm}||^2 \right\} \tag{29} $$

$$ \mathbf{v}'_l \triangleq 2\sum_{ij} \sum_{f_m \in \mathcal{F}_{ij}} \bar{\mathbf{Q}}^T_{ijml}\bar{\mathbf{r}}_{ijm}, \quad \bar{\mathbf{Q}}_{ijml} \triangleq \mathbf{K}^T\mathbf{L}({}^{l}\mathbf{q}_{ijm})\mathbf{K} \tag{30} $$

$$ \mathbf{H}_{kl} \triangleq \begin{cases} \sum_{ij}\sum_{f_m \in \mathcal{F}_{ij}} \bar{\mathbf{Q}}^T_{ijmk}\bar{\mathbf{Q}}_{ijml} & k \neq l \\ \mathbf{0}_{2\times 2} & k = l \end{cases} \tag{31} $$

$$ \mathbf{g} \triangleq \left[\mathbf{v}'^T_2 \dots \mathbf{v}'^T_N\right]^T, \quad \bar{\mathbf{H}} \triangleq \begin{bmatrix} \mathbf{H}_{22} & \dots & \mathbf{H}_{2N} \\ \vdots & \ddots & \vdots \\ \mathbf{H}_{N2} & \dots & \mathbf{H}_{NN} \end{bmatrix}. \tag{32} $$

Lastly, we employ an iterative approach by first linearizing $\bar{\mathbf{u}}$ around its current estimate, and obtain $\tilde{\boldsymbol{\psi}}$ from:

$$ \mathbf{G}^T\bar{\mathbf{H}}\mathbf{G}\tilde{\boldsymbol{\psi}} = -\mathbf{G}^T\left(\bar{\mathbf{H}}\hat{\bar{\mathbf{u}}} + \frac{1}{2}\mathbf{g}\right). \tag{33} $$

As before, the solution to (33) is used to update the orientation estimates, compute the positions, and initiate a new iteration until convergence.

3 Experimental Results

The visual and inertial data used in our tests were collected and processed online using two Google Pixel phones. Greyscale images of resolution 640×480 were saved at 30 Hz, along with consumer-grade, MEMS-based, IMU data at 200 Hz. We have divided our datasets into two categories: (i) Room-scale datasets with VICON ground-truth, and (ii) Building-scale datasets for which we have used the optimal Batch Least-Square (BLS) estimate as the ground-truth.

The visual and inertial data are first processed by the square root inverse filter [21], which operates on IMU data and feature tracks extracted from the images. The latter are produced by extracting FAST [17] corners and tracked by matching their corresponding ORB [18] descriptors in consecutive images. To detect loop-closures, we follow a bag-of-words approach using the ORB descriptors and employ the 3pt+1-RANSAC [14] for outlier-detection. As data are processed by the filter, feature tracks, loop-closures, IMU measurements, and the state estimates of the filter are passed to the BLS process.

Multiple maps are created by splitting temporally a dataset into smaller maps with equal temporal lengths. This process breaks the feature tracks spanning more than one map, but has the advantage that the common (among maps) features are known and no extra processing is required for determining them. We start by computing the BLS estimate of each map, and store the IMU-camera's trajectory and the feature positions, along with their covariances $\mathbf{P}_{im}$. The latter are approximated by inverting the 3×3 diagonal blocks of the Hessian matrix corresponding to each feature. Furthermore, we confirm the inliers among the maps' common feature pairs by passing them through the 2pt+1-RANSAC. Specifically, given the features ${}^{1}\mathbf{f}_1, {}^{1}\mathbf{f}_2$ and ${}^{2}\mathbf{f}_1, {}^{2}\mathbf{f}_2$, the minimal solver of the 2pt+1-RANSAC computes the relative yaw ${}^{1}\theta_2$ and position ${}^{1}\mathbf{p}_2$, by minimizing (1) analytically for the case of two maps.

To compute an *initial estimate* for the relative yaws, a naive solution would be to directly use the result of the 2pt+1 RANSAC in the outlier rejection step between maps 1 and $i, i = 2, \ldots, N$. There is, however, no guarantee that more than one common features exist between map 1 and every other map, hence this approach is prone to failure. Instead, we construct a graph where the nodes represent maps and the edges' weights are equal to the number of common features between the maps. We then compute the Maximum Spanning Tree (MST) of this graph following [4] which computes a tree with the maximum weights connecting *all* maps. Subsequently, for each map, we traverse the MST from the map's node to the root of the tree (*i.e.*, map 1), and compute the relative yaw by using only the relative transformation in the path to the root.

Given the inlier common feature constraints and an initial estimate for the map transformations, we only need to evaluate the common feature constraint covariance $\boldsymbol{\Omega}_{ijm}$. To do so, we assume that in each map i, the feature m's error state ${}^{i}\tilde{\mathbf{f}}_m$ is drawn from a zero-mean Gaussian distribution with covariance $\mathbf{P}_{im}$. Next, we define the noise of the cost term resulting from the observation of feature m by maps i and j as

$$\mathbf{n}_{ijm} = {}^1\mathbf{p}_j - {}^1\mathbf{p}_i + {}^1_j\mathbf{C}^j\mathbf{f}_m - {}^1_i\mathbf{C}^i\mathbf{f}_m \quad (34)$$

Where $\mathbf{n}_{ijm}$ is a zero mean Gaussian noise with covariance $\boldsymbol{\Omega}_{ijm}$. By considering the maps' transformations as (unknown) deterministic variables, $\mathbf{n}_{ijm} \simeq {}^1_i\mathbf{C}^i\tilde{\mathbf{f}}_m - {}^1_j\mathbf{C}^j\tilde{\mathbf{f}}_m$ and thus its covariance is $\boldsymbol{\Omega}_{ijm} = {}^1_i\mathbf{C}\mathbf{P}_{im}{}^1_i\mathbf{C}^T + {}^1_j\mathbf{C}\mathbf{P}_{jm}{}^1_j\mathbf{C}^T$. Note that although this approximate covariance is smaller (in the positive-definite sense) than the true covariance of the cost terms, our results show that using it significantly improves the accuracy.

Specifically, for our evaluations, we employ the proposed map-alignment process (see Sect. 2), to compute the map transformations and create a fused map along with the IMU-camera aligned trajectories. The latter are then compared to the ground truth (VICON for room-scale, BLS estimate for building-scale). Note that the errors between the fused trajectory and the ground truth are caused by (i) errors in the map transformations (*e.g.*, due to outliers or approximate covariances), and (ii) the map's local trajectory inaccuracies due to processing each map independently.

In this work, we used five datasets to evaluate our method in terms of accuracy and efficiency (see Table 1). For datasets L1 and L2, which are building size, we use as ground truth the optimal BLS solution. Datasets S1, S2, S3, on the other hand, are room size where the device is moved inside a VICON room while collecting data. A visualization of the fused trajectories against the ground truth for L1 and S1 is depicted in Figs. 1 and 2, respectively. Moreover, the position root mean square error (RMSE), the total length of the trajectory and the computation times are provided in Table 1 for the case of using covariances or without them (*i.e.*, $\boldsymbol{\Omega}_{ijm} = \sigma^2\mathbf{I}_3$), denoted by *WC* and *WOC*, respectively. To compare the accuracy and the processing time of our methods to alternative approaches that compute the 6 d.o.f. transformations between the maps, we have employed the Gauss-Newton algorithm to minimize a cost function similar to (1) where the rotations are no longer gravity-aligned. The Gauss-Newton approach is denoted by *GN* in Table 1. Additionally, we have provided the number of maps and maximum number of common features between maps. Comparing the RMSE values of GN and WC shows that our method converges to the optimal map-alignment solution.

As previously stated (Sect. 2), while our proposed methods' processing time is quartic in the number of maps, it only grows linearly in the number of common features. This is clearly shown when the number of maps grows by a factor of four, the time to compute the transformations grows by a factor of three hundred, while varying the number of common features has smaller impact. Note though, even in this case the total time is only a fraction of a second, which verifies the efficiency of our approach. Additionally, WC times are between 10%–60% faster than GN times. We have also included the times required to optimize the entire system in BLS (denoted by BLS Time), and the average time of solving each map individually (denoted by Ind. Time in Table 1) so as to highlight the advantage of using multiple maps in an online setting.

Besides the processing cost savings, Table 1 also demonstrates that our methods are, in general, more accurate than GN estimating the maps' 6 d.o.f. poses. This is due to the gravity-alignment of the maps, *i.e.*, the roll and pitch of the trajectory are accurately estimated in the visual-inertial BLS, hence estimating them in the multi-map alignment could lead to overfitting on the data and less accurate estimates. In some cases though, small errors in the roll and pitch of different maps, as in dataset S3, causes the RMSE of the 4 d.o.f. estimate to be slightly higher than that of the 6 d.o.f.

Table 1. Table of RMSE errors (cm) and the calculation times. We were unable to compute the BLS time of dataset L1 as the phone ran out of memory.

Dataset	WC-RMSE (cm)	WOC RMSE (cm)	GN-RMSE (cm)	WC-time (ms)	WOC-time (ms)	GN-time (ms)	Num. maps	Max common features	Dataset length (m)	BLS time (sec)	Ind. time (sec)
L1	**30.1**	38.1	**30.1**	371.8	**201.5**	515.2	21	276	752	–	6.59
L2	**28.3**	46.4	28.5	206.5	**133.9**	289.1	15	243	323	209	3.66
S1	**13.0**	14.6	13.1	0.7	**0.6**	1.2	5	18	112	82	25.5
S2	**5.0**	5.4	5.2	1.8	**0.9**	3.2	5	70	104	123	34.7
S3	8.0	10.0	**7.7**	1.2	**0.7**	1.9	5	28	104	102	18.9

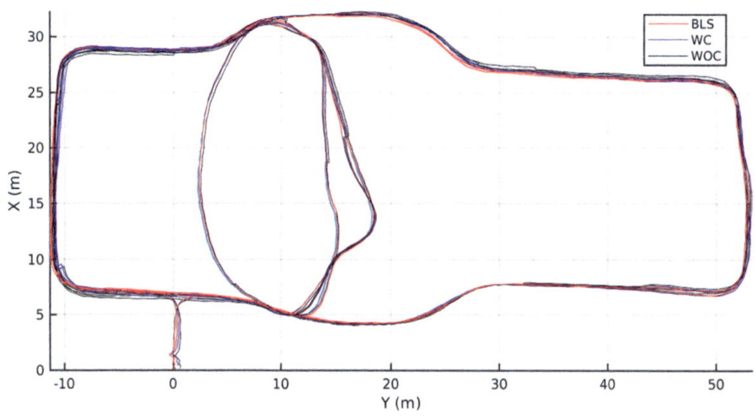

Fig. 1. Dataset L1

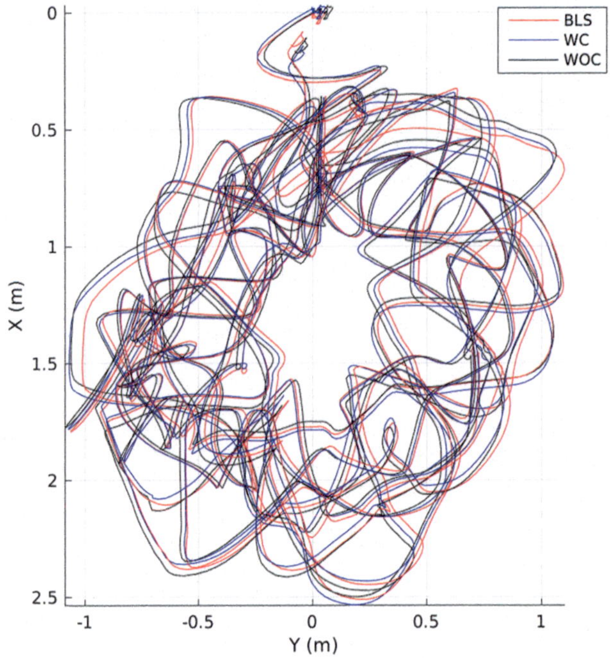

Fig. 2. Dataset S1

4 Conclusion and Future Work

In this work, we introduced an efficient method for computing the 4 d.o.f. transformation between multiple maps given 3D point feature correspondences. We cast the map alignment problem as a least squares minimization of the Mahalanobis distance between common features, and solve it efficiently by eliminating the maps' relative positions and computing their relative yaw angles first. We have experimentally evaluated our method in datasets of different scales (room-size and building-size) and have verified its accuracy and speed. As part of our future work, we seek to investigate methods that split the maps based on criteria other than time so as to ensure high accuracy. Additionally, we aim to further reduce the processing cost by efficiently selecting and reprocessing the most informative features within each, as well as across multiple maps.

References

1. Benjemaa, R., Schmitt, F.: A solution for the registration of multiple 3D point sets using unit quaternions. In: Proceedings of the European Conference on Computer Vision, Freiburg, Germany, 2–6 June 1998, pp. 34–50 (1998)
2. Chaudhury, K.N., Khoo, Y., Singer, A.: Global registration of multiple point clouds using semidefinite programming. SIAM J. Optim. **25**(1), 468–501 (2015)

3. Choudhary, S., Carlone, L., Nieto, C., Rogers, J., Christensen, H.I., Dellaert, F.: Distributed mapping with privacy and communication constraints: lightweight algorithms and object-based models. Int. J. Robot. Res. **36**(12), 1286–1311 (2017)
4. Cormen, T.H., Leiserson, C.E., Rivest, R.L., Stein, C.: Introduction to Algorithms. MIT Press, Cambridge (2001)
5. Cox, D.A., Little, J., O'Shea, D.: Ideals, Varieties, and Algorithms: An Introduction to Computational Algebraic Geometry and Commutative Algebra. Springer, New York (2007)
6. Cunningham, A., Indelman, V., Dellaert, F.: DDF-SAM 2.0: consistent distributed smoothing and mapping. In: Proceedings of the IEEE International Conference on Robotics and Automation, Karlsruhe, Germany, 6–10 May 2013, pp. 5220–5227 (2013)
7. Guo, C.X., Sartipi, K., DuToit, R.C., Georgiou, G.A., Li, R., O'Leary, J., Nerurkar, E.D., Hesch, J.A., Roumeliotis, S.I.: Large-scale cooperative 3D visual-inertial mapping in a Manhattan world. In: Proceedings of the IEEE International Conference on Robotics and Automation, Stockholm, Sweden, 16–21 May 2016, pp. 1071–1078 (2016)
8. Hesch, J.A., Kottas, D.G., Bowman, S.L., Roumeliotis, S.I.: Consistency analysis and improvement of vision-aided inertial navigation. IEEE Trans. Rob. **30**(1), 158–176 (2014)
9. Horn, B.K.: Closed-form solution of absolute orientation using unit quaternions. J. Opt. Soc. Am. **4**(4), 629–642 (1987)
10. Kim, B., Kaess, M., Fletcher, L., Leonard, J., Bachrach, A., Roy, N., Teller, S.: Multiple relative pose graphs for robust cooperative mapping. In: Proceedings of the IEEE International Conference on Robotics and Automation, Anchorage, AK, 3–8 May 2010, pp. 3185–3192 (2010)
11. Krishnan, S., Lee, P.Y., Moore, J.B., Venkatasubramanian, S.: Global registration of multiple 3D point sets via optimization-on-a-manifold. In: Proceedings of the Eurographics Symposium on Geometry Processing. SGP 2005, Aire-la-Ville, Switzerland, 4–6 July 2005 (2005)
12. McDonald, J., Kaess, M., Cadena, C., Neira, J., Leonard, J.J.: Real-time 6-DOF multi-session visual SLAM over large-scale environments. Robot. Auton. Syst. **61**(10), 1144–1158 (2013)
13. Mur-Artal, R., Montiel, J., Tardós, J.D.: ORB-SLAM: a versatile and accurate monocular SLAM system. IEEE Trans. Rob. **31**(5), 1147–1163 (2015)
14. Naroditsky, O., Zhou, X.S., Roumeliotis, S.I., Daniilidis, K.: Two efficient solutions for visual odometry using directional correspondence. IEEE Trans. Pattern Anal. Mach. Intell. **34**(4), 818–824 (2012)
15. Ni, K., Steedly, D., Dellaert, F.: Tectonic SAM: exact, out-of-core, submap-based SLAM. In: Proceedings of the IEEE International Conference on Robotics and Automation, Rome, Italy, 10–14 April 2007, pp. 1678–1685 (2007)
16. Piniés, P., Paz, L.M., Gálvez-López, D., Tardós, J.D.: CI-Graph simultaneous localization and mapping for three-dimensional reconstruction of large and complex environments using a multicamera system. J. Field Robot. **27**(5), 561–586 (2010)
17. Rosten, E., Drummond, T.: Machine learning for high-speed corner detection. In: Proceedings of the European Conference on Computer Vision, 7–13 May 2006, pp. 430–443 (2006)
18. Rublee, E., Rabaud, V., Konolige, K., Bradski, G.: ORB: an efficient alternative to SIFT or SURF. In: Proceedings of the IEEE International Conference on Computer Vision, Barcelona, Spain, 6–13 November 2011, pp. 2564–2571 (2011)

19. Williams, J., Bennamoun, M.: Simultaneous registration of multiple corresponding point sets. Comput. Vis. Image Underst. **81**(1), 117–142 (2001)
20. Williams, S.B., Dissanayake, G., Durrant-Whyte, H.: Towards multi-vehicle simultaneous localisation and mapping. In: Proceedings of the IEEE International Conference on Robotics and Automation, Washington, DC, 11–15 May 2002, vol. 3, pp. 2743–2748 (2002)
21. Wu, K.J., Ahmed, A.M., Georgiou, G.A., Roumeliotis, S.I.: A square root inverse filter for efficient vision-aided inertial navigation on mobile devices. In: Proceedings of Robotics: Science and Systems, Rome, Italy, 12–16 July 2015 (2015)

Heterogeneous Sensor Fusion via Confidence-Rich 3D Grid Mapping: Application to Physical Robots

Eric Heiden[1(✉)], Daniel Pastor[2], Pradyumna Vyshnav[3], and Ali-Akbar Agha-Mohammadi[3]

[1] Department of Computer Science, University of Southern California, 3650 McClintock Ave, Los Angeles, CA 90089, USA
heiden@usc.edu

[2] Graduate Aerospace Laboratories, California Institute of Technology, 1200 E California Blvd, Pasadena, CA 91125, USA
dpastorm@caltech.edu

[3] NASA Jet Propulsion Laboratory, California Institute of Technology, 4800 Oak Grove Drive, Pasadena, CA 91106, USA
{pradyumna.vyshnav,aliagha}@jpl.nasa.gov

Abstract. Autonomous navigation of intelligent physical systems largely depend on the ability of the system to generate an accurate map of its environment. Confidence-rich grid mapping algorithm provides a novel representation of the map based on range data by storing richer information at each voxel, including an estimate of the variance of occupancy. Capabilities and limitations are attributes of any given sensor, and therefore a single sensor may not be effective in providing detailed assessment of dynamic terrains. By incorporating multiple sensory modalities in a robot and extracting fused sensor information from them leads to higher certainty, noise reduction, and improved failure tolerance when mapping in real-world scenarios. In this work we investigate and evaluate sensor fusion techniques using confidence-rich grid mapping through a series of experiments on physical robotic systems with measurements from heterogeneous ranging sensors.

1 Introduction

Biological systems integrate information from multiple sources (*e.g.* vision, touch, etc.) to create a coherent and rich representation of the environment which allows them to reliably act in it. Robotic systems, too, largely depend on a map of their surroundings generated through various sensing instruments on-board to enable

E. Heiden and D. Pastor—Equal contribution.

Electronic supplementary material The online version of this chapter (https://doi.org/10.1007/978-3-030-33950-0_62) contains supplementary material, which is available to authorized users.

J. Xiao et al. (Eds.): ISER 2018, SPAR 11, pp. 725–736, 2020.
https://doi.org/10.1007/978-3-030-33950-0_62

robust autonomous navigation in obstacle-laden environments. Consider a robot operating in a subterranean environment, such as a cave (see Fig. 1). This robot has to take into consideration the various physical constraints such as elevation, illumination, visual occlusion, etc. Although, in some cases, perception operates independently in the context of exploration, due to the ever-changing nature of the real-world setting in question, it is imperative to make use of robotic systems with multiple sensor modalities. This provides the ability to utilize the intrinsic benefits of each sensor system based on operational requirements.

1.1 Multi-modal Mapping

In mobile robotics, occupancy grids are the most common frameworks for modeling obstacles. Compared to state-of-the-art grid-based mapping methods, *Confidence-rich Grid Mapping* (CRM) [1] not only provides an accurate representation of the map, but also estimates the consistency between the map error and the reported confidence value in each voxel. When dealing with robotic systems fusing sensor data, this form of map representation is particularly useful as it can capture noise introduced by imperfect sensors and models more accurately.

Fig. 1. Hybrid aerial-ground robot developed at NASA-JPL operating in a cave.

1.2 Heterogeneous Sensors and Sensor Fusion

Depth cameras provide rich and dense information about the scene based on stereo matching or structured light (*e.g.* Kinect v1). However, the usable range of such sensors is typically considerably limited as the measurement error increases significantly at longer distances [2]. On the other hand, although scanning lasers are very reliable and accurate, even at larger ranges, they only provide range information in the horizontal plane. As a consequence of this, significant amounts of data must be collected to obtain detailed topographical information about a region [3]. Instead of relying on singular sensor modalities, it is beneficial to leverage a variety of sensors that complement each other [4] in order to improve measurement accuracy, robustness, and spatial and temporal coverage of perception pipelines.

2 Related Work

Grid-based maps have been constructed using a variety of range sensors, including sonars [5,6], depth cameras [7], and scanning lasers [8]. Sensor fusion in occupancy grids for mobile robots was first studied by Moravec [9]. Detailed discussions on advantages of sensor fusion techniques in occupancy grids including different sensor architectures was presented by Elmenreich [4]. In 1987, Matthies and Elfes [10] combined improved versions of the sonar and stereo vision-based algorithms into a single method that builds maps integrating data from both sensors. This approach was extended to achieve robust data fusion of data from a monocular camera and a rangefinder [11]. The main technique using forward sensor models for integrating point-clouds obtained from different 3D sensors, in particular, time-of-flight sensors (Swiss-ranger, scanning laser rangefinders), and stereo vision cameras was developed by Pathak *et al.* [12]. A factor graph-based optimization framework, designed using a modular sensor-fusion system that allows for efficient and accurate incorporation of any navigation sensor of different sampling rates was discussed by Geneva *et al.* [13]. Patel *et al.* [14] propose a deep learning architecture for the sensor fusion problem that consists of two convolutional neural networks (CNNs), each consisting of a different input modality, which are fused with a gating mechanism. Deep neural network architectures that are able to fuse information generated by multiple sensors and are robust to sensor failures at runtime are demonstrated by Bohnez *et al.* [15].

3 Problem Statement

The goal of this paper is to generate a confidence-aware representation of an unknown environment using measurements obtained from a set of heterogeneous sensors mounted on vehicles with multi-modal capabilities. In particular, we adopt CRM as a spatial representation of the environment to fuse sensor data from a multitude of robots and sensors. The contributions of this paper are *(a)* the real-time implementation of distributed CRM, and *(b)* testing and analyzing CRM as a consistent, accurate and convenient mapping algorithm for heterogeneous vehicles using different ranging sensors.

4 Technical Approach

To obtain more accurate and confidence-aware maps, CRM relaxes several incorrect assumptions made by traditional mapping algorithms. First, the dependency between the voxels inside the same measurement cone is taken into account for each map update. Second, the inverse sensor model is replaced by a novel "sensor cause model" which enables a principled approach to integrating forward sensor models into the map update algorithm. Finally, besides the mean of occupancy, each voxel in the CRM stores and estimates the variance as a confidence value of occupancy. We refer to [1] for more details on CRM and provide a short summary in the following.

4.1 Confidence-Rich Representation

An occupancy map $m = [m^1, \cdots, m^n]$ is defined as a set of values over a 2D or 3D grid of n voxels. The sensor measurement and the sensor configuration at the k-th time step are given by z_k and x_k, respectively. $b_k^m = p(m|z_{0:k}, x_{0:k})$

is the probability distribution (belief) of occupancy on the map m formulated in a Bayesian framework, by compressing the information obtained from past measurements $z_{0:k} = \{z_0, \cdots, z_k\}$ and $x_{0:k} = \{x_0, \cdots, x_k\}$. In [1], we show how the map belief can be updated iteratively via a linear term α^i and a constant term β^i for each voxel i as follows:

$$p(m^i|z_{0:k}, x_{0:k}) = (\alpha^i m^i + \beta^i)p(m^i|z_{0:k-1}, x_{0:k-1}) \tag{1}$$

In our C++ implementation, the map belief b^{m^i} for each voxel i is realized as an array of particles, *i.e.* floating-point 32-bit numbers, that are updated by the α^i and β^i coefficients. We have experimentally determined the number of particles to be 50 as a sufficient trade-off between mapping accuracy, computational and storage efficiency.

4.2 Sensor Cause Model

Unlike conventional occupancy-grid algorithms, CRM maintains the interdependence between voxels through the *Sensor Cause Model* (SCM). It is a probabilistic model that reasons about a voxel c_k being cause of the measurement z_k under sensor ray x_k and is proportional to the sensor forward model $p(z_k|c_k, x_k)$, as well as the probabilities of the measurement ray bouncing off that voxel c_k and traversing through all the voxels between c_k and the sensor:

$$p(c_k|z_{0:k}, x_{0:k}) = p(c_k|b^m_{k-1}, z_k, x_k) = \eta' p(z_k|c_k, x_k)\widehat{m}^{c_k}_{k-1}\prod_{j=1}^{c^l_k-1}(1 - \widehat{m}^{g(j,x)}_{k-1}) \tag{2}$$

where c^l denotes the local index of voxel c along the ray x, i.e., $c^l = g^{-1}(c, x)$ and η' is the normalization constant.

4.3 Sensor Fusion in CRM

The Sensor Cause Model allows the CRM to embed forward models $p(z|c, x)$ for various sensors through the same update algorithm while considering the sensor's noise characteristics.

For example, for a stereo camera with focal length f and base line d_b, we can obtain the forward model $p(z|c, x) = \mathcal{N}(||G^c - x_{cam}||^{-1} f d_b, V)$ where G^c is the 3D point on the voxel grid estimated from the disparity matching between the two images recorded by the camera, x_{cam} is the camera's position and V is the variance of the Gaussian noise on the measured disparity.

The forward model provides us with its own uncertainty measurement that is embedded in the SCM such that the map update step (cf. Eq. 1) does not need to be hand-engineered for various sensors, i.e. it is *sensor-agnostic*, as opposed to conventional grid mapping algorithms. Measurements obtained from heterogeneous sensors contribute based on their accuracy to the map update, such that sensor fusion follows naturally from the Bayesian framework CRM is based on.

5 Experiments

In this section, we present experiments to analyze the performance of multiple sensors and fusion thereof on physical robotic systems. All the experiments were carried out at the Center for Autonomous Systems and Technologies (CAST[1]) located at California Institute of Technology (see Fig. 3 left). CAST has a $20 \times 12\times$ 13 m flying arena equipped with an *OptiTrack*[2] motion capturing system instrumented with 48 high-definition video cameras. We manually control the robots and use OptiTrack pose estimation throughout all our experiments.

5.1 Sensors and Vehicles

Throughout our experiments, we use the following two robotic mobility systems on which CRM was running onboard, implemented in C++ with ROS [16] integration. We provide a schematic data flow diagram of our hardware configuration in Fig. 2. While each robot independently computes the map onboard, a laptop serves as ground station to collect diagnostic information and data for visualization purposes.

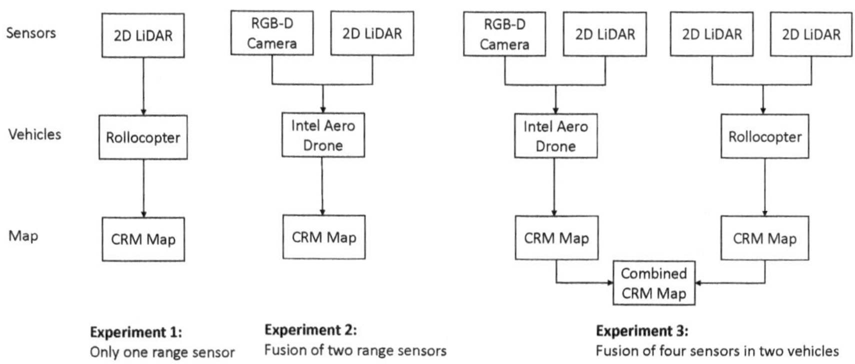

Fig. 2. Block diagram of the data flows in the three experiments that were conducted.

5.1.1 Rollocopter

Developed at NASA Jet Propulsion Laboratory (JPL), *Rollocopter* is a six degrees-of-freedom autonomous hybrid robot for both aerial and terrestrial modes (see Fig. 3 center). It is equipped with an NVIDIA Jetson TX2 microcomputer and two Hokuyo URG-04LX-UG01 LiDARs that measure ranges within a field of view of 240° and 0.36° angular resolution at a frequency of 10 Hz. Each Hokuyo sensor is rigidly mounted to the wheels which rotate as the vehicle moves, giving the planar depth sensors more coverage of the 3D space. The platform combines energy-efficient rolling on flat terrains with flying capabilities for more demanding scenarios.

[1] http://cast.caltech.edu.
[2] http://optitrack.com.

5.1.2 Intel Aero Drone

The *Intel Aero* drone is a quadcopter powered by the Intel Aero Compute Board with an Intel Atom x7-Z8750[3] CPU, equipped with an Intel RealSense R200 depth-sensing camera which records depth images at 20 Hz. In addition, we mounted a forward-facing Hokuyo URG-04-UG01 LiDAR (see Fig. 3 right) onto the vehicle.

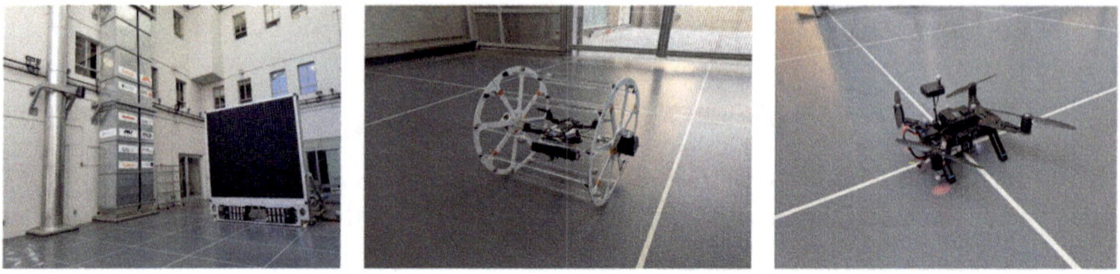

Fig. 3. Flying arena at CAST (left). Rollocopter (center) and Intel Aero Drone (right).

5.2 Experiment 1

In our first experiment, we investigate the performance of CRM compared to log-odds mapping on Rollocopter with two LiDARs. We place boxes on the floor as obstacles to be registered in the map. This presents the first time that any mapping algorithm has been implemented on this novel robot platform.

5.3 Experiment 2

We control the Intel Aero drone to take measurements with the RealSense camera and the Hokuyo LiDAR. Both sensors are aligned forward-facing in this experiment to compare how well CRM can leverage the combination of both sources of measurements in comparison to log-odds mapping. This experiment serves as our first sensor fusion application.

The depth measurements from both sensors are processed in real time by the mapping algorithms, at frequencies of 20 Hz and 10 Hz for RealSense and LiDAR, respectively (Fig. 4).

5.4 Experiment 3

In our final experiment, we combine the two vehicles, Rollocopter and Intel Aero, to exchange map updates between each other and fuse measurements from their particular sensor suites. We manually control the robots in parallel to follow paths which provide a good coverage of the obstacle course (see Fig. 6 right). The RealSense on the Intel drone is attached downward-facing to obtain depth

[3] https://software.intel.com/en-us/aero/drone-kit.

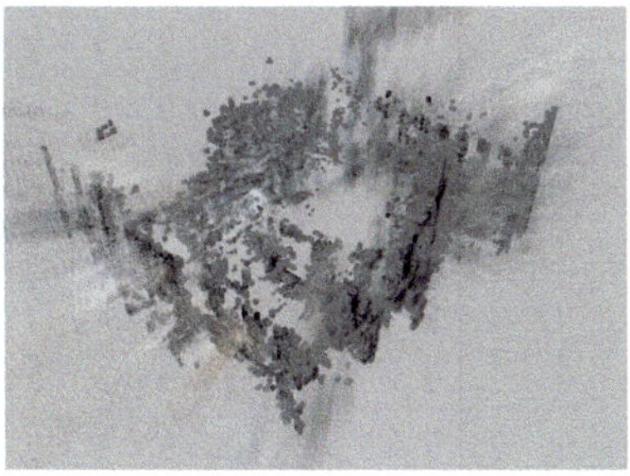

Fig. 4. Confidence-rich grid map and overlaid point cloud generated using the RealSense stereo sensor and a 2D LiDAR on the Intel Aero drone from experiment 2.

images of the floor which overlap significantly with the measurements taken by Rollocopter moving below the drone. This enables the mapping algorithm to fuse updates from both robots to obtain a more accurate model of the space that is traversable for Rollocopter. A Velodyne VLP-16 3D LiDAR is used to capture the ground-truth of the environment, as shown in Fig. 5.

With this experiment, we investigate how CRM can leverage heterogeneous sensors *and* heterogeneous robots in a multi-agent setting. We implement a procedure that allows for the exchange of map updates across different sensing agents: by broadcasting the α^i, β^i coefficients from Eq. 1 for each voxel i, the change in the CRM map can be communicated efficiently without sending the 50 particle values for each voxel. Every 10 depth measurements, the CRM instance running on each robot broadcasts these map updates in the form of ROS messages, resulting in a frequency of ca. 2 Hz (see Fig. 6 left). The receiving CRM instance listens for such update messages and applies them in accordance to the Bayesian update Eq. 1 as soon as they arrive.

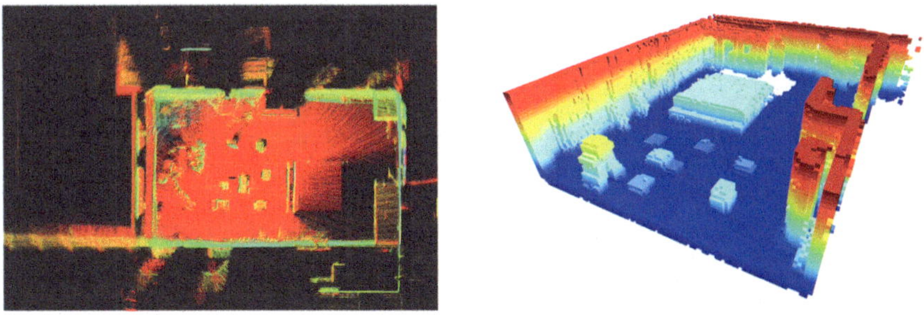

Fig. 5. *Left:* Orthographic top-down projection of point cloud of CAST captured with a Velodyne VLP-16 laser scanner. *Right:* Ground-truth map from Velodyne laser scans.

Fig. 6. *Left:* Architecture diagram for Experiment 3 with measurement rates for the different sensors (brown) and map update rates (green) between the two CRM instances (blue) running on Rollocopter and the Intel Aero Drone. *Right:* Experimental setup at CAST during Experiment 3.

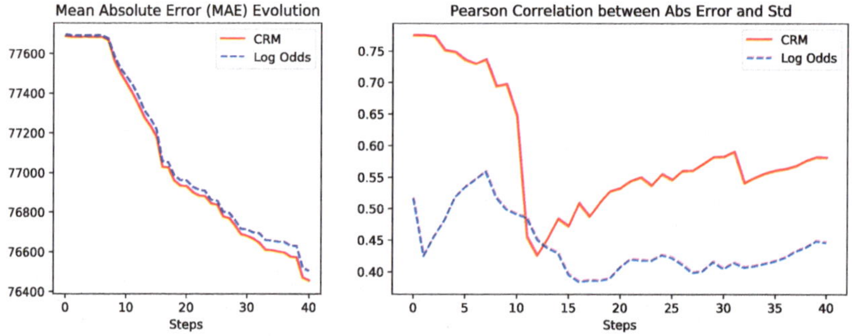

Fig. 7. Evolution of mean absolute error (left) and Pearson correlation coefficient between absolute error and estimated mapping std (right) for the LiDAR dataset obtained with Rollocopter (Experiment 1).

6 Results

In this section, we demonstrate the performance of the heterogeneous sensors when applied to multi-modal physical robotic systems. For the first experiment, using a single robot and a single range sensor, Fig. 7 shows the map mean absolute error (MAE) over time for different sensor noise std deviations. We compare CRM against log-odds-based grid mapping in the *OctoMap* framework [17]. The MAE is averaged over all voxels that were updated throughout the mapping process to best show the improvement of the affected parts of the map. As an alternative measure of consistency, we compute the *Pearson correlation coefficient* between the true error and the estimated std. deviation (see Fig. 7), which indicates that the std. deviation estimated by CRM is highly correlated with the error, hence, it reliably captures the mapping confidence.

For the second experiment, using a single robot and multiple range sensors, we compare the MAE for CRM and log-odds maps using both modalities combined and separated. Our results show that CRM yields higher accuracy than log-odds (Fig. 8) and they indicate that sensor fusion in occupancy grids benefit from our Bayesian formulation.

For the third experiment, using two robots and multiple range sensors on each robot, we show qualitative results of the resulting global map using CRM.

7 Experimental Insights

Range sensors have particular noise characteristics which we observed in the measurements obtained from the LiDAR sensors in our experiments.

Spurious Measurements: We observed that scanning lasers produce a variety of erroneous points in the vicinity of edges. These *phantom points* were usually found behind the edges close to the laser ray. This effect became more apparent at points on laser scans which were affected by a multitude of far-distanced objects. LiDARs, *e.g.* Velodyne VLP-16, that capture multiple returns for each single laser scan, are more robust to spurious measurements compared to the light-weight Hokuo LiDARs used on the robots in our experiments.

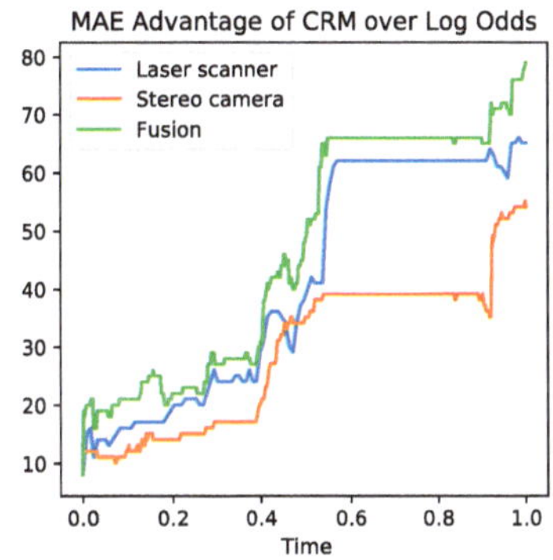

Fig. 8. Mean Absolute Error advantage of CRM over Log Odds under different sensor modalities (forward-facing LiDAR, forward-facing stereo camera, fusion of both) on the Intel Drone (Experiment 2). CRM shows an improvement in accuracy that increases with time regardless of the sensor.

Surface Reflectance: As scanning lasers rely on the beam reflected from the object surface to the receiving unit, the strength of the returning signal is affected by the reflective properties of the surface. Such specular reflection is influenced by the distance to the surface, operating frequency and the incidence angle, among others. As shown in Fig. 9, accurate depth measurements from objects with shiny surfaces are particularly difficult to obtain and result in missing or erroneous data. Filtering approaches have been shown to effectively remove echo measurements resulting from reflective and transparent objects [18].

Fig. 9. While taking laser scans from reflective surfaces, such as the floor, a distinctive group of wrong measurements appears as a halo around the non-reflected measurements. This scan was taken by rotating the LiDAR around its axis. For clarity, a simplified CAD model of the testing area is displayed.

8 Conclusion and Future Work

Our experiments have demonstrated that CRM can be used in a distributed setting where a network of robots equipped with heterogeneous sensors collaboratively construct a consistent representation of the environment in real time. While log-odds mapping achieves a similar level of accuracy as the CRM map given measurements from two equal laser scanners, the consistency in CRM is significantly higher. Furthermore, CRM has been shown to leverage the advantages of heterogeneous sensors more effectively, leading to significant improvements in accuracy.

Future research is directed towards integrating CRM with planning. Preliminary results in a simulated environment [19] have shown promising perspectives for a trajectory optimization procedure that leverages CRM's estimated confidence, besides the mean of occupancy. Such confidence-aware planners could improve the ability of mobile robots to autonomously navigate in unknown, unstructured environments. To enable true autonomy, CRM must be combined with a localization pipeline, *e.g.* visual-inertial odometry, such as Stereo MSCKF-VIO [20], or LiDAR-based odometry [21].

In some applications the communication link between the different sensor agents may be restricted or temporally blocked. Although we only send CRM updates instead of the complete map to reduce bandwidth, the issue of compressing the CRM map in an unstructured environment remains open for future research.

Acknowledgements. This research was carried out at the Jet Propulsion Laboratory under a contract with the National Aeronautics and Space Administration. U.S. Government sponsorship acknowledged.

References

1. Agha-Mohammadi, A.-A., Heiden, E., Hausman, K., Sukhatme, G.: Confidence-rich grid mapping. In: Proceedings of the International Symposium on Robotics Research, pp. 1–19 (2017)

2. Khoshelham, K.: Accuracy analysis of kinect depth data. In: ISPRS Workshop Laser Scanning, vol. 38, p. W12 (2011)
3. Castorena, J., Creusere, C.D., Voelz, D.: Modeling LIDAR scene sparsity using compressive sensing. In: 2010 IEEE International Geoscience and Remote Sensing Symposium (IGARSS), pp. 2186–2189. IEEE (2010)
4. Elmenreich, W.: An introduction to sensor fusion. Vienna University of Technology, Austria (2002)
5. Moravec, H.P., Elfes, A.: High resolution maps from wide angle sonar. In: Proceedings of the IEEE Conference on Robotics and Automation, pp. 19–24 (1985)
6. Yamauchi, B.: A frontier-based approach for autonomous exploration. In: IEEE International Symposium on Computational Intelligence in Robotics and Automation, pp. 146–151 (1997)
7. Paskin, M., Thrun, S.: Robotic mapping with polygonal random fields. In: Conference on Uncertainty in Artificial Intelligence, pp. 450–458 (2005)
8. Thrun, S.: Learning occupancy grid maps with forward sensor models. Auton. Robots **15**(2), 111–127 (2003)
9. Moravec, H.P.: Sensor fusion in certainty grids for mobile robots. AI Mag. **9**(2), 61 (1988)
10. Elfes, A., Matthies, L.: Sensor integration for robot navigation: combining sonar and stereo range data in a grid-based representataion. In: 26th IEEE Conference on Decision and Control 1987, vol. 26, pp. 1802–1807. IEEE (1987)
11. Stepan, P., Kulich, M., Preucil, L.: Robust data fusion with occupancy grid. IEEE Trans. Syst. Man Cybern. Part C (Appl. Rev.) **35**(1), 106–115 (2005)
12. Pathak, K., Birk, A., Poppinga, J., Schwertfeger, S.: 3D forward sensor modeling and application to occupancy grid based sensor fusion. In: IEEE/RSJ International Conference on Intelligent Robots and Systems 2007. IROS 2007, pp. 2059–2064. IEEE (2007)
13. Geneva, P., Eckenhoff, K., Huang, G.: Asynchronous multi-sensor fusion for 3D mapping and localization. Technical report RPNG-2017-002, University of Delaware (2017). http://udel.edu/ghuang/papers/tr_async.pdf
14. Patel, N., Choromanska, A., Krishnamurthy, P., Khorrami, F.: Sensor modality fusion with CNNs for UGV autonomous driving in indoor environments. In: International Conference on Intelligent Robots and Systems (IROS). IEEE (2017)
15. Bohez, S., Verbelen, T., De Coninck, E., Vankeirsbilck, B., Simoens, P., Dhoedt, B.: Sensor fusion for robot control through deep reinforcement learning. In: 2017 IEEE/RSJ International Conference on Intelligent Robots and Systems (IROS), pp. 2365–2370. IEEE (2017)
16. Quigley, M., Conley, K., Gerkey, B.P., Faust, J., Foote, T., Leibs, J., Wheeler, R., Ng, A.Y.: ROS: an open-source Robot Operating System. In: ICRA Workshop on Open Source Software (2009)
17. Hornung, A., Wurm, K.M., Bennewitz, M., Stachniss, C., Burgard, W.: OctoMap: an efficient probabilistic 3d mapping framework based on octrees. Auton. Robots **34**(3), 189–206 (2013)
18. Koch, R., May, S., Nüchter, A.: Detection and purging of specular reflective and transparent object influences in 3D range measurements. In: ISPRS - International Archives of the Photogrammetry, Remote Sensing and Spatial Information Sciences, pp. 377–384 (2017)
19. Heiden, E., Hausman, K., Sukhatme, G.S., Agha-Mohammadi, A.-A.: Planning high-speed safe trajectories in confidence-rich maps. In: 2017 IEEE/RSJ International Conference on Intelligent Robots and Systems (IROS), pp. 2880–2886. IEEE (2017)

20. Sun, K., Mohta, K., Pfrommer, B., Watterson, M., Liu, S., Mulgaonkar, Y., Taylor, C.J., Kumar, V.: Robust stereo visual inertial odometry for fast autonomous flight. IEEE Robot. Autom. Lett. **3**(2), 965–972 (2018)
21. Zhang, J., Singh, S.: LOAM: lidar odometry and mapping in real-time. In: Robotics: Science and Systems Conference, July 2014

FastCal: Robust Online Self-calibration for Robotic Systems

Fernando Nobre(✉) and Christoffer R. Heckman

Department of Computer Science, University of Colorado, Boulder, CO 80309, USA
{fernando.nobre,christoffer.heckman}@colorado.edu

Abstract. We propose a solution for sensor extrinsic self-calibration with low time complexity, competitive accuracy and graceful handling of often avoided corner cases: drift in calibration parameters and unobservable directions in the parameter space. It consists of three main parts: (1) information-theoretic based segment selection for constant-time estimation; (2) observability-aware parameter update through a rank-revealing decomposition of the Fischer information matrix; (3) drift-correcting self-calibration through the time-decay of segments. At the core of our FastCal algorithm is the loosely-coupled formulation for sensor extrinsics calibration and efficient selection of measurements. FastCal runs up to an order of magnitude faster than similar self-calibration algorithms (camera-to-camera extrinsics, excluding feature-matching and image pre-processing on all comparisons.), making FastCal ideal for integration into existing, resource-constrained, robotics systems.

Keywords: Self-calibration · SLAM · Real-time · Change detection · Observability · Drift

1 Introduction

Autonomous platforms destined for long-term applications equipped with multiple sensors such as cameras and Inertial Measurement Unit (IMU) have become increasingly ubiquitous. Generally these platforms must undergo sophisticated calibration routines to estimate extrinsic (sensor-to-sensor rigid body transform) parameters to high degrees of certainty before sensor data may be interpreted and fused in a Simultaneous Localization and Mapping (SLAM) context. Once fielded, calibration parameters are generally fixed for the lifetime of the platform. However, in many applications these platforms experience gradual changes in calibration parameters due to e.g. non-rigid mounting, accidental bumps and temperature dilation that can change sensor extrinsic parameters. Self-calibration addresses this by inferring extrinsic parameters pertaining to proprioceptive and exteroceptive sensors without using a known calibration target or a specific calibration routine. The motivation behind self-calibration is to remove the explicit, tedious, and sometimes nearly impossible calibration procedure from robotic applications and to enable robust long-term autonomous operation. Self-Calibration is an essential part of any long-term robotic system, as such is under

J. Xiao et al. (Eds.): ISER 2018, SPAR 11, pp. 737–747, 2020.
https://doi.org/10.1007/978-3-030-33950-0_63

constant pressure to increase its accuracy, speed and robustness. A higher speed allows its inclusion into larger systems with extensive subsequent processing (*e.g.* localization, mapping, object/activity recognition, planning) and its deployment in computationally constrained scenarios (*e.g* planetary exploration, embedded systems). A robust self-calibration system should cope with unobservable directions in the parameter space (*e.g.* due to a nonholonomic platform, measurement noise which makes unobservable parameters *appear* observable) and changes and drift in calibration parameters.

In our previous work [1] we have partially addressed constant-time self-calibration using a priority queue and informative segments. [2] addresses drift and slow changes in calibration parameters by attempting to regress the change point and retroactively correcting the state estimates and in [3] we propose an observability aware framework capable of updating only the observable directions of the parameter space, even in the presence of noisy measurements, which is then used in a reinforcement-learning framework to learn informative motions to be suggested to a human operator. In this paper we leverage some individual aspects of previous works and propose significant novel contributions:

1 A novel formulation for regressing extrinsic calibration parameters which is suitable for integration with any existing SLAM system, while being considerably faster than the system used in [1] due to a loosely coupled formulation which optimizes over relative poses instead of jointly over raw sensor measurements. We also propose a novel criteria for adding informative segments to the estimation queue which minimizes the number of times the entire segment queue needs to be optimized.
2 We handle intrinsically degenerate scenarios in noisy nonholonomic systems by only updating the observable directions of the parameter space, demonstrating the necessity and usefulness in a real-world robotic application.
3 Slow changes over arbitrary periods of time are handled by continuously renewing the segment queue by use of time-decay on measurements, this approach is shown to be much more efficient and robust compared to [2], at the cost of immediate and local accuracy.
4 Integration of the proposed self-calibration system into a real-world robotic platform operating in challenging environments for extended periods of time.[1]

1.1 Related Work

Most current techniques for vision-aided inertial navigation use filtering approaches [4–6] or a smoothing formulation. In either case the estimation is made constant-time by rolling past information into a prior distribution. Filtering methods present the significant drawback of introducing inconsistencies due to linearization errors of past measurements which cannot be corrected post hoc, particularly troublesome for non-linear camera models. Some recent work has tackled these inconsistencies; see, e.g. [7–10]. The state-of-the-art includes

[1] 3 weeks of operation, approximately 170 km driven.

methods to estimate poses and landmarks along with calibration parameters, but these approaches do not output the marginals for the calibration parameters, which are desirable for long-term autonomy applications.

The use of a known calibration pattern such as a checkerboard coupled with nonlinear regression has become the most popular method for camera calibration in computer vision during the last decade; it has been deployed both for intrinsic camera calibration [11] and extrinsic calibration between heterogeneous sensors [12]. While being relatively efficient, this procedure still requires expert knowledge to reach a discerning level of accuracy. It can also be quite inconvenient on a mobile platform requiring frequent recalibration (e.g experimental platforms which undergo constant sensor changes). In an effort to automate the process in the context of mobile robotics, several authors have included the calibration problem in a state-space estimation framework, either with filtering [13] or smoothing [14] techniques. Filtering techniques based on the Kalman filter are appealing due to their inherently online nature. However, in case of nonlinear systems, smoothing techniques based on iterative optimization can be superior in terms of accuracy [15].

2 Methodology

Algorithm 1: FastCal Algorithm

```
Data: relative pose measurements for each of the N sensors; reference sensor
      N_ref, initial guess on sensor placement.
Result: sensor-to-sensor extrinsic SE(3): Θ
Initialize Θ ← initial guess;
if num_measurements(D^candidate) < θ_meas then
    D^candidate ← new measurements;
else
    Estimate Θ|D^candidate according to (2);
    if num_segments(D^info) < θ_pq then
        D^info ← D^candidate;
    else
        Check if D^candidate should be swapped into D^info, according to 2.2 ;
        if D^info ← D^candidate then
            Estimate Θ|D^info according to (5) with TSVD;
        end
    end
end
for i ← 0 to num_segments(D^info) do
    if Time Decay for D_i^candidate according to (6) < 0.001 then
        remove D_i^candidate from D^info
    end
end
```

Table 1. FastCal parameters

Function	Symbol	Default value
TSVD threshold	$\theta_{\epsilon_{svd}}$	0.1
Maximum entropy	$\theta_{\Sigma_{max}}$	15
Number of segments in priority queue	θ_{pq}	10
Number of measurements in each candidate	θ_{meas}	10
Keyframing translation [m]	$\theta_{kf_{trans}}$	0.15
Keyframing rotation [rad]	$\theta_{kf_{rot}}$	0.1745
Time decay	θ_{λ}	0.04
Number of consecutive estimates at the same value	θ_{same}	3
Min total update	θ_{min_update}	0.008

It is common for robotic platforms to have multiple sensors, such as cameras, wheel encoders, IMU. This creates the need to obtain the relative rigid body transform between sensors so that a fused position estimate may be obtained. This is what we refer to as *calibration parameters* in this work, represented by Θ. There are other calibration parameters which can be estimated, such as camera intrinsics (*e.g.* focal length, center point, distortion parameters) but these have been found to not vary considerably even in long term operation. Sensor extrinsics however, change frequently (Sect. 3) and have considerable impact on the resulting position estimation. For these reasons we focus on estimating sensor extrinsics.

For camera intrinsic self-calibration refer to [1,16]. The proposed FastCal algorithm can be divided into three major components, summarized in Algorithm 1: (1) Selecting informative segments so as to bound the computation time. (2) Updating only the observable directions of the parameter space. (3) Considering drift in the calibration parameters over time by time-decaying measurements. We also focus on creating an algorithm which has as few tuning parameters as possible; the parameters for FastCal are summarized in Table 1, the parameters in bold are the ones with highest impact on the performance of the algorithm.

2.1 Problem Formulation

We focus on estimating the sensor-to-sensor SE(3) rigid body transform between two cameras with no co-visible features. The calibration problem can be framed as an optimization problem in a Bayesian estimation framework, by including the calibration parameters in the standard SLAM formulation:

$$\hat{\mu}_{\Theta\mathcal{X}\mathcal{L}} = \underset{\Theta\mathcal{X}\mathcal{L}}{\arg\max}\, p(\Theta, \mathcal{X}, \mathcal{L}|\mathcal{Z}) = \underset{\Theta\mathcal{X}\mathcal{L}}{\arg\min} -\log p(\Theta, \mathcal{X}, \mathcal{L}|\mathcal{Z}); \quad (1)$$

Where the estimated parameters contain the robot pose ($\mathcal{X}$), landmarks ($\mathcal{L}$) and calibration (Θ) parameters. $\mathcal{Z}$ in this context are the sensor measurements, such as landmark observations. The advantages of solving the problem in this formulation is leveraging joint information from all measurements, at the price of higher computational complexity since we must solve a larger system comprised of N poses, M landmarks in addition to the calibration parameters.

Alternatively, we can leverage the fact that most robotics systems already estimate the reduced state: $\hat{\mu}_{\mathcal{X}\mathcal{L}} = \arg\max_{\mathcal{X}\mathcal{L}} p(\mathcal{X}, \mathcal{L}|\mathcal{Z}) = \arg\min_{\mathcal{X}\mathcal{L}} -\log p(\mathcal{X}, \mathcal{L}|\mathcal{Z})$ for each camera, where $\mathcal{X} = [\mathbf{x}_{s1}, \mathbf{x}_{s2}, ..., \mathbf{x}_{sn}]$ the world position for N sensors at time t. Finding the sensor-to-sensor extrinsics can then be posed as an alignment problem, using $\mathcal{X}$ as the measurement instead of the landmark observations as in (1).

$$\hat{\Theta} = \arg\max_{\Theta} p(\Theta|\mathcal{X}) = \arg\min_{\Theta} -\log p(\Theta|\mathcal{X}); \tag{2}$$

2.2 Informative Segment Selection

We process small batches of data sequentially and decide to keep them based on their utility for the calibration. Each new batch is merged to the old ones to refine our knowledge about the calibration parameters until we reach a satisfactory level of confidence. This notion was introduced in [16] and extended to multiple sensors in [1]. We briefly explain the informative segment selection into a priority queue, to then introduce our novel metric on adding new batches of measurements. We define the current set of data samples in a priority queue as $\mathcal{D}^{info} = \{\mathbf{x}_1, ..., \mathbf{x}_n\}$ which has led to the posterior marginal density in Eq. (2), associated to the random variable $\hat{\Theta}|\mathcal{D}^{info}$. The set, $\mathcal{D}^{info}$, contains the current informative measurements for the calibration variable Θ. Our sensors, S, continuously stream new data which are used for estimating their relative positions, that we then accumulate in another batch of size ΔN denoted by

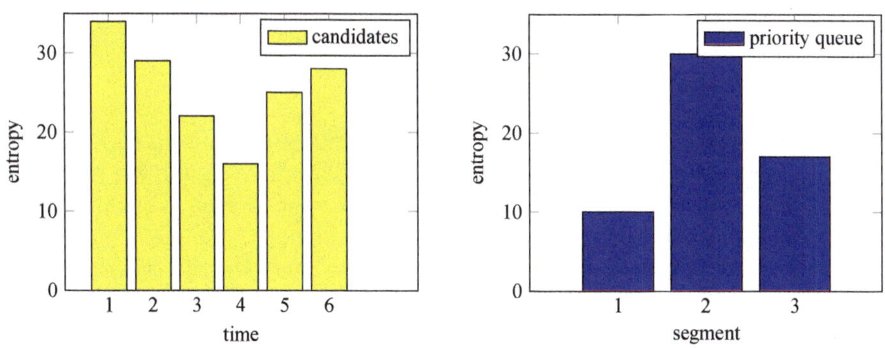

Fig. 1. Left: Entropies of rolling candidate window over time. Candidate window at time $t = 2$ could be swapped into the priority queue in place of segment 2, however by waiting until time $t = 5$ we can instead swap in the candidate window at time $t = 4$, with much more information content. Right: Entropy of three informative batches in the priority queue.

$\mathcal{D}^{candidate} = \{\mathbf{x}_{n+1}, ..., \mathbf{x}_{n+\Delta N}\}$. Intuitively, if the measurements in $\mathcal{D}^{candidate}$ are similar to those in $\mathcal{D}^{info}$, we are not improving our knowledge about Θ and we can safely discard $\mathcal{D}^{candidate}$ to keep the computation tractable. There are multiple ways of evaluating the usefulness of $\mathcal{D}^{candidate}$: Given the covariance of the estimated calibration parameters Σ_Θ, which can be obtained quickly from the solution to (2) by inverting the 6×6 information matrix, the entropy of the distribution is then given by $h = \frac{1}{2}\ln|2\pi e \Sigma_\Theta|$, where the bars denote the matrix determinant.

We propose a novel metric on adding batches to the priority queue: a candidate measurement batch is swapped into the queue iff. Its entropy beats the worst scoring segment in the priority queue *and* the rolling candidate window has achieved a local entropy minimum. This ensures that we don't needlessly swap in segments to the priority queue and ensures that we are adding the best possible local window to the queue, not just the one that beat the worst scoring segment. Figure 1 shows on the left the entropy for a rolling window of $N = 10$ candidate measurements over time, on the right the entropies for the three segments in the priority queue. By holding off on adding the candidate window to the priority queue, at time $t = 5$, we notice that the candidate entropy is increasing and as such we have reached a local minimum at $t = 4$, which is then swapped into the priority queue. This approach reduces the priority queue entropy much faster and leads to quicker convergence on Θ, as well as greatly reducing the number times the priority queue needs to be estimated.

The measurements which are added to each segment are the relative pose measurements generated independently for each sensor (each camera in our case) as shown in Fig. 2. Instead of adding each new measurement we take a keyframing based approach to self-calibration and accumulate relative pose measurements until the total relative transform reaches a threshold. In our case we use $\theta_{kf_{trans}} = 0.15\,\text{m}$ and $\theta_{kf_{rot}} = 0.1745\,\text{rad}$. This results in both much faster convergence and reduces the number of times the candidate window and priority queue need to be estimated.

2.3 Observability-Aware Estimation

We leverage our insights from [3] to use the Truncated SVD (TSVD) decomposition of the Fischer Information Matrix (FIM) and determine the observable directions in the parameter space, and only update those, even in the presence of noise. This is essential for platforms which, for example, only move on a planar surface and never excite roll, pitch or translation along the direction normal to the ground plane. In these cases the system (2) is ill conditioned and we have to either manually regularize the unobservable directions in order to make the system well conditioned or automatically detect the observable components and only update those directions. This is the approach we take in this work. The least squares solution to (2) is

$$(\mathbf{J}^T\mathbf{G}^{-1}\mathbf{J})\delta\hat{\Theta} = -\mathbf{J}^T\mathbf{G}^{-1}\mathbf{r}(\hat{\Theta}), \tag{3}$$

where J is the Jacobian matrix, G the measurement covariance matrix obtained from the first step and r the residual. There exists a solution to Eq. (3) iff the FIM is invertible, i.e. it is of full rank. The link between the rank of the FIM and observability of the parameters being estimated is well established. A singular FIM corresponds to some unobservable directions in the parameter space given the current set of observations.

Using singular-value decomposition (SVD) on the FIM we can identify a numerically rank-deficient matrix by analyzing its singular values and consequently the numerical observability of the system. The *numerical rank* r of a matrix is defined as the index of the smallest singular value σ_r which is larger than a pre-defined tolerance $\theta_{\epsilon_{svd}}$, $r = \arg\max_i \sigma_i \geq \theta_{\epsilon_{svd}}$. When the noise affecting the matrix entries has the same scale (by using column or row scaling) then the numerical rank can be determined by the singular values. The scaling matrix S can be computed as:

$$S = \text{diag}\left\{\frac{1}{||J(:,1)||}, ..., \frac{1}{||J(:,n)||}\right\} \tag{4}$$

Where $||J(:,n)||$ denotes the column norm of the Jacobian matrix, for column n. Specifically we decompose the error covariance matrix $\mathbf{G}$ from Eq. (3) into its square root form by using Cholesky decomposition, $\mathbf{G}^{-1} = \mathbf{L}^T\mathbf{L}$, we can re-write Eq. (3) in standard form: $(\mathbf{LJ})^T(\mathbf{LJ})\delta\hat{\Theta} = -(\mathbf{LJ})^T\mathbf{Lr}(\hat{\Theta})$ which are the normal equations for the linear system $(\mathbf{LH})\delta\hat{\Theta} = -\mathbf{Lr}(\hat{\Theta})$. Thus we can directly use a *rank-revealing* decomposition to estimate the numerical rank of the FIM, and consequently the numerical observability of the system. Let $(\mathbf{LJ})$ be a $m \times n$ matrix with the following SVD decomposition: $\mathbf{LJ} = \mathbf{USV^T}$, where $\mathbf{U}$ is $m \times n$ and orthogonal, $\mathbf{S} = diag(\varsigma_1, ..., \varsigma_i)$ the singular values and $\mathbf{V}$ an $n \times n$ matrix, also orthogonal. From the orthogonality of $\mathbf{U}$ and $\mathbf{V}$ we can solve (3) as $\delta\hat{\Theta} = -\mathbf{VS^{-1}ULr}(\hat{\Theta})$, Specifically, according to [17], we can efficiently obtain the update as:

$$\delta\hat{\Theta} = \sum_{i=1}^{r_\epsilon} \frac{\mathbf{u}_i^T\mathbf{r}_\Theta}{\varsigma_i}\mathbf{v}_i, \tag{5}$$

where $\mathbf{u}$ and $\mathbf{v}$ are the column vectors of $\mathbf{U}$ and $\mathbf{V}$. This allows us to only update the observable directions of the parameter space and maintain the other directions at their initial value. Establishing the value of ϵ to use is specific to the amount of noise expected in the measurements and is treated in Sect. 3.

2.4 Drift Correction

In order to correct for the inevitable drift over time on the calibration parameters due to physical shocks, maintenance, etc. We adopt a simpler strategy than what was used in [2]; We associate a exponential time-decay with each batch in the priority queue $\mathcal{D}info$. This is done by using a exponential distribution

$$p(t;\lambda) = \lambda e^{-\lambda t}, t \in [0, \infty] \tag{6}$$

Which has an expectation $E = \lambda^{-1}$, so we see that the parameter λ in (6) encodes the expected time that a set of measurements remains informative.

3 Experiments and Results

Our experimental setup consists of a robotic platform designed to transport material autonomously in warehouses, deployed in diverse real-world scenarios. This is an example of a challenging long term autonomy deployment in adverse industrial scenarios, with constrained resources and subject to physical impacts. It proves to be an excellent test-bed for the proposed *FastCal* algorithm which integrates into the SLAM module seamlessly by subscribing to sensor pose updates (Fig. 3).

The robotic platform is equipped with two pairs of global shutter stereo camera pairs, one facing forward and one facing backward, with no overlapping field of view. We focus on estimating the front-to-back camera extrinsics, given an initial rough guess obtained with a measuring tape. We implement the proposed FastCal algorithm in C++, utilizing the parameters defined in Table 1. We wish to assess how often and by how much the extrinsics parameters actually change in practice. For that end we run FastCal on the robot for a period of 14 days. In this time the robot was subject to cargo loads up to 100 kg, was transported

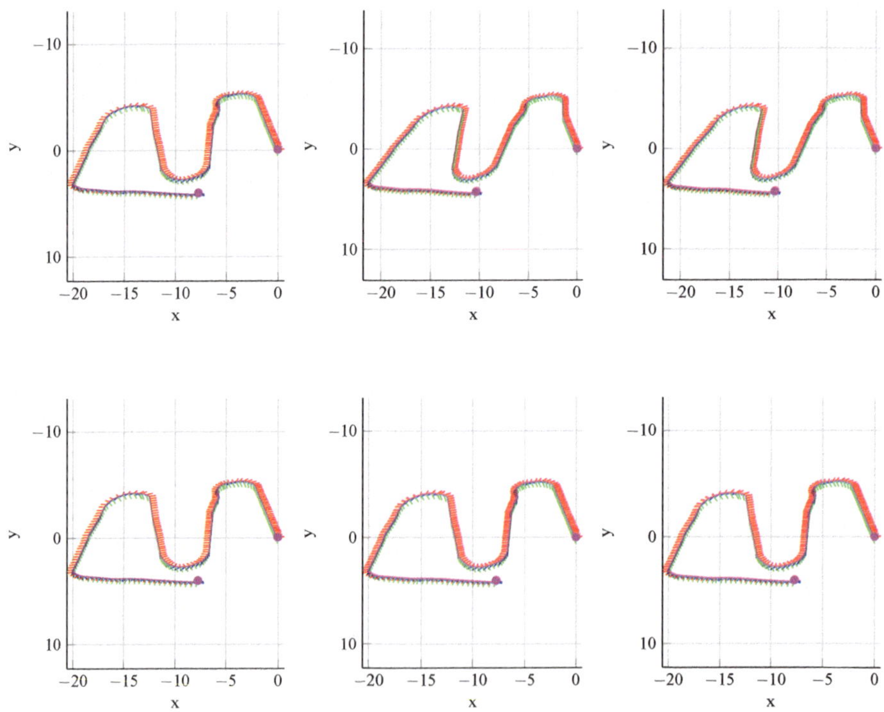

Fig. 2. Same route performed over the span of multiple weeks. Top row: no self-calibration, using initial offline calibration only. Bottom row: FastCal enabled. Each column represents a day in which that route was performed. Columns are approximately 1 week apart.

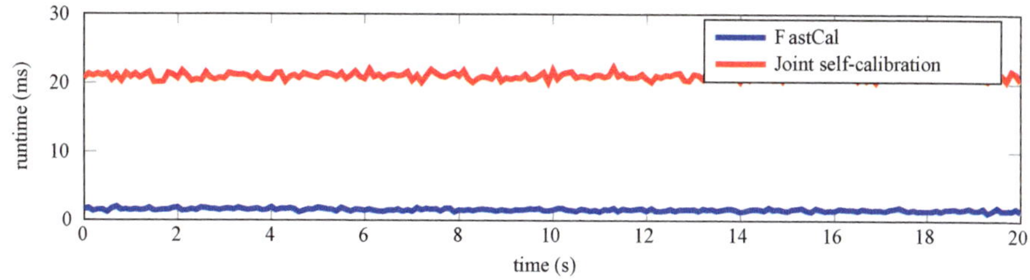

Fig. 3. Timing comparison of the proposed FastCal algorithm vs a reference implementation of tightly coupled, priority queue based self-calibration, for the same size priority queue, candidate segment, number of iterations and fair parameter settings.

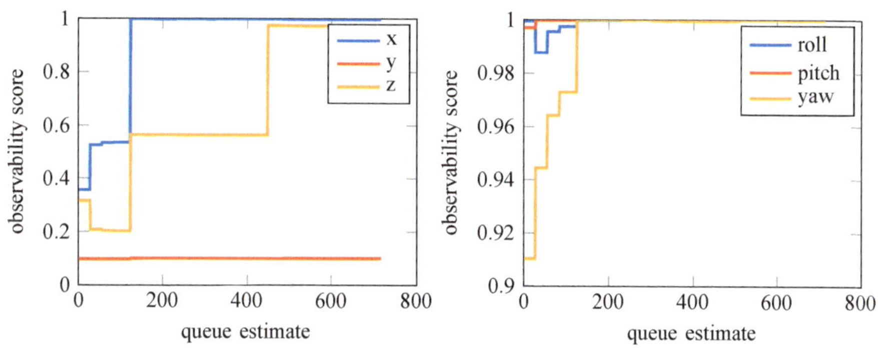

Fig. 4. Observability score for each direction of the calibration parameter (SE(3) transform between cameras). Note that due to the planar motion, the y component is completely unobservable, even in the presence of noise; the unobservable direction remains clamped at its original value as a natural consequence of FastCal, with no need for explicit regularization.

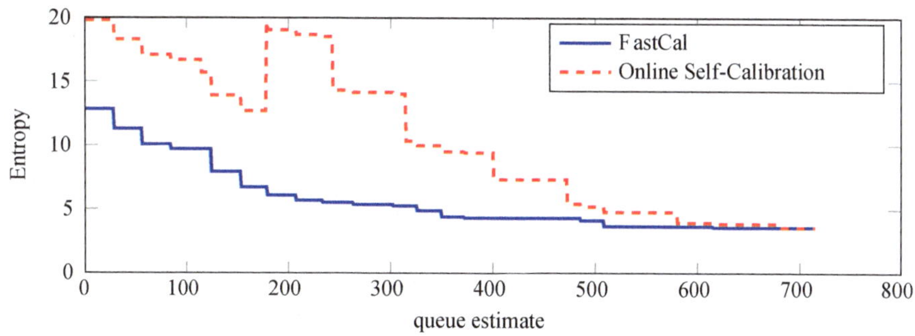

Fig. 5. Priority Queue entropy comparison of the proposed FastCal algorithm vs a implementation of [1]. The novel criteria on adding segments to the queue, coupled with the keyframing results in a much faster reduction in entropy and thus convergence on the calibration parameters.

in a truck, had it's front bumper removed and re-attached due to maintenance reasons and finally suffered one accidental head-first collision while on joystick mode. These circumstances provide valuable data points on the usefulness and necessity of a robust self-calibration algorithm (Figs. 4 and 5).

4 Conclusion

In this paper we presented a low time complexity extrinsics self-calibration algorithm that uses an information theoretic measure to add only the most locally informative measurement batches to the estimation queue, the work from [1] is improved upon by leveraging novel queue update techniques which drastically reduce the time it takes to converge on the calibration parameters. We have leveraged truncated SVD/QR decomposition to deal with unobservable directions and operate on nonholonomic platforms, similar to the algorithm presented in [3], however not applied to reinforcement learning and user motion suggestions, but to calibrating nonholonomic robotic platforms. Finally, we have incorporated time-decay as a mechanism to address drift in calibration parameters. We draw from the lessons learned from [2] to implement a simpler, more effective mechanism of dealing with drift. We have evaluated the proposed system in a variety of long term scenarios spanning 15 days and over 120 km driven and shown that it can be used as a low-overhead add-on to existing perception systems, while achieving similar accuracy to off-line calibration techniques.

References

1. Nobre, F., Heckman, C., Sibley, G.: Multi-sensor slam with online self-calibration and change detection. In: International Symposium on Experimental Robotics (ISER) (2016)
2. Nobre, F., Kasper, M., Heckman, C.: Drift-correcting self-calibration for visual-inertial SLAM. In: International Conference on Robotics and Automation (ICRA) (2017)
3. Nobre, F., Heckman, C.: Reinforcement learning for assisted visual-inertial robotic calibration. In: International Symposium on Robotics Research (ISRR) (2017)
4. Jones, E.S., Soatto, S.: Visual-inertial navigation, mapping and localization: a scalable real-time causal approach. Int. J. Robot. Res. **30**, 407–430 (2011)
5. Kelly, J., Sukhatme, G.S.: Visual-inertial sensor fusion: localization, mapping and sensor-to-sensor self-calibration. Int. J. Robot. Res. **30**, 56–79 (2011)
6. Mourikis, A.I., Roumeliotis, S.I.: A multi-state constraint Kalman filter for vision-aided inertial navigation. In: International Conference on Robotics and Automation (2007)
7. Li, M., Mourikis, A.I.: High-precision, consistent EKF-based visual-inertial odometry. Int. J. Robot. Res. **32**, 690–711 (2013)
8. Hesch, J.A., Kottas, D.G., Bowman, S.L., Roumeliotis, S.I.: Towards consistent vision-aided inertial navigation. In: Algorithmic Foundations of Robotics (2013)
9. Civera, J., Bueno, D.R., Davison, A.J., Montiel, J.M.M.: Camera self-calibration for sequential Bayesian structure from motion. In: International Conference on Robotics and Automation (2009)

10. Li, M., Yu, H., Zheng, X., Mourikis, A.I.: High-fidelity sensor modeling and self-calibration in vision-aided inertial navigation. In: International Conference on Robotics and Automation (2014)
11. Sturm, P.F., Maybank, S.J.: On plane-based camera calibration: a general algorithm, singularities, applications. In: Computer Vision and Pattern Recognition (1999)
12. Zhang, Q., Pless, R.: Extrinsic calibration of a camera and laser range finder. In: Intelligent Robots and Systems (2004)
13. Martinelli, A., Scaramuzza, D., Siegwart, R.: Automatic self-calibration of a vision system during robot motion. In: International Conference on Robotics and Automation (ICRA) (2006)
14. Kümmerle, R., Grisetti, G., Burgard, W.: Simultaneous calibration, localization, and mapping. In: Intelligent Robots and Systems (IROS) (2011)
15. Strasdat, H., Montiel, J., Davison, A.J.: Real-time monocular SLAM: why filter? In: Robotics and Automation (ICRA). IEEE (2010)
16. Keivan, N., Sibley, G.: Constant-time monocular self-calibration. In: Robotics and Biomimetics (ROBIO), pp. 1590–1595 (2014)
17. Hansen, P.C.: Rank-Deficient and Discrete Ill-Posed Problems: Numerical Aspects of Linear Inversion. SIAM, Philadelphia (1998)

Estimation and Calibration

Estimation and Calibration

Ioannis Pekleitis

This session was dedicated to state estimation and calibration. The first paper of Serlin, Sookraj, Belta, and Tron was about tracking objects observed by multiple robots and was focused on selecting the right robot for observing a moving object and then accurately tracking the identified object. The main challenge is on matching the features across multiple images to select which views contain the object. While currently the robots are in fixed positions, the next step will be to extend this work on moving observers. The second paper by Joukov, Lin, Westermann, and Kulić was about modeling the human body for tracking motions in a motion capture environment and focused on handling occlusions; in particular the problems of missing and misidentified markers. The proposed approach utilizes a Kalman filter to track the motion capture markers and label them as missing when they are occluded. The paper by Zhang and Nakamura was presented next, discussing a novel dense SLAM approach based on RGB-D data. The method is targeted to dynamic human-made environments. The main challenge is the large scene occlusions while the approach benefits by the many features and rigid structures present. The fourth paper by Marr and Kelly was focused on the calibration of 2D-LIDAR and monocular camera sensors moving in 3D. Utilizing the egomotion of the camera in 3D an approximate position of the LIDAR points is estimated. The initial point cloud is then processed with the calibration parameters as the free variables in the optimization framework. The final paper of the session was by Nissler, Durner, Marton and Triebel on enhancing the camera and camera-camera calibration by incorporating the results from a mapping module. The proposed approach is particularly beneficial when the cameras have non-overlapping fields-of-view. All the research presented in this session was focused on well-lit, well-structured, feature-rich environment demonstrating the challenging nature of the state estimation and calibration problem.

Consistent Multi-robot Object Matching via QuickMatch

Zachary Serlin(✉), Brandon Sookraj, Calin Belta, and Roberto Tron

Boston University, Boston, MA 02446, USA
zserlin@bu.edu

Abstract. In this work, we present a novel solution and experimental verification for the multi-image object matching problem. We first review the QuickMatch algorithm for multi-image feature matching and then show how it applies to an object matching test. The presented experiment looks to match features across a large number of images and features more often and accurately than standard techniques. This experiment demonstrates the advantages of rapid multi-image matching, not only for improving system performance, but also for use in new applications, such as object discovery and localization.

Keywords: Computer vision · Feature matching · Object matching

1 Motivation

In this paper, we propose a solution to the following problem: given a set of images taken from a team of robots (or camera network), match unique object features, as they enter and exit the images from multiple perspectives. This problem is fundamental to both computer vision and robotics applications, where feature matching can be used in object detection, localization, and tracking [2,21], homography estimation [15], structure from motion [6], and formation control [10]. Solutions to this problem are traditionally computationally complex, and often mismatch features when considering more than two images [2,8]. Multi-image correspondences allow for greater match reliability, and a more accurate representation of objects in the universe. The proposed solution leverages a relatively recent algorithm, QuickMatch [17], to quickly and reliably discover correspondences across multiple images. The experiments presented in this paper benchmark QuickMatch's performance by implementing an object matching framework under realistic conditions (i.e. images with clutter, repeated structures, and poor image quality); a target object is matched across a network of cameras, and then these matches are used to generate the target's trajectory.

Electronic supplementary material The online version of this chapter (https://doi.org/10.1007/978-3-030-33950-0_64) contains supplementary material, which is available to authorized users.

J. Xiao et al. (Eds.): ISER 2018, SPAR 11, pp. 751–761, 2020.
https://doi.org/10.1007/978-3-030-33950-0_64

2 Problem Statement

Given a set of images $\mathcal{I} = \{1, 2, \ldots, i, \ldots, N\}$ and a set of K_i feature vectors, x_{ik}, extracted from each image, determine matches $(x_{i_1 k_1} \leftrightarrow x_{i_2 k_2} : i_1 \neq i_2)$ between features from separate images, such that matched features represent the same point in the scene.

3 Related Work

Feature matching is a basic process in many computer vision algorithms. *Pairwise matching* is the classical approach to this task, where features between two images are compared based on a distance metric (e.g. Euclidean or Manhattan distance), and declared a match if this distance is below some threshold [8,18]. This method is used in two standard algorithms, *Brute Force* (BF) matching, and *Fast Library for Approximate Nearest Neighbors* (FLANN) matching. Pairwise matching has difficulties matching entities with repetitive structure or similar appearance (e.g. windows) because the distance metric alone does not consider the *distinctiveness* (smallest distance between features from the same image) of the features. Including distinctiveness of features during matching has been shown to be beneficial [8]. For multi-image matching, pairwise matches scale poorly with the number of images and across multiple images, match correspondences often do not belong to the same ground truth object. *Graph matching* has also been used for pairwise matching. This approach attempts to match vertices (features) and edges (matches) simultaneously to determine better pairwise matches [20], but it cannot handle the multi-image setting.

Beyond pairwise matching, a number of other approaches exist for multi-image matching (where multiple images are directly considered) that are based on optimization, graphs, and clustering. *Optimization* based approaches are based upon non-convex problems where optimization constraints must often be relaxed to reliably obtain solutions [12,20]. Moreover, these approaches require to know *a priori* the number of objects, which is often not available, and do not consider distinctiveness of the features. *Cycles in graphs* are early predecessors to the QuickMatch algorithm and have largely been used to remove inconsistent matches [7]. *Clustering* can be cast as finding clusters of similar features. Algorithms such as k-means [9] and spectral clustering [11] have been explored to this end, but also often require a predefined number of objects, and do not consider that a unique feature only occurs once in an image.

QuickMatch uses *density-based clustering* algorithms [4,19], which find clusters by estimating a non-parametric density distribution of data [13,14]. These approaches do not require prior knowledge of the number or shape of clusters, and can be modified to include feature distinctiveness by construction. This paper is an experimental extension of [17], where QuickMatch was initially introduced.

4 Contribution

The primary contribution of this paper is the testing and experimental validation of the QuickMatch algorithm under more realistic conditions, as opposed to previous evaluations using standard datasets. This experiment tests the algorithm for computational efficiency and match accuracy by employing a distributed camera network to localize a moving target.

5 Technical Approach

A two-stage, offline, centralized solution is implemented on a system of distributed ground robots and a central computer. Features are first extracted using off-the-shelf feature extraction methods (SIFT), and the features are then matched using the QuickMatch algorithm to find a given reference object. These matches are used to perform homography estimation between the reference object and the camera network to generate target trajectories.

5.1 Feature Extraction

Feature extraction aims to find and describe representative points from high dimensional data, such as an image [1,5,8,21]. Features themselves are also high dimensional vectors. In this experiment, the *scale invariant feature transform* (SIFT) feature is used, which extracts K_i 128-dimensional vectors that represent the appearance of each feature point. See [8,18] for more details on this standard feature extraction algorithm. Other feature types can be used and we also tested with Oriented FAST and Rotated BRIEF (ORB) features and Speeded-Up Robust Features (SURF), however SIFT was the most reliable.

5.2 QuickMatch

The QuickMatch algorithm is a density based clustering algorithm. It begins by calculating the distance between all features (we use Euclidean distance in our application). For each image, the minimum distance, σ_i, between any two features is used as the distinctiveness of features for that image. Recall, from above, x_{ik} is a point in the high dimensional feature space. The feature density $D(x_{ik})$ is then calculated for each point using the formula

$$D(x) = \sum_{i=1}^{N} \sum_{k=1}^{K_i} h(x, x_{ik}; \sigma_i), \tag{1}$$

$$h(x_1, x_2; \sigma) = \exp(-\frac{\|x_1 - x_2\|}{2\sigma^2}), \tag{2}$$

with kernel function h, and distinctiveness σ_i. With this feature density, the features are organized into a tree structure, with parent nodes being the nearest neighbor with a higher density.

$$\text{parent}(x_{ik}) = \underset{i'k' \in J}{\arg\min}\, d(x_{ik}, x_{i'k'}), \tag{3}$$

$$J = \{i'k' : k \neq k', D(x_{i'k'}) > D(x_{ik})\}. \tag{4}$$

Edges are directed to parents along the gradient of feature density, and ultimately toward the center of the parent cluster or to another distant cluster. Once the tree has been constructed, edges are broken if either of two criteria are met; (1) parent and child groups have nodes from the same image, or (2) the edge is larger than a user defined threshold (ρ) times σ_i. This method results in a forest of trees, where each tree is a cluster representing a unique entity in the universe. In practice, each tree represents a point that is common among images, meaning the algorithm discovers common features among very similar objects. Feature discovery will be explored further in Sect. 7.3, where groups of matching points are organized into object detections and homography transformations.

5.3 Homography and Localization

An homography is a perspective transformation between the view points of two images that can also be used to determine the relative position of an object given a reference image. Given a reference viewpoint $\widetilde{x}$, a new viewpoint can be found given the homography matrix $\overline{H}$ as $\overline{H}\widetilde{x}$. The $\overline{H}$ matrix can be estimated with a set of known relative points (or matched features) between the two images. Once $\overline{H}$ is estimated, it is possible to compare the position of objects in each image in a relative coordinate system. To improve the estimate of $\overline{H}$, random sample consensus (RANSAC) is used to remove match outliers by randomly sampling the matches, finding a fit of the data, and then removing any matches that fall outside of a user defined region [15].

Using the homography transformations between each image and a target reference image, the object can be localized up to a distance scale factor, as shown in Fig. 1(a). Given a known parameter of the target object, in this case the object's height, this ambiguity can be resolved, and the relative position can be determined. When taken together with other cameras in the network and a known global camera pose, the target object can be accurately positioned in the global reference frame, allowing for generating a target's trajectory (e.g., Fig. 4).

Homography and localization are limited by the reference images used for matching, and are prone to noisy and inaccurate measurements. Firstly, the system can only identify the known side of the object, unless the target is symmetric. To overcome this, multiple reference target images are used here. Secondly, inaccurate measurements in distance and bearing are common. These inaccuracies arise from extreme sensitivity to object height estimate errors when calculating target distance. To account for these errors in practice, multiple measurements can be used to estimate each position, and then a filter can be used to smooth the target's trajectory.

6 Experiment

The experiment consists of a team of five iRobot Create2 ground robots, each with a forward facing camera, distributed throughout the experimental area shown in Fig. 1. Each camera has a $62^\circ \times 48^\circ$ field of view, and takes a 640×480 px image at 2 Hz. Through the center of the area, the target object is driven along the trajectory shown in Fig. 1(a) over approximately thirty seconds. All cameras are triggered simultaneously and the images are sent to a central computer for feature extraction and matching. The central computer has an Intel i7-7800x 3.5 GHz processor, and runs Ubuntu 16.04 LTS and ROS Kinetic. Features are extracted using SIFT with an octave layer of 6, a contrast threshold of 0.10, an edge threshold of 15, and sigma of 1.0. The matches from QuickMatch (using $\rho = 1.1$) are used to determine which cameras observe the target object at each time step, based on the number of matches with a target image (in this experiment 10 matches are required). The matches between each reference images and the current images are used to determine the homography between them, using RANSAC with a threshold of 10.0. The homography is used to generate a bounding box around the target object using a perspective transformation on the target image corners. The relationship between pixel height of this box and distance from the camera is calibrated beforehand using an object of known size (in this experience a checkerboard pattern of know dimension). The localization points are recorded to build a target trajectory, which is then compared to ground truth measurements from an OptiTrack© motion capture system (Fig. 1(b)).

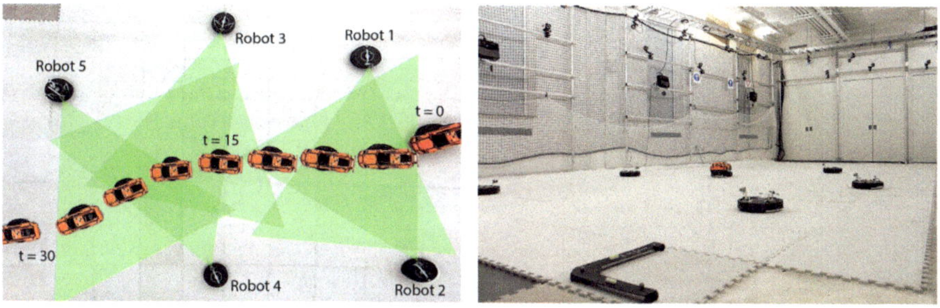

(a) (b)

Fig. 1. (a) Overhead view of experimental area with trajectory of the target object, position of the robots, and the approximate field of view for the camera network (shown in yellow). (b) Prospective view of experimental area with modified iRobot Create2 platform, target object, and overhead OptiTrack© motion capture system.

7 Results

QuickMatch is evaluated in two ways: pure matching performance, and in the context of a target localization application. The QuickMatch algorithm is first compared to standard matching algorithms in the OpenCV Software Package [2], Brute Force (BF), and FLANN. Both algorithms use the Euclidean distance metric and a threshold match distance of 0.75 [2,8]. Unlike QuickMatch, both algorithms cannot consider matches across more than two images but do have very low execution times.

QuickMatch is implemented in Python and takes 5.6 s to find matches between 6254 SIFT features (from 115 images), while BF and FLANN are both implemented in C++, and both take approximately 0.05 s to find the matches between the reference image features, and the same 6254 features. This time difference arises from two factors: the inherently slower runtime of Python compared to C++ [3], and the extra comparisons done by QuickMatch to solve the entire Multi-match problem. If BF and FLANN compared all images with all other images combinatorially (as QuickMatch implicitly does) their computation times would be ~5.75 s, which is comparable to QuickMatch's slower Python implementation. This time also does not account for the post processing time necessary to reconcile inconsistent matches from both BF and FLANN, is not required in QuickMatch.

7.1 Precision Versus Recall

Although QuickMatch is slower, it outperforms both BF and FLANN in the number of matches correctly found, and generally in terms of precision vs. recall (PR) and precision-recall area under the curve (PR AUC), which are common metrics for evaluating matching algorithms [16]. Figure 2(a) shows the precision (fraction of correctly matched images) versus recall (fraction of possible matches

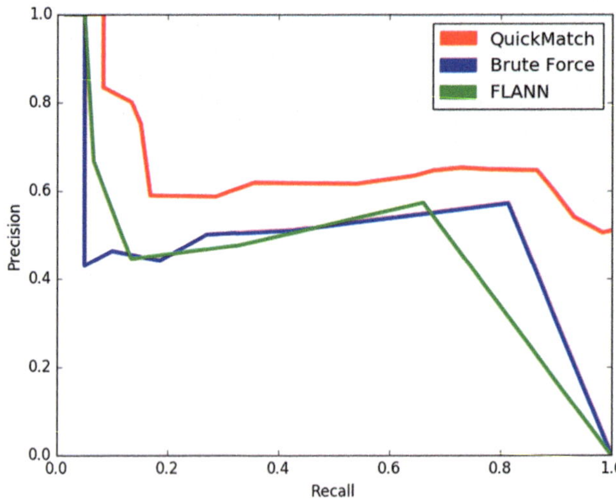

Fig. 2. (a) Precision vs. recall curves for the QuickMatch, Brute Force, and FLANN algorithms. All algorithms are run on the same feature vectors. A match is considered to exist if the number of matched features is above a threshold.

found) curves for QuickMatch, BF, and FLANN. For any recall level, QuickMatch maintains a higher precision level than either BF or FLANN. These curves are non-monotonic because mismatched features appear at a higher rate than correctly matched features at higher thresholds. PR AUC is a threshold agnostic metric used for comparing overall performance of matching algorithms [16]. In terms of PR AUC, QuickMatch achieves 0.64, while BF and FLANN reach 0.49 and 0.45 respectively. The overall increase in precision stems for QuickMatch's ability to consider more instances of the reference object, by matching cycles of features across multiple images. It is therefore able to find the reference object not only more consistently, but with many more matched features. An example of these matches is shown in Fig. 3.

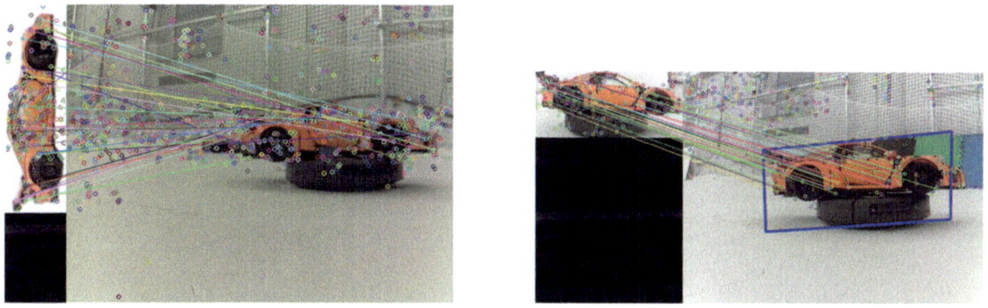

(a) (b)

Fig. 3. (a) Example image matches between the reference object image (left) and an experimental image (right): Circles represent features, and lines indicate matches. (b) Homography and localization of car with prospective transform of bounding box.

7.2 Homography and Localization

In order to further demonstrate the utility of the QuickMatch algorithm, matches were used to localize a target object in relation to the camera network, and then estimate its global trajectory. This was done using all three above algorithms with again an identical set of SIFT features. QuickMatch considers multi-image matches between the set of target images and the set of five robot images at each time step, while BF and FLANN consider matches between each target image and the robot image individually. Once feature matches are generated, RANSAC is used to estimate the homography matrix $\overline{H}$ for each pair of images while also removing outliers from the matches. The homography between the reference image and each robot image is used to generate a bounding box around the target in the robot image as shown in Fig. 3(b). This bounding box, given a known camera calibration, provides bearing and height information for the target. The target height is known and is used to find the relate distance to the target with the bounding box height. With these two values, a distance and a bearing, the object can be localized with respect to each robot.

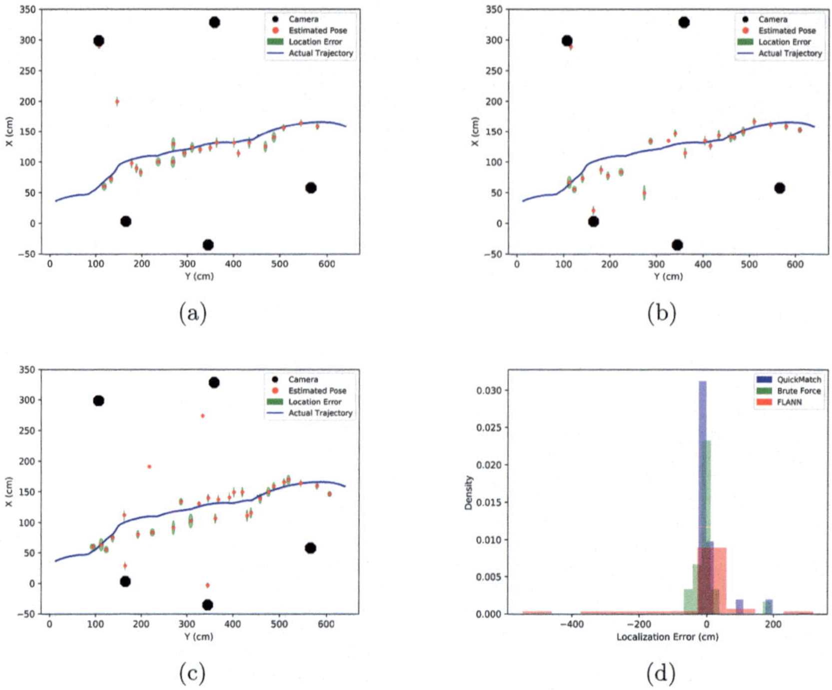

Fig. 4. (a) QuickMatch trajectory estimate. (b) BruteForce trajectory estimate. (c) FLANN trajectory estimate. (d) Histogram of estimate error for each algorithm.

The above steps are performed using the match data from each of the three above algorithms. Figures 4(a–c) show the results of the localization estimation for each algorithm. Red points are estimate target poses for each time step, blue points denote the ground truth measurements, black octagons are the camera network positions, and the green regions are the one standard deviation error between all localization estimates at each time step. The localization error was found by taking the absolute distance between the estimated and ground truth position at each time step. QuickMatch had an error of 0.2118 ± 0.4254 m, BF had an error of 0.2349 ± 0.4027 m, and FLANN had an error of 0.6232 ± 1.1722 m. QuickMatch outperforms both BF and FLANN in terms of accuracy, which is indicative of its higher match quality. BF matcher also performs well and maintains a low variance, however it is not as accurate. FLANN is the worst performing of the three, and has a number of extremely erroneous estimates. Generally, monocular camera distance measurements are very sensitive to match errors, meaning target localization error is an indirect method for testing the overall accuracy of each method. Figure 4(d) shows a histogram of the localization error, which is found by comparing the localization estimate to the ground truth pose at each time step. The histogram makes it clear that QuickMatch maintains a higher number of accurate matches and has a small number of highly erroneous estimates. In practical applications, a Kalman filter would be employed to smooth the estimates, but the values are left unaltered here to demonstrate the algorithm's output.

7.3 Feature Discovery

The QuickMatch algorithm implicitly discovers common features among images by creating clusters of similar features. These clusters correspond to specific locations in the universe, and therefore can be used to find both targets and landmarks across images. Landmarks, although not used in this paper, are points that occur commonly across all images (except when occluded), and are useful for multi-agent localization tasks. In the images collected, landmarks were the clusters with the largest number of features, because many of this images did not contain the target object. An example landmark cluster is shown in Fig. 5(a). Features belonging to the target object are generally smaller than the landmark clusters, but can still be extracted, and show key features of the target. Figure 5(b) shows one such cluster, which is the front hood of the car model. Feature discovery is one attribute of QuickMatch that does not exist in either BF or FLANN and can be useful for discerning what features are most descriptive of images from the network.

(a)

(b)

Fig. 5. (a) Landmark feature cluster. (b) Target feature cluster.

8 Conclusion

This experiment highlights the utility of QuickMatch multi-image matching for object matching. QuickMatch is able to find many more object feature matches than standard methods by considering matches across all images, not just pairwise matches. The presented experiment tests the QuickMatch algorithm in an experimental setting with realistic conditions, and shows that multi-image matching is superior to standard methods at matching the reference object (even as it enters and exits images across the entire camera network). Quickmatch is also tested with a target object localization and again outperforms both the BF and

FLANN algorithms. Beyond testing Quickmatch, this experiment also demonstrates its feature discovery ability by showing a characteristic landmark and target feature cluster from the test images. This approach is the precursor to an online, distributed, and decentralized approach. Our future work will focus on a distributed version of object discovery and localization and multi-camera homography. We also plan to use these extracted trajectories for higher level tasks. Overall, QuickMatch is shown to be a versatile multi-feature matching algorithm that outperforms standard pairwise matching algorithms.

Acknowledgements. This work was supported by the National Science Foundation under grants NRI-1734454, and IIS-1717656.

References

1. Bay, H., Ess, A., Tuytelaars, T., Gool, L.V.: Speededup robust features (SURF). Comput. Vis. Image Underst. **110**(3), 346–359 (2008)
2. Bradski, G.: The openCV library. Dobb's J. Softw. Tools **25**, 120–125 (2000)
3. Fourment, M., Gillings, M.: A comparison of common programming languages used in bioinformatics. BMC Bioinform. **9**, 82 (2008)
4. Fukunaga, K., Hostetler, L.: The estimation of the gradient of a density function, with applications in pattern recognition. IEEE Trans. Inf. Theory **21**(1), 32–40 (1975)
5. Hariharan, B., Arbelaez, P., Girshick, R., Malik, J.: Hyper-columns for object segmentation and fine-grained localization. In: IEEE Conference on Computer Vision and Pattern Recognition (2015)
6. Hartley, R., Zisserman, A.: Multiple View Geometry in Computer Vision, 2nd edn. Cambridge University Press, Cambridge (2004)
7. Huang, Q., Guibas, L.: Consistent shape maps via semidefinite programming. Comput. Graph. Forum **32**(5), 177–186 (2013)
8. Lowe, D.G.: Distinctive image features from scale-invariant keypoints. Int. J. Comput. Vis. **60**(2), 91–110 (2004)
9. MacKay, D.J.: Information Theory, Inference and Learning Algorithms. Cambridge University Press, Cambridge (2003)
10. Montijano, E., Cristofalo, E., Zhou, D., Schwager, M., Sagues, C.: Vision-based distributed formation control without an external positioning system. IEEE Trans. Robot. **32**(2), 339–351 (2016)
11. Ng, A.Y., Jordan, M.I., Weiss, Y.: On spectral clustering: analysis and an algorithm. Neural Inf. Process. Syst. **2**, 849–856 (2002)
12. Oliveira, R., Costeira, J., Xavier, J.: Optimal point correspondence through the use of rank constraints. In: IEEE Conference on Computer Vision and Pattern Recognition, vol. 2, pp. 1016–1021 (2005)
13. Parzen, E.: On estimation of a probability density function and mode. Ann. Math. Stat. **33**(3), 1065–1076 (1962)
14. Rosenblatt, M.: Remarks on some nonparametric estimates of a density function. Ann. Math. Stat. **27**(3), 832–837 (1956)
15. Szeliski, R.: Computer vision: algorithms and applications. Springer, Heidelberg (2010)
16. Ting, K.M.: Precision and recall. In: Sammut, C., Webb, G.I. (eds.) Encyclopedia of Machine Learning. Springer, Boston (2011)

17. Tron, R., Zhou, X., Esteves, C., Daniilidis, K.: Fast multi-image matching via density-based clustering. In: The IEEE International Conference on Computer Vision (2017)
18. Vedaldi, A., Fulkerson, B.: VLFeat: an open and portable library of computer vision algorithms (2008). http://www.vlfeat.org/
19. Vedaldi, A., Soatto, S.: Quick shift and kernel methods for mode seeking. In: IEEE European Conference on Computer Vision, pp. 705–718. Springer (2008)
20. Yan, J., Cho, M., Zha, H., Yang, X., Chu, S.: Multi-graph matching via affinity optimization with graduated consistency regularization. IEEE Trans. Pattern Anal. Mach. Intell. **38**, 1228–1242 (2015)
21. Zhou, X., Zhu, M., Daniilidis, K.: Multi-image matching via fast alternating minimization. In: The IEEE International Conference on Computer Vision (2015)

Real-Time Unlabeled Marker Pose Estimation via Constrained Extended Kalman Filter

Vladimir Joukov(✉), Jonathan F. S. Lin, Kevin Westermann, and Dana Kulić

University of Waterloo, Waterloo, Canada
vjoukov@uwaterloo.ca

Abstract. Marker-based based motion capture is the prevalent technique for estimating human motion. A common problem with the approach is the occlusion and mis-labeling of the markers; typically the data requires tedious manual cleaning in post processing. We present a constrained extended Kalman filter method that estimates full body human motion in real time and handles missing and mis-labeled markers. The approach is validated on two datasets and is shown to produce comparable results to using manually cleaned data. The constrained estimator ensures realistic human joint trajectories that satisfy kinematic limits.

Keywords: Motion Capture · Kalman filter

1 Introduction

Marker-based motion capture is the gold standard of capturing and analyzing human motion. A set of markers attached to the participant is tracked by multiple cameras to produce the location data of each marker. In robotics [1], motion capture is used extensively for human motion imitation, improving humanoid robot design, and human-robot interaction. However, marker occlusions, marker label swapping, and unwanted reflective surfaces all contribute to estimation error. Our goal is to create a system that can estimate full body motion using this raw marker data in real-time.

A common way to handle these problems is to fill in the missing regions using interpolation and manually correct marker labels; a time consuming task. The extended Kalman filter (EKF) is a popular approach for filling in missing markers in real time. It has been used to track individual markers [2] and rigid objects [3]. After the missing marker data is filled in, the joint angles can be estimated from the pose of each limb using inverse kinematics [4]. The aforementioned approaches

Electronic supplementary material The online version of this chapter (https://doi.org/10.1007/978-3-030-33950-0_65) contains supplementary material, which is available to authorized users.

J. Xiao et al. (Eds.): ISER 2018, SPAR 11, pp. 762–771, 2020.
https://doi.org/10.1007/978-3-030-33950-0_65

do not consider human kinematics when estimating missing marker data, potentially impacting joint angle accuracy. Meyer *et al.* [5] proposed a probabilistic approach that both builds the kinematic model and associates markers to each limb, but requires computationally expensive optimization. Steinbring *et al.* [6] used the smart sampling Kalman filter with a kinematic model utilizing the sample-based covariance estimate for marker matching, and to incorporate constraints via a non-linear mapping between the state and the joint angles.

Previously, we applied the EKF to estimate pose and used the projected covariance matrix to compute correspondence probabilities between modeled and measured markers [7]. However, the approach was computationally expensive. This previous work did not guarantee positive symmetric definite (PSD) covariance matrices, leading to invalid probabilities. In this paper, we introduce a new constrained EKF-based method for real-time full body marker-based pose estimation that can handle missing and swapping marker labels. Our method does not track each marker separately but uses a correspondence measure between the modeled and observed markers based on the error covariance of the kinematic model estimated by EKF. The constraints on the state estimate allow our approach to produce motion that satisfies human joint limits.

2 Technical Approach

We model the human body as an articulated rigid body skeleton. Adjacent bones are connected with revolute joints $\mathbf{q} = [q_1 \cdots q_N]$, with three perpendicular revolute joints used to represent spherical joints at the back, shoulder and hip. The transformation between the world frame and the base of the kinematic model is represented by 3 prismatic and 3 revolute joints. Each marker is assumed to be rigidly attached to the model (Fig. 2).

2.1 EKF Formulation

The EKF state consists of the joint positions and velocities $\mathbf{x} = [\mathbf{q}\ \dot{\mathbf{q}}]^T$. We assume a constant velocity model $\mathbf{A}$ for the state update function

$$\mathbf{x}_{k+1} = \underbrace{\begin{bmatrix} \mathbf{I} & \mathbf{\Delta t} \\ 0 & \mathbf{I} \end{bmatrix}}_{\mathbf{A}} \mathbf{x}_k + \mathbf{w}_k \tag{1}$$

where $\mathbf{x}_k = [\mathbf{q}_k\ \dot{\mathbf{q}}_k]^T$ is the vector of joint angles and velocities at time k, Δt is the sampling interval, and $\mathbf{w}_k$ is the white Gaussian noise (WGN) with zero mean and covariance $\mathbf{Q}_k$.

Our measurement vector consists of the observed 3D positions of m markers $\mathbf{z}_k = [x_1\ y_1\ z_1 \cdots x_m\ y_m\ z_m]^T + \mathbf{v}_k$ that are related to the state through non-linear forward kinematics $\mathbf{z}_k = h(\mathbf{x}_k)$. We assume that any measurement noise $\mathbf{v}_k$ from the motion capture system is WGN with covariance $\mathbf{R}_k$. Assuming that the correspondences between the measured and predicted marker positions are known, EKF can be used to recursively estimate the state [7].

Algorithm 1. EKF iteration: $\mathbf{H}_k = \frac{\partial h}{\partial x}$ is the measurement Jacobian, $\mathbf{P_k}$ is the state error covariance, $\mathbf{P_z}$ is the projected measurement error covariance, and ${}^R\mathbf{K}_k$ is the restricted optimal Kalman gain. After computing the restricted Kalman gain such that $\mathbf{x}_k < \mathbf{b}$ is satisfied we assume the joints at their limits are fixed and recompute the gain for the unrestricted joints by using the appropriate columns and rows of $\mathbf{H}_k$ and $\mathbf{P}_k^-$, $\bar{\mathbf{H}}_k$ and $\bar{\mathbf{P}}_k^-$ respectively.

1: Prediction step:
$\hat{\mathbf{x}}_k^- = \mathbf{A}\hat{\mathbf{x}}_{t-1}$
$\mathbf{P}_k^- = \mathbf{A}\mathbf{P}_{k-1}\mathbf{A}^T + \mathbf{Q}_{k-1}$
2: Update step:
$\tilde{\mathbf{z}_k} = \mathbf{z}_k - h(\hat{\mathbf{x}}_k^-)$
Compute restricted Kalman gain:
$\mathbf{P}_z = \mathbf{H}_k\mathbf{P}_k^-\mathbf{H}_k^T + \mathbf{R}_k$
$\mathbf{K}_k = \mathbf{P}_k^-\mathbf{H}_k^T\mathbf{P}_z^{-1} - (\hat{\mathbf{x}}_k^- - \mathbf{b})(\tilde{\mathbf{z}}^T\mathbf{P}_z^{-1}\tilde{\mathbf{z}})^{-1}\tilde{\mathbf{z}}^T\mathbf{P}_z^{-1}$ [8]
Recompute unrestricted Kalman gain for joints not at their limit:
$\bar{\mathbf{P}}_z = \bar{\mathbf{H}}_k\bar{\mathbf{P}}_k^-\bar{\mathbf{H}}_k^T + \mathbf{R}_k$
$\bar{\mathbf{K}}_k = \bar{\mathbf{P}}_k^-\bar{\mathbf{H}}_k^T\bar{\mathbf{P}}_z^{-1}$
Update state and covariance combining restricted and unrestricted gains:
$\mathbf{K}_k = [\bar{\mathbf{K}}_k; \mathbf{K}_k]$
$\hat{\mathbf{x}}_k = \hat{\mathbf{x}}_k^- + \mathbf{K}_k\tilde{\mathbf{z}}$
$\mathbf{P}_k = (\mathbf{I} - \mathbf{K}_k\mathbf{H}_k)\mathbf{P}_k^-(\mathbf{I} - \mathbf{K}_k\mathbf{H}_k)^T + \mathbf{K}_k\mathbf{R}\mathbf{K}_k^T$ [9]

We restrict the optimal Kalman gain [8] to satisfy human joint constraints. To ensure that the Kalman gain for the unconstrained joints remains near optimal after the restriction is applied, we assume the joints at their constraints are fixed and recompute the gain for the rest. To keep the error covariance matrix PSD, as required for our proposed marker matching, we use the Joseph form EKF covariance update [9].

2.2 Marker Matching

In practice, reflective marker-based motion capture systems often do not correctly label the markers between consecutive frames, requiring correspondences between the observed and measured markers to be identified. Furthermore, since not all of the markers may be visible in every frame, the EKF must be able to correctly change the size of the measurement vector.

In order to identify the correspondences we propose to use the inverse of the error covariance projected onto the measurement space $\mathbf{P}_z^{-1}$ to build a measure of correspondence between each observed and modeled marker. Noting that the 3×3 diagonal blocks of $\mathbf{P}_z^{-1}$ represent the covariance Σ_j^{-1} of each marker we can compute the logarithm of the normalized observation probability

$$m_{ij} = log(P(\mathbf{z}_i|\hat{\mathbf{z}}_j)) = (\mathbf{z}_i - \hat{\mathbf{z}}_j)^T\Sigma_j^{-1}(\mathbf{z}_i - \hat{\mathbf{z}}_j) \quad (2)$$

where $\mathbf{z}_i = [x_i \ y_i \ z_i]$ is the measured marker position and $\hat{\mathbf{z}}_j = [\hat{x}_j \ \hat{y}_j \ \hat{z}_j]$ is the predicted marker position using forward kinematics. We build a match matrix

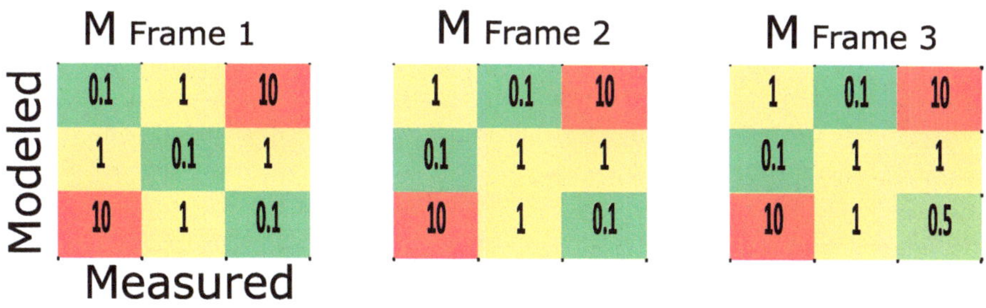

Fig. 1. Marker matching process: Consider the above **M** matrices for 3 scenarios with the matching threshold set to 0.3. (1) The proposed approach matches each marker to the modeled one since the minimum measure is on the diagonal and falls below the threshold (left). (2) A swap between marker 1 and 2 is detected (middle). (3) Marker 3 is declared missing and not used in EKF update (right).

M containing the logarithm probabilities between each measured and modeled marker and track the minimum value indices. Missing and newly matching markers are detected when the minimum value falls below a threshold. Swapping minimum indices between two modeled markers indicates a swap in the motion capture marker labeling. Finally, only values, rows and columns of $\tilde{\mathbf{z}}_k$, $\mathbf{H}_k$ and $\mathbf{K}_k$ that correspond to matched markers are used in the EKF update, respectively (Fig. 1). Note that computing $\mathbf{P}_z^{-1}$ is necessary for the EKF update step and thus our approach does not significantly add to the iteration time, allowing us to run the algorithm on-line.

3 Results

We validate our proposed approach on two different motion capture datasets.

3.1 Broad Jumping Exercises

The first dataset consists of 21 participants performing standing broad jump motions to 3 different distances. A modified Helen Hayes full-body marker set was used, consisting of 30 markers. Both the original uncleaned marker data (*raw data*), as well as a manually cleaned version (*clean data*), are available. Trajectories that could not be fully manually cleaned were excluded from this analysis, resulting in a total of 680 trajectories.

The EKF initial covariance is set to identity, process noise assumes WGN with covariance of 10^6 is present in the position, while measurement noise is set to 10 mm. A threshold of 100 m^2 was used for the correspondence measure. Joint constraints were enabled for the knee, ankle and elbow joints, using the human range limits identified in [10].

We compare three configurations: (1) the *standard raw configuration*, where the raw marker data is used as the input into the standard EKF without correspondence adjustment. Missing marker data was filled using the last known value; (2) the *standard clean configuration*, where the manually cleaned data is used as input into the standard EKF; (3) the *proposed raw configuration*, where the raw marker data is used as the input into our proposed algorithm.

We evaluate two error metrics: (1) marker position root mean square error (RMSE, ME in Table 1), where forward kinematics and the joint angles obtained from each of the three tests are used to estimate the marker locations; these estimates are compared to the cleaned data; (2) joint angle error (JE), where the joint angles estimated by the proposed method are compared to the joint angles obtained using the clean configuration. For spherical joints, a joint error metric based on rotation matrix distance (RD) [11], was used. This metric calculates the distance between two rotation matrices by taking the Frobenius norm of the $||I - R_1 R_2^T||_F$, where I is the identity matrix, and R_1 and R_2 are the two matrices being compared. This value is normalized such that 0 denotes maximum difference and 100 denotes no difference (Table 1).

In addition to the error metrics, we report the marker correspondence results of the proposed algorithm. A given marker can be identified (ID) as: (1) correct if it is associated to the marker of the same name in the observation data, (2) swapped if the names are different, (3) missing if there is no marker associated to the model marker. The absolute maximum joint velocity (MJV) is also calculated. If the MJV is high, and larger than expected for human motion it suggests that marker correspondence is not estimated accurately and thus joint estimates are changing rapidly to match to incorrectly assigned markers.

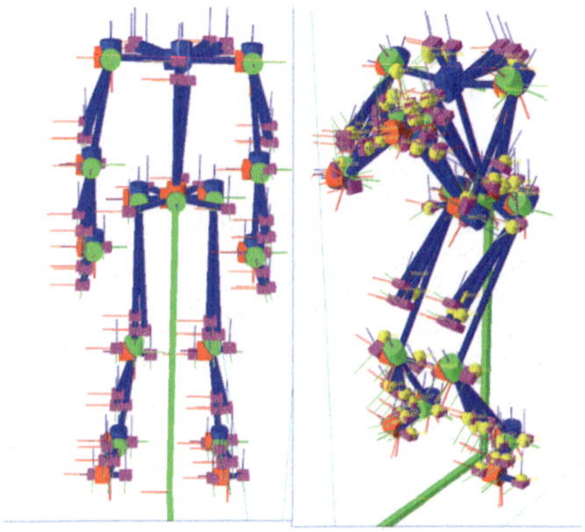

Fig. 2. Full body kinematic model with 80 markers with the participant standing upright (left) and during squat motion (right). Modeled and measured markers are drawn as pink boxes and yellow spheres, respectively. Note the numerous markers in very close proximity as the participant crosses their arms at the bottom of the squat.

The results show that RMSE is highest when using raw marker data without our proposed EKF refinements. Due to left-right symmetry, the motion capture system sometimes mislabels the left side of the body as the right side, resulting in the large RMSE observed. Our proposed approach using raw data achieves similar performance to using manually cleaned data, suggesting that the proposed algorithm is successful at detecting missing and swapping marker events, even when only a minority of the markers were deemed correctly labelled by EKF.

Table 1. The mean marker position RMSE (ME) (mm) using the standard raw (StR), standard clean (StC) and proposed raw (PrR) configurations on the broad jump dataset over different limbs, with the smallest value in each column denoted in bold. The marker identity (ID) [%] denotes if the proposed raw configuration denoted a given marker in a given frame as correct (Cor), swapped with another marker (Swa), or missing (Mis). The mean joint angle RMSE (JE) (deg) for non-spherical (Rev) joints are provided, as well as the rotation matrix distance (RD) Frobenius norm error metric [0, 100, where 100 is perfect match] for spherical joints (Sph). The absolute maximum joint velocity values (MJV) for StC (VC) and PrR (VR) are also provided. RMSE values are calculated against the original cleaned motion capture data, or against the StC EKF-derived joint angles. Note that not all body segments have both revolute and spherical joints, and are left blank in the table.

		Hip	R Leg	L Leg	Torso	R Arm	L Arm	Avg
ME	StR	51 ± 41	41 ± 65	45 ± 67	73 ± 70	165 ± 227	170 ± 225	81 ± 78
	StC	**33 ± 4**	**22 ± 8**	21 ± 3	**29 ± 6**	**12 ± 3**	**11 ± 3**	**21 ± 3**
	PpR	35 ± 3	**22 ± 8**	**20 ± 3**	30 ± 9	13 ± 5	13 ± 6	22 ± 3
JE	Rev		1 ± 12	1 ± 1	2 ± 3	5 ± 14	3 ± 13	2 ± 5
RD	Sph	99 ± 1	99 ± 1	99 ± 1		99 ± 3	99 ± 4	99 ± 2
ID	Cor	86 ± 29	90 ± 28	91 ± 28	83 ± 23	58 ± 42	52 ± 40	79 ± 21
	Swa	9 ± 27	10 ± 28	9 ± 28	14 ± 22	40 ± 42	47 ± 40	19 ± 21
	Mis	6 ± 10	1 ± 2	1 ± 2	3 ± 6	1 ± 3	1 ± 3	2 ± 2
VC	Rev		421 ± 189	413 ± 62	187 ± 220	351 ± 137	356 ± 155	344 ± 128
VR	Rev		684 ± 1515	575 ± 101	443 ± 265	1231 ± 1991	980 ± 1284	702 ± 525

Figure 3 demonstrates various joint angle tracking cases. Case 1 shows joint angle tracking with multiple missing markers. Maintaining the last known marker position, to keep the regular EKF measurement vector consistent size, causes the limb or the entire body to be pulled towards that marker. This can create a large joint angle estimation error. The proposed approach correctly identifies missing markers using the covariance based measure and excludes them from the measurement vector, maintaining a much better joint angle estimate. Since most motion capture systems do not use temporal marker information labels of markers close to each other, such as ones placed on the wrist, they are often swapped. Without the proposed swap detection regular EKF will rotate joints by 180° to match the swapped marker locations (Case 2). Our proposed approach can successfully detect swapping markers and does not rely on motion capture system labelling apart from the initial frame. When markers are occluded and only a single marker is visible on a link (Case 3), EKF will still attempt to minimize the error to the remaining marker, and may cause an unrealistic rotation. Once multiple markers are visible again the estimator with quickly snap to the correct solution, causing large velocity values when compared to cleaned MJV. These cases can be identified for manual cleaning by looking at abnormally high estimated joint velocity peaks.

3.2 Full-Body Exercises

The second dataset consists of 15 participants performing various full body motions with a total of 80 markers on the body. Due to the number and proximity of markers, occlusions were frequent and it was infeasible to manually clean and perform the same comparison as the first dataset. Therefore, joint error and rotational distance were not available due to the lack of manually cleaned data. We report the remaining error metrics in Table 2.

Table 2. The mean marker position RMSE (ME) (mm) using the proposed raw (PrR) configurations on the full-body exercise dataset. The marker identity (ID) [%] denotes if the proposed raw configuration denoted a given marker in a given frame as correct (Cor), swapped with another marker (Swa), or missing (Mis). The mean joint angle RMSE (JE) (deg) for non-spherical (Rev) joints are also provided, as well as the absolute maximum joint velocity (VR) for Rev joints. The RMSE is calculated using raw motion capture data. For time segments that the proposed algorithm identifies to have swapped or missing markers, the RMSE is not calculated. 6 of the 11 exercises are shown here for brevity: pick-and-place (PICK), arm butterfly swings (ABUT), high-knee jacks (HIGH), squats (SQUF), squats with arms crossing in front of chest (SQUC), and the corkscrew toe touch (CORK). The average (Avg) is calculated over all 11 exercise types.

		PICK	ABUT	HIGH	SQUF	SQUC	CORK	Avg
ME	PpR	25 ± 3	25 ± 3	25 ± 3	23 ± 2	24 ± 3	54 ± 103	27 ± 30
ID	Cor	75 ± 22	83 ± 16	67 ± 19	74 ± 23	71 ± 20	54 ± 41	73 ± 23
	Swa	18 ± 19	11 ± 12	22 ± 9	19 ± 23	21 ± 18	43 ± 24	19 ± 20
	Mis	7 ± 5	7 ± 5	11 ± 12	7 ± 5	8 ± 5	17 ± 8	9 ± 7
VR	Rev	206 ± 163	235 ± 176	958 ± 709	487 ± 239	549 ± 115	4872 ± 4524	1012 ± 2506

Despite multiple missing and swapping marker events, the proposed approach achieves below 30 mm RMSE accuracy (Table 2). However, these RMSE values are calculated using the raw motion capture labelling data, and not against manually cleaned ground truth. Without ground truth data, marker ID and MJV can be used to provide insight into the quality of the joint angle estimation. Simple motions like pick-and-place (PICK) and arm butterfly flutters (ABUT) appear to perform well, as is suggested by the low standard deviation of the MJV, demonstrating that marker occlusions are handled smoothly and that joint constraints keep the joints in the feasible range, eliminating the need for further post-processing (Fig. 4). These exercises also have few swapped or missing markers. Other exercises report much higher standard deviation in the MJV, such as the two squat motions (SQUF, SQUC). While this could be caused by participants moving at different velocities, it is also possible that the large velocities are caused by reconstruction error and may require some attention.

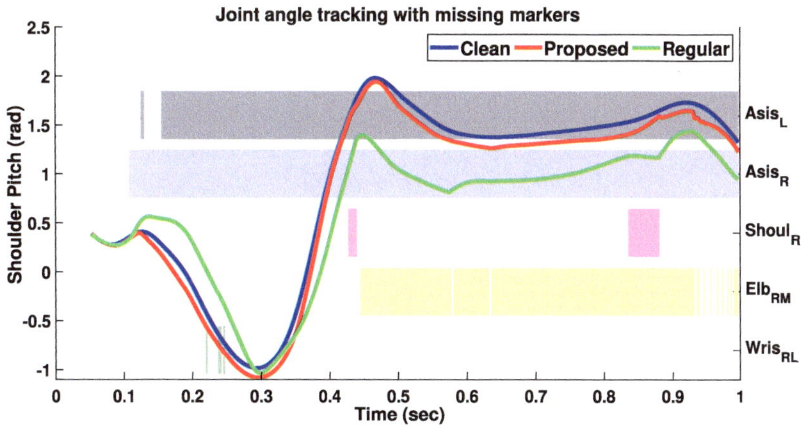

Case 1: Shoulder pitch joint angle tracking with missing markers, frames where each marker is missing are shaded. Using the last known location of a marker causes a limb or the entire body to be pulled pack during the forward jump and leads to large joint angle error. The proposed approach is able to correctly detect the missing events and ignores those markers during estimation.

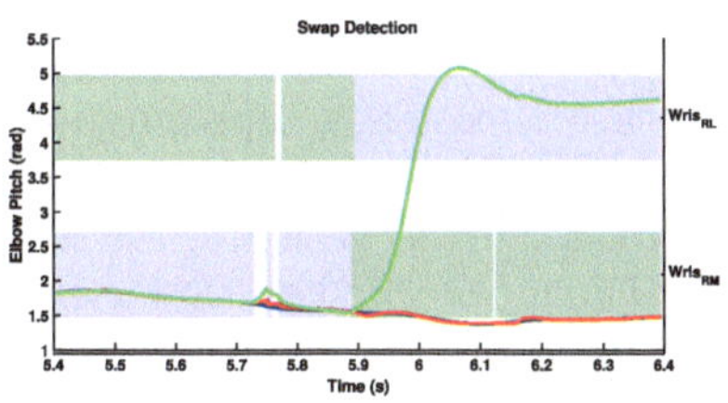

Case 2: When the labels of the two markers attached to the wrist are swapped at 5.9 seconds, regular EKF will rotate the link by 180 degrees in order to minimize the marker position error. Using the covariance metric the proposed approach correctly identifies the swapping event and re-associates the incoming measurement data. Shaded regions represent visible markers, the different colors correspond to motion capture system labels.

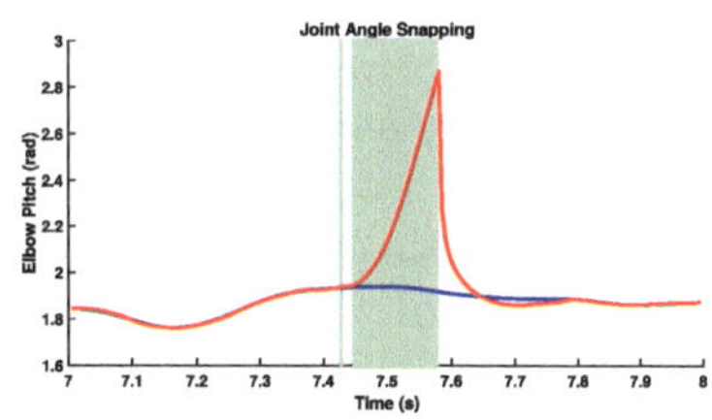

Case 3: Elbow joint angle tracking when only a single wrist marker is visible (shaded region). EKF minimizes the predicted marker position error with respect to the only visible marker causing an incorrect elbow rotation estimate. When the second wrist marker comes back into view the proposed approach accurately associates the marker and quickly snaps back to the correct solution.

Fig. 3. Joint angle tracking under different conditions

More complex motions such as squats with arms crossing at the chest (SQUC) and corkscrew toe touch (CORK) result in high MJV. Sources of error include the choice of kinematic model being too simple to capture complex spinal or shoulder movements, inaccurate kinematic model parameters due to key markers missing during model construction, or marker occlusions followed by the body moving too far from the markers for marker re-association to occur.

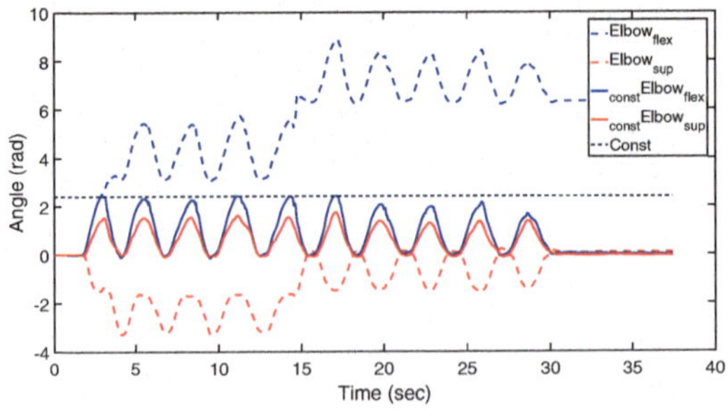

Fig. 4. Unconstrained (blue, red) and constrained (yellow, green) estimation of the left elbow flexion and supination joint angles for the constrained and unconstrained EKF. Imposing a constraint on the elbow flexion joint allows the proposed approach to maintain human like elbow trajectory. For this task our method detected 629 missing and reappearing markers and 256 marker swaps over 40 s of data.

4 Discussion and Future Work

This paper proposes an extended Kalman filter based method to perform inverse kinematics using motion capture data. At each timestep, a matching matrix is generated to map the raw maker position data to the filter measurement model allowing for missing and swapped marker labels. Furthermore, the Kalman gain is restricted to enforce realistic joint limits. The proposed algorithm was verified on a set of broad jump motions and a full body motion dataset. The marker position and joint angle RMSE of the proposed method, using raw data from the motion capture system was found to be similar to a fully cleaned version of the data, even though the raw data contain numerous sets of missing and swapped marker labels. The proposed algorithm also showed good RMSE accuracy on a set of full-body exercises with significant marker occlusions.

Several shortcomings still exist within the algorithm. The algorithm does not fully handle markers that are occluded for prolonged periods of time. When the missing marker reappears in a further location, either due to prolonged occlusion or high velocity movement, the estimate of the limb location may be too far from the marker, so that the marker is not reattached. For future work, a damping weight can be attached to markers that have disappeared, to reduce the jerking motion, until the marker reappears. Contextual spherical joint limits will also be considered, where the joint limits will be modified according to the current posture of the participant. Many complex body limbs, such as the torso and the shoulder, were presented in this paper as simple 1- or 2-dimension links, and may be unable to reach the kinematic configuration demanded by the movements of the markers, so a more complex kinematic model may be able to reduce RMSE in the future.

References

1. Kulić, D., Venture, G., Yamane, K., Demircan, E., Mizuuchi, I., Mombaur, K.: Anthropomorphic movement analysis and synthesis: a survey of methods and applications. IEEE Trans. Rob. **32**, 776–795 (2016)
2. Aristidou, A., Cameron, J., Lasenby, J.: Real-time estimation of missing markers in human motion capture. In: Bioinformatics and Biomedical Engineering, pp. 1343–1346 (2008)
3. Dorfmüller-Ulhaas, K.: Robust optical user motion tracking using a Kalman filter. Technical report, Universitat Augsburg (2007)
4. Wu, Q., Boulanger, P.: Real-time estimation of missing markers for reconstruction of human motion. In: Symposium on Virtual Reality, pp. 161–168 (2011)
5. Meyer, J., Kuderer, M., Müller, J., Burgard, W.: Online marker labeling for fully automatic skeleton tracking in optical motion capture. In: IEEE International Conference on Robotics and Automation, pp. 5652–5657 (2014)
6. Steinbring, J., Mandery, C., Pfaff, F., Faion, F., Asfour, T., Hanebeck, U.: Real-time whole-body human motion tracking based on unlabeled markers. In: IEEE International Conference on Multisensor Fusion and Integration for Intelligent Systems, pp. 583–590 (2016)
7. Joukov, V., D'Souza, R., Kulić, D.: Human pose estimation from imperfect sensor data via the extended Kalman filter. In: International Symposium on Experimental Robotics, pp. 789–798 (2016)
8. Gupta, N., Hauser, R.: Kalman filtering with equality and inequality state constraints. arXiv e-prints (2007)
9. Bierman, G.: A comparison of discrete linear filtering algorithms. IEEE Trans. Aerosp. Electron. Syst. **AES–9**, 28–37 (1973)
10. Boone, D., Azen, S.: Normal range of motion of joints in male subjects. J. Bone Joint Surg. Am. **61**, 756–759 (1979)
11. Huynh, D.: Metrics for 3D rotations: comparison and analysis. J. Math. Imaging Vis. **35**, 155–164 (2009)

PoseFusion: Dense RGB-D SLAM in Dynamic Human Environments

Tianwei Zhang(✉) and Yoshihiko Nakamura

Department of Mechano-Informatics, School of Information Science and Technology, The University of Tokyo, 7-3-1 Hongo, Bunkyo-ku, Tokyo, Japan
{zhang,nakamura}@ynl.t.u-tokyo.ac.jp

Abstract. RGB-D Simultaneous Localization and Mapping (SLAM) in indoor environments is a hot topic in computer vision and robotics communities, and the dynamic environment is a remaining problem. Dynamic environments, which are often caused by dynamic humans in indoor environments, usually lead to the camera pose tracking method failure, feature association error or loop closure failure. In this paper, we propose a robust dense RGB-D SLAM method which efficiently detects humans and fast reconstructs the static backgrounds in the dynamic human environments. By using the deep learning-based human body detection method, we first quickly recognize the human body joints in the current RGB frame, even when the body is occluded. We then apply graph-based segmentation on the 3D point clouds, which separates the detected moving humans from the static environments. Finally, the left static environment is aligned with a state-of-the-art frame-to-model scheme. Experimental results on common RGB-D SLAM benchmark show that the proposed method achieves outstanding performance in dynamic environments. Moreover, it is even comparable to the performance of the related state-of-the-art methods in static environments.

1 Introduction

The task of Simultaneous Localization and Mapping (SLAM) is to estimate the visual sensor's pose and reconstructed the three-dimensional (3D) static backgrounds at the same time. Except for some special applications, most of the SLAM approaches assume that the robot works in static environments. The dynamic environment is a challenging problem for visual SLAM since that the foreground dynamic objects occlude static background features and then result in failing or wrong features corresponding. Humans are often considered as moving obstacles in indoor environments. Particularly, they may occlude visual features during the human-robot interactions.

Generally, a SLAM framework can be divided into a front-end and a back-end part. The task of the front-end is to extract environment features, which

Electronic supplementary material The online version of this chapter (https://doi.org/10.1007/978-3-030-33950-0_66) contains supplementary material, which is available to authorized users.

J. Xiao et al. (Eds.): ISER 2018, SPAR 11, pp. 772–780, 2020.
https://doi.org/10.1007/978-3-030-33950-0_66

are used for image or point clouds alignment. The back-end part takes care of maintaining the graph structure, saving keyframe information, loop finding and dealing with camera pose drift. In [1], Saputra et al. survey the dynamic SLAM methods by the year 2016. They classify the existed dynamic SLAM methods to three types based on different application and outputs: Robust Visual SLAM, Dynamic Object Segmentation, and Joint Motion Segmentation and Reconstruction. Follow this taxonomy, we propose a novel dense RGB-D SLAM method for dynamic humans environments based on human motion segmentation and static environment reconstruction.

Our method works in multiple humans dynamic environment, it efficiently detects humans and fast reconstructs the static environments. We detect moving objects by integrating the OpenPose [2] into our front-end algorithm, which is an advanced deep learning based human detection method for a 2D color image. Dynamic point clouds are then removed by using the Min-Cut Segmentation [3]. Finally, we input the static environment point clouds to a state-of-the-art dense RGB-D SLAM framework, ElasticFusion [4] for static environment reconstruction. As a combination of OpenPose and ElasticFusion, our method is named as PoseFusion, which indicate that this SLAM framework applies human pose detection and frame-to-module schemes. Our method is tested on the well known Freiburg RGB-D SLAM dataset dynamic serials [5], and it achieved the smallest camera trajectory error compared to other state-of-the-art dynamic SLAM methods.

2 Related Works

2.1 Dynamic SLAM Methods

In dynamic SLAM survey [1], Saputra et al. divide the exited dynamic SLAM and visual odometry by their fundamental techniques, such as background/foreground initialization, deep learning, optical flow, ego-motion constraints, geometry constraints, and feature based. In this section, we only introduce recent works for RGB-D dense reconstruction, which are similar to our setting.

1. EF [4] and Kintinous [6] are extended from KinectFusion [7], which is an early dense RGB-D reconstruction framework. [4,6,7] perform great in static environments. They adopt a module maintaining scheme. Different from the others, EF is a state-of-the-art method designed for static environments within slightly dynamic scenes. It can handle small-scale environment changing since it benefits from the deformation graph based non-rigid module fusion. When a slightly changed scene occurs, EF can fuse it into the saved key-frame modules and ignore small scene changing.
2. Co-Fusion [8] (CF) is a state-of-the-art approach for tracking and reconstructing multiple moving objects using EF framework. However, CF has to first reconstruct the map in a static environment, and then they enable the dynamic object detection and tracking abilities within that reconstructed map.

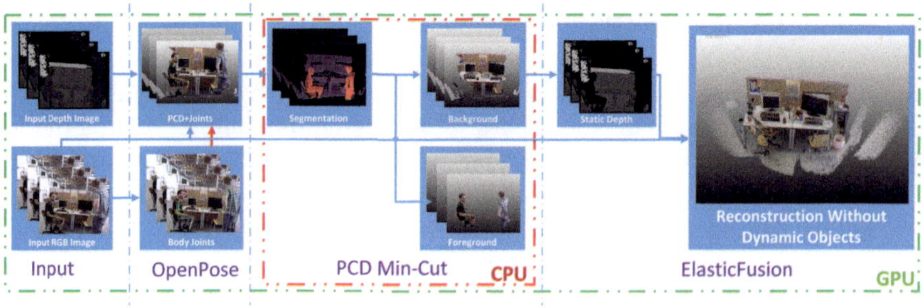

Fig. 1. PoseFusion Flowchart: Firstly input RGB images to OpenPose [2] to detect human joints. Then we project joint points to the PCD and followed by foreground removal. Finally, static backgrounds are feed into reconstruction.

3. Jaimez et al. [9] proposed an odometry method named The joint visual odometry and scene flow (SF). SF localizes the moving camera in dynamic scenes by segmenting intensity point clouds into a number of clusters (called supervoxels), and then, the clusters are divide into moving foregrounds and static backgrounds. Finally, the static background point clouds are feed into camera pose tracking.
4. Scona et al. proposed the staticFusion in [10], which extends the result of SF to environment reconstruction by combining the background segmenting and EF's dense environment reconstruction. stF achieved robust and fast dynamic segmentation and static reconstruction.

These above three SLAM methods are advanced in light dynamic environments. Same to CF and stF, the proposed method is also based on EF. Our idea is similar to SF, that is to decouple the motion of camera from the motion of moving humans. The difference is that SF tries to separate the foreground from the background by comparing the moving speeds of all segments, while we separate moving human segments by using OpenPose.

2.2 The Human Detection Methods

OpenPose [2] estimates the body joints as feature points of human beings such as wrists and shoulders in real time from a single RGB image. It is to estimate each joint position by deriving Part Confidence Maps using trained Convolutional Neural Networks. Moreover, [2] can estimate the vector fields between connected two joints, which make it robust to the occlusions no mater form self body nor multiple humans' bodies. OpenPose provide us the probable positions of human body joints in the 2D RGB image, thus we are able to quickly label the position of humans in 3D Point Clouds Data (PCD).

3 Multiple Humans Detection in Dynamic Environment

The Flowchart of proposed PoseFusion is illustrated in Fig. 1. PoseFusion take the RGB and Depth image pair as input. The RGB image is first used for body joints estimation using OpenPose, which tells the likelihood of the human joint positions on the input image. When the input image f is given, the feature map is extracted via CNN network and then output data:

$$[\boldsymbol{h} \times \boldsymbol{w}]_f \tag{1}$$

in which $\boldsymbol{h}$ is the detected human bodies, at a maximum 15, which means OpenPose can detect at maximum 15 humans within one frame. The $\boldsymbol{w}$ is the list of estimated joints, it presents a probability map on the RGB image plane which indicates the existence likelihood of at maximum 18 human joint. In matrix 1, each element has three components: u, v, p. They are the image pixel coordinates (u, v) and the existence likelihood ($p \in (0, 1]$) of the human body joint.

The estimated joint points are converted from 2D to 3D using the pinhole camera model and then, they are used for labeling humans' positions in the PCD. This processing is the red arrow in Fig. 1, and note that the green points in the point clouds stand for the projected joint positions. Then, the PCD with these green joints are inputted to Min-Cut point foreground segmentation.

Min-Cut [3] is a Graph-Cut [11] based method for segmenting objects in point clouds. Graph-cut treats every single point as a vertex and vertices are connected with their neighbors by edges. Given some vertices as foreground priors, it cuts the foreground object out of the background points by computing the weights of the edges. To apply Min-Cut, we use the human joints from Eq. 1 as foreground prior, we assign two edge weights in min-cut: the edge smooth cost C and background penalty P.

$$C = e^{-(\frac{len}{\sigma})^2} \tag{2}$$

in which len is the length of the edge, obviously, the father away the vertices are, the more is the probability the edge will be cut. The σ is a user defined parameter.

The background penalty is to weight the points connected with the foreground points. In which, for a joint point $J(J_x, J_y, J_z)$, we set an input parameter rad as the maximum horizontal (X-Y plane) radius of foreground objects. Then, for a neighbor point (x, y, z) of J, its background penalty is:

$$P = \frac{\sqrt{(x - J_x)^2 + (y - J_y)^2}}{r} \tag{3}$$

As the pose points are labeled on the human body, we set the foreground rad as 20 cm, σ as 0.25 which draws a good segmentation performance. After Min-Cut, the background segments are converted to static depth images, which are the inputs for the following static environment reconstruction, together with the original RGB image. Finally, a clean static environment reconstruction is achieved through frame-to-module map fusions.

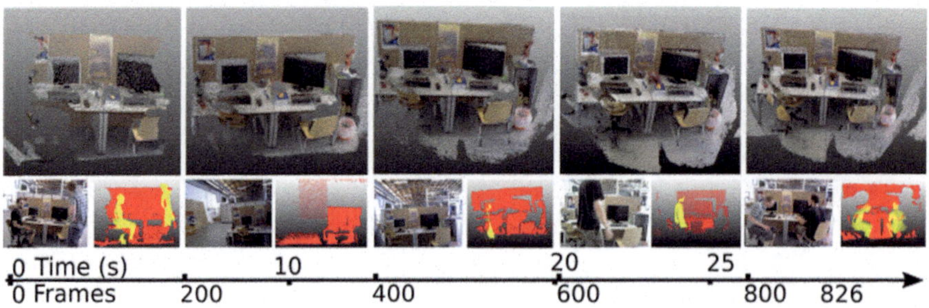

Fig. 2. Experiment Setting: RGB-D PCD frames are played as a video. The first row shows the reconstructed maps, the second row shows the respectively RGB input and foreground-background segmentations. As the frame grows, the first row reconstruction is gradually completed, the RGB viewpoint moves as the camera moves, the point clouds segmentation is also changing. There is no foreground cluster in the 4th image, the second row, since the guy walks out of the scene.

4 Experiments Setup and Results

Our method is evaluated with the Freiburg SLAM benchmark, which contains a dynamic SLAM series (fr3 series). This benchmark is wildly used for dynamic scenes comparison. The experiment is shown in Fig. 2, which take fr3/walk_xyz (in which the camera moved by a man walks in x, y, and z translations) as an example. Each fr3 dataset contains several hundred color (RGB) and Depth (D) image pairs, which can be transformed into RGB-D PCD frames. For instance, there are 827 RGB-D frames in fr3/walk_xyz, these RGB-D PCD frames can be played as a video.

In Fig. 2, the first row shows the reconstructed maps, and the second row shows the respective input RGB images and their foreground-background segmentation results. As the frame number grows, the first row reconstruction is gradually completed, the RGB viewpoint moves as the camera moves, the point clouds segmentation is also changing. In the fourth image of the second row, there is no foreground cluster since the people walk out of the camera view. These experiment setting intuitively indicate the dynamic ability of SLAM methods. One can compare the moving object detection and removal abilities throw checking when and how many moving object ghost shadows are integrated into the reconstructed scenes in the first row. One comparison with EF and PF is given in Fig. 4.

We compare our PoseFusion (PF) method with three state-of-the-art dynamic SLAM methods: Scene Flow (SF) method from [9], ElasticFusion (EF) [4] and Co-Fusion (CF) [8]. All of them are implemented from their open source repositories. StaticFusion (stF) [10] is the most recent method and its code is not opened yet.

To evaluate these four SLAM methods, we compare their absolute trajectory error (ATE) and relative pose error (RPE). ATE is well-suited for measuring the performance of visual SLAM systems. In contrast, the RPE is well-suited

Table 1. Translate ATE RMSE (cm)

Dynamic dataSet [5]	SF	EF	CF	PF
fr3/sit_static	8.67	3.18	**2.93**	3.18
fr3/sit_xyz	7.65	**0.90**	1.49	**0.90**
fr3/sit_halfsphere	50.21	30.18	27.58	**2.31**
fr3/walk_static	81.07	25.13	18.77	**6.87**
fr3/walk_xyz	65.31	67.78	37.11	**3.21**
fr3/walk_halfsphere	79.97	47.90	21.23	**4.65**

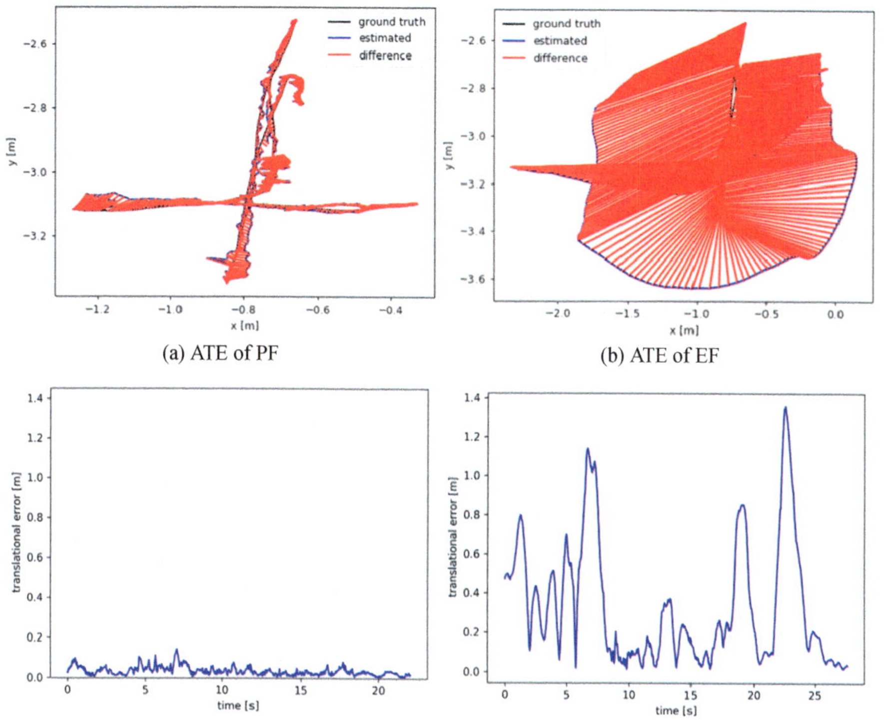

(a) ATE of PF

(b) ATE of EF

(c) RPE of PF

(d) RPE of EF

Fig. 3. Evaluation of the proposed PoseFusion compare to ElasticFusion, both are the result of fr3/walk_xyz. (a) and (c) are absolute trajectory error (ATE) and the relative pose error (RPE) of PF, while (b), (d) are ATE and RPE of EF. PF achieves very small trajectory error, average 3.21 cm (short red line segments in (a)), while EF gets big ATE, long red line segments in (b). (c) and (d) indicate PF achieves about ten times smaller RPE than EF.

Table 2. Translate RPE RMSE (cm/s)

Dynamic dataSet [5]	SF	EF	CF	PF
fr3/sit_static	2.71	1.12	**0.81**	1.19
fr3/sit_xyz	9.61	**3.90**	4.93	3.97
fr3/sit_halfsphere	41.10	29.19	23.80	**5.19**
fr3/walk_static	27.77	20.33	18.79	**3.77**
fr3/walk_xyz	35.32	21.78	37.11	**2.11**
fr3/walk_halfsphere	69.74	77.91	31.32	**4.50**

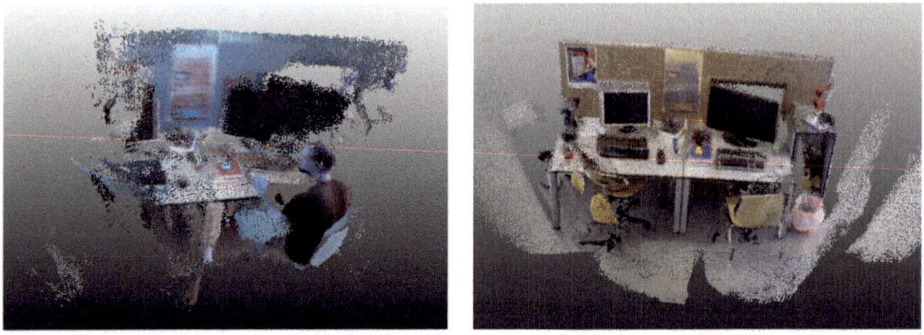

(a) ElasticFusion Reconstruction (b) PoseFusion Reconstruction

Fig. 4. EF and PF Scene Reconstruction of fr3/walk_xyz. (a) is the result of EF, 67.78 cm absolute trajectory error makes an obvious wrong reconstruction. The wall, the table, and the desktops are not aligned and multiple people shadow corrupt the final map. While (b) is reconstructed by PF, which is well aligned with only 3.21 cm error. The dynamic performance of proposed is as good as the EF's static performance. Note that, the left chair's ghost shadow in PF reconstruction is not a wrong alignment, it is because the people moved that chair to that place for several seconds, which can be observed in our supplement video.

for measuring the drift of a visual odometry system, for example, the drift per second. The ATE directly measures the difference between points of the ground truth and the estimated trajectory. The RPE computes the error in the relative motion between pairs of timestamps.

Tables 1 and 2 show the root-mean-square error (RMSE) of translate ATE (cm) and RPE (cm/s). All of these six datasets are dynamic scenes, but the first three: fr3/sit_static; sit_xyz and sit_halfsphere are low dynamic scenes, while the other three are high dynamic scenes. From these two Tables, one can obviously find that our PF method achieved the smallest estimation errors in highly dynamic situations. In the light dynamic scenes, EF achieved the best performances, since it is designed for static environments.

Figures 3 and 4 show the comparison between PF and EF run on the fr3/walk_xyz dataset. Inside the former figure set, (a) and (c) are ATE and RPE of PF, while (b), (d) are ATE and RPE of EF. PF achieves very small trajectory error, average 3.21 cm (difference of the same time-stamp are shown as the short red line segments in (a)), while EF gets big ATE, which can be seen as the long red line segments in (b). (c) and (d) indicate that PF achieves around ten times smaller RPE than EF.

Figure 4 intuitively indicate the performances of EF and PF. Big trajectory error of EF results in obvious wrong map reconstruction. The wall, the table, and the desktops are not aligned, as well as people shadows corrupt the final map of (a). While (b) is reconstructed by PF, which is well aligned. The dynamic performance of PF is as good as the EF's static performance. Note that, the left chair's ghost shadows in PF reconstruction is not wrong alignments, they are reconstructed since that the people moved that chair to that places for several seconds, which can be observed in our supplement video.

5 Conclusions

In this paper, we propose a novel dense RGB reconstruction approach for the dynamic human environments. We combine the advanced deep learning based human body detection method and static dense SLAM approach to deal with the dynamic environment problem using dense RGB-D data. We compare the proposed method with EF [4], CF [8], and SF [9]. Our approach significantly decreases the alignment error and archive 3.21 cm average ATE in TUM benchmark, while EF results in a 67.78 cm ATE.Moreover, We not only propose one approach for a particular dynamic environment setting but also provide a framework for solving SLAM "Type II Error", which means the front-end part doesn't "reject" the wrong features of the moving objects so that the moving shadows are integrated to key-frames on the back-end, thus breakdown loop closing. According to the proposed, one can integrate advanced learning-based object recognition methods to handle dynamic environments. For instance, involving vehicle recognition methods for high way scenes, and including semantic labeling methods for service robots in house and office scenes.

Acknowledgement. This work was supported by JSPS Grants-in-Aid for Challenging Research (Pioneering) Grant Number JP17H06291.

References

1. Saputra, M.R.U., Markham, A., Trigoni, N.: Visual SLAM and structure from motion in dynamic environments: a survey. ACM Comput. Surv. (CSUR) **51**(2), 37 (2018)
2. Cao, Z., Simon, T., Wei, S.E., et al.: Realtime multi-person 2D pose estimation using part affinity fields. In: IEEE Conference on Computer Vision and Pattern Recognition (CVPR), vol. 1, no. 2, p. 7 (2017)

3. Golovinskiy, A., Funkhouser, T.: Min-cut based segmentation of point clouds. In: 2009 IEEE 12th International Conference on Computer Vision Workshops (ICCV Workshops) (2009)
4. Whelan, T., Leutenegger, S., Salas-Moreno, R.F., Glocker, B., Davison, A.J.: ElasticFusion: dense SLAM without a pose graph. In: Robotics: Science and Systems (RSS) (2015)
5. Sturm, J., Engelhard, N., Endres, F., Burgard, W., Cremers, D.: A benchmark for the evaluation of RGB-D SLAM systems. In: IEEE/RSJ International Conference on Intelligent Robots and Systems (IROS) (2012)
6. Whelan, T., Kaess, M., Fallon, M., et al.: Kintinuous: spatially extended kinectfusion. CSAIL Technical report (2012)
7. Newcombe, R.A., Izadi, S., Hilliges, O., et al.: KinectFusion: real-time dense surface mapping and tracking. In: 10th IEEE International Symposium on Mixed and Augmented Reality (ISMAR) (2011)
8. Rünz, M., Agapito, L.: Co-fusion: real-time segmentation, tracking and fusion of multiple objects. In: IEEE International Conference on Robotics and Automation (ICRA) (2017)
9. Jaimez, M., Kerl, C., Gonzalez-Jimenez, J., Cremers, D.: Fast odometry and scene flow from RGB-D cameras based on geometric clustering. In: IEEE International Conference on Robotics and Automation (ICRA) (2017)
10. Scona, R., Jaimez, M., Petillot, Y.R., et al.: StaticFusion: background reconstruction for dense RGB-D slam in dynamic environments. In: Institute of Electrical and Electronics Engineers (2018)
11. Boykov, Y., Funka-Lea, G.: Graph cuts and efficient ND image segmentation. Int. J. Comput. Vis. **70**(2), 109–131 (2006)

Unified Spatiotemporal Calibration of Monocular Cameras and Planar Lidars

Jordan Marr(✉) and Jonathan Kelly

Space and Terrestrial Autonomous Robotic Systems Laboratory,
Institute for Aerospace Studies, University of Toronto, Toronto, Canada
{jordan.marr,jonathan.kelly}@robotics.utias.utoronto.ca

Abstract. Monocular cameras and planar lidar sensors are complementary. While monocular visual odometry (VO) is a relatively low-drift method for measuring platform egomotion, it suffers from a scale ambiguity. A planar lidar scanner, in contrast, is able to provide precise distance information with known scale. In combination, a monocular camera-2D lidar pair can be used as a performance 3D scanner, at a much lower cost than existing 3D lidar units. However, for accurate scan acquisition, the two sensors must be spatially and temporally calibrated. In this paper, we extend recent work on a calibration technique based on Rényi's quadratic entropy (RQE) to the unified spatiotemporal calibration of monocular cameras and 2D lidars. We present simulation results indicating that calibration errors of less than 5 mm, 0.1°, and 0.15 ms in translation, rotation, and time delay, respectively, are readily achievable. Using real-world data, in the absence of reliable ground truth, we demonstrate high repeatability given sufficient platform motion. Unlike existing techniques, we are able to calibrate in arbitrary, target-free environments and without the need for overlapping sensor fields of view.

1 Introduction

Multiple sensors, and sensing modalities, are often used together to enable more robust and reliable robot task execution. Sensors are typically selected to be complementary, leveraging the benefits of the combination while mitigating individual limitations and drawbacks. A monocular camera and a planar lidar unit, in particular, are complementary: the camera is able to provide six degrees-of-freedom (6-DOF) motion estimates with low drift, but suffers from a scale ambiguity; the lidar, in turn, provides accurate, metric distance information, but only within the unit's single scanning plane. Together, the sensors are able to operate as a low-cost 3D lidar scanner, generating dense, registered 3D point clouds (for a bargain price).

Despite the benefits of multimodal, multisensor fusion, there is a caveat: in any multisensor system, calibration is necessary to ensure correct data fusion.

Electronic supplementary material The online version of this chapter (https://doi.org/10.1007/978-3-030-33950-0_67) contains supplementary material, which is available to authorized users.

J. Xiao et al. (Eds.): ISER 2018, SPAR 11, pp. 781–790, 2020.
https://doi.org/10.1007/978-3-030-33950-0_67

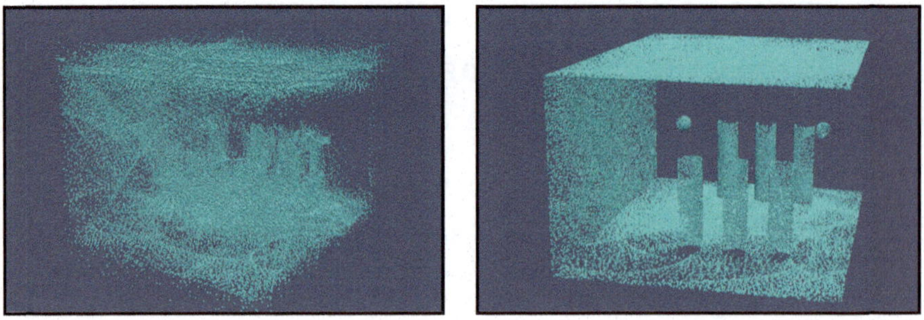

Fig. 1. Results from one calibration simulation. The initial calibration errors, per-axis, were 6 cm and 3.5°, respectively, and the time delay error was 25 ms. Prior to calibration (left), the point cloud is clearly disordered and has a high entropy; after optimization (right), the calibration errors have been dramatically reduced and the structure in the environment is clearly visible.

Accurate and precise knowledge of the spatial transform between the sensors is needed to properly integrate the individual measurements into a single reference frame. A second factor of critical importance is temporal calibration [15], that is, the alignment of the sensor clocks and identification of any inherent processing delays. Naïvely fusing data based on raw sensor timestamps can result in substantial estimation errors.

We seek to develop a unified camera-lidar spatiotemporal calibration procedure that (1) solves for spatial and temporal calibration parameters simultaneously, and (2) works in arbitrary environments, absent of any specific calibration targets or known scene structure. Our technique builds on the methods described in [10,17], and [9]. These calibration algorithms all determine the extrinsic transform between one or more sensor pairs (including at least one lidar) by minimizing the Rényi Quadratic Entropy (RQE) [16] of a composite lidar point cloud. Our novel contributions include: (1) a problem formulation that unifies spatial and temporal calibration, (2) the use of a theoretically well-founded method for interpolating poses for sensors that update at different frequencies, (3) the incorporation of a fast Bayesian optimization procedure that accelerates the calibration process, and (4) an experimental demonstration that spatiotemporal calibration can be carried out for camera and lidar sensors in arbitrary configurations and in environments with unknown structure.

2 Technical Approach and Related Work

Canonical methods for camera-lidar spatial calibration almost always impose undesirable constraints on both the requisite calibration environment and the procedure itself. For example, it is often necessary to ensure that the sensors have overlapping fields of view [14], that each observes a scene with known structure [4], or that a known calibration object is available [7,18]. Other, more general calibration method exist, but can require that each sensor is able to measure its own egomotion [5,6,8], which is not possible for a planar lidar scanner moving in three dimensions. Moreover, the bulk of the existing calibration literature either ignores temporal calibration, or assumes that the temporal offsets (delays) have already been identified by some other means.

The key idea in our proposed calibration approach (as in [10,17], and [9]) is that, when lidar scan data are registered (given a set of corresponding camera poses), the calibration parameters affect the entropy of the final lidar point cloud. Inaccurate calibration causes a 'blurring' effect (see Fig. 1), with separate lidar scans affected in different ways. The ultimate result (with incorrect calibration) is a point cloud with a greater entropy. As shown in Fig. 2, each calibration parameter affects the point cloud entropy in a different way, with the rotational and trajectory scale parameters having the largest impact. Extensive experimental work by ourselves and others has shown that the overall point cloud entropy is minimized by the ground truth calibration parameters. We therefore solve for the optimal parameters by minimizing a cost function derived from the RQE. In our case, the monocular camera serves as an egomotion sensor only—the method is therefore extensible to platforms that utilize other sensors, such as stereo cameras, IMUs, or GPS-INS units, in place of the single camera.

2.1 Point Cloud Generation

Our goal is to estimate the set of (constant) parameters that define the transform from the lidar frame $\underrightarrow{\mathcal{F}}_L$ to the camera frame $\underrightarrow{\mathcal{F}}_C$,

$$\Xi = \begin{bmatrix} x_L & y_L & z_L & \phi_L & \theta_L & \psi_L & s & t_d \end{bmatrix}^T. \tag{1}$$

The six spatial parameters (the translation parameters x_L, y_L, and z_L, and the rotation parameters ϕ_L, θ_L, and ψ_L)) are used to construct the 4×4 rigid body transformation matrix $\mathbf{T}_{C,L}$, from the lidar frame to the camera frame. The scale factor s is applied to the translational component of the camera poses (i.e., to the top-right elements of the transform $\mathbf{T}_{G,C}$ from the camera frame to the global frame). In practice, we have found that including an additional range bias parameter, as is done in [15], does not substantially improve (and sometimes negatively impacts) the calibration results.

We account for the inter-sensor measurement time delay by adopting a continuous time representation of the sensor platform trajectory,

$$\mathbf{T}_{G,L}(t) = \mathbf{T}_{G,C}(t + t_d)\,\mathbf{T}_{C,L}. \tag{2}$$

The transform $\mathbf{T}_{G,L}(t)$ is evaluated at time steps $t = t_1, t_2, \ldots, t_N$ to obtain a corresponding lidar pose for all N lidar scans along a trajectory. We obtain $\mathbf{T}_{G,C}(t + t_d)$ by interpolating between (unsynchronized) camera poses. If the delay value t_d is positive, the camera clock leads the lidar clock (and vice-versa for a negative t_d). Interpolation is carried out using the STEAM model [2], which permits a principled estimation of the uncertainty of an interpolated pose [1,3]. The camera pose uncertainties are used to weight the contribution of individual lidar scans within the entropy calculation.

At time t_n, the lidar produces a scan consisting of K measurements $(x_{L_n}^{(k)}, y_{L_n}^{(k)})$, $k = 1, \ldots, K$, each representing the coordinates of a measured point in the scan plane. We transform these points into the global frame using our current estimate of the lidar pose,

$$\hat{\mathbf{p}}_{G_n}^{(k)} = h^{-1}(\mathbf{p}_{L_n}^{(k)} \mid \mathbf{T}_{G,C_n}, \boldsymbol{\Xi}) = \mathbf{T}_{G,L}(t_n) \begin{bmatrix} x_{L_n}^{(k)} \; y_{L_n}^{(k)} \; 0 \; 1 \end{bmatrix}^T. \tag{3}$$

In the equation above, we have followed the notation used in [10] (viz. the inverse sensor model). To simplify the notation, we omit the homogeneous component in the remainder of our derivation, such that $\hat{\mathbf{x}}_{G_n}^{(k)} \leftarrow \hat{\mathbf{p}}_{G_n}^{(k)}$, as follows,

$$\hat{\mathbf{x}}_{G_n}^{(k)} = \begin{bmatrix} x_{G_n}^{(k)} \; y_{G_n}^{(k)} \; z_{G_n}^{(k)} \end{bmatrix}^T \leftarrow \begin{bmatrix} x_{G_n}^{(k)} \; y_{G_n}^{(k)} \; z_{G_n}^{(k)} \; 1 \end{bmatrix}^T = \hat{\mathbf{p}}_{G_n}^{(k)}. \tag{4}$$

We use the Jacobian of the model (3) and the camera pose covariance to obtain a position covariance estimate for each lidar point in the global frame, according to

$$\boldsymbol{\Sigma}_n^{(k)} = \mathbf{J}_n^{(k)} \mathbf{Q}_n \mathbf{J}_n^{(k)T}, \quad \mathbf{J}_n^{(k)} = \frac{\partial h^{-1}(\mathbf{x}_{L_n}^{(k)} | \mathbf{T}_{G,C_n}, \boldsymbol{\Xi})}{\partial \mathbf{T}_{G,C_n}}, \tag{5}$$

where $\mathbf{T}_{G,C_n}$ is the interpolated 6-DOF camera pose, that is, $\mathbf{T}_{G,C}(t_n + t_d)$, and $\mathbf{Q}_n$ is the associated camera pose covariance matrix. Note that the translational uncertainty of the camera pose must be scaled to match the (metric) scale of the trajectory, and that $\mathbf{Q}_n$ is the covariance matrix of the interpolated camera pose, rather than the covariance matrix of any of the measured camera poses.

2.2 Rényi's Quadratic Entropy Computation

Grouping all the lidar scans together, we model the full set of lidar points, $\hat{\mathbf{X}}$, as a Gaussian mixture. There are $NK = M$ points in total, $\hat{\mathbf{x}}^{(m)}$, for $m = 1, \ldots, M$, with the associated point covariance matrices $\boldsymbol{\Sigma}^{(m)}$, $i = 1, \ldots, M$. The coordinates of all points are expressed in the global frame $\underrightarrow{\mathcal{F}}_G$ (henceforth we drop the subscripted frame identifier). The RQE of the point cloud can be computed as

$$H[\hat{\mathbf{X}}] = -\log \left(\frac{1}{M^2} \sum_{i=1}^{M} \sum_{j=1}^{M} \mathcal{N} \left(\hat{\mathbf{x}}^{(i)} - \hat{\mathbf{x}}^{(j)}, \boldsymbol{\Sigma}^{(i)} + \boldsymbol{\Sigma}^{(j)} + 2\sigma^2 \mathbf{I}_3 \right) \right). \tag{6}$$

The term $\sigma^2 \mathbf{I}_3$ (where $\mathbf{I}$ is the 3×3 identity matrix) is included to represent the uncertainty of the lidar range measurements. We omit the details of this derivation and refer readers to [9,10] for more information.

Given that the logarithm is a monotonic function and that the $\frac{1}{M^2}$ term is constant, we can minimize the RQE and find the optimal calibration parameters, $\boldsymbol{\Xi}^*$, by solving

$$\boldsymbol{\Xi}^* = \underset{\boldsymbol{\Xi}}{\operatorname{argmin}} \; - \sum_{i=1}^{M} \sum_{j=1}^{M} \mathcal{N} \left(\hat{\mathbf{x}}^{(i)} - \hat{\mathbf{x}}^{(j)}, \boldsymbol{\Sigma}^{(i)} + \boldsymbol{\Sigma}^{(j)} + 2\sigma^2 \mathbf{I}_3 \right). \tag{7}$$

The RQE value quantifies the 'structuredness' of the 3D environment in a manner that is partially robust to outliers as a result of the Gaussian weighting. Clouds that contain many points clustered together on surfaces will tend to have a lower RQE compared to clouds that contain points scattered in space. Therefore, a point cloud produced with the correct camera-lidar calibration should have a low RQE (see Fig. 1). This result holds even in environments that lack large planar surfaces (such as calibration targets).

Following [9,10], it is possible to speed up the nominally $O(M^2)$ RQE computation by ignoring points that contribute negligible entropy. Introducing a tuning parameter, k, one can then make the approximation

$$\mathcal{N}\left(\hat{\mathbf{x}}^{(i)} - \hat{\mathbf{x}}^{(j)}, \boldsymbol{\Sigma}^{(i)} + \boldsymbol{\Sigma}^{(j)} + 2\sigma^2\mathbf{I}\right) \approx 0 \;\text{ if}$$
$$\left\|\hat{\mathbf{x}}^{(i)} - \hat{\mathbf{x}}^{(j)}\right\| \geq 2k\left(\max\left(\lambda_1(\boldsymbol{\Sigma}^{(i)}), \lambda_1(\boldsymbol{\Sigma}^{(j)})\right) + \sigma^2\right), \tag{8}$$

where $\lambda_1(\boldsymbol{\Sigma})$ is the largest eigenvalue of the matrix $\boldsymbol{\Sigma}$. A larger value of k will reduce the computation time at the expense of accuracy. A k-d tree can be used to efficiently determine which point pairs do and not have a significant impact on the RQE calculation [10].

2.3 Bayesian Optimization

Although the simplification made in (8) results in a tractable runtime for the algorithm, it also introduces an additional difficulty. Because different sets of points are selected on each iteration of the optimization, the cost function no longer has continuous analytical derivatives (see Fig. 2) and therefore cannot be optimized easily by Newton's method. Despite the 'jagged' micro-scale behaviour of the cost function, it is still (very nearly) smooth at a macro-scale. Further, in the neighbourhood of the true calibration values, the function appears (nearly) convex.

Given the nature of the cost function (i.e., the expense of its evaluation and its discontinuous derivatives), an alternative minimization approach involves Bayesian optimization [12]. This method samples the parameter space to build up a surrogate model of the cost function as a Gaussian process. The optimizer then efficiently queries the cost function at points in the parameter space that will add the most information (or produce the greatest potential improvement). We use the BayesOpt library [11] to solve for the calibration parameters; this has allowed us to achieve an eightfold or greater speedup in computation when compared with [9].

3 Experimental Results

We validated the proposed calibration procedure by running a series of simulation studies and by processing several real-world calibration datasets. Our goals were to evaluate the performance of the calibration algorithm under a variety of conditions, to determine the amount of data required to calibrate reliably, and to assess what type of sensor platform motion was needed.

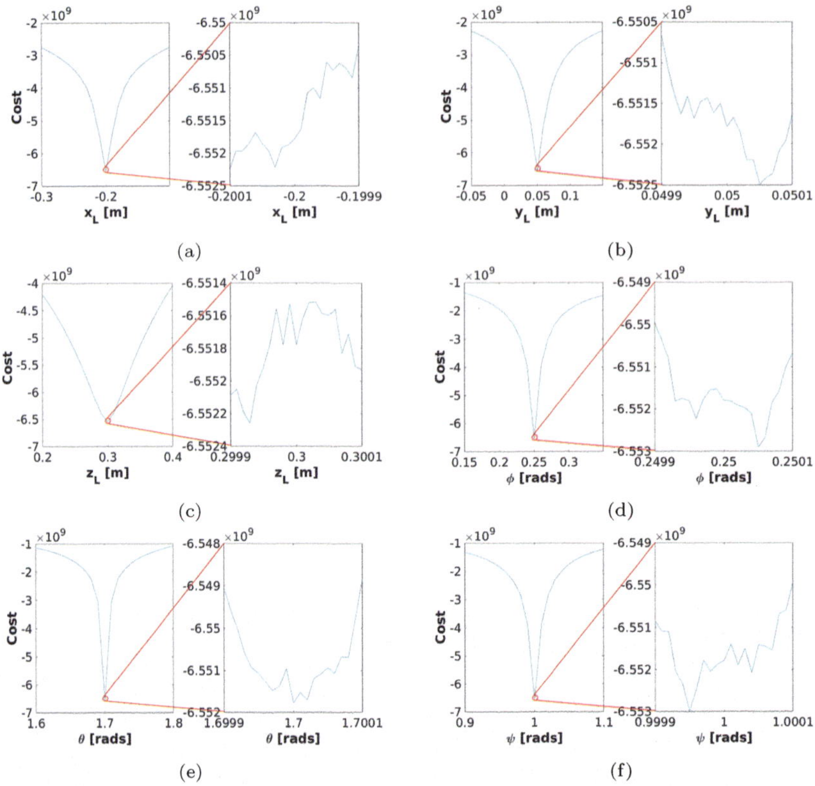

Fig. 2. Simulation results showing the macro- and micro-scale behaviour of the RQE-based cost function for the spatial transform parameters. To generate each plot, the remaining seven calibration parameters were held constant at their ground truth values.

3.1 Simulation Studies

We began by performing several simulations in five different environments with varying numbers of planar, cylindrical, and spherical surfaces. The simulation environments were:

1. **Simple Room**, composed of six planes arranged as a rectangular prism,
2. **Circular Room**, composed of a single cylindrical volume with two planes making up the floor and ceiling,
3. **Plane City**, composed of 11 planes (including the floor and ceiling) arranged as a rectangular prism, with the additional interior planes at irregular angles,
4. **Parking Garage**, similar to the simple room environment, but containing six cylindrical 'pillars' and two spherical targets (see Fig. 1), and
5. **Forest**, the most challenging environment, composed of a 'wavy' ground surface and a series of cylindrical 'trees' with attached spherical 'leaves'; the trees and leaves were placed randomly.

During each simulation trial, a virtual camera-lidar pair followed a sinusoidal trajectory (in both position and orientation) that allowed the lidar to repeatedly scan most of the surfaces in the environment from multiple, distinct vantage points. We evaluated the effects of noisy camera pose estimates by adding uncorrelated, zero-mean Gaussian noise with a standard deviation of 5 mm (position) and 0.5° (orientation) to each simulated camera pose. Scale uncertainty was simulated by uniformly scaling the translational component of each camera pose. Both sensors generated data at 40 Hz, but the updates were offset in time to simulate realistic delays between the sensor clocks. Each simulation dataset was 90 s in length, with the lidar typically scanning between two and four million points.

The initial simulation results are summarized in Table 1. In most cases, we were able to produce calibration estimates with errors of less than 5 mm for translation, 0.1° for orientation, and 0.15 ms for time delay. One exception, however, is calibration in the very challenging ‘Forest’ environment, where the lidar scanned far fewer points (simply because the environment itself is sparse). We believe that the calibration accuracy in this environment could be improved substantially by modifying the sensor platform trajectory such that the lidar primarily scans the ground and the simulated ‘trees’, rather than free space.

We also tested the effects of longer sequences on the achievable calibration accuracy by generating several new simulation datasets, each 180 s in length. The accuracy of the results was comparable, suggesting that there is a diminishing return in terms of the overall trajectory length. Accurate calibration requires the lidar to scan the scene from many different viewpoints in order to effectively constrain the calibration parameters—however, at some stage, most surfaces will have been scanned repeatedly (resulting in a high point density), and adding more data may not substantially improve the calibration results.

Table 1. Average absolute error over five unique trajectories in each of five different environments.

Environment	Average absolute error - $\lvert\mu\rvert$							
	x [mm]	y [mm]	z [mm]	ϕ [°]	θ [°]	ψ [°]	s $[\times 10^{-3}]$	t_d [ms]
Simple room	1.07	1.68	2.21	0.012	0.025	0.012	0.013	0.020
Circular room	0.96	3.12	3.54	0.026	0.050	0.008	0.055	0.135
Plane city	2.45	3.37	2.80	0.010	0.038	0.009	0.126	0.029
Parking garage	2.45	4.33	3.69	0.028	0.064	0.014	0.178	0.081
Forest	12.9	11.4	13.9	1.06	0.548	0.235	35.8	4.76

3.2 Real-World Experiments

We validated our procedure in the real world by carrying out several tests with a Hokuyo UTM-30LX lidar and a PointGrey Flea3 monocular camera (rigidly attached and, notably, configured with non-overlapping fields of view) in our cluttered laboratory environment. The lidar acquired data at 40 Hz, while the camera operated at a higher frequency (enabling us to interpolate the camera poses). One large dataset was collected and then divided into three separate, non-overlapping subsets, each 100 s in length. There were approximately three million lidar scan points in each subset. We determined the trajectory of the camera (up to an unknown scale factor) using ORB-SLAM2 [13].

Calibration results are shown in Table 2. The optimized parameters are consistent across each calibration trial, with few exceptions (e.g., the z translation in Dataset 2). We attribute these differences to the specific platform trajectory; this highlights the necessity of exciting all degrees of freedom of movement to sufficiently constrain the full set of calibration parameters.

Table 2. Calibration results for three real-world datasets.

	Estimated calibration parameters							
	x [mm]	y [mm]	z [mm]	ϕ [°]	θ [°]	ψ [°]	Scale	t_d [s]
Dataset 1	50.5	0.2	−211.0	181.7	−1.1	−88.8	0.216	0.016
Dataset 2	55.1	7.7	−234.2	186.1	1.0	−89.4	0.212	0.018
Dataset 3	47.8	−0.4	−224.2	184.2	−1.5	−89.7	0.214	0.018

4 Discussion

Our experience with the RQE-based spatiotemporal calibration technique has produced several useful insights. We have discovered that a key to obtaining accurate calibration is to use a large-scale sensor trajectory. In our simulation studies, we observed that accuracy improved substantially when the sensor platform moved through a larger portion of the simulation environment; the same was true in our real-world experimental trials. We believe one reason for the improvement is that larger/longer trajectories effectively reduce the impact of camera pose estimation errors (in terms of the relative magnitudes of the pose errors compared to the overall trajectory). Although increased trajectory length implies that lidar scans are taken farther apart (in space and time), scanning surfaces from widely disparate viewpoints has been shown by us and by others to better constrain the calibration parameters. When neighbouring scans are captured farther apart, there are fewer valid calibration transforms that will collect lidar points on common surfaces (regardless of the exact shape of the surfaces involved).

At present, the calibration algorithm considers all points within a single lidar scan to have been captured simultaneously. This is a reasonable approximation when the sensor platform is moving slowly, but ignores the possibility of motion distortion when the platform is rotating or translating quickly. Our previous work [9] showed that incorrect time delay estimates (on the order of 20 ms) can result in the miscalibration of spatial transform parameters by as much as

1.3 cm and 0.4°. Given the sensitivity of calibration to timing errors, a potential improvement to our approach would be to switch to a continuous-time trajectory representation (where each lidar point has a specific and unique time stamp). However, it is not immediately clear how to do so within the RQE framework while maintaining computational tractability.

As a final note, we emphasize the importance, within this framework, of calibrating both the spatial and temporal transform simultaneously. Although the relative time delay between sensor data streams is not correlated with the spatial transform between the sensors (and could be potentially be determined in some other manner), the minimum of the RQE-based cost function is only found when the spatial and temporal calibration parameters are at (or very close to) their ground truth values.

5 Conclusions and Future Work

We have presented a unified spatiotemporal calibration algorithm for monocular cameras and planar lidar scanners, based on the minimization of the entropy of a composite lidar point cloud. Our calibration procedure allows for estimation of both the temporal and spatial offsets between a camera and lidar in arbitrary environments, without the constraints imposed by existing methods. In particular, we do not rely on overlapping sensor fields of view, or on any known structure or calibration objects within the scene. Our approach compares favourably to state-of-the-art methods, given that we remove these constraints.

There are several directions for future work. We are investigating ways to parameterize the camera trajectory and each lidar scan together in continuous time, to better account for potential motion distortion and improve calibration accuracy (while limiting computational complexity). We are also exploring the use of the RQE-based calibration technique with other sensor pairs, including inertial measurement units and planar and 3D lidars.

References

1. Anderson, S.: Batch continuous-time trajectory estimation. Ph.D. thesis, University of Toronto, Toronto, Ontario, Canada (2017)
2. Anderson, S., Barfoot, T.D.: Full STEAM ahead: exactly sparse Gaussian process regression for batch continuous-time trajectory estimation on SE(3). In: Proceedings of the IEEE/RSJ International Conference on Intelligent Robots and Systems (IROS 2015), Hamburg, Germany, pp. 157–164, September 2015
3. Anderson, S., Barfoot, T.D., Tong, C.H., Särkkä, S.: Batch nonlinear continuous-time trajectory estimation as exactly sparse Gaussian process regression. Auton. Robot. **39**(3), 221–238 (2015)
4. Bok, Y., Choi, D., Vasseur, P., Kweon, I.S.: Extrinsic calibration of non-overlapping camera-laser system using structured environment. In: Proceedings of the IEEE/RSJ International Conference on Intelligent Robots and Systems (IROS 2014), Chicago, Illinois, USA, pp. 436–443, September 2014

5. Brookshire, J., Teller, S.: Automatic calibration of multiple coplanar sensors. In: Proceedings of the Robotics: Science and Systems (RSS 2011), Los Angeles, California, USA, June 2011
6. Brookshire, J., Teller, S.: Extrinsic calibration from per-sensor egomotion. In: Proceedings of the Robotics: Science and Systems (RSS 2012), Sydney, Australia, July 2012
7. Guo, C.X., Roumeliotis, S.I.: An analytical least-squares solution to the line scan lidar-camera extrinsic calibration problem. In: Proceedings of the IEEE International Conference on Robotics and Automation (ICRA 2013), Karlsruhe, Germany, pp. 2943–2948, May 2013
8. Kelly, J., Roy, N., Sukhatme, G.S.: Determining the time delay between inertial and visual sensor measurements. IEEE Trans. Robot. **30**(6), 1514–1523 (2014)
9. Lambert, J., Clement, L., Giamou, M., Kelly, J.: Entropy-based sim(3) calibration of 2D lidars to egomotion sensors. In: Proceedings of the IEEE International Conference on Multisensor Fusion and Integration for Intelligent Systems (MFI 2016), Baden-Baden, Germany, pp. 455–461, September 2016
10. Maddern, W., Harrison, A., Newman, P.: Lost in translation (and rotation): rapid extrinsic calibration for 2D and 3D lidars. In: Proceedings of the IEEE International Conference on Robotics and Automation (ICRA 2012), Saint Paul, Minnesota, USA, pp. 3096–3102, May 2012
11. Martinez-Cantin, R.: BayesOpt: a Bayesian optimization library for nonlinear optimization, experimental design and bandits. J. Mach. Learn. Res. **15**, 3915–3919 (2014)
12. Mockus, J.: Bayesian Approach to Global Optimization – Theory and Applications. Mathematics and Its Applications, vol. 37. Kluwer Academic Publishers (1989)
13. Mur-Artal, R., et al.: ORB-SLAM: a versatile and accurate monocular SLAM system. IEEE Trans. Robot. **31**(5), 1147–1163 (2015)
14. Pandey, G., McBride, J., Savarese, S., Eustice, R.: Automatic extrinsic calibration of vision and lidar by maximizing mutual information. J. Field Robot. **32**(5), 696–722 (2014)
15. Rehder, J., Siegwart, R., Furgale, P.: A general approach to spatiotemporal calibration in multisensor systems. IEEE Trans. Robot. **32**(2), 383–398 (2016)
16. Rényi, A.: On measures of entropy and information. In: Proceedings of the Fourth Berkeley Symposium on Mathematical Statistics and Probability, Volume 1: Contributions to the Theory of Statistics, pp. 547–561. University of California Press (1961)
17. Sheehan, M., Harrison, A., Newman, P.: Automatic self-calibration of a full field-of-view 3D n-laser scanner. In: Khatib, O., Kumar, V., Sukhatme, G. (eds.) Experimental Robotics. Springer Tracts in Advanced Robotics, vol. 79, pp. 165–178. Springer, Heidelberg (2010)
18. Zhang, Q., Pless, R.: Extrinsic calibration of a camera and laser range finder (improves camera calibration). In: Proceedings of the IEEE/RSJ International Conference on Intelligent Robots and Systems (IROS 2004), Sendai, Japan, vol. 3, pp. 2301–2306, September 2004

Simultaneous Calibration and Mapping

Christian Nissler[1(✉)], Maximilian Durner[1], Zoltán-Csaba Márton[1], and Rudolph Triebel[1,2]

[1] German Aerospace Center (DLR), Institute of Robotics and Mechatronics, Münchner Str. 20, Oberpfaffenhofen, 82234 Weßling, Germany
{Christian.Nissler,Maximilian.Durner,Zoltan.Marton, Rudolph.Triebel}@dlr.de

[2] Department of Computer Science, Technical University of Munich (TUM), Boltzmannstrasse 3, 85748 Garching, Germany

Abstract. We present evaluation experiments of a hand-eye calibration and camera-camera calibration method, which is applicable to cases where classical calibration methods fail. As described in our earlier works, the calibration works by performing rotational movements with the robot and estimating the rotational axes by tracking fiducial markers or other static parts of the environment (if the camera is moved with the robot, as in this experiment; otherwise tracking the robot itself or markers on it with a static camera). We extend our earlier work by virtually increasing the field of view of the cameras by using a mapping approach. We compare our results with an extended classical approach for the challenging case of calibrating a compliant humanoid robot having cameras with non-overlapping fields of view.

1 Motivation

Cameras are the most commonly used sensors for robotic perception due to their excellent combination of data density and cost and energy efficiency. Therefore, important robotic perception tasks such as navigation, localization, or object detection are usually addressed by processing images from cameras that are mounted on the robot. This requires an accurate computation of the transformation between the camera coordinate frame and the robot or end effector frame, as well as the pairwise transformation between the cameras. The former is usually denoted *hand-eye calibration*, and it can be done by estimating the unknown transformations from the forward kinematics of the robot [12] or by employing an external tracking system [5]. The latter is known as *camera-to-camera calibration*, and it is usually performed by observing a known calibration pattern with both cameras [13] and minimizing a set of error equations from pairs of corresponding features. This can also be done for other sensors such as 3D laser scanners [1].

However, those classical calibration methods can fail, e.g. if no (precise) kinematic information of the robot is available, or if there is not enough overlap

J. Xiao et al. (Eds.): ISER 2018, SPAR 11, pp. 791–800, 2020.
https://doi.org/10.1007/978-3-030-33950-0_68

between the cameras and no external tracking is available (e.g. for recalibration during deployment). In this study, we show how calibration can be done in those challenging situations. In particular, we perform rotational movements with the robot and estimate the rotational axes by tracking fiducial markers (AprilTags [10]) in the surrounding. Note that in principle any feature or object detection providing 6D poses would work for our approach. If only 3D positions are given, as in [14], the mapping part can be skipped, as explained below.

This paper builds on our previous work [8,9] and extends it two ways: First, we address here the limitation of cameras with relatively small opening angles and without any overlap, which can lead to unstable results. In the improved method presented here, we additionally build a *map* of relative transformations between markers in the surrounding to virtually increase the field of view.

Second, we show the applicability of this new approach by calibrating a humanoid robot using this map. Before, only hand-eye calibration could be done in this particular case, but now we also provide an accurate camera-to-camera calibration.

2 Technical Approach

2.1 Mapping

If several markers are seen in a camera image, this additional information can be used. A graph of connected markers is built, refer to Fig. 1. Here, each node is a seen marker, and each edge is a vector of relative transformations. By averaging multiple detections in this connected graph, an optimal solution for the relative transformations between makers is obtained. In order to do this, a weight/quality value for each edge of the graph is computed. For the rotational part, a Bingham distribution [3] is fitted to all relative rotations of this marker pair. The concentration term of the Bingham distribution describes the spread of the distribution and can therefore be used as a quality term for the rotation. For the translational part, a Principal Component Analysis (PCA) is performed. The Eigenvalue of the PCA describes the quality of the translational distribution. The translational and rotational quality value are then combined using the SRT distance [11]. The optimal paths are found using the Dijkstra algorithm, resulting in the "best" connections (depending on the quality values) of all markers to one chosen parent marker.

2.2 Calibration

By obtaining the relative transformations between the markers, a map of the environment is known. This can be used for calibration by transforming all markers to a common coordinate system. The robot is firstly rotated around one axis. During this motion, all visible markers are tracked. By fitting the detections with a Bingham distribution (for the orientation) and a set of concentric circles (for the centroid), an axis of rotation can be obtained. Following this, the robot is tilted around one consecutive axis and the same rotation is repeated. For a more detailed description of the procedure please see [8,9].

3 Experiments

Two experiments are presented: a hand-eye calibration and camera-camera of a humanoid robot and a hand-eye calibration of a robotic arm of a mobile platform. In Fig. 3 the humanoid robot be seen during the experiment, in Fig. 2 the experiment using the mobile platform is depicted.

3.1 Hand-Eye and Camera-Camera Calibration of Humanoid Robot

The goal of the experiment was to calibrate the humanoid robot Rollin'Justin [2, 7]. For this robot, no reliable (forward) kinematics are available due to its compliant joints and it employs cameras with very little to no overlap, which makes a classical calibration challenging. See also Fig. 3 for a picture of the robot and its cameras, which are attached at the front, left, right and back of the platform of the robot.

In a first step, a map of surrounding markers is built by moving the robot and tracking markers from several angles (see Sect. 2). By optimizing the graph, all tracked markers can be transformed to a common frame of reference. This is then used in the actual calibration procedure: while rotating the robot around one axis, AprilTags in the surrounding are tracked and transformed into a common frame of reference (found before). The robot is tilted around another axis and the same rotation like before is repeated (e.g. rotation 1 is performed leaned back, rotation 2 is performed leaned to the front), at the same time tracking AprilTags again. The goal of the calibration is to find the rotation axis in space, around which the robot's rotation occurred. The estimation of the rotation axis can be split up in two parts: first, the vector around which the rotation occurs is estimated by fitting a Bingham distribution to the relative rotations of the transformed markers. Afterwards, the point around which the rotation occurs is estimated by fitting a circle to all detected marker positions. More details about this procedure can be found in [9]. After finding the rotation axes for

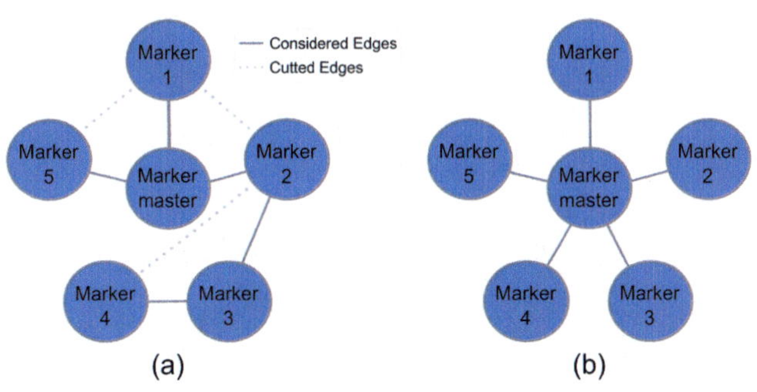

Fig. 1. (a) The initial graph, showing connected marker pairs; (b) the optimized graph, in which the optimal pose between a chosen master marker and all seen markers is computed

several tilts, the intersection of those axes are estimated. Based on this, the transformation of all cameras relative to each other and to the robot's frame can be estimated. In total three different robot orientations are used in the experiment: a position of the robot tilted back relative to its upright axis (called orientation1 in the following), an upright position (orientation2) and leaning to the front (orientation3). In the next section we evaluate the quality of the found camera-camera calibration regarding the different configurations.

3.2 TCP Modeling

In addition to the presented marker graph generation, we are able to replace the master April Tag with an external coordinate frame. Therefore, the corresponding pose of the desired coordinate frame to each April Tag detection is recorded and treated as the master marker. In this experiment the marker graph is modeled with respect to the Tool Center Point TCP of a robotic arm which results in a marker graph representing the relative pose of each April Tag with respect to the TCP (see $T_{tcp}^{marker_x}$ in Fig. 2). In other words, by observing at least one of the related AprilTags the TCP pose can be estimated visually, which is an essential part of vision-based robot control. Furthermore, given the relative TCP-to-April Tag poses the Hand-Eye calibration (Camera-to-TCP) can be evaluated by comparing the visual TCP pose estimation and the one obtained by the forward kinematics.

The robotic system used in this experiment, consists of a mobile platform, additionally equipped with a robotic arm as well as a stereo-camera mounted on a pan tilt unit (see Fig. 4). To generate the TCP-to-April Tag graph the robotic arm is placed in various (here five) arm joint configurations and rotates the TCP stepwise (in this set-up 60 steps) around its z-axis.

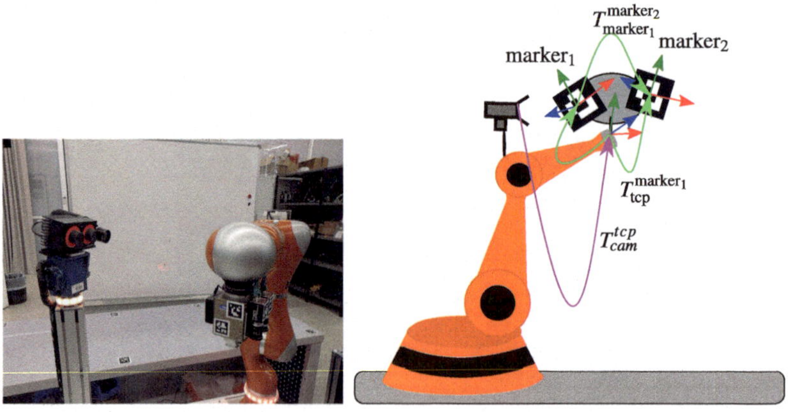

Fig. 2. (left) a picture of the used robotic arm and camera attached to a mobile platform, (right) an overview of the transformations, where the transformation colored in pink is assumed static and provided to the algorithm, and the transformations colored in green are estimated by the algorithm

Besides the images, which are recorded by the stereo-camera directed towards the gripper, we also log the TCP poses obtained by the forward kinematic for each step. This results in 60 images of the robot's end effector and corresponding TCP-logs for each of the five arm joint configurations.

The gathered data is then divided by its joint configurations into five data splits. In a 5-fold cross-validation manner the marker graph is modeled on four data splits and evaluated on the remaining one.

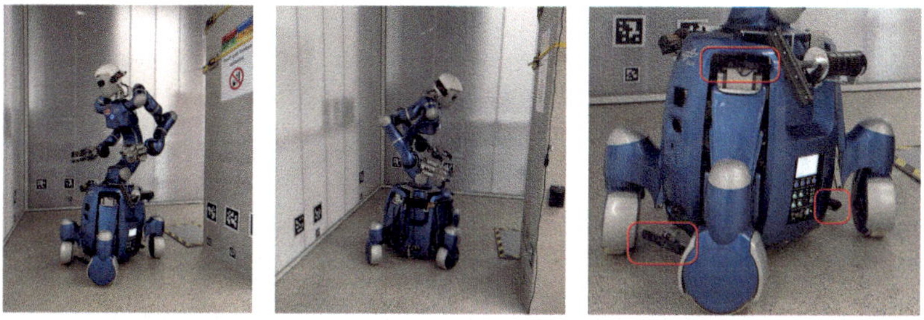

Fig. 3. The robot used in the experiment during the rotational motion around its up axis, tracking AprilTags in the surrounding (placed at the walls). The image on the right shows a close-up of the robot with its used cameras (marked red)

4 Results

In order to evaluate the quality of the resulting relative marker transformations described in Sect. 2, ground truth was recorded by measuring a test object equipped with several AprilTags, by using a measurement arm (FaroArm Quantum M). The differences to the result of our mapping yield a translational error of 0.44 mm $\pm$ 0.36 mm and a rotational error of 0.65° $\pm$ 0.34° (mean $\pm$ STD based on pairwise tag detections in 15 frames).

The result of this obtained map is applied to the tracked markers by transforming them to a common frame of reference, the result of this can be seen in Fig. 4. In Fig. 4 (left) the individual markers seen by a camera are shown, whereas the localization of the common frame of reference yields much larger segments (Fig. 4 right).

By estimating the axis of rotation based on the transformed markers instead of the individual markers, a stable estimation of the rotation axis can be obtained. The result of this can be seen in Fig. 5 (left), which shows two rotation axes as seen by two different cameras. Those two rotation axes are describing the same rotation, which can be seen by overlaying one on another (Fig. 5 (right)): they match perfectly.

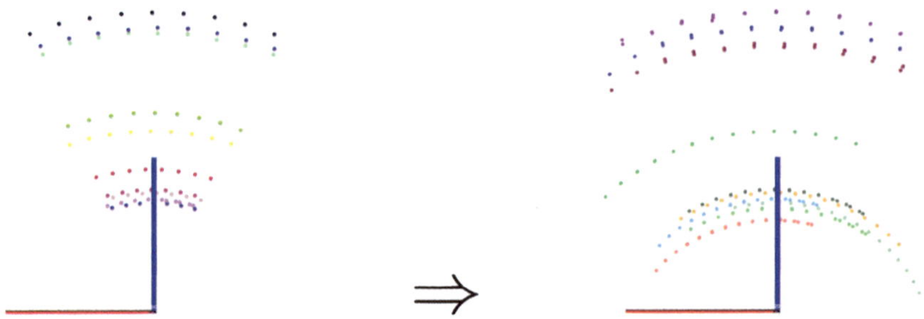

Fig. 4. (left) The tracked AprilTags during a rotation seen by one camera, the colors representing connected components of the graph, i.e. markers seen together; (right) By changing the detections to the mapped common frame of reference the field of view gets virtually increased, facilitating a better circle fit

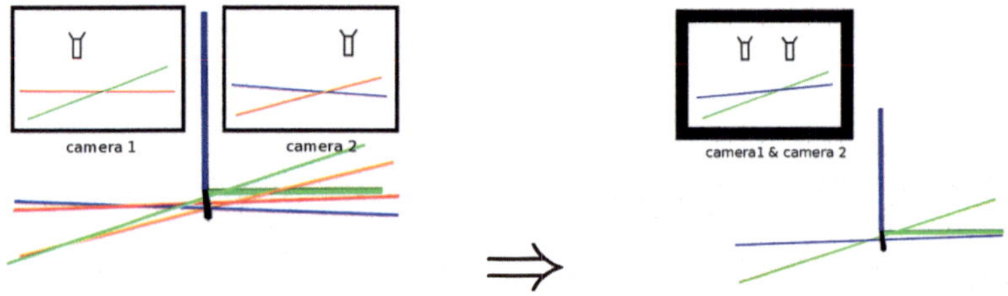

Fig. 5. (left) Two rotation axes seen by two cameras, by overlaying one onto the other we can find the relative transformations between the cameras (right)

4.1 Calibration Evaluation

In Table 1 we are evaluating the quality of the found rotation axis in several ways: Firstly, we estimate both the line-to-line distance (d) for each found axis pair and the estimated angle (α) between the two rotation axes. Secondly, we calculate the area spanned by the estimated rotation axes intersection. The are of this triangle, for which each edge consists of an intersection point of an axis pair gives an estimate of how good three axes intersect. The result can be seen in Table 3 for the case without the marker modeling part described before for all combinations of orientations.

Table 2 shows the corresponding results after marker modeling. It can be seen that the line-to-line distances are reduced and the area of the spanned triangle reduced drastically. Note that a small line-to-line distance d doesn't necessarily mean a better axis estimation, because (non-parallel) lines in space always produce a point with a minimal distance. However, the opposite is true: if the distance d is very high, the found axes intersection is also of poor quality. The spanned area of the triangle is a good estimate of how well the axis intersect. By repeating the experiment for different orientations (more than 3) this would allow to find the optimal combination.

Table 1. Intersection error of estimated axes (d) of each camera and estimated tilt (α) of the robot without marker modeling

	Orientation1–3		Orientation1–2		Orientation2–3		All orientations
	$d(mm)$	$\alpha(°)$	$d(mm)$	$\alpha(°)$	$d(mm)$	$\alpha(°)$	$A(m^2)$
Front	10.09	8.07	9.54	4.40	2.89	3.67	0.0050
Left	113	6.84	5.94	4.29	76.52	3.07	0.0267
Right	34.77	8.50	5.94	4.29	27.17	4.22	0.0011
Back	6.16	8.03	6.26	4.26	1.12	3.77	0.0117

Table 2. Intersection error of estimated axes (d) of each camera and estimated tilt (α) of the robot after marker modeling

	Orientation1–3		Orientation1–2		Orientation2–3		All orientations
	$d(mm)$	$\alpha(°)$	$d(mm)$	$\alpha(°)$	$d(mm)$	$\alpha(°)$	$A(m^2)$
Front	2.08	8.00	0.47	4.13	3.17	3.83	0.0008
Left	13.06	7.62	12.18	4.06	0.85	3.56	0.0007
Right	7.18	8.14	13.50	4.20	5.75	3.94	0.0002
Back	4.76	8.15	1.99	4.24	2.42	3.91	0.0024

In Table 3 we compare our results with calibration results performed by an extension of classical hand-eye calibration and camera-camera calibration, using the upper cameras at the head of Justin to estimate the transformation between non-overlapping cameras.

Without the marker modeling the results produce relatively large rotational and translational errors, the camera pairs left to right were even not possible to calibrate at all. Table 4 shows the results after marker modeling, which reduces

Table 3. Results of camera calibration compared to classical calibration, without marker modeling, showing both translational (e_{trans}) and rotational errors (e_{rot}) for all camera combinations (rows) and robot orientations (columns), as well for a solution based on averaging all found calibrations (*mean*)

	Orientation1–3		Orientation1–2		Orientation2–3		Mean	
	e_{trans}	e_{rot}	e_{trans}	e_{rot}	e_{trans}	e_{rot}	e_{trans}	e_{rot}
Left - right	—	—	—	—	—	—	—	—
Back - left	1.69	27.03	0.55	25.21	1.06	25.7	1.19	11.14
Back - right	0.35	7.87	0.18	5.96	0.27	6.73	0.31	6.73
Back - front	0.22	2.84	1.21	2.84	0.58	2.76	0.68	2.8
Front - left	1.81	11.51	0.65	9.17	0.5	14.41	0.53	11.69
Front - right	0.46	11.7	1.08	10.83	0.33	11.01	0.32	11.11

the rotational and translational error in all cases. It can also be seen that by averaging over all found calibrations, a calibration can be found which deviates from the classical solution by a few cm and degrees.

Note however that in that case the "classical" calibration cannot be considered ground truth, but also introduces errors by relying on the kinematics of the head cameras.

Table 4. Results of camera calibration compared to classical calibration, after marker modeling, showing both translational (e_{trans}) and rotational errors (e_{rot}) for all camera combinations (rows) and robot orientations (columns), as well for a solution based on averaging all found calibrations (*mean*)

	Orientation1-3		Orientation1-2		Orientation2-3		Mean	
	e_{trans}	e_{rot}	e_{trans}	e_{rot}	e_{trans}	e_{rot}	e_{trans}	e_{rot}
Left - right	0.03	9.36	0.08	9.85	0.09	9.45	0.04	9.45
Back - left	0.24	2.36	0.05	2.85	0.21	1.88	0.16	2.35
Back - right	0.25	5.48	0.1	5.33	0.15	5.95	0.16	5.51
Back - front	0.17	1.46	0.3	1.5	0.36	1.47	0.08	1.46
Front - left	0.08	3.06	0.32	3.59	0.16	2.64	0.1	3.04
Front - right	0.09	9.4	0.39	9.37	0.21	9.6	0.1	9.43

4.2 TCP Modeling

In Fig. 6 the translational errors (right) and rotational errors (left) are shown for five different configurations of the robot. Generally the errors are in the range of less than 1 cm and 2°, which is consistent to the ground truth estimates shown before. However, there are configurations of the robot which lead to a slightly higher error, like e.g. fold 4 in Fig. 6. This can be explained by the relatively

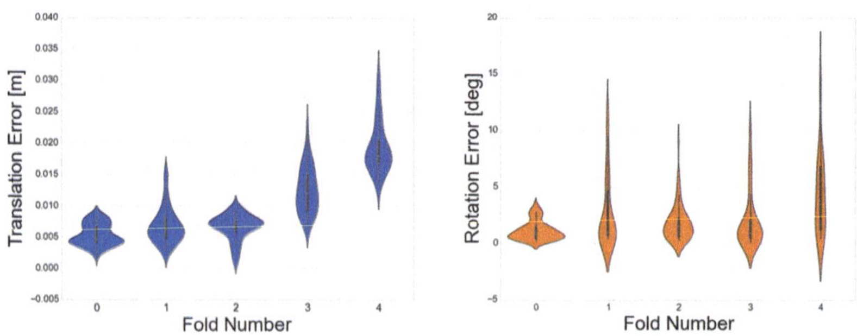

Fig. 6. Pose estimation error per fold. The average over all folds results in a translation error of 0.0089 ± 0.0052 m and a rotation error of $2.24 \pm 2.64°$

imprecise angle encoders and therefore poor kinematics of the robot. However, if the TCP is visible in the camera image, this can be accounted for by employing the found transformations of our method.

5 Discussion

Based on the results of the experiments, we have shown two things:

First, we built a map of markers in the environment by tracking markers during movement of the robot. By building a graph and optimizing it, we obtain the relative transformations (rotations and translations) between all seen markers.

This map can be used to calibrate the robot: another point of view is to regard this map as one large calibration object. In contrast to classical calibration methods the idea here is not to estimate the transformations based on the forward kinematics of the robot, but by estimating rotation axes around which the robot rotates based on tracked markers in the surrounding.

We previously [9] compared our method to state-of-the art, classical hand-eye and camera-camera calibration methods. In this study we improved on previous limitations of our method by utilizing a parallel running mapping algorithm. We experimentally show the application of this to the camera-camera and hand-eye calibration of a humanoid robot, which could not be calibrated in a stable manner with our method before. The robot does not offer forward kinematic information and its cameras have no overlapping field of view, which hinders the application of classical methods. We show the comparison of our method to a method based on external tracking using the robot's head cameras.

We also show another application of this method: By observing markers attached to a robot's TCP, we are able to both estimate the transformations in between the markers and the transformations between the robot's TCP frame to this markers, thereby finding a model describing the TCP.

One limitation of our method is the dependency on the angle of view of seen markers. This can make the estimation of the centroid of rotation unstable, resulting in relatively large translational errors compared to classical calibration methods. We showed in this study that we can mitigate this effect by building a map of the environment. However, this is a direction we want to pursue further. In the future we want to implement methods from the SLAM community like GTSAM [4] and error redistribution (e.g. g2o [6]), enhancing the quality of the rotation centroid estimation and reducing the translational error.

References

1. Andreasson, H., Triebel, R., Lilienthal, A.: Vision-based interpolation of 3D laser scans. In: Proceedings of the International Conference on Autonomous Robots and Agents (ICARA) (2006)
2. Bäuml, B., Schmidt, F., Wimböck, T., Birbach, O., Dietrich, A., Fuchs, M., Friedl, W., Frese, U., Borst, C., Grebenstein, M., et al.: Catching flying balls and preparing coffee: humanoid rollin' justin performs dynamic and sensitive tasks. In: International Conference on Robotics and Automation (ICRA). IEEE (2011)

3. Bingham, C.: An antipodally symmetric distribution on the sphere. Ann. Stat. **2**, 1201–1225 (1974)
4. Dellaert, F.: Factor graphs and GTSAM: a hands-on introduction. Technical report, Georgia Institute of Technology (2012)
5. Figueroa, N.B., Schmidt, F., Ali, H., Mavridis, N.: Joint origin identification of articulated robots with marker-based multi-camera optical tracking systems. Robot. Auton. Syst. **61**(6), 580–592 (2013)
6. Kümmerle, R., Grisetti, G., Strasdat, H., Konolige, K., Burgard, W.: G^2o: a general framework for graph optimization. In: International Conference on Robotics and Automation (ICRA). IEEE (2011)
7. Leidner, D., Borst, C., Dietrich, A., Albu-Schaeffer, A.: Classifying compliant manipulation tasks for automated planning in robotics. In: International Conference on Intelligent Robots and Systems (IROS) (2015)
8. Nissler, C., Marton, Z.-C.: Robot-to-camera calibration: a generic approach using 6D detections. In: 1st International Conference on Robot Computing (IRC). IEEE (2017)
9. Nissler, C., Marton, Z.-C., Kisner, H., Thomas, U., Triebel, R.: A method for hand-eye and camera-to-camera calibration for limited fields of view. In: International Conference on Intelligent Robots and Systems (IROS). IEEE (2017)
10. Olson, E.: AprilTag: a robust and flexible visual fiducial system. In: International Conference on Robotics and Automation (ICRA). IEEE (2011)
11. Pham, M.T., Woodford, O.J., Perbet, F., Maki, A., Stenger, B., Cipolla, R.: A new distance for scale-invariant 3D shape recognition and registration. In: 2011 International Conference on Computer Vision, pp. 145–152, November 2011
12. Strobl, K.H., Hirzinger, G.: Optimal hand-eye calibration. In: International Conference on Intelligent Robots and Systems (IROS). IEEE (2006)
13. Strobl, K.H., Hirzinger, G.: More accurate camera and hand-eye calibrations with unknown grid pattern dimensions. In: International Conference on Robotics and Automation (ICRA), pp. 1398–1405. IEEE (2008)
14. Teichman, A., Miller, S., Thrun, S.: Unsupervised intrinsic calibration of depth sensors via SLAM. In: Proceedings of Robotics: Science and Systems, Berlin, Germany, June 2013

Author Index

J. Xiao et al. (Eds.): ISER 2018, SPAR 11, pp. 801–804, 2020.
https://doi.org/10.1007/978-3-030-33950-0

Printed by Printforce, the Netherlands